Science and Technology Against Microbial Pathogens

Research, Development and Evaluation

Science and Technology Against Microbial Pathogens

Research, Development and Evaluation

Proceedings of the International Conference on Antimicrobial Research (ICAR2010)

Valladolid, Spain 3 – 5 November 2010

editor

A. Mendez-Vilas

Formatex Research Center, Spain

World Scientific

NEW JERSEY · LONDON · SINGAPORE · BEIJING · SHANGHAI · HONG KONG · TAIPEI · CHENNAI

Published by

World Scientific Publishing Co. Pte. Ltd.
5 Toh Tuck Link, Singapore 596224
USA office: 27 Warren Street, Suite 401-402, Hackensack, NJ 07601
UK office: 57 Shelton Street, Covent Garden, London WC2H 9HE

British Library Cataloguing-in-Publication Data
A catalogue record for this book is available from the British Library.

The cover images are courtesy of C. Diaz, P.L. Schilardi, and M.A. Fernandez, Lorenzo de Mele, Scanning Probe Microscopies and Surface Physical Chemistry, Laboratory and Biomaterials Laboratory, INIFTA, UNLP-CONICET, La Plata, Argentina.

SCIENCE AND TECHNOLOGY AGAINST MICROBIAL PATHOGENS
Research, Development and Evaluation
Proceedings of the International Conference on Antimicrobial Research (ICAR2010)

ISBN-13 978-981-4354-85-1
ISBN-10 981-4354-85-6

Printed in Singapore by B & Jo Enterprise Pte Ltd

Introduction

This book contains a compilation of works which were presented at the International Conference on Antimicrobial Research (ICAR2010), which took place in Valladolid (Spain), during November 3rd-5th 2010 (http://www.formatex.org/icar2010). This inaugural edition was attended by more than 500 delegates from about 60 countries.

Plenary Lectures were given by Paul Williams, from the Centre for Biomolecular Sciences, Nottingham University (UK), who talked about "Quorum sensing as an antibacterial target"; Stephen Bentley, from the Wellcome Trust Sanger Institute, Cambridge (UK), with a speech entitled "Viewing evolution and global spread of drug resistant bacterial pathogens through whole genome sequencing"; and Thierry Jouenne, from the Facility of the European Institute for Peptide Research, University of Rouen (France), who delivered the speech "The Biofilm proteome: the holy grail".

This conference series will provide a new forum in Europe for the presentation, exchange and dissemination of information and experiences on anti-microbe strategies (against bacteria, fungi or protozoans), in biotic or abiotic environments, in planktonic or adhered states, in biologically specific or unspecific ways, *in vitro* or *in vivo*, in a general context marked by the threat posed by the increasing antimicrobial resistance of pathogenic microorganisms. "Anti" is here taken in a wide sense as "against cell cycle, adhesion, or communication", when harmful for human health, industry or economy (infectious diseases, chemotherapy, food, biomedicine, agriculture, livestock, biotechnology, water systems...). It will include topics on antimicrobial resistance, (early) microbial and resistance detection, enhancement of innate defences against pathogens, as well as methods & techniques.

The topics discussed at the conference were:

Antimicrobial chemistry (experimental and computational). Analytical detection of antibiotics in complex samples.

Synthesis and screening of novel chemical compounds for antimicrobial action. Natural, synthetic and semi-synthetic antibiotics. Analogs. Structural determination. *In-silico/ab-initio/de-novo* antimicrobials discovery. New targets for antimicrobials. Rational design of antimicrobials. Bioinformatics and comparative genomics for the identification of antimicrobial targets…

Antimicrobial natural products.

Antimicrobial substances from terrestrial and marine organisms. Antimicrobial peptides. Antimicrobial enzymes. Essential oils. Bioactive phytochemicals. Plant/Herbal extracts. Purification. Structural determination…

Antimicrobials mechanisms of action.

Methods and Techniques.

Antimicrobial resistance. Superbugs. Multi-resistant strains. Emerging and re-emerging pathogens.

Microbial resistance to antibiotics and biocides. Molecular mechanisms. Resistance genes. Prevention of resistance. Surveillance & statistics. Genetics and Proteomics. Emerging and re-emerging bacteria and fungi in humans, animals, and plants. Methicillin-resistant *Staphylococcus aureus* (MRSA), Vancomycin Intermediate/Resistant *Staphylococcus aureus* (VISA/VRSA), *Clostridium difficile*, *Mycobacterium tuberculosis*, Vancomycin-resistant *enterococcus* (VRE), *Cryptosporidium*, *Plasmodium parasite*, *Plasmodium falciparum*, *Leishmania* species, *Klebsiella pneumoniae*, *Streptococcus pneumoniae*, *Acinetobacter baumannii*, *Cryptococcus*, *Escherichia coli* O157:H7, *Helicobacter* spp., *Enterobacter sakazakii*, *Serratia* spp., Fluoroquinolone-Resistant *Pseudomonas aeruginosa* (FQRP)...

Antimicrobial microbes.

Microbial-derived toxins. Bacteriocins (colicins, microcins, lantibiotics…). Archaeocins. Biocontrol approach to microbial invasions (probiotics, lactic acid bacteria…). Biosynthesis of antibiotics. Genetic and metabolic engineering. Gene regulation…

Antimicrobial viruses.

Bacteriophages. Phage therapy and biocontrol in humans, animals (agriculture-farm animals, aquaculture), plants, food industry… Materials functionalization with bacteriophages. Using bacteriophages for microbiological detection…

Antimicrobial materials science and surface chemistry. Biofilms.

Antimicrobial, anti-adhesive surfaces & coatings. Microbial adhesion to surfaces. Biofouling. Biofilm formation, control and eradication. Novel characterization techniques. Physical and chemical (inorganic (e.g. silver, copper compounds) and organic) surface modification. Cationic surfaces. Functionalization strategies for polymers, metals, metal oxides, ceramics. Drug-eluting concepts. Biofilms susceptibility to antimicrobials. Antibiotic resistance of microorganisms in biofilms. Genomics and Proteomics...

Antimicrobials in consumer products.

Textiles (hygienic clothing, activewear, medical textiles…), paper industry, active packaging (food industry…), public buildings (hospitals, schools, restaurants, day care centers, nursing homes…). Safety and toxicological aspects…

Antimicrobial physics.

Exploitation of physical properties for killing/inactivating microbes: surface tension (nano-emulsions), radiation, ultrasounds, temperature, specific properties of nano-materials (nano-particles, nano-tubes/wires, nano-crystals, nano-grained materials…). Resistance to physical agents…

Non-antibiotic biocides. Hygiene and Sterilizing.

Disinfectants, antiseptics, preservatives… Mechanism of action. Resistance to non-antibiotic biocides. Combination of physical and chemical treatments. Hygiene and Sterilizing. Sanitizers. Regulatory issues. Good practices…

Techniques and Methods.

Susceptibility Testing. Rapid microbial and resistance detection. Detection of antibiotics in environmental samples. Microscopy, microanalysis & spectroscopy, single-cell studies, high-throughout studies, nanomechanical studies, microfluidics, lab-on-a-chip concepts, miniaturized science, analysis of microbial surfaces, heterogeneity, statistics. Interaction of antimicrobial drugs with model membranes. Analytical techniques…

The Intelligent war.

Interfering microbe-microbe communication (quorum sensing) as antimicrobial strategy.

Strengthening of innate immune system as antimicrobial strategy.

Immunotherapy, immunomodulating agents, cytokines (interleukins, colony-stimulating factors, interferons…), hormones… Novel vaccines for preventing or treating disease…

Antimicrobials evaluation. Pre-clinical and clinical trials.

Public awareness, learning & teaching, influence on policy-makers. Regional regulatory frameworks and experiences on antimicrobials.

Organizing Committee:

Scientific Advisory Committee:

David Wareham, Queen Mary University London, Centre for Immunology and Infectious Disease, Blizard Institute of Cell and Molecular Science, Barts & The London School of Medicine and Dentistry, London, UK

Debora B Barbosa, Department of Dental Materials and Prosthodontics, Araçatuba Dental School, São Paulo State University (UNESP), Araçatuba, SP, Brazil

Karl Lohner, Institute of Biophysics and Nanosystems Research, Austrian Academy of Sciences, Graz, Austria

Gill Diamond, Department of Oral Biology, UMDNJ-New Jersey Dental School, Newark, NJ, USA

Francieso Xavier Malcata, Centre of Biotechnology and Fine Chemicals, School of Biotechnology, Portuguese Catholic University, Porto, Portugal

Manuel V. Simões, LEPAE, Department of Chemical Engineering, Faculty of Engineering, University of Porto, Porto, Portugal

Sun Chul Kang, Department of Biotechnology, College of Engineering, Daegu University, Kyoungsan, Kyoungbook, Republic of Korea

Robert A. Bonomo, Department of Medicine, Pharmacology, Molecular Biology and Microbiology, Case Western Reserve University School of Medicine, Cleveland, Ohio, USA, and Research Service, Louis Stokes Cleveland Department of Veterans Affairs Medical Center, Cleveland, Ohio, USA

Mahendra Rai, Department of Biotechnology, Sant Gadge Baba Amravati University, Amravati, Maharashtra, India

Pascal M. Dohmen, Department of Cardiovascular Surgery, Charité Hospital, Medical University Berlin, Berlin, Germany

Francois Niyonsaba, Atopy (Allergy) Research Center, Juntendo University School of Medicine, Hongo, Bunkyo-ku, Tokyo, Japan

Roland Pieters, Department of Medicinal Chemistry and Chemical Biology, Utrecht Institute for Pharmaceutical Sciences, Utrecht University, Utrecht, The Netherlands

Geraint Rogers, Molecular Microbiology Research Laboratory, Pharmaceutical Science Division, Franklin-Wilkins Building, King's College London, London, UK

Octávio Luiz Franco, Department of Biology, Federal University of Juiz de Fora, Juiz de Fora, MG, Brazil

Fiona Regan, National Centre for Sensor Research, School of Chemical Sciences, Dublin City University, Glasnevin, Dublin, Ireland

Miguel Castanho, Centre of Chemistry & Biochemistry, Faculty of Sciences, University of Lisbon, Lisbon, Portugal

Pilar García Suárez, Department of Technology and Biotechnology of Dairy Products, Dairy Products Institute of Asturias (IPLA-CSIC), Asturias, Spain

Valerie Edwards-Jones, Manchester Metropolitan University, Manchester, UK

Jürgen Reichling, Institute of Pharmacy and Molecular Biotechnology, Department of Biology, University of Heidelberg, Heidelberg, Germany

Neil Marsh, Department of Chemistry, University of Michigan, Ann Arbor, MI, USA.

Paul M. Tulkens, Cellular and Molecular Pharmacology Unit, Catholic University of Louvain, Brussels, Belgium

Kevin Laupland, O'Brien Centre for the Bachelor of Health Sciences Program, Health Sciences Centre, Faculty of Medicine, University of Calgary, Alberta, Canada

Martti Vaara, Northern Antibiotics Ltd, Helsinki, Finland.

Paavo Kinnunen, Helsinki Biophysics and Biomembrane Group, Medical Biochemistry, Institute of Biomedicine, University of Helsinki, Finland

Ray Dawson, DSTO (Defense Science & Technology Organisation), Melbourne, Australia

Tim Maisch, Antimicrobial PDT, Clinic and Polyclinic for Dermatology, Regensburg University Hospital, Germany

Angel Concheiro, Department of Pharmacy and Pharmaceutical Technology, University of Santiago de Compostela, Santiago de Compostela, Spain

Tatsushi Mogi, Department of Biomedical Chemistry, Graduate School of Medicine, The University of Tokyo, Tokyo, Japan

Juana Fernández-López, Agrofood Technology Department, Miguel Hernández University, Alicante, Spain

Quintín Rascón-Cruz, Faculty of Chemical Sciences, Autonomous University of Chihuahua, Chihuahua, Mexico

Xiaoqiang Wang, Plant Biology Division, Samuel Roberts Noble Foundation, Ardmore, OK, USA

Maria Luisa Mangoni, Istituto Pasteur-Fondazione Cenci Bolognetti, Department of Biochemical Science and Unit of Advanced Molecular Diagnostic, II Faculty of Medicine and Surgery, Rome, Italy

PJ Cullen, School of Food Science and Environmental Health, Dublin Institute of Technology, Dublin, Ireland

Enrique Querol, Biotechnology and Biomedicine Institute, Autonomous University of Barcelona, Barcelona, Spain
Shunmugaperumal Tamilvanan, Department of Pharmaceutical Sciences, University of Antwerp, Antwerp, Belgium
Arvind Bansal, Department of Computer Science, Kent State University, Kent, Ohio, USA
Maria do Carmo Bastos, Institute of Microbiology Prof. Paulo de Góes, UFRJ, Rio de Janeiro, Brazil
Axel Dalhoff, Institute for Infection Medicine, University Hospital Schleswig-Holstein, Kiel, Germany
Kevin J Saliba, Medical School & Research School of Biology, College of Medicine, Biology and Environment, The Australian National University, Canberra, Australia
Richard W. Scott, PolyMedix, Inc., Radnor, PA, USA
Amparo Estepa, IBMC, Miguel Hernández University, Elche, Spain
Elrefaie Sobhy Kenawy, Polymer Research Group, Chemistry Department, Faculty of Science, Tanta University, Tanta, Egypt
Isidra Recio, Institute of Industrial Fermentation (CSIC), Madrid, Spain
Emilio Montesinos, Institute of Food and Agricultural Technology, University of Girona, Girona, Spain
Jagat R. Kanwar, Institute for Technology & Research Innovation (ITRI), Deakin University, Geelong, Australia
Amlan K. Patra, West Bengal University of Animal & Fishery Sciences, Kolkata, India
Ahmed Yousef, The Ohio State University, USA
Hao Zhang, School of Food Science and Technology, Jiangnan University, China
Gregory N. Tew, Department of Polymer Science & Engineering, University of Massachusetts, Amherst, USA

Venue
Congress Center of Valladolid (Feria Valladolid), Valladolid, Spain.

centrodecongresos deValladolid

Media Sponsors

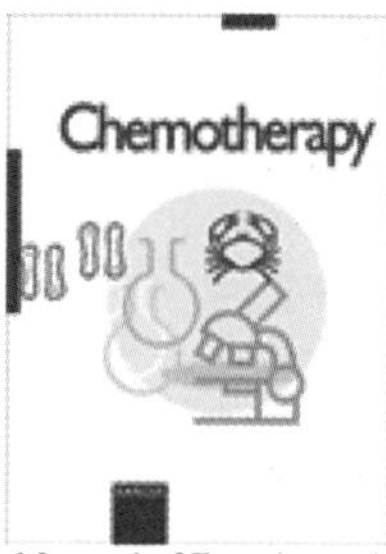

International Journal of Experimental and Clinical Chemotherapy

Archives of Clinical Microbiology

DANISCO
First you add knowledge...

viruses
Viruses-Open Access-Journal

OMYA

DSM

STERIS

Contents

Introduction v

Antimicrobial Peptides

A new class of Scots pine antimicrobial proteins, which act by binding β-glucan
Sanjeewani Sooriyaarachchi, Adrian Suárez Covarrubias, Wimal Ubhayasekera, Frederick O. Asiegbu and Sherry L. Mowbray 3

Antimicrobial aza-β^3-peptides: Structure-activity relationship?
B. Legrand, M. Laurencin, C. Zatylny-Gaudin, J. Henry, A. Bondon and M. Baudy Floc'h 7

Differential antimicrobial activities of Human Beta-Defensins against Methicillin Resistant (MRSA) and Methicillin sensitive (MSSA) *Staphylococcus aureus*
N. D. S. Herathge, J. T. George and D. A. Rowley 9

Isolation of antimicrobial peptides from New Zealand spinach *Tetragonia tetragonioides*
T. Neubauerová, M. Macková, T. Macek, M. Šanda, M. Králová, I. Doležílková and P. N. Holmsgaard. 13

Lysostaphin: molecular changes that preserve staphylolytic activity
S. Becker, J. Foster-Frey, A. Powell, H. Mohammadi, D. E. Kerr and D. M. Donovan 18

Purification and characterization of antimicrobial peptides from fleshfly larvae haemolymph
T. Neubauerová, M. Macková, T. Macek, M. Šanda and Z. Vobůrka 23

Structural and functional diversity of natural antimicrobial oligopeptides
Aexander A. Zamyatnin 28

The role of Gram-negative envelope LPS on the bactericidal properties of proteins and peptides: the case of eosinophil cationic protein
D. Pulido, M. Torrent, M. V. Nogués and E. Boix 34

Non-antibiotics Biocides

Evaluation of biocidal activity of Evolyse, a disinfectant based on hydrogen peroxide and silver nitrate
M. Barbara Pisano, V. Altana, M. Elisabetta Fadda, L. Mura, M. Deplano and S. Cosentino 41

Increased resistance to detergent in *Enterococcus faecalis*
Jacqueline Keyhani and Ezzatollah Keyhani 46

Legionella pneumophila isolation rate in a Spanish hospital pre- and post-installation of an electro-chemical activation system for potable water disinfection
Jose-Maria Rivera, Juan-Jose Granizo, Jose-Maria Aguiar, Ana Vos-Arenilla, Maria-Jose Giménez and Lorenzo Aguilar 51

Antimicrobial Evaluation: Clinical and Pre-clinical Trials

Adherence to ART and its associated factors among HIV Aids Patients in Addis Ababa
Ezra Muluneh 57

Effectiveness and safety of miconazole with hydrocortisone (Daktacort) feminine care cream in the treatment of vulvar candidiasis
J. Perez-Peralta and G. Balaccua 67

Natural Products: Terrestrial and Marine Organisms

Analysis of the 2-Phenylethyl isothiocyanate present in *Brassica* leaves and their potential application as antimicrobial agent against bacteria strains isolated from Human and Pig gastrointestinal tracts
A. Aires, C. Dias, R. N. Bennett, E. A. S. Rosa and M. J. Saavedra 75

Antimicrobial effect of carvacrol on *Escherichia coli* K12 growth at different temperatures
C.M. Belda-Galbis, A. Martínez and D. Rodrigo 80

Bacteriostatic effect of cocoa powder rich in polyphenols to control *Cronobacter sakazakii* proliferation on infant milk formula
M. C. Pina-Pérez, D. Rodrigo and A. Martínez-López 85

Curcumin: a natural antibiofilm agent
Maya Moshe, Jonathan Lellouche and Ehud Banin 89

Evaluation of *in vitro* and *in vivo* antibacterial and antifungal activity of "Camelyn M"
Benedikte Maglakelidze, Guguli Abashidze, Inga Dadeshidze, Vakxtang Mshvildadze, Andre Pichete, Vincent Perreten, Shota Tsanava, Nata Shubladze and Koba Gurielidze 94

In vitro antifungal activity of several essential oils from aromatic plants of Aragón (NE, Spain)
S. Lorán, O. Segman, C. Rota, P. Conchello and A. Herrera 99

Influence of some natural compounds on freshwater microfoulants
O. A. Cuzman, C. Faraloni, T. Turchetti, M. Camaiti and P. Tiano 104

Natural antimicrobial agents against the microbiota associated with insoles
N. Cuesta Garrote, M. M. Sánchez Navarro, F. Arán Aís and C. Orgilés Barceló 109

Plant essential oils and their components for the control of phytopathogen and mycotoxigenic fungi in crops
C. Morcia and V. Terzi 114

Possibility of using *Rosmarinus officinalis, Myrtus communis* and *Origanum sp.*'s essential oil as fungicide in pickling and tanning processes
Eser Eke Bayramoglu 118

Screening of aqueous methanol plant extracts for their antibacterial activity
Kashif Nauman and Muhammad Arshad 123

Antimicrobial Surfaces. Biofilms. Quorum Sensing. Consumer Products.

Antimicrobial active packaging films based on sorbic acid
C. Hauser, J. Wunderlich and G. Ziegleder 131

Bacteriophages actions on *Salmonella* Enteritidis biofilm
A. A. Ferreira, R. C. S. Mendonça, H. M. Hungaro, M. M. Carvalho and J. A. M. Pereira 135

Biocompatibility and antibacterial property of cold sprayed ZnO/Titanium composite coating
Noppakun Sanpo, Chen Hailan, Kelvin Loke, Koh Pak Keng, Philip Cheang, C. C. Berndt and K. A. Khor 140

Effect of penicillins on the acidification of yogurt made from ewe's milk during the storage
M. I. Berruga, M. C. Beltrán, B. Novés, A. Molina and M. P. Molina 145

Exploring thermostable quorum quenching lactonases to counteract bacterial infections in cystic fibrosis
L. Mandrich, E. Porzio, L. Merone, F. Febbraio, R. Nucci and G. Manco 150

Furanones and Thiophenones in Control of *Staphylococcus epidermidis* Biofilm Infections?
J. Lönn-Stensrud, T. Benneche and A. A. Scheie 155

Innovative High Surface Area CuO Pretreated Cotton Effective in Bacterial Inactivation under Visible Light
A. Torres, C. Ruales, C. Pulgarin, A. Aimable, P. Bowen, V. Sarria and J. Kiwi 160

Is synergy of antimicrobials the effective way of management of resistance among cosmetically significant skin microflora?
B. K. Mohanty, S. Gokul Shankar, M. S. Ranjith and M. Prabhamanju 164

Microencapsulation of *Melaleuca alternifolia* (tea tree) oil as biocide for footwear applications
M. M. Sánchez-Navarro, N. Cuesta-Garrote, F. Arán-Aís and C. Orgilés-Barceló 170

Plasma polymerisation and retention of antibacterial properties of terpinen-4-ol
Kateryna Bazaka, Mohan V. Jacob, Russell J. Crawford and Elena P. Ivanova 175

Preparation of silver or zinc loaded nanocapsules with core-shell architecture and their application as metal-ion release agents in plastics leading to antibacterial and fungicidal surface properties
M. Gladitz, S. Reinemann, J. Bauer and H. J. Radusch 180

Structure-reactivity relations for DC-magnetron sputtered Cu-layers during *E. coli* inactivation in the dark and under light
C. Castro, R. Sanjines, C. Pulgarin, P. Osorio, S. A. Giraldo and J. Kiwi 186

Study of submerged and biofilm fermentation of *Bacillus subtilis* using fish protein for production of lipopeptide antibiotic iturin A
Umme Salma Zohora, Abdul Wahab Khan, Masahiro Okanami, Takashi Ano and Mohammad Shahedur Rahman 190

The influence of DC air plasma and cellulase enzyme on the antimicrobial activity of Azadirachtin (neem leaf extract) treated cotton fabric
E. Nithya, S. Jayakumar, K. Vaideki and R. Rajendran 196

Urolithins, metabolites produced by human colonic microflora, act as *Quorum Sensing* inhibitors of *Yersinia enterocolitica* affecting its gene expression
J. A. Giménez, P. Truchado, M. Larrosa, J. C. Espín, F. Tomás-Barberán, M. T. García-Conesa and A. Allende 202

Use of lactic acid bacteria biofilms as biocontrol agents
Emel Ünal, Selin Kalkan and Zerrin Erginkaya 207

Methods and Techniques. Mechanisms of Action. Physics.

A new approach for detection of bacterial contamination in cooling lubricants
D. Oberschmidt, A. Spielvogel, C. Hein, J. E. Langbein, D. Lorenz, U. Stahl and E. Uhlmann 213

Development of a liquid-medium assay for screening antimicrobial natural products against marine bacteria
M. Geiger, J. Dupont, O. Grovel, Y. F. Pouchus and P. Hess 219

Experimental planning can help to optimize the selective photoinactivation of microorganisms
J. R. Perussi, P. L. Fernandes, C. Bernal and H. Imasato 225

Fungal evaluation on oolong tea irradiated with different water activities
G. B. Fanaro, R. C. Duarte, F. T. Rodrigues, B. Corrêa and A. L. C. H. Villavicencio 230

Gamma radiation effects on bacteria and fungi in coffee (*Coffea arabica* L.)
F. T. Rodrigues, R. C. Duarte, G. B. Fanaro and A. L. C. H. Villavicencio 233

Irradiation effect on antifungal potential of clove essential oil
R. C. Duarte, G. B. Fanaro, F. T. Rodrigues and A. L. C. H. Villavicencio 236

Microarray for rapid detection of microbial resistance genotypes
Tomasz A. Leski, Gary J. Vora, Brian Barrows, Guillermo Pimentel and Chris R. Taitt 241

Photosensitization as human and environmentally friendly antimicrobial tool
Zivile Luksienė 246

Rapid diagnosis of neonatal bacteremia using polymerase chain reaction
Amany El-Sharif and Raghdaa Ali 249

Shelf life extension of semi-dried fish Nile Tilapia (*Oreochromis niloticus*) by ultraviolet and infrared irradiations
Narumol Matan 254

Study on the effect of RF plasma pretreatment on the antimicrobial efficacy of neem leaf extract processed cotton fabric
K. Vaideki, S. Jayakumar, E. Nithya and R. Rajendran 259

Resistance and Susceptibility

A 3-year review on the profile of multidrug-resistant Gram-negative in a tertiary teaching hospital in Malaysia
H. Habsah, Z. Z. Deris, M. Zeehaida, A. R. Zaidah, H. Siti Asma'and I. Nabilah 271

Antimicrobial susceptibility in clinical isolates of *Staphylococcus aureus* harbouring of *mec*A and *luk*FS-PV genes in Northern Portugal
N. Silva, C. Prudêncio, C. Tomaz and R. Fernandes 275

Antimicrobial susceptibility profile and effect of stem bark extracts of *Curtisia dentata* on multi-drug resistant verotoxic *Escherichia coli* and *Acinetobacter* spp. isolates obtained from water and wastewater samples
Hamuel James Doughari, Patrick Alois Ndakidemi, Izanne Susan Human and Spinney Benade 281

Antimicrobial utilization in Intensive care units of a private tertiary care hospital
Pramil Tiwari, Vani Yadav and Shilpi Singh 288

Bacterial clearance from blood in mice infected by *S. pneumoniae* (penicillin MIC = 16 μg/ml) presenting specific IgG (non-protective levels) and treated with sub-therapeutic regimens of cefditoren (a highly bound cephalosporin)

Fabio Cafini, Jose Yuste, Maria-Jose Giménez, David Sevillano, Lorenzo Aguilar, Luis Alou, Martha Torrico, Natalia González, Ernesto García, Jose Prieto and Pilar Coronel 292

Characterisation of methicillin resistant *Staphylococcus aureus* isolates from hospitalised patients

Vladimir Kmet, Daniela Ohlasova and Milan Niks 295

Characterization of methicillin-resistant coagulase-negative Staphylococci Isolates from blood cultures in a Brazilian University Hospital

Valéria Cataneli Pereira and Maria de Lourdes Ribeiro de Souza da Cunha 298

Control of bacterial contamination in boar semen doses

J. M. Morrell and Margareta Wallgren 303

Diffusion of extended-spectrum β-lactamase producing *Enterobacter cloacae* in a kidney transplantation unit

S. Hammami, I. Boutiba-Ben Boubaker, A. Kamoun, R. Ghozzi, M. Saidani, A. Slim and S. Ben Redjeb. 309

Effect of antifungal agents on non-*Candida albicans Candida* species enzymatic activity

M. Negri, T. Lorenço, S. Silva, M. Henriques, J. Azeredo and R.Oliveira 313

Effect of chitosan, nisin and storage time on the growth of *Listeria innocua* and *Shewanella putrefaciens* in fish homogenates

L.I. Schelegueda, M.F. Gliemmo and C.A. Campos 318

ESBL-producing Enterobacteriaceae in the northern portugal – antimicrobial susceptibility and molecular epidemiology

R. Fernandes and C. Prudêncio 323

Observations on the antimicrobial susceptibility of *Staphylococcus pseudintermedius* following the introduction of cefovecin for clinical use in Europe

Y. Chaudhry, A. Robinson and K. S. Godinho 329

Oxacillin resistance among *Staphylococcus aureus* isolated from peritoneal dialysis related peritonitis

C. H. Camargo, T. M. C. Moraes, M. L. R. S. Cunha, J. Caramori, A. L. Mondelli, A. C. Montelli and P. Barretti 334

Resistance detection and susceptibility profile in *Staphylococcus* spp. isolated from patients with Urinary Tract Infection (UTI)

Adriano Martison Ferreira, Mariana Fávero Bonesso, Alessandro Lia Mondelli and Maria de Lourdes Ribeiro de Souza da Cunha 338

Resistance distribution profile of MBL, ESBL and multidrug resistant Gram negatives isolated at a tertiary care hospital in India

K. H. Bhutada and V. R. Shende 343

Study on selection and transfer of antimicrobial resistant *Escherichia coli* from broiler breeders to their progeny

G. Manfreda, F. Pasquali, A. Lucchi, A. De Cesare, D. Giovanardi, M. Stonfer and A. Franchini 347

Survey of *Fusarium* species infecting potato and their resistance to fungicides used in Canada

K. Lugosch, R. D. Peters, T. Barasubiye and K. Drake 352

Vancomycin Resistant Enterococci (VRE) in equine-faecal samples

Mohamed O. Ahmed, Peter D. Clegg, Nicola J. Williams, Keith E. Baptiste and Malcolm Bennet 357

Chemistry

Antimicrobial cyclic pseudopeptides including Aza-β^3-amino acids
M. Laurencin, B. Legrand, L. Mouret, A. Bondon, Y. Fleury and M. Baudy Floc'h 365

Effect of paracetamol on the pharmacokinetics of cephalexin in dogs
N. A. Afifi, M. Atef, K. Abo-El-Sooud and N. El-Mokadem 367

Importance of the C9 absolute configuration for the antifungal activity of natural and semisynthetic sesquiterpenes
M. Derita, M. Di Liberto and S. Zacchino 373

Specificity determinants dramatically reduce hemolytic activity in amphipathic α-helical antimicrobial peptides: antimicrobial activity against Gram-negative pathogens, *Acinetobacter baumannii* and *Pseudomonas aeruginosa*
Ziqing Jiang, Adriana I. Vasil, Lajos Gera, Michael L. Vasil and Robert S. Hodges 376

Studies to find a better high level disinfectant for heat-sensitive endoscopes
Norman Miner, Valerie Harris, Towanda Ebron and Natalie Lukomski 381

Towards the rational design of antimicrobial peptides: Recent developments in computational tools
M. Torrent, D. Andreu, M. V. Nogués and E. Boix 386

Antimicrobial Microbes and Viruses. Biosynthesis of Antibiotics

Antimicrobial properties of *Lactobacillus plantarum* Tensia (DSM 21380) and Inducia (DSM 21379)
M. Rätsep, P. Hütt, R. Avi, M. Utt and E. Songisepp 393

Cell growth control by tRNase ribotoxins from bacteria and yeast
Eyemen Kheir, Christian Bär, Daniel Jablonowski and Raffael Schaffrath 398

Comparison of anti-listerial effect spectrum of bacteriocins
Selin Kalkan, Emel Ünal, Zerrin Erginkaya 403

Isolation and screening of novel antibiotic producing *Streptomyces* from southwest Turkey soils
A. Ugur and O. Ceylan 406

Production and characterization of enterocin from *Enterococcus faecium* IJ-06 and IJ-21 isolated from indigenous dairy products
Imran Javed and Safia Ahmed 411

Structural model for a holin-like protein Tmp3 exhibiting broad spectrum antibacterial activity
Thangamani Rajesh and Paramasamy Gunasekaran 416

The hospital environment as a source of new antimicrobial substances: antimicrobial activity of *Bacillus amyloliquefaciens* strain HNA3 isolated in a surgery room
M.Guida, R. A. Nastro, M. Inglese, M. Trifuoggi, S. Scherillo, R. Gesuele, V. Di Onofrio and G. Liguori 421

The Inhibitor activity of *Lactobacillus rhamnosus* against rope-forming *Bacillus* strains
Selçuk Arslan, Zerrin Erginkaya, Emel Ünal 425

Uncovering the mechanism of resistance against type III bacteriocin, enterolysin A
L. Maliničová, M. Piknová, P. Pristaš and P. Javorský 428

Antimicrobial Peptides

A new class of Scots pine antimicrobial proteins, which act by binding β-glucan

Sanjeewani Sooriyaarachchi[1], **Adrian Suárez Covarrubias**[2], **Wimal Ubhayasekera**[1§], **Frederick O. Asiegbu**[3] **and Sherry L. Mowbray**[1, 2*]

[1] Department of Molecular Biology, Swedish University of Agricultural Sciences, Box 590, Biomedical Center, SE-751 24, Uppsala, Sweden
[2] Department of Cell and Molecular Biology, Uppsala University, Box 596, Biomedical Center, SE-751 24, Uppsala, Sweden
[3] Department of Forest Ecology, University of Helsinki, Box 27, FI-00014 Helsinki, Finland
[§] Present address: MAX-lab, Lund University, Box 118, S-221 00 Lund, Sweden and Institute of Medicinal Chemistry, University of Copenhagen Universitetsparken 2, DK-2100 Copenhagen Ø, Denmark

*Corresponding author

Scots pine (*Pinus sylvestris*) produces several small, highly homologous, disulfide-rich proteins (Sp-AMPs) in response to fungal pathogenic attacks. We report here the expression, structure and function of these proteins. One of the Sp-AMPs was cloned into and over-expressed in *Pichia pastoris*. The purified protein shows antifungal activity against *Heterobasidion annosum*, causing morphological changes in spores and hyphae. Binding studies revealed that it binds to soluble and insoluble β-(1,3)-glucans, major components of the fungal cell wall, with high affinity. Homology modeling studies suggest a Greek-key-β-barrel fold having a conserved patch on the surface that can accommodate at least 4 sugar units. We conclude that these proteins represent a new class of antimicrobial proteins that can be classified as pathogenesis related (PR) protein family 18.

Keywords antimicrobial proteins, β-(1,3)-glucan, binding, *Heterobasidion annosum*, homology modelling, inhibition, pathogen, *Pinus sylvestris*

1. Introduction

Crop plants such as conifers are highly abundant in the Northern hemisphere, where they are threatened by the highly destructive pathogen *Heterobasidion annosum*, the causative agent of root and butt rot disease. The main host plants for this pathogen are Scots pine, Norway spruce and Douglas fir. This fungus acts as a parasite as well as a saprophyte due to its ability to degrade both cellulose and lignin components of wood. *H. annosum* infection in forest plantations has serious economic consequences, and so fungicides, biological control agents and silvicultural methods are currently used for disease management. However, it is an urgent goal to investigate more effective and environmentally friendly methods for controlling the disease.

Antimicrobial proteins are one of the main components of the nature's tool kit in the defences of plants against pathogens. These are a structurally and functionally diverse group of proteins, which are designated as antimicrobial proteins (AMPs) solely on the basis of the common feature that they counter the attacks of microbes [1]. The expression of AMPs may in fact be triggered by several factors, including pathogenic attacks, drought, wounding, stress and heavy metals [2]. According to their properties, AMPs have been classified into a number pathogenesis related (PR) protein families; there are currently 17 PR families [3, 4].

Studies using gene expression analysis in Scots pine (*Pinus sylvestris*) have uncovered a new class of antimicrobial proteins (designated Sp-AMPs) produced after infection with *H. annosum* [5]. It has been observed that non-pathogenic fungi can also promote AMP expression in the plants [6]. Expression of at least 5 AMP genes was ultimately suggested, where four of them (sequence identity 93-100%) being deposited in the sequence databases (Sp-AMP1-4).

The present study describes the heterologous expression of one of the Sp-AMPs (Sp-AMP3), together with binding as well as inhibition studies, and homology modelling.

2. Material and methods

2.1 Cloning, expression and purification of Sp-AMP3

Sp-AMP3 (Gene Bank accession number AF410954) was cloned into *Pichia pastoris* with and without a poly-His tag at the C-terminus. The AMP expression was induced with methanol. The expressed proteins were precipitated with ammonium sulfate, then resuspended in 20 mM HEPES, pH 7.0, and dialysed against the same buffer. The amino acid sequence was confirmed by MALDI-TOF mass spectroscopy. Affinity, chelating and size exclusion chromatography were used for purification. Protein samples were stored in pH 7.0, 20 mM HEPES buffer at -20 °C.

2.2 Binding studies

Binding studies with insoluble sugars were carried out with chitin, chitosan and β-(1,3)-D-glucan (Sigma-Aldrich). 2 mg of each carbohydrate polymer was incubated at 4 °C for 6 h with 30 μl of a 0.75 mg/ml Sp-AMP3 solution and then centrifuged. The pellet was washed with 30 μl of 20 mM HEPES buffer, pH 7.0, followed by 30 μl of 1 M NaCl in 10 % acetic acid. The supernatant and the wash fractions were analyzed by SDS-PAGE.

Binding of soluble sugars to Sp-AMP was investigated using changes in protein tryptophan fluorescence in an SPF-500 spectrofluorometer (Aminco) at 25 °C. 2 mg/ml sugar solution (glucose, laminari-sugars up to laminarihexose, laminarin, cellobiose, cellotriose and cellopentose, all obtained from Sigma-Aldrich) in 10 mM HEPES, pH 7.0, was added to Sp-AMP samples (1 ml of 0.5 μM protein in the same buffer), mixed well and equilibrated at room temperature. The emission spectra were recorded at 285 nm, using an excitation wavelength of 390 nm.

2.3 Antifungal activity

Inhibition of *H. annosum* growth on Hagem agar plates in the presence of Sp-AMP3 was observed. The mycelial growth was tested by placing an agar plug containing the fungus in the centre of a Petri dish containing 20 ml of the medium. The plates were incubated in the dark at room temperature until the mycelial growth zone had a diameter of ~3 cm. Filter paper disks, each soaked with 10 μl of 0.75 mg/ml Sp-AMP3, were placed at the growth front. Protein storage buffer was used as the control.

Similarly, spore suspension-spread Hagem agar plates, with sterilized filter paper disks containing Sp-AMP3 added, were used to test the inhibition of spore germination. These plates were incubated in the dark at room temperature for a week with repeated addition of protein (10 μl of 0.75 mg/ml) and control solutions every 24 h.

2.4 Scanning electron microscopic studies

H. annosum hyphae were incubated with 20 μl of 0.75 mg/ml Sp-AMP3 for 16 h at room temperature. The hyphae were then blotted and observed using a Hitachi TM-1000 tabletop scanning electron microscope.

2.5 Homology modeling

The *Macadamia integrifolia* AMP (MiAMP1) structure solved by NMR (PDB entry 1C01 [7]) was used as the template to build a homology model of Sp-AMP3 after BLAST [8] searches. The pair-wise sequence alignment of Sp-AMP3 and MiAMP in CLUSTALW [9] was used to generate the model in the program SOD [10]. The model was adjusted in O [11] to improve the packing. Figures were prepared in O, Molscript [12] Molray [13] and Pymol (http://www.pymol.org/).

3. Results and discussion

We wished to explore the means by which several highly identical Sp-AMPs are used by plants as a defence strategy to combat a range of fungal invaders. As a first priority, we required protein for functional studies. Although the yield is very low, sufficient quantities were obtained from a *Pichia* expression system. Bacterial systems produced only inclusion bodies, and all the refolding trials were unsuccessful. The correct sequence of the expressed protein was confirmed with mass spectroscopic analysis.

Antifungal activity

H. annosum hyphal growth and spore germination were inhibited by Sp-AMP3. This effect was greater as the concentration of the protein increased (Fig. 1). Control samples inhibited neither mycelial growth nor spore germination. The poly-His tag attached to protein sample did not interfere with the activity, this construct giving comparable inhibition.

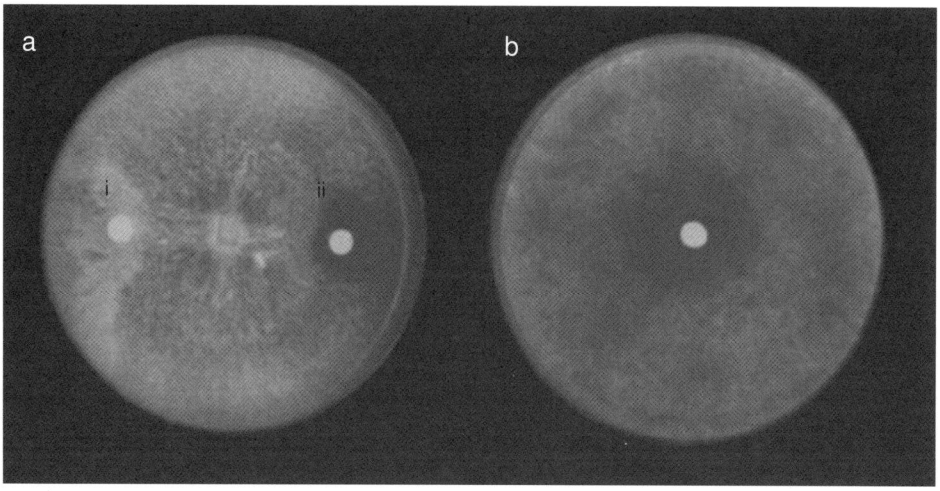

Fig. 1 Sp-AMP3 inhibits *H. annosum* growth and spore germination. In panel A, (i) is the control, i.e. protein storage buffer, while (ii) is Sp-AMP3. Panel B shows inhibition of spore germination by Sp-AMP3.

Microscopic observation

Scanning electron microscopic observations show that Sp-AMP3 pre-treatment had severely affected both hyphae and spores by shrinking and flattening.

Binding studies

Components of the fungal cell wall were tested to find the target carbohydrate for Sp-AMP. SDS-PAGE results of binding experiments clearly show that Sp-AMP3 binds to curdlan (an insoluble β-(1,3)-D-glucan) but not to chitin and chitosan. The supernatant or the wash fractions did not contain the protein, but the NaCl/acetic acid denatured fraction included most.

Soluble sugar binding to Sp-AMP3 was tested using the fluorescence change of at least one tryptophan residue of the protein upon binding. Sugars did not themselves fluoresce in the absence of Sp-AMP3. Significant binding was observed for β-(1,3)-glucan containing sugars (including laminari oligosaccharides up to laminarihexose and laminarin).

Homology model

Sp-AMP3 may safely be said to represent all the other Sp-AMPs, due to the high amino acid sequence identity in the group. Therefore, observations based on the homology model of that protein are generally applicable to other Sp-AMPs. The model predicts that Sp-AMP will exhibit a Greek-key-β-barrel fold (Fig. 2A). A patch of conserved residues probably forms the glucan-binding site, which can accommodate at least 4 sugar units Fig 2. There are no clefts or pockets suggested by the model (Fig. 2B), thus ruling out the possibility of enzymatic activity. The content of negatively charged residues is very low, giving the protein a relatively high pI (~8.5 or more). Three disulfide bonds help to maintain the shape of the molecule. There are polar residues as well as two aromatics (Tyr46 and Phe76) on the putative binding surface (Fig. 2A). The only tryptophans in Sp-AMPs (Trp6 and Trp72) are not exposed, suggesting that there may be a global change of the structure upon soluble sugar binding, to give the fluorescence change observed.

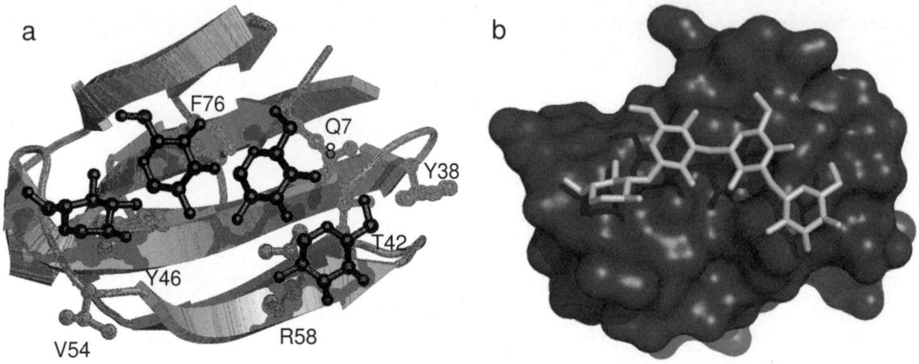

Fig. 2 Sp-AMP3 Homology model. A. Ribbon cartoon of Sp-AMP3 model, showing the conserved residues on the surface believed to be important for sugar binding in dark gray. Manually modelled laminaritetraose is shown in black. B. Surface of the Sp-AMP3 model with modelled laminaritetraose.

Antimicrobial properties, and increased expression of Sp-AMPs in response to pathogenic attacks, suggest a role of these proteins in pathogen defence. Sp-AMPs are not similar to any other known pathogenesis related protein family, but fulfil the requirements to be classified as a new family, PR family 18.

4. Conclusions

Sp-AMPs are proposed to represent a new class of antimicrobial proteins that act by binding β-glucans, a major component of fungal cell walls.

The inhibitory effects of Sp-AMP3 on spores and hyphae development of *H. annosum*, as well as the discovery of their binding to particular fungal cell wall components, suggests the potential for use of Sp-AMPs or small-molecule compounds with similar properties as biocontrol agents.

Acknowledgements Authors would like to acknowledge the Swedish Research Council for the Environment, Agricultural Sciences and Spatial Planning (FORMAS) for a travel grant (SS) and project grant (SLM), and the Swedish Research Council (VR) for a project grant. We especially thank Prof. Ulf Hellman, Ludwig Institute for Cancer Research, Uppsala, Sweden for mass spectroscopic analysis, and Gulaim A. Seisenbaeva, Department of Chemistry, Swedish University of Agricultural Sciences for scanning electron microscopic analysis.

References

[1] Montesinos, E., Antimicrobial peptides and plant disease control. FEMS Microbiol Lett, 2007. 270(1): p. 1-11.

[2] Broekaert, W.F. Antimicrobial peptides from plants. Critical Reviews in Plant Sciences, 1997. 16(3): p. 297-323.

[3] van Loon, L.C. and E.A. Van Strien, The families of pathogenesis-related proteins, their activities, and comparative analysis of PR-1 type proteins. Physiol. Mol. Plant Pathol, 1999. 55: p. 85-97.

[4] Okushima, Y., et al., Secreted proteins of tobacco cultured BY2 cells: identification of a new member of pathogenesis-related proteins. Plant Mol Biol, 2000. 42(3): p. 479-88.

[5] Asiegbu, F.O., et al., Isolation of a novel antimicrobial peptide gene (Sp-AMP) homologue from *Pinus sylvestris* (Scots pine) following infection with the root rot fungus *Heterobasidion annosum*. FEMS Microbiol Lett, 2003. 228(1): p. 27-31.

[6] Adomas, A., et al., Comparative analysis of transcript abundance in *Pinus sylvestris* after challenge with a saprotrophic, pathogenic or mutualistic fungus. Tree Physiol, 2008. 28(6): p. 885-97.

[7] McManus, A.M., et al., MiAMP1, a novel protein from *Macadamia integrifolia* adopts a Greek key beta-barrel fold unique amongst plant antimicrobial proteins. J Mol Biol, 1999. 293(3): p. 629-38.

[8] Altschul, S.F., et al., Gapped BLAST and PSI-BLAST: a new generation of protein database search programs. Nucleic Acids Res, 1997. 25(17): p. 3389-402.

[9] Thompson, J.D., D.G. Higgins, and T.J. Gibson, Clustal-W - Improving the sensitivity of progressive multiple sequence alignment through sequence weighting, position-specific gap penalties and weight matrix choice. Nucleic Acids Research, 1994. 22(22): p. 4673-4680.

[10] Kleywegt, G.J., et al., Around O. Rossmann, M.G., et al. (Eds.). International Tables for Crystallography, Vol. F. Crystallography of Biological Macromolecules, 2001: p. 353-356.

[11] Jones, T.A., et al., Improved methods for building protein models in electron density maps and the location of errors in these models. Acta Crystallogr A, 1991. 47 (Pt 2): p. 110-9.

[12] Kraulis, P.J., Molscript - a program to produce both detailed and schematic plots of protein structures. Journal of Applied Crystallography, 1991. 24: p. 946-950.

[13] Harris, M. and T.A. Jones, Molray - a web interface between O and the POV-Ray ray tracer. Acta Crystallographica Section D-Biological Crystallography, 2001. 57: p. 1201-1203.

Antimicrobial aza-β[3]-peptides: Structure-activity relationship?

B. Legrand[2], M. Laurencin[1], C. Zatylny-Gaudin[3], J. Henry[3], A. Bondon[2] and M. Baudy Floc'h[1]

[1] Université de Rennes 1. ICMV, UMR CNRS 6226 Sciences Chimiques de Rennes, 263 Avenue du Général Leclerc, 35042 Rennes Cedex, France.
[2] Université de Rennes 1. RMN-ILP, UMR CNRS 6026, CS 34317, Campus de Villejean, 35043 Rennes cedex, France
[3] Université de Caen Basse Normandie, Equipe Physiologie et Ecophysiologie des Mollusques Marins, LBBM, UMR IFREMER, 14032, Caen Cedex, France.

Keywords pseudopeptides; structure-activity relationships

1. Introduction

Designing antimicrobial molecules based on pseudopeptides to increase the activity, selectivity and bioavailability of natural peptides is now widespread. Recently, numerous peptidomimetics have been developed for biological applications including azapeptides, β-peptides, peptoids, oligoureas In this context, aza-β^3-aminoacids were used as new blocks to compose antimicrobial peptide sequences.
From a natural antimicrobial peptide, H-ALSGDAFLRF-NH_2 (AD), depending on the aza-β^3 residue insertions, the modifications can result either in inactive pseudopeptides or in a drastic enhancement of the antimicrobial activity without cytotoxicity. We present here, the first NMR solution structures of peptides containing aza-β^3 amino acids to study their structure-activity relationship.

2. Results and Discussion

In SDS micelles, the global fold of AD and AK displays disordered N-terminal regions and well-defined amphipatic helical C-terminal moieties (5-10) (Fig 1). The Aβ^3K peptide is based on AK, substituting the lysine in position 5 by an aza-β^3-lysine (Aβ^3K). Its global fold is very different from the two previous peptides and do not exhibit a helical C-terminal part. A typical aza-β^3 residues hydrazino turn, previously described in crystals, can be also observed in solution.[1-3]Despite the two possible conformations of the aza-β^3 lysine nitrogen stereocenter, only one set of signals is observable on the NMR spectra and the NOESY spectrum have numerous NOEs. The configuration of the stereocenter of the aza-β^3-Lys5 was not defined in the CNS topology and parameter files, and two set of conformations can be obtained from the NMR restraints according to the configuration R or S of the $N_\alpha$stereocenter. Based on the NMR spectra, one cannot discriminate if only one conformation is present in solution or if two configurations are in fast exchange.

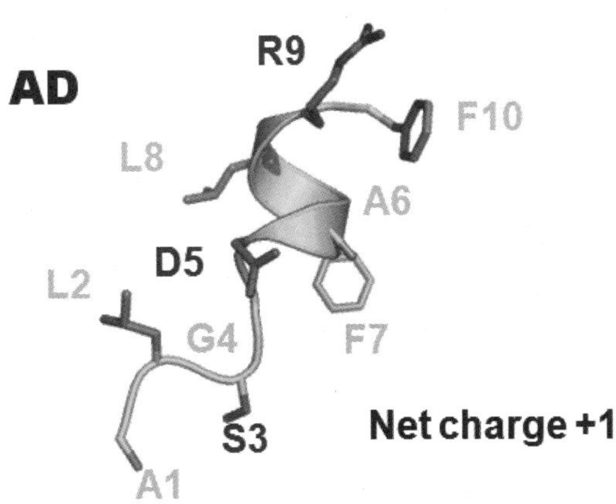

Fig. 1 AD natural cuttlefish peptide NMR structure.

Table 1 Sequences and MIC activity of the natural peptide and of five synthetic analogues. *(MIC : Minimum Inhibitory Concentration in μM; na: not active; nd: not determined)*

Name	Primary sequence	*Staphylococcus aureus*	*Escherichia coli*
AD	H-ALSGDAFLRF-NH_2	na	584
AK	H-ALSG**K**AFLRF-NH_2	262	65
A$\boldsymbol{\beta}^{3}$**K**	H-ALSG-**aza-**$\boldsymbol{\beta}^{3}$**K**-AFLRF-NH_2	na	na
K-2Nal7	H-ALSGKA-**aza-**$\boldsymbol{\beta}^{3}$**-1Nal**-LRF-NH_2	nd	268
K-1Nal	H-ALSGKA-**aza-**$\boldsymbol{\beta}^{3}$**-1Nal**-LR-**aza-**$\boldsymbol{\beta}^{3}$**-1Nal**-NH_2	15	30
K-2Nal	H-ALSGKA-**aza-**$\boldsymbol{\beta}^{3}$**-2Nal**-LR-**aza-**$\boldsymbol{\beta}^{3}$**-2Nal**-NH_2	30	118

Its global fold is very different and do not exhibit a helical C-terminal part as the two previous peptides anymore. A typical aza-β^3 residues hydrazino-turn, previously described in crystals with a characteristic geometry, can be also observed in solution (Fig 2).

Two new peptides, named K2Nal7 and K2Nal, were designed with subsequent substitution of the phenylalanine by aza-β^3-naphtylalanine amino acids (aza-β^3Nal). As AD and AK, only their C-terminal moeities are ordered but not fold in helix. They share a similar turn which approach the two hydrophobic aromatic residues in position 7 and 10. Nevertheless, in each case, they are too far to pack each other (~9 Å) and no NOEs can be detected between the residues 7 and 10 aromatic protons. The K2Nal7 and K2Nal dominant structures do not have a hydrazino-turn, indeed the hydrazino-turn NOEs marker previously noticed on the Aβ^3K NOESY spectrum cannot be detected for these peptides.

The incorporation of aza-β^3 residues broke the helical structures to induce different peptide folds. The Aβ^3K structure is quite rigid and shows a hydrazino-turn surrounded by two β-turn (Fig 2). However, despite the presence of aza-β^3 amino-acids the K-2Nal7 and K-2Nal appears to be quite flexible, their structures do not have a stable hydrazino-turn and only the C-terminal part is well defined.

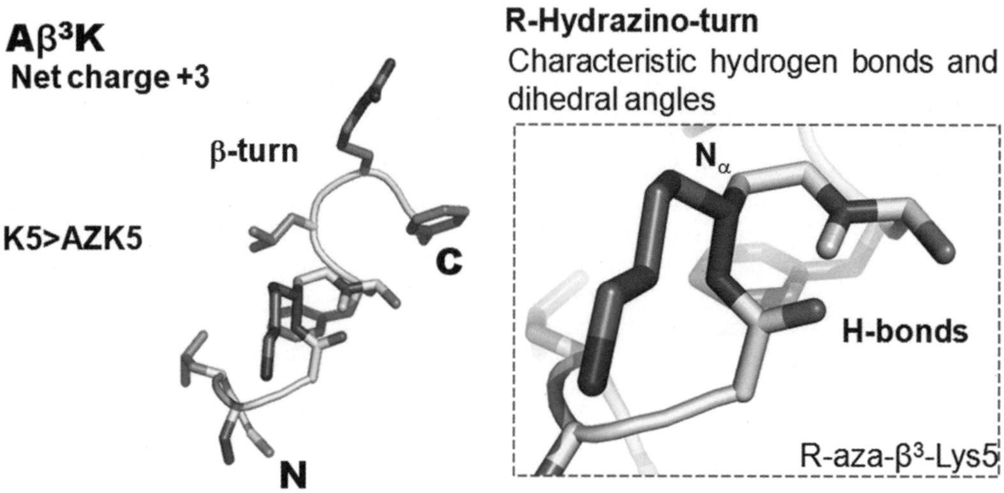

Fig. 2 Aβ^3K peptide NMR Structure.

3. Conclusion

Structural parameters, such as peptide helicity, hydrophobicity, hydrophobic moment, peptide charge and the size of the hydrophobic/hydrophilic domain, on membrane activity and selectivity could increase the antibacterial activity and improve the prokaryotic selectivity of natural and peptides analogues.

References

[1] 1. Cheguillaume, A., et al. *J. Org. Chem* **66**, 4923-4929 (2001).
[2] Salaün, A., Potel, M., Gall, P., Le Grel, P. *J. Org. Chem.* **70**, 6499-6502 (2005).
[3] Le Grel, P., Salaün, A., Potel, M., Le Grel, B., Lassagne, F. *J. Org. Chem.* **71**, 5638-5645 (2006).

Differential antimicrobial activities of Human Beta-Defensins against Methicillin Resistant (MRSA) and Methicillin sensitive (MSSA) *Staphylococcus aureus*

N.D.S. Herathge[1], J.T. George[1] and D.A. Rowley[1]

[1] School of Health and Bioscience, University of East London, Stratford Campus, Water Lane, London, E15 4LZ

Human beta-defensins (hBDs) are an intrinsic part of the innate immune response. These peptides have been shown to possess varying potencies against a wide spectrum of bacteria, viruses and fungi. Due to increasing resistance to traditional antimicrobials, interest is being shown in these peptides as possible therapeutic molecules. In this study, the bactericidal activity of two peptides (human beta defensin-2: hBD-2, and human beta defensin-3: hBD-3) was assessed against a range of *Staphylococcus aureus* (*S. aureus*) strains, including MRSA, using a standard bactericidal assay. The antibiotic sensitive strains of *S. aureus* were more susceptible to killing and exhibited a dose response to the test peptides reaching 99% kill at the highest concentration. In comparison the MRSA clinical isolates although still susceptible to the peptides showed lower percentage kill at all the concentrations examined, reaching a maximum at 68% kill. Importantly, here we show for the first time a clear and significant difference in bactericidal activity of defensins towards Methicillin sensitive and Methicillin resistant strains of *S. aureus*.

Keywords: MRSA, HBD-2, HBD-3, Bactericidal activity.

Introduction

Human epithelial and endothelial surfaces represent the first barrier against invading micro-organisms. The initial response by cells on these surfaces during infection is therefore crucial to disease outcome. As well as forming a physical barrier, it has been demonstrated that these cells can produce proteins or peptides with antimicrobial activities. One such class of peptides is the human beta-defensins (hBD's). These are small (4-5kDa) positively charged molecules which contain three di-sulphide bridges, which contribute to a very stable structure (1).The antimicrobial action of human β-defensins is initiated via interaction of the negatively charged bacterial lipid bilayer through an electrostatic attraction with the positively charged arginines and lysines in the peptide (2). These peptides either cause target cell depolarization followed by permeabilization, and/or translocate across the membrane into the cytoplasm. While the first mechanism most likely leads to irreversible damage of the target membrane potentially followed by leakage of cytoplasmic contents, peptides translocate across the membrane and accumulate intracellularly. Once inside the cell these peptides target a variety of essential cellular processes to mediate cell killing including inhibition of nucleic acid, protein, cell wall synthesis and induction of autolytic enzymes (3).

S aureus infections range from minor skin infections to life threatening infections. This bacterium has evolved different mechanisms to evade innate immune barriers such as cationic antimicrobial peptides (CAMs), complement system and phagocytosis (4, 5). *S. aureus* CAM resistance mechanisms have evolved to reduce their net negative charge and cellular hydrophobicity (6). In this study, the antibacterial activities of hBD-2 and hBD-3 against *S. aureus* including MRSA were evaluated.

Materials and Methods

Bacterial strains and growth conditions. Methicillin sensitive *S. aureus* (MSSA) strains used in this study were *S. aureus* NCTC 10655, NCTC 8331, NCTC 8511, NCTC 10970, NCTC 9717 and one freshly isolated strain from human skin. MRSA clinical isolates (MRSA 4, 7, 12, 98, 99 and 100) were obtained from NHS London. All strains were grown in isosensitest broth under aerobic conditions at 37°C.

hBD-2 and hBD-3 peptides. Recombinant human BD-2 (300-49) and human BD-3 (300-52) were obtained from PeproTech Ec Ltd. U.K.

Antimicrobial assays.
A dilution series of hBD-2 (300-49) or hBD-3 (300-52) at 3μg/ml (10^{-6}M), 0.3 μg/ml (10^{-7}M) and 0.03μg/ml (10^{-8}M) were prepared in 500μl 10mM potassium phosphate buffer (pH 7.4) (7). Bacterial strains were grown for 3 hours to the semilogarithmic stage. 1×10^5 CFU/ml cell number was obtained by diluting the 3 hours incubated culture. Then, 10 μl of 1×10^5 /ml of Staph. *aureus* was added into each of the peptide concentrations.

Control experiments were prepared for each strain using potassium phosphate buffer only. Control and samples were incubated in an orbital shaker at 37°C for 20 minutes. After incubation 100 µl from control and samples were spread on isosensitest agar plates and incubated at 37°C overnight. Colony counts were taken and antimicrobial activities of different peptide concentrations were evaluated. The results were expressed as an average of triplicate experiments.

Results

The effect of hBD-2 and hBD-3 on the *S. aureus* strains is shown in figures 1 and 2 respectively. Statistical analysis of the results at 0.3µg/ml (which is the closest to physiological concentrations) using t-Test: Two-Sample Assuming Equal Variance reveals a significant difference in the result for MRSA vs MSSA strains with p values <0.00000005 for hBD-2 and <0.0005 for hBD-3.

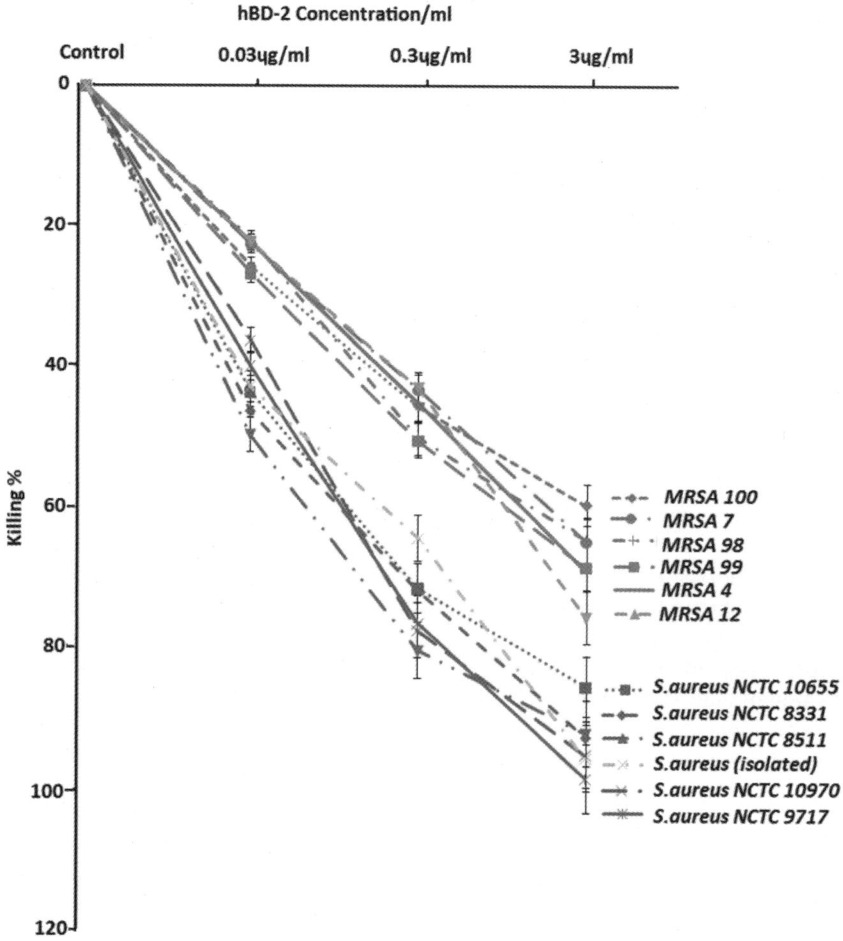

Figure 1. Antimicrobial activity of hBD-2 against different *S. aureus* strains.

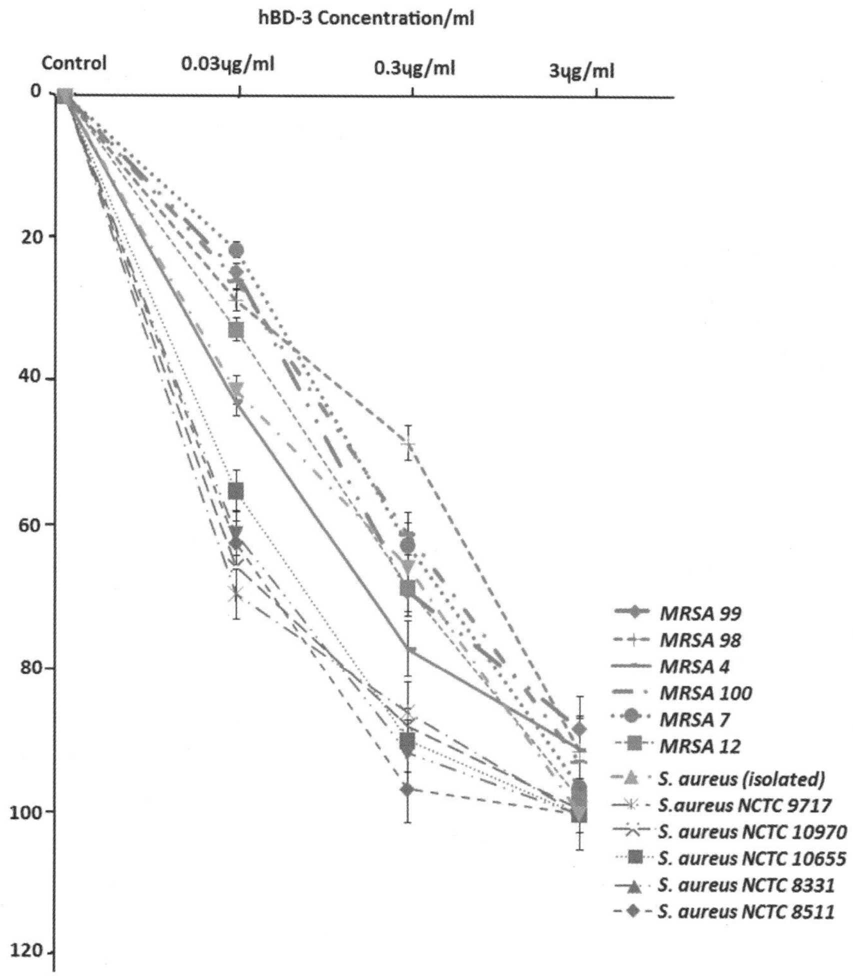

Figure 2. Antimicrobial activity of hBD-3 against different *S. aureus* strains.

Discussion

Despite the fact that beta defensins exhibit killing activity against MRSA strains it is significantly less potent than that observed against the Methicillin sensitive isolates of *S. aureus* (Fig 1 and Fig 2). Similar findings have been reported before for hBD-2 however, this is the first time such a distinct difference has been observed (8). In addition, MRSA isolates have also demonstrated decreased susceptibility to other antimicrobial peptides (9). The mechanism of reduced susceptibility may be due to different cell wall envelope alterations such as charge differences and thicknesses (10) or, the activation of an antimicrobial peptide sensor (aps) system by the Gram-positive bacterium. Cationic antimicrobial peptides initiate the gene regulatory responses via aps, leading to D-alanylation of teichoic acids by the *dlt* system and lysinylation of phospholipids by MprF in the cytoplasmic membrane, which result in a reduction of net negative charge of the membrane. The peptides are then either expelled or imported for proteolytic inactivation (11, 12). However, the exact cause of the differing sensitivities to the beta-defensins is currently under investigation. This finding obviously has implications for the potential use of such molecules in clinical treatment.

To conclude, for the first time we report a significant difference in the susceptibility of *S. aureus* strains to the beta defensin antimicrobials particularly hBD-2. MRSA strains being significantly more resistant to their killing activity. The mechanism underlying the lower sensitivity is of obvious importance and is currently under investigation.

References

[1] Harder J, Siebert R, Zhang Y, Matthiesen P, Christophers E, Schlegelberger B, Schröder JM. Mapping of the gene encoding human beta-defensin-2 (DEFB2) to chromosome region 8p22-p23.*Genomics.* 1997; 46:472-475.

[2] Ganz T. Defensins: Antimicrobial peptides of innate immunity. *Nature Rev. Immuno.* 2003; 3:710-719.

[3] Jenssen H, Hamill P. Hancock RE. Peptide antimicrobial agents. *Clin. Microbio. Rev.* 2006; 19:491-511.

[4] Rooijakkers SH, van Kessel KP, van Strip JA. Staphylococcal innate immune evasion. *Trends in Microbiology.* 2005; 13: 596-601.

[5] Bobin-Dubreux S, M-E, Reverdy ME, Nervi C, Rougier M, Bolmstrom A, Vandenesch F, Etienne J. Clinical isolate of vancomycin-heterointermediate *Staphylococcus aureus* susceptible to methicillin and in vitro selection of a vancomycin-resistant derivative. *Antimicrob. Agents Chemother.* 2001 ;45:349–352.

[6] Fedtke I, Gotz F, Peschel A. Bacterial evasion of innate host defences - the *Staphylococcus aureus* lesson. *International Journal of Medical Microbiology.* 2004; 294:189-194.

[7] Hamanaka Y, Nakashima M, Wada A, Ito M, Kurazono H, Hojo H, Nakahara Y, Kohno S, Harder J, Meyer-Hoffert U, Teran LM *et al.* Mucoid Pseudomonas aeruginosa, TNF-alpha, and IL-1beta, but not IL-6, induce human beta-defensin-2 in respiratory epithelia. *Am J Respir Cell Mol Biol.* 2001; 22: 714– 721.

[8] Midorikawa K, Ouhara K, Komatsuzawa H, Kawai T, Yamada S, Fujiwara T, Yamazaki K, Sayama K, Taubman MA, Kurihara H, Hashimoto K, Sugai M. Staphylococcus aureus susceptibility to innate antimicrobial peptides, beta-defensins and CAP18, expressed by human keratinocytes. *Infection and Immunity.* 2003;71:3730-3739.

[9] Clarke SR, Mohamed R, Bian L, Alexander F. Routh AF, Kokai-Kun JF, Mond JJ, Tarkowski A, Foster SJ. The *Staphylococcus aureus* Surface Protein IsdA Mediates Resistance to Innate Defenses of Human Skin. *Cell host and Microbe.* 2007; 1:199-212.

[10] Cui L, Ma X, Sato K., Okuma K, Tenover FC, Mamizuka EM, Gemmell CG, Kim MN, Ploy MC, El-Solh N, Ferraz V, Hiramatsu K. Cell wall thickening is a common feature of vancomycin resistance in *Staphylococcus aureus. J. Clin. Microbiol.* 2003; 41:5-14.

[11] Li M, Cha DJ, Lai Y, Villaruz AE, Sturdevant DE, Otto M. The antimicrobial peptide-sensing system asp of *Staphylococcus aureu. Molecular Microbiology.* 2007; 66:1136-1147.

[12] Peschel A, Collins LV. Staphylococcal resistance to antimicrobial peptide of mammalian and bacterial origin. *Peptides.* 2001; 22: 1651-1659.

Isolation of antimicrobial peptides from New Zealand spinach *Tetragonia tetragonioides*

T. Neubauerová[1,2], M. Macková[1,2], T. Macek[1,2], M. Šanda[1,3], M. Králová[1], I. Doležílková[1,2], P. N. Holmsgaard[4]

[1]Department of Biochemistry and Microbiology, Institute of Chemical Technology, Technická 5, 166 28, Prague 6, CZ
[2]Joint Laboratory of ICT and IOCB, Institute of Organic Chemistry and Biochemistry, Czech Academy of Sciences, Flemingovo nam. 2, 166 28 Prague 6, CZ
[3]Institute of Organic Chemistry and Biochemistry, Czech Academy of Sciences, Flemingovo nam. 2, 166 28 Prague 6, CZ
[4]Dept. of Biology, University of Copenhagen, Solvgade 83H, Copenhagen, Denmark

Corresponding author: email: neubauet@vscht.cz

The number of resistant pathogenic microbial species has increased in the last years. In the future it seems, there will be fewer possibilities to cure multiresistant infections by existing antibiotics. This can become a great problem for public health. One of the potential solutions seems to be short cationic, mainly hydrophobic peptides synthetised as a part of innate immunity of all living organisms. In our project, leaves of New Zealand spinach (*Tetragonia tetragonioides*) were chosen as a source of these peptides. We used several separation techniques based to gain fractions containing low molecular peptides. Antimicrobial activity of these peptide fractions was tested against gram-positive bacteria Staphylococcus aureus (clinical isolate) and gram-negative bacteria *Pseudomonas aeruginosa* DBM 3081. Mass spectrometry analysis of antimicrobial active fractions proved presence of low molecular peptides with molecular masses to 5 kDa.

Keywords: antimicrobial peptides; isolation; *Tetragonia tetragonioides*

1. General remarks

Overloading of antibiotic drugs in both human and veterinary medicine during the last years led to emergence of the resistant food-borne pathogens and pathogens causing various nosocomial infections. So far resistance to almost all available antibiotics has developed. The alarming occurrences of resistance among bacteria led the World Health Organisation (WHO) to announce antimicrobial drug resistance as a main public health concern[1]. To bacteria, which cause thousands deaths and million cases of food-borne illnesses, belongs *Staphylococcus aureus* resistant to methicilin and vancomycin (MRSA and VRSA), *Streptococcus* and *Enterococcus* resistant to vancomycin and linezolid, *Salmonella*, *Campylobacter* and *Listeria monocytogenes*. In the hospitals danger rising of multidrug resistant strains like *Pseudomonas aeruginosa*, *Acinetobacter baumanii*, *Escherichia coli* O157:H7 and *Klebsiella pneumoniae*[2,3] was recorded. Therefore, contamination and proliferation of food pathogens are of a great concern for food safety and public health. New Zealand spinach *Tetragonia tetragonioides* belongs to the family *Aizoaceae*, genus T*etragonia*, and vegetates at temperate and subtropical regions mainly at south hemisphere like New Zealand, Australia, southern Africa and South America. Main part of this plant forms leaves that have similar flavour and texture properties as spinach (*Spinacea oleracea*), a consumable vegetable. The plant is an annual with weak, spreading stems 2 to 4 feet long that are covered with dark green leaves 2 to 4 inches long. The leaves are smaller and fuzzier than those of regular spinach. When it's cooked, the two are virtually indistinguishable. Currently, this crop has negligible economic importance in many countries in Europe and throughout the world. However, as the investigations by Kmiecik and Jaworska (1999)[4] show, it is worth propagating because of its high and uniform yields and value, recommending it for the food processing industry. *Tetragonia tetragonioides* has been used for treatment of gastric cancer in East Asia countries. Recent studies suggest, that this plant effectively inhibits the ulcer formation induced by sedative drugs[5]. *Tetragonia tetragonioides* contains a number of active compounds: two of cerebrosides have the antiulcerogenic activity[6], the polysaccharides isolated from the leaves also have anti-inflammatory effects[7] and a water extract also showed inhibition of tryptic activity which leads to significant inhibition of trypsin-induced TNF-α (Tumor necrosis factor α) in HMC-1 cells (mast cells)[8]. But no antimicrobial peptide has been isolated from New Zealand spinach *Tetragonia tetragonioides* yet.

2. Materials and methods

2.1 Extraction of peptides and proteins from spinach leaves

Approximately 800 grams of spinach leaves were used for peptide extraction. The leaves homogenized using liquid nitrogen were extracted by three hours of incubation in an extraction buffer containing a protease inhibitors (Sigma). Residual plant material was removed by centrifugation (6 000 x g, 15 min, 4°C). The extract was precipitated by ammonium sulphate. First ammonium sulphate was added to a concentration of 30% to precipitate impurities and large proteins (16 000 x g, 15 min, 4°C), the pellet was discarded, and 90% ammonium sulphate concentration was used to precipitate the peptides and small proteins (16 000 x g, 15 min, 4°C). The salt was removed from the pellet containing the proteins by dialysis in a Tris-HCl buffer (2mM Tris-HCl, pH 7,5) at 4°C. The buffer was changed three times over the course of 1,5 days. The sample was concentrated by evaporation in a speed-vac to a final volume of about 17 ml.

2.2 Purification of crude extract

The sample was filtered through a 50kDa Amicon filter (5000 rpm, 30 min, 4°C) to remove impurities and proteins larger than 50 kDa. The protein concentration was determined to 23 mg/ml by the Bradford method using bovine serum albumin as a standard (Bradford, 1976) prior to separation on a solid phase extraction (SPE) column with C18 filling. 1 ml sample was added to the SPE column per run. SPE cleans the sample by binding the proteins to the column enabling the washout of salts and other impurities and the proteins was afterwards separated into a hydrophilic and hydrophobic fraction by elution with MilliQ water solution of 0,1% trifluoroacetic acid (TFA) and 0,1% TFA with 80% acetonitrile (ACN) respectively. The eluted factions were concentrated by speed-vac evaporation and the final volume of the hydrophobic fraction was approximately 4 ml divided into three portions, while the volume of hydrophilic fraction was approximately 1,5 ml. The hydrophobic portions contained 1,672 mg/ml, 3,345 mg/ml and 1,166 mg/ml of protein, while the hydrophilic fraction contained 0,015 mg/ml of protein.

2.3 Reverse-phase chromatography purification (RP-HPLC)

Before use of RP-HPLC, the sample were lyophilized and dissolved in MilliQ water solution of 0,1% trifluoroacetic acid (TFA) to concentration 1mg/ml. Samples were separated using HPLC system Hewlett Packard model 1100, computer operated system with Chemstation program. HPLC semipreparative procedure employed column Discovery® BIO Wide Pore C8 250 mm x 10 mm x 5 μm (Sigma). Elution was performed at 3 ml/min using gradient of acetonitril from 0,8 to 80% in 0,1% TFA (v/v). The time of run of hydrophilic part after SPE column was 45 minutes and of hydrophobic part was 75 minutes. The absorbance was monitored at 218 nm. Fractions were manually collected, evaporated on speed-vac, lyophilized and reconstituted in 100μl of MilliQ water before analysis of antimicrobial activity.

2.4 Microorganisms and antimicrobial assay

Antimicrobial activity was assayed by modified radial diffusion method. For the antimicrobial assay were chosen gram-negative bacteria: *Escherichia coli* and *Pseudomonas aeruginosa*, gram-positive bacteria: *Staphylococcus aureus* and *Bacillus megaterium*. Bacteria were cultivated in LB medium in thermal shaker for 8 hours at 37°C. In the exponential phase of growth were diluted to the absorbance 0.1 at 550 nm. The suspension of microscopic fungi was prepared in distilled water to the absorbance 0.2 at 550 nm. Aliquot of 100 μl of prepared microbial suspension was poured on the Mueller-Hinton agar plate and tested sample was transferred into pit in agar. The clear zones of inhibition were measured after incubation 24 hours at 37°C.

2.5 Antimicrobial activity

Antimicrobial activity was assayed by modified radial diffusion method. Gram-positive bacterial pathogen *Staphyloccocus aureus* was clinical isolate from Dept. of Clinical Microbiology, University Hospital Bulovka (Prague), and gram-negative bacterial pathogen *Pseudomonas aeruginosa* is from collection of bacterial strains of Department of Biochemistry and Microbiology of ICT Prague (DBM 3081). Bacteria were cultivated in LB medium under shaking for 8 hours at 37°C. In the exponential phase of growth the cell suspensions were diluted to the absorbance 0,1 at 550 nm. Aliquot of 100 μl of prepared microbial suspension was poured on the Mueller-Hinton agar plate and tested sample was transferred into pit in agar made by cork borer. The diameter of clear zones representing intensity of inhibition was measured after incubation of Petri dishes for 24 hours at 37°C. As positive controls commercial antibiotics were used: vancomycin against *Staphylococus aureus* and kanamycin against *Pseudomonas aeruginosa* (Table 1).

2.6 Mass spectrometry analysis

Mass spectra of the isolated fractions were recorded using a hybride FT-HR mass spectrometer LTQ Orbitrap XL (Thermo). Samples dissolved in acetonitrile/0.1% formic acid in water (1 : 1, v/v) were injected into the mobile phase of the same composition at a flow rate of 100 μl/min. The electrospray ion source was operated in positive ion mode with ESI voltage and capillary voltage at 4500 and 30V, respectively. Nitrogen was used as the sheath, auxiliary, sweep and collision gas. The capillary temperature was 280°C. In the MS mode, the collision energy was set at 10 V. Fractions were MS denovo sequenced using CID fragmentanion. Fragmentation mass spectra were measured using by in FT mass spectrometry mode with resolution 100 000. Fragmentation spectra were processed by "peak" sofware (Thermo).

3. Results

3.1 Isolation of antimicrobial peptides from spinach

For isolation of antimicrobial fractions from *Tetragonia tetragonioides* 800 grams of leave biomass were homogenized and treated as described above. The yield after filtration and SPE purification was 8,24 mg proteins of hydrophobic portion and 0,02 mg proteins of hydrophilic portion. These were analysed by HPLC on C8 column. 28 fractions of hydrophobic and 8 of hydrophilic portions were finally collected.

3.2 Antimicrobial assay

All these fractions after RP-HPLC were tested for antimicrobial activity on both bacteria *Staphylococcus aureus* and *Pseudomonas aeruginosa.* Nine fractions of hydrophopic portion (A) and 7 of hydrophilic portion (B) showed inhibitory effect. Fractions 1A, 5A and 4B showed significant inhibition on both bacteria. Fractions 7A, 8A and 9A showed significant inhibition on bacteria Staphylococcus aureus. Fractions 6A and 5B showed significant inhibition on bacteria *Pseudomonas aeruginosa.* Other fractions showed moderate antimicrobial effect, these fractions were 4A on bacteria *Staphylococcus aureus* and 2A,3A, 1B, 2B, 3B, 6B and 7B on bacteria *Pseudomonas aeruginosa* (Table 1).

3.3 Characterization by MS

All active fractions were analysed by mass spectrometry Orbitrap, but structure of only several fractions was successfully analysed. In fractions 1A, 4B and 5B only low molecular compounds were identified. MS analysis proved presence of short antimicrobial peptides of molecular weight between 1 932,9 and 4 894,0 Da in fractions 5A, 6A, 7A and 8A. In peptide fraction 6A partial aminoacid sequence of LPV was determined (Table 1). The first way of isolation steps were separations on the SPE and RP-HPLC.

4. Discussion

We used biochemical techniques based on predicted properties of AMP´s like hydrophobic and amphipatic properties (SPE and RP-HPLC), and low molecular masses (membrane centrifugation) to isolate low molecular substances with explicit hydrophobic character. Antimicrobial assays performed with isolated fractions showed, at several cases, antimicrobial activity. Characterization of these fractions led to several conclusions. Fractions 1A and 5B, which had antimicrobial effect against bacteria *Staphylococcus aureus* and/or *Pseudomonas aeruginosa*, contained mixture of low molecular substances. Also fraction 4B, which inhibited the growth of *Staphylococcus aureus*, contained substance of low molecular weight 112,05 Da. These fractions probably contain other antimicrobial substances not of peptide origin. These can belong to phenolic and polyphenolic substances, terpenoids, flavonoids and essencial oils, alkaloids and others, but only part of them can be extracted into water phase – anthocyanins, tannins, saponins and terpenoids[9]. That research is not in the scope of this article.

Mass spectrometry analysis of four antimicrobially active fractions showed molecular masses of 1 932,9, 3 101,3, 3 455,4 and 4894,0. Such bigger molecular masses could point to various polysaccharides, but during purification step with the SPE column all these fractions were eluted by 80% of acetonitril, polysaccharide would be eluted by the water phase. All these fractions were absorbing at wavelength 218 and 280 nm. A wavelength of 218 nm is rather selective for peptide bond and 280 nm is selective for aromatic compounds like aromatic aminoacids. All these findings lead to conclusion of antimicrobial peptide presence in these four fractions.

Mass spectrometry analysis of fraction with molecular mass 1 923,9 affirmed partial inside aminosequence LPV. This aminosequence has several similarities in the antimicrobial peptide database (Wang, 2004). Next to frog AMPs (brevinins and temporins), insect AMPs (ceratotoxins, lebocins and cecropins) and one bird beta defensin (ostricacin-4), most of similarities were with plant cyclotides: isolated from *Oldenlandia affinis* (kalata peptides) and from a *Viola genus* (varv peptides, vibi peptides, cycloviolacins, violapeptide 1 and vhl-2), unfortunately at these cyclotides has not been any antimicrobial activity published yet.

Cyclotides are a recently characterized but rapidly growing family of head-to-tail macrocyclic plant-derived peptides. Typically containing 28–37 amino acids, they are defined by their unique knotted disulfide topology linking six conserved cysteine residues and their characteristic cyclic backbone (cyclic cystine knot) motif[10,11]. This structure is base for the remarkable thermal, chemical and enzymatic stability that has made them a source of interest. To date, more than 80 cyclotides have been characterized from the plant families *Rubiaceae, Violaceae* and *Cucurbitaceae*, with a recent estimating there may be in excess of 9000 different cyclotide sequences present in nature[12].

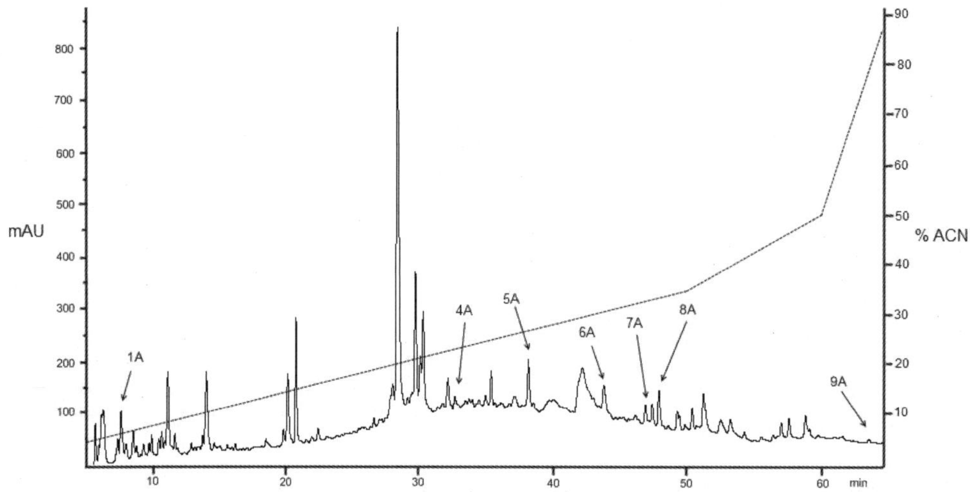

Fig. 1 RP-HPLC of hydrophobic part, antimicrobialy active fractions are assigned by numbers.

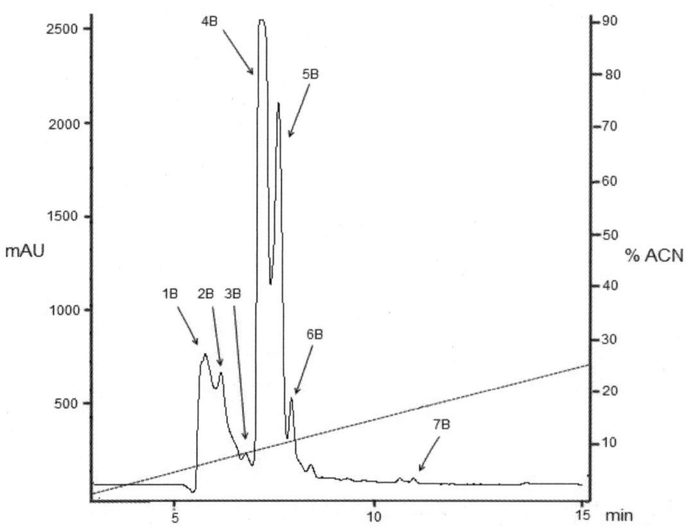

Fig. 2 RP-HPLC of hydrophilic part, antimicrobialy active fractions are assigned by numbers.

Table 1 Characterization of fractions. Table at bottom presents controls of antibiotics.

Fraction	M_W [Da]	Partial sequence	Inhibition zone [mm] *Staphylococcus aureus*	*Pseudomonas aeruginosa*
1A	Mixture of low M_W substances		9	5
5A	3 101,3		5	7
6A	1 932,9	-LPV-	-	6
7A	3 455,4		7	-
8A	4 894,0		8	-
9A			6	-
4B	112,05		7	7
5B	Mixture of low M_W substances		-	7
4A			5	-
2A, 3A, 1B, 2B, 3B, 6B, 7B			-	5

Antibiotic	Concentraction [μg/ml]	Inhibition zone [mm] *Staphylococcus aureus*	*Pseudomonas aeruginosa*
Vancomycin	**10**	**7**	**-**
Kanamycin	**20**	**-**	**8**

Acknowledgements This work was sponsored by the grants MSM 6046137305, Z 405 505 06, GACR 305/09/H008, GACR 525/ 09/ 1693 1693 and financial support from specific university research MSMT no. 21/2010.

References

[1] (WHO) World Health Organisation.. The use of essencial drugs. *Sixth Report of the WHO expert committee, WHO Tech. Rep.*, 1995,*850*.

[2] Levin A.S., Barone A.A., Penco J., Santos M. V., Marinho I. S., Arruda E. A. G., Manrique E. I.,. Costa S. F. Intravenous colistin as therapy for nosocomial infections caused by multidrug-resistant *Pseudomonas aeruginosa* and *Acinetobacter baumannii. Clin. Infect. Dis.* 1999, *28*,1008-1011.

[3] Wiener J., Quinn J. P., Bradford P. A., Goering R. V., Nathan C., Weinstein R., A. Multiple Antibiotic-Resistant Klebsiella and Escherichia coli in Nursing Homes. *JAMA* , *281*, 517 -523

[4] Kmiecik, W. & Jaworska, G. Effect of growing methods of New Zealand spinach on its yield and pattern of harvests. *Folia Horticulturae*, 1999, 11, 75–85.

[5] Okuyama E. and Yamazaki M. The principles of *Tetragonia tetragonioides* having anti-ulcerogenic activity. II. Isolation and structure of cerebrosides. *Chem. Pharm. Bull.*, 1983. *31*, 2209-2219.

[6] Cambie R. C. and Ash J. Fijian Medicinal plants, CSIRO: Melbourne, Australia, 1994, 65p., ISBN 0643054049.

[7] Kato M., Takeda T., Ogihara Y., Shimizu M., Nomura T. and Tomita Y. Studies in the structure of the polysacharides from *Tetragonia tetragonioides*. *Chem. Pharm. Bull. Jpn.*, 1985, 33, 3675–3680.

[8] Kang O., Choi Y., Park H., Tae j., Kang Ch., Lee D., Kim J. and Lee Y. Inhibitory effect of *Tetragonia tetragonioides* water extract on the production of TNF-α and tryptase in trypsin-stimulated human mast cells. *Natural Product Sciences*, 2005, 11, 207-3907.

[9] Cowan, M. M. Plant products as antimicrobial agents. *Clin. Microbiol. Rev.*, 1999,12, 564 -582.

[10] Craik D. J., Daly N. L., Mulvenna J., Plan M. R., Trabi M. Discovery, structure and biological activities of the cyclotides. *Curr. Protein Pept. Sci.* 2004;5:297–315.

[11] Craik D. J., Daly N. L., Bond T., Waine C. Plant cyclotides: a unique family of cyclic and knotted proteins that defines the cyclic cystine knot structural motif. *J. Mol. Biol.* 1999;294:1327–1336.

[12] Simonsen S., Sando L., Ireland D. C., Colgrave M. L., Craik D. J. A continent of plant defense peptide diversity: cyclotides in Australian hybanthus (Violaceae). *Plant Cell.* 2005;17:3176–3189.

Lysostaphin: molecular changes that preserve staphylolytic activity

S. Becker[1], J. Foster-Frey[1], A. Powell[1], H. Mohammadi[1], D. E. Kerr[2] and D. M. Donovan[1]

[1]Animal Biosciences and Biotechnology Lab, ANRI, ARS, USDA, Bldg 230, Room 104, BARC-East, 10300 Baltimore Ave, Beltsville, MD 20705-2350

[2]Department of Animal Science, University of Vermont, Burlington, VT

Lysostaphin is a potent staphylolytic enzyme with known resistance mechanisms. To avoid resistance development, this lab is developing lysostaphin-LysK fusion proteins harboring three, unique, synergistic lytic activities. Some of the molecular changes to lysostaphin and their effects on staphylolytic activity are described.

Keywords: antimicrobial, peptidoglycan hydrolase, *Staphylococcus aureus*

1. Introduction

Staphylococcus aureus is a renowned pathogen successful at acquiring antibiotic resistance traits, with known strains resistant to virtually every antibiotic. There is a need for novel antimicrobials to overcome this trend. Bacteriophage (phage) endolysins are natural bacteriocides expressed at the end of the phage lytic cycle to help mature phage escape from the bacterial cell. Gram positive cell walls are degraded by these PG hydrolases when exposed externally, allowing the internal osmotic pressure to lyse the cell. These peptidoglycan (PG) hydrolases are potential sources of new candidate antimicrobials,[1] with numerous patents awarded for antimicrobial applications[2].

Lysostaphin is a PG hydrolase bacteriocin, expressed and secreted (mLyso) by *S. simulans* to lyse *Staphylococcus aureus*. It, also, has numerous antimicrobial uses, including recent fusion to carbon nanotubes for use in antimicrobial surface coatings e.g. paint[3]. mLyso acts by degrading the glycyl-glycine inter-peptide bridge of the cell wall PG, and has been studied extensively in animal models of staphylococcal disease. Lysostaphin is able to kill *S. aureus* in biofilms, is effective against drug resistant staphylococci, and is synergistic with β-lactam antibiotics, antimicrobial peptides, and phage lytic enzymes. However, known mLyso resistance mechanisms exist, making mLyso less attractive as a stand alone antimicrobial, as illustrated by mLyso-resistant strains of methicillin resistant *S. aureus* (MRSA)[4]. mLyso consists of an N-terminal peptidase domain and a C-terminal SH3b cell wall binding domain http://www.sanger.ac.uk/Software/Pfam/.

In an effort to thwart *S. aureus* resistance, our lab is creating triple-acting fusion PG hydrolases with three unique, synergistic, lytic activities[5]. It is believed that no bacterium can evade three simultaneous lytic activities. Fusions with mLyso and the synergistic staphylococcal phage K endolysin, LysK, have been shown to kill MRSA[6].

It is important, when making fusion constructs, to understand the types of mLyso modifications that can be tolerated, while retaining lytic activity. A common modification is amino acid tags at the N- or C-terminus e.g. His tags, for aid in purification. Another modification is from the work of Kerr et al., where amino acid residues 125 and 232 were altered from N to Q to prevent glycosylation in mammalian expression systems[7]. Finally, there are reports of removal or insertion of amino acids at the N-terminus of mLyso or in designing our fusion constructs. To test the role of these sequences, we created a series of deletion mutants that remove up to 31 N-terminal residues (preceding the endopeptidase domain), and examined the effects on staphylolytic activity in multiple assays.

2. Results and Discussion

2.1 Effect of N- vs. C-terminal 6 x His tag on mLyso lytic activity

To assess the effect of a His-tags, mLyso constructs were subcloned and expressed in pET-21a (Novagen) and pQE-2 (Qiagen) vectors which express a C- and N-terminal 6xHis tag fusion, respectively. The fusions are created through subcloning at unique, six base pair, restriction enzyme (RE) recognition sites, such that two amino acid codons are introduced into the primary mLyso sequence (Genbank YP_003505772.1) at the site of fusion. The C-terminal amino acid of mLyso is K, the two inserted codons at the engineered XhoI RE site in pET-21a code for LE (Fig. 1). The N-terminal mLyso amino acid is A, while the recombinant version (Sigma) initiates with an M. The two amino acid codons resulting from the engineered NdeI RE site in pQE-2 code for

H and M. Control experiments to remove the N-terminal His tag via diamino-peptidase (DAPase; Qiagen), required an R codon engineered in the N-terminal sequences (Fig. 1). In light of a recent report that a small N-terminal deletion of mLyso increased its lytic activity [patent # WO/2003/082184], it was also decided to remove three N-terminal amino acids (AAT) and include this deleted version of mLyso (Δ3 mLyso) in these experiments.

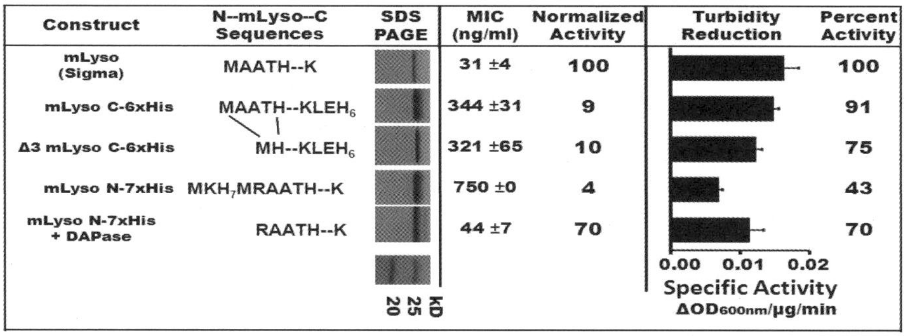

Fig. 1. N- and C-terminal His tag effects on mLyso antimicrobial activity. SDS PAGE analysis of commercial mLyso (Sigma) and NiNTA (Qiagen) purified constructs. Minimum inhibitory concentrations (MICs) and turbidity reduction assay with commercial mLyso and NiNTA purified mLyso-His fusions. Normalized MIC Activity = mLyso (Sigma) MIC/ Construct MIC x 100. Normalized turbidity reduction activity = specific activity construct/specific activity mLyso x 100. All assays were performed with *S. aureus* strain Newman. Results represent triplicate data from 20 μg of each construct, in a minimum of 8 experiments, from a minimum of 2 protein preparations. Error bars = SEM.

SDS PAGE analysis indicates that both the commercial mLyso and all N- and C-terminal His tag preparations of nickel chromatography purified (Qiagen)[8] proteins yield single protein products at >95% purity (Fig. 1). To examine the effect of C- and N-terminal His tags on antimicrobial activity, the minimum inhibitory concentration (MIC) assay was performed (as described previously[6]). The MIC assay determines the amount of enzyme required to inhibit the growth of *S. aureus* strain Newman at 20 hours post inoculation of 5 x 10^5 cells/ml in Tryptic Soy Broth media. The non-His-tagged mLyso (Sigma) required 31 ± 4 ng/ml to inhibit growth of the inoculum. Addition of the C-terminal His tag to mLyso, generating mLyso C-6xHis, reduced the MIC activity ~10X requiring 344 ± 31 ng/ml to inhibit growth, while deletion of the N-terminal 3 amino acids (AAT) of the C-terminal 6 x His tag construct (Δ3 mLyso C-6xHis) had little additional effect (321 ±65 ng/ml). The N-tagged construct (mLyso N-7xHis) was determined to have an MIC of 750 ± 0 ng/ml, indicating that the N-terminal His tag had a much greater negative effect on the activity than the C-terminal His tag. When the N-terminal His tag is removed via DAPase treatment, the lytic activity increased ~17X compared to the untreated N-terminally tagged construct, approaching the activity of mLyso. The DAPase treated samples consist of 48 hour digests with DAPase and the DAPase activator cysteamine. MIC assays performed with DAPase and cysteamine or DAPase buffer alone yielded no apparent inhibition of growth (data not shown). Also, DAPase treatment did not alter the migration of either mLyso or C-tagged mLyso in the SDS PAGE (data not shown).

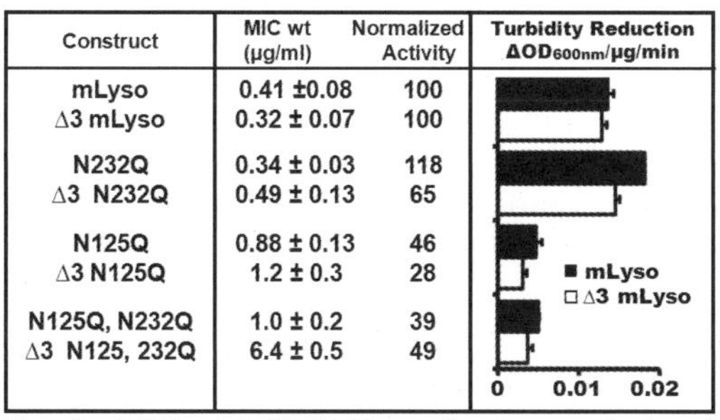

Fig. 2. MIC and Turbidity reduction analysis of mLyso C-6xHis tagged site-directed mutant constructs on *S. aureus* strain Newman. MIC data is the result of a minimum 4 experiments. Turbidity reduction data represents a minimum of 2 experiments with triplicate replicates. Normalized activity is calculated by division of construct MIC/ mLyso MIC (or Δ3 mLyso) x100. Error bars = SEM.

The His-tag mLyso variants and commercially prepared non-His-tagged mLyso were also tested in turbidity reduction assays with live cells (as previously described[8];Fig. 1), using the same protein preparations as the MIC assay. 20μg of protein in the turbidity reduction assay indicate that the C-terminal His-tagged mLyso has 91% of the non-tagged mLyso activity while the N-terminally tagged version shows 43% of the non-tagged activity level. Treatment of the full length N-terminally tagged version with DAPase shows a trend toward full activity recovery (70% of the non-tagged mLyso activity levels).

```
                                 ▼   ▼
Lysostaphin           26   GINGGMHYGVDFFMNIGTPVKAISSGKIVEAGWSNYGGGNQIGLIENDGV
ALE-1                144   GINGGNHYGVDFFMNVGTPVRAISDGKIVEAGWTNYGGGNEIGLVENDGV
LytM                 204   YHGGGAHYGVDAAMPENSPVYSLTDGTVVQAGWSNYGGGNQVTIKEANSN
S. sap. peptidase     71   SEFDGKHYGIDYHLPEDTPIKAAADGKVTRT-FDDDLGGKVIQIAESNGE

                                 ▼                               ▼ ▼
Lysostaphin           76   HRQWYMHLSKYNVKVGDYVKAGQIIGWSGSTG-YSTAPHLHFQRMVNSFS
ALE-1                194   HRQWYMHLSKFNVKVGDRVKAGQIIGWSGSTG-YSTAPHLHFQRMTNSFS
LytM                 254   NYQWYMHNNRLTVSAGDKVKAGDQIAYSGSTG-NSTAPHVHFQRMSGGIG
S. sap. peptidase    120   YHQWYMHLNEFKVEVGDDVKAGDTIALSGNTGEQTTGAHLHFQRMEGGVG

                           *
Lysostaphin          125   NSTAQDPMPFLKSAGYGKAGGTVTPTPNTGWKTNKYGTLYKSESASFTPN
ALE-1                243   NNTAQDPMPFLKSAGYGSN--STSSSNNNGYKTNKYGTLYKSESASFTAN
LytM                 303   NQYAVDPTSYLQSR------------------------------------
S. sap. peptidase    170   NDYAIDPKDYVEDLPNGEE-----------------SLFEVE-------
```

Fig. 3. Endopeptidase domain sequence alignments of staphylococcal LAS metalloproteases. Arrows indicate conserved H and D residues involved in zinc binding and protein function of LytM (L77184) and ALE-1 (D86328). N125 of the mature lysostaphin (X06121) is conserved in all four proteins and noted with an asterisk. *S. saprophyticus*. peptidase is a 194 aa putative peptidase (YP_300378).

These results indicate that although nickel column chromatography is an expedient method for rapid isolation of lytic enzymes, the addition of His tags appears to reduce the lytic activity of mLyso. In both the MIC and turbidity reduction assays, the C-terminal fusions are less deleterious than N-terminal His-tag fusions. The MIC assay is much more sensitive to a His-tag (N- or C-) than the turbidity reduction assay.

DAPase removal of the N-terminal His tag appears to at least partially retrieve the lost activity but this protocol is not likely a viable alternative for isolating lysins as antimicrobials, due to the expense of DAPase and the negative effect on stability and overall activity that a 48 hour DAPase digestion would likely have on the activity of most enzymes. The lack of full mLyso activity retrieval after DAPase treatment may be due to the engineered R amino acid remaining at the N-terminus after DAPase treatment (necessary for limiting the progressive DAPase digestion), or an effect of the required 48 hour digestion protocol. When interpreting these results it is important to be mindful that mLyso yields disparate quantitative results between multiple quantitative PG hydrolase assays [9].

2.2 Activity changes resulting from N125Q and N232Q mLyso mutations.

In a prior study, the asparagine residues 125 and 232 of mLyso were conservatively changed to glutamine residues in order to prevent glycosylation of mLyso in the mouse mammary epithelial cells [7]. The resultant mLyso double mutant, as a transgene, protected both mice and cows from *S. aureus* infection but showed a five to 10 fold reduced activity compared to wild type mLyso[10]. The contribution of each mutant residue to the reduction in activity was never determined. The mutation N125Q is located in the endopeptidase domain and the N232Q mutation is located in the C-terminus, near the SH3b cell wall binding domain, such that both residues were candidates for contributing to the reduced activity of the double mutant. It is known that a truncated version of mLyso, lacking the C-terminal SH3b binding region was reported to be enzymatically active but lost its ability to target just *S. aureus*[11]. In other works, the histidine residues of ALE-1 (a glycyl-glycine PG hydrolase >80% identical to mLyso) were mutated to demonstrate their essential role in the endopeptidase active site[12]. In that same study, ALE-1 mutation W358A (similar to mLyso N232Q, located C-terminal to the SH3b domain) reduced cell wall binding of a GFP fusion to the 92 amino acid C-terminus (including the SH3b domain) by 50%. To determine the contribution of each of the N125Q and N232Q mutations to the reduced activity, we created and characterized the activity of mLyso C-6xHis constructs with the mutations on separate mLyso constructs.

The site-directed mutations described here were created using standard PCR cloning techniques. The resultant constructs were sequence verified and tested in turbidity reduction assays (Fig. 2). The single mutant N125Q shows >50% reduction in activity compared to the mLyso C-6xHis. The single mutant N232Q, shows wild type levels of activity. The double mutant N125,232Q shows >50% reduction in activity, nearly identical to the single mutant N125Q. This suggests that in this assay, the N125Q mutation is the only mutation having an effect, and the N232Q mutation is virtually silent. This series of mutations was also created in the Δ3 mLyso C-6xHis version described in Fig. 1 and the results were nearly identical (Fig. 2). This work suggests that in order to maintain full activity, and avoid glycosylation of mLyso in mammalian systems, it might be prudent to avoid the N residues and consider mutating neighboring glycosylation recognition residues and test for their effect on glycosylation and lytic activity.

There are no previous site-directed mutagenesis studies reported for mLyso to help explain how the N125Q mutation might function at the molecular level. However, LytM[13] and ALE-1[12](two closely related metallo-proteases) have been subjected to crystallography and mutagenesis, respectively, and the residues involved in zinc binding identified (summarized in Fig. 3). Lysostaphin N125 is located in the endopeptidase domain and is perfectly conserved among the closely related proteins, LytM, ALE-1, and a putative *S. saprophyticus* peptidase (Fig. 3). N125, or its equivalent in either LytM[13, 14] or ALE-1[15], is not thought to be involved in zinc binding. However, information on the binding of lysostaphin to substrates can be inferred from mLyso's homology to the crystallographically characterized LytM [although LytM crystallization was not performed with a true substrate or product]. Firczuk et al. noticed that the crystal contacts in two crystal forms of in-vitro activated LytM involved the insertion of a glycine rich loop of one LytM molecule in the putative active site cleft of another LytM molecule and therefore suggested that these glycine rich loops might mimic a substrate [13]. In one case, the glycine-rich loop comes so close to LytM N303, the equivalent of mLyso N125, that replacement of asparagine with glutamine (one additional carbon in the side chain) would lead to a clash. Hence, the reduced activity of the mLyso variant N125Q is consistent with the crystallographic data (M. Bochtler, personal communication).

2.3 N-terminal deletions of mLyso between residue 1 and 31.

There is an apparent conflict between a recent patent application [patent # WO/2003/082184] where an N-terminal 2 amino acid deletion in mLyso yielded a 2X increase in mLyso activity, and our N-terminal 3 amino acid deletion that does not appear to have a significant effect on activity levels (Figs. 1 and 2). We decided to examine activity from a mini-deletion series in the N-terminus of mLyso (mLyso C-6xHis).

For the N-terminal deletions in Fig. 4, PCR cloning was first used to insert a SalI RE site at the second amino acid codon of mLyso, resulting in the addition of two new codons encoding VD [+SalI(VD)]. This SalI site harboring mLyso vector became the parental vector for PCR cloning of the deletion series fragments. Deletion fragments were created with multiple forward primers and one reverse primer. The forward primer series was designed to contain a SalI site positioned at 5 codon intervals as illustrated in Fig. 4. The single reverse primer spanned the 3' region of the mLyso gene, including the XhoI RE site (not shown). The deletion fragments were SalI, XhoI digested and subcloned into similarly digested pET21a. The protein construct harboring just the two inserted residues (+Sal VD) and one deletion construct (Δ2-26) were insoluble and could not be isolated in sufficient quantities for further analysis. All other deletion constructs (Δ2-6, Δ2-11, Δ2-16, Δ2-21, Δ2-31) yielded a single band in SDS PAGE analysis, and in zymogram analysis [8] only the longest deletion Δ2-31 (of the purifiable constructs) was inactive and lacked a prominent zone of clearing. In the MIC, the Δ2-31 construct was below the range of detectable MIC activity and showed only 1% mLysoC-6xHis activity in the turbidity reduction assay. Plate lysis assays were performed as described previously [8] with 10μl of enzyme solution at various concentrations spotted onto a lawn of freshly prepared mid-log phase cells (Fig. 4).

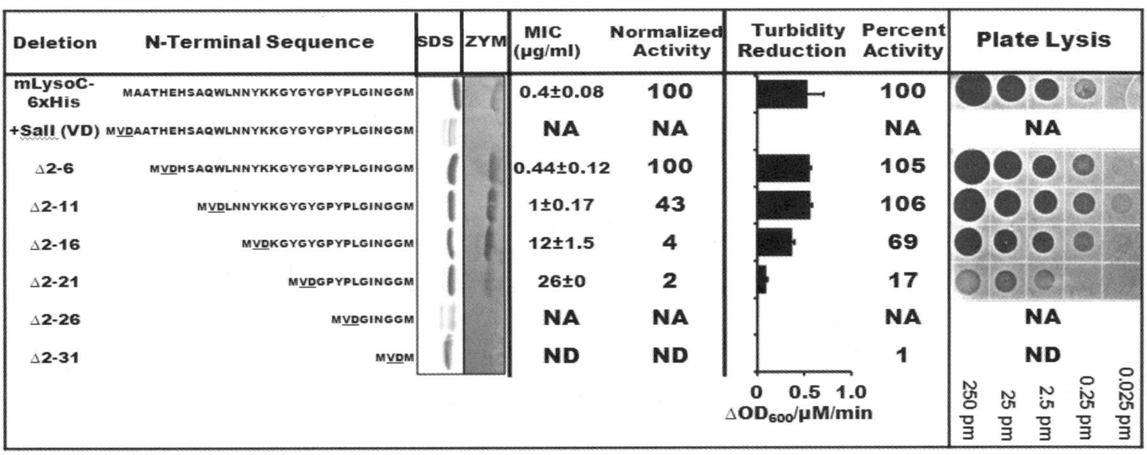

Fig. 4. Mini-deletion series of the N-terminal 31 amino acids of mLyso C-6xHis. All deletion constructs harbor 2 inserted residues (VD; underlined). SDS PAGE and zymogram analysis of 5 μg of NiNTA purified constructs. MIC and turbidity reduction assay normalization are as described in Fig. 1. Turbidity reduction assay were performed with 1 μM NiNTA purified construct. Normalized turbidity reduction activity = mLyso C-6xHis MIC/ Construct MIC x 100. Plate lysis assays were performed by spotting 10 μL containing defined picomoles (pm) amounts of protein onto a lawn of *S. aureus*. All assays were performed with *S. aureus* strain Newman. Results represent data from a minimum of 6 replicates, comprised of at least 2 independent protein preparations. Turbidity reduction experiments are performed in triplicate. Error bars = SEM. NA (non-assayable due to insolubility); ND (not determined, activity is below detection in the assay).

In Fig. 4, the constructs amenable to purification show a correlation between a loss of activity and the increasing size of the N-terminal deletion. The removal of the first 6 mLyso residues does not appear to have an effect on activity. However, the interpretation of this result needs to be qualified by the fact that the parent construct harboring MVD at the N-terminus (prior to any deleted residues) was insoluble, so not assayable. Despite this caveat, the first deletion product that was assayable (Δ2-6) shows an identical activity in all assays as the parental mLysoC-6xHis, suggesting that the small insertion (VD) does not significantly alter activity. However, there is a sharp drop in MIC activity when residues 6 to 11 are removed. The turbidity reduction assay shows a similar drop when residues 11- 16 are removed. The activities continue to drop with each successively larger deletion until virtually all activity lost in the MIC assay at Δ2-21 and in both the turbidity reduction and plate lysis assay at Δ2-31. Another caveat that needs to be considered when interpreting these results is the fact that all of these constructs also harbor a C-terminal 6 x His tag. Experiments in Fig. 1 indicate that the C-terminal tag does reduce the activity to ~80% of non-tagged mLyso levels, such that there needs to be consideration given to the unknown source of that reduction and its potential interactions with these N-terminal deletions. The results of Fig. 2 should also be considered where the activity of the 3 amino acid deleted version (Δ3mLysoC-6xHis) is identical to mLysoC-6xHis in a series of experiments with mutations that significantly alter activity.

Conclusions

Most minor changes to mLyso are tolerated with minimal effect on activity. None improved mLyso activity. Both N- and C-terminal His tags reduced activity with N-terminal being the most deleterious. N125Q reduces mLyso activity up to 50%. Deletions in the N-terminal 6 amino acids did not significantly alter activity.

Acknowledgments This work was supported in part by NIH grant1RO1AI075077-01A1; NRI grant 2007-35204-18395 and US State Dept funds, all awards to DMD. Mention of trade names or commercial products in this article is solely for the purpose of providing specific information and does not imply recommendation or endorsement by the U.S. Department of Agriculture.

References

[1] Loessner MJ. Bacteriophage endolysins--current state of research and applications. Curr Opin Microbiol 2005 August;8(4):480-7.
[2] Donovan DM. Bacteriophage and peptidoglycan degrading enzymes with antimicrobial applications. Recent Patents in 2Biotechnology 2007 June 1;1(2):113-22.
[3] Climo MW, Ehlert K, Archer GL. Mechanism and suppression of lysostaphin resistance in oxacillin-resistant *Staphylococcus aureus.* Antimicrob Agents Chemother 2001 May;45(5):1431-7.
[4] Donovan DM, Becker SC, Dong S, Baker JR, Foster-Frey J, Pritchard DG. Peptidoglycan hydrolase enzyme fusions for treating multi-drug resistant pathogens. Biotech International 2009 April;21(2):6-10.
[5] Becker SC, Foster-Frey J, Donovan DM. The phage K lytic enzyme LysK and lysostaphin act synergistically to kill MRSA. FEMS Microbiol Lett 2008 October;287(2):185-91.
[6] Pangule RC, Brooks SJ, Dinu CZ et al. Antistaphylococcal nanocomposite films based on enzyme-nanotube conjugates. ACS Nano 2010 July 27;4(7):3993-4000.
[7] Kerr DE, Plaut K, Bramley AJ et al. Lysostaphin expression in mammary glands confers protection against staphylococcal infection in transgenic mice. Nat Biotechnol 2001 January;19(1):66-70.
[8] Donovan DM, Foster-Frey J. LambdaSa2 prophage endolysin requires Cpl-7-binding domains and amidase-5 domain for antimicrobial lysis of streptococci. FEMS Microbiol Lett 2008 July 30;287(1):22-33.
[9] Kusuma C, Kokai-Kun J. Comparison of four methods for determining lysostaphin susceptibility of various strains of *Staphylococcus aureus.* Antimicrob Agents Chemother 2005;49(8):3256-63.
[10] Baba T, Schneewind O. Target cell specificity of a bacteriocin molecule: a C-terminal signal directs lysostaphin to the cell wall of *Staphylococcus aureus.* EMBO J 1996 September 16;15(18):4789-97.
[11] Fujiwara T, Aoki S, Komatsuzawa H et al. Mutation analysis of the histidine residues in the glycylglycine endopeptidase ALE-1. J Bacteriol 2005 January;187(2):480-7.
[12] Wall RJ, Powell A, Paape MJ et al. Genetically enhanced cows resist intramammary Staphylococcus aureus infection. Nat Biotechnol 2005 April;23(4):445-51.
[13] Firczuk M, Mucha A, Bochtler M. Crystal structures of active LytM. J Mol Biol 2005 December 2;354(3):578-90.
[14] Odintsov SG, Sabala I, Marcyjaniak M, Bochtler M. Latent LytM at 1.3A resolution. J Mol Biol 2004 January 16;335(3):775-85.
[15] Kajimura J, Fujiwara T, Yamada S et al. Identification and molecular characterization of an N-acetylmuramyl-L-alanine amidase Sle1 involved in cell separation of *Staphylococcus aureus.* Mol Microbiol 2005 November;58(4):1087-101.

Purification and characterization of antimicrobial peptides from fleshfly larvae haemolymph

T. Neubauerová[1,2], M. Macková[1,2], T. Macek[1,2], M. Šanda[1,3] and Z. Vobůrka[3]

[1]Department of Biochemistry and Microbiology, Institute of Chemical Technology, Technická 5, 166 28, Prague 6, CZ
[2]Joint Laboratory of ICT and IOCB, Institute of Organic Chemistry and Biochemistry, Czech Academy of Sciences, Flemingovo nam. 2, 166 28 Prague 6, CZ
[3]Institute of Organic Chemistry and Biochemistry, Czech Academy of Sciences, Flemingovo nam. 2, 166 28 Prague 6, CZ

Corresponding author: email: neubauet@vscht.cz

In 1995, when World Health Organisation announced the multiresistances against classic antibiotics as a great health problem, microbiological research led to find new terapeutics against infections caused by resistant microorganisms that are not sensitive to classical antibiotic treatments. Classical antibiotics mainly act by interfering in a specific target, there are several ways how bacteria resist - preventing the antibiotic to enter to the organism, changing permeability of the membrane, producing an enzyme that inactivates the antibiotic and/or change intramolecular targets. Pathogens with multiresistance cause infections, reaching several thousands of deaths in the world annually[1]. The appearance of such multiresistant microorganisms is exponentially growing in hospitals and clinical environments. Such infections are predominantly caused by pathogens such as *Staphylococcus aureus*, *Escherichia coli*, *Pseudomonas aeruginosa*, *Klebsiella pneumoniae* and *Enterococcus* sp.

Keyword: antimicrobial peptide; isolation; fleshfly

1. General remarks

One of the solutions potentially solving health problem with antibiotic resistance of bacteria seems to be substances with new mechanism of activity - antimicrobial peptides. Pathogens has not developed efficient system of resistance against these AMPs. These peptides are synthetised as a part of innate immunity of all living organisms as a quick weapon against infection with non-specific mechanism of action. They have broad spectrum of antimicrobial activity (gram-positive and gram-negative bacteria, fungi, viruses, protozoa and cancer cells), immunomodulatory abilities and show synergistic effects with classical antibiotics. Most of these peptides are linear short and cationic with portion of hydrophobic aminoacids, which confers the ability to adopt amphipatic helixes in the presence of biological membrane. To kill bacteria antimicrobial peptides incorporate into the biological membrane to form pores or micelles and affect the homeostasis. Some antimicrobial peptide act also on intramolecular targets (synthesis of DNA, proteosynthesis, peptide folding).[2] So far there are over 1 400 of antimicrobial peptides in the databases, most of them were isolated from insect, amphibian and plants.[1]

2. Materials and methods

2.1 Isolation of larval haemolymph

In our project larvae of the fleshly *Neobellieria bullata* at the third instar were chosen as a source of these peptides. The larvae were manipulated on the ice. Next to control larvae, in some larvae was the production of these peptides induced by pinprick. After 22 hours of incubation at room temperature the haemolymph was squeezed from induced and non-induced larvae working on ice.

2.2 Purification of antimicrobial peptides

To remove large cells the haemolymph was centrifuged at 1000 x g for 10 min at 4°C. The supernatant was precipitated in four times volume of 90% of methanol with 1% of acetic acid and centrifuged (13 000 x g for 30 min at 4°C). Supernatant was evaporated by the speed-vac and lyophilized. This crude extract was dissolved in deionised water with 0,1% trifluoroacetic acid (TFA) and separated on SPE (solid phase extraction) column

Chromabond C_{18} (Machery-Nagel) to hydrophilic part and hydrophobic part eluted by 80% of acetonitril (ACN) with 0,1% TFA.

2.3 Ion-exchange and reversed-phase liquid chromatography (Ionex-FPLC, RP-HPLC)

Hydrophobic part was lyophilized and dissolved in application buffer (50 mM ammonium acetate pH 5) and filtered through 0,22 µm pores to remove impurities and contaminants.

Ion-exchange Fast protein liquid chromatography (FPLC) analysis was performed with Biologic DuoFlow system (Bio-Rad). Sample was dissolved in application buffer (ammonium acetate, pH 5), degassed under vacuum and separated on weak katex column with CM Sepharose™ FF (ligand carboxymethyl, GE Healthcare). Elution was made by gradient of NaCl, monitored wavelength was 218 nm. The fractions after FPLC were evaporated by speed-vac, lyophilized and purified on reversed-phase liquid chromatography system (RP-HPLC).

Before use of RP-HPLC, the samples were lyophilized and resuspended in solution of 0,1%°TFA (v/v) and pure water (Millipore system DirectQ UV – MilliQ water) to final concentration of protein/peptide 2 mg/ml.

Samples were separated using HPLC system Hewlett Packard model 1100, computer operated system with Chemstation program. HPLC semipreparative procedure employed column Discovery BIO Wide Pore C8 (250 mm x 10 mm x 5 µm, Sigma). Elution was performed at 3 ml/min using gradient of acetonitril from 0.8 to 80% in 0.1% TFA (v/v). The time of hydrophobic part run after SPE column was 75 minutes. The absorbance was monitored at 218 and 280 nm. Fractions were manually collected, evaporated, lyophilized.

Isolated fractions are reconstituted in 100µl of MilliQ water before analysis of antimicrobial activity and other analysis.

2.4 Microorganisms and antimicrobial assay

Antimicrobial activity was assayed by modified radial diffusion method. For the antimicrobial assay gram-negative bacteria were chosen: *Escherichia coli* and *Pseudomonas aeruginosa*, gram-positive bacteria: *Staphylococcus aureus* and *Bacillus megaterium* and fungi: *Candida scotii, Aspergillus ochraceus* and *Mucor* species.

Bacteria were cultivated in LB medium in thermal shaker for 8 hours at 37°C. In the exponential phase of growth bacterial suspensions were diluted to the absorbance of 0.1 at 550 nm. The suspension of microscopic fungi was prepared in distilled water to the absorbance of 0.2 at 550 nm. Aliquots of 100 µl of prepared microbial suspension was poured on the Mueller-Hinton agar plate or Sabouraud agar in case of fungi and tested sample was transferred into pit in agar. The clear zones of inhibition were measured after incubation for 24 hours at 37°C.

2.5 Tricine gel electrophoresis and electroblotting

Method for tricine gel electrophoresis[3] was optimized for our system, the electrophoresis was performed with two gels - 4% and 16% gel and reducing sample buffer. After electrophoresis the gels were stained by Coomassie Brilliant Blue (CBB) or silver. For the N-terminal sequencing the unstained gel was blotted to PVDF (polyvinylidendifluoride) membranes and stained by CBB. The molecular weight markers that we used are Mark12™ Unstained Standard (Invitrogen) and Color Marker Ultra Low Range (Sigma) for blotting.

2.6 Mass spectrometry analysis

Mass spectra of the isolated fractions were recorded using a hybride FT-HR mass spectrometer LTQ Orbitrap XL (Thermo). Samples dissolved in acetonitrile/ 0.1% formic acid in water (1:1, v/v) were injected into the mobile phase of the same composition at a flow rate of 100 µl/min. The electrospray ion source was operated in positive ion mode with ESI voltage and capillary voltage at 4500 and 30V, respectively. Nitrogen was used as the sheath, auxiliary, sweep and collision gas. The capillary temperature was 280°C. In the MS mode, the collision energy was set at 10 V. Fractions were MS denovo sequenced using CID fragmentanion. Fragmentation mass spectra were measured using by in FT mass spectrometry mode with resolution 100 000. Fragmentation spectra were processed by "peak" sofware (Thermo).

3. Results and discussion

As a first alternative of isolation we chose separations on the SPE and RP-HPLC. The chromatograms show a lot of peaks, that were additionally tested for antimicrobial activity.

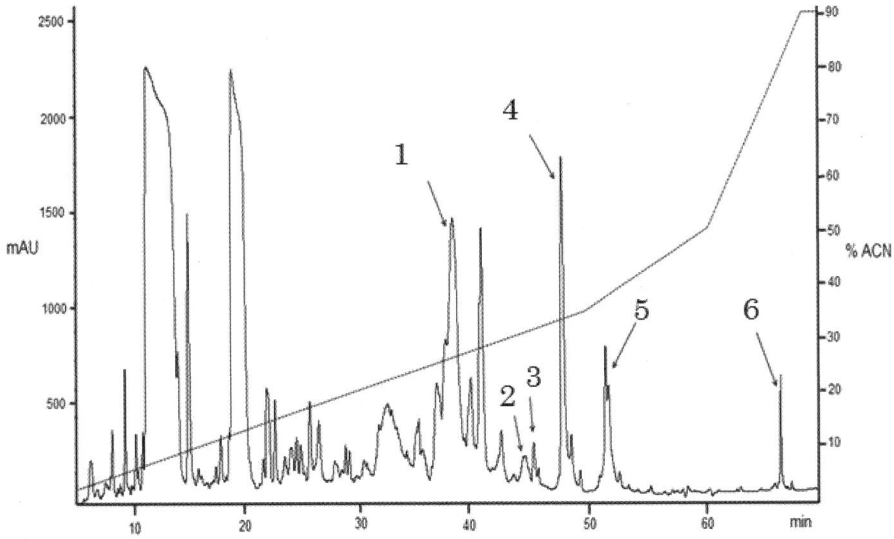

Fig. 1 RP-HPLC of hydrophobic part, antimicrobially active fractions are assigned by numbers.

Constitutive antimicrobial dipeptide, earlier identified as β-AlaTyr[4], we found in 15 minute. Three fractions **1, 2** and **3** inhibited the growth of bacteria *Bacillus megaterium*. Other fractions showed differences between activities. Fractions **4** and **6** inhibited the growth of bacteria *Escherichia coli* and *Staphylococcus aureus*. The widest spectrum of antimicrobial activity showed fraction 5. Results of antimicrobial activity are in table 1. The tested concentrations, based on the lyophilized weight, were about 0.5 mg/ml. If the activity was positively detected peptides were further characterized.

Table 1 Characterization of fractions, the letters in bold correspond to the database peptide.

Microorganism	Fraction					
	1	2	3	4	5	6
Escherichia coli				+		+
Pseudomonas aeruginosa					+	
Staphylococcus aureus				+	+	+
Bacillus megaterium	+	+	+		+	
Candida scotii					+	

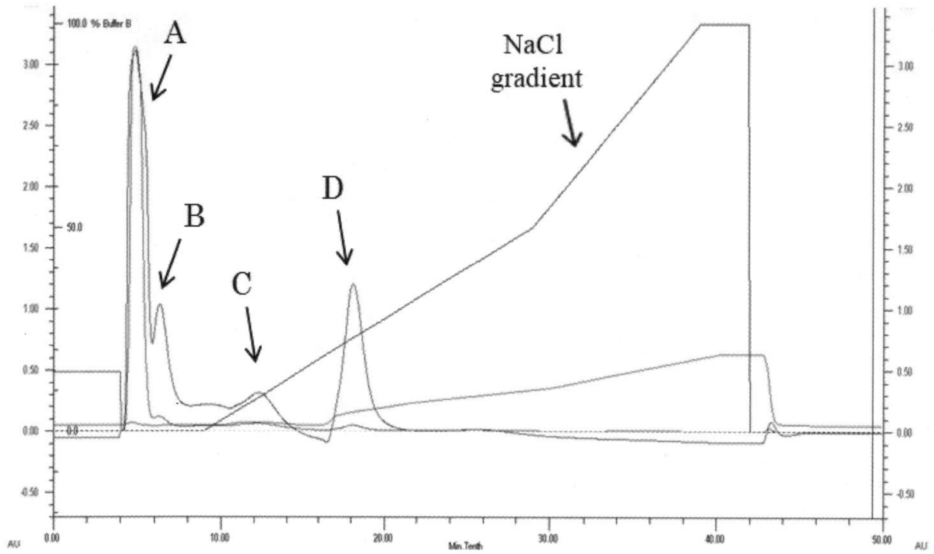

Fig. 2 FPLC of hydrophobic part after SPE separation.

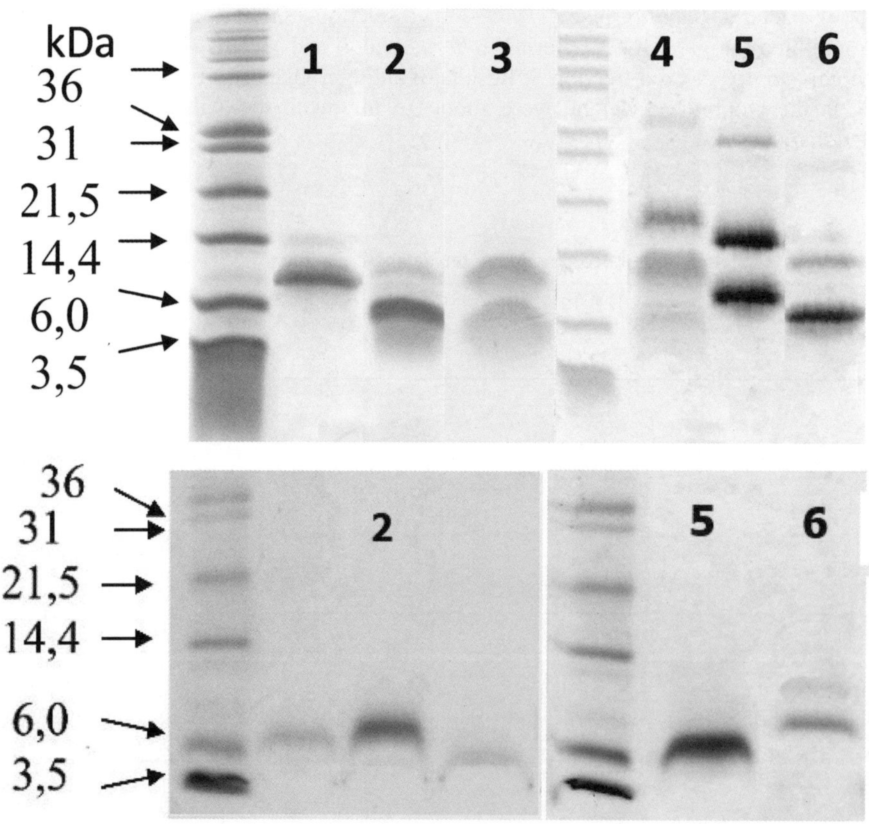

Fig. 3 Tricine electrophoresis of antimicrobially active fractions. Picture above shows these fractions separated only by RP-HPLC and picture down represents peptide fractions purified by both RP-HPLC and ionex FPLC. Fraction 2, and 5 were this way obtained homogenic.

Tricine chromatography showed presence of small peptides in all these active fractions (Fig. 3). Addition of weak katex FPLC before RP-HPLC obviously helped to separate the peptide to homogeneity. For example peak of the retention time of the fraction 2 was detected in 3 peaks on FPLC separation.

Mass spectroscopy was able to determine several molecular masses, but the method of fingerprinting using tryptic digest did not get any results. N-terminal sequencing exhibited structures present in mainly insect and amphibian antimicrobial peptides, but corresponding no more than in 40%.[5]

Table 2 Characterization of fractions.

Fraction	M_w-PAGE [kDa]	M_w-MALDI TOF [kDa]	Partial N-terminal sequence	Similarity[5]
1	8	7,600	KKKTRKRTLPKN	
2	5	6,276	SRDDXPVQQVPFXXP	
2b	12	9,621	AIAMAR	
3	8; 14	6,820	DAGTVAKHPS	35.71% - Temporin-1CSd
4a	12	6,825	DHYEXGXNXESX	
4b	16	9,761	DLSVGRLHPEYENK	36 %- Caerin 1.6 34.78 %- Ocellatin-5
5a	8	13,223	EDEWTVKNGEQIKQI	
5b	16	16,018	ANIQLSDEQKAKV	35.29 % - Lasioglossin LL-II
6	6	6.820	DAPASNPIDDLATI	

Acknowledgements This work was sponsored by the grants MSM 6046137305, GACR 305/09/H008, GACR 522/09/1693 and financial support from specific university research MSMT no. 21/2010.

References

[1] Neubauerová T., Macková M., Macek., Koutek B.: Kationické antimikrobiální peptidy, *Chem. Listy* 2009, 103, 460-468.

[2] Mcphee J.B., Hancock R.E.W.: Function and therapeutic potential of host defence peptides, *J. Pept. Sci.* 2005, *11,* 677-687

[3] Schägger H.: Tricine-SDS PAGE, *Nat. Protoc.* 2006, 1, 16-22.

[4] Ciencialová A. Neubauerová T., Šanda M., Šindelka R., Cvačka J., Vobůrka Z, Buděšínský M., Kašička V.,Sázelová P., Šolínová V., Macková M., Koutek B., Jiráček J.: Mapping the peptide and protein immune response in the larvae of the fleshfly *Sarcophaga bullata, J. Pept. Sci.* 2008, 14, 670–682.

[5] Wang G., Li X. and Wang Z.: APD2: the updated antimicrobial peptide database and its application in peptide design. *Nucleic Acids Res.* 2009, 37, 933-937.

Structural and functional diversity of natural antimicrobial oligopeptides

Aexander A. Zamyatnin[*,1,2]

[1] Computer Biochemistry Group, A. N. Bach Institute of Biochemistry, Russian Academy of Sciences, Leninsky prosp. 33, Moscow 119071, Russian Federation
[2] El Centro Cientifico Tecnologico de Valparaiso, Departamento de Informática, Universidad Técnica Federico Santa Maria, Avda. España 1680, Valparaiso, Chile

*Corresponding author: e-mail: aaz@inbi.ras.ru or alexander.zamyarnin@usm.cl, Phone: +7 4959543066 or +56 322652986

Antimicrobial peptides have been isolated and characterized from tissues and organisms representing virtually every kingdom and phylum, ranging from prokaryotes to humans. These molecules are small, relatively compact compounds, which are structurally and functionally rather diverse. This consideration is focussed on the structural and functional properties of the natural antimicrobial oligopeptides that are ribisomally synthesised without extensive posttranslational modifications. They are cationic molecules as the molecules are usually rich in lysine, and arginine. They are also amphipathic, containing hydrophobic and hydrophilic region. Natural antimicrobial oligopeptides have a broad spectrum of antimicrobial activity. They serve as an ancient defence mechanism against pathogenic microorganisms that easily come into contact with the host through the environment. These molecules are considered as part of the innate immune system of all species. Lastly, the review discusses the potential applications in health care products.

Keywords antimicrobial oligopeptide; structure; function; location; fragment; fragmentomics, fragmentome; EROP-Moscow database

1. Introduction

Regulatory oligopeptides generally do not exceed ~50 amino acid residues [1, 2]. For more than a century, natural oligopeptides have attracted scientific attention as biochemical regulators. The very first such oligopeptide, carnosine (β-Ala-His), was discovered by Gulevitch and Amiradzhibi in 1900 in Russia [3]. Since that time, thousands oligopeptide regulators have been described, and now almost 600 new natural oligopeptides emerge annually, out of a literature of more than 20,000 publications each year.

It is known that natural oligopeptides are involved in practically all regulatory processes [4, 5]. Chemical structures are now known for over 10000 oligopeptides found in more than 2000 living organisms representing all biological kingdoms [5]. The spectrum of their functional activity is very broad. Among them more than 2200 natural antimicrobial (antiviral, antifungal, antibacterial, antitumour, antiparasitic, and hemolytic) oligopeptides were described.

The term 'antimicrobial peptides' refers to a large number of peptides first characterized on the basis of their antibiotic and antifungal activities. In addition to their role as endogenous antibiotics, antimicrobial peptides, also called host defence peptides, participate in multiple aspects of immunity (inflammation, wound repair, and regulation of the adaptive immune system) as well as in maintaining homeostasis. The possibility of utilizing these multifunctional molecules to effectively combat the ever-growing group of antibiotic-resistant pathogens has intensified research aimed at improving their antibiotic activity and therapeutic potential, without the burden of an exacerbated inflammatory response, but conserving their immunomodulatory potential [6].

Since oligopeptides play an important role in various types of regulation (e.g. innate immunity [7]), they are of great interest for biotechnology [8], and obtaining new information about their structures and functions is therefore important. Keeping this in mind, we carried out the structural–functional consideration of diversity of all known antimicrobial oligopeptides.

2. Structure

Most natural oligopeptides formed by ribosomal synthesis are fragments formed in an organism from specialized precursors due to splitting off of the signal peptide and pre-peptide(s). This process is specific to large protein structures and small oligopeptide regulators [9]. Primary structure of the first such natural antimicrobial oligopeptides circulin A and circulin B discovered in tropical tree *Chassalia parviflora* were elucidated in 1965 [10] and 1968 [11]. They were cyclic oligopeptides, consisted of 30 and 31 amino acid residues respectively, and demonstrated high homology:

circulin A	**G IPCGESCV**W**IPCIS**AA**LGCSCKNKVCYRN**
circulin B	**G**V**IPCGESCV**F**IPCIS**TL**LGCSCKNKVCYRN**

These oligopeptides had antibiotic activity, inhibited the cytopathic effects and replication of the human immunodeficiency virus, and was active against both Gram-positive and Gram-negative bacteria [12].

At the same time, more and more facts appear showing that there are also oligopeptides in different organs and tissues of living organisms that are not formed from specialized precursors, but are rather natural fragments of well studied proteins [13-16]. Functional role in an organism is not established for most. Studying functions of some of such fragments and of those obtained by experimental proteolysis has shown that they can serve as oligopeptide regulators. Thus, a number of fragments of milk proteins (casein [17-21], α-lactalbumin, β-lactoglobulin, and lactoferrin [22, 23]) as well of cytochrome *c* [24, 25] and hemoglobin [26-29] are able to exhibit opioid activity, i.e. they are neuropeptides. It is also known that some lactoferrin [30] and hemoglobin [31-33] fragments also exhibit antimicrobial activity.

A lot of natural antimicrobial oligopeptides were discovered in all kingdoms (viruses, fungi, bacteria, plants, and animals). The data of their primary structures, functional properties, and other characteristics were inputted into EROP-Moscow (**E**ndogenous **R**egulatory **O**ligo**P**eptide) database (http://erop.inbi.ras.ru [34]) which permitted to present general consideration of different structural and functional groups of natural oligopeptides including antimicrobial oligopeptides.

The number of decoded amino acid sequences of antimicrobial oligopeptides is rapidly increasing. As seen from Fig. 1, to date the rate of decoding is now more than 800 sequences per five years. They are cationic molecules as the molecules are rich in histidine, lysine, and arginine. E.g., antimicrobial oligopeptide misgurin obtained from the loach *Misgurnus anguillicaduatus* and consisted of 21 amino acid residues contains five arginyl and four lysyl residues [35]. Antimicrobial oligopeptides are also amphipathic, containing both hydrophobic and hydrophilic regions.

Based on the structural size and conformational structure, they are classified into four categories [36]: (1) cysteine-rich amphiphilic β-sheet peptides; (2) amphiphilic α-helical peptides; (3) cysteine-disulfide ring peptides; and (4) linear peptides with one or two predominant amino acids.

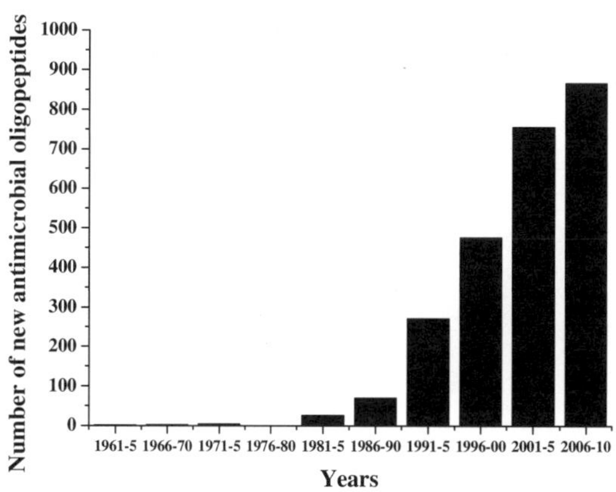

Fig. 1 Increase in the number of decoded amino acid sequences of natural antimicrobial oligopeptides from year 1961. It is seen that at present almost 200 primary structures a year are decoded (the data of 2010 are not complete).

The biodiversity of all known potential natural antimicrobial oligopeptides can be seen from Fig. 2. They were found in viruses (2), fungi (42), bacteria (130), plants (317), and animals (1699).

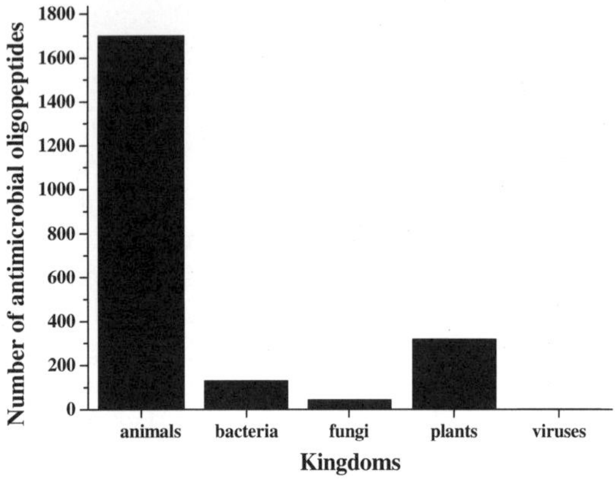

Fig. 2 Natural antimicrobial oligopeptides found in species of different kingdoms.

Antimicrobial properties were found for the most small oligopeptide molecules consisting of two amino acid residues only. It can be seen in Fig. 3. Their primary structure can exceed 50 residues but such oligopeptides are out of range of EROP-Moscow,

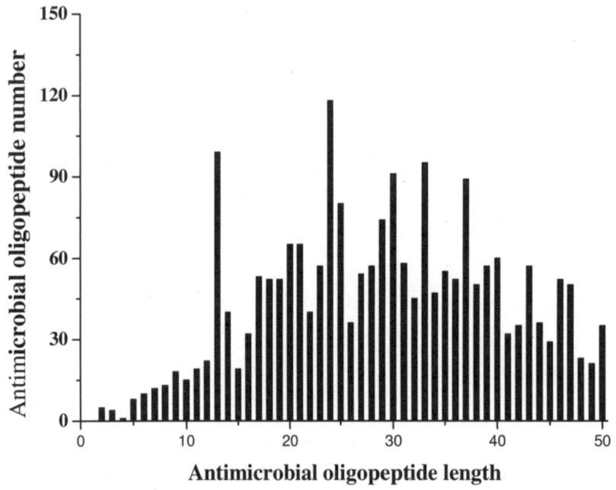

Fig. 3 Distribution of known natural antimicrobial oligopeptides by amino acid residue number.

There is a great diversity of their location. E.g., according to EROP-Moscow data different human antimicrobial oligopeptides were found in salivary gland and saliva, gastrointestinal tract (small intestine, ileum, colon), parotid gland, sweat gland, skin (epidermis, keratinocytes), tracheal epithelium, lung epithelial cells, bone marrow, skeletal muscles, liver, kidney, male reproductive system (epididymis, testis), female reproductive system (vagina, placenta), blood (erythrocytes, neutrophils, granulocytes, lymphocytes), airway secretion, urea, etc. It was found almost 50 different oligopeptide structures in human blood and about 30 fragments of α- and β-chains of hemoglobin were among them.

3. Functions

Natural antimicrobial oligopeptides have a broad spectrum of antimicrobial activity. They serve as an ancient defence mechanism against pathogenic microorganisms that easily come into contact with the host through the environment. These molecules are considered as part of the innate immune system of all species. Multiple antimicrobial peptides are simultaneously present at the same site, and they are thought to work in concert, to effectively fight infection.

A large body of information about different types of antimicrobial activity does not permit to consider all these types in detail. EROP-Moscow contains information on more than 2200 potential antimicrobial oligopeptides. Only half of them was obtained experimentally and their functions were studied in different test systems (Table 1). It has been shown in numerous publications that they could possess antimicrobial (895), antifungal (353), antiviral (78), antitumour (14), antiparasitic (25) or hemolytic (196) activity (Fig. 4). Most of theses oligopeptides exhibit more than one type of antimicrobial activity. E.g., four brevinins 1Vla, 1VLb, 1VLc, and 1VLd, obtained from skin secretion of Central American frog *Lithobates vaillanti*, were active against Gram-negative bacterium *Eschericha coli* (MIC=25 mM), Gram-positive bacterium *Staphylococcus aureus* (MIC =3 mM), fungus *Candida albicans* (MIC =3 mM), and possessed cytolytic activity (LC50=4-8 mM) against human erythrocytes and hepatoma-derived cells [37].

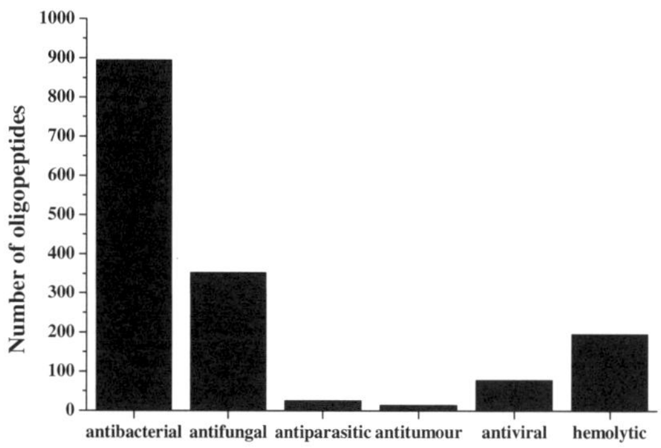

Fig. 4 Natural antimicrobial oligopeptides possessing different types of functional activity.

Primary structures of the other half were elucidated by translation from nucleotide sequences only and were not tested. The belonging to antimicrobial substances was postulated by the similarity of their primary structures with functionally characterized antimicrobial oligopeptides.

Table 1 Examples of different kinds of antimicrobial activity of natural antimicrobial oligopeptides (analysis of EROP-Moscow database data).

ANTIBACTERIAL	MIC, μM	ANTIFUNGAL	MIC, μM
Gram-negative bacteria		*Aspergillus niger*	2.9 ÷ 90
Escherichia coli	1.8 ÷ 2500	*Candida albicans*	3 ÷ 3000
Pseudomonos aeruginosa	6.8 ÷ 10	*Pichia pastoris*	16.5
Citobacter freundii	75	*Schizosaccharomyces pombe*	11
Gram-positive bacteria		*Trichodema viride*	1.6
Bacillus subtilis	0.8 ÷ 40	*Fisarium oxysporum*	25
Staphylococcus aureus Kal B1	0.2 ÷ 70	**ANTIVIRAL**	
Micrococcus luteis	1.3 ÷ 34.4	HIV	1.5 ÷ 10
Listeria monocytogenes	34.4 ÷ 90.9	**ANTITUMOUR**	**LC50, μM**
Bacillus subtilis	0.8 ÷ 40	Hepatoma cells	30 ÷ 50
Sarcina lutea	1.9	**HEMOLYTIC**	
Lactococcus lactis	25	Human blood erythrocytes	4 ÷ 200
Streptococcus uberis	50	**ANTIPARASITIC**	
Leuconostic lactis	32.4 ÷ 72.9	Trypanosoma cruzi	6

4. Potential applications

Antimicrobial peptides, being cationic and amphiphilic in nature, have been able to penetrate bacterial wall and defence membrane, causing death of bacterial cells. This mechanism of killing is different from that of the classical antibiotics. With the development of antibiotic resistance by some bacteria, there is an urgent need for the development of new therapeutic drugs and antimicrobial peptides can be the answer to the need for new antibiotics. Many antimicrobial peptides have shown potential for development into therapeutic drugs and topical products because of their broad range of antimicrobial activities. Up to now, there have been some animal derived antimicrobial peptides, such as protegrin [38], histatin [39], human lactorferricin [40], which have undergone pre-clinical or phase III trials for commercial developments as topical products. They are used for antibacterial, antifungal, antiviral, and to combat other infections caused by open wounds [41].

Enrichment of food products with preliminary prepared fragments of food (and, possibly, inedible) proteins is an another example of such a practical application. Protein fragments are safe, nontoxic, and free from side effects. Elaboration of the technology of such an enrichment and its manufacturing application should result in enhancement of the immune status of the consumer. In addition, fragments with antimicrobial effects can be considered as natural preservatives of food products. Their addition will increase the shelf life of these products without any dangerous side effects due to their natural origin.

Active natural fragments of proteins are promising for hygiene and cosmetics. For example, their use as components of hygienic and cosmetic preparations could result in the creation of effective toothpastes and creams of a new generation [13].

Acknowledgements This work was supported by the Program of Fundamental Investigations of the Presidium of the Russian Academy of Sciences and by the Chilean National Research Foundation FONDECYT (grant No. 1080504).

References

[1] Zamyatnin AA. Amino acid, peptide, and proten volume in solution. *Annual Review of Biophysics and Bioengineering*. 1984;13:145-165.

[2] Zamyatnin AA. EROP-Moscow specialized data bank for endogenous regulatory oligopeptide. *Protein Sequence and Data Analysis*. 1991;4:49-52.

[3] Gulewitch VS, Amiradzhibi S. Ueber das Carnosin, eine neue organische Base des Fleischextraktes. *Berichte der Deutsche Chemische Gesellschaft*. 1900;33:1902-1903.

[4] Sewald N, Jakubkem HD. *Peptides: Chemistry and Biology*. Weinheim, Wiley-VCH Verlag GmbHp; 2002.

[5] Zamyatnin AA. Structural and functional diversity of endogenous regularory oligopeptides. In: Rolka K, Rekowsky P, Silberring J, eds. *Peptides 2006*. Geneva, Kenes International; 2007:384-385.

[6] Auvynet C, Rosenstein Y. Multifunctional host defense peptides: antimicrobial peptides, the small yet big players in innate and adaptive immunity. *FEBS Journal*. 2009;276:6497-6508.

[7] Ganz T. Defensins: antimicrobial peptides of innate immunity. *Nature Reviews. Immunology*. 2003;3:710-720.

[8] Tavares LS, Santos MO, Viccini LF, Moreira JS, Miller RNG, Franco OL. Biotechnological potential of antimicrobial peptides from flowers. *Peptides*. 2008;29:1842-1851.

[9] Zamyatnin AA. Biochemical problems of regulation by oligopeptides. *Biochemistry (Moscow)*. 2004;69:1565-1573.

[10] Fujikawa K, Suketa Y, Hayashi K, Suzuki T. Chemical structure of circulin A. *Experientia*. 1965;21:307-308.

[11] Hayashi K, Suketa Y, Suzuki T. Chemical structure of circulin B. *Experientia*. 1968;24:656-657.

[12] Gustafson KR, Sowder RC II, Henderson LE, Parsons IC, Kashman Y, Cardellina JH II, MacMahon JB, Buckheit RW, Pannel LK, Boyd MR. Circulins A and B: Novel HIV-inhibitory macrocyclic peptides from thr tropical tree Chassalia parvifolia. *Journal of the American Chemical Society*. 1994;116:9337-9338.

[13] Zamyatnin AA, Borchikov AS. Enzymatic fragments as possible regulatory oligopeptides. *Neurochemical Journal*. 2007;1:188-195.

[14] Zamyatnin AA. Fragmentomocs of proteins and natural oligopeptides. *Biophysics*. 2008;53:329-335.

[15] Zamyatnin AA. Hemoglobin as a potential source of natural regulatory oligopeptides. *Biochemistry (Moscow)*. 2009;74:201-208.

[16] Zamyatnin AA. Fragmentomics of natural peptide structures. *Biochemistry (Moscow)*. 2009;74:1575-1585.

[17] Brantl V, Teschemacher H, Henschen A, Lottspeich F. Novel opioid peptides derived from casein (beta-casomorphins). I. Isolation from bovine casein peptone. *Hoppe-Seyler's Zeitschrift für Physiologische Chemie*. 1979;360:1211-1216.

[18] Henschen A, Lottspeich F, Brantl V, Teschemacher H. Novel opioid peptides derived from casein (beta-casomorphins). II. Structure of active components from bovine casein peptone. *Hoppe-Seyler's Zeitschrift für Physiologische Chemie*. 1979;360:1217-1224.

[19] Brantl V, Teschemacher H. Opioids in milk *Trends Pharmacological Sciences*. 1983;4:193-193.

[20] Brantl V. Novel opioid peptides derived from human beta-casein: human beta-casomorphins. *European. Journal of Pharmacology*. 1984;106:213-214.

[21] Koch G, Wiedemann K, Teschemacher H. Opioid activities of human beta-casomorphins. *Naunyn-Schmiedeberg's Archives of Pharmacology*. 1985;331:351-354.

[22] Teschemacher H, Koch G, Brantl V. Milk protein-derived opioid receptor ligands. *Biopolymers. Peptide Science*. 1997;43; 99-117.

[23] Meisel H. Biochemical properties of regulatory peptides derived from milk proteins *Biopolymers. Peptide Science*. 1997;43: 119-128.

[24] Brantl V, Gramsch C, Lottspeich F, Henschen A, Jaeger KH, Herz A. Novel opioid peptides derived from mitochondrial cytochrome b: cytochrophins. *European Journal of Pharmacology*. 1985;111:293-294.

[25] Zadina JE, Kastin AJ, Ge LJ, Brantl V. Hemorphins, cytochrophins, and human beta-casomorphins bind to antiopiate (TYR-MIE-1) as well as opiate binding sites in rat brain. *Life Sciences*. 1990;47:PL25-30.

[26] Brantl V, Gramsch C, Lottspeich F, Mertz R, Jaeger KH, Herz A. Novel opioid peptides derived from hemoglobin: hemorphins *European Journal of. Pharmacology*.1986;125:309-310.

[27] Zhao Q, Garreau I, Sannier F, Piot JM. Opioid peptides derived from hemoglobin: hemorphins. *Biopolymers. Peptide Science*. 1997;43:75-98.

[28] Nyberg F, Sanderson K, Glamsta EL. The hemorphins: a new class of opioid peptides derived from the blood protein hemoglobin *Biopolymers. Peptide Science*. 1997;43:147-156.

[29] Ivanov VT, Karelin AA, Philippova MM, Nazimov IV, Pletnev VZ. Hemoglobin as a source of endogenous bioactive peptides: the concept of tissue-specific peptide pool *Biopolymers. Peptide Science*. 1997;43:171-188.

[30] Ueta E, Tanida T, Osaki T. A novel bovine lactoferrin peptide, FKCRRWQWRM, suppresses *Candida* cell growth and activates neutrophils. *Biopolymers. Peptide Science*. 1997;43:240-249.

[31] Froidevaux R, Krier F, Nedjar-Arroume N, Vercaigne-Marko D, Kosciar E, Ruckebusc C, Dhulster P, Guillochon D. Antibacterial activity of a pepsin-derived bovine hemoglobin fragment. *FEBS Letters*.2001;491:159-163.

[32] Fogaca AC, da Silva PI Jr, Miranda MT, Bianchi AG, Miranda A, Ribolla PE, Daffre S. Antimicrobial activity of a bovine hemoglobin fragment in the tick Boophilus microplus. *Journal of Biological. Chemistry*. 1999;274:25330-25334.

[33] Ivanov VT, Karelin AA, Mikhaleva II, Vas'kovskii BV, Sviryaev VI, Nazimov IV. Isolation, structure and properties of novel endogenous peptides *Bioorganicheskaya Khimia.*1992;18:1271-1310.
[34] Zamyatnin AA, Borchkov AS, Vladimirov MG, Voronina OL. The EROP-Moscow oligopeptide database. *Nucleic Acids Research.* 2006;34:D261-D266.
[35] Park CB, Lee JH, Park IY, Kim MS, Kim SC. A novel antimicrobial peptide from the loach, *Misgurnus anguillicaudatus. FEBS Letters*, 1997;411:173-178.
[36] Martin E, Ganz T, Lehrer RI. Defensins and other endogenous peptide antibiotics of vertebrates. *Journal of Leukocyte Biology.* 1995;58:128-136.
[37] Conlon JM, Raza H, Coquet L, Jouenne T, Leprince J, Vaudry H, King JD. Purification of peptides with differential cytolytic activities from the skin secretions of the Central American frog, Lithobates vaillanti (Ranidae). *Comparative Biochemistry and Physiology*, 2009;150C:150-154.
[38] Kokryakov VN, Harwig SSL, Panyutich EA, Shevchenko AA, Aleshina GM, Shamova OV, Korneva HA, Lehrer RI. Protegrins: Leucocyte antimicrobial peptides that combine features of corticostatic defensins and tachyplesins. *FEBS Letters*, 1993;327:231-236.
[39] Oppenheim FG, Yang Y-C, Diamond RD, Hyslop D, Offner GD, Troxler RF. The primary structure and functional characterization of the neutral histidine-rich polypeptide from human parotid secretion. *Journal of Biological Chemistry*, 1986;261:1177-1182.
[40] Odell EW, Sarra R, Foxworthy M, Chapple DS, Evans RW. Antibacterial activity of peptides homologous to a loop region in human lactoferrin. *FEBS Letters.* 1996;382:175-178.
[41] Sugarito H, Yu PL. Avian antimicrobial peptides: the defense role of β-defensins. *Biochemical and Biophysical Research Communications.* 2004;323:721-727.

The role of Gram-negative envelope LPS on the bactericidal properties of proteins and peptides: the case of eosinophil cationic protein

D. Pulido*, M. Torrent, M.V. Nogués and E. Boix

Department of Biochemistry and Molecular Biology, Universitat Autònoma de Barcelona, 08193 Cerdanyola del Vallès, Spain.

*Corresponding author: e-mail: david.pulido@campus.uab.cat, Phone: +34 935811707

The human Eosinophil Cationic Protein (ECP) is an antimicrobial protein involved in host defense. The protein presents a high affinity to lipopolysaccharides (LPS) and can specifically agglutinate Gram-negative bacteria cells. We have identified the ECP antimicrobial domain that can reproduce most of the protein activity, and partly retain the cell aggregation properties. With the aim to reduce the antibacterial predicted region we have synthesized a peptide only containing the two α-helices of the antimicrobial domain linked by and aminohexanoic acid. To better understand the ECP antimicrobial mechanism of action and evaluate the contribution of the LPS components, we have assayed a battery of *E. coli* K12 strain variants that present distinct chemotypes (*Ra* to *Re* chemotypes), which are defective in some of the genes involved in the LPS synthesis and display progressively truncated LPS at their outer membrane. The data correlates the LPS structure with the bacteria viability and reveals that both the length and charge of the LPS alters the bacteria cell susceptibility to the protein cytotoxicity.

Keywords: Eosinophil cationic protein; *E. coli* K12; Lipopolisaccharide

1. General remarks

The Gram-negative cell structure contains two main layers: an outer layer, which is an asymmetric membrane mainly composed of lipopolysaccharides (LPS) and an inner membrane, constituted mostly by phospholipids. The periplasmic space contains an extra layer of peptidoglycan (PG) much thinner than those found in Gram-positive bacteria. The LPS molecule has a lipid backbone, which is extended with a carbohydrate chain as it is shown in Fig. 1.

It has been shown that Gram-positive bacteria are much more susceptible to specific antibiotics due to the absence of the LPS outer layer [1, 2]. Thus, the LPS layer seems to be crucial for the reduced permeability of Gram-negative bacteria. Several studies pointed out different hypothesis to explain the role of LPS on cell permeability. Hancock and coworkers suggested that LPS is structured by intermolecular cross bridges mediated by divalent cations; [3] an hypothesis that could explain the enhanced uptake of ß-lactams in *E. coli* in the presence of a chelator agent such as EDTA [4]. In contrast, Parker and coworkers proposed another hypothesis where LPS molecules could be tightly linked by hydrogen bonding between sugar groups without the assistance of divalent cations [5]. It has also been suggested that the LPS carbohydrate chain could act as a hydrophilic barrier against the internalization of hydrophobic molecules [6]. Several studies have shown that the outer membrane permeability is well correlated with the thick of the LPS saccharide moiety thus supporting the aforementioned hypothesis [7].

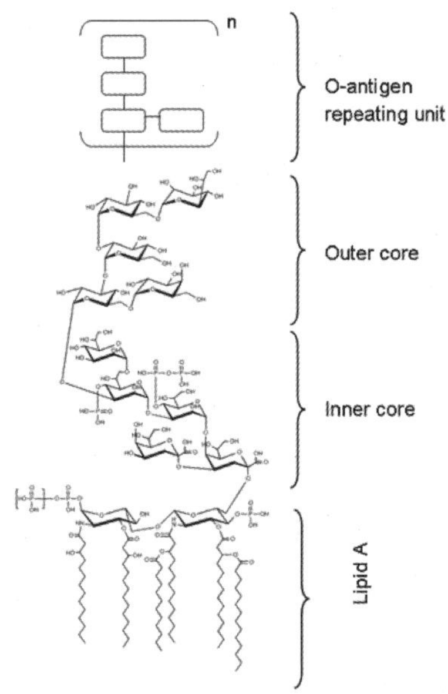

Fig. 1. *Lipopolysaccharide structure.* The core and Lipid A structures are taken from a LPS portion of *E. coli* K-12. Reproduced from [14] with permission.

In contrast with what one would expect, some antimicrobial cationic proteins and peptides such as lysozyme or lactoferrin are shown to be more active in strains displaying longer LPS core molecules. This is also the case of the eosinophil cationic protein (ECP). ECP is a cationic protein released by eosinophils during the inflammation process. ECP presents a high antimicrobial activity (below 1μM) against both Gram-negative and Gram-positive bacteria and can agglutinate

E. coli cells before the cell membrane permeation occurs. This behaviour has also been observed for model membranes, i.e. ECP promotes the vesicle aggregation before its leakage. The ECP agglutinating activity in *E. coli* cells has suggested being dependent on the high protein binding affinity for LPS [8, 9, and 10]. In subsequent studies, it has been described that the ECP N-terminus can almost reproduce the antimicrobial properties of the whole protein, including the LPS binding affinity [11]. Further investigation to refine the N-terminus antimicrobial domain has lead to the synthesis of a shorter peptide with two α-helical key segments linked by an aminohexanoic acid (Fig. 2) that represents the minimum template that can retain the antimicrobial properties of ECP.

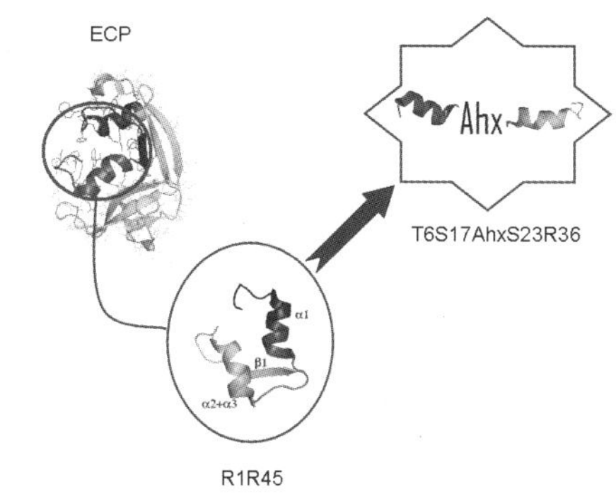

Fig. 2. *Pharmacophore refinement of ECP.* The figure displays the eosinophil cationic protein and its N-terminus derived peptides: R1R45 and T6S17AhxS23R36.

In order to evaluate the contribution of LPS on the bactericidal properties of ECP and their derived peptides, a wide battery of *E. coli* K12 strain variants has been selected. These mutant strains can present two different phenotypes (rough and deep rough), which are defective in some genes involved in the LPS synthesis and display progressively truncated LPS molecules at their outer membrane [12]. These two phenotypes include different chemotypes. Fig. 3 illustrates the tested strain variants: the Ra chempotype (D21 strain) that possesses the entire LPS core except the O-antigen, the Rc chemotype (D21e7 and D21e19) that does not display any sugar moiety after the first glucose residue and the Rd chemotype (D21f1 and D21f2) that only contains the KDO and heptoses residues [13].

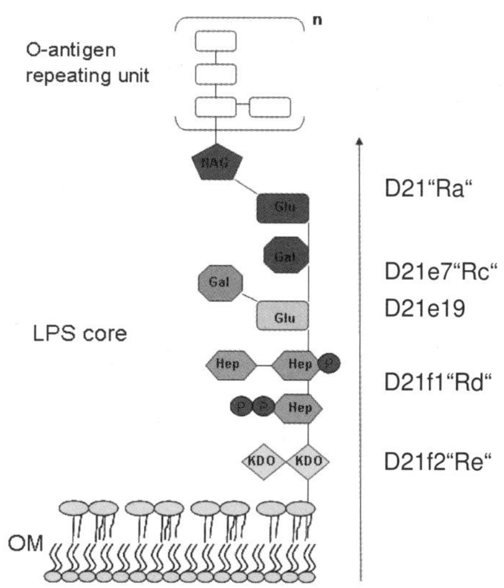

Fig. 3. *LPS structure of the E. coli mutant strains.* Representation of the lipopolisaccharide core and O-antigen of *E. coli* and the different strains from D21 to D21f2. Hep is heptose, Glu is glucose, Gal is galactose and KDO is 3-deoxy-d-manno-octolunosonic acid.

2. Results and discussion

In previous studies we determined the antimicrobial properties of ECP using both model membranes and bacteria cell cultures [8, 9, and 10]. We also identified the first 45 N-terminal residues of the protein as the antimicrobial domain of ECP. This domain almost retains all the antimicrobial properties of the whole protein [11]. In order to evaluate how the LPS of the cell envelope of Gram-negative bacteria could affect on the antimicrobial properties of the protein and their derived peptides we have assessed the protein and peptide activities on the D21 to D21f2 *E. coli* K12 mutants by using several methodologies.

2.1 Antimicrobial activity

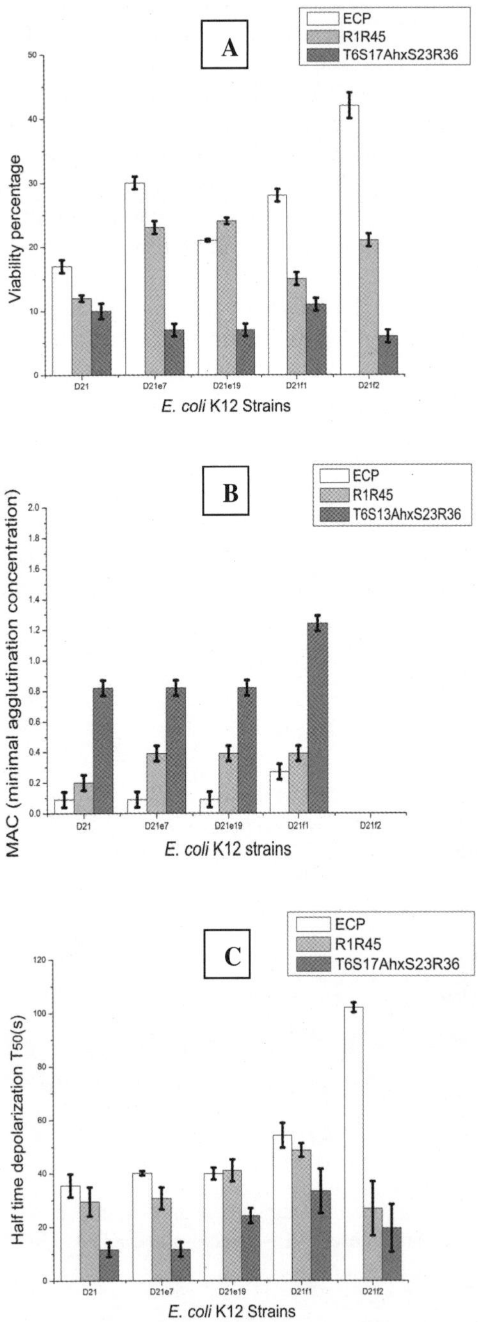

Fig. 4. *Antibacterial assays* A) The viability percentage was determined after incubation of mid-log phase grown *E.coli* K12 cultures with 5 μM of protein and peptides. The percentage of viability after incubation with ECP, R1R45 and T6S17AhxS23R36 is shown on the bar graph. B) The agglutination activity was evaluated by the Minimal Agglutination Concentration (MAC). The results for ECP, R1R45 and T6S17AhxS23R36 are shown on the bars graph. D21f2 strain does not present agglutination below concentration of 10 μM. C) Membrane depolarization activity on *E. coli* K12 cells was followed using the $DISC_3(5)$ assay using a 5 μM protein concentration. The time needed to achieve a 50% depolarizing effect is depicted for ECP, R1R45 and T6S17AhxS23R36.

At a first approach to better understand the role of the LPS outer layer structure on the ECP antimicrobial activity and their derived peptides we have evaluated the antimicrobial activity using the Live/Dead fluorescent viability assay. The cell viability percentage after the treatment with 5μM ECP was higher in those strains with a shorter LPS structure (42±2 % for the D21f2 strain), while strains carrying longer LPS cores showed a reduced viability (17±1 % for the D21 strain). Our data suggest that ECP needs to bind with LPS in order to promote the cell death. On the other hand, the ECP-derived peptides R1R45 and T6S17AhxS23R36 do not display significant differences in the viability assay for the different strains tested, suggesting that these peptides could present a different mechanism of action, apparently unrelated with the LPS core length.

2.2 Minimal agglutination activity (MAC)

The minimal agglutination activity, calculated as the minimal peptide concentration that can induce agglutination in bacteria, has also been tested for ECP and ECP-derived peptides. The protein shows a high affinity for LPS and can agglutinate *E. coli* K12 cells below 0.1 μM (Fig. 4B). On the other hand, R1R45 and T6S17AhxS23R36 peptides need a higher concentration in order to promote bacteria cell agglutination, perhaps due to a lower affinity for LPS. Thus we conclude that the LPS core seems to be crucial for the protein interaction and cell agglutination properties. The hypothesis is supported by the results obtained on the D21f2 strain, which lacks the LPS core as neither ECP nor ECP-derived peptides can induce agglutination at concentrations below 10 μM.

2.3 Depolarization activity

Two main steps characterize the ECP antimicrobial mechanism of action; first, the protein triggers the bacteria cell agglutination and then, promotes the cell permeation [9].

It has been shown that ECP can promote the cell membrane depolarization of both Gram-positive and Gram-negative bacteria. Here, we have analyzed the depolarization activity of ECP and the ECP-derived peptides. As it can be seen in Fig. 4C, the ECP depolarization activity is slowed down as a function of the LPS length core. On the other hand, R1R45 and T6S17AhxS23R36 peptides show a similar half-depolarization time for all the strains tested except for the D21f2 strain, which shows a lower half-depolarization time. One plausible explanation for the higher susceptibility of the D21f2 truncated strain is that it is unable to correctly assemble the outer membrane porins. In order to fill the corresponding gaps at the outer membrane the strain could display some phospholipid patches on the LPS outer layer that could, increase the cell permeability to antimicrobial agents [12, 13]. These data also suggest a different mechanism of action for the ECP-derived peptides in comparison with the previously described process for protein [9].

2.4 Scanning electron microscopy (SEM)

With the aim to study the cell population behavior, the cell morphology and the bacteria cell surface details of ECP treated cultures were visualized by scanning electron microscopy (Fig. 5). For all *E. coli* K12 mutants, cell damage and agglutination were observed after incubation with 10μM of ECP. In contrast, the characteristic baton-shaped cell morphology was maintained. Moreover, we can highlight that the D21f2 aggregates were rare and significantly less crowded than the aggregates observed for the D21 strain.

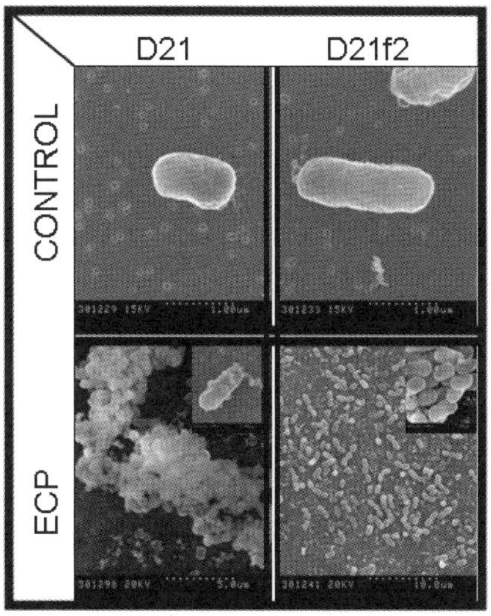

Fig. 5. Scanning electron micrographs of *E. coli* K12 D21 and D21f2 strains incubated with 10μM ECP during 4h. The figure shows the cell population agglutination and the surface cell wall damage.

Acknowledgements D.P. is the recipient of a predoctoral grant from Universitat Autònoma de Barcelona (Spain).The work was supported by the Ministerio de Ciencia e Innovación [grant BFU2009-09371].

References

[1] Nikaido, H. Prevention of drug access to bacterial targets: permeability barriers and active efflux. *Science;* 1994 264, 382-388.
[2] Zimmermann, W, and Rosselet, A. *Antimicrob. Agents Chemoter.* 1997; 12, 368-372.
[3] Hancock, R. E. W. Alterations in outer membrane permeability. *Annu. Rev. Microbiol.* 1984; 38, 237-264.
[4] Stan-Lotter, H, Gupta, M, and Sanderson, K. E, The influence of cations on the permeability of the outer membrane of Salmonella typhimurium and other gram-negative bacteria. *Can. J. Microbiol.* 1979; 73, 2615-2629.
[5] Parker, C. T, Kloser, A. W, Schnaitman, C. A, Stein, M. A, Gottesman, S, and Gibson, B.W. *J. Bacter.* 1992; 174, 2525-2538.
[6] Nikaido, H, and Vaara, Molecular basis of bacterial outer membrane permeability. M. *Microbiol. Rev.* 1985; 49, 1-32.
[7] Hiruma, R, Yamaguchi, A, and Sawai, T. The effect of lipopolysaccharide on lipid bilayer permeability of beta-lactam antibiotics. *FEBS Lett.* 1984; 170, 268-272.
[8] Torrent, M, Sánchez, D, Buzón, V, Nogués, M.V, Cladera, J, and Boix. E, *BBA Biomembranes.* 2009; 1788, 1116-1125.
[9] Torrent, M, Badia, M, M. Moussaoui, M.V. Nogues, and E . Comparison of the membrane interaction mechanism of two antimicrobial RNases: RNase 3/ECP and RNase 7. *FEBS J.* 2010; 277, 1713.
[10] Torrent, M, S. Navarro, M. Moussaoui, M.V. Nogues, and E. Boix. Eosinophil cationic protein high-affinity binding to bacteria-wall lipopolysaccharides and peptidoglycans. *Biochemistry.* 2008; 47, 3544-3555.
[11] Torrent M., Torre B., Nogués V., Andreu D., Boix E. Bactericidal and membrane disruption activities of the eosinophil cationic protein are largely retained in an N-terminal fragment. *Biochem J.* 2009; 421, 425-34.
[12] Snyder. D.S., Mc Intosh. T.J., The lipopolysaccharide barrier: correlation of antibiotic susceptibility with antibiotic permeability and fluorescent probe binding kinetics. *Biochemistry.* 2000; 38,11777-87.
[13] Roantree, R. J., Kuo, T., and MacPhee, D. G. The effect of defined lipopolysaccharide core defects upon antibiotic resistances of Salmonella typhimurium. *J. Gen. Microbiol.*1977; 103, 223-234.
[14] Boix E, Nogués V. Mammalian antimicrobial proteins and peptides: overview on the RNase A superfamily members involved in innate host defence. *Mol. BioSyst.* 2007; 3:317-35.

Non-antibiotics Biocides

Evaluation of biocidal activity of Evolyse, a disinfectant based on hydrogen peroxide and silver nitrate

M. Barbara Pisano[*,1], **V. Altana**[2], **M. Elisabetta Fadda**[1], **L. Mura**[2], **M. Deplano**[1] and **S. Cosentino**[1]

[1] Department of Experimental Biology, Section of Hygiene, Cittadella Universitaria, 09042 Monserrato, Cagliari, Italy
[2] ALL Side, Via Forlanini 11, 07100 Sassari, Italy

* Corresponding author: e-mail: barbara.pisano@unica.it, Phone: +39 0706754145

The *in vitro* biocidal activity of a novel disinfectant, Evolyse, against a wide spectrum of microorganisms, including bacteria, spores and fungi was analysed. The antimicrobial efficacy of disinfectant delivered by a nebulising device was also evaluated under practical conditions, in different hospital environments by microbiological monitoring of air and surfaces, before and after disinfection treatment. The disinfectant exhibited bactericidal activity against all bacterial strains tested after 5 min contact at room temperature. Sporicidal activity against *B. cereus* ATCC 11178 spores was obtained after 45 min contact at room temperature. The disinfectant also showed fungicidal activity against *A. flavus* ATCC 46283 and an *A. niger* DBS strain after 15 min contact at room temperature and against *C. albicans* ATCC 10231, *S. cerevisiae* ATCC 2601 and *P. chrysogenum* ATCC 9179 after 30 min contact time. Overall, disinfection treatment with Evolyse was able to reduce up to 70% of the microbial growth of air and surfaces of the different hospital environments analysed. The results of this study showed the good biocidal activity of Evolyse and offer promising perspective for the use of this product in hospital environments.

Keywords Evolyse; antimicrobial activity; hospital environments

1. Introduction

Indoor pollution by microbial contaminants is increasingly receiving attention as a public health problem [1]. Exposure to microbial pollutants is, in fact, related to many negative consequences, such as infectious disease, toxic effects, allergies and asthma [1, 2]. The main source of microbial pollutants are people and air conditioning systems, which allow the survival and multiplication of microorganisms [3]. Indoor air and surfaces disinfection is considered the most important measure in preventing the spread of infectious disease, especially in hospitals and in those settings where the absence of pathogenic microorganisms is required [4]. The control of indoor biological contamination is particularly important, as various pathogens may be transmitted from these environments to patients [5].

The increased usage of products containing the commonly used biocides, such as phenols, aldehydes, biguanides and chlorine and its derivatives has raised some concerns about their overall efficacy, but also about the possible production of by-products that are harmful to humans or have a corrosive effect on surfaces. Given the good disinfection efficacy, disinfection agents based on peroxides such as hydrogen peroxide and its compounds may well replace older biocides including substances such as chlorine-releasing agents [7]. Evolyse is a new peroxygen based disinfectant containing hydrogen peroxide (5.9 %) and silver nitrate (10 ppm). The aim of this study was to test the *in vitro* biocidal activity against a wide spectrum of microorganisms, including bacteria, spores and fungi and to evaluate its antimicrobial efficacy under practical conditions in different hospital environments.

2. Materials and Methods

2.1 Quantitative suspension test

Homogeneous suspensions of each bacterial and yeast test strain (table 1), containing about 1 - 5 x 10^8 CFU/mL were prepared in Nutrient and Sabourad broth (Microbiol, Cagliari, Italy), respectively. Conidial suspensions of *Aspergillus* and *Penicillium chrysogenum* strains (about 5×10^6 CFU/mL) were prepared according to the simplified procedure reported by Tortorano et al. [8]. The spore suspension of *Bacillus cereus* ATCC 11178 was prepared from a culture on nutrient agar 10 days old at 30° C. The culture was harvested by centrifugation, washed 4 times, resuspended in a final volume of 10 mL sterile distilled water, then heated at 75° C for 10 min to inactivate vegetative cells.

In vitro activity against the test strains was determined using the quantitative suspension tests described by the European Committee for Standardization [9, 10, 11]. Briefly, one millilitre of bacterial, fungal or spore suspension and 1 mL of interfering solution, bovine albumin 3%, (Sigma, Milan Italy) to simulate dirty conditions, were added to 8 mL disinfectant. After the desired contact time, (see table 1), 1 mL test solution was transferred to 8 mL neutralizing solution (catalase 0.05%) and 1 mL distilled water. After a further 5 min neutralization time, 1 mL of the neutralized test mixture was transferred, in duplicate, into empty dishes and covered with melted 20 mL of Tryptic Soy Agar (TSA, Microbiol, Cagliari, Italy) or Potato dextrose Agar (PDA, Microbiol). TSA and PDA plates were incubated at 37°C for 48 h, and 30° C for 4- 5 days, respectively. The suspensions test containing *Legionella pneumophila* ATCC 33152 were spread on Buffered Charcoal Yeast Extract Agar (BCYE, Microbiol) plates which were incubated at 37° C for 5 days. The effectiveness of the disinfectant agent was expressed as log reduction factor (RF) by the formula RF= log N_i-log N_r where N_i is the initial cells suspension count and N_r is the count after exposure to disinfectant. Reduction of 5 $\log_{10}$, 4 $\log_{10}$ and 3 $\log_{10}$ were regarded as denoting bactericidal, fungicidal and sporicidal activity, respectively.

2.2 Test under practical conditions

The antimicrobial activity of Evolyse was analysed under practical conditions, in the following hospital environments: operating room, sterilization room, dental office. The disinfectant was nebulised in the environments at the concentration of 1 mL/m^3 by means of a nebulizer. The efficacy of this disinfection system was evaluated by microbiological monitoring of air and surfaces, before and after treatment. Microbial sampling, after disinfection treatment, was carried out 30 min after nebulisation.

Microbial air contamination was assessed by active sampling using the Surface Air System (SAS, PBI International) to determine the number of CFU/m^3. Plate count agar (PCA, Microbiol, Cagliari, Italy) and PDA supplemented with chloramphenicol 0.05% (w/v) were used for detection of total bacterial count and fungal count, respectively.

Contact plates (PBI International) containing different agar media were used for enumeration of microorganisms on surfaces. Three horizontal surfaces (floor, table, chair/bed/table trolley) were tested.

The following microbiological parameters were determined: Total Bacteria Count in PCA, Staphylococci in Mannitol Salt Agar (MSA, Microbiol) *Enterobacteriaceae* in Violet Red Bile Glucose Agar (VRBGA, Microbiol), yeasts and moulds in PDA. Incubation conditions were 48h at 30° C for total bacterial count, 48 h at 37° C for Staphylococci, 24 h at 30° C for *Enterobacteriaceae*, 120 h at 25° C for yeasts and moulds. Results, before and after treatment, were expressed as CFU/24 cm^2 and percentage reduction (%) values were calculated.

3. Results

The biocidal activity of Evolyse was tested on a large panel of microorganisms including Gram-positive, Gram-negative, spore-forming bacteria and fungal strains belonging to different yeast an mould species.

As can be observed in table 1, the reduction in viability higher than 5 log units, required by the EN 1276 norm to qualify the product as being an efficient disinfectant, was obtained by the non-diluted product, in experiments simulating dirty conditions (bovine albumin 3%), after 5 min contact time at room temperature for all bacterial species tested.

Table 1 – Bactericidal, sporicidal and fungicidal activity of Evolyse in suspension tests under dirty conditions (bovine albumin 3%) .

Test microorganisms	Initial inocula ($\log_{10}$ cfu/mL)	European Standard test	Contact time (min)	$\log_{10}$-reduction§ factor
Staphylococcus aureus ATCC 25923	8.54	EN 1276	5	> 6.84
Staphylococcus aureus DBS collection	8.94	EN 1276	5	> 5.00
Listeria monocytogenes ATCC 7644	8.60	EN 1276	5	> 6.18
Pseudomonas aeruginosa ATCC 27853	8.84	EN 1276	5	> 6.39
Escherichia coli ATCC 25922	8.40	EN 1276	5	> 6.40
Salmonella enteritidis ATCC 13076	8.85	EN 1276	5	> 6.85
Legionella pneumophila ATCC 33152	8.00	EN 1276	5	> 5.84
Bacillus cereus ATCC 11178 spores	8.28	EN 13704	45	3.00
Candida albicans ATCC 10231	8.40	EN 1650	30	> 4.97
Saccharomyces cerevisiae ATCC	7.80	EN 1650	30	> 5.80
Aspergillus flavus ATCC 46283	6.70	EN 1650	15	> 4.70
A. niger DBS collection	6.30	EN 1650	15	> 4.30
Penicillium chrysogenum ATCC 9179	6.75	EN 1650	30	> 4.60

*methicillin-resistant strain

§ For colony counts under the limit of detection (< 1.0×10^2), this value was used in the calculation of the reduction

Additionally, reduction in viability higher than 4 log units, required by the EN 1650 norm to qualify the product as being an efficient fungicidal disinfectant, was obtained after 15 min contact time at room temperature for *A. flavus* ATCC 46283 and *A. niger* DBS strain, and after 30 min contact time for *C. albicans* ATCC 10231, *S. cerevisiae* ATCC 2601 and *P. chrysogenum* ATCC 9179. Sporicidal activity against *B. cereus* ATCC 11178 spores was obtained after 45 min contact at room temperature (reduction of the initial count by 3 log_{10} units, according to EN 13704).

The efficacy of Evolyse was also examined under practical conditions by evaluating its ability to reduce bio-contamination of air and surfaces in different hospital environments. In figure 1 the mean percentage reductions in bacterial and fungal counts in the air are reported. As can be seen, there was a mean reduction in bacterial and fungal count, respectively, of 73.5% and 65.3% for dental office, 83.5% and 56.5% for sterilization room. In the operating room the percentage reduction in microbial count was 100% in all samples analysed.

Fig. 1 Efficacy of Evolyse under field tests: Percentage reductions in microbial counts of the air in the different hospital environments analysed.

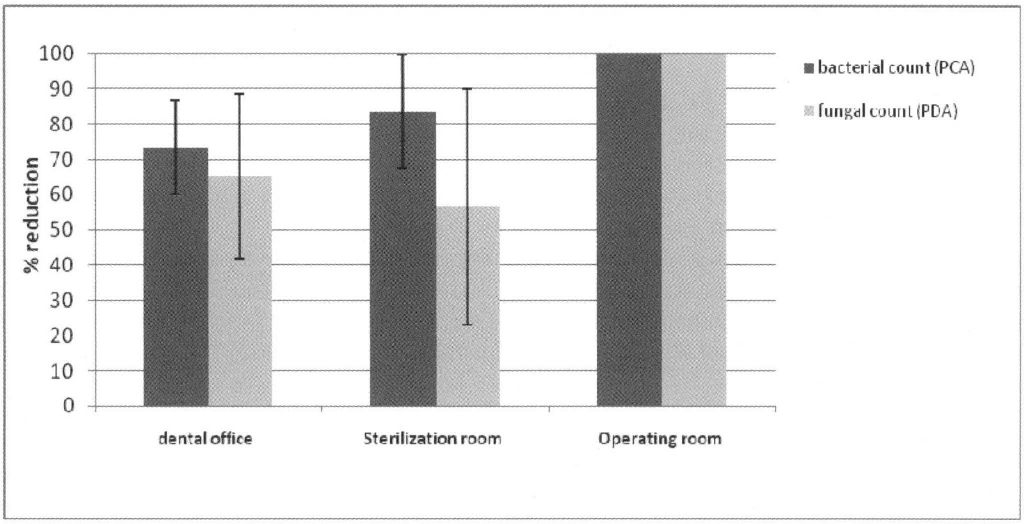

Fig. 2 Efficacy of Evolyse under field tests: Percentage reductions in microbial counts on the surfaces of the different hospital environments analysed.

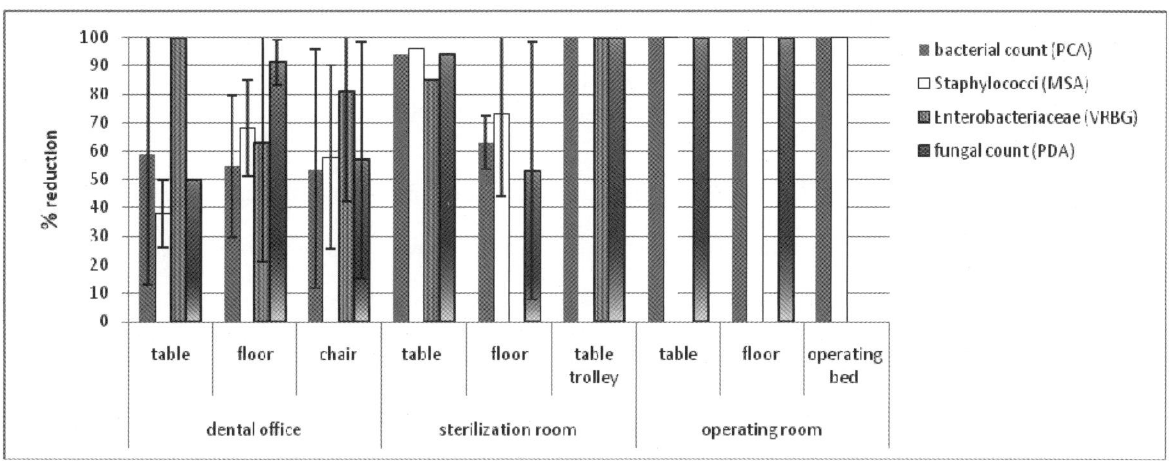

The percentage reductions in microbial counts on surfaces in the different hospital environments are shown in figure 2. Disinfection treatment with Evolyse resulted in 38-100 % reduction of microbial growth on surfaces of dental office. Twenty-six of 44 (59%) surface samples carried out in this environment showed percentage reduction values higher than 50%. The percentage reductions in the surfaces of the sterilization room ranged from a minimum of 53% to a maximum of 100%, values obtained in the floor and in the table trolley, respectively. Overall, the reductions was higher than 70% in more than 50% of microbial samples (66.6%) analysed in this room. Finally, it was noticed that in the operating room, before treatment, *Enterobacteriaceae* were not detected in any of sampled surfaces. For the other microbiological parameters, a decrease of 100% was observed in all sampled surfaces, after disinfection treatment.

4. Discussion and Conclusions

In this study the effect of a new biocidal product, Evolyse, on viable microorganism was examined by *in vitro* tests as well as in a field study.

In the suspension tests, the undiluted product caused a high reduction in viability of the tested microorganisms. The spectrum of microorganisms used in this study is wide and covers pathogens or opportunists for man and animals. In particular, it included species traditionally considered to be nosocomial pathogens, resistant to antibiotics, found in hospital water distribution system.

Evolyse was able to inactivate more than 5.00 $\log_{10}$ CFU/ml of all the studied bacterial strains, within 5 min contact time, under dirty conditions (BSA 3%). In addition, a reduction factor higher than 4.00 $\log_{10}$ CFU/ml was observed against the two *Aspergillus* species tested within 15 min contact time, under dirty conditions, but a delayed fungicidal activity against the two yeast species *C. albicans* ATCC 10231 and *S. cerevisiae* ATCC 2601 and the mould species *P. chrysogenum* ATCC 9179 was observed. A contact time of 30 min was necessary to inactivate more than 4.00 $\log_{10}$ CFU/ml of these fungal species. Other studies confirmed that reductions of studied microorganisms increased with the time of exposure and the disinfectant concentrations [12, 13]. This product also demonstrated activity against the spores of *B. cereus* ATCC 11178 strain.
These results are in line with other studies that demonstrated the bactericidal, fungicidal and sporicidal effects of H_2O_2 and other peroxygens [12, 14, 15, 16].

The biocidal effect of Evolyse was evaluated in a field study in three different hospital environments, dental office, sterilization room and operating room. The results showed that the disinfection treatment with Evolyse was to able to reduce microbial pollution in the air and surfaces of all three hospital environments. As for the surfaces, the percentage reduction of all microbiological parameters were close to 100% in the sterilization room and 100% in the operating room. The microbial counts in the surfaces of the dental office decreased, after disinfection treatment, with percentages lower than those recorded for surfaces of the other two environments. This minor reduction could have been due to the fact that doors were not kept closed during the sampling time. The results of this study showed the good biocidal activity of Evolyse and offer promising perspective for the use of this product in hospital environments. On the other hand, a further field study could allow to better evaluate product utility and efficacy in reducing the incidence of hospital-acquired infections.

References

[1] Douwes J, Thorne P, Pearce N, Heederik D. Bioaerosols health effects and exposure assessment: progress and prospect. *The Annals Occupational Hygiene.* 2003;47:187-200.

[2] World Health Organization: WHO guidelines for indoor quality: dampness and mould. Copenhagen 2009.

[3] Uduman SA, Farrukh AS, Nath KN, Zuhair MY, Ifrah A, Khawla AD, Sunita P. An outbreak of *Serratia marcescens* infection in a special-care baby unit of a community hospital in United Arab Emirates: the importance of the air conditioner duct as nosocomial reservoir. *Journal of Hospital Infection.* 2002;52:175-180.

[4] Rutal WA, Weber DJ. Control: the role of disinfectants and sterilization. *Journal of Hospital Infection.* 1999; 43:43-45.

[5] Kramer A, Schwebke I, Kampf G. How long do nosocomial pathogens persist on inanimate surfaces? A systematic review. *BMC Infectious Disease.* 2006;6:130.

[7] Dettenkofer M, Block C. Hospital disinfection: efficacy and safety issues. *American Journal of Infectious Control.* 2001;29:259-261.

[8] Tortorano AM, Viviani MA, Biraghi E, Rigoni AL, Prigitano A, Grillot R, EBGA Network. *In vitro* testing of fungicidal activity of biocides against *Aspergillus fumigatus. Journal of Medical Microbiology.* 2005;54:955-957.

[9] European Committee for Standardization. *European Standard EN 1276: chemical disinfectants and antiseptics. Quantitative suspension test for the evaluation of bactericidal activity of chemical disinfectants and antiseptics used in food, industrial, domestic, and institutional areas.* Test methods and requirements (phase 2, step 1). Brussels 1997.

[10] European Committee for Standardization. *European Standard EN 1650: chemical disinfectants and antiseptics. Quantitative suspension test for the evaluation of fungicidal activity of chemical disinfectants and antiseptics used in food, industrial, domestic, and institutional areas.* Test methods and requirements (phase 2, step 1). Brussels 1997.

11] European Committee for Standardization. *European Standard EN 13704: chemical disinfectants. Quantitative suspension test for the evaluation of sporicidal activity of chemical disinfectants used in food, industrial, domestic, and institutional areas.* Test methods and requirements (phase 2, step 1). Brussels 2002.

[12] Briñez WJ, Roig-Sagués AX, Hernández Herrero MM, López-Pedemonte T, Buenaventura G. Bactericidal efficacy of peracetic acid in combination with hydrogen peroxide against pathogenic and non pathogenic strains of *Staphylococcus* spp., *Listeria* spp. and *Escherichia coli*. *Food Control*. 2006;17:516-521.
[13] Aarnisalo K, Salo S, Miettinem H, Suihko ML, Wirtanen G, Autio T, et al. Bactericidal efficiencies of commercial disinfectant against *Listeria monocytogenes* on surfaces. *Journal of Food Safety*. 2000;20:237-250.
[14] Khadre MA, Yousef AE. Sporicidal action of ozone and hydrogen peroxide: a comparative study. *International Journal of Food Microbiology*. 2001;71:131-138.
[15] Majcher MR, Bernard KA, Sattar S A. Identification by quantitative carrier test of surrogate spore-forming bacteria to assess sporicidal chemicals for use against *Bacillus anthracis*. *Applied and Environmental Microbiology*. 2008;676-681.
[16] García de Lomas J, Lerma M, Cebrián L, Esteban E, Giménez MJ, Aguilar L, Domínguez V, Randez JJ. Evaluation of the in-vitro cidal activity and toxicity of a novel peroxygen biocide: 2-butanone peroxide. *Journal of Hospital Infections*. 2008; 68:248-254.

Increased resistance to detergent in *Enterococcus faecalis*

Jacqueline Keyhani* and Ezzatollah Keyhani

Laboratory for Life Sciences, Saadat Abade, Sarve Sharghi 56, 19979 Tehran

* Corresponding author: e-mail: keyhanius2002@yahoo.com, Phone: +98-21-66956974

Enteroccocus faecalis, a gram-positive bacterium ubiquitous in the environment and normal inhabitant of the human gastrointestinal tract, has emerged as a major health problem exacerbated by its ability at surviving adverse conditions, including prolonged stay on environmental surfaces. Here its tolerance to detergents that are routine household chemicals was investigated and compared to that of *E. coli*. A clinical isolate of *E. faecalis* with plasmid-mediated antibiotics resistance grown in the presence of increasing concentrations of sodium N-lauroylsarcosinate (sarkosyl) and sodium dodecyl sulphate (SDS) exhibited a sarkosyl minimum inhibitory concentration (MIC) of 0.1% (10% for *E. coli*) and a SDS MIC of 0.05% (6% for *E. coli*). A mutant with accrued sarkosyl resistance (MIC = 0.6%) was isolated after growth in the presence of 0.06% sarkosyl. While there was no cross-resistance to SDS, other mutant characteristics gave it advantages over the original strain, emphasizing the potential emergence of *E. faecalis* with increasing tolerance to common household detergents.

Keywords *Enteroccocus faecalis*; *E. coli*; sarkosyl; SDS; antibiotics resistance; plasmids

1. Introduction

A gram-positive commensal bacterium inhabiting the gastrointestinal tracts of humans and other mammals, *Enterococcus faecalis* is also found in water, soil, vegetation and food [1-5]. In addition, the bacterium has been used in probiotic cultures such as cheese and other milk product starters, and it is among the main constituents of some probiotic food supplements [6, 7]. Unfortunately, although considered harmless to human health for a long time, *E. faecalis* has since emerged as a major health problem. It has been identified as a leading nosocomial pathogen and as an infectious agent in endocarditis, meningitis, endophthalmitis and other serious diseases [8-11]. Furthermore, *E. faecalis* has been recognized as a prevalent cause of endodontic infections [12].

The emergence of *E. faecalis* as a major health problem can be attributed primarily to the robust nature of the organism. The bacterium grows at temperatures ranging from 10 to 45°C, in hypotonic, hypertonic, acidic, or alkaline conditions and it is able to withstand detergents, oxidative stress and desiccation [13]. Besides its sturdiness and high adaptability to environmental changes, *E. faecalis* often displays resistance to various antimicrobial agents, either intrinsic or acquired [2, 3, 6, 8, 10, 14]. It has also been reported to survive prolonged stay on environmental surfaces, a property that multiplies the chances of spreading infections [15].

Gram-positive enteric bacteria are reputedly very sensitive to detergents, especially compared to gram-negative enteric bacteria [16], and this would provide a way of controlling pathogen spread, at least on environmental surfaces. However, *E. faecalis* is known to become less sensitive to normally lethal levels of detergents such as bile salts or sodium dodecyl sulphate (SDS) after brief exposure to homologous sublethal concentrations, with strong cross-resistance observed with bile salts- and SDS-adapted cells [17]. This prompted us to further investigate *E. faecalis* tolerance to detergents, specifically sodium N-lauroylsarcosinate (sarkosyl) and sodium dodecyl sulfate (SDS), two anionic detergents that are routine household chemicals. While sarkosyl is a foaming and cleansing agent in shampoo, shaving foam and foam wash products, SDS is a common component in household cleaning products.

A clinical isolate of *E. faecalis* that exhibited plasmid-mediated resistance to tetracycline and kanamycin was used in this study. The strain was grown in the presence of increasing concentrations of sarkosyl or SDS and its sensitivity to these two detergents was compared to that of a gram-negative member of the bacterial community in the gastrointestinal tracts of humans and other mammals, namely *Escherichia coli.*

2. Materials and Methods

The *Escherichia coli* strains HB101, HB101(pBR322) and HB101(pBR325) were provided by Prof. N.O. Keyhani, Department of Microbiology and Cell Sciences, University of Florida at Gainesville, FL. The *E. faecalis* strain used in this study was a clinical isolate provided by Prof. S. Lauwers from the Microbiolgy Department of the Vrije Universiteit Brussel University Hospital, Brussels, Belgium. pBR322 is a 4.3 kb plasmid carrying resistance to ampicillin and tetracycline [18]; pBR325 is a 5.7 kb plasmid carrying resistance to ampicillin, tetracycline and chloramphenicol [19].

Luria-Bertani (LB) broth and LB-agar plates were used throughout this study. All media, detergents and antibiotics were prepared according to [20]. Detergent sensitivity and antibiotic sensitivity were determined by the minimum inhibitory concentration (MIC) method where MIC was defined as the minimum concentration of detergent/antibiotic required to observe no growth for up to 30 h culture at 37°C, with aeration, in LB medium.

E. faecalis was tested for the presence of catalase by adding a few drops of 30% H_2O_2 to 0.1 ml bacterial suspension and watching for bubble formation; *E. coli* strains were used as catalase-positive controls. *E. faecalis* was also tested for growth in the presence of 6.5% NaCl, growth at pH 9.6, and growth on bile-aesculin agar with black discoloration of the agar. In addition, cells were stained according to the Gram-stain procedure

Plasmid DNA was extracted from *E. faecalis* plasmid-carrying strain by a novel method specifically developed for *E. faecalis* as described in [21], and from *E. coli* plasmid-carrying strains by the alkaline lysis method [22]. Plasmid DNA preparations were electrophoresed in 1% agarose gels as described in [20], except that ethidium bromide was added to the electrophoresis buffer only, at a final concentration of 0.5 μg/ml. The agarose was type-II, low-endo-osmotic agarose. Gels were 8x10 cm.

3. Results

3.1 Comparative sensitivity to detergents of *E. faecalis* and *E. coli*

Figure 1 shows *E. faecalis* (clinical isolate used in this study) and *E.coli* (strain HB101(pBR322)) growth curves obtained by monitoring the absorbance at 550 nm at timed intervals during culture at 37°C, with aeration, in LB broth containing various concentrations of sarkosyl or SDS. *E. faecalis* was much more sensitive than *E. coli* to both detergents. No growth was observed for *E. faecalis* in the presence of 0.1% sarkosyl (Fig. 1a, closed diamonds) while *E. coli* growth remained unaffected even in the presence of 0.5% sarkosyl (Fig. 1b, open triangles). Duration of the lag and log phases progressively increased for *E. faecalis* as sarkosyl concentrations

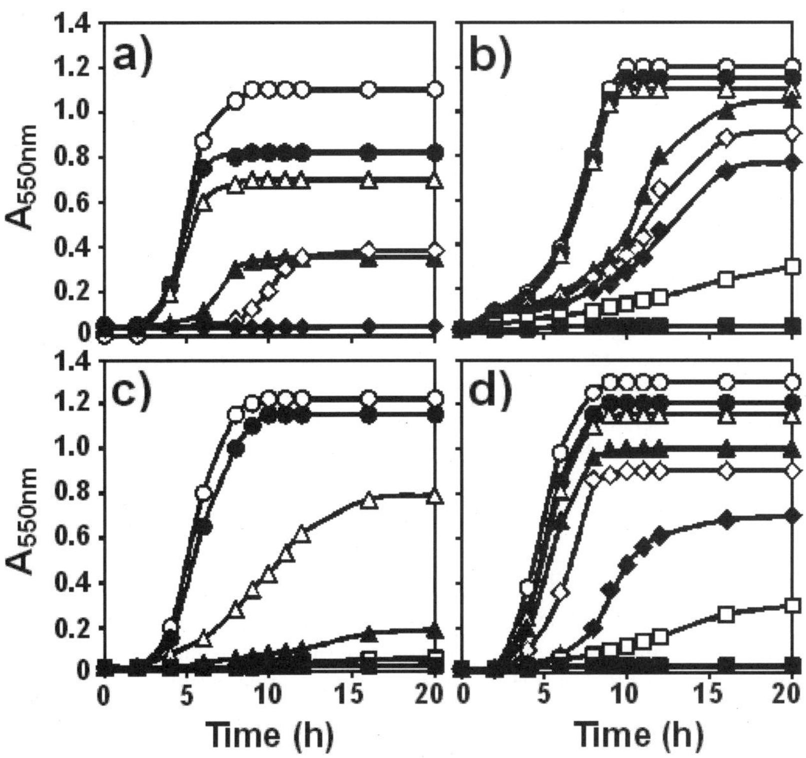

Fig. 1 Growth in LB medium at 37°C with aeration of: a) *E. faecalis* in the absence (○) and presence of sarkosyl 0.003% (●), 0.006% (△), 0.02% (▲), 0.06% (◇), 0.10% (◆); b) *E. coli* in the absence (○) and presence of sarkosyl 0.25% (●), 0.5% (△), 1% (▲), 3% (◇), 5% (◆), 7% (□), 10% (■); c) *E. faecalis* in the absence (○) and presence of SDS 0.003% (●), 0.006% (△), 0.03% (▲), 0.05% (□), 0.25% (■); d) *E. coli* in the absence (○) and presence of SDS 0.25% (●), 0.5% (△), 1% (▲), 2% (◇), 4% (◆), 5% (□), 6% (■).

increased to 0.02% and 0.06%, with considerable reduction in cell density at the stationary phase (Fig 1a, closed triangles and open diamonds). For *E. coli*, extension of the lag and log phases was not observed until sarkosyl concentration reached 1% (Fig. 1b, closed triangles), a concentration already 10 times higher than the detergent MIC for *E. faecalis*. Similarly, while no growth was observed for *E. faecalis* in 0.05% SDS (Fig. 1c, open squares), growth was still observed in 5% SDS for *E. coli* (Fig. 1d, open squares).

The sarkosyl MIC that was 0.1% for *E. faecalis* was 10% for *E. coli*, and the SDS MIC that was 0.05% for *E. faecalis* was 6% for *E. coli*. Cultures were maintained for 30 h; no further change in absorbance was observed for any of the growth curves presented in Figure 1. Growth curves obtained with *E. coli* strain HB101(pBR325) were similar to those obtained for *E. coli* strain HB101(pBR322) presented in Figure 1b and d.

3.2 Emergence of an *E. faecalis* mutant strain with accrued resistance to sarkosyl

Growth in the presence of 0.06% sarkosyl led to the isolation of a mutant *E. faecalis* strain that exhibited accrued resistance to the detergent. Figure 2 shows the growth curves of the original strain (Fig. 2a) and of the mutant strain (Fig. 2b) of *E. faecalis* in the absence and in the presence of sarkosyl at detergents concentrations equal to or higher than the original MIC value of 0.1%. The mutant strain grew well in the presence of 0.25% sarkosyl (Fig. 2b, open triangles), a concentrations 2.5 times the detergent MIC and would still grow in the presence of 0.5% sarkosyl (Fig. 2b, closed diamonds). The detergent MIC for this mutant strain was 0.6%, increasing it by a factor of 6 compared to the MIC identified for the original strain.

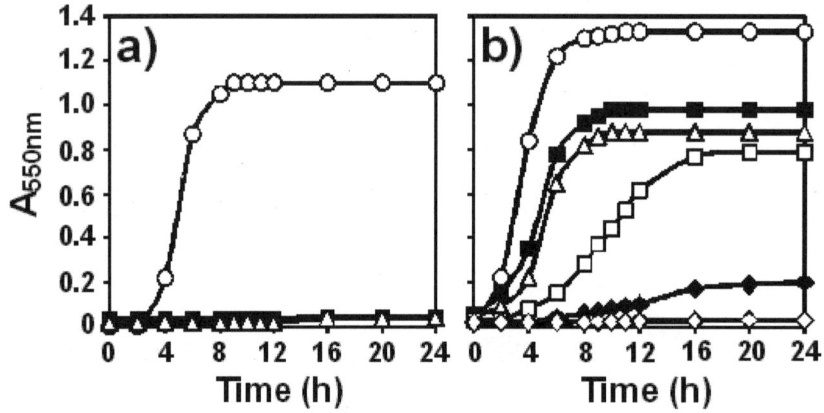

Fig. 2 Comparative sensitivity to sarkosyl of the original *E. faecalis* strain used in this study and the isolated *E. faecalis* mutant strain. a) Growth curves obtained during growth of *E. faecalis* (original strain) in the absence (○) and presence of 0.1% (■) and 0.25% (△sarkosyl. b) Growth curves obtained during growth of the mutant strain in the absence (○) and presence of 0.1% (■), 0.25% (△), 0.4% (□), 0.5% (◆), and 0.6% (◇) sarkosyl.

The mutant strain did not exhibit accrued resistance to SDS; no growth was observed after up to 30h culture in the presence of 0.05% SDS, a concentration identified as MIC against the original *E. faecalis* strain. However, the mutant strain grew faster than the original strain in the absence of detergent, featuring a lag phase reduced by half and a log phase reduced by 30%, and reaching a higher cell density at stationary phase (Fig. 2a and b, open circles). Progressive increase in the duration of the lag phase and, eventually, of the log phase was observed as the concentration of sarkosyl in the culture medium increased Fig. 2b). Cultures were maintained for 30 h; no further change in absorbance was observed for any of the growth curves presented in Figure 2.

3.3 Further characteristics of the *E. faecalis* mutant strain

Table 1 summarizes the properties of the isolated mutant strain compared to the original strain. Both strains presented the characteristics of *E. faecalis* strains, being gram-positive, exhibiting no catalase activity, being able to grow in the presence of 6.5% NaCl as well as at pH 9.6, and growing on bile-esculin agar with black discoloration of the agar. As indicated in Table 1 and as further developed below, both strains were resistant to kanamycin and tetracycline; however, they were both sensitive to ampicillin and chloramphenicol, with MICs of 5 μg/ml (not shown in the table).

Table 1 Characteristic features of the original *E. faecalis* strain and of the mutant strain

Characteristics	Original strain	Mutant strain
Gram staining	Gram positive	Gram positive
Catalase activity	Not detectable	Not detectable
Growth in 6.5% NaCl	+	+
Growth at pH 9.6	+	+
Growth on bile-esculin agar	Black discoloration of agar	Black discoloration of agar
MIC SDS (%)	0.05	0.05
MIC sarkosyl (%)	0.10	0.60
MIC kanamycin (µg/ml)	2000	500
MIC tetracycline (µg/ml)	50	50

3.4 Antibiotics resistance in the *E. faecalis* mutant strain

The original strain used in this study exhibited resistance to tetracycline and kanamycin associated with the presence of two small plasmids (~ 7 and ~ 5.7 kb) [21]. The mutant strain retained the antibiotics resistance although its resistance to kanamycin was reduced compared to the original strain, with a MIC dropping from 2000 µg/ml to 500 µg/ml. This was not due to plasmid loss; plasmid DNA preparations from the mutant were identical to those of the original strain after electrophoresis in agarose gels. Figure 3a shows pictures of agarose gels after electrophoresis of plasmid DNA preparations from the original and from the mutant *E, faecalis* strain; both preparations showed two sets of bands well resolved for the linear and supercoiled forms, indicating the presence of two plasmids as previously identified in the original *E. faecalis* strain. In Figure 3b, a plasmid preparation from the original *E. faecalis* strain that was electrophoresed for a longer time shows the 2 sets of bands well resolved for each of the DNA form (relaxed circular, linear and supercoiled). In Figure 3c, a plasmid

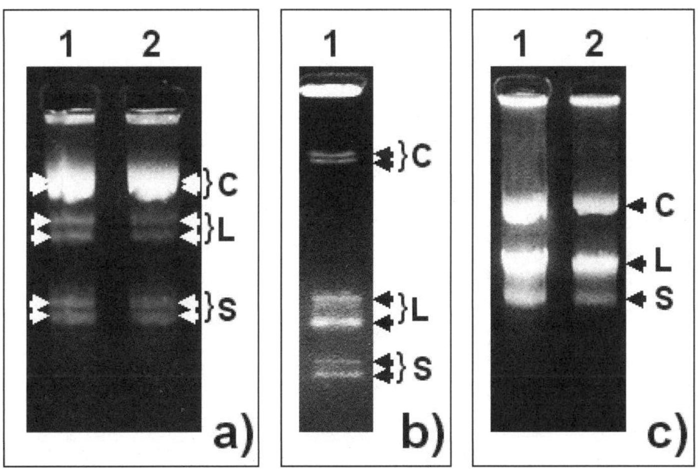

Fig. 3 Agarose gel electrophoresis of plasmid DNA preparations. a) Plasmids isolated from *E. faecalis* strains - lane 1: original strain; lane 2: mutant strain. b) Plasmids isolated from *E. faecalis* original strain. c) pBR322 isolated from HB101(pBR322) – lane 1: 30 µl-aliquot; lane: 15 µl-aliquot. Letters refer to different topological forms in plasmid preparations - C: relaxed circular; L: linear; S: supercoiled.

preparation from HB101(pBR322) shows the expected set of three single bands corresponding to the relaxed circular, linear and supercoiled form of a single plasmid DNA [23].

Similarly to those from the original *E. faecalis* strain, plasmid DNA preparations from the mutant strain conferred resistance to kanamycin and/or tetracycline to *E. coli* strain HB101 after transformation.

4. Discussion

The propensity of *E. faecalis* to propagate diseases is largely enhanced by its ubiquity, its capability of surviving numerous environmental challenges and its ability to acquire resistance to multiple antibiotics. Furthermore, multi-resistant strains have acquired virulence genes resulting in hospital-adapted clones, triggering the search for alternative therapy [10, 14].

On the other hand, high sensitivity to detergents as displayed by *E. faecalis* would provide for a way to control the pathogen spread, at least on environmental surfaces. But the bacterium's high adaptability to environmental conditions includes the ability to survive lethal concentrations of SDS after brief exposure to sublethal concentrations of that detergent or of bile salts [17]. Furthermore, here we show the occurrence of a mutant strain with accrued resistance to sarkosyl, another anionic detergent, after growth in half the detergent MIC. While SDS is a common component in household cleaning products, sarkosyl is used as a foaming and cleansing agent in shampoo, shaving foam and foam wash products. Although the mutant did not exhibit cross-resistance to SDS, it displayed characteristics that gave it advantages over the original strain. Interestingly the mutant retained the two plasmids present in the original strain although we had previously shown that sarkosyl is a potential curing agent for the strain [21]. The findings reported here further emphasize the remarkable capacity of *E. faecalis* to resist harsh treatment and the potential emergence of strains with increasing tolerance to detergents commonly used for cleaning purposes in households. This, in turn, further increases the challenge of finding ways to control the spread of *E. faecalis*-propagated infections.

Acknowledgements Both authors equally contributed to this work.

References

[1] Gilmore M. *The Enterococci: Pathogenesis, Molecular Biology, and Antibiotic Resistance.* Washington, DC: American Society for Microbiology Press; 2002.

[2] Silva N, Igrejas G, Figueiredo N, Gonçalves A, Radhouani H, Rodrigues J, Poeta P. Molecular characterization of antimicrobial resistance in enterococci and Escherichia coli isolates from European wild rabbit (Oryctolagus cuniculus). *Sci Total Environ.* 2010;408:4871-4876.

[3] Poeta P, Costa D, Igrejas G, Rodirgues J, Torres C. Phenotypic and genotypic characterization of antimicrobial resistance in faecal enterococci from wild boars (Sus scrofa). *Vet Microbiol* 2007;125:368-374.

[4] Marcinek H, Wirth R, Muscholl-Silberhorn A, Gauer M. *Enterococcus faecalis* gene transfer under natural conditions in municipal sewage water treatment plants. *Appl Environ Microbiol* 1998;64:626-632.

[5] Fifadara N, Radu S, Hassan Z, Beuchat LR, Rusul G. Hemolytic and nonhemolytic vancomycin-resistant *Enterococcus faecalis* isolated from beef imported to Malaysia. *J Food Prot* 2003;66:1845-1850.

[6] Cocconcelli PS, Cattivelli D, Gazzola S. Gene transfer of vancomycin and tetracycline resistance among *Enterococcus faecalis* during cheese and sausage fermentation. *Int J Food Microbiol* 2003;88:315-323.

[7] Eaton TJ, Gasson MJ. Molecular screening of *Enterococcus* virulence determinants and potential for genetic exchange between food and medical isolatess. *Appl Environ Microbiol* 2001;67:1628-1635.

[8] Simjee S, Fraise AP, Gill MJ. Plasmid heterogeneity and identification of a Tn*5281*-like element in clinical isolates of high-level gentamicin-resistant *Enterococcus faecium* isolated in the UK. *J Antumicrob Chemother* 1999;43:625-635.

[9] Nallapareddy SR, Singh KV, Sillanpää J, Garsin DA, Höök M, Erlandsen SL, Murray BE. Endocarditis and biofilm-associated pili of *Enterococcus faecalis. J Clin Investig* 2006;116:2799-2807.

[10] Sava IG, Heikens E, Huebner J. Pathogenesis and immunity in enterococcal infections. *Clin Microbiol Infect* 2010;16:533-540.

[11] Forestier F, Salvanet-Bouccara A, Mohand-Said M, Dublanchet A, Emond JP. Recurrent acute Enterococcal faecalis endophthalmitis following cataract surgery: a case report with a favorable outcome. *J Fr Ophtalmol* 2006;29:181-183.

[12] Stuart CH, Schwartz SA, Beeson TJ, Owartz CB. Enterococcus faecalis: its role in root canal treatment failure and current concepts in retreatment. *J Endod* 2006;32:93-98.

[13] Kayaoglu G, Ørstavik D. Virulence factors of *Enterococcus faecalis*: relationship to endodontic disease. *Crit Rev Oral Biol Med* 2004;15:308-320.

[14] Uchiyama J, Rashel M, Maeda Y, Takemura I, Sugihara S, Akechi K, Muraoka A, Wakiguchi H, Matsuzaki S. Isolation and characterization of a novel Enterococcus faecalis bacteriophage phiEF24C as a therapeutic candidate. *FEMS Microbiol Lett* 2008;278:200-206.

[15] Noskin GA, Stosor V, Cooper I, Peterson LR. Recovery of vancomycin-resistant enterococci on fingertips and environmental surfaces. *Infect Control Hosp Epidemiol* 1995;16:577-581.

[16] Kramer VC, Nickerson KW, Hamlett NV, O'hara C. Prevalence of extreme detergent resistance among the enterobacteriaceae. *Can J Microbiol* 1984;30:711-713.

[17] Flahaut S, Frere J, Boutibonnes P, Auffray Y. Comparison of the bile salts and sodium dodecyl sulfate stress responses in *Enterococcus faecalis. Appl Environ Microbiol* 1996;62:2416-2420.

[18] Bolivar F, Rodriguez R, Greene PJ, Betlach M, Heyneker HL, Boyer HW, Crosa J, Falkow S. Construction and characterization of new cloning vehicles. II. A multipurpose cloning system. *Gene* 1977; 2:95-113.

[19] Bolivar F. Construction and characterization of new cloning vehicles. III. Derivatives of plasmid pBR322 carrying unique *Eco*RI sites for selection of *Eco*RI generated recombinant molecules. *Gene* 1978;4:121-136.

[20] Sambrook J, Fritsch EF, Maniatis T. *Molecular Cloning – A Laboratory Manual.* 2nd ed. Cold Spring Harbor, NY: Cold Spring Harbor Laboratory Press; 1989.

[21] Keyhani J, Keyhani E, Attar F, Haddadi A. Sensitivity to detergents and plasmid curing in *Enterococcus faecalis. J Ind Microbiol Biotechnol* 2006;33:238-242.

[22] Birnboim HC, Doly J. A rapid alkaline extraction procedure for screening recombinant plasmid DNA. *Nucleic Acids Res* 1979;7:1513-1523.

[23] Summers DK. *The Biology of Plasmids.* Oxford: Blackwell Science; 1996.

Legionella pneumophila isolation rate in a Spanish hospital pre- and post-installation of an electro-chemical activation system for potable water disinfection

Jose-Maria Rivera[1,2], Juan-Jose Granizo[3], Jose-Maria Aguiar[4], Ana Vos-Arenilla[5], Maria-Jose Giménez[6*] and Lorenzo Aguilar[6]

[1]Preventive Medicine Department, School of Medicine, Universidad Complutense, Avda. Complutense s/n, 28040 Madrid, Spain;
[2]Bioseguridad Ambiental (BSA), Doña Juana I de Castilla 52, 28027 Madrid, Spain;
[3]Grana Datos, Isla de Arosa 11, 28223 Pozuelo de Alarcón, Madrid, Spain;
[4]Control Microbiológico, Ctra. Coruña Km. 23,200, Edificio Las Rozas 23, 28230 Las Rozas, Madrid, Spain;
[5]Nursery Department, Nursery School, Universidad Complutense, Avda. Complutense s/n, 28040 Madrid, Spain;
[6]Microbiology Department, School of Medicine, Universidad Complutense, Avda. Complutense s/n, 28040 Madrid, Spain

*Corresponding author: Fax no: +34 91 3941511; email: prism-ag@hotmail.com

A retrospective analysis of Legionella isolation (1/January/2005 to 31/December/2009) in the routinely monthly surveillance in a Spanish hospital with a water disinfectant apparatus installed in the inflow of water-heating tanks (on 23/September/2007) was performed. Data of all 521 samples (250 pre- and 271 post-installation) collected from unfiltered taps from the potable hot water system was considered. Chlorine levels were higher ($p<0.001$) post-installation (0.55 ± 0.41 vs. 1.21 ± 0.38 ppm). A total of 28.2% (147/521) samples were positive for *L. pneumophila*; 112 (76.2%) of them serogroups 2-14. Post-installation, the isolation rate was lower (46.8% vs. 11.1%; $p<0.001$) due to reduction in *L. pneumophila* serogroups 2-14 from 38.4% to 5.9% ($p<0.001$). In the step-wise multivariate analysis, the installation of the system was the main variable associated with the reduction of Legionella isolation ($p<0.001$).

Keywords: Legionella; water disinfection; hospital potable water supply; electro chemical activation system

INTRODUCTION

Since the discovery in 1985 that hospital-acquired legionnaires' disease usually arises through the potable water supply [1,2], nearly all hospital outbreaks have been linked to potable water. Hospitals whose water supplies are not colonised by Legionella have not cases of nosocomial legionellosis [3,4]. Low chlorine levels in potable water systems are associated with high isolation rates [5], being serogroups 2–14 responsible for seasonality in isolation rates in these systems [5,6]. This may be important because not all nosocomial legionellosis are aetiologically assigned because the Legionella urinary antigen test is specific for serogroup 1 [4]. Disinfectant methods increasing chlorine levels may prevent seasonality isolation waves.

The aim of this study was to explore the efficacy of a commercialised electro-chemical activation system for Legionella disinfection of hospital potable water systems through a retrospective analysis in a Spanish general public hospital of the Legionella isolation rate in the pre-installation period compared with the rate in the post-installation period.

MATERIALS AND METHODS

A retrospective analysis of data on Legionella isolation obtained in the routine monthly surveillance performed at Residencia Cantabria (Santander, Spain) from 1 January 2005 to 31 December 2009 by BSA was performed. A water disinfectant apparatus (Hydrostel; Idrovital S.r.l, Empoli, Fi, Italy) was installed in the inflow of water-heating tanks on 23 September 2007 and consisted of a completely automatic system for the production of an aqueous solution of NaOCl (sodium hypochlorite) from water and NaCl (99.5% purity) through an electrolysis procedure (Electro Chemical Activation System) [7,8]. Data from all 521 samples collected over the study period from unfiltered taps from the potable hot water system were considered. Water was collected by opening the hot water tap (at peripheral points with two taps) or opening the valve in the hot position (at peripheral points with mixer valves). Sample collection and processing was performed on a routine basis as described previously [5]. For microbiological processing, water samples were concentrated 100-fold immediately on arrival at the laboratory. A 0.1 ml aliquot was plated onto GVPC (glycine, vancomycin, polimixin B, cyclohexamide) selective agar medium (Oxoid Ltd., Basingstoke, Hampshire, UK) and incubated at 36°C for 10 days. Colonies morphologically consistent with Legionella species were plated onto buffered charcoal yeast extract (BCYE) agar (Oxoid Ltd.) and blood agar (Oxoid Ltd.), and incubated for 48h. Colonies growing on BCYE agar but not on blood agar were definitively identified as *Legionella* using a commercially available latex agglutination test (Oxoid Ltd., DR0800) that distinguishes *L. pneumophila* serogroup 1, *L. pneumophila* serogroups 2–14, and other *Legionella* species. Comparisons of means were made using the t -test or the Mann–Whitney U -test if normality was not shown. Differences between percentages were assessed by the $\chi 2$ test or Fisher's exact test when necessary. Crude odds ratios and 95% confidence intervals were calculated.

A step-wise logistic regression analysis was performed to predict *Legionella* isolation using a predictive model (including the installation of the system, chlorine levels and temperature as independent variables). In order to explore the influence of the installation of the system on the seasonality of isolation rates, data were grouped by seasons in the pre- and post-installation periods, using medians of isolation rates in each season. Seasonal trends were adjusted to different models of linear and non-linear regressions, selecting the model with the highest R^2.

RESULTS AND DISCUSSION

Table 1 shows per-season mean ± SD of chlorine levels (ppm) and temperature (ºC), marking with a black line the time of the installation of the disinfectant apparatus. As shown, temperature and chlorine levels were higher ($p<0.001$) post-installation vs. pre-installation (51.6 ± 3.2 vs. 42.8 ± 10.6ºC and 0.55±0.41 vs. 1.21±0.38 ppm). Table 2 shows yearly and seasonal distribution of samples collected, no. of samples positive to *L. pneumophila* serogroup 1 (S1) and serogroups 2-14 (S2-14), total positive samples and percentage of positive samples with respect to total samples collected., marking with a black line the time of the installation of the disinfectant apparatus. As shown, post-installation the isolation rate was lower (46.8% vs. 11.1%; $p<0.001$; OR=0.14; 95%CI=0.09-0.22) due to the significant reduction in *L. pneumophila* serogroups 2-14 from 38.4% to 5.9% ($p<0.001$; OR=0.10; 95%CI=0.06-0.18), but not significant for serogroup 1 (from 9.6% to 5.2%; $p=0.052$; OR=0.51; 95%CI=0.26-1.02), due to the scarce number of serogroup 1 isolates. The logistic regression analysis [$p<0.001$, R2 (Cox)=0.335] showed a significant association ($p<0.001$, OR=0.26, 95%CI=0.15-0.46) between the reduction in *Legionella* isolation rate and the installation of the electrolytic device.

Figure 1 shows the fit by cubic regression of seasonal trends before and after installation of the electrolytic device. By comparing isolation rates by season before and after the installation of the system, the decrease in *Legionella* isolation was significant ($p<0.001$) in spring, summer and autumn. The electrochemical activation system was efficacious in reducing *Legionella* isolation rates (countering its increase in spring, summer and autumn) to a minimum (as in winter pre-installation), by significantly increasing chlorine levels.

Figure 1. Fit by cubic regression of seasonal trends before (black columns) and after (grey columns) installation of the electrolytic device

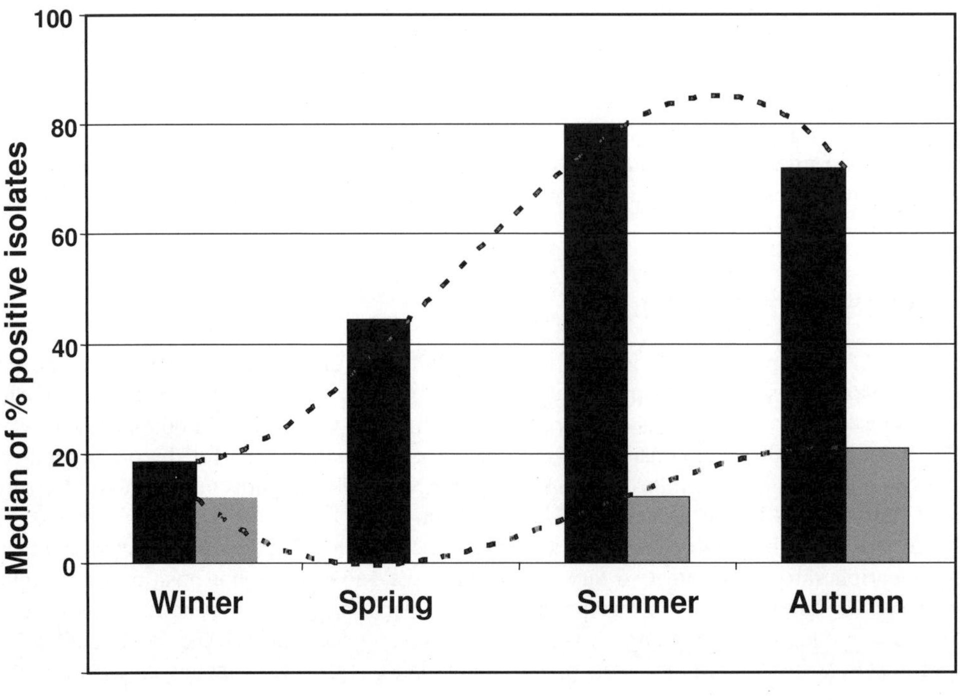

Table 1. Yearly and seasonal mean ± SD chlorine levels (ppm) and temperature (ºC). The black line shows the time of the installation of the apparatus. (W=winter, SP=spring, SU=summer, A= autumn).

	2005				2006				2007				2008				2009			
	W	**SP**	**SU**	**A**	**W**	**SP**	**SU**	**A**	**W**	**SP**	**SU**	**A**	**W**	**SP**	**SU**	**A**	**W**	**SP**	**SU**	**A**
Chlorine (ppm)	0.27 (0.30)	0.27 (0.30)	0.24 (0.13)	0.80 (0.17)	0.76 (0.26)	0.30 (0.23)	0.42 (0.30)	0.75 (0.26)	1.13 (0.35)	0.65 (0.37)	0.80 (0.27)	1.01 (0.67)	1.31 (0.40)	1.32 (0.13)	1.18 (0.28)	1.45 (0.21)	1.18 (0.28)	1.43 (0.24)	1.01 (0.35)	1.10 (0.25)
Temp. (ºC)	37.0 (8.1)	40.9 (8.3)	41.6 (11.7)	42.7 (10.4)	42.4 (11.1)	39.6 (11.6)	42.9 (11.3)	44.1 (10.7)	45.2 (12.2)	45.5 (10.8)	51.8 (2.7)	51.7 (2.7)	52.7 (1.8)	52.4 (1.1)	50.9 (8.0)	51.4 (4.0)	51.3 (1.9)	52.2 (1.7)	50.9 (0.8)	50.8 (0.9)

Table 2. Yearly and seasonal distribution of samples collected, no. of samples positive to *L. pneumophila* serogroup 1 (S1) and serogroups 2-14 (S2-14), total positive samples and percentage of positive samples with respect to total samples collected. The black line shows the time of the installation of the apparatus. (W=winter, SP=spring, SU=summer, A= autumn).

	2005				2006				2007				2008				2009			
	W	**SP**	**SU**	**A**	**W**	**SP**	**SU**	**A**	**W**	**SP**	**SU**	**A**	**W**	**SP**	**SU**	**A**	**W**	**SP**	**SU**	**A**
No. Samples	30	30	20	16	27	27	25	8	23	17	27	35	17	32	29	24	34	35	32	33
Positive S1	0	2	4	4	2	0	1	0	3	6	2	3	1	0	2	4	4	0	0	0
Positive S2-14	5	8	12	3	3	12	20	8	8	10	7	9	1	0	5	1	0	0	0	0
Total positive	5	10	16	7	5	12	21	8	11	13*	9	12	2	0	7	5	4	0	0	0
% positive	16.7	33.3	80.0	43.8	18.5	44.4	84.0	100	47.8	76.5	33.3	34.3	11.8	0.0	24.1	20.8	11.8	0.0	0.0	0.0

*3 samples yielded *L. pneumophila* S1 and S2-14.

REFERENCES

[1] Johnson JT, Yu VL, Best MG, Vickers RM, Goetz A, Wagner R, Wicker H, Woo A. Nosocomial legionellosis in surgical patients with head-and-neck cancer: implications for epidemiological reservoir and mode of transmission. *Lancet* 1985;2:298-300.

[2] Sabrià M, Yu VL. Hospital-acquired legionellosis: solutions for a preventable infection. *Lancet Infectious Diseases* 2002;2:368-373.

[3] Sabrià M, Mòdol JM, Garcia-Nuñez M, Reynaga E, Pedro-Botet ML, Sopena N, Rey-Joly C. Environmental cultures and hospital-acquired Legionnaires' disease: a 5-year prospective study in 20 hospitals in Catalonia, Spain. *Infection Control and Hospital Epidemiology* 2004;25:1072-1076.

[4] Stout JE, Yu VL. *Legionella* in the hospital water supply: a plea for decision making based on evidence-based medicine. *Infection Control and Hospital Epidemiology* 2001;22:670-672.

[5] Rivera JM, Aguilar L, Granizo JJ, Vos-Arenilla A, Giménez MJ, Aguiar JM, Prieto J. Isolation of *Legionella* species/serogroups from water cooling systems compared with potable water systems in Spanish healthcare facilities. Journal of Hospital Infection 2007;67:360-366.

[6] Rivera JM, Granizo JJ, Aguilar L, Giménez MJ, Aguiar JM, Prieto J. Is there a relationship between monthly rainfall and *Legionella* isolation in potable water systems in Spanish health-care facilities?. Infection Control and Hospital Epidemiology 2009;30:306-308.

[7] Franzin L, Cabodi D, Bonfrate N, Tortorelli F, Bernini L, Giorgione N. Evaluation of two disinfection systems for *Legionella* eradication from a hospital water supply. Abstracts of the 23rd European Working Group for Legionella Infections, May 11-13, 2008, Madrid (Spain). Abstract P8.03.

[8] World Intellectual property organization. WO/2007/057940. Available at: www.wipo.int/pctdb/en/wo.jsp.

Antimicrobial Evaluation:

Clinical and Pre-clinical Trials

Adherence to ART and its associated factors among HIV Aids Patients in Addis Ababa

Ezra Muluneh

May 2010 Haramaya University

Background: Customary practice; fasting and holly water use is very common in Ethiopia, and most people belong to one of religious denomination, this study was undertaken to investigate the influence of the socio-cultural factors among HIV AIDS patients in Addis Ababa.
Objective: To determine adherence levels and identify demographic and socio-cultural factors that influence Adherence to ART among HIV AIDS Patients.
Methods: A cross sectional survey design was used to conduct the study..
Result: High spirituality was found to have significant association **2.781 [1.230 – 6.287]** P< 0.05 with non-adherence..
Conclusion and Recommendation: High spirituality was found to be the main challenges for appropriate adherence. Therefore, strengthen the existing adherence-counseling strategy, involving religious leaders in designing and implementing continues patient education tools and operational research are recommended.

Keywords HIV/AIDS; Adherence; Antiretroviral therapy

1. Introduction

Antiretroviral therapy (ART) for the treatment of HIV disease, which specify the use of at least 3 different type of drugs (multiple individual drugs) taken once, twice, or 3 times a day [1, 2], which became a constant challenge for patients, especially in light of the treatment guidelines for HIV disease; treatment non-adherence was found to be mainly a response to the burdens of complicated drug regimens that disrupt one's daily routine. Moreover, HIV-infected patients who are already taking an extraordinary number of antiretroviral pills often have co-morbid conditions that require additional medications [3].

Ethiopia has made tremendous efforts in scaling up ART service; there are a number of studies conducted on HIV AIDS treatment adherence and possible factors contributing to ARV non-adherence in Ethiopia. These Studies have described a range of factors affecting ARV treatment adherence at various levels, i.e. at individual, community and health facility levels, individual factor such as being spiritual is very common in developing countries, and in Ethiopia as well. No study has been found to indicate the impact of spiritual believes on patient adherence. Therefore these studies aimed at socio-cultural factors that contribute to adherence are identified and proper structures for support or improving adherence are developed. The specific objectives were *i.* To determine the adherence level of HIV AIDS patients to ART. ii. To assess the socio-cultural factors influence adherence to ART among HIV AIDS Patients and iii. To identify the socio-demographic factors that affect adherence to ART among HIV AIDS patients.

2. Methods

The study was conducted in Addis Ababa city, the study carried out from January, 2010 to April, 2010. The study employed a cross-sectional design. The source population was all HIV positive individuals who were on ART and live in Addis Ababa. The study population was individuals who were on ART attending the health center by the time of visit.

Studies, conducted in Addis Ababa reported 81.2% [using self report] and Ethiopian Adherence survey indicate that full Adherence was 83% [using retrospective document review] [4]. the sample size calculated based on the above mentioned Studies, giving any particular outcome to be with in 2.5% marginal error and 95% confidence interval of certainty [alpha =0.05], 867, and adding 10% response rate, a total 954 interviewees required. Considering the proportion of health centers to communities, and to ensure equal representation among the inhabitants, ten health centers selected using random sampling techniques.

The study employed a two stage sampling method. The number of samples in each health centers was determined based on proportion of patients on ART. Where patient was not interested being included in the study, he/she was considered as non respondent. Dependent variable was adherence to ART. Independent variables were: sex, age, marital status, having child, level of education, occupation, income, social drug use, knowledge about objective of treatment, stigma, disclosure, community support, and spirituality. Patient exit interview was undertaken in twenty two working days; the period was chosen to get sufficient number of participants. A five question index was constructed to measure spirituality [Table 1], as holly water use was not found to be common spiritual characteristics among most religious participants, excluded from measurement index of spirituality. The field test of data collection instruments was carried out at hospital of Addis Ababa, so that logical flow of the questions, understanding by respondents, missed points and other relevant issues were assessed. Based on the findings of the pre-test, the instruments were finalized and duplicated for use.

Table 1. Measures of Spirituality

#	Descriptions	Response	Value
1	• Religious Place Attendance		
	• Questions		
	• Do you attend religious place frequently?	Yes, No	1, 0
2	• Fasting		
	• Questions		
	• Do you conduct fasting program in your religion?	Yes, No	1, 0
3	• Some people Believe that one can cure with faith the sick person		
	• Question		
	• Do you share this view?	Yes, No	1, 0
4	• People Believe that HIV AIDS is a result of sin / punishment		
	• Question		
	• Do you share this view?	Yes, No	1, 0
5	• Some people Believe that People contract HIV AIDS because they were not following the word of god		
	• Question		
	• Do you share this view?	Yes, No	1, 0

The coding of data, entry and cleaning was carried out using Epi 6.04 and analysis part was carried out accordingly.

Binary logistic regression was undertaken to explore the factors associated with adherence at $p < 0.05$, to provide odds ratio [OR] and 95% confidence intervals [CI]. First assessed bi-variate associations between the dependent and independent variable, and seven covariates not significantly associated [$P > 0.10$] with the outcome variable were dropped from further consideration in modeling. The remaining candidate covariates were entered in to a multiple logistic regression model, adjusted for the other remaining covariates.

For analysis purpose age, income and education were categorized in to groups.

3. Result

3.1 Background characteristics

A total of 954 Participants were invited to the study, of which 54 refused to accept invitation, and the response rate was 94.34%, 28 had missing Adherence data and omitted from analysis.

The median age of respondent's was 34 years, IQR [28 - 40], with a peak at age group between 30-44 years. Most of the respondent's 60% [523] were females. Twenty four percent [209] of respondents were married,

40.3% [351] were single, and these findings showed that majority of respondents were either married or single. Majority of respondents, 58.8% [513] said that they have child. Most respondents 56.8% [495] had attained Secondary education; the respondents who had attained college education were 15.5 % [135].

Most of the respondent's, 72.5 percent of the respondents were employed. Majority of respondent's 45.4% [396] said that their income were less than 350.00 [Eth Birr], and 10.9% [95] said their income were more than 1000.00 [Eth Birr]. 43.46% [379] of respondent's knew that ARV treatment reduces the viral load and therefore prevent progression to AIDS, 44.84% [391] said ARV treatment reduces pain. 43.6% [354] of respondent's said that they had emotional and practical support from community.

Majority of respondent's 55.4 % [483] said that they disclose their HIV status to friends or relatives. 80.7 % [704] said that they did not avoid friends, 78.9% [688] respondent's said that they were not using social drugs.

The background characteristics of 872 Participants in the study shown in Table 2

3.1.1 Spiritual characteristics

All respondents had denomination to religion, majority of them 71.8% [626] were orthodox Christian. Majority of respondent's 83.7% [730] said that they regularly attend religious place, 51.1% [446] said that they have practiced fasting. Spiritual believes; 45.3% [395] said that they partially believe that one can cure with faith the sick person, 47.5% [414] believe that HIV AIDS is a result of sin/punishment, 35.3% [308] believe that they have contracted HIV AIDS because they were not following the word of god. Further spirituality measured according to spiritual characteristics, those who score 0, categorized as non-spiritual, scored 1 – 2, Moderate spiritual, and scored 3 – 5, high spiritual, accordingly, 13.1% [114] were non-spiritual, and 57 % [497] were high spiritual.

3.1.2 Factors associated with adherence

3.1.2.1 Demographic factors

On bi-variate analysis sex, marital status, education, employment, Knowledge about use of ARV, were not found to be associated with non-adherence. Aged 31–44 was associated with non-adherence **2.127 [1.021 – 4.431]** at P< **0.05**. Table 3 depicts demographic and socio-cultural factors associated with non- adherence.

Table 2. Background characteristics

Characteristics	Frequency	%
Sex		
Male	349	40
Female	523	60
Age		
< 30	336	38.5
31 – 44	407	46.7
>45	129	14.8
Religion		
Orthodox Christian	626	71.8
Protestant Christian	105	12
Catholic Christian	16	1.8
Muslim	125	14.3

Characteristics	Frequency	%
Marital Status		
Married	209	24
Div / Sep / widowed	312	35.8
Single	351	40.3
Educational Status		
< 4	242	27.8
5 – 12	495	56.8
>12	135	15.5
Occupation		
Employed	632	72 .5
Unemployed	240	27.5
Monthly Income		
< 350	396	45.4
351 – 1000	381	43.7
>1000	95	10.9
Having child		
Yes	513	58.8
No	359	44.2
Community support		
Yes	354	40.6
No	518	59.4
Knowledge		
Yes	379	43.5
No	493	56.5
Social drug use		
Yes	184	21.1
No	688	78.9

Characteristics	Frequency	%
Stigma		
Yes	168	19.3
No	704	80.7
Disclosure		
Yes	483	55.4
No	389	44.6

Table 3. Association of demographic and socio-cultural factors with adherence to ARV in Addis Ababa, May, 2010

Study variables	Adherent N	Non Adherent N	Crude Odds ratio [95% CI]
Sex			
Male	303	46	1.15 [0.764 - 1.731]
Female	462	61	1
Age			
< 30	294	42	1.905 [0.899 – 4.035]
31 – 44	351	56	2.127 [1.021 – 4.431]
>45	120	9	1
Marital Status			
Married	176	33	1.213 [0.749 – 1.964]
Div/ Sep/ Wid	285	27	0.613 [0.372 – 1.010]
Single	304	47	1
Educational Status			
< 4	217	25	1.08 [0.53 – 2.19]
5 – 12	426	69	1.52 [0.813 – 2.84]
>12	122	13	1
Occupation			
Employed	555	77	1
Unemployed	210	30	1.03 [0.656 – 1.616]

Monthly Income			
< 350	361	35	0.56 [0.288 – 1.091]
351 – 1000	323	58	1.039 [0.552 – 1.955]
>1000	81	14	1
Having child			
Yes	458	55	1.41 [0.94 – 2.116]
No	307	52	1
Community support			
Yes	310	44	1
No	455	63	0.976 [0.647 – 1.472]
Knowledge			
Yes	341	38	1
No	424	69	1.46 [0.959 – 2.224]
Social drug use			
Yes	156	28	1.384 [0.869 – 2.204]
No	609	79	1
Stigma			
Yes	151	17	0.768 [0.444 – 1.329]
No	614	90	1
Disclosure			
Yes	427	56	1
No	338	51	1.151 [0.767 – 1.726]
Spirituality			
Non Spiritual	107	7	1
Moderate Spiritual	236	25	1.619 [0.671 – 3.861]
Highly Spiritual	422	75	2.717 [1.217 – 6.065]

3.1.2.2 Socio-cultural factors

On bi-variate analysis Stigma, disclosure, Community support, social drug use were not found to be associated with non-adherence. High Spirituality was associated with non- adherence 2.717 [1.217 – 6.067] at P< 0.05. Variables which had P < 0.05 in the bi-variate binary logistic regression analysis were also reevaluated again with adherence as a dependent variable and controlling for the effect of independent variables (age, marital status, having child, income, knowledge and spirituality). Variables found to be associated with adherence on the multivariate analysis; age became non-significant at p<0.05, the others, marital status, having child, income, knowledge remained non-significant, on multivariate analysis high Spirituality 2.781 [1.230 – 6.287] at P< 0.05 times more likely to report non-adherence, having a relatively increased OR. Table 4 depicts demographic and socio-cultural factors associated with non- adherence.

Table 4. Association of demographic and socio-cultural factors with adherence to ARV in Addis Ababa, May, 2010

Study variables	Adherent N	Non Adherent N	Crude Odds ratio [95% CI]	Adjusted* Odds ratio [95% CI]
Age				
< 30	294	42	1.905 [0.899 – 4.035]	1.554 [0.706– 3.422]
31 – 44	351	56	2.127 [1.021 – 4.431]	1.913 [0.905 – 4.045]
>45	120	9	1	
Marital Status				
Married	176	33	1.213 [0.749 – 1.964]	1.126 [0.624 – 2.031]
Div/ Sep/ Wid	285	27	0.613 [0.372 – 1.010]	0.646 [0.386 – 1.081]
Single	304	47	1	
Monthly Income				
< 350	361	35	0.56 [0.288 – 1.091]	0.612 [0.306 – 1.224]
351 – 1000	323	58	1.039 [0.552 – 1.955]	1.116 [0.582 – 2.141]
>1000	81	14	1	
Having child				
Yes	458	55	1.41 [0.94 – 2.116]	0.76 [0.455 – 1.292]
No	307	52	1	1
Knowledge				
Yes	341	38	1	1
No	424	69	1.46 [0.959 – 2.224]	1.508 [0.978 – 2.324]
Spirituality				

Study variables	Adherent N	Non Adherent N	Crude Odds ratio [95% CI]	Adjusted* Odds ratio [95% CI]
Non Spiritual	107	7	1	1
Moderate Spiritual	236	25	1.619 [0.671 – 3.861]	1.639[0.677 – 3.968]
Highly Spiritual	422	75	2.717 [1.217 – 6.065]	2.781 [1.230 – 6.287]

After adjusting for; age [grouped], marital status, income [categorized], having child and knowledge and spirituality.

3.2 Discussion

3.2.1 Level of Adherence

The result shows higher level of adherence in a large proportion of HIV AIDS patients, the rate of overall adherence was better to those reported from previous studies in Addis Ababa and sub-Saharan and developed countries [5, 6]. The study revealed that they were two types of adherence, Adhering to drugs and food on time and adhering to drug on time only, of which customary fasting practice is the main reason for sub optimal adherence.

Most patients with sub-optimal adherence report that they were taking ARV with out food on time, and some also reported that they took ARV with food but used to shift time. These practices alter the pharmacokinetic and Pharmaco-dynamic effect of ARV, especially those with narrow therapeutic index and consequently their effectiveness and tolerability. These may exacerbate the occurrence of virologic failure and drug resistance.

Some practice from chronic illness like Diabetes mellitus, hypertension poor adherence to prescribed drugs during fasting period may lead to therapeutic failure [7, 8].

The possible occurrence is using drugs with short half life, two doses daily; dosing may lead to sub therapeutic level during customary fasting practice. The practice like adhere to drug on time and dietary instruction in case of participants, who used to shift time, may ingesting high heavy fatty meals predawn and after sun set.

Food considerably reduce the absorption of protease inhibitors , and fatty meals may reduce the absorption and bioavailability of thymidine analogues, zidovudine [9]. Therefore fasting practice in all religious denomination make it difficult to take drugs that should be taken with food, particularly for drugs dosed twice daily.

Customary fasting practice may make difficult to predict blood levels of drug concentration, and the concentration may be in sufficient to suppress viral population.

3.2.2 Factors associated with adherence

3.2.2.1 Demographic factors

Though common predictors of non-adherence include socio-demographic factors, this study did not report association between socio-demographic factors; age, marital status, having child and income, and adherence. Literatures reported that certain socio-demographic variables have influence over adherence to HAART; however, others showed no association [10] [11-13].

3.2.2.2 Socio-cultural factors

The study suggests that while belong to religious denomination plays significant role in people's attitude about HIV, More consistent associations are found between socio-cultural factors; high spirituality and adherence. The finding corroborate those from studies that showed the greater level of spirituality were associated with health outcomes [14-16].

Previous studies corroborate the obtained finding that African Americans tend to be more spiritual and utilize more religious coping [both positive and negative] than their Caucasian counter part. [17].

Though common predictors of non-adherence include social drug use, lack of community support, and HIV status disclosure, and Knowledge, this study did not report association between aforementioned factors and adherence [18-20].

In this study in contrast to others most socio-demographic and socio-cultural factors had no association with non-adherence, a number of factors might explain the difference: the difference in the assessment methodologies, the cut off points, the study setting, the time of the study and others, and also this may as a result of dynamic patient behaviors toward adherence [21], which are influenced by multiple factors varying over time.

3.3 Limitations of the study

Data on adherence was collected by self report and there might a probability for information bias, mainly recall and social desirability biases. [22]

3.4 Conclusion and Recommendation

Since being spiritual is very common characteristics among most HIV AIDS patients, and high spirituality shows significant association with outcome of interest, thus will have serious implication for emergence of resistance strains and eventually clinical failure. Therefore, strengthen the existing adherence-counseling strategy, involving religious leaders in designing and implementing continues patient education tools and operational research are recommended

Abbreviation: AIDS: Acquired Immune Deficiency Syndrome; ART: Antiretroviral therapy; ARV: Antiretroviral drugs used for the treatment of HIV infection; HAART: Highly Active Anti-retroviral Therapy; HIV: Human immunodeficiency virus; OR: Odds Ratio; Eth Birr: Ethiopian Currency

Acknowledgment: My special thanks are extended to my advisors **Dr. Negussie Deyessa** and **Mr. Temesgen Workayehu** for their technical support and constructive comments.

REFERENCE

[1] Eric, S.e.a., *Improving Adherence to Antiretroviral Therapy.* . Cliggott Publishing, Division. AIDS, 2003: p. 13(2):81-90.

[2] Monforte, A., L. Testa, and F.e.a. Adorni, Clinical outcome and predictive factors of failure of highly active antiretroviral therapy in antiretroviral-experienced patients in advanced stages of HIV-1 infection. . AIDS 1998: p. 12:1631-7.

[3] Paterson, D.L., S. Swindells, and J.e.a. Mohr, Adherence to protease inhibitor therapy and outcomes in patients with HIV infection. 2000: p. 133:21-30.

[4] INRUD and MOH, Survey of Adherence to Antiretroviral Medicine in Ethiopia. 2007: p. 13.

[5] Millis, E.J., J.B. Nachenga, and D.R.e.e. Bangsberg, Adherence to HAART: A Systematic Review of Developed and Developing Countries. 2006.

[6] Ivers, L., D. Kendric, and K. Doucette, Efficacy of antiretroviral therapy programs in resource poor settings: A meta analysis of the published litrature. HIV AIDS clinical Infection diseases 2005. 41(217 - 224).

[7] Aadil, N., I.E. Houti, and S. Moussamih, *Drug intake during ramadan* Britheish medical journal, 2004. 329: p. 778 - 782.

[8] Benaji, B., N. Mounib, and R.e.e. Rocky, *Diabetes and Ramadan : Reveiw of the literature* Diabetes resereach and clinical practice 2006. 73: p. 117 - 125.

[9] Habib, A. and M.e.e. Shepherd, Adherence to antiretroviral therapyduring muslim ramadan fasting. Springer science 2008.

[10] John Schneider, H. Kaplan Sherrie, and e.e. Sheldon Greenfield, Better Physician-Patient Relationships Are Associated with Higher Reported Adherence to Antiretroviral Therapy in Patients with HIV Infection. Gen Intern Med, 2004. 19:1096-1103.

[11] Nachega Jean, B., Michael Hislop, and W.e.e. Dowdy David, Adherence to Highly Active Antiretroviral Therapy Assessed by Pharmacy Claims Predicts Survival in HIVInfected South African Adults. J Acquir Immune Defic Syndr, 2006. 43(1):1-7.

[12] Carballo, E., C. Cadarso-Suarez, and I.e.e. Carrera, Assessing relationships between health-related quality of life and adherence to antiretroviral therapy. Quality of Life research 2004. 13(3):587-99.

[13] Berg Karina, M., A. Demas Penelope, and A.e.e. Howard Andrea, Gender Differences in Factors Associated with Adherence to Antiretroviral Therapy. J Gen Intern Med, 2004. 19:1111-1117.

[14] Coleman, C. and W. Holzemer, Spirituality, Psycological well being and HIV symptoms for African Americans living with HIV AIDS. J Assoc Nurses AIDS Care, 1999. 10; 42 - 50

[15] Pargament, k., S. McCarthy, S, and P.e.e. haash, Religion and HIV; a review of the literature and clinical implications. S Med J, 2004. 97; 1201 - 9
[16] Truck, I., N. MeCain, and R. ELswick, Spirituality and Psycho-social factors in persons living with HIV AIDS. J Adv Nurs, 2001. 33; 776 - 83
[17] Peterman, A., G. Fitchett, and M.e.e. Brady, Measuring spiritual well-being in people with cancer: the functional assessment of chronic illness therapy-Spiritual Well-being Scale (FACIT-Sp). Ann Behav Med, 2002. 24:49-58.
[18] Bruno, S. and S.e.e. Marc, Adherence to highly active antiretroviral therapies (HAART) in HIV-infected patients: from a predictive to a dynamic approach. Soc Sci Med, 2002. 54:1481-1496.
[19] Stirratt Michael, J., H. Remien Robert, and e.e. Anna Smith, The Role of HIV Serostatus Disclosure in Antiretroviral Medication Adherence. AIDS Behav, 2006. 10(5):483-93.
[20] Kumarasamy, N., A. Safren Steven, and R.e.e. Raminani Sudha, Barriers and facilitators to antiretroviral medication among HIV positive patients in Chennai, India: a qualitative study. AIDS Patient Care and STDs, 2005. 19(8):526-537.
[21] Carrier, P. and V.e.e. Cailleton, The dynamics of adherence to highly active antiretroviral therapy: results from the french national APROCO cohort, A two year follow up. Acquire Immune Deficency syndrome 2001. 28(232 - 239).
[22] Chesney, M., Factors affecting adherence to antiretroviral therapy. Clinl Infec Dis, 2000. 30:S171-S76.

Effectiveness and Safety of Miconazole with Hydrocortisone (Daktacort) Feminine Care Cream in the Treatment of Vulvar Candidiasis

J. Perez-Peralta*[1] and G. Balaccua[2]

[1]Quirino Memorial Medical Center, Katipunan Rd., Project 4, Quezon City, Philippines
[2] Janssen Pharmaceutica, Edison Road, Barrio Ibayo, Paranaque City, Philippines

* Corresponding author: e-mail: rxjudypmd@yahoo.com. Phone: +632 7184609

This is a multicenter, open label, single arm prospective phase IV clinical trial to determine the effectiveness and safety of miconazole + hydrocortisone (Daktacort) feminine care cream in the treatment of vulvar candidiasis. Female patients 20 – 60 years old with confirmed vulvar candidiasis were enrolled, treated with miconazole + hydrocortisone cream twice daily and followed-up after 14 days and if necessary, 28 days. A total of 115 patients were enrolled. Out of these, pruritus relief was achieved in in 110 subjects (96%) within 1 hour. Time to itch relief was as fast as 3 minutes in 5% of subjects and 9 minutes in 57%. Seventy-four followed up until day 14 and seven until day 28, while 43 did not follow-up. Itch relief profile was similar between those who followed-up and those who did not. Cure rate was 88% at day 14 and 97% at day 28. There was a significant improvement in quality of life based on Modified Itch Severity Scale scores. There were 2 non-serious (pruritus) and no serious adverse events. It is concluded that miconazole + hydrocortisone cream is effective and safe for vulvar candidiasis.

Keywords: miconazole + hydrocortisone cream; vulvar candidiasis

1. Introduction

Vulvovaginal candidiasis occurs in 75% of healthy women at least once in their life [1]. Of all the causes of non-venereal vulvovaginal infections, perhaps the most common culprit is *Candida*. Candidiasis is diagnosed through a KOH mount, which has a sensitivity of 50-80% [2, 3] and its characteristic signs and symptoms. About 40% to 75% of sexually active women have experienced the symptoms of vulvovaginal candidiasis: itching, irritation, soreness, pain with intercourse, burning on urination, and a white, curd-like discharge. Chronic vulvovaginal candidiasis ranks 3rd (10%) among the most common causes of chronic vulvar symptoms referred to a dermatologist [4]. Though not life-threatening, vulvar candidiasis causes pruritus, genital discomfort, reduced sexual pleasure, and psychological distress leading to a decrease in productivity and impaired quality of life [5].

In the past, gentian violet 1% solution, sitz baths with sodium bicarbonate or Burrows solution have been used for pruritic relief, but with limited effectivity. Boric acid is 90% effective, however patients using it have often noted a slight watery discharge. It is also toxic when taken orally [3].

According to the Omaha Classification of vulvovaginal candidiasis (VVC), Genesis I or primary VVC is a spectrum ranging from predominant vulvitis to vaginitis. Genesis III is systemically influenced VVC. It is recommended that a topical antifungal cream may be used alone in Genesis I VVC or as an adjunct to oral and intravaginal antifungals in the treatment of Genesis III VVC. Oral or intravaginal antifungals are recommended for the treatment of Genesis II or antibiotic-associated VVC. However, in cases where vulvitis is clinically significant, simple eradication of the vaginal reservoir will not afford prompt relief of symptoms. Total reliance on intravaginal medications in cases with significant vulvitis is associated with frequent therapeutic failures. The physician must treat for vulvitis and this is best achieved thru direct application of medication to the perineum and vulva [6].

Women with a history of repeated infections require extended therapy with either topical or oral treatment. Extended topical therapy is also indicated for pregnant women who were not cured by single-application, high-dose topical therapy [2]. No particular adverse effect in the mother or fetus was noted with miconazole. The CDC recommendations for pregnancy are limited to 7 day treatment with topical azoles, with miconazole among the list of those with most effective results [3].

The vulva is a neglected area of the skin, hidden away by ignorance, embarrassment, and taboo with the result that women with vulvar diseases, desperate for help, suffer needlessly [7]. Approximately 75% of patients use over-the counter medications [8] such as miconazole, clotrimazole or povidone-iodine. The use of over-the counter medicines in women with chronic vulvovaginal symptoms has progressively increased over the years primarily because no known effective treatment exists for the management of recurrent vaginal yeast infections [9].

With many of these preparations available over the counter, one must be aware that self-diagnosis of yeast vaginitis is unreliable. Incorrect self-diagnosis results in overuse of topical antifungal agents, with subsequent risks of contact or irritant vulvar dermatitis and delayed diagnosis of other conditions [3]. Self- medication with over the counter medications is recommended only for women who previously have been diagnosed with vulvovaginal candidiasis and who have recurrence of the same symptoms. The CDC guidelines recommend that any woman whose symptoms persist after using an over the counter preparation or has recurrence of symptoms within 2 months should seek medical care. Culture should also be done for confirmation [10].

This is a Phase IV study, which aims to determine the effectiveness and safety of miconazole with hydrocortisone (Daktacort) feminine care cream in the treatment of patients with vulvar candidiasis. Study results will help guide clinicians and consumers on the correct usage of this "over-the-counter" drug.

2. Study Objectives

2.1. To determine the effectiveness of minconazole + hydrocortixone cream for the treatment of vulvar candidiasis based on cure rate, pruritus relief and quality of life improvement using the Modified Itch Severity Scale.

2.2. To determine the safety of miconazole + hydrocortisone cream for the treatment of vulvar candidiasis based on the occurrence of adverse events and incidence of discontinuation due to adverse events.

3. Methods

3.1 Study Design

This was a multicenter, open-label, single arm prospective phase IV clinical trial conducted in Quirino Memorial Medical Center and Dr. Jose Fabella Memorial Hospital in the Philippines.

3.2 Inclusion and Exclusion Criteria

Patients may be included in the study if they are female, 18 to 60 years old; with positive KOH mount of vulvar scrapings; have vulvar candidacies with at least mild pruritus; and have signed the informed consent to participate.

Patients are excluded from the study if they are pregnant or lactating; diagnosed with vaginal candidias; with a history of sexually transmitted infections; with allergy to miconazole and/or hydrocortisone; with prutitus due to other etiologic agents aside from fungi; or with tuberculous skin infection, varicella, or vulvar malignancy.

3.3 Treatment and Evaluations

After obtaining informed consent, patients who passed screening based on inclusion/exclusion criteria, were enrolled. Baseline past history and demographic data were taken as well as the Modified Itch Severity Scale scores.

Subjects were given a 2 week supply (2 tubes of 3.5g each) of study drug. Time for pruritus relief was noted after the first application before the patient left the clinic. Each subject was instructed to apply the study drug topically twice daily until follow up after 14 (± 3) days for study evaluations. A patient was considered cured if repeat KOH mount of vulvar scrapings tested negative. If repeat KOH mount remained positive, the patient was given another 2-week supply of study drug and asked to follow up after 28 (± 3) days from baseline. Persistence of KOH mount positivity at 28 days was considered as treatment failure and the patient was referred to an obstetrician-gynecologist for further management.

3.4 Evaluation Criteria

The effectiveness of miconazole + hydrocortisone cream was evaluated using cure rate, pruritus relief, and quality of life. Cure was based on clinical evaluation and confirmation with KOH mount and was assessed at day 14 and day 28. Time to pruritus relief was determined as the number of minutes from the first application to pruritus relief. Quality of life was determined using the Modified Itch Severity Scale, which employs a 6-section questionnaire evaluating degree of pruritus based on frequency, extent and intensity of itch as well as the effect of itch on the patient's mood, sexual desire and function, and sleep to obtain a modified itch severity score. The modification of the scale consists in assuming a 1% extent for body surface area involved for all patients, corresponding the the estimated surface area of the vulvar area.

The safety of miconazole + hydrocortisone cream was determined using the frequency and severity of adverse events/reactions as well as the frequency of discontinuation due to adverse events.

3.5 Sample Size and Statistical Treatment of Data

Estimated sample size was 114 based on 80% power, alpha of 0.05, assumed 80% cure rate, 10-point margin of error and adjusting for an assumed 50% drop-out rate. Cure rate was computed as the percentage of patients cured at days 14 and 28. Time to pruritus relief was analyzed using Kaplan Meier Plot. Changes in the individual itch severity scale items from baseline to study end were analyzed using chi-square test, while the Modified Itch Severity score change from baseline to study end was analyzed using paired t-test. Analysis of the modified itch severity items was done on intent-to-treat principle. Frequency of adverse events and adverse event discontinuations was also determined.

4. Results

4.1 Patient Profile

A total of 115 patients were enrolled in the study. Of these, 74 subjects followed-up at day 14 and 41 did not. Among these 74 subjects, 65 were cured, 7 continued treatment and followed-up at day 28 and 2 did not. There was a total of 43 patients (37%) who did not follow-up. All patients who followed-up at day 28 were cured.

Table 1 shows the demographic profile of the enrolled subjects. Mean age was 37.9 years and mean body mass index (BMI) was 23.7 kg/m^2. Most (87.9%) were sexually active and on the average those with children have around 3.

Table 1. Demographic profile of subjects[1]

Demographic characteristic	Value
Age (years)	
Mean ± standard deviation (Range)	37.8 ± 6.4 (20.0 – 50.0)
Body mass index, BMI (kg/m^2)	
Mean ± standard deviation (Range)	23.7 ± 4.5 (15.1 – 38.0)
Sexually active	87.9 %
Number of children[2]	
Mean ± standard deviation (Range)	2.8 ± 1.7 (1.0 – 9.0)

[1]n = 115; [2] n = 85 (30 had no child)

4.2 Concomitant Medications and Diseases

Around one- sixth (17%) of the patients used medication for vulvar itch. Almost one- third (29%) of had used various forms of antifungals prior to enrollment, yet still continued to complain of vulvar pruritus. A similar number had prior intake of antibiotics, most of which had broad-spectrum coverage. Among the most common concomitant medications were hypertension (7 patients) and diabetes (3 patients).

4.3 Effectiveness

The Kaplan Meier Plot of time to itch relief among enrolled subjects is shown in Figure 1. Mean time to itch relief was around 13 minutes. Majority of subjects (57%) experienced relief in ≤ 9 minutes, and 5% in as fast as 3 minutes. Most patients (96%) achieved itch relief in 1 hour or less. Times achieved for pruritus relief in the drop-out group were similar.

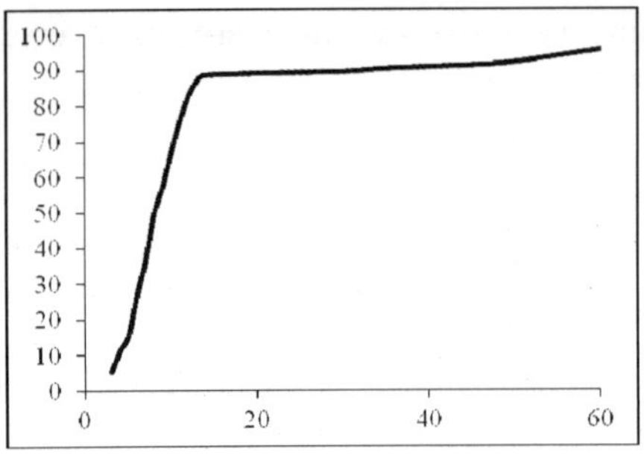

Figure 1. Time to itch relief, all subjects (n = 115). Most patients (96%) achieved itch relief in ≤ 1 hour.

Figure 2 shows the cure rate of subjects in the study, which was around 88 % at day 14, as 65 out of the 74 subjects had a negative KOH mount result by day 14. An additional 7 subjects achieved cure on day 28, increasing the number to 72 out of 74 subjects with a negative KOH mount result, giving a cumulative cure rate of around 97 %. The 2 patients with a positive KOH mount result on day 14 failed to follow up on Day 28.

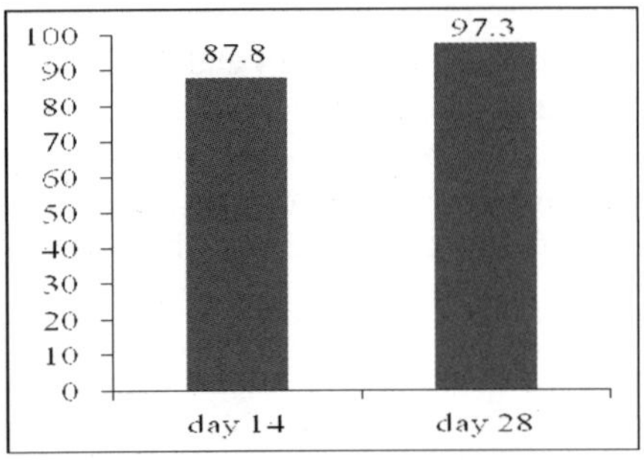

Figure 2. Percentage cure rate of subjects. KOH mount confirmed cure rate was around 88% at day 14 and 97% at day 28.

For the Modified Itch Severity Scale items, there was a significant decrease in pruritus frequency, extent and intensity over time. There was also a significant decrease in mood changes, and improvement in sleep and sexual desire and function. Figure 3 shows the change in the Modified Itch Severity Scale score from baseline to study end, showing a highly significant decrease from 7.2 to 4.3 ($p < 0.0001$). Based on this, quality of life was improved due to treatment with miconazole + hydrocortisone feminine care cream.

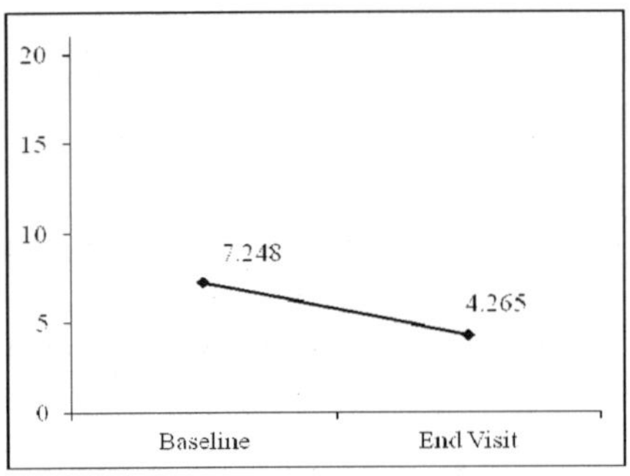

Figure 3. Modified Itch Severity Scale score. Score considers frequency, extent and intensity of itch as well as its effect on mood change, sexual desire and function, and sleep. Score significantly decreased from around 7.2 at baseline to 4. 3 at study end ($p<0.0001$).

4.4 Safety

The safety analysis included the complete data set on all patients and all visits. Only 2 (1.7%) of patients experienced adverse events. This would translate to 98% of patients having no adverse event during the study period. Both patient experience pruritus after application. One patient (<1%) dropped-out due to adverse event. The patient recovered after change in medication. One patient continued medication and the event resolved. This patient also achieved clinical cure after 14 days of treatment. No serious adverse event was noted.

5. Discussion

Most patients in this study were in the childbearing age. Recurrent vulvovaginal infection is a prevalent disease especially among women of child-bearing age [11]. However, the results show that this may also be true for vulvar candidiasis. Normal and overweight to obese patients were both affected by vulvar candidiasis. This may be explained that obesity is commonly noted to be a predisposing feature to yeasts in the vagina rather than cases of simple vulvitis [3].

In cases where vulvitis is clinically significant, simple eradication of the vaginal reservoir will not afford prompt relief of symptoms. Total reliance on intravaginal medications in cases with significant vulvitis is associated with frequent therapeutic failures. This study's results support this by showing that 29% of the subjects had used various forms of antifungal prior to enrollment, yet still continued to complain of vulvar pruritus.

Previous intake of antibiotics suppress normal bacterial population and allow opportunistic colonization by yeasts [3]. This is supported by this study's results showing that 29% had prior intake of antibiotics, most of which had broad-spectrum coverage.

Seven subjects had hypertension while three subjects had diabetes mellitus. Diabetics are known to be high risk for vulvovaginal candidiasis development. Calcium channel blockers used for hypertension and oral hypoglycemic used for diabetes mellitus treatment may have clinically important interactions with oral ketoconazole or fluconazole [3]. These interactions may not be likely in topical use of miconazole because all subjects who completed treatment were able to achieve cure.

The cure rate of miconazole + hydrocortisone cream was 88% to 97%. This study's cure rate range was actually higher than the 80-90% cited in medical literature [3]. The drop-out rate of 37% was analyzed to see if a possible cause for this was early dissatisfaction with the study drug. Results were similar between the entire group and those who did not follow-up. The mean time of itch relief was 13 minutes among the subjects in the entire group, compared with 12 minutes among the drop-out group. Itch relief within one hour was achieved in 96% of the subjects in the entire group, compared to 98% in the drop-out group. Itch relief within 3 minutes was achieved in 5.2% in the entire group, compared with 4.7% in the drop-out group. The median time was 9.0 min among those in the entire group while it was 8.5 minutes among the drop-out group. Based on this, it could be inferred that failure to follow up was unlikely due to disappointment with the drug. Perhaps like many patients who have already achieved relief, the drop-out group did not see further need for a return visit. This study succeeded in showing how an effective drug can improve quality of life by quantifying and qualifying the vulvar pruritus experienced by an individual. Up until this time, a scientific and concrete way of doing this has eluded clinicians and patients. This study's application of the Modified Itch Severity Scale score demonstrates how evidence can be gathered to show that there was a highly significant decrease in the frequency, extent, intensity of the vulvar pruritus, over time. There was also improvement of mood, sleep changes and sexual desire and function due to relief from vulvar pruritus. Overall, there was a highly significant decrease in the Itch Severity Scale Score over time. Based on this, there was improvement in quality of life due to the treatment with miconazole + hydrocortisone cream.

Only 2 (1.7%) of patients experienced adverse events. Both patients experienced pruritus after application. There were no serious adverse event with the treatment of miconazole + hydrocortisone cream. On occasion, the topical application results in increased local symptoms, such as burning and irritation. Relief of these local symptoms may be achieved by application of low-potency topical corticosteroids [3].

6. Conclusion

Miconazole + hydrocortisone (Daktacort) feminine care craeam is effective and safe in curing vulvar candidiasis with rapid resolution of vulvar itch, leading to improved quality of life.

Acknowledgment Grant support for this work was provided by Janssen Pharmaceutica. We thank Dr. Guillerma Panay of Dr. Jose Fabella Memorial Hospital. We also thank Dr. Rosalinda Arandia, Dr. Hedeliza Bellosillo, Dr. Emmaline Masangkay of Quirino Memorial Medical Center. We also acknowledge the technical support provided by MacGerald Cueto, Janus Relos, Jean Baria, and Ariane Dayag

References

[1] Calero I, Centeno C. Folia Clinica en Obstetrica y Ginecologia. 2005; 53; 32.
[2] Boardman LA, Boardman LA. *Women Health Primary Care* 2000; 3(12):857-862.
[3] Sweet RL, Gibbs RS, Eds. *Infectious Diseases of the Female Genital Tract.* Philadelphia: Woterrs Kluwer/ Lipincott Williams & Wilkins. 2009.
[4] Go F. Australas J Dermatol. 1996; 37(1); 12-18.
[5] Eschenbach DA. *The New England Journal of Medicine.* 2004. 351 (9): 851 – 852.
[6] Monif GRG, Baker DA. *Infectious Diseases in Obstetrics & Gynecology.* 6th Ed. Ohama, NE: IDI Publications. 2008.
[7] Margesson, LJ. *Dermatol Clin. Vulvar Disease Pearls.*2006; 24 (2); 145-155.

[8] Sobel JD, Faro S, Force RW, *et al. American Journal of Obstetrics & Gynecology*. 1998; 178 (2); 203-211.
[9] *Patient Care. Mississauga*. 2005; 16 (2); 131.
[10] Centers for Disease Control and Prevention. *Guidelines for treatment sexually transmitted diseases.* MMWR 2006; 55: 54-55.
[11] Giraldo P, von Nowaskonski A, Gomes FAM, *et al. Obstetrics & Gynecology*. 2000; 95 (3); 413-416.
[12] Welsh BM, Berzins KN, Cook KA, Fairley CK. *Med J Aust.* 2003; 178(8):391-395.
[13] Majeski CJ, Johnson JA, Davison SN, *et al. British Journal of Dermatology, Canada*/ 2007; 156; 667-673.

Natural Products:

Terrestrial and Marine Organisms

Analysis of the 2-Phenylethyl isothiocyanate present in *Brassica* leaves and their potential application as antimicrobial agent against bacteria strains isolated from Human and Pig gastrointestinal tracts

A. Aires[*,1], C. Dias [1,2], R. N. Bennett[1], E.A. S. Rosa[1] and M. J. Saavedra[2]

[1]CITAB/UTAD-Centre for the Research and Technology for Agro-Environment and Biological Sciences. Agronomy Department, University of Trás-os-Montes and Alto Douro, , P.O. Box 1013, 5000-801 Vila Real, Portugal.

[2]CECAV/UTAD-Veterinary and Animal Science Research Centre. Department of Veterinary Science, University of Trás-os-Montes and Alto Douro, P.O. Box 1013, 5000-801 Vila Real, Portugal.

* Corresponding author: e-mail: alfredoa@utad.pt, Phone: +351 259350435.

This work aims to evaluate the effect of 2-Phenyethyl isothiocyanate (PEITC), a predominant isothiocyanate derived from glucosinolate named glucoranturtiin, particularly important in *Brassica rapa* (turnip-inflorescences, leaves, stems and roots) and *Nasturtion officinalis* (watercress), against pathogenic Gram-negative and Gram-positive bacteria, isolated from Human (clinical) and animal gastrointestinal segments. Also standard bacteria strains were sued in the study. Using a disc diffusion method, different doses of PEITC were tested. The compound was dissolved in dimethyl sulfoxide (DMSO) and tested at 5 different concentrations using a standard disk-diffusion assay (0.015, 0.15, 0.75, 1.50, and 3.00 μmoles). Positive (commercial standards antibiotic) and negative controls (DMSO only) were always included in the tests. All the experiments were conducted in triplicate. The antibacterial activity was also assessed by the application of the percentage of relative inhibition zone (%RIZD) in relation to the antimicrobial effect of commercial standard antibiotics. The results showed that, there were clear concentration differences ($P < 0.05$) with respect to the *in vitro* growth inhibition effects as well as differences in the sensitivities of the individual bacteria. The antimicrobial activity of PEITC was always a dose-dependent. The results also showed that antimicrobial activity of PEITC was higher in Gram-positive and lower in Gram-negative. Our results are useful and suggest that PEITC might be useful in controlling Human and animal pathogenic bacteria's growth and therefore could be considered as alternative to, or in combination with, the current antibiotic-based controls used for treating pathogenic bacteria. However, further studies are needed to determine their MIC values, to evaluate their commercial and economical effectiveness, the eventual synergistic effects with others compounds, as found in the consumption of the *Brassica* foods; toxicity bioassays must as well be performed.

Keywords: Phytochemycals; isothiocyanates; PEITC; antimicrobial activity; health.

1. Introduction

The sulphur secondary plant metabolites named glucosinolates occurs in plants simultaneously with an endogenous β-thioglucosidase enzyme commonly named myrosinase (E.C. 3.2.1.14.7), which is responsible, after tissue disruption (e.g. cut, ground or chewed), for their hydrolysis into diverse biologically active products such as isothiocyanates (ITC), nitriles, thiocyanates and various indole compounds from the indolylglucosinolates [1]. Over the past 30 years, it has been shown that many classes of secondary metabolites (phytochemicals) including glucosinolates and specifically their hydrolysis products, are important bioactives in the diet and they can have an important role as a therapeutic tool [2]. Nonetheless their biological effects are dependent on the levels and classes of glucosinolates present [3]. Sulforaphane (SFN), derived from 4-methylsulfinylbutylglucosinolate (glucoraphanin), has been the focus of many researchers because of its potent anticarcinogenic activity [4], and recently has been the focus of antimicrobial studies with *Helicobacter pylori,* a potential causal agent of stomach cancer [5-7]. Various microorganisms, including some intestinal bacteria, are known to be capable of metabolizing glucosinolates including *Escherichia coli* [8], *Aspergillus niger* [9], *Geotrichum candidum* [10], *Enterobacter cloacae* [11], *Lactobacillus agilis* R16 [12] and *Bacteroides thetaiotaomicron* II8 [13]. Somes studies [14] have showed that a mixed population of bacteria isolated from human faecal material quantitatively converted sinigrin to allylamine, and glucotropaeolin (benzylglucosinolate) to benzylamine and more recently anointer study [15] using gnotobiotic rats, have shown that both plant myrosinase and microbial myrosinase influence the in vivo catabolism of sinigrin (to AITC) and benzylglucosinolate (to BITC) but that the microbial population can significantly reduce the availability of ITC produced; suggesting either sequestration or further metabolism of the ITC. In the past other studies [16] have shown that volatile compounds from two *Brassicacea* species, including isothiocyanates, were 25% as effective against E. coli, 33% as effective against *Bacillus subtilis*, and equally effective against *Streptococcus lactis* compared with the antibiotic controls. SFN has been shown to effectively inhibit *H. pylori*, the bacteria responsible for human gastritis and stomach cancer [5-7]. SFN can be safely delivered in the diet, particularly from broccoli sprouts, and could be a powerful instrument in the treatment of *H. pylori*-associated gastric diseases. Despite these promising results, fewer studies have been conducted with other hydrolysis products against other human pathogenic or gastrointestinal tract bacteria and lesser involving bacteria strains isolated from animal food sources. Thus, the current *in vitro* study was set in order to evaluate the role of 2-Pheneylethylisothiocyanate, an important ITC derived from the aromatic glucosinolate gluconasturtiin (2-Phenetylethylglucosinolate) largely present in plants like turnips and watercress (largely used daily in human diet), on aerobic bacteria isolated from the human and animal intestinal tract.

2. Materials and Methods

2.1 Glucosinolate hydrolysis product

The isothiocyanate 2-phenylthyl (PEITC) was obtained from Sigma-Aldrich. The compound was tested at 0.015, 0.15, 0.75, 1.5 and 3.0μmoles. The concentrations were prepared using dimethyl sulfoxide (DMSO, Sigma-Aldrich).

2.2 Isolation, identification and growth of bacterial strains

The bacterial strains used in this study were isolated from human faecal material supplied by the Hospital Center of Vila Real / Peso da Régua, North of Portugal. Also animal gastrointestinal segments were collected from slaughters houses in North of Portugal-Vila Real region and from them were isolated different bacterias. Each sample was grown for 24 h at 37 °C, in an enriched medium of Selenit Broth (Oxoid CM395) and isolation was performed in plates with agar SS (Oxoid CM99) and CPSID (BioMerieux 43211) followed by several re-isolation steps. When necessary the samples were recovered with Brain heart infusion broth. Identification of the strains was performed using standard biochemical classification techniques [17] using API 20E, API 20NE, API Staphy and API Step (BioMerieux), according to specific recommendations [18], followed by genetic identification through 16S rRNA sequencing by Dr. Antonio Martinez-Murcia [Molecular Diagnostics Center (MDC), Biomolecular Technologies, S.L. Crta. Ncnal. 340, km. 29, Apdo. 169, Orihuela (Alicante), Spain; http://www.mdc-bt.com]. After identification, the strains were kept in aliquots of brain-heart infusion medium (BHI medium, Oxoid) containing 15% (v/v) glycerol at -70°C. With this procedure, 18 aerobic bacterial strains from clinical source were obtained; Gram-negative: *Acinetobacter baumanii*, *Citrobacter freundii*, *Enterobacter asburiae*, *Enterobacter cloacae*, *Enterobacter hormaechei*, *Echerichia* coli (two strains 1 and 2), *Hafnia alvei*, *Klebsiella oxytoca*, *Klebsiella pneumoniae*, *Morganella morganii*, *Proteus mirabilis*, *Pseudomonas aeruginosa*, *Salmonella typhi* and *Stenotrophomonas maltophilla*; Gram-positive: *Enterococcus faecalis*, *Staphylococcus aureus* and *Staphylococcus saprophyticus*. From animal source were obtained 4 strains two gram-nevative: *Escherichia coli*$_1$, *Escherichia coli*$_5$; and two gram-positive: *Enterococcus faecalis* $_{49}$ and *Enterococcus faecalis* $_{52}$. In addition were added 4 commercial bacterial strains, also two gram -negative: *Escherichia coli* CETC 434 *and Pseudomonas aeruginosa* ATCC 10145; and two gram-positive: *Staphylococcus aureus* CETC 976 and *Listeria monocytogenes* ATCC 15313.

2.3 Screening of antibacterial effects using a disc diffusion method

Isolated colonies from pure strains were transferred from the BHI solid medium grown overnight and inoculated into 4.0 mL of 0.9% NaCl solution. From these stocks, the suspensions used for the bioassays were prepared by adjusting the turbidity to 0.5 McFarland standard units. Antibacterial activity was tested using a modification of the disc diffusion method [19]. A loop of bacteria from the agar-slant stock was cultured in nutrient broth overnight and spread with a sterile cotton swab into Petri dishes (90 mm of diameter) containing 20 ml of Mueller–Hinton Agar (Oxoid). Sterile filter paper discs (6 mm in diameter) (Oxoid) impregnated with 15 ll of the test chemicals were placed on the agar plate seeded with the test microorganism; the plate was incubated in an inverted position overnight at 37 °C. All test compounds were applied, in dimethyl sulfoxide (DMSO, Sigma-Aldrich), at five different concentrations, 0.0, 0.015, 0.15, 0.75, 1.50 and 3.0 μmoles per disc. In all experiments, a negative control (15 μl of solvent, DMSO) and positive controls consisting of standard commercial antibiotics of vancomicin (30 μg) (Oxoid), gentamicin (10 μg) (Oxoid) were used. After overnight incubation, the diameter in mm of the inhibitory or clear zones around the discs was recorded. The antibacterial activity was expressed as the mean of inhibition zone diameters (mm). The antibacterial activity was also assessed by the application of the equation (adapted from a previous study [20]: %RIZD = [(IZD sample-IZD negative control)/IZD antibiotic standard] x100. RIZD is the percentage of relative inhibition zone diameter and

IZD is the inhibition (mm). According to the authors, the current equation compensates for the possible effect of the solvent (blank) other than water on the IZD. Also, the resulting IZD of the samples was either higher or equal to the IZD blanks, and thus the obtained percentages were positive.

2.4 Statistical analysis

All experiments were performed in triplicate. The data were analyzed using One-Way anova. The differences between the mean values were separated using Duncan's comparison test. The results were presented as the mean ± SEM (standard error of the mean). Significance level for the separation was set at $P < 0.05$. Statistical analyses were performed using the statistical program Superanova ver. 1.11 software (Abacus Concepts, Berkeley, CA, USA).

3. Results and Discussion

The results achived with thr current study are presents in Tables 1and 2. The antimicrobial activity of PEITC were proportionate to the dose applied ($P < 0.05$). Higher doses led to higher antibacterial activity. In general, the lowest dose applied (0.015 µmoles) had no effect, whilst the major dose applied (3.0 µmoles) exhibited the major inhibition effect, varying in its intensity according to the bacteria tested ($P<0.05$) (Table 1).

Table 1 Inhibition zone diameter (in mm) of 2-Phenylethylisothiocyanate and respective controls.

Bacteria strain	Positive control [a,b]	Negative control [b]	Concentration (µM)[b]				
			0.015	0.15	0.75	1.5	3.0
Acinetobacter baumanii	16.0 ± 0.0	-	-	-	-	-	9.3±1.5
Citrobacter freundii	18.0 ± 0.0	-	-	-	-	-	-
Enterobacter asburiae	16.3 ± 0.7	-	-	-	-	8.3±0.3a	9.0±0.6a
Enterobacter cloacae	15.0 ± 0.0	-	-	-	-	-	
Enterobacter hormaechei	18.0 ±0.0	-	-	-	-	-	8.3±0.3
Escherichia $coli_1$	17.0 ± 0.6	-	-	-	7.7±0.7a	8.7±0.7ab	9.7±0.7b
Escherichia $coli_2$	15.7 ± 0.3	-	-	-	8.0±0.0a	9.0±0.6ab	9.7±0.7b
Hafnia alvei	19.0 ± 1.0	-	-	-	7.7±0.3a	9.3±0.3b	10.3±0.3c
Klebsiella oxytoca	12.0 ± 0.0	-	-	-	-	7.0±0.0a	7.3±0.3a
Klebsiella pneumoniae	14.7 ± 0.3	-	-	-	-	-	7.7±0.3
Morganella morganii	17.0 ±0.0	-	-	-	7.7±0.7a	11.3±0.3b	17.0±1.6c
Proteus mirabilis	17.0 ±0.0	-	-	-	8.0±0.0a	8.7±0.3a	10.7±0.3b
Pseudomonas aeruginosa	16.7 ± 0.3	-	-	-	-	-	7.3±0.3
Salmonella typhi	14.0± 0.6	-	-	-	-	8.0±0.7a	8.7±0.7a
Stenotrophomonas maltophilia	18.3 ± 0.9	-	-	-	-	8.0±0.0a	11.0±0.6b
Enterococcus faecalis	14.7 ± 0.3	-	-	-	7.0±0.0a	10.0±0.0b	14.3±0.3c
Staphylococcus aureus	16.3 ± 0.3	-	-	-	-	29.0±0.0a	44.0±0.0b
Staphylococcus saprophyticus	16.7 ± 0.3	-	-	-	22.7±1.6a	37.3±0.3b	49.0±0.6c
Escherichia $coli_1$	17.0 ± 0.0	-	-	-	-	-	-
Escherichia $coli_5$	17.0 ± 0.0	-	-	-	-	-	-
Enterococcus faecalis $_{49}$	17.0 ± 0.0	-	-	-	-	-	10.0±0.0
Enterococcus faecalis $_{52}$	15.0 ± 0.0	-	-	-	-	-	11.0±0.0
Escherichia coli CETC 434	19.0 ± 0.0	-	-	-	-	-	10.0±0.0
Pseudomonas aeruginosa ATCC 10145	16.0 ± 0.0	-	-	-	-	-	-
Staphylococcus aureus CETC 976	17.0 ± 0.0	-	-	-	-	Without growth	Without growth
Listeria monocytogenes ATCC 15313	20.0 ± 0.0	-	-	-	-	Without growth	Without growth

[a]The commercial antibiotics used were Gentamicin for Gram (-)ve and Vancomicin for Gram (+)ve.
[b]The symbol (-) means that compound have no effect on bacteria strain in vitro growth.

For all bacteria tested, as expected the negative control performed with the solvent (DMSO) used in preparation of the compounds was ineffective in the suppression of the bacterial growth ($P < 0.05$) and the positive antibiotic controls were effective ($P < 0.001$), varying according to the antibiotic used and the antibiotic sensitivity of the bacteria tested. For the three Gram-positive bacteria the *S. aureus* and *S. saprophyticus* were strongly inhibited by PEITC. For the Gram-negative bacteria there was a considerable variation in sensitivity and PEITC had low inhibitory activity particularly for the *E. hormaechei*, *E. coli* strains, *K. oxytoca*, *M.*

morganii, *P. aeruginosa* and *S. maltophilla*, which may be related to its lipohilic nature and slower diffusion through the agar.

The results of antibacterial activity expressed as % RIZD (Table 2) can often reveal additional information on the differential effects of the test compounds vs the antibiotic controls. For example, a test compound giving an RIZD of 50% is half as effective as the antibiotic at the concentration tested, whereas a test compound of >100% is more effective than the antibiotic. It is also clear that some of the bacteria, such as *C. freundii* and *P. aeruginosa* (Table 2) were less sensitive to the ITC.There is considerable concern about the development of antibiotic resistance in pathogenic bacteria and this is clearly evident in *E. coli* O157:H7, Methicillin Resistant *S. aureus* (MRSA) and several other bacterial pathogens of humans and animals. Also, there is an increasing occurrence of enteric pathogens on raw vegetable and salad products [21, 22]. Therefore it is important to identify new sources of effective antimicrobial compounds for food products and within the diet.

Table 2 Antimicrobial activity expressed as percentage of relative inhibition zone diameter (RIZD %) for the 3.0 μmoles dose.

Bacteria strain	Concentration (μmoles)[b]
	3.0
Acinetobacter baumanii	58.3
Citrobacter freundii	-
Enterobacter asburiae	55.0
Enterobacter cloacae	-
Enterobacter hormaechei	46.3
Escherichia coli$_1$	56.8
Escherichia coli$_2$	61.7
Hafnia alvei	54.6
Klebsiella oxytoca	61.1
Klebsiella pneumoniae	52.4
Morganella morganii	100.0
Proteus mirabilis	62.8
Pseudomonas aeruginosa	44.0
Salmonella typhi	60.4
Stenotrophomonas maltophilia	62.5
Enterococcus faecalis	97.7
Staphylococcus aureus	269.6
Staphylococcus saprophyticus	294.2
Escherichia coli$_1$	-
Escherichia coli$_5$	-
Enterococcus faecalis $_{49}$	58.8
Enterococcus faecalis $_{52}$	73.3
Escherichia coli CETC 434	52.6
Pseudomonas aeruginosa ATCC 10145	-
Staphylococcus aureus CETC 976	Without growth
Listeria monocytogenes ATCC 15313	Without growth

There is evidence that many classes of plant compounds, specifically phytochemicals, are a potentially useful source of novel antimicrobials [23, 24]. The results of the present study show that the antibacterial activity of PEITC might be useful in controlling human and animal pathogenic bacteria. It is known that conventional antimicrobial products are more effective against Gram-positive bacteria compared with Gram-negative bacteria [25]. Consequently, antimicrobial products with significant activity against Gram-negative bacteria can be a valuable source of therapeutic agents to control infections specifically caused by those bacteria. This *in vitro* study provides data indicating the potential for these natural dietary chemicals as an alternative to, or in combination with, current antibiotic therapies. However, more studies are needed to evaluate their effectiveness *in vivo*, determining their MIC values, evaluating synergistic effects of hydrolysis product mixtures, and effects against biofilms.

Acknowledgements The authors are grateful do Dra. Ana Paula Castro, from the Central Hospital of Trás-os-Montes e Alto Douro, North of Portugal who provided the clinical bacteria strains.The authors are also grateful to Dr. Antonio Martinez-Murcia from Molecular Diagnostics Center (MDC), Biomolecular Techionolgies, Orihuela (Alicante), Spain (http://www.mdc-bt.com) for the genetic identification through 16S rRNA sequencing of the clinical and animal bacteria strains used in this study. Also, the authors acknowledge the financial support provided by the Portuguese Foundation for Science and Technology (A. Aires, SFRH/BD/18586/2004 and SFRH/BPD/65029/2009; C. Dias-SFRH/BGCT/33354/2008).

References

[1] Bones AM, Rossiter JT. The enzymic and chemically induced decomposition of glucosinolates. *Phytochemistry*. 2006; 67: 1053–1067.

[2] Johnson IT. Phytochemicals and cancer. *Proceedings of the Nutrition Society*. 2007;66:207-215.

[3] Griffiths DW, Birch ANE, Hillman JR. Antinutritional coumponds in the Brassicaceae: Analysis, biosynthesis, chemistry and dietary effects. Review article. *Journal of Horticultural Science & Biotechnology*. 1998;73:1-18.

[4] Juge N, Mithen RF, Traka M. Molecular basis for chemoprevention by sulforaphane: a comprehensive review. *Cellular and Molecular Life Sciences*. 2007;64:1105–1127.

[5] Fahey JW, Haristoy X, Dolan PM, Kensler TW, Scholtus I, Stephenson KK, Talalay P, Lozniewski A. Sulforaphane inhibits extracellular, intracellular, and antibiotic-resistant strains of Helicobacter pylori and prevents benzo[a]pyreneinduced stomach tumor. *Proceedings of the National Academy of Sciences of the United States of America*. 2002;99:7610-7615.

[6] Haristoy X, Duprez KADA, Lozniewski A. Efficacy of Sulforaphane in eradicating Helicobacter pylori in human gastric xenografts implanted in nude mice. *Antimicrobial Agents and Chemotherapy*. 2003;47:3982–3984.

[7] Haristoy X, Fahey JW, Scholtus I, Lozniewski A. Evaluation of the antimicrobial effects of several isothiocyanates on Helicobacter pylori. *Planta Medica*. 2005;71:326-330.

[8] Oginsky EL, Stein AE, Greer MA. Myrosinase activity in bacteria as demonstrated by the conversion of progoitrin to goitrin. *Proceedings of the Society for Experimental Biology and Medicine*.1965;119: 360-364.

[9] Ohtsuru M, Hata T. General characteristics of the intracellular myrosinase from Aspergillus niger. *Agricultural Biology and Chemistry*. 1973;37:2543-2548.

[10] Staron, T. Obtention des proteines a partir des grains oleagineuses par des methodes microbiologiques. *Proteins & Graines Oleagineuses* 1975;11-12:579-589.

[11] Tani N, Ohtsuru M, Hata, T. Isolation of myrosinase producing microorganism. *Agric Biol Chem*. 1974;38:1617-1622.

[12] Llanos Palop M, Smiths JP, ten Brink B. Degradation of sinigrin by *Lactobacillus agilis* strain R16. *International Journal of Food Microbiology*. 1995;26: 219-229.

[13] Elfoul L, Rabot S, Khelifa N, Quinsac A, Duguay A, Rimbault A. Formation of allylisothiocyanate from sinigrin in the digestive tract of rats monoassociated with a human colonic strain of *Bacteroides thetaiotaomicron*. *FEMS Microbiology Letters*. 2001;197;99–103.

[14] Combourieu B, Elfoul L, Delort A-M, Rabot S. Identification of new derivatives of sinigrin and glucotropaeolin produced by the human digestive microflora using 1H NMR spectroscopy analysis of in vitro incubations. *Drug Metabolism and Disposition*. 2001;29:1440-1445.

[15] Rouzaud G, Rabot S, Ratcliffe B, Duncan AJ. Influence of plant and bacterial myrosinase activity on themetabolic fate of glucosinolates in gnotobiotic rats. *British Journal of Nutrition*. 2003; 90:395-404.

[16] Hashem FA, Saleh MM. Antimicrobial components of some Cruciferae plants (Diplotaxis harra Forsk. and Erucaria microcarpa Boiss.). *Phytotherapy Research*. 1999;13:329-332.

[17] Murray PR, Boron EJ, Pfaller MA, Tenover FC, Yolken RH, eds. *Manual of Clinical Microbiology*, 7th edn. Washington, DC: ASM Press; 1999.

[18] Jorgensen JH, Turnidge JD, Washington JA. Antibacterial susceptibility tests: dilution and disk diffusion methods. In: Murray, PR, Baron, EJ, Pfaller, MA, Tenover FC, Yolken R, eds. *Manual of Clinical Microbiology*., 7th edn. pp.. Washington, DC: ASM Press; 1999:1526-1543.

[19] Bauer AW, Kirby MDK, Sherria JC, Turck M. Antibiotic susceptibility testing by standard single disc diffusion method. *American Journal of Clinical Pathology*. 1966;45:493-496.

[20] Rojas J, Ochoa JV, Ocampo SA, Muñoz JF. Screening for antimicrobial activity of ten medicinal plants used in Colombian folkloric medicine: a possible alternative in the treatment of non-nosocomial infections. *BMC Complementary and Alternative Medicine*. 2006; 6:1-6.

[21] Rangel JM, Sparling PH, Crowe C, Griffin PM, Swerdlow DL. Epidemiology of Escherichia coli O157:H7 outbreaks, United States 1982–2002. *Emerging Infectious Diseases*. 2005;11:603–609.

[22] Brandl MT. Fitness of human enteric pathogens on plants and implications for food safety. *Annual Review of Phytopathology*. 2006;44:367-392.

[23] Ahmad I, Aqil F. *In vitro* efficacy of bioactive extracts of 15 medicinal plants against ESbL-producing multidrug-resistant enteric bacteria. *Microbiological Research*. 2007;162:264-275.

[24] Vattem DA, Mihalik K, Crixell SH, McLean RJC. Dietary phytochemicals as quorum sensing inhibitors. *Fitoterapia*. 2007;78:302-310.

[25] McDonnell G, Russell AD. Antiseptics and disinfectants: activity, action, and resistance. *Clinical Microbiology Reviews*. 1999;12:147–179.

Antimicrobial effect of carvacrol on *Escherichia coli* K12 growth at different temperatures

C.M. Belda-Galbis[1], **A. Martínez**[1] and **D. Rodrigo**[*,1]

[1] Food Preservation Group, Department of Preservation and Food Quality, Instituto de Agroquímica y Tecnología de Alimentos, CSIC, Avda. Agustín Escardino 7, 46980 Paterna, Valencia, Spain

* Corresponding author: e-mail: lolesra@iata.csic.es, Phone: +34 963900022

A study was carried out to evaluate the antimicrobial activity of carvacrol on *Escherichia coli* K12 (CECT 433) at different incubation temperatures (30 and 37 ºC). Regardless of the temperature, both the maximum specific growth rate (μ_{max}) and the lag phase (λ) showed to be dependent on the carvacrol concentration but in different ways. While λ was directly proportional to the antimicrobial concentration, μ_{max} showed to be inversely proportional to it, being 0.14 µl ml^{-1} a lethal concentration, and existing a log-linear relationship between that parameter and the tested concentrations ($R \geq 0.900$). Comparing the results for both temperatures for each concentration, no significant differences were detected in μ_{max} but λ is always longer at 30 ºC than at 37 ºC ($P_{value} \leq 0.05$). These results indicate that carvacrol is active against *E. coli* K12 growth, so it could be used to extend the shelf life and improve the safety margins of minimally processed foods.

Keywords essential oils; carvacrol; antimicrobial activity; *Escherichia coli*; kinetic growth parameters

1. Background

In spite of the improvements that have been achieved in materia of hygiene and food safety, both are increasingly important issues. Governments all over the world are intensifying their efforts to improve them because illnesses caused by the consumption of foods contaminated with pathogens have a wide economic and public health impact. In 2005 alone, for example, 1.8 million people died from diarrhoeal diseases and a great proportion of these cases can be attributed to pathogens presence (and/or their toxins) in food and water [1].

One of the most dangerous microorganisms that can be transmitted by food is *Escherichia coli* O157:H7, which has emerged over the last decades [1]. Controlling its concentration and growth remains an important objective for food industry since it is the major cause of hemorrhagic colitis, haemolytic uremic syndrome, and thrombotic thrombocytopenic purpura in humans, especially in infants, children and elderly, by consumption of undercooked ground beef, raw dairy products or vegetables, that have been in contact with bovine manure [2].

Till now, the use of chemical preservatives, thermal processing, vacuum packaging or modified atmosphere and/or refrigeration, have been the main technologies used by food industry to achieve food preservation and extent their shelf-life. In response to the increasing consumer demand for fresh and natural food, minimally processed, alternative technologies without relying neither on heat nor in chemical preservatives, are being developed [3, 4]. In this context, essential oils derived from medicinal plants, herbs and spices are gaining interest for their potential as preservative ingredients given they are generally recognised as safe (GRAS) [4, 5] and possess antibacterial, antifungal and antioxidant properties that have been known since antiquity [4].

The aim of the present work was to evaluate the effect of the oregano-derived volatile carvacrol on *E. coli* K12 (CECT 433) growth, used as surrogate for *E. coli* O157:H7, to assess its potential as a natural antimicrobial product to be used for food preservation at different temperatures.

2. Materials and methods

2.1 Bacterial strain and growth conditions

Glycerinated cryovials of *E. coli* cells with around 2.41×10^9 cfu ml^{-1} were obtained from lyophilized strain *E. coli* K12 (CECT 433) provided by the Spanish type culture collection, using the method described by Pina-Pérez *et al.* [6].

2.2 Carvacrol

Carvacrol ≥ 98% (5-Isopropyl-2-methylphenol) was obtained from Sigma-Aldrich Chemie GmbH (Steinheim, Germany). To test its antibacterial properties, before each experiment repetition, a stock solution was freshly prepared diluting it in dimethyl sulfoxide ACS reagent ≥ 99.9% (DMSO) [7, 8], obtained from Sigma-Aldrich Chemie GmbH (Steinheim, Germany).

2.3 Antimicrobial assay

To asses the antimicrobial activity of carvacrol, bacterial growth curves were determined by absorbance based assays using an automated microtiter plate reader (POLARstar Omega, BMG LABTECH GmbH, Offenburg, Germany).

The assay reaction mixtures consisted on different carvacrol concentrations (0, 0.04, 0.06, 0.08, 0.10, 0.12 and 0.14 µl ml^{-1}) diluted in NB (Nutrient Broth, Scharlab S.L., Spain), and microbial cells from stock culture, to achieve an initial bacterial inoculum of around 10^7 cfu ml^{-1}.

Samples without bacteria, with or without carvacrol, were used as blanks and negative controls. Sample contained only sterile growth media and *E. coli* K12 were used as positive control.

A polystyrene microtiter plate (Deltalab S.L., Barcelona, Spain) was filled with 250 µl of sample per well in quadruplicate, after vigorous shaking. The plate was incubated with double orbital shaking (500 rpm). The bacterial growth was monitored until the population reached the stationary phase. The absorbance was recorded at 600 nm every 30 min. The experiment was repeated at two different incubation temperatures (30 and 37 °C) in triplicate.

2.4 Modelization of *E. coli* K12 growth

Experimental data from optical density were transformed into natural log by means of absorbance *versus* log(cfu ml^{-1}) calibration curves that were previously obtained. Those data at the different incubation temperature and carvacrol concentration combinations were fitted to two mathematical primary non-linear models to obtain growth kinetic parameters (μ_{max} and λ) by using the Statgraphics Centurion XV software (StatPoint Technologies, Inc., Warrenton, VA, USA).

The Baranyi-Roberts model [9]:

$$y(t) = y_0 + \mu_{max} A(t) - \ln\left(1 + \frac{e^{\mu_{max} A(t)} - 1}{e^{(y_{max} - y_0)}}\right) \tag{1}$$

where $y(t)$ is ln(cfu ml^{-1}) of cell concentration at time t; y_0 is the initial cell concentration (ln(cfu ml^{-1})); y_{max} is the maximum cell concentration (ln(cfu ml^{-1})); μ_{max} is maximum specific growth rate in terms of ln(cfu ml^{-1} h^{-1}); and $A(t)$ is defined as:

$$A(t) = t + \frac{1}{v}\ln(e^{-vt} + e^{-h_0} - e^{(-vt-h_0)}) \tag{2}$$

where v is rate of increase of the limiting substrate, assumed to be equal to μ_{max}; and $h_o = \mu_{max} \times \lambda$.

The modified Gompertz equation [10]:

$$\log_{10}(N) = A + Ce^{-e^{-B(t-M)}} \tag{3}$$

where N is the number of microorganisms at time t in terms of log(cfu ml^{-1}); A is the asymptote when time decreases indefinitely (approximately equivalent to the logarithm of initial cell number) (log(cfu ml^{-1})); C is the increase in the logarithm of the number of microorganisms when time increases indefinitely (number of cycles of growth) (log(cfu ml^{-1})); B is relative growth rate at $t = M$ (log(cfu ml^{-1}) h^{-1}); and M is the time required to reach μ_{max} (h). A, C, B, and M are four parameters that combinated were used to derive μ_{max} [10] and λ [11].

$$\mu_{max} = \frac{BC}{e} \tag{4}$$

$$\lambda = M - \left(\frac{1}{B}\right) + \frac{\log_{10}(N_0) - A}{BC/e} \tag{5}$$

where e = 2.718 and N_0 is the log10 count at $t = 0$ in terms of log(cfu ml^{-1}).

A secondary model for μ_{max} as a function of carvacrol concentration was also developed in order to describe its effect on this parameter. Specifically, the natural log of μ_{max} estimations at each carvacrol concentration were fitted to a linear equation based on the Bigelow model [12] using the Microsoft Office Excel 2003 software (Microsoft Corporation, USA).

The goodness of fit of experimental data to the primary models was obtained by calculating the adjusted correlation coefficient (adjusted R^2), mean square error (MSE) and the standard error of the fit (SE of the fit).

The secondary models were also externally validated by using the coefficient of determination value (R^2).

2.5 Statistical data analysis

The statistical analysis was performed with Statgraphics Centurion XV software (StatPoint Technologies, Inc., Warrenton, VA, USA). This analysis consisted in an ANOVA to detect significant differences between μ_{max} and λ average values at different carvacrol concentrations and temperatures. If it was necessary, a multiple range test was also applied to determine which levels of each factor were perceptibly different ($P_{value} \leq 0.05$).

3. Results and discussion

3.1 Growth curves

The effect of different carvacrol concentrations (0, 0.04, 0.06, 0.08, 0.10, 0.12 and 0.14 μl ml^{-1}) on *E. coli* K12 growth at 37 °C and 30 °C was studied. Figure 1 shows the growth curves obtained in each case. Carvacrol clearly shows a bacteriostatic effect on *E. coli* K12 growth when its concentration is less than 0.14 μl ml^{-1}, and a bactericidal effect when it is equal or higher than 0.14 μl ml^{-1}. Since 0.14 μl ml^{-1} is the lowest concentration inhibiting visible growth of test microorganism, it can be considered as the minimum inhibitory concentration (MIC) [13]. This value is very close to what have been previously reported by other researchers working with *E. coli* in liquid media [5, 14-19].

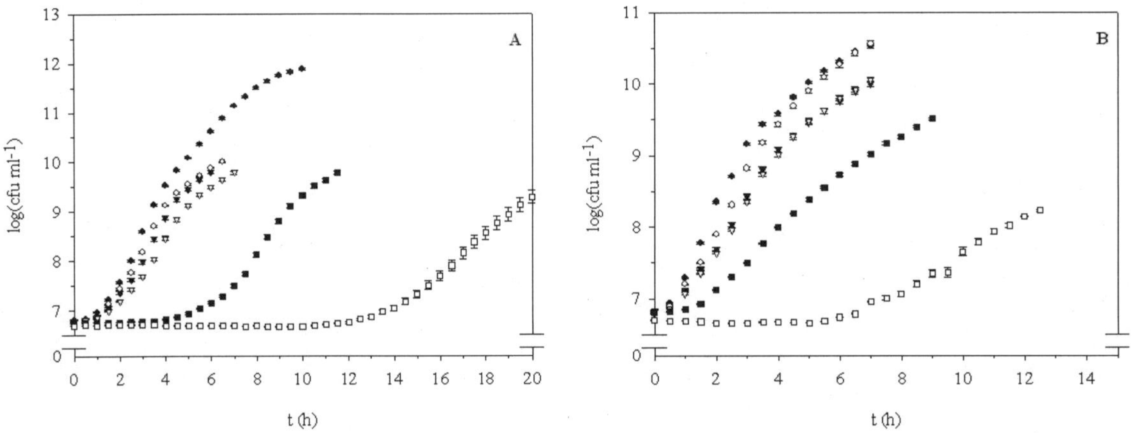

Fig. 1 *E. coli* K12 growth curves at 30 °C (A) and 37 °C, depending on the carvacrol concentration (μl ml^{-1}): 0 (●), 0.04 (○), 0.06 (▼), 0.08 (▽), 0.10 (■), and 0.12 (□). The variation coefficient was expressed by errors bars.

3.2 Modeling growth and determination of kinetic parameters

Experimental data from each repetition were fitted to the modified Gompertz and Baranyi-Roberts equations in order to obtain μ_{max} and λ values (Tables 1 and 2). Both were chosen due to the sigmoid behavior of *E. coli* K12 growth curves. In all cases the goodness of fit was evaluated calculating adjusted R^2, MSE and SE of the fit. Adjusted R^2 value was always greater than 99%, and both MSE and SE of fit values were less than 0.01 and 0.13, respectively. Therefore, both nonlinear models are equally effective for describing sigmoidal curves of carvacrol antimicrobial effect when modeling *E. coli* growth.

Table 1 Mean and standard deviation (SD) values for the μ_{max} and λ obtained from experimental data fitted to the Baranyi-Roberts model.

Carvacrol (μl ml^{-1})	μ_{max} ± SD (log(cfu ml^{-1}) h^{-1})		λ ± SD (h)	
	37ºC	30ºC	37ºC	30ºC
0	1.001 ± 0.068	0.978 ± 0.100	0.519 ± 0.101	1.192 ± 0.167
0.04	0.851 ± 0.091	0.841 ± 0.072	0.655 ± 0.025	1.226 ± 0.164
0.06	0.744 ± 0.125	0.790 ± 0.106	0.771 ± 0.102	1.380 ± 0.226
0.08	0.708 ± 0.112	0.698 ± 0.108	0.807 ± 0.112	1.644 ± 0.279
0.10	0.447 ± 0.058	0.614 ± 0.065	1.345 ± 0.180	5.980 ± 0.453
0.12	0.411 ± 0.079	0.508 ± 0.126	7.734 ± 1.993	14.609 ± 2.317

Table 2 Mean and standard deviation (SD) values for the μ_{max} and λ obtained from experimental data fitted to the modified Gompertz equation.

Carvacrol (μl ml^{-1})	$\mu_{max} \pm$ SD (log(cfu ml^{-1}) h^{-1})		$\lambda \pm$ SD (h)	
	37ºC	30ºC	37ºC	30ºC
0	0.954 ± 0.065	1.058 ± 0.091	0.474 ± 0.090	1.309 ± 0.109
0.04	0.861 ± 0.103	0.911 ± 0.078	0.668 ± 0.073	1.325 ± 0.163
0.06	0.762 ± 0.115	0.844 ± 0.079	0.805 ± 0.154	1.488 ± 0.257
0.08	0.739 ± 0.125	0.760 ± 0.086	0.820 ± 0.109	1.788 ± 0.320
0.10	0.416 ± 0.021	0.619 ± 0.071	1.339 ± 0.015	5.926 ± 0.457
0.12	0.467 ± 0.021	0.522 ± 0.120	7.951 ± 1.624	14.743 ± 2.388

3.3 Effect of carvacrol concentration and incubation temperature on *E. coli* K12 growth

From μ_{max} and λ values obtained from each assay repetition, an ANOVA was performed to detect significant differences on parameters average value at different concentrations of carvacrol under isothermal conditions, and at different incubation temperatures for each carvacrol concentration.

A shown in tables 1 and 2, regardless of the temperature, both μ_{max} and λ showed to be dependent on the carvacrol concentration. There are statistical evidences to conclude that an increase in the antimicrobial concentration produces a decrease in μ_{max} ($P_{value} \leq 0.05$), although there were not detected significant differences between all factor levels, especially at low doses, maybe because differences between tested concentrations are quite low. λ is prolonged with increasing the antimicrobial concentration but there were only significant differences ($P_{value} \leq 0.05$) between the reference sample and those with higher carvacrol concentration (0.12 μl ml^{-1} at 37ºC, and 0.10 or 0.12 μl ml^{-1} at 30ºC). These results indicate that there is threshold concentration below which no significant differences exist between λ values. This threshold concentration is lower at 30ºC than at 37ºC probably due to 37ºC is the optimal growth temperature for *E. coli*. However, this difference maybe also is due to that there is an antimicrobial synergistic between incubation temperature and carvacrol concentration.

When μ_{max} and λ values obtained at different temperatures for each carvacrol concentration were compared (Tables 1 and 2), no significant differences were detected in μ_{max} but λ is always longer at 30ºC than at 37ºC ($P_{value} \leq 0.05$), whatever the concentration of carvacrol, probably because 37ºC is the optimal temperature for *E. coli* growth, as already mentioned above.

3.4 Effect of carvacrol concentration on μ_{max}

With the aim of determine the effect of carvacrol concentration on μ_{max}, the establishment of a secondary model based on the Bigelow model [12] was carried out. As shown in figure 2, a log-linear relationship between that parameter and the concentration of carvacrol can be found ($R \geq 0.900$) at a 95% confidence level, at each incubation temperature. That permits interpolation in the region of observations and provide information on interactions between this two factors [20].

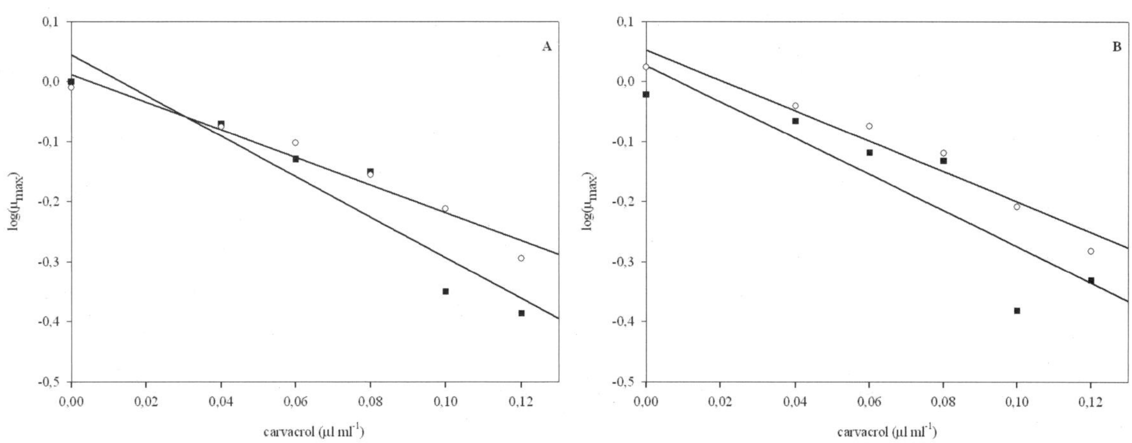

Fig. 2 Effect of carvacrol concentration on μ_{max} at different temperatures. Data correspond to those obtained with the Baranyi-Roberts (A) and the modified Gompertz (B) models at 30 ºC (○, dotted line) and at 37 ºC (■, solid line).

4. Conclusion

The results obtained in the present study demonstrate that carvacrol is active against *E. coli* K12 so it could potentially be used to extend the shelf life and improve the safety margins of minimally processed foods, as primary or additional microbial control measure, independently or combinated with low storage temperature. However further studies *in vitro* and in food models are required to assess the efficacy of this compound as a preservative, especially when it is in contact with the food matrix. It would also be important to know if it really exits a synergistic effect between carvacrol and temperature, or whether it might take place between carvacrol and another physical or chemical agent.

Acknowledgements This study was carried out with funds from European project NAFISPACK (FP7-KKB-2007-1). C. M. Belda-Galbis is especially grateful to the CSIC for providing her a JAE-predoctoral grant.

References

[1] WHO. Food safety and foodborne illness. World Health Oraganization Fact Sheet 237, Geneva; revised March 2007.

[2] Codex Alimentarius Commission. Risk profile for enterohemorragic *E. coli* including sprouts, ground beef and pork Report nr. CX/FH 03/5_Add. 4. FAO and WHO Orlando, USA; 2002.

[3] Raso J, Barbosa-Cánovas GV. Nonthermal preservation of foods using combined processing techniques. *Critical Reviews in Food Science and Nutrition.* 2003;43(3):265-285.

[4] Burt S. Essential oils: their antibacterial propierties and potential applications in food – A review. *International Journal of Food Microbiology.* 2004;94:223-253.

[5] Kim J, Marshall MR, Wei CI. Antibacterial activity of some essential oil components against five foodborne pathogens. *Journal of Agricultural and Food Chemistry.* 1995;43:2839-2845.

[6] Pina-Pérez MC, García-Fernández MM, Rodrigo D, Martínez-López A. Monte Carlo simulation as a method to determine critical factors affecting two strains of *Escherichia coli* inactivation kinetics by high hydrostatic pressure. *Foodborne Pathogens and Disease.* 2010;7(4):459-466.

[7] Iscan G, Kirimer N, Kurkcuoglu M, Baser KHC, Demirci F. Antimicrobial screening of *Mentha piperita* essential oils. *Journal of Agricultural and Food Chemistry.* 2002;50: 3943-3946.

[8] Vardar-Unlu G, Candan F, Sokmen A, Daferera D, Polissiou M, Sokmen M, Donmez E, Tepe B. Antimicrobial and antitoxic activity ofthe essential oil and methanol extracts of *Thymus pectinatus* Fisch. et Mey. var. *pectinatus* (Lamiaceae). *Journal of Agricultural and Food Chemistry.* 2003;51:63-67.

[9] Baranyi J, Roberts TA. A dynamic approach to predicting bacterial growth in food. *International Journal of Food Microbiology.* 1994; 23:277-294.

[10] Gibson AM, Bratchell N, Roberts TA. Predicting microbial growth: growth responses of salmonellae in a laboratory medium as affected by pH, sodium chloride and storage temperature. *International Journal of Food Microbiology.* 1988;6:155-178.

[11] McMeekin TA, Olley JN, Ross T, Ratkowsky DA. *Predictive Microbiology - Theory and Application.* Taunton, Somerset, United Kingdom, Research Studies Press; 1993.

[12] Bigelow WD. The logarithmic nature of thermal death time curves. *The Journal of Infectious Diseases.* 1921;29:528-536.

[13] Delaquis PJ, Satnich K, Girard B, Mazza G. Antimicrobial activity of individual and mixed fractions of dill, cilantro, coriander and eucaliptus essential oils. *International Journal of Food Microbiology.*2002;74:101-109.

[14] Helander IM, Alakomi HL, Latva-Kala K, Mattila-Sandholm T, Pol I, Smid EJ, Gorris LGM, Von Wright A. Characterization of the action of selected essential oil components on gram-negative bacteria. *Journal of Agricultural and Food Chemistry.* 1998;46:3590-3595.

[15] Cosentino S, Tuberoso CIG, Pisano B, Satta M, Mascia V, Arzedi E, Palmas F. *In vitro* antimicrobial activity and chemical composition of Sardinian *Thymus* essential oils. *Letters in Applied Microbiology.* 1999;29:130-135.

[16] Juliano CA, Mattana A, Usai M. Composition and *in vitro* antimicrobial activity of the essential oil of *Thymus herba-barona* Loisel growing wild in Sardinia. *Journal of Essential Oil Research.* 2000;12:516-522.

[17] Olasupo NA, Fitzgerald DJ, Gasson MJ, Narbad A. Activity of natural antimicrobial compounds agaisnt *Escherichia coli* and *Salmonella enterica* serovar Typhimurium. *Letters in Applied Microbiology.* 2003;36:448-451.

[18] Burt SA, Vlielander R, Haagsman HP, Veldhuizen EJA. Increased activity of essential oil components carvacrol and thymol against *E. coli* O157:H7 by addition of food stabilizers. *Journal of Food Protection.* 2005;68:919-926.

[19] Si W, Gong J, Tsao R, Zhou T, Yu H, Poppe C, Johnson R, Du Z. Antimicrobial activity of essential oils and structurally related synthetic food additives towards selected pathogenic and beneficial gut bacteria. *Journal of Applied Microbiology.* 2006;100:296-305.

[20] Dong Q, Tu K, Guo L, Li H, Zhao Y. Response surface model prediction of growth parameters from spores of *Clostridium sporogenes* under different experimental conditions. *Food Microbiology.* 2007;24:624-632.

Bacteriostatic effect of cocoa powder rich in polyphenols to control *Cronobacter sakazakii* proliferation on infant milk formula

M.C. Pina-Pérez, D. Rodrigo*, and A. Martínez-López

CSIC. Institut of Agrochemistry and Food Technology. Food preservation and quality department. 46980 Paterna, Valencia (SPAIN).

*Corresponding author: lolesra@iata.csic.es

Public health authorities, researchers and manufacturers are urged to develop alternative preservatives methods for milk and dairy derivative products (infant milk formula, dairy beverages, etc) specifically to avoid or reduce foodborne illness among children [1]. Taking into account that epidemiological studies have implicated *C. sakazakii* as an occasional contaminant of powder infant products, specifically infant formulas, with severe consequences, it could be relevant to propose alternative control measures to avoid growth and infection by this microorganism. The possible inhibitory/bacteriostatic effect of cocoa powder to control *Cronobacter sakazakii* in milk formulas (MF) was studied at three concentrations: 1%, 2.5% and 5 % (w/v).

The growth study was carried out at 25ºC in MF and supplemented MF at different concentrations using an initial *C. sakazakii* concentration of 10^{2} cfu/mL. The data results were fitted to Gompertz equation in order to obtain growth biological parameters.

According to present study results, there are no significant differences between supplemented/not supplemented milk formula specific growth rate values (μ, h^{-1}), which means that there is not an inhibitory effect of cocoa on *C. sakazakii* growth at 25ºC. Related to lag phase values (λ, h) it was observed an extension of λ duration when MF was supplemented with cocoa at 5% (w/v). The results of the present study point out that a bacteriostatic but not inhibitory effect, on *C. sakazakii* cells, could be attributed to cocoa powder at 5% (w/v) concentration when it is added to milk formula and remains in contact with cells at ambient temperature (25ºC). These results would be applied to a future pasteurized milk formula product supplemented with cocoa powder rich in polyphenols enhancing its shelf life.

Keywords *Cronobacter sakazakii*, cocoa powder, antimicrobial effect, bacteriostatic effect, preservation methods.

1. Introduction

Following recent food poisoning cases, reported worldwide and sometimes involving the death of the individuals concerned, many of whom are children, international public health organizations have focussed their attention and efforts on preventing and avoiding potential food hazards, especially in infants and young children. Since 2004, the Food and Agriculture Organization (FAO) and the World Health Organization (WHO) have taken note of the hazards associated with the potential presence of pathogens in powdered foods, such as infant formula, which highlight the presence of *Cronobacter sakazakii* [2]. From the food-safety standpoint public authorities, researchers and dairy-product manufacturers are working together to develop alternative methods to thermal treatment, and thus find a better way of preserving the nutritional properties of milk and dairy products. Various technologies associated with the addition of complementary ingredients to generate innovation of an assortment of flavors, textures, and functional attributes are being applyed in milk derivatives. In this sense, several studies reveal results on the preservative action of these natural ingredients such as e.g. cinnamon, anise, cocoa, or vanillin. In this respect, the availability and recent interest in certain minimally processed cocoa derivatives/ concentrates and flavonoid-rich extracts, are encouraging the development of a variety of products with functional properties. Cocoa (*Theobroma cacao*) has multiple uses in the food industry nowadays. Recently, there is growing interest in cocoa flavonoids due to their antioxidant capacity and associated benefits for human health [3]. These flavonoid-rich products demonstrate high bioavailability in humans when administered with milk [4, 5]. These ingredients may be formulated as liquids or powders (dehydrated) and in the dairy industry can be added to milk or water at different stages in the process, and could be considered as additional hurdles to control microbial presence and microorganism proliferation on the final product.

2. Objective

The main objective of this study is to determine the potential effect of cocoa powder on inhibition or control of *Cronobacter sakazakii* growth in infant milk formula at ambient temperature. The aim is to contribute to a better understanding of hurdle technology.

3. Material and Methods

3.1 Microorganism culture

The pure culture of *C. sakazakii* (equivalent to strain 29544 ATCC) was supplied lyophilized by the Spanish Type Culture Collection (CECT 858), and reconstituted to a final concentration of 10^7 cfu/mL that remained stored at -80ºC. The *Cronobacter sakazakii* culture vials were used to inoculate infant milk formula supplemented/not supplemented with cocoa

3.2 Cocoa powder

In this study, we used a cocoa powder marketed as CocoanOX 12% (CCX), a polyphenol-rich cocoa powder developed using an industrial process that maintains a significant fraction of the original cocoa flavonoids intact (Naturex S.L., Spain).

3.3 Substrate of inoculation

In this study we developed a milk formula (MF), mixture of reconstituted powdered infant formula, to which cocoa powder (CCX) was added at three different concentrations: 1% (w/v) (MF-1%); 2.5% (w/v) (MF-2.5%) and 5% (w/v) (MF-5%). A MF without CCX was used as a control.
Using an initial inoculum of 10^2 cfu/mL, it was carried out a study of *Cronobacter sakazakii* cells growth into different MF at 25ºC. Three replicates of growth curves were obtained for each substrate.

3.4 Viable *C. sakazakii* cell counts

Cell suspensions of each treated sample were serially diluted in sterile peptone water 0.1% and then plated on tryptose-soy agar (TSA, Scharlab Chemie Barcelona, Spain) in duplicate. The plates were incubated at 37°C for 24 hours and Colony Forming Units (cfu) were counted.

3.5 Modelling *C. sakazakii* growth

Gompertz equation (equation 1) was used to analyze the *C. sakazakii* growth behavior at 25 ºC in MF and in MF with different CCX concentrations [6]:

$$N = No + Ce^{-e((2.718*\mu)/(\lambda - t))+1} \qquad (1)$$

where N, is the number of microorganisms (cfu/mL) at time t; N_0 is the initial number of microorganisms (cfu/mL); C is the difference between initial and final cell numbers; λ is the delay before growth (hours); μ is the maximum specific growth rate.

4. Results and Discussion

The growth curves of *C. sakazakii* in MF, supplemented/not supplemented with CCX at different concentrations (1, 2.5 and 5% (w/v)), were obtained at 25ºC (Figure 1)

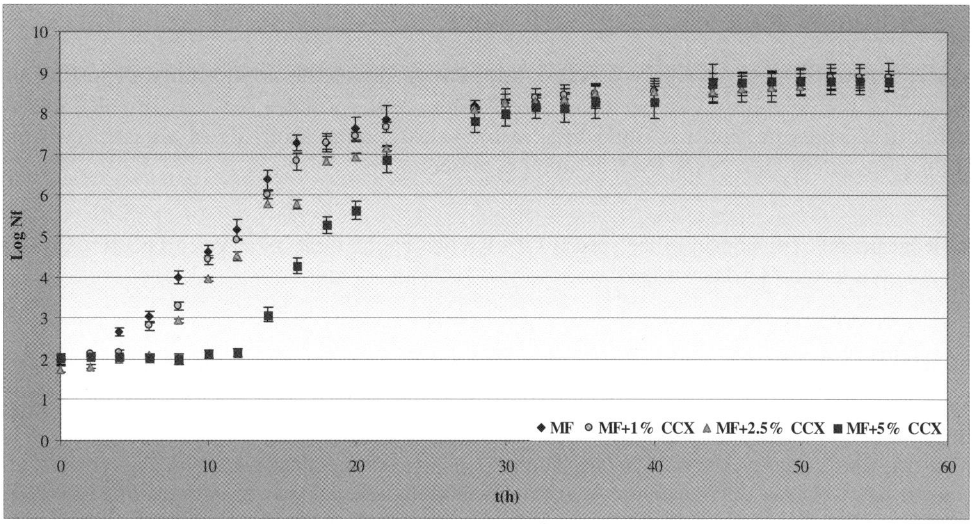

Fig. 1 *Cronobacter sakazakii* growth curves at 25ºC in milk formula (MF) supplemented / not supplemented with different cocoa powder rich in polyphenols concentration: 1, 2.5 and 5% (w/v).

Gompertz equation was used to model the *C. sakazakii* growth behavior at this temperature. Table 1 shows the growth parameter values: log N*f* (Log(CFU/mL)), specific growth rate ($\mu_{máx,}$ (h^{-1})) and lag time (λ, h).

Table 1. The Gompertz biological parameters for *Cronobacter sakazakii* growth curves at 25ºC in milk formula (MF), and supplemented MF with three different concentrations of cocoa powder rich in polyphenols: 1%, 2.5%, and 5% (w/v).

		Gompertz parameters			
MFs		**Log N_0**	**Log N_f**	μ **(h^{-1})**	λ **(h)**
MF		2.112 ± 0.415	8.662 ± 0.368	0.384 ± 0.037	3.242 ± 0.609
MF-CCX	**1%**	2.001 ± 0.212	8.087 ± 0.245	0.376 ± 0.023	3.051 ± 0.808
MF-CCX	**2.5%**	2.086 ± 0.132	8.035 ± 0.095	0.374 ± 0.010	3.444 ± 0.631
MF-CCX	**5%**	1.945 ± 0.110	7.942 ± 0.080	0.350 ± 0.015	4.789 ± 0.122

An ANOVA analysis was carried out in order to prove the differences between substrates on *C. sakazakii* growth at 25ºC. No significant differences ($P < 0.05$) between μ_{max} values for not supplemented/supplemented SM-CCX beverages at different CCX concentration were observed. However, from the data in table 1, it can be observed that the CCX concentration contributed to extend lag phase duration being significantly different for 5% (w/v) ($P < 0.05$). The higher the lag phase, the higher the bacteriostatic effect of the substance.

On the other hand, the Log N*f* achieve on MF was significantly higher than the one achieved on supplemented MFs at different CCX concentration, being no significant different between 1, 2.5 and 5% (w/v) results.

There is no an inhibitory effect of CCX on *C. sakazakii* growth in different beverages, but a significantly bacteriostatic effect was observed with 5% supplemented CCX beverage according to the growth parameters values obtained. According to our results, Gabis and Langlois [8] reported that the growth of several test organisms in milk was retarded by the presence of cocoa powder. Toshio et al. [9] studied the antimicrobial activities of aqueous cacao mass extract, which showed an inhibitory effect against enterohemorrhagic *Escherichia coli* (EHEC) O157:H7 strain. In addition, the production of verotoxins of EHEC O157:H7 strain was significantly inhibited by 8.0% cocoa extract. These results demonstrate that cocoa mass has antimicrobial effects on EHEC O157:H7. Furthermore, Pina et al. [10] obtained a lethal and bacteriostatic effect of cocoa powder (CCX) to face *Bacillus cereus* growth at 5, 20, and 37 ºC by addition at 2.5% (w/v) in skim milk substrate.

5. Conclusions

The present research work provides valuable information about the applicability of tasting ingredients with antimicrobial effect to control *Cronobacter sakazakii* into milk formulas specially oriented to infant population. According to the present results it could be possible to extend the shelf life of a pasteurized product from the microbiological safety view point, even at ambient temperature.

Acknowledgements: The support of the special I+D+I action by the Generalitat Valenciana (GV/2010/064) to emergent research groups is gratefully acknowledged.

References

[1] OMS (2008) "WHO Surveillance Programme for Control of Foodborne Infections and Intoxications in Europe. 7th Report, Spain". Disponible en www.who.it

[2] Urmenyi AMC, and Franklin AW. Neonatal death from pigmented coliform infection. *Lancet* 1961; 1:313–5.

[3] Cooper KA, Donovan JL, Waterhouse AL, et al. Cocoa andhealth: a decade of research. *Br J Nutr* 2008;99:1–11.

[4] Cienfuegos-Jovellanos, E.; Pasamar, M.A.; Fritz, J.; Arcos, J.; Ramón, D.; Castilla, Y. 2007. Method for obtaining polyphenol-rich cocoa powder with a low fat content and cocoa thus obtained. *WO* 2007/096449.

[5] Tomas-Barberan FA, Cienfuegos-Jovellanos E, Marin A, et al.A new process to develop a cocoa powder with higher flavonoid monomer content and enhanced bioavailability inhealthy humans. *Journal of Agricultural Food Chemistry* 2007;55:3926–3935.

[6] Gibson, A., Bratchell, N. y Roberts, T. 1988. Predicting microbial growth : growth responses of salmonellae in a laboratory medium affected by pH, sodium chloride and storage temperature. *International Journal of Food Microbiology*, 6, 155-178.

[8] Busta E.F. and Peck M.L. 1968. Antimicrobial effect of Cocoa on Salmonellae. *Applied Microbiology*, 424-425.

[9] Toshio T., Haruhiko T., Hiroyuki Y., Takako O., Susumu S., Masanori K., Shuich H., Shigeru K. 1999. Antibacterial Effects of Cacao Mass on Enterohemorrhagic *Escherichia coli* O157:H7. *Journal of the Japanese Association for Infectious Diseases* 73 (7): 694-701

[10] Pina-Pérez M.C., A.B. Silva-Angulo, D. Rodrigo, A. Martínez-López. 2009. Synergistic effect of Pulsed Electric Fields and CocoanOX 12% on the inactivation kinetics of Bacillus cereus in a mixed beverage of liquid whole egg and skim milk. *International Journal of Food Microbiology* 130: 196–204

Curcumin: a natural antibiofilm agent

Maya Moshe[1,2], Jonathan Lellouche[1,2] and Ehud Banin[1,2*]

[1]The Mina and Everard Goodman Faculty of Life Sciences, [2]The Institute for Nanotechnology and Advanced Materials, Bar-Ilan University, Ramat Gan, 52900, Israel.

* Corresponding author: e-mail: banine@mail.biu.ac.il, Phone: +972-3-5317288, Fax: +972-3-5317008

Biofilms resistance to killing prompts the need to discover new agents that inhibit and/or eradicate these microbial communities. Screening natural compounds for antibiofilm activity is becoming a promising approach. In this study we characterized the antibiofilm activity of curcumin. Curcumin is an extract of turmeric known to exhibit antimicrobial properties, however, its antibiofilm activity has yet to be thoroughly investigated. We find that curcumin reduces biofilm formation by *Staphylococcus aureus* and *Escherichia coli* and in addition, can effect removal and / or killing of mature biofilms. Preliminary data indicates that curcumin imposes membrane damage in *S. aureus* and morphological changes in *E. coli* cell wall which may possibly mediate the mechanism by which curcumin exerts its anti-biofilm activity. In addition, curcmin induced expression of oxidative stress related genes in *E. coli,* which may further imply on a possible mode of action. Our findings highlight the potential of curcumin to serve as a natural antibiofilm pharmacological agent.

Keywords biofilm; curcumin; antimicrobial

1. Introduction

Biofilms are surface-associated microorganisms embedded in a self-produced extracellular polymeric matrix [1, 2]. A phenotypic consequence of biofilm growth with profound clinical implications is markedly enhanced resistance to host immune killing and also to antimicrobial agents [2, 3]. This has made biofilms a major cause of medical infections [1, 2] and highlights the urgent need to develop innovative treatments for biofilm-associated infections. One approach for searching novel antibiofilm agents is to examine compounds isolated from natural sources that are commonly used in alternative medicine for treating bacterial infections [4]. Ancient cultures recognized that certain spices and herbs help preserve foods and have medicinal value. Accordingly, since the late 19th century scientists have been validating experimentally that some spices, herbs, and their components exhibit antimicrobial properties [5]. However, the antibiofilm activity of such compounds has not been addressed rigorously. Recently, antibiofilm properties have been attributed to several natural phenolic compounds such as epigallocatechin gallate (EGCg), the main polyphenol component of green tea [4]. Here we investigate whether another natural phenolic compound, curcumin, possesses antibiofilm activity.

Curcumin is a natural yellow colored product extracted from the Indian herb, *Curcuma longa.* This low molecular weight polyphenol is the major curcuminoid in turmeric, a rhizome used in India for centuries as a spice, for coloring, and as a medicinal agent [6]. Recent studies have demonstrated that curcumin exhibits a wide range of beneficial pharmacological activities including, anti-inflammatory, antioxidant, antiviral, antiangiogenic and antitumorigenic effects (reviewed in [7]). Notably, several studies have reported that curcumin can act as an antimicrobial agent against a wide range of pathogenic bacteria [8-11]. The aforementioned studies focused mainly on the activity of curcumin against free-living (i.e. planktonic) cultures and our knowledge on curcumin's antibiofilm activity is therefore fairly limited [12, 13]. To address this we have examined the capacity of curcumin to inhibit biofilm formation and eradicate mature biofilms using three model biofilm forming organisms: *Staphylococcus aureus*, *Escherichia coli* and *Pseudomonas aeruginosa*.

2. Material and methods

2.1 Bacterial strains, reagents and growth conditions

The strains studied were: *E. coli* C600, *S. aureus* 8325-4, *P. aeruginosa* PA01 and *P. aeruginosa* PA14. *E. coli* and *P. aeruginosa* strains were grown in tryptic soy broth medium (TSB). *S. aureus* 8325-4 strain was grown in B medium (1% tryptone, 0.5% yeast extract, 0.1% K_2HPO_4, 0.1% glucose and 0.5% NaCl). All experiments were carried out at 37°C. Curcumin (Sigma Chemical Co.) was dissolved in dimethyl sulfoxide (*DMSO*) and added to the media at various concentrations (0-100 µg/ml) as indicated. The final concentration of DMSO was adjusted to 1% in all experiments. For the planktonic growth inhibition assay the initial inoculum was 1.0 x 10^5

CFU/ml and growth was evaluated by taking samples for viability counts at various time points. Samples for transmition electron microscopy (TEM) were prepared as described previously [14].

2.2 Static and flow-cell biofilm assays

Static biofilm assay was carried out using the crystal violet (CV) staining assay [14]. Briefly, after the dilution of an overnight culture, cells were inoculated into 96-well plates and incubated for 18 h at 37°C without shaking in the presence and absence of various concentrations of curcumin. The biofilm biomass was determined using crystal violet (CV) staining [14]. To study the effect of curcumin on established biofilms we used a flow cell biofilm system described previously [15]. The biofilms were pregrown for three days on 10% TSB or B medium to obtain mature biofilms. On the third day, 50 µg/ml curcumin was added to the growth medium and the biofilms were treated for another 24 h. Viability was determined using the Live/Dead staining kit (Molecular Probes, Invitrogen). Propidium iodide and SYTO9 fluorescence were detected using excitation wavelengths of 488 nm and emission wavelengths of >660 nm and >500 nm, respectively. To image biofilms, we used TCS SPE Leica (Leica Microsystems INC) confocal scanning laser microscopy (CSLM). Images were analyzed using the Imaris (Bitplane AG, Switzerland) image analysis package. The PHLIP biofilm analysis software was used to determine the biovolume and ratio between the live and dead cells within the biofilm and was based on the proportion of green (i.e. live) versus red (i.e. dead) pixels in the image [16]. The SD is calculated from 5 fields viewed in three independent experiments.

2.3 *sodA* and *micF* promoter fusion expression assays

To determine the stress response of the bacteria following exposure to curcumin we utilized *E. coli* reporter strains carrying either a sodA-*lux* or mic*F*-*lux* promoter fusions as described previously [17]. These constructs emit light as a response to oxidative stress and superoxide damage (*sodA*) and superoxide or hyperosmotic stress (*micF*). For assaying luminescence, overnight cultures were diluted to approximately 10^7 cells per ml in fresh TSB (with ampicillin), and were allowed to grow at 37 °C for 6 h with and without various concentrations of curcumin (0-100 µg/ml). Luminescence was monitored using a temperature-controlled (37 °C) microtiter plate luminometer (Synergy 2, BioTek Instruments, Winoosky). Luminescence values are presented as relative luminescence units vs. growth (RLU/ OD_{595}).

3. Results and discussion

3.1 Curcumin inhibits *E. coli* and *S. aureus* but not *P. aeruginosa* biofilm formation

First, we performed the static biofilm formation assay to determine whether curcumin could inhibit biofilm formation. Curcumin reduced biofilm formation by *E. coli* by 86% and effectively blocked biofilm formation by *S. aureus* (Fig. 1 a-b) at the highest concentration tested (100 µg/ml). Interestingly, curcumin had no inhibitory effect on *P. aeruginosa* biofilm formation (Fig. 1 c).

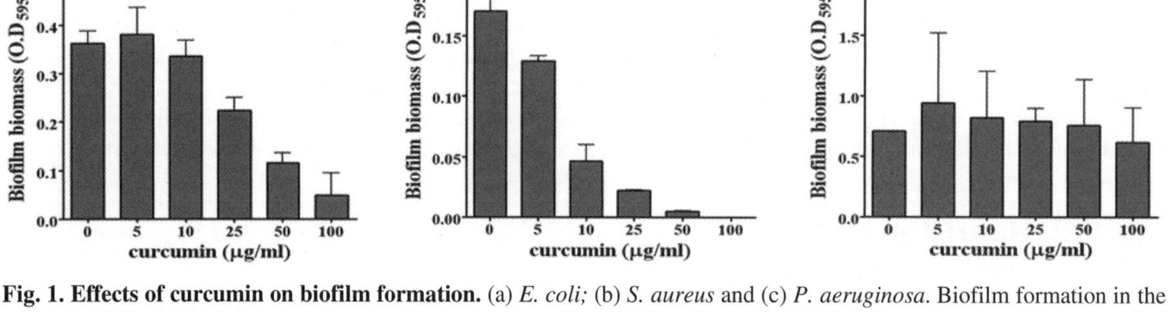

Fig. 1. Effects of curcumin on biofilm formation. (a) *E. coli;* (b) *S. aureus* and (c) *P. aeruginosa*. Biofilm formation in the presence of the indicated curcumin concentrations was evaluated using the static biofilm formation assay and quantified using crystal violet. The data represented are the mean ± SD derived from three independent experiments.

To delineate further the mechanism by which curcumin affects biofilm formation, we investigated whether curcumin influences the planktonic growth of *E. coli*, *S. aureus* and *P. aeruginosa*. We found that in the concentration range tested (25-100 µg/ml) curcumin had no influence on *E. coli* (Fig. 2 a) or *P. aeruginosa* (Fig. 2 c) growth and only slightly delayed *S. aureus* growth (Fig. 2 b); the delay was only evident in the first 12 h.

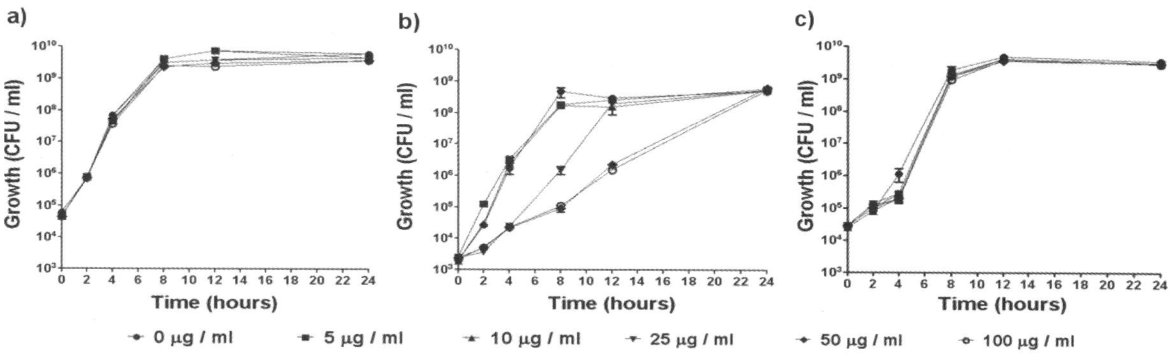

Fig. 2. Effects of curcumin on planktonic growth. (a) *E. coli*, (b) *S. aureus*, (c) *P. aeruginosa*. Bacterial growth in the presence of the indicated concentrations of curcumin (µg/ml) was monitored using viable counts. The data represented are the mean ± SD derived from three independent experiments.

These findings support that inhibition of *S. aureus* and *E. coli* biofilm formation by curcumin is not due to a bactericidal effect but rather results from inhibiting the process of biofilm formation itself. In agreement, Tajbakhsh and colleagues reported that only at a concentration of 187.5 µg/ml curcumin inhibited *S. aureus* growth while having no antibacterial activity against *E. coli* or *P. aeruginosa* [8]. Notably, contradictory to our data it has been reported that curcumin exhibits antibacterial and antibiofilm effects against a *P. aeruginosa* clinical isolate when used at 10-30 µg/ml and 1.5-3 µg/ml, respectively [13]. Microarray analyses demonstrated that curcumin down-regulated *P. aeruginosa* genes involved in quorum sensing and biofilm formation, such as *lasI* and *rhlA*, and this effect on expression was proposed to underlie curcumin antibiofilm activity [13]. We double checked planktonic growth and biofilm formation of two additional *P. aeruginosa* strains routinely used for biofilm studies, *PAO1* and *PA14*, in the absence and presence of even higher concentrations of curcumin (5-100 µg/ml) and found no inhibitory affect of curcumin (data not shown). In light of the conflicting data, we suggest that the activity of curcumin against *P. aeruginosa* could be strain specific. Given this, future work should address what confers resistance to curcumin in the two *P. aeruginosa* strains studied here.

3.2 Curcumin affects established *E. coli* and *S. aureus* biofilms

Having evidenced an inhibitory effect of curcumin on *E. coli* and *S. aureus* biofilm formation, we examined whether curcumin (50 µg/ml) can eradicate biofilms of *S. aureus* and *E. coli* that have already been established. To this end, we utilized the flow cell biofilm model system whereby biofilms are pregrown for 48 h before being exposed to curcumin for an additional 24h and then are assessed using CSLM imaging and analyses. We found that curcumin treatment reduced the biomass of *S. aureus* biofilms by 99% but had little effect on viability. In contrast, curcumin had little effect on the biomass of *E. coli* biofilms but killed cells within as indicated by propidium iodide red staining (Fig. 3 a). Indeed, PHLIP analysis confirmed that 89% of the cells were not viable in the curcumin treated *E. coli* biofilm as compared to 44% in the untreated control (Fig. 3 b). This increased sensitivity of *E. coli* biofilm cells to curcumin detected in the flow-cell system in contrast to that observed in the planktonic cultures may be due to the fact that in the flow-cell system the bacteria are constituently challenged with fresh media containing curcumin throughout the experiment. Taken together, the data indicates that curcumin exhibits antibiofilm activity against both *E. coli* and *S. aureus*, but that the mechanisms underlying this outcome are most likely different.

a)

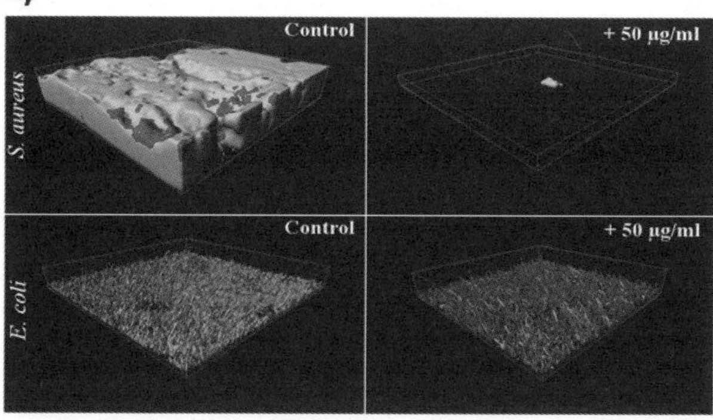

b)

	S. aureus		*E. coli*	
Curcumin	-	+	-	+
Biomass (µm³)	336,731± 193,383	5,657± 6,202	39,231± 1,861	27,663 ±4,808
Viable cells (%)	44 %	44 %	89 %	44 %
Dead cells (%)	56 %	56 %	11 %	56 %

Fig. 3. Curcumin activity against established *E. coli* and *S. aureus* biofilms. (a) Biofilms were grown in flow cells at 37°C for 3 days before the addition of curcumin (final concentration 50 g/ml). After an additional 24 h of growth, cell

viability was assessed using Live/Dead staining. Live cells are represented in green and dead cells in red. (b) Biovolume and viability were quantified using the PHLIP software as described in the Materials and Methods.

3.3 Effect of curcumin on *E. coli* and *S. aureus* morphology

To investigate in more detail the mechanisms underlying curcumin activity, planktonic *E. coli* and *S. aureus* were grown for 24 h with and without 100 µg/ml of curcumin and then their morphology examined. Transmission electron microscopic analysis of *E. coli* samples revealed that curcumin causes a slight reduction in bacterial size (data not shown). In addition, the images indicate that the cell wall becomes amorphous and in parallel, unspecified filaments appear on and around the cells (Fig. 4, marked with arrow). We speculate that these morphological changes serve to prevent *E. coli* biofilm formation. Curcumin treated *S. aureus* cells showed substantial cell wall damage and leakage (Fig. 4). In addition, although we saw no effect on cell size, treated *S. aureus* cells appeared to contain several chromosomes instead of one (Fig. 4, marked with arrows). This phenotype could indicate curcumin is deregulating *S. aureus* cell division. In this regard, it is noteworthy that another Gram-positive bacterial strain, *B. subtilis*, was reported recently to suppress cell division in response to curcumin. The proposed mechanism involved inhibition of FtsZ assembly in the Z ring [9].

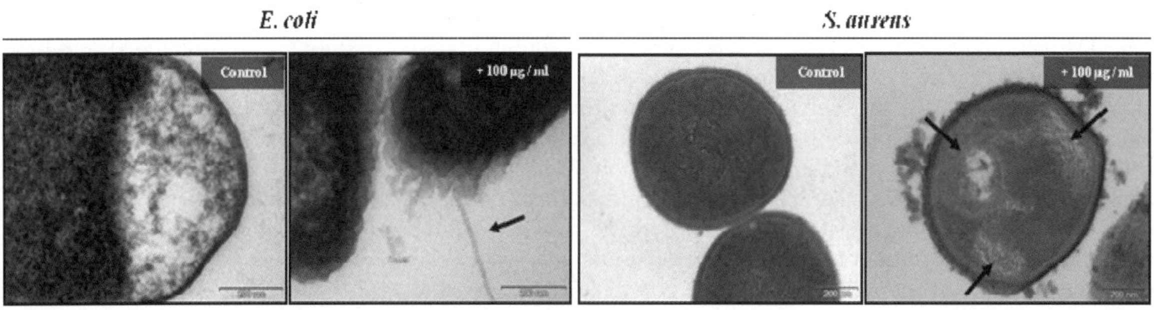

Fig. 4. Curcumin affects *E. coli* and *S. aureus* cell morphology. TEM images of planktonic cultures grown with or without curcumin (100 µg/ml) for 24 h.

3.4 Curcumin induces stress related genes in *E. coli*

To further characterize the physiological response upon exposure to curcumin we utilized *E. coli* biosensor strains that are indicators for various bacterial stress responses [17]. Curcumin treatment caused overexpression of the *micF* gene by 4 fold and the *sodA* gene by 7 fold (Fig. 5). The *micF* gene is an antisense RNA that represses the expression of *ompF*, a major outer membrane porin required for passive entry of antibiotics and nutrients [18]. Thus, curcumin most-likely reduces expression of *ompF* through the induction of *micF*. Interstingly, *OmpF* is known to be overexpressed in biofilms [19], therefore, the anti-biofilm activity of curcumin against *E. coli* may be due to its negative affect on *ompF* expression. Curcumin also induced the expression of *sodA* (manganese superoxide dismutase) which responds to elevated amounts of superoxide radicals. There has been several reports on the pro-oxidant activity of curcumin and its ability to induce reactive oxidant species (ROS) [6, 20]. These results implicate a primary mode of superoxide-type stress toxicity due to curcumin treatment.

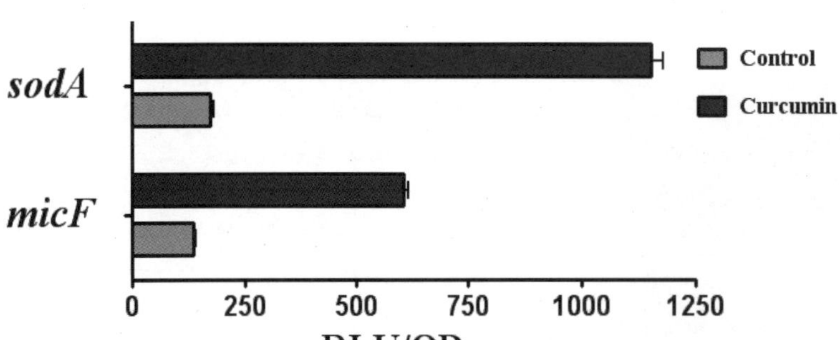

Fig. 5. Overexpression of stress related genes in the presence of curcumin. For assaying luminescence, cells were grown at 37 °C for 6 h without (grey) and with (black) curcumin (100 µg/ml). Luminescence values are presented as relative luminescence units vs. growth (RLU/OD_{595}).

4. Conclusions

The contribution of the biofilm lifestyle to the refractory nature of many infections is widely accepted and highlights the urgent need to discover new antibiofilm agents. Our findings evidence the potential of curcumin to serve as such an agent. The lack of toxicity of curcumin together with its recently improved aqueous solubility [21], underscore the conclusion that curcumin should be considered for further development as an antibiofilm pharmacological agent.

Acknowledgements The work was supported by the Bar-Ilan University nanotechnology startup grant (E.B.).

References

[1] Hall-Stoodley L, Costerton JW, Stoodley P. Bacterial biofilms: from the natural environment to infectious diseases. *Nat Rev Microbiol.* 2004;2:95-108.
[2] Costerton JW, Stewart PS, Greenberg EP. Bacterial biofilms: a common cause of persistent infections. *Science.* 1999;284:1318-1322.
[3] Davies D. Understanding biofilm resistance to antibacterial agents. *Nat Rev Drug Discov.* 2003;2:114-122.
[4] Blanco AR, Sudano-Roccaro A, Spoto GC, Nostro A, Rusciano D. Epigallocatechin gallate inhibits biofilm formation by ocular staphylococcal isolates. *Antimicrob Agents and Ch.* 2005;49:4339-4343.
[5] Cowan MM. Plant products as antimicrobial agents. *Clin Microbiol Rev.* 1999;12:564-580.
[6] Aggarwal BB, Sundaram C, Malani N, Ichikawa H. Curcumin: The Indian solid gold. *Adv Exp Med Biol.* 2007;595:1-15.
[7] Strimpakos AS, Sharma RA. Curcumin: Preventive and therapeutic properties in laboratory studies and clinical trials. *Antioxid Redox Sig.* 2008;10:511-545.
[8] Tajbakhsh S, Mohammadi K, Deilami I, Zandi K, Fouladvand M, Ramedani E, et al. Antibacterial activity of indium curcumin and indium diacetylcurcumin. *Afr J Biotechnol.* 2008;7:3832-3835.
[9] Rai D, Singh JK, Roy N, Panda D. Curcumin inhibits FtsZ assembly: an attractive mechanism for its antibacterial activity. *Biochem J.* 2008;410:147-155.
[10] Mahady GB, Pendland SL, Yun G, Lu Z. Turmeric (Curcuma longa) and curcumin inhibit the growth of *Helicobacter pylori*, a group 1 carcinogen. *Anticancer Res.* 2002;22:4179-4181.
[11] Kumar S, Narain U, Tripathi S, Misra K. Syntheses of curcumin bioconjugates and study of their antibacterial activities against beta-lactamase-producing microorganisms. *Bioconjugate Chem.* 2001;12:464-469.
[12] Pattiyathanee P, Vilaichone RK, Chaichanawongsaroj N. Effect of curcumin on *Helicobacter pylori* biofilm formation. *Afr J Biotechnol.* 2009;8:5106-5115.
[13] Rudrappa T, Bais HP. Curcumin, a known phenolic from curcuma longa, attenuates the virulence of *Pseudomonas aeruginosa* PAO1 in whole plant and animal pathogenicity models. *J Agr Food Chem.* 2008;56:1955-1962.
[14] Lellouche J, Kahana E, Elias S, Gedanken A, Banin E. Antibiofilm activity of nanosized magnesium fluoride. *Biomaterials.* 2009:5969-5978.
[15] Banin E, Brady KM, Greenberg EP. Chelator-induced dispersal and killing of *Pseudomonas aeruginosa* cells in a biofilm. Applied And Enviromental Microbiology 2006;72:2064-2069.
[16] Beyenal H, Donovan C, Lewandowski Z, Harkin G. Three-dimensional biofilm structure quantification. *J Microbiol Met.* 2004;59:395-413.
[17] Belkin S, Smulski DR, Vollmer AC, VanDyk TK, LaRossa RA. Oxidative stress detection with *Escherichia coli* harboring a katG'::lux fusion. *App Environ Microb.* 1996;62:2252-2256.
[18] Delihas N, Forst S. MicF: An antisense RNA gene involved in response of *Escherichia coli* to global stress factors. *J Mol Biol.* 2001;313:1-12.
[19] Schembri MA, Kjaergaard K, Klemm P. Global gene expression in *Escherichia coli* biofilms. *Mol Microbiol.* 2003;48:253-267.
[20] Haukvik T, Bruzell E, Kristensen S, Tonnesen HH. Photokilling of bacteria by curcumin in different aqueous preparations. Studies on curcumin and curcuminoids XXXVII. *Pharmazie.* 2009;64(10):666-673.
[21] Bisht S, Feldmann G, Soni S, Ravi R, Karikar C, Maitra A, et al. Polymeric nanoparticle-encapsulated curcumin ("nanocurcumin"): a novel strategy for human cancer therapy. *J Nanobiotechnol.* 2007;5:3-10.

Evaluation of *in vitro* and *in vivo* antibacterial and antifungal activity of "Camelyn M"

Benedikte Maglakelidze[1], Guguli Abashidze[1], Inga Dadeshidze[2], Vakxtang Mshvildadze[3], Andre Pichete[3], Vincent Perreten[4] Shota Tsanava[5], Nata Shubladze[6], Koba Gurielidze[1*]

1. Scientific-Pharmaceutical Company "Camelyn", 0127, Kindzmarauli str. 7, Tbilisi, Georgia.
2. Kutateladze Institute of Pharmacochemistry, 0159, sarajishvili str. 36, Tbilisi, Georgia.
3. Laboratoire LASEVE, Universite du Quebec a` Chicoutimi, 555, Boulevard de l' Universite, Chicoutimi, Que., Canada G7H 2B1.
4. Institute of Veterinary Bacteriology,University of Berne,Postfach Länggass-Strasse 122, CH-3001 Berne,Switzerland.
5. National Center for Disease Control & Medical Statistics, 0177, Asatiani 9, Tbilisi, Georgia.
6. National Center for Tuberculosis and Lung Diseases, 0101, Maruashvili 50, Tbilisi, Georgia.

*Corresponding author: Scientific-Pharmaceutical Company "Camelyn", 0127, Kindzmarauli str. 7, Tbilisi, Georgia. E-mail: kgurielidze@yahoo.com

The minimum inhibitory concentration (MIC) of "Camelyn M" active compound obtained from special sort of honey in one of region of Georgia, was determined both by agar and broth dilution methods against some of strains of bacteria and fungy. The antibacterial action of "Camelyn M" was further tested in animal models. "Camelyn M" was seen to possess powerful inhibitory action (0,012-0,150 µg/mL) against most test bacteria in *in vitro* studies. *In vivo* studies showed that the drug offered significant protection ($p<0.001$) to mice challenged with a virulent bacterium.

"Camelyn M" exhibited potent in vitro activities against fluconazole-resistant strains of *Candida albicans* , *Candida glabrata*, *Candida tropicalis*, *Candida parapsilosis* and *Candida krusei*, with MICs at which 90% of isolates were inhibited of 0.012 µg/ml, respectively.

Keywords: "Camelyn M", honey , antimicrobial, Candida, MIC.

Introduction

Antibiotics are one of our most important weapons in fighting bacterial and fungal infections and have greatly benefited the health-related quality of human life since their introduction. However, over the past few decades these health benefits are under threat as many commonly used antibiotics have become less and less effective against certain illnesses not only because many of them produce toxic reactions but also due to emergence of drugresistant bacteria and fungy(2,3,4,5,6,7,8).

"Camelyn M" were isolated from special species of honey from one of region of Georgia (1). "Camelyn M" is a mixture of different biologically active compounds: aldehydes, ketones, bioorganyc acides, (6). Different physicochemical parameters of Camelin's compounds, such as solubility, primary structure, molecular weight, and other properties play a significant role in the biological activities of "Camelyn M" (6). Our studies demonstrated that "Camelyn M" has strong antibacterial and antifungal activity in a wide variety of bacterial and fungal species. "Camelyn M" acts by as detergent on the bacterial and fungal cell membranes and selectively inhibiting wearies of living mechanism of bacterial and fungal cells.

The present paper describes the detailed *in vitro* and *in vivo* antibacterial and antifungal activity of "Camelyn M".

Materials and Methods

Media

Liquid media used for this study were peptone water [PW; Oxoid brand bacteriological peptone 1%(w/v) plus Analar NaCl 0.5%(w/v)], nutrient broth (NB; Oxoid), Mueller Hinton broth (MHB; Difco). Solid media were peptone agar (PA), bromothymol blue lactose agar media (BLA), nutrient agar (NA) and Mueller Hinton agar (MHA), obtained by solidifying the liquid media with 1.2%(w/v) agar (Oxoid No.3).

Antibacterial and Antifungal agents. "Camelyn M" was isolated from honey as described previously (1) for the *in vitro* and *in vivo* studies. For the in vitro study, fluconazole (FLC) and itraconazole (ITC) were extracted from commercial preparations purchased from PSP Pharmaceuticals, Inc. (Tbilisi, Georgia.)

Organisms:

*Bacteria and fungy:*A total of strains of bacteria and fungy were obtained from the American Type Culture Collection .

Media

Liquid media used for this study were peptone water [PW; Oxoid brand bacteriological peptone 1%(w/v) plus Analar NaCl 0.5%(w/v)], nutrient broth (NB; Oxoid), Mueller Hinton broth (MHB; Difco). Solid media were peptone agar (PA), bromothymol blue lactose agar media (BLA), nutrient agar (NA) and Mueller Hinton agar (MHA), obtained by solidifying the liquid media with 1.2%(w/v) agar (Oxoid No.3). In case of BLA.

In vitro tests. The MICs for the test organisms were determined by the broth microdilution method described in NCCLS document M27-A2 (3,4,5). The MICs of "Camelyn M" were defined as the lowest concentration that resulted in slight growth (approximately 90% inhibition) or the absence of growth at 48 h.

In vivo tests

Swiss strain of male white mice weighing 18-20g were used for the *in vivo* studies. The virulence of the test strain *S.typhimurium* NCTC 74 was exalted by repeated mouse passage and the median lethal dose (MLD or LD50) of the passaged strain corresponding to 0.95x109 CFU/ mouse suspended in 0.5mL NB served as the challenge dose17 for all the groups of animals. Reproducibility of the challenge dose was ensured by standardization of its optical density in a Klett-Summerson colorimeter at 640nm and determination of the CFU count in NA.

To determine the toxicity of "Camelyn M", 40 mice were taken, 20 of which were injected 60g of the drug, and the rest 20 received 30g of "Camelyn M". They were kept under observation upto 100 hours. Two groups of mice, 20 animals per group (each mouse weighing about 20g) were kept in separate cages. Group I was intraperitoneally administered 30g "Camelyn M" per mouse (0.1mL from 300g/mL solution of "Camelyn M"), and group II was given 60g of the drug per mouse (0.1mL from 600g/mL solution of "Camelyn M"). After 3 hours, each group I and II was challenged with 50 MLD of *S.typhimurium* NCTC 74. A control group of 60 mice was also injected similarly with the same bacterial strain, and 0.1mL sterile saline instead of "Camelyn M". The protective capacity of the drug was determined by recording the mortality of the mice in different groups upto 100 hours of the treatment, and statistically by $x2$ test. In another experiment, 4 groups of mice, 5 animals per group, were taken. Groups 1 and 3 were administered 60 g of "Camelyn M", while groups 2 and 4 were given 0.1mL sterile saline. After 3 hours, all the groups were given a 50 MLD challenge of *S.typhimurium* NCTC 74. After 2 hours, groups 1 and 2 were sacrificed. Their heart blood was collected aseptically; their livers and spleens were removed aseptically and homogenised in tissue homogenisers. CFU counts of the individual organs were determined separately. The same procedure was applied on groups 3 and 4, 18 hours after the challenge. Statistical analysis of the *in vivo* data was done by Student's t-test. The concentration of "Camelyn M" in mouse blood was assayed by measuring the diameter of the inhibition zones by serum-soaked filter paper discs (6 mm diameter, 3 mm thick, Millipore, absorbing 0.0 3mL volume) on a lawn flooded with bacteria from an 18 hours broth culture of *S. typhimurium* 74 on peptone agar. The drug concentrations in the sera were determined by referring these values to a standard calibration curve prepared with known concentrations of the drugs.

Results

All the bacterial strains tested were found to be resistant to many antibiotics. However, "Camelyn M" showed powerful antimicrobial action against all the bacteria"Camelyn M" at a range of concentration of 0,156-3.0 μg/ml inhibited all bacterial strains:

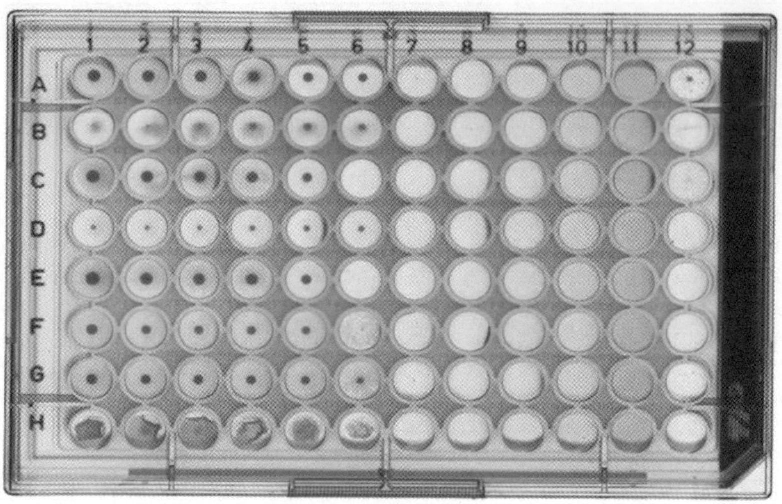

Fig 1. MICs of “Camelyn M” for **specific antibiotic resistant bacterias** as determined by broth dilution in Müller-Hinton broth.

In vertical- strains of:

A) - *E. faecium* SF11770
B) - *B. antracis ChadA8*
C) - *S. aureus* BM3318
D) - *B. antracis JFS854*
E) - *S.typhimurium* NCTC 74
F) - *S. sciuri* CSLA3
G) - *S. haemolyticus* VPS617
H) - *B. cereus* AND934

In horizontal- MICs of “Camelyn M”

1)- 0.002
2)- 0.004
3)- 0.009
4)- 0.019
5)- 0.039
6)- 0.078
7)- 0.156
8)- 0.310
9)- 0.62
10)- 1.25
11)- 2.50

A B

Fig 2. MICs of “Camelyn M” for specific antibiotic resistant bacterias *of Bacilus anthracis*.
A-after 4 hours inoculation and B-after 4 hours inoculation.

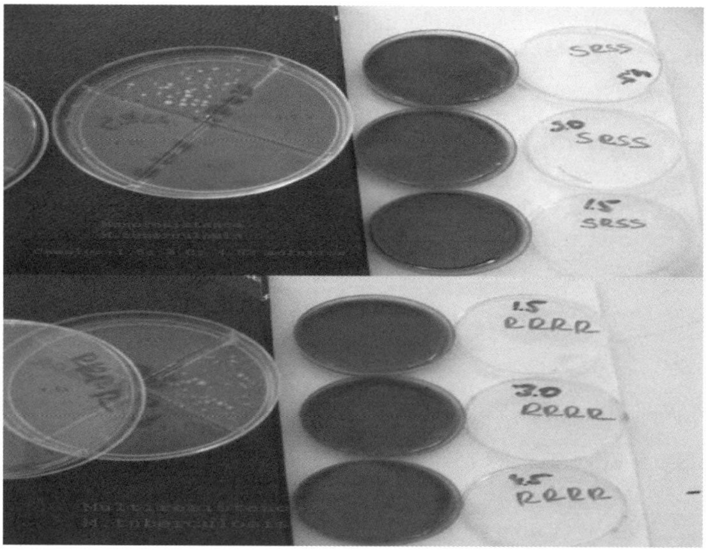

Fig 3. MICs of "Camelyn M" for **specific multi and monoresistent bacterias** ***of M.tuberculosis***.

***In vitro* antifungal activity.**
Table 1 shows the spectrum of activities of "Camelyn M" and other reference against various fungal strains. "Camelyn M" exhibited potent activities against *C. albicans*, *C. glabrata*, *C. tropicalis*, *C. guilliermondii*, and *C. neoformans*, with MICs ranging from 0.012 µg/ml for all the above listed cases. Table 2 shows the MICs of "Camelyn M" and other reference agents for clinical yeast isolate of *C. albicans*. Table 2 shows the results for *C. albicans* separately for FLC-susceptible (FLC-S) strains (FLC MICs, ≤8 µg/ml) and FLC-susceptible dose-dependent (FLC-S-DD) and FLC-resistant (FLC-R) strains (FLC MICs, ≥16 µg/ml), according to the guidelines of NCCLS document M27-A2 (14). "Camelyn M" exhibited potent activities against *C. albicans* (FLC-S), with MICs at which 90% of isolate are inhibited (MIC_{90}s) of 0.012 µg/ml. "Camelin M" also exhibited potent activities against the FLC-S-DD and FLC-R strains of *C. albicans* (MIC range, 0.012 µg/ml).

Table 1. Antifungal spectrum of "Camelyn M".

Organism and strain	MIC (µg/ml)			
	Camelyn M	FLC	ITC	AMB
Candida albicans ATCC 24433	0.012	0.25	0.016	0.12
Candida glabrata ATCC 90030	0.012	4.00	0.12	0.12
Candida tropicalis ATCC 750	0.012	2.00	0.06	0.25
Candida parapsilosis ATCC 22019	0.012	2.00	0.03	0.5
Candida krusei TIMM3378	0.012	32.00	0.06	0.25
Candida guilliermondii ATCC 9390	0.012	2.00	0.03	0.06

Table 2. *In vitro* antifungal activity of "Camelyn M" against clinical isolate of *Candida albicans*.

Organism (clinical isolates) and agent	MIC (µg/ml)	
	50%	90%
Candida albicans		
"Camelyn M"	0.012	0.012
FLC	0.25-8	1.00
ITC	0.016	0.03
AMB	0.12	0.12

"Camelyn M" belongs to the natural, antibacterial and antifungal agents with a complex mechanism of action. The fact that "Camelyn M" comprises biologically very active phenolic compounds is very interesting. This compounds may be met in leaves and rootstocks of many plants and ensure of resistance of these plant with respect to different pathogens. Natural phenolyc compoumds is characterized with strong detergent properties and at occurrence in small size organisms cause destruction of the cell membrane. Resulting from the mentioned, it may be presumed that the biological activity of "Camelyn" includes the above mechanisms, which of cause is a subject of deeper investigations.

As it is seen from Tables 1 and 2 "Camelyn M" displays strong in vitro activity with respect to all the test strains in the same concentration -MIC (μg/ml)=0.012. This concentration is more low than all the antibiotics tested by us.

In conclusion, the results of the present study suggest that “Camelyn M” is a promising compound, in particular, for the treatment of disseminated or bacterial and mucosal infections induced by *C. albicans*, including FLC-resistant strains.

References

1. B.Maglakelidze. “Antitumoral properties of “Camelyn” and mechanism of activity” Monograf, Tbilisi, 2004 (in Georgian)
2. Berdy, J., 1967. Recent developments of Antibiotic Research and Classification of Antibiotics According to Chemical Structure, Advances in Applied Microbiology, 309.
3. Bergey’s manual of determinative bacteriology, 2000. Actinomycetales. 9th edition.
4. Hugo, W.B., Russel, A. D., 1983. Pharmaceutical Microbiology, 3rd edition, Blackwell Scientific publications.
5. Barchiesi, F., A. M. Tortorano, L. F. Di Francesco, M. Cogliati, G. Scalise, and M. A. Viviani. 1999. In-vitro activity of five antifungal agents against uncommon clinical isolates of *Candida* spp. J. Antimicrob. Chemother. 43:295-299.
6. Kamai, Y., M. Kubota, T. Fukuoka, Y. Kamai, N. Maeda, T. Hosokawa, T. Shibayama, K. Uchida, H. Yamaguchi, and S. Kuwahara. 2003. Efficacy of CS-758, a novel triazole, against experimental fluconazole-resistant oropharyngeal candidiasis in mice. Antimicrob. Agents Chemother. 47:601-606.
7. Herreros, E., M. J. Almela, S. Lozano, F. Gomez de las Heras, and D. Gargallo-Viola. 2001. Antifungal activities and cytotoxicity studies of six new azasordarins. Antimicrob. Agents Chemother. 45:3132-3139.
8. Herreros, E., C. M. Martinez, M. J. Almela, M. S. Marriott, F. Gomez de las Heras, and D. Gargallo-Viola. 1998. Sordarins: in vitro activities of new antifungal derivatives against pathogenic yeasts, *Pneumocystis carinii*, and filamentous fungi. Antimicrob. Agents Chemother. 42:2863-2869.

In vitro antifungal activity of several essential oils from aromatic plants of Aragón (NE, Spain)

S. Lorán[*,1], O. Segman, C. Rota[1], P. Conchello[1] and A. Herrera[1]

[1] University of Zaragoza, Department of Animal Production and Food Science, C/ Miguel Servet 177, 50013 Zaragoza, Spain

* Corresponding author: e-mail: sloran@unizar.es

In this study we have evaluated the antifungal activity of thirteen essential oils (EOs), collected in four different geographical areas of Aragón, north eastern Spain, against a selection of three strains of the genus *Fusarium* (*Fusarium verticilloides* CECT 2987, *Fusarium verticilloides* CECT 2982 *and Fusarium graminearum* CECT 2150). In vitro antifungal activity was evaluated by the disc diffusion method. Three of the EOs assayed (*Thymus vulgaris* Local, *Thymus vulgaris* French and *Satureja montana*) showed an important activity against the three strains studied by producing a zone diameter of inhibition ≥ 19 mm. In addition, *Fusarium graminearum* was the most sensitive strain against the essential oils studied. However, in those cases in which no inhibition was observed, it could be seen a lower growth density of the moulds inoculated than in the corresponding negative control. Therefore, more precise data about the antifungal activity of these EOs are needed.

Keywords essential oils; Aragón (NE, Spain); *Fusarium verticilloides*; *Fusarium graminearum*; antifungal activity

1. Introduction

Spoilage and poisoning of foods by fungi reduces their shelf life and market value. In addition, the growth of toxigenic fungi and mycotoxins produced in food and grains stored for long period of time present a potential hazard to human and animal health [1, 2].

For many years, a variety of different chemical and synthetic compounds have been used as antimicrobial agents to inhibit the plant pathogenic fungi. However, the widespread indiscriminate use of chemical preservatives has led to a number of ecological and medical problems which in addition to the economic considerations, make it necessary to adopt strategies that are accessible, simple in application, and nontoxic to humans and plants [3]. In consequence, some research has been made to find natural products that could safely be used as substitutes for fungicides to partially or completely inhibit the growth of fungi as well as mycotoxin biosynthesis.

Among these natural products, essential oils (EOs) from aromatic and medicinal plants have received particular attention as potential natural agents for food preservation [4, 5]. In fact, EOs and their main components are becoming increasingly important in the food industry because of their reported wide spectrum of antibacterial activity as well as antimycotic, antiparasitic and insecticidal properties along with other biological activities. Thus, their use seems to be particularly important for public health in developing countries.

Essential oils are volatile mixture of organic compounds usually obtained from non-woody parts of the plant through steam or hydro-distillation [4, 6]. They are characterized by two or three major components at quite high concentrations (up to 80%) compared to other components present only in trace amounts [4, 5]. At present, approximately 3000 essential oils are known, 300 of which are commercially important, especially for the pharmaceutical, agronomic, food, sanitary, cosmetic and perfume industries [5].

Fungi of the genus *Fusarium* are widely distributed in nature and cause deterioration of food and feeds [7, 8]. In Aragón (Spain), maize diseases caused by *Fusarium* have been reported since the early 1980's [9]. Several species from this genus are economically relevant because, apart from their ability to infect and cause tissue destruction, predominantly in corn and other cereals as constituents of food and feed products, they may produce mycotoxins on the crops in the field and in storage [10-12]. The presence of mycotoxins in grains creates severe problems for animals, given that they may perform poorly or become sick and die, and for human health [10,12]. Therefore, the threat of mycotoxins entering the human food supply through contaminated grain products and exposed animals is also a concern for regulatory officials.

A wide variety of EOs and many of their main components, mainly mono and sesquiterpenes, have been shown to have antifungal properties against several strains of the genus *Fusarium* [8, 11, 12]. What's more, a direct relationship exists between inhibitory effects of EOs on fungal growth and fusariotoxins production [8, 11, 12].

This study was undertaken to investigate the *in vitro* antifungal activity, and so their possible use as food preservatives, of thirteen essential oils extracted from selected plants of Aragón (NE Spain) against spoilage and mycotoxin producing fungi (*Fusarium verticilloides* and *Fusarium graminareum*).

2. Material and methods

2.1 Plant material and isolation of essential oils

Plants were collected from experimental fields located in four different geographical areas of Aragón, a region in north-eastern Spain with a dry continental climate. The essential oils investigated were obtained from *Thymus vulgaris* Local (L), *Thymus vulgaris* French (F), *Satureja montana*, *Rosmarinus officinalis*, *Lavandin abrialis*, *Lavandin super*, *Lavandin grosso*, *Lavandula latifolia*, *Lavandula angustifolia*, *Lavandula lavandulifolia*, *Hissopus officinalis*, *Salvia officinalis* and *Salvia sclarea.*

Pulverized aerial parts of the dried plants were subjected to steam distillation for 1 h using a Clevenger-type apparatus. The essential oil was collected and stored in sealed glass vials at 4°C in the dark until used. Identification and purity of samples was guaranteed all the time.

The gas chromatography characterization of these essential oils has been previously reported [13].

2.2 Test microorganism and growth conditions

In this study three mycotoxigenic strains of the genus *Fusarium* were used: *Fusarium verticilloides* CECT 2982, *Fusarium verticilloides* CECT 2987 and *Fusarium graminareum* CECT 2150. Both *Fusarium. verticilloides* and *Fusarium graminareum, also known as Gibberella moniliformis and Gibberella zeae respectively,* are included in the number of species of the genus *Fusarium* most frequently isolated from maize in temperate climates. All strains were obtained from the Spanish Collection of Type Cultures (Colección Española de Cultivos Tipo, CECT, Valencia, Spain).

Fungal strains were grown in Saboreaud broth (Merck, Darmstadt, Germany) and incubated at 25 ºC/7 days. Stock cultures of the molds were maintained on Potato Dextrose Agar (PDA; Merck, Darmstadt, Germany) at 5 ºC.

2.3 Antifungal assay (disk diffusion assay)

The preliminary *in vitro* antifungal activity was determined by the disc agar diffusion technique. In so doing, cultures of the moulds were grown on PDA slants for 7 days at 25 ºC. Inoculums were prepared by adding about 20 ml of sterile 0.5% Tween 20 (Merck-Schuchardt, Hohenbrunn, Germany) in distilled water. Spores were loosened by gentle brushing of the conidiophores with a sterile inoculating loop. The spore suspension was filtered through four layers of sterile cheese cloth to remove mycelial debris and then diluted to a final concentration of 10^6 spores per ml. The number of spores per ml was determined by direct microscopic count.

Then, each plate of PDA was inoculated with 0.1 ml of the spore suspension which was spread on the agar surface and allowed to dry for approximately 2 h. Next, 15 μl of each essential oil were dropped on 6 mm diameter filter paper disks (Whatman No.1) and placed on the agar surface. The plates were incubated for 7 days at 25 ºC. Every day plates were observed but only after 7 days of incubation inhibition zone diameters around each of the discs (diameter of inhibition zone plus diameter of the disc) were measured in millimetres. The scale of measurement was the following (disk diameter included): ≥20 mm zone of inhibition = strongly inhibitory; <20–12 mm zone of inhibition = moderately/mildly inhibitory; <12 mm = not inhibitory.

3. Results and discussion

The antimicrobial activity of the EOs assayed, against the fungal strains tested is summarized in Table 1. As it can be seen in the data, using the filter paper disc agar diffusion technique, the different essential oils inhibited growth to variable extents, depending on the oil and the strain assayed.

Higher antifungal activity was found in the EOs of *Thymus vulgaris* F. and *Satureja montana*, which were strong inhibitors in all cases, *and Thymus vulgaris* L. (moderate to strong inhibitory activity depending on the *Fusarium* strain tested). These three EOs showed a zone diameter of inhibition ≥ 19 mm in all cases. However, inhibition zones lower than 12 mm, which was considered as no inhibition of the moulds, were exhibited after seven days of incubation by the other EOs assayed.

The different levels of antifungal activities expressed by the EOs assayed may be explained by the variations in their chemical composition. At this concern, the most effective EOs *Thymus vulgaris* F. and *Satureja montana,* both have as mayor component carvacrol (43.7% and 56.7%, respectively) followed by thymol

(7.5%), *p*-cymene (6.7%), linalool (5.5%) and 1,8-cineole (1.8%) (*Thymus vulgaris* F.) and by *p*-cymene (17.2%), borneol (2.4%) and γ-terpinene (2.2%) (*Satureja montana*). In fact, carvacrol, since it contains both a hydroxyl group and a system of delocalised electrons, is a well-known phenolic compound that has demonstrated to possess high antifungal activity against phytopathogenic fungi [14, 15]. Several authors have demonstrated high antifungal activity of thymol (also present at significant concentration in *Thymus vulgaris* F.), in different fungal species including *Fusarium verticilloides* [11, 15] while *p*-cymene (at concentrations of 17.2% in *Satureja montana* EO) has been reported to be less fungitoxic than thymol and carvacrol [15]. However, similar activities have been obtained by both *Thymus vulgaris* F. and *Satureja montana* EOs.

The chemical profile of *Thymus vulgaris* L., with 1,8-cineole (20.8%), linalool (14.2%), carvacrol (10%) and *p*-cymene (7.5%) as major constituents, was rather different than the showed by the more effective *Thymus vulgaris* F. The genus *Thymus* has numerous species and varieties and their essential oil composition has been studied. The composition of the essential oils is known to vary not only between species and chemotypes but also with the geographical region, climate conditions, age of the plant, methods of drying and extraction. Then again, the content of antimicrobial compounds of these EOs explains the differences in the inhibitory activity against both the bacterial [16] and fungi (present study) strains tested. Actually, previous reports showed that 1,8-cineole possessed weak and/or no antifungal activity against phytopathogenic fungal species [17, 18].

What's more, nor carvacrol, neither thymol were present at quantifiable concentrations in the other EOs assayed. All in all, maybe we shouldn't attribute the entire activity of a complex mixture to a single or particular constituent and it should be considered that minor components in essential oils, as well as synergistic and/or antagonistic interactions between the volatile components could also affect the antifungal properties of the essential oils.

Table 1 Inhibiton zone (mm) exhibited by the essential oils studied against *Fusarium verticilloides* 2982, *Fusarium verticilloides* 2987 and *Fusarium graminareum* 2150.

Essential oils	Inhibition zone (mm)		
	Fusarium verticilloides 2982	*Fusarium verticilloides* 2987	*Fusarium graminareum* 2150
Thymus vulgaris (L)	19.5	19	30
Thymus vulgaris French (F)	37.5	38	c.i.
Satureja montana	35	41	c.i.
Rosmarinus officinalis	n.i.	n.i.	n.i.
Lavandin abrialis	n.i.	n.i.	n.i.
Lavandin super	n.i.	n.i.	n.i.
Lavandin grosso	n.i.	n.i.	n.i.
Lavandula latifolia	n.i.	n.i.	n.i.
Lavandula angustifolia	n.i.	n.i.	n.i.
Lavandula lavandulifolia	n.i.	n.i.	n.i.
Hissopus officinalis	n.i.	n.i.	n.i.
Salvia officinalis	n.i.	n.i.	n.i.
Salvia sclarea	n.i.	n.i.	n.i.

n.i.: not inhibited (inhibition zone < 12 mm); c.i.: whole inhibition (no visible growth of the inoculated moulds)

These results are in accordance with earlier studies that have reported the elevated antifungal activity of *Satureja montana* [19] and *Thymus vulgaris* EOs, being the latest even better than most of the synthetic fungicides that were assayed at the same time [20]. However, the comparison of the different data reported is not easy due to the different methodologies employed to asses the antifungal activity of the EOs.

At this concern, some considerations should be taken into account in the evaluation of the antifungal activity of the EOs studied with the employed disk diffusion assay technique. So, during the incubation period, the inoculated plates were observed every day and the inhibition zones around the discs were attempted to be measured. However, due to the heterogeneous growth of the moulds, only the measured carried out after seven days of incubation was considered in the end. Nonetheless, these observations let us to know that, in general, the inhibition zones around the discs were getting lower from the third to the seven day. Therefore, a number of EOs (*Rossmarinus officinalis*, *Lavandin grosso* or *Lavandula latifolia*) that at the end of the incubation time were discharged due to their low anti-*Fusarium* activity showed moderate activity (inhibition zone > than 12 mm) the earlier incubation days. What's more, in some cases in which very low inhibition zones were observed, it could be seen a lower growth density of the moulds inoculated than in the corresponding negative control (filter paper disk without EO added) (Figure 1). This perceptible inhibition of the growth of the moulds, which was probably due to the effect of volatile compounds, couldn't be evaluated with this methodology.

Thus, the use of other methodologies as the agar dilution method, broth dilution method or in vitro contact assay [18; 21] should be employed in order to obtain more precise data about the anti-*Fusarium* activity of the EOs studied.

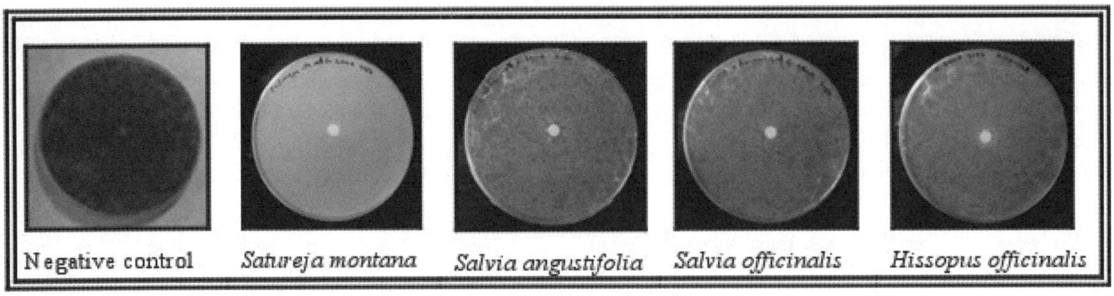

Fig. 1 Micelial growth of inoculated plates (*Fusarium graminearum 2150*) in which some of the essential oils assayed were added.

Nonetheless, the species of *Fusarium verticilloides* and *Fusarium graminareum* have been previously reported to be inhibited by the EOs of *Zataria multiflora* and *Cuminum cyminum* [22]; *Ocimum basilicum* and *Ocimum gratissimum* [12] or lemongrass, clove, cinnamon, oregano and palmarosa oils [8] In addition, it has been demonstrated that there is a direct association between inhibitory effects of EO on fungal growth and fusariotoxins production [8, 12].

Phenolic compounds eugenol and carvacrol were present as major components in a number of these essential oils. The antimicrobial activity of the EOs that, similarly to our results, were rich in phenols can be attributed to the presence of an aromatic nucleus and a phenolic OH group that is known to be reactive and to form hydrogen bonds with active sites of target enzymes [23].

Besides, no differences were observed in the level of inhibition of the two different strains of *Fusarium verticilloides* studied. However, *Fusarium graminearum* was the most sensitive strain with inhibition zones of 30 mm (*T. vulgaris* L.) and whole inhibition (no visible growth of the inoculated moulds) for *Satureja montana* and *Thymus vulgaris* F EOs. Conversely to these results, previous studies have reported that *Fusarium graminareum* was a more resistant strain to the effect of EOs than other strains of the genus *Fusarium* when they were studied all together [21].

Whatever the case may be, the EOs studied in this work have been previously investigated for their in vitro antimicrobial activity. In accordance to our results, the MIC and MBC values revealed that *Satureja montana* and *Thymus vulgaris* (F) were the most inhibitory EOs examined against the gram-negative and gram-positive bacterial strains assessed [16]. Similarly to our results, the EO of *Thymus vulgaris* L. possessed less antimicrobial properties than *Thymus vulgaris* F.

In summary, the results presented here contribute to the knowledge of antifungal activities of the tested EOs obtained from aromatic plants growing in Aragón (NE, Spain). The most effective EOs studied against the mycotoxigenic strains of *Fusarium verticilloides* and *Fusarium graminareum* were *Thymus vulgaris* (Local and French) and *Satureja montana*. The effectiveness of these EOs is mostly due to the presence of phenols such as carvarcrol and/or thymol. These EOs may be potential antifungal compounds and could be considered as suitable alternatives to chemical additives for future plant disease management programs eliminating fungal spread and therefore mycotoxin production. Thus, more studied are needed to better evaluate this antifungal activity. In addition, it would be also necessary to evaluate the efficacy of essential oils in real agricultural substrates.

Acknowledgements This research was supported by the Government of Aragón (Spain)-Fondo Social Europeo through Project DGA/Grupo de Investigación Consolidado (Project AO1). Thanks are also given to Dr. J. Burillo (Servicio de Investigación Agraria, Zarogoza, Spain) who kindly provided the essential oils.

References

[1] Guo BZ, Russin JS, Brown RL, Cleveland TE, Widstrom NW. Resistance to aflatoxin contamination in corn as influenced by relative humidity and kernel germination. *Journal of Food Protection*. 1996;59:276–281.

[2] Soliman KM, Badeaa RI. Effect of oil from some medicinal plants on different mycotoxigenic fungi. *Food and Chemical Toxicology*. 2002;40:16691675.

[3] Shukla R, Kumar A, Prasad CS, Srivastava B, Dubey NK. Antimycotic and antiaflatoxigenic potency of *Adenocalymma alliaceum* Miers. on fungi causing biodeterioration of food commodities and raw herbal drugs. *International Biodeterioration & Biodegradation*. 2008;62:348–351.

[4] Burt S. Essential oils: their antibacterial properties and potential applications in foods – a review. *International Journal of Food Microbiology*.2004;94:223-253.

[5] Bakkali F, Averbeck S, Averbeck D, Idaomar M. Biological effects of essential oils – A review. *Food and Chemical Toxicology*. 2008;46:446–475.

[6] Batish DR, Singh HP, Kohli RK, Kaur S. Eucalyptus essential oil as a natural pesticide. *Forest Ecology and Management*. 2008;256:2166–2174.

[7] SCOOP (ScientiWc Cooperation Task 3.2.10 of the European Commission). Collection of occurrence data of *Fusarium* toxins in food and assessment of dietary intake by the population of EU Member States. Final report. 2003. *Directorate – General Health and Consumer Protection*. European Commission, Brussels, Belgium.

[8] Velluti A, Marín S, Gonzalez P, Ramos AJ, Sanchis V. Initial screening for inhibitory activity of essential oils on growth of *Fusarium verticillioides*, *F. proliferatum* and *F. graminearum* on maize-based agar media. *Food Microbiology*. 2004;21:649-656.

[9] Palazón C F, Palazón ML. Estudio patológico de problemas de nascencia de maíz en la zona de Ejea de los Caballeros (Zaragoza).*Información Técnica Económica Agraria.* 1982;46:47–53.

[10] Martins HM, Almeida I, Marques MF, Guerra MM. Fumonisins and deoxynivalenol in corn-based food products in Portugal. *Food and chemical toxicology*. 2008;46:2585-2587.

[11] Dambolena JS, López AG, Cánepa MC, Theumer MG, Zygadlo JA, Rubinstein HR. Inhibitory effect of cyclic terpenes (limonene, menthol, menthone and thymol) on *Fusarium verticillioides* MRC 826 growth and fumonisin B1 biosynthesis. *Toxicon*. 2008;51:37-44.

[12] Dambolena JS, Zunino MP, López AG, Rubinstein HR, Zygadlo JA, Mwangi JW, Thoithi GN, Kibwage IO, Mwalukumbi JM, Kariuki ST. Essential oils composition of *Ocimum basilicum* L. and *Ocimum gratissimum* L. from Kenya and their inhibitory effects on growth and fumonisin production by *Fusarium verticillioides*. *Innovative Food Science and Emerging Technologies*. 2010;11:410–414.

[13] Burillo J, García-Vallejo MC. 2003. Investigación y Experimentación de Plantas Aromáticas y Medicinales en Aragón. Cultivo, transformación y analítica. Departamento de Agricultura, Dirección General de Tecnología, Gobierno de Aragón. Zaragoza, Spain.

[14] Vázquez BL, Fente CCM, Franco CM, Vázquez MJ, Cepeda A. Inhibitory effects of eugenol and thymol on Penicillium citrinum strains in culture media and cheese. *International Journal of Food Microbiology*. 2001;67:157-163.

[15] Kordali S, Cakir A, Ozer H, Cakmakci R, Kesdek M, Mete E. Antifungal, phytotoxic and insecticidal properties of essential oil isolated from Turkish Origanum acutidens and its three components, carvacrol, thymol and *p*-cymene. *Bioresource Technology*. 2008;99:8788-8795.

[16] Rota, C, Carramiñana, JJ, Burillo, J, Herrera, A. In Vitro antimicrobial activity of essential oils from aromatic against selected foodborne pathogens. *Journal of Food Protection*. 2004; 67(6): 1252-1256.

[17] Pitarakili D, Tzakou O, Loukis A, Harvala C. Volatile metabolites from *Salvia fruticosa* as antifungal agents in soil-borne pathogens. *Journal of Agricultural and Food Chemistry*. 2003;51:3294–3301.

[18] Kordali S, Kotan R, Cakir A. Screening of *in-vitro* antifungal activities of 21 oxygenated monoterpenes as plant disease control agents. *Allelopathy Journal*. 2007;19:373–392.

[19] Fraternale D, Giamperi L, Bucchini A, Ricci D, Epifano F, Genovese S,Curini M. Chemical composition and antifungal activity of the essential oil of *Satureja montana* from central Italy. *Chemistry of Natural Compounds*. 2007; 43: 622-624.

[20] Kumar A, Shukla R, Singh P, Prasad CS, Dubey NK. Assessment of Thymus vulgaris L. essential oil as a safe botanical preservative against post harvest fungal infestation of food commodities. *Innovative Food Science and Emerging Technologies*. 2008;9:575–580.

[21] Kordali S, Cakir A, Akcin TA, Mete E, Akcin A, Aydin T, Kilic H. Antifungal and herbicidal properties of essential oils and n-hexane extracts of *Achillea gypsicola* Hub-Mor. and *Achillea biebersteinii* Afan. (Asteraceae). *Industrial crops and products*. 2009;29:562–570.

[22] Naeini A, Ziglari T, Shokri H, Khosravi AR. Assessment of growth-inhibiting effect of some plant essential oils on different Fusarium isolates. *Journal de Mycologie Médicale*. 2010;20:174-178.

[23] Farag R.S, Daw ZY, Abo-Raya SH. Influence of some spice essential oils on *Aspergillus parasiticus* growth and production of aflatoxins in a synthetic medium. *Journal of Food Science*. 1989; 54:74-76. (del articulo de Velutti et al., 2004).

Influence of some natural compounds on freshwater microfoulants

O. A. Cuzman[*,1], **C. Faraloni**[2], **T. Turchetti**[3], **M.Camaiti**[1] and **P. Tiano**[1]

[1] ICVBC-CNR, Institute for the Conservation and Promotion of Cultural Heritage, Via Madonna del Piano 10, 50019 Sesto Fiorentino (FI), Italy

[2] ISE-CNR, Institute of the Ecosystems Study, Via Madonna del Piano 10, 50019 Sesto Fiorentino (FI), Italy

3 IPP-CNR, Plant Protection Institute, Via Madonna del Piano 10, 50019 Sesto Fiorentino (FI), Italy

The antifouling properties of five natural compounds (poly-alkyl pyridinium salts – pAPS, Ceramium botryocarpum extract – CBE, zosteric acid – ZA, capsaicin – CS and cinnamaldehyde – CI) were tested against mixed and isolated freshwater biofilm microbial components, in laboratory conditions. These active natural agents can be extracted from marine (*Reniera sarai, Zostera marina, Ceramium botryocarpum*) or terrestrial (chili pepper, cinnamon) natural sources and for some of them (ZA, CS and CI), the corresponding chemical copounds can be synthesized as well. The evaluation of their efficacy in liquid and solid cultural medium, using serial dilution and diffusion method, was performed. The most efficient antibiofouling agents (ABAs) seems to be pAPS and CI at all tested concentrations. ZA exhibited a contrary behavior with respect to the others inhibitors, being less efficient at the highest concentrations. However, most of the ABAs tested are efficient against cyanobacteria rather than algae, and only CI showed a inhibitor effect against fungi as well.

Keywords biofilm; natural antifoulants;

1. Introduction

Microbial colonization occurs especially on submerged or wet surfaces both in terrestrial and aquatic environments, being a common phenomenon with ecological and industrial implications. The colonizers are usually associated in a mixed microbial community belonging to phototrophic and heterotrophic fractions. The term biofilm is often associated only with the bacterial fraction, while a phototrophic biofilm includes micro-autotrophs and the heterotrophs as well, all the microorganisms being embedded in the extracellular polymeric substances (EPS), that enhance its resistance against the stressing external factors [1, 2]. Biofouling of an inanimate substratum by unsought micro/macroorganisms can cause serious technical and/or economical inconveniences (e.g. water systems clogging, ships' hull fouling and biocorrosion) [3, 4]. These problems can be reduced by mechanical removal and using antifouling coatings. A new approach, based on the antibiofouling strategies in the natural environment [5, 6, 7], is to use natural products extracted from various marine or terrestrial organisms that might repel or inhibit fouling organisms [8, 9]. Many secondary metabolites, such as halogenated furanones produced by the red alga *Delisea pulchra* [10], poly-alkyl-pyridinium salts isolated from a marine sponge *Reniera sarai* [11, 12, 13, 14, 15], zosteric acid from eealgrass *Zostera marina* [16, 17, 18, 19, 20], capsaicin from *Capsicum* sp. [21, 22, 23], cinnamaldehyde from cinnamon [24, 25, 26], aaptamine and analogs from *Aaptos* sp. marine sponge [27], are reported for their ability to inhibit the microbial growth.

The five natural antibiofouling agents selected and tested in this research for inhibition of freshwater microorganisms development are: **(i) poly-alkyl-pyridinium salts (pAPS)**[†] (*3-alkylpyridinium* active compound); **(ii) zosteric acid (ZA)** (*p-sulphoxy-cinnamic acid*); **(iii) extract of *Ceramium botryocarpum* (CBE)**[‡] a dichloromethane extract of a marine rhodophyta, containing a salt of natural digeneaside [28] that have been successfully tested against microorganisms involved in the microfouling process, such as marine bacteria, diatoms and microalgae (*Ulva* sp.) [29, 30]; **(iv) Capsaicin (CS)** (*8-methyl-N-vanillil-6-nonenamide*) and **(v) cinnamaldehyde (CI)** (*trans-3-phenyl-2-propenal*).

2. Materials and methods

2.1. Natural compounds

Three of the tested antibiofouling agents (ABAs) are present in marine environment (pAPS, ZA and CBE) while the other two have a terrestrial origin (CI and CS). The biocide Algophase (A) was used as a positive control. The chemical structures of these compounds are presented in **Figure 1**.

[*] Corresponding author: e-mail: cuzman@icvbc.cnr.it, Phone: +39 0555225473

† kindly provided by Sepčić K (Univ. of Ljubliana, Biotechnical Faculty, Slovenia)

‡ kindly supplied by A.Bazes (Univ. of South-Bretagne, LBCM, France)

(i) Poly-alkyl-pyridinium salt (pAPS)* (3-alkylpyridinium active compound) was used as water solution (0.0114% w/v). The concentrations used for serial dilution experiment were: 50 µg mL^{-1} (dil 0), 5 µg mL^{-1} (dil 1), 0.5 µg mL^{-1} (dil 3), while for diffusion experiment were: 100 µg mL^{-1}, 50 µg mL^{-1}, 30 µg mL^{-1}.
(ii) Zosteric acid (ZA) was synthesised according to previously reported procedures [31] in this case an incomplete purification of the active product was obtained by treatment with diethyl ether and methanol. The synthesised compound was characterised by MS† and FT-IR and the resulting spectra were in agreement with the presence of zosteric acid. ZA was used as water solution (3% w/v). The concentrations used for serial dilution experiment were: 5 mg mL^{-1} (dil 0), 500 µg mL^{-1} (dil 1), 50 µg mL^{-1} (dil 3), while for diffusion experiment were: 10 mg mL^{-1}, 5 mg mL^{-1}, 2 mg mL^{-1}.
(iii) Extract of *Ceramium botryocarpum* (CBE)‡ was used as water solution (2% w/v). The concentrations used for serial dilution experiment were: 5 mg mL^{-1} (dil 0), 500 µg mL^{-1} (dil 1), 50 µg mL^{-1} (dil 3), while for diffusion experiment were: 10 mg mL^{-1}, 5 mg mL^{-1}, 2 mg mL^{-1}.
(iv) Capsaicin (CS) was purchased from Aldrich Chemical Co. with a composition of 65% capsaicin and 35% dihydrocapsaicin. It was used as methanol solution (0.7% w/v). The concentrations used for serial dilution experiment were: 800 µg mL^{-1} (dil 0), 80 µg mL^{-1} (dil 1), 8 µg mL^{-1} (dil 3), while for diffusion experiment were: 1600 µg mL^{-1}, 800 µg mL^{-1}, 200 µg mL^{-1}.
(v) Cinnamaldehyde (CI) were purchased from Aldrich Chemical Co. It was used as methanol solution (10% v/v). The concentrations used for serial dilution experiment were: 1.5% v/v (dil 0), 0.15% v/v (dil 1), 0.015% v/v (dil 3), while for diffusion experiment were: 3% v/v, 1.5% v/v, 0.5% v/v.
(vi) Concentrate Algophase (A)§ (N-butyl-1,2 benzisothiazolin-3-one) is a biocide with a large spectrum of action (bacteria, cyanobacteria, algae, fungi). This product was developed by PHASE and has been validated by Chia *et al.* for its antimicrobial efficiency against bacterial and fungal strains [32]. The concentrations used for serial dilution experiment were: 5% v/v (dil 0), 0.5% v/v (dil 1), 0.05% v/v (dil 3), while for diffusion experiment were: 15% v/v, 5% v/v, 3% v/v.
(vii) Methanol (MeOH) used as solvent for CI and CS was tested for serial dilution experiment at the following concentrations: 15% v/v (dil 0), 1.5% v/v (dil 1), 0.15% v/v (dil 3), while for diffusion experiment were: 30% v/v, 15% v/v, 5% v/v.

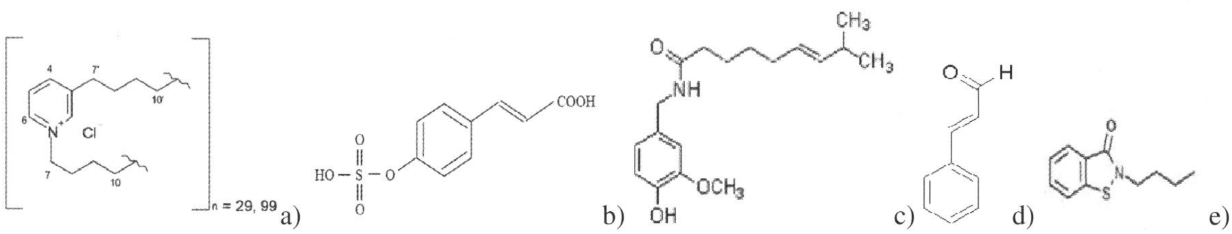

Fig. 1 Chemical structrures of (a) poly-alkyl pyridinium salt – pAPS, (b) zosteric acid – ZA; (c) capsaicin – CS; (d) cinnamaldehyde – CI and (e) algophase – A.

2.2. Serial dilution method

Two sets of 9 tubes each were prepared for each antibiofouling agent, containing triplicates of the three different concentrations (see 2.1.) of aqueous dilutions. In each tube was added 100 µl of biological inoculum (≈ 0.8 mg) which consisted in a mixed natural biofilms with the following main genera: *Scenedesmus* sp., *Chlorella* sp., *Aphanocapsa* sp., *Gloeocapsa* sp. and *Navicula* sp. All the tubes were therefore kept at 27 ºC, under low artificial lighting conditions with continuous white fluorescent light at photosynthetic photon flux density of 10 µmol photon m^{-2}.s^{-1}. Biological growth was evaluated by "in vivo" measurements of chlorophyll *a* (680 nm) and optical density (750 nm) using a Beckman DU 640 spectrophotometer. The estimation of pigment content was corrected by subtraction of residual absorbance at 750 nm [33]. Media and standard deviation of three measurements for each dilution was considered. Small aliquots (0.75ml) from each tube were diluted with 0.75 ml H_2O, sonicated in the cuvette for 10 seconds in continuous pulse at 10 watts, using a MISONIX Sonicator, in order to disperse clumps and chains [34], and therefore immediately measured. Some ABAs (CS, CBE, CI, A) induced an opaqueness or a bit yellowing of the water at the start concentrations, in this case the spectrophotometric measurements could not be considered, being out of range. The presence of the phototrophs was evaluated by microscopical observations as well, using a Nikon Eclipse E600 microscope.

The inhibition of the phototrophic growth was expressed by a percentage of inhibition and was calculated:

* kindly provided by Sepčić K (Univ. of Ljubliana, Biotechnical Faculty, Slovenia)
† GC-MS analysis was made by Prof. M. Bambagiotti of the Department of Pharmacology, University of Florence, Italy
‡ kindly supplied by A.Bazes (Univ. of South-Bretagne, LBCM, France)
§ kindly provided from Phase - Proind Srl. Italy

% inhibition = ($chla_c$ - $chla_t$) / $chla_c$ x 100, where $chla_c$ is the mean concentration of chlorophyll *a* of the controls and $chla_t$ is the mean concentration of chlorophyll *a* in the test samples.

2.3. Diffusion method

For each ABA a set of three Petri dishes were used for testing the efficiency against 3 types of microorganisms isolated from the collected biofilms from monumental fountains. The selected microorganisms (a cyanobacteria – *Chroococcus* sp.; a green alga – *Apatococcus* sp.; and a fungus – *Torula* sp.) have been grown on specific cultural media (BG11; BG11 modified and diluted 1:1; and PDA respectively), using 0.25ml inoculum with a concentration of $\approx 5.7x10^6$ cells/ml for cyanobacteria, $\approx 3.8x10^6$ cells/ml for algae and $\approx 1.5x10^6$ cells/ml for fungi. The Petri dishes with phototrophs were incubated in continuous low light at 28°C, while the one with fungi at 25°C, in darkness. A visual inspection at different time intervals (7, 24 and 75 days) was performed.

3. Results and discussion

3.1. Serial dilution method

The chlorophyll *a* absorbance of the phototrophs slightly varied for all the ABAs in the first 20 days. Clearly differences of the in vivo absorbance at 680 nm have been noticed after 40 days, suggesting some changes in the phototrophic community behavior in presence of ABAs. The absorbance of chlorophyll *a* measured in vivo generally increased with dilution of concentration for all tested ABAs, as can be seen in Figure 2a for pAPS, where the lower concentration (dil 3) has an absorbance closer to the reference one. Only the ZA solutions exhibited a contrary behavior for the chlorophyll *a* absorbance with respect to all others ABAs. The highest absorption was in fact recorded at the highest concentration (dil 0) as can be seen in Figure 2b. This behavior can be attributed to the presence of intermolecular hydrogen bonding between the –COOH and $-SO_2OH$ groups. The activity of the sulphate esteric group may be reduced at higher concentrations. CI and A solutions presented the lowest values for the chlorophyll *a* absorbance even at the less concentrate solution tested (dil 2).

The microscopical observations of the ABAs dilutions revealed the absence of the phototrophs in the CI (dil 0 and dil 1), CS (dil 0), A (dil 0, dil 1 and dil 2) solutions, which seem to have completely reduced the phototrophic development, having a toxic behavior rather than an inhibitory one. Cyanobacteria was the group less abundant and diatoms were that one more present.

The effect of ABAs on the phototrophic growth is presented in Table 1. It can be clearly seen that most of the ABAs possess an inhibitory effect at higher tested concentrations (dilution 0). The most efficient ABAs were pAPS and CI at all tested concentrations that reduced the growth by 50%. The solvent MeOH showed an inhibitory effect only at the highest concentration tested (15%) which means that the positive effect of CI and CS can be assigned to their actions. The control A was efficient at all tested concentrations.

Table 1 Inhibition activity (%) of the tested ABAs and controls after 60 days.

Tested formulates	dil 0	dil 1	dil 2
pAPS	98	70	49
ZA	22	40	53
CBE	> 45	45	0
CI	>84	84	77
CS	>69	69	5
A (biocide)	>88	>88	88
MeOH (solvent)	75	42	9

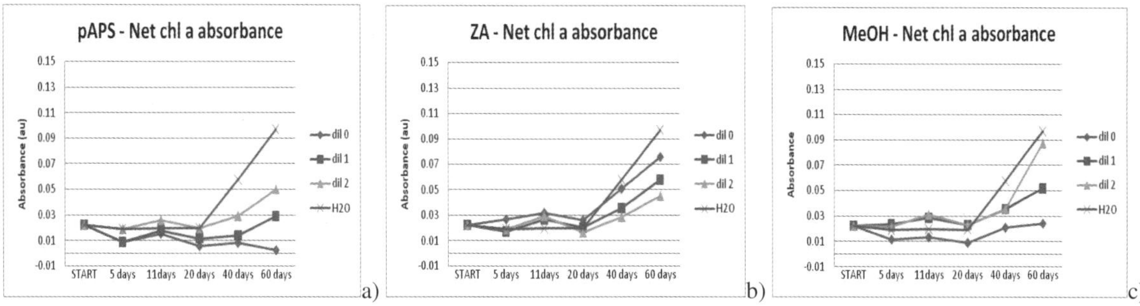

Fig. 2 In vivo fluorescence of chlorophyll *a* detected for phototrophic development in aqueous solutions containing antibiofouling agents: (a) pAPS; (b) ZA. and (c) MeOH.

3.2. Diffusion method

At the end of experiments, the results of the diffusion methods, at tested concentrations, revealed that most of the ABAs are efficient against cyanobacteria rather than algae. Only CI and CBE showed an inhibitor effect against fungi, at tested concentrations as can be seen in Table 2. No fungal development was present in the Petri dish with the tested biocide. CS showed a certain efficacy against fungi only in the first four days and only at the most two concentrated solution tested. In the Petri dish containing the pAPS small drops of exudates were observed on the surface of the fungus, and therefore pAPS may induce a specific reaction of the fungal strain.

Table 2 The efficiency of the ABAs, A and MeOH against tested microbial strains. Key to symbols: + = efficient; - = no efficient; +/- = uncertain.

		ZA (mg mL^{-1})	pAPS (μg mL^{-1})	CBE (mg mL^{-1})	CS (μg mL^{-1})	CI (% v/v)	A (% v/v)	MeOH (% v/v)
Concentration	C1	10	100	10	1600	3	15	30
	C2	5	50	5	800	1.5	5	15
	C3	2	30	2	200	0.5	3	5
Alga	C1	+/-	+/-	+/-	+/-	+	+	+/-
	C2	+/-	+/-	+	+/-	+	+	-
	C2	+/-	+/-	+	-	+	+	-
Cyanobacterium	C1	+	+	+/-	+	+	+	+
	C2	+	+	+/-	+	+	+	+/-
	C3	+	+	+	+	+	+	-
Fungus	C1	+/-	-	+	-	+	+	+
	C2	+/-	-	+	-	+	+	+/-
	C3	+/-	-	+	-	+/-	+	-

4. Conclusion

The results obtained in this research showed a good efficiency for all tested ABAs, especially against cyanobacteria, rather than against algae and fungi. pAPS and CI were the most efficient at all tested concentrations, while ZA exhibited a contrary behavior with respect to the others inhibitors, being less efficient at the highest concentrations.

Acknowledgements This work has been made in the frame of the EPISCON Project (MEST-CT -2005-020559 - European PhD in Science for Conservation).

References

[1] Roeselers G, van Loosdrecht, Muyzer. Heterotrotrophic pioneers facilitate phototrophic biofilm development. *Microbial Ecology.* 2007;54:578-585.

[2] Sekar R, Venugopalan, Nandakumar K., Nair KVK, Ro VNR. Early stages of biofilm succession in a lentic freshwater environment. *Hydrobiologia.* 2004;512:97-108.

[3] Coester SE, Cloete TE. Biofouling and biocorrosion in industrial water systems. *Critical Reviews in Microbiology.* 2005;31:213-233.

[4] Schultz MP, Swain GW. The influence of biofilm on skin friction drag *Biofouling*. 2000;15:129-139.
[5] Evans LV, Clarkson N. Antifouling strategies in the marine environment. *Journal of Applied Bacteriology. Symposium Supplements*. 1993; 74:119S-124S.
[6] Yebra DM, Kiil S, Dam-Johansen K. Antifouling technology – past, present and future steps towards efficient and environmentally friendly antifouling coatings, *Progress in Organic Coatings*. 2003; 50:75-104
[7] Fusetani N. Biofouling and antifouling. *Natural Product Reports*. 2004; 21:94-104.
[8] de Nys R, Steinberg PD. Linking marine biology and biotechnology. *Environmental Biotechnology*. 2002; 13:244-248.
[9] Omae I. General aspects of tin-free antifouling paints, *Chemical Review*. 2003;103:3431-3448.
[10] Givskov M, de Nys R, Manefield M, Gram L, Maximilien R, Eberl L, Molin S, Steinberg PD, Kjelleberg S. Eukaryotic interference with homoserine lactone-mediated prokaryotic signaling, *Journal of Bacteriology*. 1996; 178:6618-6622.
[11] Sepčić K, Guella G, Mancini I, Pietra F, Dalla Serra M, Manestrina G, Tubbs K, Maček P, Turk T. Characterization of anticholinesterase-active 3-alkylpyridinium polymers from the marine sponge *Reniera sarai* in aqueous solutions. *Journal of Natural Products*. 1997; 60:991-996.
[12] Faimali M, Sepčić K, Turk T, Geraci S. Non-toxic antifouling activity of polymeric 3-alkylpyridinium salts from the Mediteranean sponge *Reniera sarai* (Pulitzer-Finali). *Biofouling*. (2003; 19:47-56.
[13] Garaventa F, Faimali M, Sepčić K, Geraci S. Laboratory analysis of antimicrofouling activity of Poly-APS extracted from *Reniera sarai* (Porifera: Demospongiae). *Biol. Mar. Mediter.* 2003; 10:565-567.
[14] Sepčić K, Turuk T. 3-Alkylpyridinium compounds as potential non-toxic antifouling agents. *Progress in Molecular and Subcellular Biology*. 2006; 42:105-124.
[15] Eleršek T, Kosi G, Turk T, Pohleven F, Sepčić K. Influence of polymeric 3-alkylpyridinium salts from the marine sponge *Reniera sarai* on the growth of algae and wood decay fungi. *Biofouling*. 2008; 24:137-143.
[16] Zimmerman RC, Alberte RS, Todd JS, Crews P. Phenolic acid sulphate esters for prevention of marine biofouling, *US Patent 5384176*. 1995
[17] Todd JS, Zimmerman RC, Crews Ph, Alberte RS. The antifouling activity of natural and synthetic phenol acid sulphate esters. *Phyto*1993 *chemistry*. 1993; 34:401-404.
[18] Callow ME, Callow JA. Attachment of zoospores of the fouling alga Enteromorpha in the presence of zosteric acid. *Biofouling*. 1998; 13:87-95.
[19] Stanley MS, Callow ME, Rerry R, Alberte RS, Smith R, Callow JA. Inhibition of fungal spore adhesion by zosteric acid as the basis for a novel, nontoxic crop protection technology. *Phytopatology*. 2002; 92:378-383.
[20] Newby BZ, Cutright T, Barrios CA, Xu Q. Zosteric acid – an effective antifoulant for reducing fresh water bacterial attachment on coatings. *Journal of Coatings Technology and Research*. 2006; 3:69-76.
[21] Cichewicz RH, Thorpe PA. The antimicrobial properties of chile peppers (Capsicum species) and their use in Mayan medicine. *Journal of Ethnopharmacology*. 1996; 52:61-70.
[22] Molina-Torres J, Garcia-Chávez A, Ramírez-Chávez E. Antimicrobial properties of alkamides present in flavouring plants traditionally used in Mesoamerica: affinin and capsicin, *Journal of Ethnopharmacology*. 1999; 64:241-248.
[23] Xu Q, Barrios CA, Cutright T, Newby BZ. Assessment of antifouling effectiveness of two natural product antifoulants by attachment study with freshwater bacteria. *Environmental Science and Pollution Research*. 2005; 12:278-284.
[24] Bowled BL, Sackitey SK, Williams AC. Inhibitory effects of flavor compounds on Staphylococcus aureus WRRC B124. *Journal of Food Safety*. 1995; 15:337-347.
[25] Heandler IM, Alakomi HL, Latva-Kala K, Mattila-Sandholm T, Pol L, Smid EJ, Gorris AGM, von Wright A. Characterization of the action of selected essential oil compounds on Gram-negative bacteria. *Journal of Agriculture and Food Chemistry*. 1998; 46:3590-3595.
[26] Niu C, Gilbert A, Gilbert ES. Subinhibitory concentrations of cinnamaldehyde interfere with quorum sensing. *Letters in Applied Microbiology*. 2006; 43:489-494.
[27] Herlt A, Mander L, Rombang L, Rumampuk R, Soemitro S, Steglich W, Tarigan P, von Nussbaum F. Alkaloids from marine organisms. Part 8: isolation of bisdemethylaaptamine and bisdimethylaaptamine-9-O-sulfate from an Indonesian *Aaptos* sp. marine sponge. *Tetrahedron*. 2004; 60:6101-6104.
[28] Claude A, Bondu S, Michaud F, Bourgougnon N, Deslandes E. X-ray structure of a sodium salt of digeneaside isolated from red alga *Ceramium botryocarpum*. *Carbohydrate Research*. 2009; 344:707-710.
[29] Gross EM. Allelopathy of aquatic autotrophs. *Critical reviews in plant sciences*. 2003; 22(3,4):313-339.
[30] Bazes A, Silkina A, Defer D, Bernède-Bauduin C, Quéméner E, Braud JP, Bourgougnon N. Active substances from *Ceramium botryocarpum* used as antifouling products in aquaculture. *Aquaculture*. 2006; 258:664-674.
[31] Alexandratos SD. Synthesis and purification of zosteric acid. *US Patent 5990336*. 1999.
[32] Chia LL, Warwick EF. Antimicrobial composition containing N-(N-butyl)-1,2,benzisothiazolin-3-one. *US Patent 0197366*. 2005
[33] Lee TY, Tsuzuki M, Takeuchi T, Yokoyama K, Karube I. Quantitative determination of cyanobacteria in mixerd phytoplankton assamblages by an in vivo fluorometric method. *Analytica Chimica Acta*. 1995; 302:81-87.
[34] Petersen FC, Pecharki D, Scheie AA. Biofilm mode of growth of Streptococcus intermedius favored by a competence-stimulating signaling peptide. *Journal of. Bacteriology*. 2004; 186:6327-6331.

Natural antimicrobial agents against the microbiota associated with insoles

N. Cuesta Garrote*, M.M. Sánchez Navarro, F. Arán Aís and C. Orgilés Barceló

INESCOP, Footwear Research Institute. Pol. Industrial Campo Alto s/n. 03600 Elda (Alicante). Spain

* Corresponding author: ncuesta@inescop.es ; +34 965395213

A tea tree oil (TTO) and a lemon essential oil preparation of defined chemical composition was studied as natural antimicrobial agents, using two different microbroth methods, for its in vitro activity against four strains described as microbiota associated to feet skin (*Escherichia coli* SG13009, *Bacillus subtilis* 168, *Klebsiella pneumoniae* and *Staphylococcus aureus*). The bacterial strains were from established type collections.
While negative controls have reached optical densities close to 0.9 and 1, inocula with tea tree or lemon essential oil have decreased close to 0.2 after 24 hours, indicating that the natural agents have greatly inhibited bacterial growth.
Both the lemon and TTO oils inhibited only *Bacillus subtilis* (0.5 and 0.7 cm) and *Staphylococcus aureus* (0.3 and 0.7 cm) growth.
Overall, the obtained data confirm the effectiveness of the TTO and the lemon essential oil against the bacteria associated with leather insoles, in particular against *Bacillus subtilis* and *Staphylococcus aureus*.

Keywords footwear industry; natural agents; microbiota; DGGE; antimicrobial

1. Introduction

The environment our feet are in shows favorable conditions for bacteria and fungi growth. Once the microbial growth is completed, some of these microorganisms produce secondary metabolites that contribute to bad odor [1] generated in the feet. One way to prevent or reduce the bad smell, apart from the foot care and hygiene measures would be through the use of antimicrobial agents.

The interest in natural products and essential oils has grown in response to the increasing resistance of antibiotics, synthetic disinfectants and germicides. In this sense, the essential oil of *Melaleuca alternifolia* (Australian tea tree oil, TTO) has been successfully employed for its germicidal activity to control wound infections [2], as well as feet infections cause by dermatophytes such as *Microsporum gypseum* or *Trichophyton interdigital* [3]. In the same way, the lemon essential oil (Citrus sp.) has been used as a traditional medicine for severe atopic and bacterial dermatitis cause by human skin bacteria such as *Staphylococcus* or *Streptococcus* [4].

Therefore, the aim of this study is the analysis of the microbiota associated with leather insoles worn by healthy users. Furthermore, the effect of two natural essential oils (tea tree oil and lemon extract) as antimicrobial agents against some of the microorganisms isolated from these insoles was analyzed.

2. Experimental

2.1 Materials

Strains associated to feet skin [5,6]: *Escherichia coli SG13009 (QIAGEN), Bacillus subtilis 168 ATCC 23857, Klebsiella pneumoniae CECT 141 and Staphylococcus aureus CECT 239.*
Culture media: Nutrient Broth medium (meat extract (Cultimed), peptone bacteriological (Cultimed), Luria Bertani medium Lennox (MP Biomedicals LCC) and saline 0, 9 % (NaCl (Panreac)
Natural agents: Almond oil (Guinama S.L.U), lemon oil (Guinama S.L.U), tea tree oil (Guinama S.L.U)

Primers and polymerase: **rRNA 16S Archaea** (907r 5' CCGTCAATTCMTTTGAGT 3'; 344(GC)f 5' CGCCCGCCGCGCCCCGCGCCGGTCCCGCCGCCCCCGCCCGACGGGGYGCAGGCGCGA 3'), **rRNA 16S Bacteria** (907r 5' CCGTCAATTCMTTTGAGT 3'; 341 (GC)f 5' CGCCCGCCGCGCCCCG CGCCGGTCCCGCCGCCCCCGCCCGCC 3') and **rRNA 18S Eukarya** (EUK1Ar 5' CTGGTTGATC CTGCCAG 3'; EUK156GCf 5' CGCCCGCCGCGCCCCGCGCCGGTCCCGCCGCCCCCGCCCGACG GGACCAGACTTGGCCTCC 3'). **DNA Taq polymerase** (ROCHE).

2.2 DNA extraction

The starting material leather insoles had been used continuously for several users during 15 days. For total acid nucleic extraction, the leather insoles surface was previously scraped by using sterile swabs (Vidrafoc) soaked in 2 mL of saline 0.9 % (NaCl (Panreac). This step was repeated three times so as to make sure as much material as possible was obtained. After that, DNA extraction throughout the "Masterpure DNA purification kit (Epicentre) MCD85201" was carried out.

Once the DNA extraction was completed, the DNA concentration (ng/µl, as absorbance measured at 260 nm) and purity (ratio 260 nm/280 nm) were checked using a Nanodrop Spectrophotometer ND – 10000 device. DNA extracted from unused templates was used as a negative control.

2.3 rRNA 16S amplification and DGGE gel

In order to obtain a high number of copies of the phylogenetically conserved 16S and 18S fragments, PCRs have been carried out using specific primers for Bacteria, Archaea and Eukarya. The PCR products were visualized in an agarose gel 1% using the documentation system Gel Doc XR (BIORAD).

Once the amplification products were obtained, microbial composition was analyzed on a denaturing gradient gel electrophoresis (DGGE) that is a molecular fingerprinting method that separates polymerase chain reaction (PCR)-generated DNA products. Each of the bands obtained was further amplified again on a second DGGE in order to check the purity and get enough DNA products for advanced sequencing. Furthermore, the bands pattern was analysed by means of FPQuest software to determine the degree of similarity between them.

2.4 Antimicrobial assays with natural essential oils

The antimicrobial effect of two natural essential oils (tea tree oil and lemon essential oil) was tested throughout measuring the absorbance (optical density) at 600 nm in liquid medium and measuring the halos inhibition by agar diffusion. In both assays, *Bacillus subtilis*, *Staphylococcus sp.* (both identified by DGGE) and two additional species described as microbiota associated to foot skin [2] (*Escherichia coli* and *Klebsiella pneumoniae*) were tested.

In **liquid medium**, three tubes for each species with 4 mL of medium, were inoculated and the natural biocides studied were poured into each of them: 0.5 mL of almond oil (allegedly innocuous), 0.5 mL of lemon extract oil and 0.5 mL of tea tree oil (antimicrobial agents). Moreover, an oil-free tube was inoculated as a negative control. The inocula were incubated at 37 ° C and 200 rpm in an orbital incubator Gallemkamp, and after that, the culture growth (absorbance at 600 nm) was measured at 8 and 24 hours using the spectrophotometer CECIL CE3021.

The evaluation of **the inhibition halos assay** by seeding four plates (60 mm diameter) for each type of bacteria with 250 µL of inoculum was carried out. A disc of cloth, which had been impregnated with 125 µL of the aforementioned conditions in the liquid medium, was placed on each of the seeded plates: almond oil, lemon extract, tea tree oil and an oil-free cloth disc (control). After 24 and 48 hours growing the plates at 37 ºC in an incubator INE 200 -800 Memmert, the inhibition halos were measured.

3. Results and Discussion

3.1 DNA extraction

Once the DNA extraction using the Masterpure DNA purification kit was completed, the DNA concentration (ng /µL, as absorbance measured at 260 nm) and purity (ratio 260 nm/280 nm) obtained was suitable, to continue with further experiments.

Table 1 DNA concentration and purity of the samples.

	concentration (ng /µL)	purity (ratio 260 nm/280 nm)
Sample 1	156.2	1.75
Sample 2	178.4	1.89

3.2 rRNA 16S amplification and DGGE gel

After obtaining the amplification products of Bacteria and having analyzed the microbial composition in a DGGE gel, the banding pattern of the bacteria community associated with the templates of both users is shown in Figure 1.

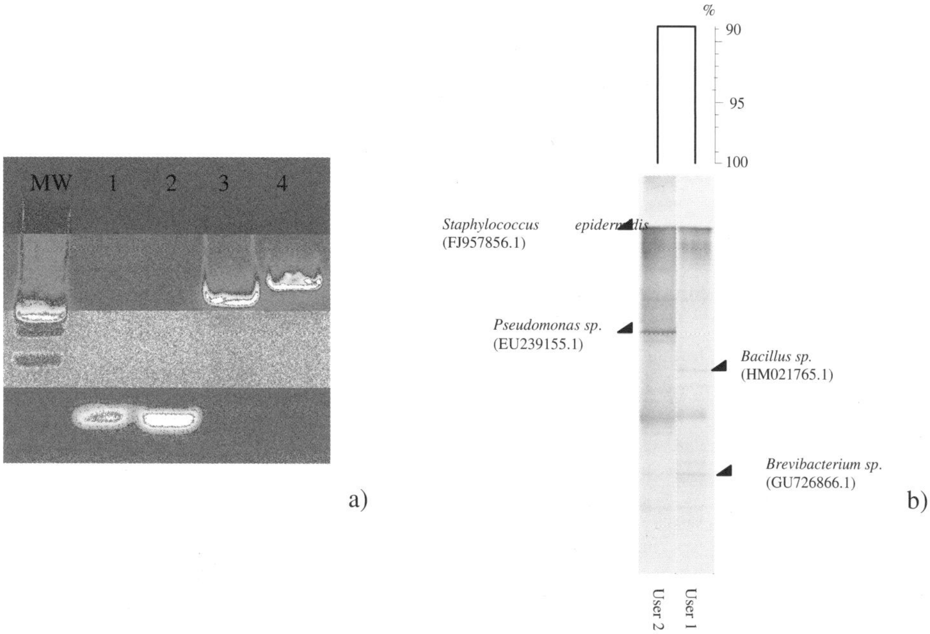

Fig. 1 a) *PCR RNA 16S* amplification samples (MW: molecular weight; 1: PCR negative control; 2: DNA negative control; 3: DNA sample 1; 4: DNA sample 2. b) *DDGE Gel.* Degree of similarity between the microbiota bands pattern obtained in the DGGE gel and the software FPQuest.

For amplification of 16S and 18S rRNA gene with primers respectively for Archaea and Eukarya no bands were obtained in test conditions, so new amplification conditions should be optimised. Amplification products with Bacteria specific primers were obtained, corresponding with the partial sequences of the rRNA 16S gen of: *Bacillus sp.* (HM021765.1) and *Brevibacterium sp.* (GU726866.1) for user 1 and *Staphylococcus epidermidis* (FJ957856.1) and *Pseudomonas sp.* (EU239155.1) for user 2.

3.3 Antimicrobial assays with natural essential oils

The antimicrobial effect of two natural essential oils (tea tree oil and lemon extract) were tested against *Bacillus subtilis*, *Staphylococcus sp.* (both identified by DGGE) and two additional species described as microbiota associated to foot skin (*Escherichia coli* and *Klebsiella pneumoniae*) in liquid medium and measuring the halos inhibition by agar diffusion.

Measurement of optical density in liquid medium

As a negative control, an oil – free tube was used, so as to determine the 100 % of the microbial growth. Moreover, as a positive control almond oil was added into one of the four tubes for each species, that has been described as allegedly innocuous. In this case, the possible inhibitory effect after the addition of the oil over the bacterial growth was analyzed in Figure 2.

Table 2 shows that while negative controls have reached optical densities close to 0.9 and 1, inocula with tea tree or lemon essential oil have decreased close to 0.2 after 24 hours, indicating that the natural agents have greatly inhibited bacterial growth.

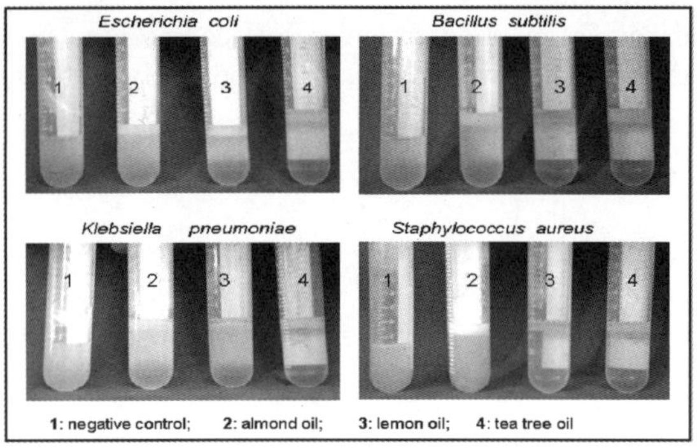

Fig. 2 Bacterial growth cultures in liquid medium for different oils in four bacterial strains.

Bacterial growth measuring the culture absorbance at 600 nm was evaluated. This is an indirect way of analysing growth throughout the medium turbidity.

Table 2 Absorbance measurements at 600 nm at 8 and 24 hours in different conditions.

	conditions	8 hours	24 hours
E. coli	control	0.711	0.956
	almond	0.886	0.888
	lemon	1.034	0.682
	tea	0.54	0.219
B. subtilis	control	0.904	0.916
	almond	0.897	0.83
	lemon	0.954	0.254
	tea	0.598	0.096
K. pneumoniae	control	0.777	0.968
	almond	0.822	0.922
	lemon	0.6	0.783
	tea	0.352	0.114
S. aureus	control	0.665	0.956
	almond	0.655	1.014
	lemon	0.628	0.51
	tea	0.431	0.273

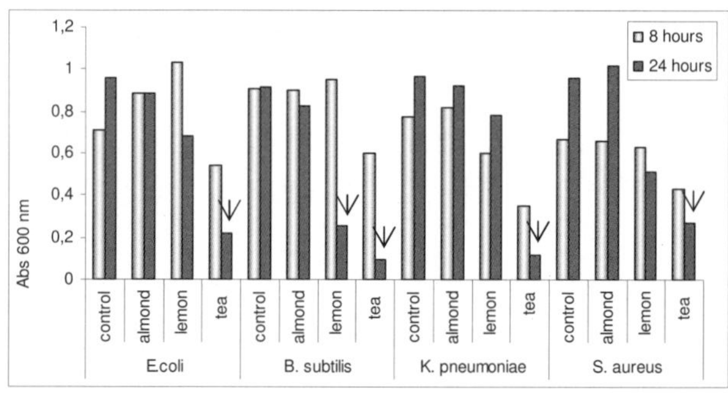

Fig. 3 Absorbance measurements plots at 8 and 24 hours for the four bacterial strains tested.

Figure 3 shows that lemon oil has not had any inhibitory effect over *E. coli* and *K. pneumoniae* growth. The arrows indicate the samples in which the growth inhibition was significant.

Measurement of inhibition halos in agar diffusion

Once the inocula were grown till 10^8 ufc / mL, they were spread into four plates of solid medium. Over each of the four seeded plates a disc of cloth soaked in each of the four conditions mentioned above in the liquid medium were placed: almond oil, lemon oil, tea tree oil and a plate without oil (control).

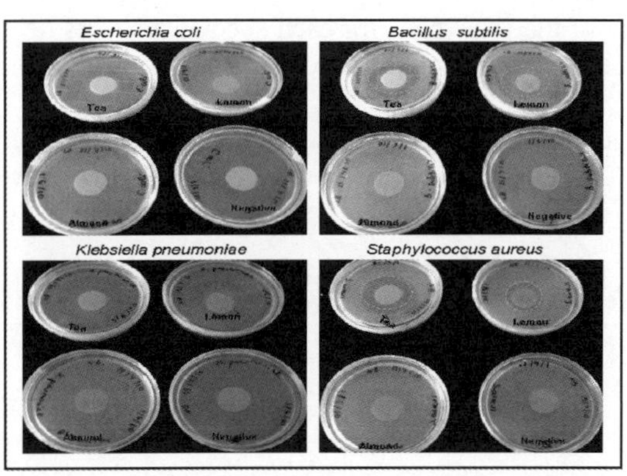

Fig. 4 Inhibition halos assay in agar medium plates for each type of natural oil.

Figure 4 shows that natural oils tested did not modify *Klebsiella pneumoniae* and *Escherichia coli* growth. In the same way as in the previous assays, the positive control (almond oil) of the four species grew properly throughout the plate. In the case of plates in which the lemon essential oil and the tea tree oil were added, the existence of halos of inhibition is shown.

Figure 5 shows the inhibition halos measurements for each species. Table 3 shows that both the lemon and TTO inhibited only *Bacillus subtilis* (0.5 and 0.7 cm) and *Staphylococcus aureus* (0.3 and 0.7 cm) growth.

Table 3 Inhibition halos measurements at 24 and 48 hours in different conditions.

	conditions	Inhibition halos (cm)	
		24 hours	48 hours
E.coli	control	0	0
	Almond oil	0	0
	Lemon oil	0	
	Tea tree oil	0	0
B.subtilis	control	0	0
	Almond oil	0	0
	Lemon oil	0.7	0.5
	Tea tree oil	0.6	0.7
K. pneumoniae	control	0	0
	Almond oil	0	0
	Lemon oil	0	0
	Tea tree oil	0	0
S. aureus	control	0	0
	Almond oil	0	0
	Lemon oil	0.3	0.3
	Tea tree oil	0.6	0.7

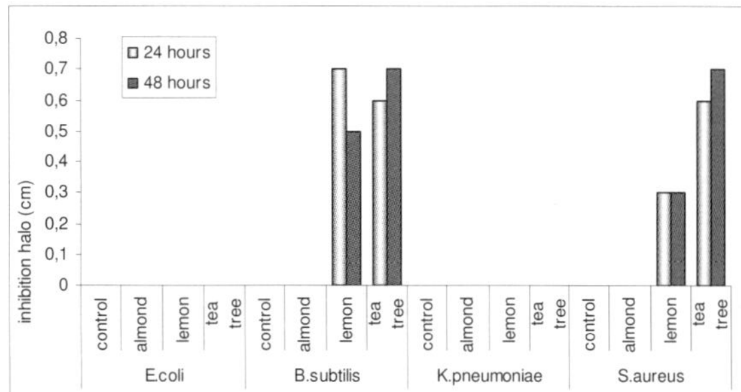

Fig. 5 Inhibition halos measurements plots at 24 and 48 hours for the four bacterial strains tested.

4. Conclusions

Among the microbiota associated with pig leather insoles worn, the amplification and identification of four bacteria strains were obtained: *Bacillus subtilis, Staphylococcus epidermidis, Pseudomonas sp. and Brevibacterium sp.* On further experiments the Eukarya species will be analysed.

According to the results obtained, PCR and DGGE are suitable techniques for the identification of the microbiota associated to worn leather insoles.

Both natural essential oils tested showed an inhibitory effect on most of the microbial species tested. Nevertheless, lemon oil did not show any inhibitory effect on *E. coli* and *K. pneumoniae* cultures.

Both natural antimicrobial agents tested have proved their suitability as possible candidates to be used on different components of footwear.

In further research the inhibitory minimal concentration will be determined.

Acknowledgements The Research was partially funded by The Institute of Small and Medium Industry of the Generalitat Valenciana and FEDER, through Action 1 of its R&D Programme for Technological Institutes 2010 (project IMIDIC/2010/106).

References

[1] Ara, K., Hama, M., Akiba, S., Koike, K., Okisaka, K., Hagura, T., Kamiya, T., Tomita, F. Foot odour due to microbial metabolism and its control. *Can. J. Microbil.* 2006. 52 (4): 357 – 364.

[2] Syed TA, Qureshi ZA, Ali SM, Ahmad S, Ahmad SA. Treatment of toenail onychomycosis with 2% butenafine and 5% *Melaleuca alternifolia* (tea tree) oil in cream. *Trop Med Int Health.* 1999; 4: 284 – 287.

[3] Hsin Chieh Li, Jean-Philippe Bouchara, Mark Ming-Long Hsu, Richard Barton, Shuli Su and Tsung Chain Chang. Identification of dermatophytes by sequence analysis of the rRNA gene internal transcribed spacer regions *Journal of Medical Microbiology* 2008, 57: 592–600.

[4] Baik, Jong Seok, Sang-Suk Kim, Jung-A Lee, Tae-Heon Oh, Ji-Young Kim, Nam Ho Lee1, and Chang-Gu Hyun. Chemical Composition and Biological Activities of Essential Oils Extracted from Korean Endemic Citrus Species. *J. Microbiol. Biotechnol.* 2008; 18 (1): 74 – 79.

[5] Gao Z, Tseng CH, Pei Z, Blaser MJ. Molecular analysis of human forearm superficial skin bacterial biota. *Proc Natl Acad Sci U SA* .2007; 104::2927 – 2932.

[6] Sedlock DM, Bailey DM. Microbiocidal activity of octenidine hydrochloride, a new alkanediylbis [pyridine] germicidal agent. *Antimicrob Agents Chemother.* 1985; 28(6): 786 – 790.

Plant essential oils and their components for the control of phytopathogen and mycotoxigenic fungi in crops

C. Morcia and **V. Terzi***

CRA-GPG, Genomic Research Centre, Via San Protaso 302, 29017-Fiorenzuola d'Arda, Italy

*Corresponding author: e-mail: valeria.terzi@entecra.it Phone: +39 0523983758

The demand for reduction in the use of pesticides in agriculture increases interest in natural substances for the control of plant diseases as alternative measures to the use of synthetic fungicides that can result in problems like environmental pollution, phytotoxicity and selection of resistant pathogen populations. The potential role of essential oils to control plant pathogens have been evaluated in several recent works. In particular the antimicrobial activity of Tea Tree Oil and its components have been evaluated against three different classes of fungal pathogens that are responsible for relevant diseases that affect cereal production.

Keywords crop protection; essential oils

1. Introduction

It is worldwide recognized that the full yield potential of crops cannot be expressed, because limited by abiotic and biotic stresses. In particular, plant pathogen fungi can cause relevant losses in terms of yield, quality and safety. Moreover, some phytopathogenic fungi produce different classes of mycotoxins, that can persist along the whole agri-food chains, with a strongly negative impact on human and animal health. Starting from these considerations, it is clear that research activities focused on the development of sustainable strategies for crop protection are needed because emerging, re-emerging and endemic plant pathogens continue to challenge our ability to safeguard plant health worldwide. Today, chemical control remains the main measure to reduce the incidence of plant diseases. However, an important problem against the use of these chemicals is the development of resistant pathogen population [1-3]. Besides the application of chemicals increases the risk of high level toxic residues in the products and causes a range of human health problems [4]. Because the future of crop protection should lie in a combination of several different tools and strategies (Integrated Pest and Disease Management, IPDM), alternative measures to synthetic pesticides have been developed, including biological agents, mineral salts and plant extracts [5]. Plants produce in fact an enormous variety of different natural products, called "secondary metabolites", to protect themselves from pathogens (fungi, viruses, bacteria), insects and herbivores [6,7]. Many secondary metabolites are undoubtedly the result of an adaptive evolutionary process, and confer to plants defence function against herbivores (toxic materials), or able them to a chemical communication towards the same or other species (volatile attractants or warning pigments for pollinators and/or natural insect predators, and allelopathic substances). Some of these compounds have a traditional use in crop protection, others have been characterised as the basis of chemical synthesis programmes to obtain new products and some others have been used for semi-synthetic derivatives [8].

2. Essential oils for fungal control

Great interest is now focused on the evaluation of plant pathogen control strategies based on the application of essential oils (EOs). Both lipophilic character of the hydrocarbon skeleton of the essential oil constituents, as well as the hydrophilic character of their functional groups, are key factors for the antimicrobial action: the presence of phenolic compounds such as carvacrol, eugenol, isoeugenol and thymol ensure strong antifungal activities, as reported in several works. In Table 1 is reported a survey of the most recent works focused on the use of EOs for fungal control in crops.

Table 1. The concentrations of some essential oils found effective after *in vitro* and *in vivo* assay for the control of crop pathogens are reported.

Essential oil source	***Plant pathogen fungus***	***Plant disease***	***Oil concentrations***	***Ref.***
Origanum syriacum L. var. bevanii, Lavandula stoechas L. var. stoechas, Rosmarinus officinalis L	*Botrytis cinerea*	Grey mould disease of tomato	Volatile phase effects: 0.2 µg/ml air (*O. syriacum*), 1.6 µg/ml air (*L. stoechas, R. officinalis*) Contact phase: 12.8 µg/ml (*O. syriacum*), >25.6 µg/ml (*L. stoechas, R. officinalis*)	[9]
Asarum heterotropoides var. mandshuricum	*Alternaria humicola, Colletotrichum gloeosporioides, Rhizoctonia solani, Phytophthora cactorum*	Various	IC50 values: 0.42 µg/ml (*C. gloeosporioides, R. solani*) IC50 values: 0.32 µg/ml (*A. humicola*), IC50 values: 0.073 µg/ml (*P. cactorum*)	[10]
Pelargonium graveolens L. Her, Ocimum basilicum var basilicum L., Cuminum cyminum L.	*Fusarium oxysporum, Fusarium solani, Fusarium moniliforme, Fusarium. dimerum, Fusarium equiseti, Fusarium lateritium*	Root rot disease of cumin	Treatment of cumin seeds with 4% of oils resulted in protection of cumin seedlings.	[11]
Clove, cedar wood, lemongrass, peppermint, citronella, nutmeg	*Phomopsis azadirachtae*	Die-back disease of neem	1500 ppm: complete inhibition (citronella oil), 85,5% inhibition (pippermint oil), 80,6% inhibition (cedar wood oil), 2000 ppm: complete inhibition (lemongrass oil) 2500 ppm: 79.1% inhibition (nutmeg oil), 78.8% inhibition (clove oil)	[12]
Cestrum nocturnum L.	*Botrytis cinerea, Colletotrichum capsici, Fusarium oxysporum, Fusarium solani, Phytophthora capsici, Rhizoctonia solani, Sclerotinia sclerotiorum*	Gray mold rot, vascular wilt, fruit rot, water soaked spot	MIC: 62.5 µg/ml (*P. capsici*), 125 µg/ml (*B. cinerea*), 250 µg/ml (*F. solani*) 250 µg/ml (*S. sclerotiorum*), 500 µg/ml (*C. capsici, F. oxysporum*)	[13]
Astronium fraxinifolium, Inga marginata, Malva sylvestris, Matayba elaeagnoides, Miconia argyrophylla, Myrcia fallax, Ocimum gratissimum, Origanum vulgare, Rollinia emarginata, Siparuna arianeae, Styrax pohlii, Tabebuia serratifolia and Trichilia pallid	*Colletotrichum lindemuthianum*	Anthracnose of common bean	MIC: 1.45 mg/mL (*A.fraxinifolium, M. elaeagnoides, M. argyrophylla, O. gratissimum, O. vulgare, R. emarginata, S. pohlii, T.pallida*), 0.72 mg/mL (*I. marginata, M. sylvestris, M. fallax, S. arianeae, T.serratifolia*).	[14]

Tagetes minuta, Lippia javanica, Amaranthus spinosus, Vigna unguiculata	*Fusarium verticillioides, Fusarium proliferatum*	Diseases in grains, peanuts and cassava	MIC : 0.02-0.32 mg/ml (*V. unguiculata, A. spinosus*), 0.02-2.5 mg/ml (*T. minuta, L. javanica*)	[15]
Salvia officinalis	*Plasmopara viticola*	Diseases in grapevine	50 ml of sage extract/liter of water (Each plant received approximately 40 ml of treatment solution, sage extract exhibited an efficacy of 79.9-95.8%)	[16]
Metasequoia glyptostroboides	*Fusarium oxysporum, Fusarium solani, Phytophthora capsici, Colletotrichum capsici, Sclerotinia sclerotiorum, Botrytis cinerea and Rhizoctonia solani.*	Gray mold rot, vascular wilt, fruit rot, water soaked spot	MIC: 62.5-1000 μg/ml (oil), 500-4000 μg/ml (extract). 500 μg/ml of oil has a100% antifungal effect against *P. capsici* in "*in vivo*" experiment	[17]
Spiraea alpina Pall	*Rhizoctonia solani, Fusarium graminerum, Pyricularia oryzea, Exserohilum turcicum*	Diseases in solanaceae and cereals	500 μg/ml: complete inhibition of *R.solani* 1000 μg/ml: 36.4-59.5% inhibition of *F.graminearum*, 20.3-33.3% inhibition of *P.oryzea* and 44-100% inhibition of *E.turcicum*	[18]
Melodorum fruticosum	*Collectotrichum asianum, Collectotrichum siamense, Collectotrichum gloeosporioides, Collectotrichum fruiticola, Collectotrichum acutatum, Collectotrichum sp., Trichoderma reesei, Lasiodiplodia theobromae.*	Various	MIC: 250-1000 μg/ml (oil), 125-250 μg/ml (dichloromethane extract), 250-500 μg/ml (hexane extract), 500-1000 μg/ml (methanol extract)	[19]

3. Tea Tree Oil for fungal control in cereals

The activity of *Melaleuca* essential oil (Tea Tree Oil, TTO) and of its major components has been evaluated for the control of three different classes of plant pathogen fungi that affect small grain cereals widely cultivated in Mediterranean environments [20]. The fungi considered (*Fusaria, Pyrenophora graminea, Blumeria graminis*) are characterized by different life cycle, pathogenicity behavior, transmission mode and effects on plants and derived agro-food products. The antifungal activities of TTO and of single oil components have been evaluated *in vitro*, determining the impact of different concentrations of the substances on *Fusaria* and *Pyrenophora graminea* mycelia growth. Moreover, the effects of TTO and of single oil components have been studied *in vivo*, on cereal leaves colonized by *Blumeria graminis* under controlled conditions of environment and infection. The results obtained have demonstrated that different species of *Fusaria* and *Pyrenophora* are susceptible to TTO, with MIC50 ranging from 1 to 2%, but TTO components have strongest activity, with fungicide effects against these fungal species at concentrations that are not phytotoxic. On the contrary, TTO is particularly effective in *Blumeria graminis* control at concentration equal or higher than 0.5% that did not affect seed germinability nor adult plant health.

The conclusions are that TTO and its components are promising tools for the control of wide range of phytopathogens.

Acknowledgements The support of *Riduzione rischio contaminazione da micotossine nei cereali,* Emilia-Romagna project is gratefully acknowledged.

References

[1] Moorman, GW, Lease RJ. Benzimidazole and dicarboximide-resistant *Botrytis cinerea* from Pennsylvania greenhouse. *Plant Disease*. 1992;76:477-480.
[2] Diaenz F, Santos M, Blanco R, Tello JC. Fungicide resistance in *Botrytis cinerea* isolate from strawberry crops in Huelva (southwestern Spain). *Phytoparasitica*. 2002;30:529–534.
[3] Yoon CS, Ju EH, Yeoung YR, Kim BS. Survey of fungicide resistance for chemical control of *Botrytis cinerea* on paprika. *The Plant Pathology Journal*. 2008;24:447-452.
[4] Abhilash PC, Singh N. Pesticide use and application: an indian scenario. *Journal of Hazardous Materials*. 2009;165:1–12.
[5] Dayan FE, Cantrell CL, Duke SO. Natural products in crop protection. *Bioorganic & Medicinal Chemistry*. 2009;17:4022–4034.
[6] Dixon RA. Natural products and plant disease resistance. *Nature*. 2001;411:843-847.
[7] Schlaeppi K, Mauch F. Indolic secondary metabolites protect Arabidopsis from the oomycete pathogen *Phytophthora brassicae*. *Plant Signaling & Behavior*. 2010;5:1099-1101.
[8] Copping LG, Duke SO. Natural products that have been used commercially as crop protection agents. *Pest Management Science*. 2007;63:524–554.
[9] Soylu EM, Kurt S, Soylu S. *In vitro* and *in vivo* antifungal activities of the essential oils of various plants against tomato grey mould disease agent *Botrytis cinerea*. *International Journal of Food Microbiology*. 2010;143:183–189.
[10] Dan Y, Liu H-Y, Gao W-W, Chen S-L. Activities of essential oils from *Asarum heterotropoides* var. *mandshuricum* against five phytopathogens. *Crop Protection*. 2010;29:295–299.
[11] Hashem M, Moharam AM, Zaied AA, Saleh FEM. Efficacy of essential oils in the control of cumin root rot disease caused by *Fusarium* spp. *Crop Protection*. 2010;29:1111-1117.
[12] Prasad MN, Shankara Bhat S, Sreenivasa MY. Antifungal activity of essential oils against *Phomopsis azadirachtae*- the causative agent of die-back disease of neem. *Journal of Agricultural Technology*. 2010;6(1):127-133.
[13] Al-Reza SM, Rahman A, Ahmed Y, Kang SC. Inhibition of plant pathogens *in vitro* and *in vivo* with essential oil and organic extracts of *Cestrum nocturnum* L. *Pesticide Biochemistry and Physiology*. 2010;96:86–92.
[14] Pinto JMA, Souza EA, Ferreiera Oliveira D. Use of plant extracts in the control of common bean anthracnose. *Crop Protection*. 2010;29:838-842.
[15] Thembo KM, Vismer KM, Nyazemal NZ, Gelderblom WCA, Katerere DR. Antifungal activity of four weedy plant extracts against selected mycotoxigenic fungi. *Journal of Applied Microbiology*. 2010;109:1479–1486.
[16] Dagostin S, Formolo T, Giovannini O, Pertot I. *Salvia officinalis* extract can protect grapevine against *Plasmopara viticola*. *Plant Disease*. 2010;94:575-580.
[17] Bajpai VK, Kang SC. Antifungal activity of leaf essential oil and extracts of *Metasequoia glyptostroboides* Miki ex Hu *Journal of American Oil Chemical Society*. 2010;87:327–336.
[18] Teng Y, Yang Q, Yu Z, Zhou G, Sun Q, Jin H, Hou T. *In vitro* antimicrobial activity of the leaf essential oil of *Spiraea alpina* Pall. *World Journal of Microbiology and Biotechnology*. 2010;26:9–14.
[19] Pripdeevech P, Chukeatirote E. Chemical compositions, antifungal and antioxidant activities of essential oil and various extracts of *Melodorum fruticosum* L. flowers. *Food and Chemical Toxicology*.2010;48:2754–2758.
[20] Terzi V, Morcia C, Faccioli P, Valè G, Tacconi G, Malnati M. In vitro antifungal activity of the tea tree (*Melaleuca alternifolia*) essential oil and its major components against plant pathogens. *Letters in Applied Microbiology*. 2007;44:613–618.

Possibility of using *Rosmarinus officinalis, Myrtus communis* and *Origanum sp.*'s essential oil as fungicide in pickling and tanning processes

Eser[1] Eke Bayramoglu

[1]Ege University, Faculty of Engineering, Department of Leather Engineering, 35100 Bornova-IZMIR – TURKIYE eser.eke@ege.edu.tr

The objective of this study was to examine the applicability of *Rosmarinus officinalis, Myrtus communis* and *Origanum sp.* essential oil as fungicide against fungus that grows on leather during pickling and tanning processes. In the study, 2-n-Octyl-4-isothiazolin-3-one containing commercial fungicides were also used for control. During the microbiologic tests (ASTM D 4576-86/ 91), the growth of mould species like *Aspergillus niger, Alternaria alternata, Penicillium chrysogenum* and *Trichoderma auroviride* that cause problems in leather industry were also investigated against these essential oil and fungicide.

As a result of this study, it has been determined that oregano essential oil has most effective antifungal activity.

Keywords Leather industry, Pickling, Tanning, Essential oil, *Rosmarinus officinalis, Myrtus communis* and *Origanum sp.*

1. Introduction

Fungi are a group of eukaryotic microorganisms distinguished from plants by their lack of chlorophyll, differences in their cell walls, and by the fact that they are not truly multicellular. Fungi spores exist in the air we breathe and sometimes in the water we drink. Together with the bacteria, many fungi play a major role as decomposers serving to bring about the rotting and decay of dead matter, and the recycling of this matter in the environment [1]. This can cause serious damages for pickled pelts and chrome tanned leathers during leather production, since leather is an organic material that consists of many nutrients for the fungi, which acquire the necessary nutrients for growth through absorbtion. The moulds and unicellular yeasts secrete enzymes into the surrounding environment breaking down (hydrolyze) complex organic compounds into simpler ones. As a result of this extracellular digestion, simpler compounds such as glucose and amino acids can be absorbed. Since the moulds are composed of numerous hyphae, the mycelium forms a tremendously large surface area for nutrient absorbtion[1]. Fungi also produce chromophores that can stain the grain surface [2]. The pigments produced by some moulds, which are virtually imposible to remove, can create problems with dye uptaken and cause fatty spues[3-5]. They may also generate bad smell.

Fungal growth is influenced by many factors in the environment. Pickled and chrome tanned leather storage under humid conditions, high storage temparature and also the low pH value are the factors causing the growth of mould, besides the organic material of leather. Therefore, it has been considered that it would be appropriate to use fungicides in the above mentioned stages[4-6]. However, the chemicals used for this purpose are generally harmful to human health and nature. A number of widely used fungicides may not be employed for wet blue in the future due to regulations or other restrictions [7]. Thus, leather market necessitates new biocides harmless to human health and natural life. In this study, a completely natural product 'essential oils' were used as a fungicide in different percentages and it was compared with commercial fungicides. An essential oil is a volatile material derived by physical process from odorous plant material of a single botanical form and species with which it agrees in name and odor. The antimicrobial activity of essential oils and their derivatives was recognized long time ago. Ibn-i Sina was the first man, who invented the essential oils by steam distillation techniques and used (980-1037) [8]. As essential oils and their components are natural products, they are becoming increasingly popular, and such antimicrobial agents are being used in many different sectors such as pharmacy, food industry, bakery and also leather industry[9-13]. In this study, essential oil of *Origanum sp., Rosmarinus officinalis* and *Myrtus communis* were used.

The objective of this study was to determine whether the essential oil of Origanum sp**.** Rosmarinus officinalis *and* Myrtus communis can be used as an alternative to fungicides in the pickling and wet blue process, and to compare it with one of widely used commercial fungicide.

2. Materials and Methods

2.1. Materials

2.1.1. Raw skin

In this study, dry salted Turkish domestic sheep skins were used in order to obtain pickled and chrome-tanned leather.

2.1.2. Fungicide

In the experiments, N-octyl-isothiazolinone (N-OITZ) based fungicide was used. The studies at lower rate have been conducted with fungicide. The reason for this was to compare the results with the commercially available materials, which are reported to have the same effect at the same dosage. The objective of this study was not to examine the MIC value of the commercially available product. Therefore, the amount indicated in the catalogue was used after the necessary contact with the experts of the company were made, and one material was selected for the study. A special attention has been paid in order to purchase the recently produced material containing active material.

2.1.3. Moulds

Species of mould used in this study are ***Aspergillus niger***, ***Alternaria alternata***, ***Penicillium chrysogenum*** and ***Trichoderma auroviride***.

2.1.4. Essential oil

All essential oil's chemical structure and the percentage of its constituents are given in Table I,II,III. This information was obtained by the company 'Türer Tarım'.

Table I-Chemical constituents of the *essential* oil of ***R. officinalis*** [14].

R. officinalis	
Constituents	%
1,8 - cineole	25
α- pinene	19
camphor	19
p- cymene	17
camphene	9
β- pinene	5
borneol	2
unidentified	4

Table II-Chemical *constituents* of the *essential* oil of ***Origanum sp.*** [14].

Origanum sp.	
Constituents	%
carvacrol	78.84
thymol	8.90
β - Bisabolene	2.47
p- cymene	2.03
γ- Terpinen	1.36
caryophylene	1.31
terpinen 4-ol	1.09
(+)- Borneol	0.75
B- Phellandrene	0.40
linalool	0.35
α- pinene	0.28
unidentified	1.48

Table III-Chemical constituents of the essential oils of ***Myrtus communis*** [14].

Myrtus communis	
Constituents	%
1,8 cineole	24.8
α- pinene	23.5
limonene	14.2
linalool	7.5
myrtenol acetat	6.7
α- terpined/terpenil acetate	4.5
nerole/myrtenol	2.9
linalyl acetate	2.8
geranile acetate	1.9
iso-butyl iso-butyrate	1.4
carvacrol	1.1
geraniol	1.1

2.1.5. Media

Malt Extract Agar (M.E.A) was used in keeping stocked cultures and obtaining fresh cultures in the experiments.

2.2. Methods

2.2.1. Obtaining the leather samples used in the experiment

Raw skins were processed so as to obtain two different groups. One group was processed in order to obtain pickled pelts and the other group was processed for obtaining chrome-tanned leathers, which were the main materials of the study. For all leather production, one standard garment leather processing method was applied.

The group of pickling pelts was obtained during preserved pickling process. The features of the pelts are as follows:

1- Preserved pickled pelt without fungicide (pH:1.5)
2-Preserved pickled pelt obtained by using N-octyl-isothiazolinone (N-OITZ)-based fungicide (0.02%, the amount advised by the chemical company)
3- Preserved pickled pelt obtained by using ***Origanum sp., Myrtus communis*** *and* ***R. officinalis*** essential oils (ratio: 2%).
4- Preserved pickled pelt obtained by using ***Myrtus communis+Origanum sp., Myrtus communis+ R. officinalis*** essential oils (ratio: 1%+1%)
*The second group of the leathers, which would be chrome-tanned, was primarily processed through a normal pickling process (pH:3).
5- Chrome tanned leather without fungicide.
6- Chrome tanned leather obtained by using N-octyl-isothiazolinone (N-OITZ) based fungicide (0.02 %, the amount advised by the chemical company)
7- Chrome tanned leather obtained by using ***Origanum sp., Myrtus communis*** *and* ***R. officinalis*** essential oils (ratio:2%).
8- Chrome tanned leather obtained by using ***Myrtus communis+Origanum sp., Myrtus communis+ R. officinalis*** essential oils (ratio: 1%+1%)

The leather samples used in the experiments were cut in a sterile room under sterile conditions. The laboratory punch was used to punch 3 circular specimens, measuring 3 cm in diameter, out of the leather samples to be tested. All equipments and tools were cleaned with ethanol and were flamed before and after changing samples. The leathers were processed with essential oil or biocide for 4 hours after being cut. Every trial was duplicated and repeated twice.

2.2.2. Activating mould species

Mould species used in the experiments were horizontally prepared and separately incubated in the tubes with M.E.A., and fresh cultures were obtained through incubation at 27 °C for a week.

2.2.3. Microbioiogical test standards

ASTM D 4576-86 (Reapproved 1991) microbiological test method is used for mould growth resistance of wet blue and other leather samples. In this study, this test standard was applied [15]. Specimens of the leather were placed in petri-dishes on inoculated nutrient agar-medium and incubated for four weeks. Following this period, the specimens were detected and evaluated respectively in terms of the mould growth.

3. Results and Discussion

It was obtained that 2 % of essential oil of ***Origanum sp.*** was found to be the best antifungal effect even better than the effect of commercial fungicide, which are commonly used in quantities on the market. Even though this amount was reported to be commercially appropriate, the results of the experiments revealed the contradictory especially for contaminant fungi.

When all the test fungi behaviour against antifungal agent was observed, it was found that each one of them displayed different attitudes. ***Trichoderma auroviride*** *was* the most resistant during the study.

We observed that both pickled and wet blue leathers were not sterile normally so contaminant fungi can also grow easily on the leathers but when we applied 2% Origanum sp.'s essential oil to the leather it was effective also these fungi.

4. Conclusion

As a result of this study, it was concluded that ***Origanum sp.*** essential oil can be used as a new fungicide that does not give harm to nature and human health due to its completely natural structure.

The optimum antifungal effect of oregano essential oil on leathers was obtained at 2% based on ASTM D 4576/86 standards.

The studies indicated that the commercial fungicide used at suggested amounts did not provide enough protection against contaminant fungi. At this point, it is possible to state that the companies should continue working on research and development processes to provide enough protection.

Acknowledgements The authores would like to express their gratitude to Hartmut Rehbein for the information he has provided and for the interest he has shown in the study, to Prof. Dr. İsmail Karaboz who have provided fungi cultures. The authores' heartfelt thanks are also due to Türer Tarım for providing essential oil and also Meriç Özbek and Altuğ Akça for their help in lab. study.

References

[1] Pommerville, J.C., Alcamo's Fundamentals Of Microbiology, Seventh Edition, Jones And Bartlett Publishers, USA, 2004

[2] Buckman notes,Fungicides for the Leather Industry, 9 pages, USA, 1999.

[3] Bayramoğlu,E.E., Small Moulds Can Cause Big Problems, World Leather, Volume:23, Number 4,p:42-43,Aug/Sept 2010

[4] Karaboz, I., Gülümser, G., Bayramoğlu, E.E., Tabakhanelerde Depolama Sırasında Gelişen Bazı Fungusların Koruma Piklesi Ve Kromla Tabaklama Asamasinda Deride Olusturduklari Pigmentasyonun Incelenmesi,Ege Univ. Ziraat Fak. Derg., 40(3):129-136,2003

[5] Anonymus, Mould-Cause And Prevention, Leather International, Volume: 203, No:4708, UK. ,February 2001

[6] Yapici,B.M., Karaboz,I.,The Effect Of Various Antifungal Compounds In The Growth Of Molds That Frequently Appear On Tanned Leather, Jalca,Vol.92(2),38-45,1997.

[7] Adminis, U.,Huynh, C.,Money, CA, The Need For Improved Fungicides For Wet-Blue,International Union Of Leather Technologists And Chemical Societies 16 Congress, Cape Town, 2001

[8] Akgül A.,Baharat Bilimi Ve Teknolojisi; Gida Teknolojisi Dergisi,Yayin No:15, Ankara,329,1993

[9] Bayramoğlu E.E., Unique Biocide For The Leather Industry: Essential Oil of Oregano,JALCA, 102(11), 347-353, 2007.

[10] Bayramoğlu E.E, Gülümser G., Karaboz İ., Ecolological and Innovative Fungicide For Leather Industry: Essential Oil of Origanum minutiflorum, JALCA, 101(3),96-104, 2006.

[11] Bayramoglu, E.E., Natural and Enviroment Friendly New Bactericide For Leather Industry:Essential Oil Of ***Origanum minutiflorum,*** Ansinet, Journal Of Biological Sciences, Vol 5 No 4: 455-457 , 2005.
[12] Bayramoğlu E.E.E, Antibacterial Activity of *Myrtus communis* Essential Oil Used in Soaking, JSLTC,Volume 90,Number 5, 2006.
[13] Bayramoğlu E.E., Soaking with Storax- Possibility Of Using Siğla Tree (*Liquidambar Orientalis* Mill. var *Orientalis*) Storax As Bactericide In The Soaking Float, JALCA, Vol.105, 2010.
[14] Türel Tarım –Raports of essential oil constituents from TBAM ,2005
[15] ASTM D 4576-86 (Reapproved 1991),Test Method For Mould Growth Resistance of Wet Blue And Other Leather Samples.

Screening of aqueous methanol plant extracts for their antibacterial activity

Kashif Nauman[1] and Muhammad Arshad[2]

[1] Department of Meat Technology, University of Veterinary and Animal Sciences, Lahore,54100 Pakistan
[2] Department of Veterinary Microbiology, University of Agriculture Faisalabad,38000 Pakistan

The purpose of this experimental study was to observe the antibacterial effect of aqueous methanol extracts of ten local Pakistani plants against two gram negative bacteria, *(Escherichia coli, Pasturella multocida)* and three gram positive bacteria *(Bacillus cereus, Staphylococcus aureus, Corynebacterium bovis)* by using disc diffusion method. The minimum inhibitory concentration (MIC) was determined by agar well diffusion method and agar dilution method. All the bacteria were susceptible to different plant extracts. *Azadirachta indica* , *Peganum harmala* and *Swertia ciliate* showed antibacterial activity against all the tested bacteria. The extract of *Piper nigrum* was most effective antibacterial activity of the ten plant extracts used. *Bacillus cereu* and *Pasteurella multocida* were the most sensitive bacteria against most of the plant extracts.While *Nicotiana tabacum, Punica granatum, Ferula asafetida ,Fumeria parviflora, Morus alba* and *Areca cutechu* different effect on different bacteria. The present study will be helpful for the local community, traditional healers and those who involved in the study of ethnomedicine.

Keywords: Plant extracts; Pakistani medicinal plants; Antibacterial activity.

INTRODUCTION

The microorganisms have developed resistence to many antibiotics because of indiscriminate use of antimicrobial drugs that create a big problem in the treatment of infectious diseases (Davis. 1994). Plants have been shown to be a potential source for the new anti-microbial agents (Mitscher et al., 1987). The antimicrobial activity of essential oils is assigned to a number of small terpenoids and phenolic compounds (thymol, carvacrol, eugenol), which also in pure form demonstrate high antibacterial activity (Conner, 1993).Plants still continue to be almost the exclusive source of drugs for the majority of the world's population (Hamburger and Hostettmann, 1991). The use of medicinal plants as traditional medicines is well known in rural areas of many developing countries (Sandhu and Heinrich. 2005; Gupta *et al.*, 2005). Traditional healers claim that their medicine is cheaper and more effective than modern medicine. In developing countries, low-income people such as farmers, people of small isolate villages and native communities use folk medicine for the treatment of common infections (Rojas *et al.*, 2006; Hammer *et al.*, 1999). In Western medicine substances derived from higher plants constitute 25% of prescribed medicines and 74% of the 121 bioactive plant-derived compounds currently in worldwide use were identified through research based on leads from ethnomedicine (Farnsworth et al., 1985). The screening of plant extracts has been of great interest to scientists for the discovery of new drugs effective in the treatment of several diseases (Dimayuga and Garcia 1991)

This study was done to carry out the antibacterial activity of Plants like *Nicotiana tabacum* (Tobacco), *Swertia ciliate* (Cheretta), *Punica granatum* (Pomegranate), *Ferula asafetida* (Hing), *Azadirachta indica* (Neem), *Piper nigrum* (Black Pepper), *Peganum harmala* (Harmal), *Fumeria parviflora* (Shahtra), *Morus alba* (Mulberry) and *Areca cutechu* (Areca nut).

MATERIAL AND METHODS

Plant materials and Preparation of Plant Extract

Ten plant species were selected for screening of antibacterial activity. Diffrent parts of The Nicotiana tabacum, Swertia ciliate, Punica granatum, Ferula asafetida, Piper nigrum, Peganum harmala, Fumeria parviflora and Areca cutechu were procured from local market whereas Azadirachta indica and Morus alba were obtained from the local area. Different parts of the plants were dried under shade and then ground with the help of grinder. These grinded materials were socked in solvent having 70% methanol and 30% distilled water and shaken up to three days daily. The whole material was filtered through cotton cloth and the filtrate was collected and the remaining was socked again in the same solvent. The filtration was done thrice. The collected filtrate having the plant extract was kept for evaporation in rotary evaporator. Further evaporation was done in vacuum oven for dryness and the final weighed was calculated.

Microorganisms

The microorganisms that were used in the present study consisted of two gram-negative bacteria Pasturella multocida, Escherichia coli (ATCC 29922) and three gram-positive bacteria Bacillus cereus, Corynebacterium bovis and Staplococcus aureus (ATCC 29923). The Escherichia coli (ATCC 29922) and Staphylococcus aureus (ATCC 29923) were procured from Department of Clinical Medicine and Surgery, (CMS) University of Agriculture Faisalabad. Pasteurella multocida was isolated from bone marrow of HS infected cattle. Bacillus cereus was isolated from soil and Corynebacterium bovis was isolated from mastitis milk of buffalo. All these bacteria were identified on the basis of culture, colony, and morphological characteristics and biochemical tests.

The antibacterial effect of plant extracts was tested after the inoculation of bacterial species on an agar media plate for 18 to 24 hours at 37C. A single colony was mixed in 5 ml of sterile normal saline solution to adjust the density of bacterial culture to 0.5 McFarland.

Antibacterial Activity

The antimicrobial susceptibility was carried out by Disc diffusion method (Nostro *et al.,* 2000). Spread the bacterial culture (that was adjusted to 0.5 McFarland standard) uniformly on the surface of Muller Hinton agar plates. The plates were dried for 15 min before placing paper discs (10 mg/disc) that had been impregnated with solution of plant extracts (200mg/ml). Standard commercial antibiotics were used as positive control while pure DMSO and Methanol were used as a negative control. Antibiotic discs of Amoxicillin (25 ug Oxid U.K) and Teramycine (25 ug Oxid U.K) were used as positive control. The negative control were (DMSO) 100% and Methanol 70%. The plates were then incubated at 37C for 18 to 24 hours. After the incubation, the plates were examined for the zone of inhibition and then measured the diameter of zone of inhibition.

Determination of Minimum Inhibitory Concentration

The MIC of the extracts was measured by agar dilution method and agar well diffusion method (Nostro *et al.,* 2000 ; Basu *et al.,* 2005.

Agar Dilution Method

Those plant extracts that showed inhibition zone diameter in disc diffusion test were assayed the determination of MIC. A stock solution of each plant extract was prepared in pure DMSO that was serially diluted two fold in pure DMSO. 1 ml of each dilution was incorporated in 20 ml of melted Muller Hinton agar at 45-50 C. Plates contain only 1%(V/V) DMSO was used as control. The bacterial culture was inoculated on the surface of agar plates in the form of spotting that contained different concentrations of plant extracts. Agar plates were incubated for 24-36 hrs and observed for the presence or absence of visible bacterial growth. The MIC was defined as the lowest concentration of the plant extract that inhibited the visible growth of each micro-organism.

Agar Well Diffusion Method

First prepared the Bacterial culture suspension in 5 ml normal saline and its turbidity was adjusted to that of 0.5 McFarland (contained approximately 1.0×10^7 CFU/ml). The bacterial suspension was mixed @ of 1 ml/20 ml of Muller-Hinton agar at 48 C. Wells were prepared in agar containing bacterial culture with the help of well cutter. Various concentrations of plant extracts that were prepared in pure DMSO were introduced in wells and plates were incubated at 37 C for 24-36 hrs. After incubation, the plates were observed for the presence or absence of bacterial growth around wells containing various concentrations of plant extracts. MIC was defined as lowest concentration of plant extract that inhibited the visible growth of bacteria in the form of zone of inhibition around wells containing various concentrations of plant extracts.

RESULTS

The results of the zone of inhibition determined by disc diffusion test are presented in table 1. It is evident from the results that all the bacteria were susceptible to different plant extracts. However, only *piper nigrum* showed antibacterial activity against all the tested bacteria. While the remaining plants showed varying degree of activity against different bacteria. *Swertia ciliate*, *Ferula asafetida* and *Fumeria parviflora* showed no activity against Staphylococcus aureus while showed antibacterial activity against all other tested bacteria. The extract of *Nicotiana tabacum* showed antibacterial activity against *Coryne. Bovis, E. coli, Staph. aureus* and *B. cereus* but no activity against *Past. multocida*. The extract of *Punica granatum* was effective against *Coryne. Bovis, Staph. aureus Pasteurella multocida* and *Bacillus cereus*. The extracts of *Peganum harmala* showed no antibacterial activity against *Pasteurella multocida* while showed activity against *Coryne. Bovis, Staph. aureus, E. coli* and *Bacillus cereus*. *Morus alba* showed no effect against *E. coli* but showed some effect against all other tested bacteria.

DISCUSSION

It is evident from the results of the present studies that the medicinal plants are very effective against bacteria, even that, only a few reports are available regarding the antibacterial activities of plants used in the present studies such as (Toshimasa et al., 2005) reported the antibacterial activity of *Santalum album* against *Helicobacter pylori*. The antibacterial activity of *Impactien balsamina* was reported by (Habbal et al., 2005) against *Staph. aureus, E. coli* and *Pseudomonas aeruginosa.* (Phulan and Khullar, 2004) also reported the antibacterial activity of *Embelia ribes.* (Usman et al., 2003) reported the antibacterial activities of ethanolic extracts of fruits, leaves, stems and roots of *Citrulus colocynthi.* (Taylor et al., 1996) reported antibacterial and antifungal efficacy of *Mallotus phillppensis* (Euphorbiaceae) and *Terminalia alata* (Combretaceae) and observed significant antibacterial activity. Some of the previous research (Mahato and Chaudhary, 2005) has been conducted just to find out the zone of inhibition against microorgainms by plant extracts but the present studies have been conducted to find out both the zone of inhibition and minimum inhibitory concentration values of effective plant extracts.

The results of the present studies regarding the antibacterial activity of *Ferula asafetida* are in link with the recent studies conducted by Kumar et al. (2006) in Pakistan neighboring country India. Like that of the results of the present studies, the antimicrobial activity of the leaves of *Lawsonia inermis* has been reported against *Staph. aureus, Streptococcus* spp., *Pseudemonas aeruginosa, Candida albicans, Fusarium oxysporum*, and *Aspergillus niger* by (Muhammad and Muhammad, 2005) in Nigeria. A number of reports (Taylor et al., 1996; Muhammad and Muhammad, 2005; Toshimasa et al., 2005; Kumar et al., 2006) are available regarding the antimicrobial activity of plants in different parts of the world but in Pakistan, such type of studies have been conducted for the first time. However, there is need to conduct further research on the chemical analysis of the used plant extracts to find out the active compounds.

Research into the effects of local medicinal plants is expected to boost the use of these plants in the therapy against disease caused by the test bacterial species and other microorganisms. It is possible that better therapy for many microbial diseases may be found in the bark, leaves etc. of these plants. The preliminary results of this investigation appear to indicate that a number of medicinal plants used in the present studies particularly *Piper ngrum*, *Swertia ciliate*, *Ferula asafetida* and *Fumeria parviflora* have high potential of antimicrobial activity.

Table 1. Antibacterial activity of various plants by Disc diffusion test (mm).

Botanical name with family	Parts used	Diameter of zone of inhibition (mm)				
		E. coli	*B. cereus*	*S. aureus*	*P. multocida*	*C. bovis*
Nicotiana tabacum	Leaves	14 ± 0.05	19 ± 0.02	15± 0.01	NA	26 ± 0.0
Swertia ciliate	Bulb	17 ± 0.11	13 ± 0.31	NA	19 ± 0.09	25 ± 0.51
Punica granatum	Fruit	NA	20 ± 0.5	12 ± 0.02	12 ± 0.45	26 ± 0.21
Ferula asafetida	Seeds	20 ± 0.04	11± 0.23	NA	NA	25 ± 0.18
Fumeria parviflora	*Fruit*	NA	12 ± 0.35	NA	13 ± 0.0	19 ± 0.13
Piper nigrum	Seeds	11 ± 0.33	15 ± 0.03	11 ± 0.61	13 ± 0.47	20 ± 0.26
Peganum harmala	*Fruit*	27 ± 0.31	15 ± 0.09	12 ± 0.0	NA	30 ± 0.21
Azadirachta indica	Leaves	NA	23 ± 0.47	18 ± 0.13	16 ± 0.33	30 ± 0.08
Morus alba	*Bark*	NA	24 ± 0.02	17 ± 0.05	NA	28 ± 0.12
Areca cutechu	*Fruit*	13 ± 0.41	19 ± 0.07	12 ± 0.12	NA	30 ± 0.07
Amoxicillin		18 ± 0.03	21 ± 0.0	20 ± 0.12	21 ± 0.16	22 ± 0.01
Teramycin		20 ± 0.0	22 ± 0.0	21 ± 0.08	24 ± 0.0	23 ± 0.02
Blank (DMSO)		NA	NA	NA	NA	NA
Methanol		NA	NA	NA	NA	NA

Table 2. Minimum inhibitory concentrations (MIC) of aqueous methanolic extracts of plants by agar dilution method (mg ml^{-1} of agar).

Plants/Organisms	*Staphylococcus aureus*	*Pasteurella multoceda*	*Crynobacterium bovis*	*Escherichia col*	*Bacillus cereus*
Nicotiana tabacum	12.5	3.12	12.5	12.5	6.25
Swertia ciliate	6.25	12.5	6.25	6.25	12.5
Punica granatum	NA	12.5	12.5	12.5	6.25
Ferula asafetida	NA	25	6.25	12.5	25
Fumeria parviflora	NA	25	12.5	25	12.5
Azadirachta indica	NA	NA	25	NA	25
Peganum harmala	NA	25	NA	NA	12.5
Piper nigrum	25	12.5	25	12.5	25
Morus alba	25	25	NA	25	NA
Areca cutechu	NA	12.5	NA	25	25

Table 3. Minimum inhibitory concentrations (MIC) of aqueous methanolic extracts of plants by agar well diffusion method (mg ml^{-1}).

Plants/Organisms	*Staphylococcus aureus*	*Pasteurella multoceda*	*Crynobacterium bovis*	*Escherichia col*	*Bacillus cereus*
Nicotiana tabacum	6.25	6.25	25	25	12.5
Swertia ciliate	6.25	12.5	6.25	6.25	12.5
Punica granatum	NA	25	25	12.5	6.25
Ferula asafetida	NA	25	12.5	25	50
Fumeria parviflora	NA	50	25	50	25
Piper nigrum	NA	NA	50	NA	50
Peganum harmala	NA	50	NA	NA	25
Azadirachta indica	50	25	50	25	25
Morus alba	50	50	NA	50	NA
Areca cutechu	NA	25	NA	25	50

LITERATURE CITED

1. Basu. S., A. Ghosh and B. Hazara, 2005. Evaluation of the antibacterial activity of *Ventilago madraspatana* Gaertn., Rubia cordifolia Linn. And Lantana camara Linn.: Iolation of Emodin and Physcion as active antibacterial agents. *Phytotherapy Research*, 19:888-894.
2. Conner, D. E, 1993. Naturally occurring compounds. *Journal of Antimicrobials in Foods*, 4: 441–468.
3. Davis, J., 1994. Inactivation of antibiotics and the dissemination of resistance genes. Science 264, 375–382.
4. Dimayuga, R. E and S. K. Garcia, 1991. Antimicrobial screening of medicinal plants from Baja California Sur. Mexico. *Journal of Ethnopharmacol*, 31: 181–192.
5. Farnsworth, N. R., O. Akerele, A. S. Bingel, D. D. Soejarto, and Z. G. Guo, 1985. Medicinal, plants in therapy. Bull. WHO, 63: 965–9871.
6. Gupta, M. P., P. N. Solis, A. I. Calderon, S. F. Guionneau, M. Correa, C. Galdames, C. Guerra, A. Espinosa, G. I. Alvenda, G. Robles and R. Ocampo, 2005. Medical ethnobotany of the Teribes of Bocas del Toro, Panama. *Journal of Ethnopharmacology* , 96: 389-401.
7. Habbal, O. A., A. A. Al-Jabri, A. H. El-Hag, Z. H. Al-Mahrooqi and N. A. Al-Hashmi, 2005. In-vitro antimicrobial activity of Lawsonia inermis Linn (henna). A pilot study on the Omani henna. *Saudi Medicinal Journal*, 26: 69-72
8. Hamburger, M and K. Hostettmann, 1991. The link between phytochemistry and medicine. Phytochemistry, 30: 3864–3874.

9. Hammer, K. A., C. F. Carson and T. V. Riley, 1999. Antimicrobial activity of essential oils and other plant extracts. *Journal of Applied Microbiology*, 86: 985-990.
10. Kumar, V. P., S. C. Neelam, P. Harish and M. Rajani , 2006. Search for antibacterial and antifungal agents from selected Indian medicinal plants. *Journal of Ethnopharmacology* 107: 182–188
11. Mitscher, L. A., S. Drake, S. R. Golloapudi and S. K. Okwute, 1987. A modern look at folkloric use of anti-infective agents. *Journal of National Products*, 50: 1025–1040.
12. Muhammad, H. S and S. Muhammad, 2005. The use of *Lawsoni*a *inermis* linn. (henna) in the management of burn wound infections. *African Journal of Biotechnology*, 4: 934-937
13. Nostro, A. M., P. Germano, V. D. Angelo, A. Mariko and M. A. Cannatelli, 2000. Extraction methods & bioautography for evaluation of medicinal plant antimicrobial activity. *Letters in Applied Microbiology*, 30: 379-384.
14. Phulan. R and N. Khullar, 2004. Antimicrobial evaluation of some medicinal plants for their anti-enteric potential against multi-drug resistant *Salmonella typhi. Phytotherapy Research,*18: 670 – 673
15. Rojas, J. J., V. J. Ochoa, S. A. Ocampo and J. F. Munoz, 2006. Screening for antimicrobial activity of ten medicinal plants used in Colombian folkloric medicine: A possible alternative in the treatment of non-nosocomial infections. *BMC Compelmentary and Alternative Medicine*, 6: 2-8
16. Sandhu, D. S and M. Heinrich, 2005. The use of health foods, spices and other botanicals in the Sikh community in London. *Phytotherapy Research*, 19: 633-642
17. Taylor, R. S. L., F. Edel, N. P. Manandhar and G. H. N. Towers, 1996. Antimicrobial activities of southern Nepalese medicinal plants. *Journal of Ethnopharmacology,* 50: 97-102
18. Toshimasa, O., H. Shibata, T. Higuti, K. Kodama, T. Kusumi, and Y. Takaishi, 2005. Anti-*Helicobacter pylori* Compounds from *Santalum album. Journal of Nat. Prod,* 68: 819 -824
19. Usman, M., A. H. Brohi, S. W. Ahmed, I. Azhar and H. Bang 2003. Antibacterial screening for Citrullus Colocynthis. *Pakistan journal of pharmaceutical sciences*, 16: 1-6

Antimicrobial Surfaces.

Biofilms.

Quorum Sensing.

Consumer Products.

Antimicrobial active packaging films based on sorbic acid

C. Hauser[*,1,2], **J. Wunderlich**[1] **and G. Ziegleder**[1]

[1] FRAUNHOFER Institute for Process Engineering and Packaging IVV, Department Food Quality, Giggenhauser Str. 35, 85354 Freising, Germany

[2] Friedrich-Alexander University Erlangen Nürnberg, Henriette Schmidt-Burkhardt Chair of Food Chemistry, Schuhstr. 19, 91052 Erlangen, Germany

* Corresponding author: e-mail: carolin.hauser@ivv.fraunhofer.de

Common packaging films were coated with an antimicrobial active polyvinyl acetate lacquer containing sorbic acid as active agent. The antimicrobial efficacy was determined against the test strains *Escherichia coli* DSM 1567, *Staphylococcus aureus* DSM 346 and *Saccharomyces cerevisiae* DSM 70449 using the Japanese Industrial Standard Method JIS Z 2801:2000. In addition, storage tests were performed with Gouda cheese and pork loin. The Gouda cheese was inoculated with 5 x 10^2 colony forming units (cfu) of the mould *Penicillium roqueforti* and covered with a lacquered PE-film with and without sorbic acid. After 7 days of storage a visible difference between the two samples was found. On the pork a count reduction to one fifth of the initial inoculum could be observed after 9 days of storage. These findings affirm that antimicrobial active packaging films based on sorbic acid can make a significant contribution to the quality and safety of packed food.

Keywords active packaging; sorbic acid; antimicrobial films

1. Introduction

Consumer demand for fresh and minimally processed foods is growing rapidly. On the other side, retailers and distributors are asking for food products that stay fresh and – especially microbiologically – safe for a very long time in order to get more logistic flexibility.

Therefore, packaging material with a passive barrier layer is no longer sufficient. For this reason, principally in the Japanese market, commercially available active packaging material has been introduced [1]. Since 2009 there has been a new EU regulation on active and intelligent packages that facilitates the employment of this new packaging technology in the European Union [2].

At the Fraunhofer Institute for Process Engineering and Packaging IVV, Germany, an antimicrobial active packaging film has been newly developed. First trials have shown that this active film is very effective against various bacteria and moulds. It is also transparent and sealable [3]. The activity of this film is based on sorbic acid. Sorbic acid is a conventionally used preservative that represents no health or allergenic hazard [4, 5]. It is approved according to food law for many different kinds of foods [6]. This film releases very low amounts of sorbic acid to the surface of the packaged food. In this way it protects the food at the contact area from microbial contamination. Thus longer shelf life and safer food can be achieved. The common way of preserving foodstuff by adding the preservative to the whole food matrix can be avoided.

2. Materials and Methods

2.1 Film preparation

Polyvinyl acetate was dissolved in ethyl acetate by stirring at room temperature. After the polyvinyl acetate was solved, sorbic acid was added to the lacquer. The lacquer was applied in a lacquering and laminating plant on a common polyethylene packaging film and a polyethylene polyamide composite film respectively. The lacquer was dried by an airstream in the drying tunnel and the film was winded afterwards.

2.2 Test for antimicrobial efficacy

To show the antimicrobial efficacy of the developed active films the Japanese Industrial Standard Method (JIS Z 2801:2000) was applied.

The used test strains were *E.coli* DSM 1576, *Staphylococcus aureus* DSM 346 and *Saccharomyces cerevisiae* DSM 70449. All test strains were obtained from "Deutsche Sammlung von Mikroorganismen und Zellkulturen

GmbH (DSM)". An aliquot of an overnight culture was added to a 500-fold diluted nutrient broth in order to adjust a concentration of at least 1 x 10^5 cells/ml.

Three samples of each test film were cut to a size of 5 x 5 cm, inoculated with 400 µl of the test inoculum and covered with a sterile PET-film. Two uncoated films were used in triplicates as reference. One for the determination of the initial cell count directly after inoculation, the other reference was treated like the test samples. All samples were incubated for 24 hours at the specific growing temperature of the germs (*E.coli* at 37 °C, *Staphylococcus aureus* at 30 °C, *Saccharomyces cerevisiae* at 25 °C) and 90 % relative humidity. A colony count after 24 h showed the number of the surviving germs.

2.3 Storage test of Gouda cheese

Pre-sliced and packed Gouda cheese was cut to a size of 7 x 7 cm and inoculated with 5 x 10^2 spores of *Penicillium roqueforti*. Finally they were covered with the lacquered PE-film. For reference samples, inoculated cheese slices were covered with a non-lacquered PE-film. The cheese samples were stored in a sterile petri-dish at 23 °C. Each inoculation was done in triplicates. The evaluation was done visually.

2.4 Storage test of pork

Pork loin was cut in a sterile environment into pieces of 5 x 5 cm and a thickness of approx. 2 cm. These test pieces were inoculated with 1 x 10^3 cfu of *E. coli* and finally vacuum packed in the lacquered composite films of polyethylene and polyamide. For reference samples, inoculated meat pieces were packed in a non-lacquered composite film. The samples were stored at 8 °C up to 9 days. The viable count of *E. coli* was determined on violet red bile dextrose agar (VRBD) after a certain time of storage (0, 1, 3, 7, 9 days). Each inoculation and the colony count were done in triplicates.

3. Results and Discussion

3.1 Test for antimicrobial efficacy

For *E. coli* DSM 1567 a count reduction of 6.5 log, for *Staphylococcus aureus* DSM 346 a count reduction of 6.2 log and for *Saccharomyces cerevisiae* DSM 70449 a count reduction of 5.8 log was observed in comparison to the reference samples (Figure 1). This proved that the developed antimicrobial film is effective against pathogen food-infecting microorganisms like *E. coli* and *Staphylococcus aureus* as well as food spoiling contaminants like yeasts.

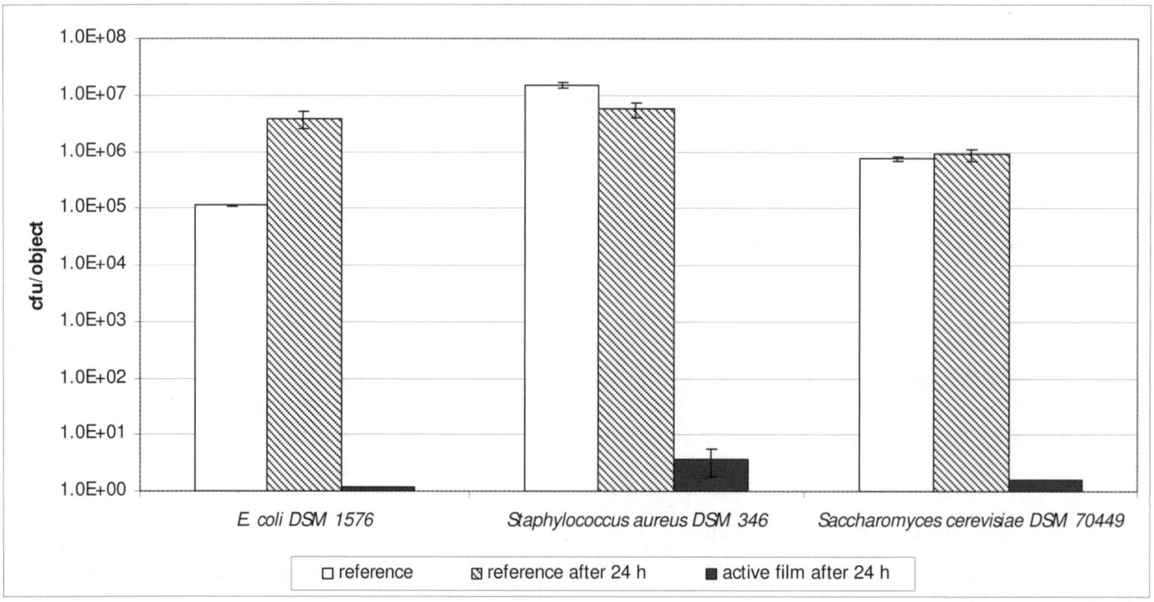

Fig. 1 Antimicrobial efficacy according to JIS Z 2801:2000.

3.2 Storage test of Gouda cheese

A positive effect on the shelf life of cheese could be demonstrated in storage tests with the mould *Penicillium roqueforti*. *Penicillium roqueforti*, which is often used for the production of blue-cheese, is a known contaminant of cheese [7]. It can be carried onto the Gouda and other cheese surfaces e.g. by cutting instruments or the spores in the air [8]. A visible difference to the reference covered with an unlacquered film without sorbic acid could already be observed after 7 days of storage. The cheese covered with the lacquered film containing sorbic acid showed no difference to the initial state, while on the reference several green to blue spots were detectable underneath the film due to the growing mould.

3.3 Storage test of pork

Figure 2 shows that a count reduction to one fifth of the initial inoculum could be observed, while the bacterial count of *E. coli* on the reference samples increases to almost 10^5 colony forming units after 9 days of storage. Freshly slaughtered meat, that is nearly sterile, generally becomes contaminated with microorganisms during subsequent cutting and processing stages e.g. by faeces, hides and skins, germs on tools, hands or rinsing water [9, 10]. So it is not random that especially surfaces of meat pieces can get contaminated with pathogens like *E.coli*.

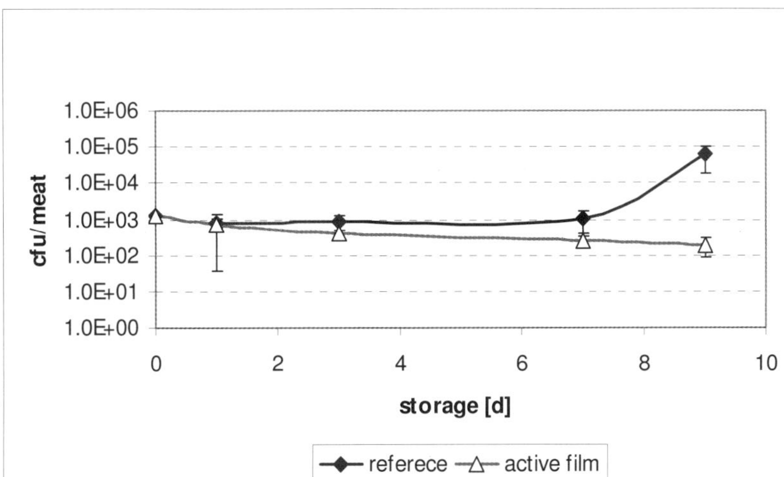

Fig. 2 *Development of E. coli on inoculated pork loin in a period of 9 days at 8 °C.*

4. Summary

It could be shown that the antimicrobial active packaging film, developed at Fraunhofer IVV, is effective against various microorganisms. It can prevent and even reduce the growing of pathogens such as *E.coli* on food surfaces. In this way it make a significant contribution to the quality and safety of packed food.

Acknowledgements This work was partially funded by the German Federal Ministry of Economics and Technology (BMWi) via the German Federation of Industrial Research Associations (AiF) under contract number AiF Nr. 14954N

References

[1] Brody AL, Strupinsky ER, Kline LR. Antimicrobial Packaging. In: *Active Packaging for Food Applications.* Boca Raton, FL: CRC Press; 2001:131.

[2] Commission Regulation No 450/2009/ EC of 29 May 2009. *Active and intelligent materials and articles intended to come into contact with food.* Official Journal of the European Union; 2009; L135:3-11.

[3] Hauser C, Rodler N. Antimikrobiell active Folien. *LVT Lebensmittelindustrie.* 2009;5-6:28-30.

[4] Sofos JN. *Sorbate food preservatives.* Boca Raton, FL: CRC Press; 1989:205.

[5] Walker R. Toxicology of sorbic acid and sorbates. *Food Additives and Contaminants.* 1990;7:671-676.

[6] European Parliament and Council Directive No 95/2/EC of 20 February 1995. *Food additives other than colours and sweeteners.* am. by Commission Directive 2010/69/EU of 22 October 2010. Official Journal of the European Union;2010;L279:22-31.
[7] Fraunhofer IVV. Internal results.
[8] Kure CF, Wasteson Y, Brendehaug J, Skaar I. Mould contaminants on Jarlsberg and Norvegia cheese blocks from four factories. *International Journal of Food Microbiology.* 2001;70:21-27.
[9] Hauser C, Langowski HC, Rieblinger K. Verbesserter Qualitätserhalt von Käseschnittflächen in optimierten transparenten Verpackungen. *Deutsche Milchwirtschaft.* 2010;5:157-160.
[10] Warriss PD. *Meat Science: An introductory text.* 2nd ed. Oxfordshire, UK: CABI; 2010:131.
[11] Gracey J, Collins DS, Huey R, eds. *Meat Hygiene.* 10th ed. London, UK: Harcourt Brace and Company; 1999:339.

Bacteriophages actions on *Salmonella* Enteritidis biofilm

A. A. Ferreira[1], R. C. S. Mendonça[1*], H. M. Hungaro[1], M. M. Carvalho[1] and J. A. M. Pereira[1]

[1] Bacteriophages group, Food Technology Department, Viçosa Federal University, P.H. Rolfs Avenue, ZIP CODE: 36570-000, University Campus, Viçosa, Minas Gerais, Brazil.

* Corresponding author: email: rmendonc@ufv.br

The aim of this study was to produce biofilms of *Salmonella* Enteritidis in stainless steel coupons and evaluate its reduction by the use of a pool of lytic bacteriophages. It was observed an increase in the number of cells in the biofilm according to the time of cultivation, in which the count of *Salmonella* spp. was 6.1 log $UFC \cdot cm^{-2}$ to 4 days, 6.9 log $UFC \cdot cm^{-2}$ with 8 days and 7.7 log $UFC \cdot cm^{-2}$ with 12 days. The largest reductions by the action of bacteriophages were observed in treatments with 10^7 $PFU \cdot mL^{-1}$ of bacteriophages, 35 min of contact with biofilm at 8 days of age. Bacteriophages that were used were able to reduce the *Salmonella* Enteritidis biofilm in coupons of stainless steel of different ages, and are an alternative to control this pathogen in the food industry.

Keywords: bacteriophages, biofilm, *Salmonella* spp.

1. Introduction

Infections caused by *Salmonella* spp. are responsible for the high incidence of diarrhea in humans and constitute a public health problem [1, 2]. The consumption of contaminated food of animal origin, particularly eggs, meat and dairy products are considered the main source of salmonellosis in humans [3]. Like other microorganisms, *Salmonella* sp. may be involved in processes of adhesion and formation of biofilm on surfaces of equipment and utensils in the food industry, which may compromise the microbiological quality of food and cause health risks to consumers [4, 5]. The adhesion of bacterial cells in food processing equipment may lead to a variety of problems such as corrosion of metal surfaces and cross-contamination of processed foods [6].

Due to various reasons, the use of bacteriophages is presented as a promising strategy for the control of biofilms, since the organisms growing in biofilms exhibit extreme tolerance to antimicrobial agents commonly used in food industry [7]. Bacteriophages offer an alternative because they are ubiquitous organisms incapable of infecting human cells and therefore have great potential as biocontrol agents in food [8].

Studies show the potential application of bacteriophages to control of *Escherichia coli* biofilm in PVC coupons [9], in the biofilms of *Pseudomonas aeruginosa*, *Acinetobacter johnsonii* and *Bacillus subtilis* on ultrafiltration membranes for wastewater treatment [10] and reduction of *Pseudomonas fluorescens* on stainless steel coupons [11].

The objective of this work was to produce biofilms of *Salmonella* Enteritidis in stainless steel coupons and evaluate its reduction through the use of a pool of lytic bacteriophages.

2. Material and Methods

This study was conducted at the Laboratory of Microbiology of Foodborne and Water Pathogens (LAMPOAH) of Viçosa Federal University, Minas Gerais, Brazil.

2.1. Microorganisms

In this study it was used a strain of *Salmonella enterica* subsp. *enterica* serovar Enteritidis (ATCC 13076). Five lytic bacteriophages were used and they were obtained from the database of microorganisms from LAMPOAH, isolated from feces of extensively farmed birds in different farms [12].

2.2. Surface evaluated

For the assessment of bacterial adhesion, stainless steel coupons 1 cm^2 (AISI 304) were used. The coupons were brushed and washed with detergent, immersed in distilled water and alcohol at 70 %, and then were attached to aluminum wires, wrapped in craft paper and sterilized in an autoclave at 121 °C for 15 minutes.

2.3. Formation of *Salmonella* Enteritidis biofilm on stainless steel coupons

The coupons were immersed in erlenmeyer flasks containing Tryptic Soy Broth (TSB - DifcoTM) diluted in water at 1/20 of original concentration, recommended for their common use, and adding to 10 % inoculum of *Salmonella* Enteritidis previously activated. The coupons were transferred to a new flask contend a diluted inoculated medium every two days of incubation in order to keep nutrient content to ensure the cells viability and increase biofilm formation. This proceed was repeat at complete 12 incubations days. The flask was keep at all times at 25 °C.

2.4. Phage infection and quantification of adhered cells

The coupons subjected to the formation of biofilm were removed from the cell suspension and washed in 0.85 % saline solution to remove non-adherent cells. The washed coupons were immersed in a mixture of five bacteriophages at concentrations of 10^5, 10^7 and 10^9 $PFU{\cdot}mL^{-1}$ for 10, 35 and 60 minutes. After these treatments, the coupons were washed again with 0.85 % saline solution for the removal of bacteriophages and lysed cells. Then they were immersed in test tubes containing 10 mL of 0.85 % saline and subjected to the action of ultrasound (Quimis Q-335D) for 30 minutes to remove the remaining adherent cells. The quantification of *Salmonella* Enteritidis was performed in Tryptone Soy Agar (TSA - Acumed ®) by the technique of micro-droplets [13]. In order to evaluate the effect of the treatments with bacteriophages, a control treatment was performed, submitting the coupons to the same conditions of treatment without the addition of bacteriophages.

2.5. Evaluation of the action of bacteriophages in biofilms of *Salmonella* Enteritidis by photomicrography

The biofilm formation and action of bacteriophages were examined by photomicrograph by electronic microscopy scanning. The preparation of coupons of observation was carried out using the methods described by [14]. The coupons were observed in a LEO 1430 VP Electroinc Scanning Microscope, with increases of 800 X to 4500 X.

2.6. Experimental design

It was used a Box-Behnken design with three factors at three levels of concentration of bacteriophages (10^5, 10^7 and 10^9 $PFU{\cdot}mL^{-1}$), contact times (10, 35 and 60 min) and biofilm age (4, 8, 12 days) with three replications at the central point.

The statistical analysis were performed in the MINITAB® Statistical Software program version 13, by means of methodology in response surface.

3. Results and Discussion

3.1. *Salmonella* biofilm formation on stainless steel coupons

It was observed an increase in the number of adhered cells in the control treatment over the incubation time at 25 °C. The average count of adhered cells of *Salmonella* Enteritidis was 6.1 ± 0.05 log $UFC{\cdot}cm^{-2}$, 6.9 ± 0.10 log $UFC{\cdot}cm^{-2}$ and 7.7 ± 0.08 log $UFC{\cdot}cm^{-2}$ after 4, 8 and 12 days of incubation, respectively. The results indicate that the incubation period was sufficient for the cells of Salmonella Enteritidis to be adhered to the surface of stainless steel coupons, and the range of 8 days of incubation from the initial assessment led to an increase of 1.6 log UFC cm^{-2}. Similar results were obtained by [15], who observed the count of 3 x 10^5 $UFC{\cdot}cm^{-2}$ in the evaluation of biofilm formation by *Salmonella* Weltevreden in stainless steel coupons. In another study, Hurrell et al. (2009) observed the formation of biofilms of different strains of *Salmonella* in PVC, with an average variation from 5.3 to 6.6 Log $UFC{\cdot}cm^{-2}$, and in polyurethane, ranging from 6.0 to 6.9 log $UFC{\cdot}cm^{-2}$.

The biofilm formation of *Salmonella* in this study, despite of having been held in culture medium, simulated actual conditions of the food industry, such as temperature, presence of nutrients, contact time, inadequate sanitary conditions, which influence the cell adhesion and establishment of the biofilm.

The cell concentrations in biofilm formation observed in this study are sufficient to cause damage in relation to contamination by spoilage microorganisms and pathogens, corrosion of surfaces, decrease in the heat transfer and mass and resistance to chemical and physical agents [4, 15, 17,18].

3.2. Effect of bacteriophage on biofilms

The effect of the suspension of bacteriophages in reducing the number of cells in biofilm of *Salmonella* Enteritidis on stainless steel coupons following the model Y = - 0.59 to 0.15 A + 0.57 B - 0.02 C -0.03 B2 + 0,

0035 BC, where A: concentration of bacteriophages, B: time of the biofilm and C: contact time. This model explained 83 % of the variations in the results of reduction in the count of Salmonella on surfaces of stainless steel coupons and showed acceptable fit (pf_{aj} = 0.33). The largest reductions of Salmonella of nearly 1 log UFC·cm^{-2} were observed in treatments with 107 PFU·mL^{-1} of bacteriophages, 35 min contact with biofilm at 8 days of age (Figure 1).

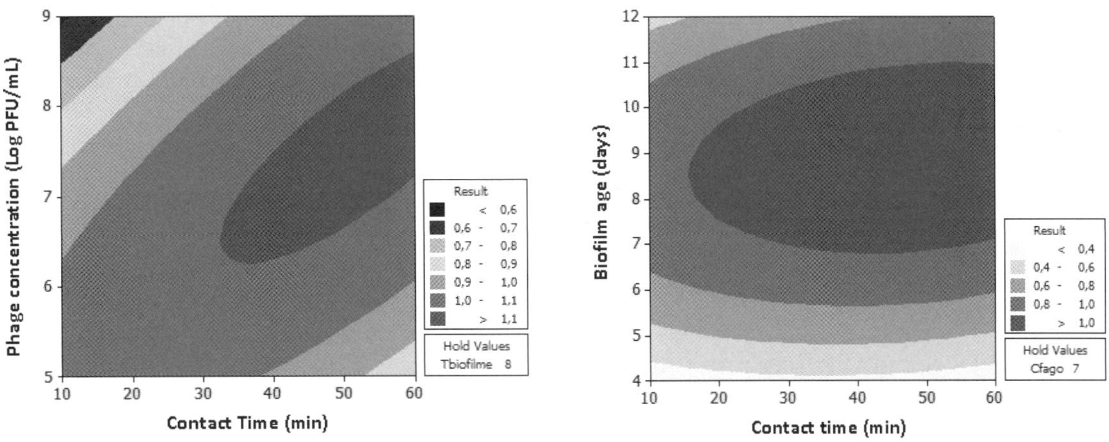

(A) **(B)**

Figure 1. Reduction of the number of adhered cells of *Salmonella* Enteritidis (log UFC·mL^{-1}) in relation to the treatment conditions that were used. A - Influence of contact time and concentration of bacteriophage in the reduction *Salmonella* counts. B - Influence of biofilm age and concentration of bacteriophage in the reduction Salmonella counts.

In the microscopic evaluation was observed a reduction of adherent cells treated with bacteriophage at a concentration of 10^9 and 107 PFU·mL^{-1} and little reduction of treatment with 10^5 PFU·mL^{-1} compared with the control treatment (Figure 2).

(A) (B) (C) (D)

Figure 2. Visualization of biofilms of *Salmonella* spp. on coupons of stainless steel submitted a bacteriophages treatment (A (control – original biofilm), B (after contact with 10^5 PFU mL^{-1} for 35 min), C (after contact with 10^7 PFU·mL^{-1} for 35 min) and D (after contact with 10^9 PFU·mL^{-1} for 35 min.)).

Results than those found in this study, reduction of microorganisms in biofilms were obtained by other authors, possibly due to the increased contact time used in these studies. Sillankorva et al. (2008) observed reductions of 3-5 log cycles in biofilms of *Pseudomonas fluorecens* on coupons of stainless steel by using bacteriophage suspension in the concentration of 10^7 PFU·mL^{-1}, for 4 hours of exposure. Tail et al. (2002) obtained the total eradication of biofilms of *Enterobacter cloacae* by exposing the biofilms for 24 hours to a solution containing three different types of bacteriophages.

The infection of cells in the biofilm by bacteriophages is highly dependent on the chemical composition and environmental factors such as temperature, phase and growth medium and concentration of bacteriophages [21, 22, 23]. The integrity of the biofilm can be quickly destroyed when the phage has the exopolysaccharide-degrading enzymes, or when there is considerable cell lysis by the action of bacteriophage [23, 24].

The treatment with bacteriophages shows improved efficiency in newly formed biofilms, whereas older biofilms exhibit cellular debris and dead cells that can immobilize the bacteriophages treatments [19]. Possibly, this could be the explanation for the best effect of bacteriophages on biofilm of 8 days of age compared with the biofilm of 12 days of age obtained in this study.

The bacteriophages have the capacity to act in different layers of the biofilm, including the basal layer, by being able to circulate through the channels and pores that were formed in the biofilm [25], this factor can influence the release of cells into the medium and planktonic cells were lysed by the free bacteriophages.

Other authors showed that the pre-treatment of surfaces with bacteriophages might reduce the adhesion of bacteria and the formation of biofilm, due to the fact that bacteriophages can adhere to contact surfaces and prevent cell adhesion [26].

The use of bacteriophages may be an alternative in the elimination and prevention of biofilm formation in contact surfaces. Recent reviews have discussed the use of bacteriophage in the control and elimination of bacteria in various areas [27, 28] and examples of applications have been published, especially compared to the control of *Salmonella* sp. [29, 30, 31, 32, 33, 34], *Campylobacter* [30, 34, 35], *Listeria monocytogenes* [37, 38, 39], *Enterobacter sakazakii* [40] and *Staphylococcus aureus* [8].

Despite the low reduction of *Salmonella* cells in the biofilm obtained in the present study, the pool of bacteriophages that was used showed potential to be applied in the controll of this pathogen in the food industry, including the formation of biofilms.

References

[1] FERNANDES, S.A.; TAVECHIO, A.T.; GHILARDI, A.C.R.; DIAS, A.M.G.; ALMEIDA, I.A.Z.C.; MELO, L.C.V. *Salmonella* sorovars isolated from humans in São Paulo State, Brasil, 1996-2003. *Revista do Instituto de Medicina Tropical de São Paulo*, v. 48, p. 179-184, 2006.

[2] ZAIDI, M.B.; MCDERMOTH, P.F.; CRAY-FERDOGA, P.; LEON, V.; CANCHE, C.; HUBERT, S.K.; ABBOTT, J.; LEON, M.; ZHAO, S.; HEADRICK, M.; TOLLEFSON, L. Nontyphoidal *Salmonella* from human clinical cases asymptomatic children and raw retail meats in Yucatan, Mexico. *Clinical Infectious Disease*, v. 42, p. 21-28, 2006.

[3] CARVALHO, A.C.B.; CORTEZ, A.L.L. *Salmonella* spp. em carcaças, carne mecanicamente separada, lingüiça e cortes comerciais de frango. *Ciência Rural,* v. 35, n. 6, p. 1-5, 2005.

[4] POULSEN, L. V. Microbial biofilm in food processing. *Lebensmittel-Wissenschaft und-Technologie*, v.32, p.321-326,1999.

[5] ANDRADE, N.J. Higiene na indústria de alimentos: avaliação e controle da adesão e formação de biofilmes bacterianos. São Paulo: Varela, 2008, 412p.

[6] POMPERMAYER, D. M. C.; GAYLARDE, C. C. The influence of temperature on the adhesion of mixed cultures of *Staphylococcus aureus* and *Escherichia coli* to polypropylene. *Food Microbiology*, v.17, p.361-365, 2000.

[7] DONLAN, R. M. Controlling clinically relevant biofilms using bacteriophages. *Biofilm Perspectives* n.2, April 2006.

[8] GARCÍA, P.; MADERA, C.; MARTÍNEZ, B.; RODRÍGUEZ, A. Biocontrol of *Staphylococcus aureus* in curd manufacturing processes using bacteriophages. *International Dairy Journal*, v.17, p.1232-1239, 2007.

[9] Doolittle, M.M.; Cooney, J.J.; Caldwell, D.E.*Canadian Journal of Microbiology*, v.41, p.12-18, 1995.

[10] GOLDMAN, G.; STAROSVETSKY, J.; ARMON, R. Inhibition of biofilm formation on UF membrane by use of specific bacteriophages. *Journal of Membrane Science*, v.342, p.145-152, 2009.

[11] SILLANKORVA, S.; NEUBAUER, P.; AZEREDO, J. *Pseudomonas fluorescens* biofilms subjected to phage phiIBB-PF7A. *BMC Biotechnology,* v.8, p. 79, 2008.

[12] HÚNGARO, H. M. *Biocontrole de Salmonella enterica em pele de frango de corte.* Dissertação de mestrado em Ciência e Tecnologia de Alimentos, UFV, Viçosa, MG, 2010. 89p.

[13] SILVA, N.; JUNQUEIRA, V. C. A.; SILVEIRA, N. F. A. *Manual de métodos de análise microbiológica de alimentos.* São Paulo: Varela, 2010.

[14] SILVEIRA, M. Preparo de amostras biológicas para microscopia eletrônica de varredura. In: SOUZA, W.; HADDAD, A.; SESSO, A. (Eds). *Manual sobre técnicas básicas em microscopia eletrônica.* Sociedade Brasileira de Microscopia Eletrônica, v.1, p. 71-79, 1989.

[15] JOSEPH, B.; OTTA, S.K.; KARUNASAGAR, INDRANI; KARUNASAGAR, I. Biofilm formation by *Salmonella* spp. on food contact surfaces and their sensitivity to sanitizers. *International Journal of Food Microbiology*, v.64, p. 367–372, 2001.

[16] HURRELL, E.; KUCEROVA, E.; LOUGHLIN, M.; CAUBILLA-BARRON, J.; FORSYTHE, S.J. Biofilm formation on enteral feeding tubes by *Cronobacter sakazakii*, *Salmonella* serovars and other Enterobacteriaceae. *International Journal of Food Microbiology*, v. 136, p.227–231, 2009.

[17] GIBSON, H.; TAYLOR, J.H.; HALL, K.E.; HOLAH, J.T.; Effectiveness of cleaning techniques used in the food industry in terms of the removal of bacterial biofilms. *J. Appl. Microbiol.*, v.87, p.41–48, 1999.

[18] STEPANOVIC, S.; CIRKOVIC´, I.; RANIN, L.; SVABIC´-VLAHOVIC, M. Biofilm formation by *Salmonella* spp. and *Listeria monocytogenes* on plastic surface. *Letters in Applied Microbiology,* v. 38, p.428–432, 2004.

[19] SILLANKORVA, S.; NEUBAUER, P.; AZEREDO, J. *Pseudomonas fluorescens* biofilms subjected to phage phiIBB-PF7A. *BMC Biotechnology*, v.8, p. 79, 2008.

[20] TAIT, K.; SKILLMAN, L.C.; SUTHERLAND, I.W. The efficacy of bacteriophage as a method of biofilm eradication. *Biofouling*, v.18, p.305-311, 2002.

[21] SILLANKORVA, S. M. *Utilização de Bacteriófagos no Controle de Células Suspensas e Biofilmes de Pseudomonas fluorecens.* Dissertação de mestrado em Tecnologia do Ambiente, Universidade do Minho. Escola de Engenharia, Departamento de Engenharia Biológica. Guimarães, Portugal, 2004. 111p.

[22] CHAIGNON, P., SADOVSKAYA, I., RAGUNAH, CH., RAMASUBBU, N., KAPLAN, J. B.; JABBOURI, S. Susceptibility of *Staphylococcal* biofilms to enzymatic treatments depends on their chemical composition. *Applied Microbiology and Biotechnology*, v.75, p.125–132, 2007.

[23] SIMÕES, M.; SIMÕES, L. C.; VIEIRA, M. J. A review of current and emergent biofilm control strategies. *LWT - Food Science and Technology*, v.43, p.573–583, 2010.

[24] LU, T.K.; COLLINS, J.J. Dispersing biofilms with engineered enzymatic bacteriophage, *Proceedings of the National Academy of Sciences U.S.A*, v.104, p.11197-11202, 2007.

[25] LAWRENCE, J.R; KORBER, D.R.; HOYLE, B.D.; COSTERTON, J.W.; CALDWELL, D.E. Optical sectioning of microbial biofilms. *Journal of Bacteriology*, v.173, p.6558-6567, 1991.

[26] CURTIN, J.J.; DONLAN, R.M. Using bacteriophages to reduce formation of catheter-associated biofilms of *Staphylococcus epidermidis*. *Antimicrob Agents Chemother*, v.50, p.1268-1275, 2006.

[27] HAGENS, S.; LOESSNER, M.J. Application of bacteriophages for detection and control of foodborne pathogens. *Applied Microbiology and Biotechnology*, v. 76, n. 3, p. 513-519, 2007.

[28] GARCÍA, P.; MARTÍNEZ, B.; OBESO, J.M.; RODRÍGUEZ, A. Bacteriophages and their application in food safety. *Letters in Applied Microbiology*, v. 47, p. 479–485, 2008.

[29] LEVERENTZ, B.; CONWAY, W.S.; ALAVIDZE, Z.; JANISIEWICZ, W.J.; FUCHS, Y.; CAMP, M.J.; CHIGHLADZE, E.; SULAKVELIDZE, A. Examination of bacteriophage as a biocontrol method for *Salmonella* on fresh-cut fruit: a model study. *Journal of Food Protection*, v. 64, p. 1116–1121, 2001.

[30] GOODE, D.; ALLEN, V.M.; BARROW, P.A. Reduction of experimental *Salmonella* and *Campylobacter* contamination of chicken skin by application of lytic bacteriophages. *Applied and Environmental Microbiology*, v. 69, p. 5032–5036, 2003.

[31] WHICHARD, J.M.; SRIRANGANATHAN, N.; PIERSON, F.W. Suppression of *Salmonella* growth by wild-type and large plaque variants of bacteriophage Felix O1 in liquid culture and on chicken frankfurters. Journal of Food Protection, v. 66, p. 220–225, 2003.

[32] FIORENTIN, L.; VIEIRA, N.D.; BARIONI, W.JR; BARROS, S. *In vitro* characterization and in vivo properties of *Salmonellae* lytic bacteriophages isolated from free-range layers. *Brazilian Journal of Poultry Science*, v. 6, n. 2, p. 121 - 128, 2004.

[33] BIELKE, L.R.; HIGGINS, S.E.; DONOGHUE, A.M.; DONOGHUE, D.J.; HARGIS, B.M.; TELLEZ, G. Use of wide-host-range bacteriophages to reduce *Salmonella* on poultry products. *International Journal of Poultry Science*, v. 6, n. 10, p. 754-757, 2007.

[34] BIGWOOD, T.; HUDSON, J.A.; BILLINGTON, C.; CAREY-SMITH, G.V.; HEINEMANN, J.A. Phage inactivation of foodborne pathogens on cooked and raw meat. *Food Microbiology*, v. 25, p. 400–406, 2008.

[35] ATTERBURY, R.J.; CONNERTON, P.L.; DODD, C.E.R.; REES, C.E.D.; CONNERTON, I.F. Application of host-specific bacteriophages to the surface of chicken skin leads to a reduction in recovery of *Campylobacter jejuni*. *Applied and Environmental Microbiology*, v. 69, n. 8, p. 6302–6306, 2003.

[36] O'FLYNN, G.; ROSS, R.P.; FITZGERALD, G.F.; COFFEY, A. Evaluation of a cocktail of three bacteriophages for biocontrol of *Escherichia coli* O157:H7. *Applied and Environmental Microbiology*, v. 70, p. 3417–3421, 2004.

[37] DYKES, G.A.; MOORHEAD, S.M. Combined antimicrobial effect of nisin and a listeriophage against *Listeria monocytogenes* in broth but not in buffer or on raw beef. *International Journal of Food Microbiology*, v. 73, p. 71–81, 2002.

[38] LEVERENTZ, B.; CONWAY, W.S.; CAMP, M.J.; JANISIEWICZ, W.J.; ABULADZE, T.; YANG, M.; SAFTNER, R.; SULAKVELIDZE, A. Biocontrol of *Listeria monocytogenes* on fresh-cut produce by treatment with lytic bacteriophages and a bacteriocin. *Applied and Environmental Microbiology*, v. 69, p. 4519–4526, 2003.

[39] CARLTON, R.M.; NOORDMAN , W.H.; BISWAS, B.; DE MEESTER, E.D.; LOESSNER, M.J. Bacteriophage P100 for control of *Listeria monocytogenes* in foods: genome sequence, bioinformatic analyses, oral toxicity study, and application. *Regulatory Toxicology and Pharmacology*, v. 43, p. 301–312, 2005.

[40] KIM, K.P.; KLUMPP, J.; LOESSNER, M.J. *Enterobacter sakazakii* bacteriophages can prevent bacterial growth in reconstituted infant formula. *International Journal of Food Microbiology*, v. 115, p. 195–203, 2007.

Biocompatibility and Antibacterial property of Cold Sprayed ZnO/Titanium Composite Coating

Noppakun Sanpo[*1], **Chen Hailan**[2], **Kelvin Loke**[3], **Koh Pak Keng**[4], **Philip Cheang**[5], **C.C.Berndt**[6] **and K.A. Khor**[7]

[1,6] Industrial Research Institute Swinburne, Swinburne University of Technology, Melbourne, Australia
[2] School of Chemical and Biomedical Engineering, Nanyang Technological University, Singapore
[3] Singapore Technologies Kinetics Ltd, Singapore
[4,5] School of Science & Technology, SIM University, Singapore
[7] School of Mechanical and Aerospace Engineering, Nanyang Technological University, Singapore

*Corresponding author: e-mail: Nsanpo@groupwise.swin.edu.au, Phone: +61 466654558

The deposition using cold gas dynamic spraying of Zinc oxide/Titanium (ZnO/Ti) composite have been fabricated on Al 6061 substrate, their biocompatibility and antibacterial properties were studied. Osteoblast cell was used for the biocompatibility test while Escherichia coli (DH5α) was employed in antibacterial study. The ratios of ZnO to Ti in their composite powders were 20:80, 50:50 and 80:20 (weight %). The ZnO/Ti coatings were successfully deposited using cold spraying parameters of 13-15 Bars under helium gas, temperature of between 300-400°C. The results obtained showed that the ZnO/Ti coatings exhibited significant antibacterial effects against E. coli. Moreover, biocompatibility results reported that cell viability decreased with increasing of ZnO concentration in coating composition.

Keywords; Cold spray, Zinc oxide, Titanium, Biocompatibility, Antibacterial property

1. Introduction

Cold Spray is a relatively new coating technology that produces deposits by high velocity impact of feed stock powders at low temperature onto the substrate. The particle and substrate temperatures are low during Cold Spray processes so that oxidation and phase transformation of the sprayed material and substrate are negligible. The solid particle and substrate need to be plastically deformed enough to dissipate the particle kinetic energy on the impact of particle on the substrate [1]. This implies that the particle velocity has to exceed a critical value to plastically deform sufficiently and obtain good bonding strength between the particle and substrate. Only the particles that attained a velocity higher than the critical velocity can be deposited to produce the coating [2].

In Cold Spray, the particles involved are typically micro-structured powder particles (approximately 15-50 μm). The preheating and the de Laval nozzle allow for a velocity of the powder above its critical velocity and aids in the formation of the coating. Most metals such as Cu, Al, Ni, Ti and Ni-based alloys can be deposited by cold spray [3], and even cermets and ceramics can be embedded into a substrate to form thin layered coatings. A potential advantage of this process over thermal spray is the ability to generate dense coatings while retaining their feedstock material chemistry and phase composition.

Recently, the antibacterial activity of ceramic powders has been pointed out and much attention has been directed towards them as a new substitute for conventional organic agents, because organic agents can contain noxious materials harmful to humans. Ceramic powders of zinc oxide (ZnO), calcium oxide (CaO) and magnesium oxide (MgO) have been found to show a marked antibacterial activity without the presence of light. Moreover, the use of these ceramics has the advantage of containing mineral elements that are essential to the human body as well as a strong antibacterial activity, even when present in a small amount without the irradiation of light. In previous works, Al-ZnO coating were earlier deposited on glass at room temperature using cold spraying parameters [4]. The results indicated that the antibacterial activity increased with increasing ZnO concentration in the composite powder feedstock and cold sprayed coating.

Medical implants are important to modern medicine and play an important role in the replacement and repair of human parts, ranging from organs to limbs. Metallic implants made from titanium and its alloys, stainless steel, cobalt-based alloys and so on, still dominate the implant market on account of their high mechanical properties and ease of machining [5].

It is well known that the success and long-term durability of these implants are impacted by the presence of bacteria in the vicinity of the implants. Bacterial infection after implant placement can cause significant complications thereby increasing medical costs, morbidity, and patient dissatisfaction [6].

The ideal medical implant should possess both antibacterial function and excellent cell biocompatibility. Therefore, means to mitigate bacterial infection and enhance biocompatibility on coating surfaces have aroused interest.

In this study, we demonstrate a novel approach to fabricate biocompatible composite coatings consisting of ZnO and titanium by cold spray technology. The aim of this study, the deposition characteristics of cold sprayed ZnO/Ti coating was examined using microstructural analysis of as-sprayed coatings comprising various ratios of ZnO/Ti composite powders. The biocompatibility of ZnO/Ti coating was observed by cell culture test. Moreover, the biofunctionality of ZnO/Ti coating was also evaluated using bacteria assay to ascertain the killing rate against E coli.

2. Experimental Procedure

2.1 Synthesis of ZnO powder and Cold Spray ZnO/Ti Powder Preparation

Zinc oxide nanopowder was synthesized from zinc acetate. This was added into sodium hydroxide solution for about an hour. The precipitate was washed with Deionised water, followed by ethanol, before drying in a 37 °C oven. Zinc oxide nanopowder was analyzed with FESEM and EDX for compositional analysis as shown in Fig. 1a. The commercially available titanium powder (-45 mesh) was used as an initial material as shown in Fig. 1b. Both powders that were used to generate the coatings were sieved to less than 45 micron before blending in a ball-mill for about 24 hours. The ZnO and Ti powder were mixed in a ratios of 20:80, 50:50 and 80:20 (wt.%).

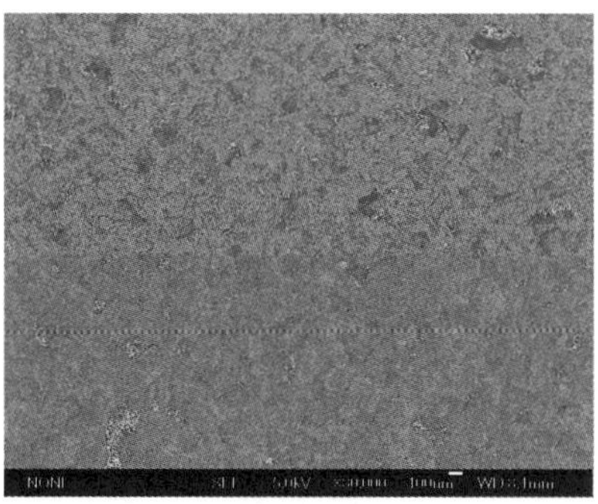
a)

b)

Fig. 1 SEM micrographs of a) ZnO powder and b) Ti powder

2.2 Cold Spray Parameters ZnO/Ti Powders

The cold spray system was obtained from the Russian Academy of Science, Siberia and co-developed with ST Kinetics (Singapore). An ABB Robot (IRB2400) was used in conjunction with the cold spray system. Al 6061 will be cleaned with acetone and dried with pure air before using as coating substrate. Helium was used as the accelerating and carrier gas operating at pressures of 13 to 15 Bars in the pre-chamber. The gas was preheated to a temperature of 300 to 400°C. The standoff distance from the gun exit was 15 mm. The traverse speed of the gun relative to the substrate was 60 mm/s. The microstructure of the ZnO/Ti coating was analyzed using a Field Emission scanning electron microscope (JEOL JSM-6700F).

2.3 Tests of Antibacterial Activity

The antimicrobial experiment was modified from Standard Test Method for Determining the Activity of Incorporated Antimicrobial Agent(s) in Polymeric or Hydrophobic Materials (ASTM E 2180), which is commonly used in School of Biological Science (SBS), Nanyang Technological University Singapore, to estimate the antimicrobial abilities of antimicrobial samples.

2.3.1 Bacterial Qualitative Test

To perform qualitative analysis of the anti-bacterial properties of the ZnO/Ti coating, LB agar solution was poured onto Petri dishes after the cold sprayed sample was placed perpendicular to the base of the Petri dish. Only half the sample was covered by the LB agar. The E. coli was introduced onto the LB agar surface by using a micro-pipette to spread the bacteria over the agar surface. The agar plates were placed upside down in a 37 °C incubator for 24 hrs before further analysis.

2.3.2 Bacterial Quantitative Test

The E. coli stock was stored inside micro-centrifuge tubes at -80 ^{o}C. Using a sterile wire loop, a single colony was extracted and streaked on to another agar plate. The plates were further incubated upside-down at 37 ^{o}C for another 24 hrs. Using a sterile wire loop, an isolated colony was placed into a test tube containing 10 ml of broth, and vortexed for 60s. 100µl of solution was plated on to agar dishes using a spreader. The plates were again incubated at 37 ^{o}C (upside down) for 24 hrs. Choose and count the number of colonies in plates containing 30 to 300 colonies per plate. One colony represents a colony-forming unit (CFU).

Determine the geometric mean of the number of E. Coli from the triplicate incubation period control and incubation period treated samples by the following equation:

$$\text{Geometric mean} = (\text{Log}_{10}X1 + \text{Log}_{10}X2 + \text{Log}_{10}X3) / 3$$

where: X = number of organisms recovered from the incubation period control (substrate without coating) or incubation period treated samples (substrate with coating)

Percent E. Coli Reduction—Use the following equation to calculate the percent reduction:

$$\%\ \text{E. Coli reduction} = [(a-b) \times 100] / a$$

where: *a and b* = the antilog of the geometric mean of the number of E. Coli from the incubation period control and treated samples repectively.

2.4 Biocompatibility Test

Human osteoblast cell lines MG63 was used to assess the biocompatibility of the titanium and zinc oxide coating. Osteoblast cells were grown in a complete culture medium consisting of Dulbecco's Modified Eagle Medium: Nutrient Mixture F-12 (DMEM/F12), supplemented with 2 mM L-glutamine, 100 IU/ml penicillin, 100 µg/ml streptomycin, and 10% heat-inactivated fetal bovine serum (PAA). All cells were maintained at 37°C in a saturated-humidity atmosphere of 95% air and 5% CO_2.

2.4.1 Cell Viability Assay

Osteoblast cells were seeded on the sterilized coating surface at a density of 5000 cells/ml and cultured for 24 hours. 3-(4,5-dimethylthiazole-2-yl)-2,5-diphenyl tetrazolium bromide (MTT; Sigma) cell proliferation assays was performed to asscess the viability of cells. At least three samples for each condition were used. Cells grow in the well without adding anything was used as negative control and 1mg/ml Triton X-100 was used as positive control in all biocompatibility tests.

3. Results and Discussion

3.1 Microstructure of ZnO/Ti Cold Sprayed Coating

SEM pictures of ZnO/Ti coating surface are shown in Fig. 2. These images presented that the surface of cold sprayed coating consisted of ZnO powder embedded in a continuous Ti matrix. The coating thicknesses of all samples were in the same range of about 60-80 micron.

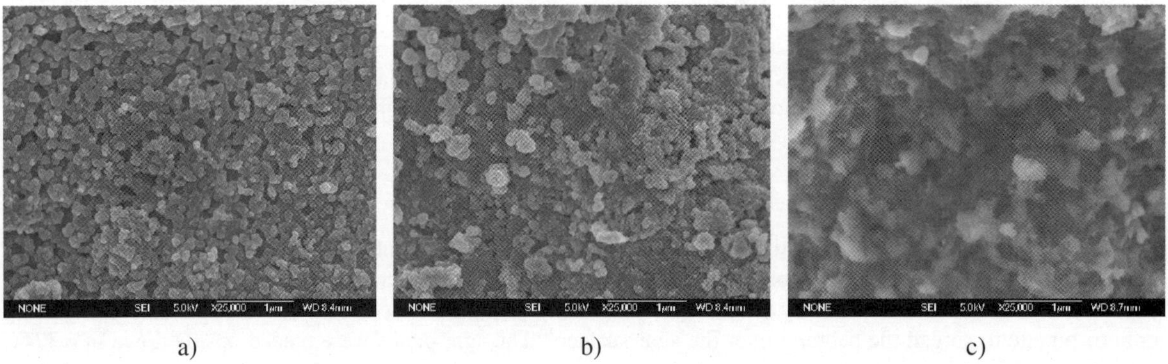

Fig. 2 SEM micrographs of ZnO /Ti coating with different composition: (a) ZnO 20/ Ti 80, (b) ZnO 50/ Ti 50 and (c) ZnO 80/ Ti 20

According to the EDX results of all ZnO/Ti coatings, it can be stated that the composition of the ZnO of the cold sprayed coating was slightly higher than that of the starting feedstock powder while the amount of Ti was slightly lower than starting feedstock powder as shown in Table 1.

Table 1 The average of ZnO/Ti contents (wt. %) in the powder feedstock and coatings

Powder composition	Powder Feedstock		Coating	
	ZnO	Ti	ZnO	Ti
ZnO 20/Ti 80	20	80	9.45	90.55
ZnO 50/Ti 50	50	50	33.41	66.59
ZnO 80/Ti 20	80	20	53.78	46.22

3.2 Antibacterial Results of ZnO/Ti Cold Sprayed Coating

Figure 3 shows that all cold sprayed samples have a very distinctive killing effect on E. coli. A lack of E.Coli colonies growing in the region surrounding the ZnO/Ti coated samples was observed and appeared as clearance zone. The area of the clearance zone increased with increasing ZnO powder concentration in the composite powder feedstock and cold sprayed coating, showing that the killing effect on E.Coli increased with increasing amount of ZnO.

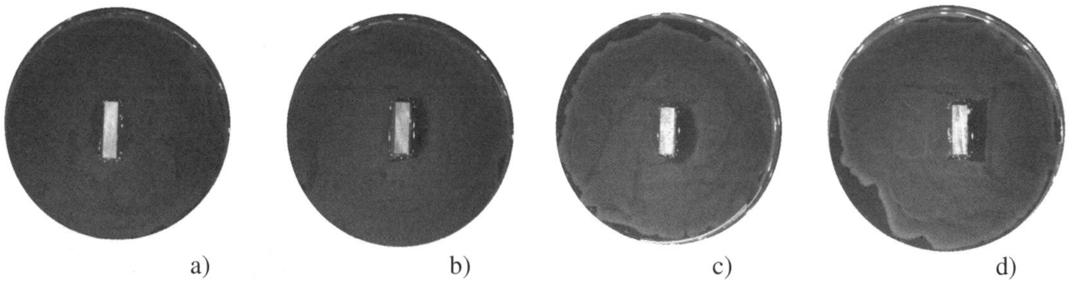

Fig. 3 Qualitative analysis on the anti-bacterial properties of ZnO/Ti coatings. A clearance zone (absence of E.Coli colonies) surrounding the coated samples was observed on samples (b) ZnO 20/ Ti 80, (c) ZnO 50/ Ti 50 and (d) ZnO 80/ Ti 20 in an increasing manner compared with Al 6061 without coating (a)

Figure 4 shows that all cold sprayed samples displayed an even more pronounced killing effect on E. coli. The killing rate increased with increasing ZnO powder concentration in the composite powder feedstock and cold sprayed coating.

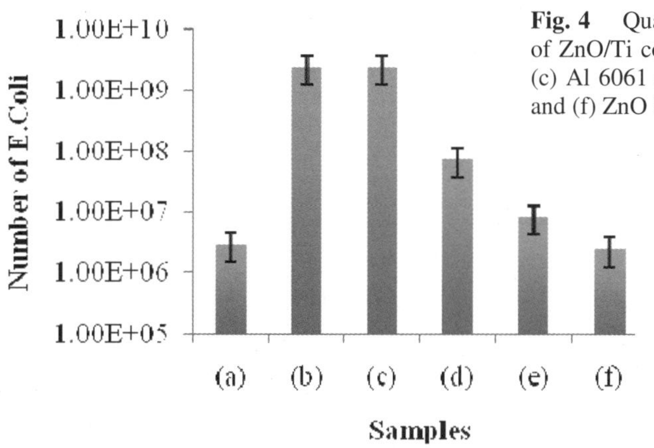

Fig. 4 Quantitative analysis on the anti-bacterial properties of ZnO/Ti coatings : (a) E. coli at 0 hr, (b) E. coli at 24 hrs, (c) Al 6061 substrate, (d) ZnO 20/ Ti 80, (e) ZnO 50/ Ti 50 and (f) ZnO 80/ Ti 20

Table 2 presents the percent reduction of E. Coli of ZnO 20/ Ti 80, ZnO 50/ Ti 50 and ZnO 80/ Ti 20 coatings which are 13.34%, 25.38% and 32.06%, respectively. It can be clearly seen that the percent reduction of E. Coli decreased with increasing ZnO composition on coating samples.

Table 2 The calculation of percent reduction of E. Coli for ZnO/Ti coatings

Samples	Calculation	
	Geometric Mean (X)	Percent Reduction of E. Coli (%)
Control sample (without coating)	9.132	-
ZnO 20/Ti 80	7.914	13.34
ZnO 50/Ti 50	6.814	25.38
ZnO 80/Ti 20	6.204	32.06

3.3 Biocompatibility Results of ZnO/Ti Cold Sprayed Coating

After culturing osteoblast cells on the samples for 24 hours, MTT assay was used to assess the viability of cells. The negative control was normalized as 100% while other conditions were presented as percent scale relative to negative control. From Fig.5, cells with substrate without coating (c) showed the viability at the same level as ZnO20/Ti80 (d) which resulted about 80% of negative control. The viability of cells on ZnO20/Ti80 was higher than those of ZnO50/Ti50 and ZnO80/Ti20 samples, thus it was proven that the cell viability decreased with increasing of ZnO concentration in coating composition.

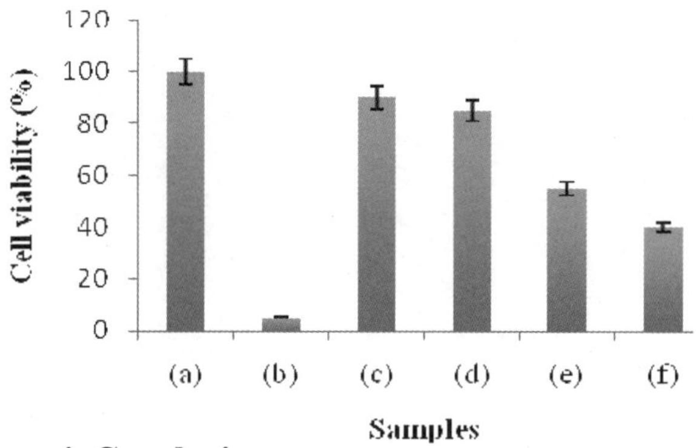

Fig. 5 Cell viability of osteoblast exposure to ZnO/Ti coating:
(a) Negative control, (b) Positive control, (c) Al 6061 substrate, (d) ZnO20/Ti80, (e) ZnO50/Ti50 and (f) ZnO80/Ti20.

4. Conclusions

ZnO/Ti coatings were successfully deposited on Al 6061 substrate at room temperature using cold spraying parameters of 13-15 Bars at pre-heated helium gas. EDX analysis verified composition of the Ti of the cold sprayed coating was slightly higher than that of the starting feedstock powder while the amount of ZnO on coating was slightly lower than starting feedstock powder. For the biocompatibility results, it was clearly seen that the cell toxicity of coating samples increased with increasing ZnO/Ti powder concentration in the composite powder feedstock and cold sprayed coating. These as-sprayed coatings of varying ZnO/Ti composition retained their inherent antibacterial property as clearly verified from bacteria assay. The results indicated that the antibacterial activity increased with increasing ZnO/Ti powder concentration in the composite powder feedstock and cold sprayed coating.

Acknowledgements We wish to acknowledge the funding support for this project (RG1/04) from Nanyang Technological University, and the joint research collaboration with ST Kinetics (Singapore).

References

[1] Hirotaka Fukanuma, Naoyuki Ohno, Bo Sun and Renzhong Huang. In - flight particle velocity measurements with DPV - 2000 in cold spray. *Surface and Coatings Technology*. 2006;201:1935-1941.

[2] Hyung-Jun Kim, Chang-Hee Lee and Soon-Young Hwang. Fabrication of WC-Co coatings by cold spray deposition. *Surface and Coatings Technology*. 2005;191:335-340.

[3] R.C McCune, W.T. Donlon, E.L. Cartwright, A.N. Papyrin, E.F. Rybicki, J.R. Shadley. *Thermal Spray: Practical Solutions for Engineering Problems. ASM International*. Materials Park; OH;1996:397-403.

[4] Noppakun Sanpo, Saraswati, Tan Meng Lu, Philip Cheang. Anti-Bacterial Property of Cold Sprayed ZnO-Al Coating. *BMEI*. 2008;1:488-491.

[5] Y.Z. Wan, G.Y. Xiong, H. Liang, S. Raman, F. He, Y. Huang. Modification of medical metals by ion implantation of copper. *Applied Surface Science*. Elsevier Science, Amsterdam; 2007;253: 9426-9429.

[6] Mahan J, Seligson D, Henry S L, Hynes P, Dobbins J. Factors in pin tract infections. *Orthopedics*;1991;14: 305-308.

Effect of penicillins on the acidification of yogurt made from ewe's milk during the storage

M.I. Berruga[1*] , M.C. Beltrán[2], B. Novés[3], A. Molina[1] and M.P. Molina[2]

[1]Departamento de Ciencia y Tecnología Agroforestal y Genética, Sección de Calidad Alimentaria, IDR, ETSIA; Universidad de Castilla-La Mancha, 02071 Albacete, Spain
[2]Instituto de Ciencia Animal, Universidad Politécnica de Valencia, Camino de Vera, 14, 46071 Valencia, Spain
[3]Instituto Técnico Agronómico Provincial, Diputación de Albacete, Albacete, Spain

*Corresponding author: mariaisabel.berruga@uclm.es, Phone: +34 967599363

The effects of three penicillins (penicillin G, ampicillin and amoxicillin), at concentrations close to the Maximum Residue Limits (MLRs) regulated by the European Union as safe for consumers, have been studied in yoghurts made from ewe's milk during the storage in order to know if these concentrations were also apt for the correct development of the fermentation processes. Milk for yoghurts was fortified with three different concentrations: 2, 4 (MRL) and 6 μg/kg of each antibiotic and the acidification and microbial evolution were evaluated until 21 days of refrigeration storage. Results showed that concentrations of some of the penicillins, equal or inferior to MRL, could inhibit the normal growth of lactic acid bacteria and provoke some alterations in the acidity parameters by causing damage on the production of the isomer L(+) lactic acid, wich is easier to assimilate by consumers.

Keywords yogurt; ewe milk; penicillins; acidification; microbial growth

1. Introduction

The protection of consumers against the antibiotic contamination of milk or dairy products has been regulated by the European Union by means of the Maximum Residue Limits (MRLs) which guarantee food safety at a high level [1]. For controlling the presence of these inhibitors in cow milk a vast supply of screening methods have been commercialized during the last few years [2]. The availability of these tests for ewe milk analysis is not as abundant as for cow milk, and only several methods based on the bacterial growth inhibition have been validated for ewe milk [3-5]. Most of them have good sensitivity for the detection of β-lactams, one of the antibiotic groups more widely used in ovine therapy [6]. Generally, official and dairy industry laboratories use these microbial screening tests to control the occurrence of antibiotics due to the easy use and low cost. However, sometimes these tests are not sensitive enough to detect some penicillins like ampicillin or amoxicillin at the MRL concentrations [3-5].On the other hand, milk which contains penicillins under the maximal legal concentration (MRL) can be used for elaborating dairy products even though it contains antibiotics, and this amount of antibiotics could have some effect on the fermentation process.

Another important aspect regarding consumer protection is the observation of withdrawal periods by farmers, but on some occasions the separation within a non-violative and a violative sample is too thin and antibiotics could be present above the MRL even if producers carry out these security times [7], and milk could have levels of antibiotics high enough to cause inhibition effects. If milk contaminated with antibiotics, at allowed or violative levels, reaches the dairy industry already fermented this could provoke serious problems during the coagulation phases or affect the final product. Although most of time milk destined to dairy products is pasteurized, it has been demostrated that penicillins are not completely inactivated after these heat treatments and could continue provoking a microbial inhibition effect [8].

Controls for avoiding the presence of antibiotics above the MRLs guarantee food safety for consumers, but it is not completely clarified if these measures are enough to preserve dairy fermentation processes [9]. In fact, the few previous studies carried out observed slight acidification delays in some dairy products at antibiotic concentrations very close to the MRLs [10-12]. The aim of this present study was to establish the effect of the presence of low concentrations of some penicillins, even inferior to the MRL, on acidity parameters and microbial growth in yoghurts made with ewe milk during the storage at refrigeration conditions.

2. Materials & Methods

2.1. Ewe milk, antibiotics & manufacture of yoghurts

Ewe's milk used in yoghurt manufacture was collected from the bulk tank of the Manchegan-breed flock of the experimental farm of the ITAP (Diputación de Albacete, Spain). The flock did not receive any drugs during the experiments. The chemical composition (%) of milk was as follows: total solids 18.48 ± 1.01; fat 7.01 ± 0.62 and protein 5.96 ± 0.48. Somatic cell counts (SCC) and bacterial counts were 593,000 ± 159,000 somatic cells/ml and 101,000 ± 59,000 cells/ml respectively; and the pH was 6.74 ± 0.02.

Three antibiotic concentrations were assayed corresponding to 0.5 x MRL, MRL and 1.5 x MRL. Thus were as follows: penicillin "G", amoxicillin, ampicillin: 2, 4, 6 μg/kg; (all these antibiotics were obtained from Sigma-Aldrich). The antimicrobial solutions were prepared as described Montero *et al.* [5].

Three lots of 40 yoghurts were made for each antibiotic and each concentration, and nine lots for controls. Ewe milk was pasteurised at 80°C for 1 min, cooled to 45°C and inoculated with a 2 IU/1000 L of commercial yoghurt culture comprised of *Streptococcus thermophilus* and *Lactobacillus delbrueckii* subsp. *bulgaricus* (DELVO-YOG® TY-17A DSL, DSM Food Specialties, Delft, The Netherlands). Aliquots of 100 ml dispensed in sterile plastic containers were incubated at 43°C until the pH decreased to about 4.65-4.60. Then the containers were stored at 2°C for 24 h and after that at 12°C for 2, 14 and 21 days analysis.

2.2.Chemical analysis

The titratable acidity (Dornic method) and D(-) /L(+) lactic acid (Boehringer Mannheim/R-Biopharm enzymatic assay, ref.1112821) were analysed in triplicate.

2.3. Microbial analysis

L. delbrueckii subsp. *bulgaricus* and *S. thermophilus* counts were estimated on MRS pH 4.6 agar and M 17 agar (both from Scharlau Chemie, Barcelona, Spain) incubated at 37°C (72 and 24 h, respectively). The evolution of moulds and yeasts and *Enterobacteriaceae* were also determined on Rose Bengal agar (incubated at 25°C, 96 h) and Violet Red Bile Dextrose (VRBD) agar (incubated at 32°C, 48 h), respectively.

2.4. Data & statistical analysis

The data obtained were analysed using analysis of variance. When the differences among concentrations were significant ($P < 0.05$), Tukey's test was carried out to determine the differences between pairs of groups. Statistical analyses were performed using SPSS version 13.0 for Windows statistical software.

3. Results & Discussion

In a previous work, the authors observed delays on the time of fermentation for ampicillin when this antibiotic was at the MRL or superior and for penicillin G when was at the MRL [13], but all yoghurts assayed were stored when fermentation finished at pH between to 4.65-4.60. Tables 1 to 3 show the parameters related to the acidification and microbial evolution of refrigerated yoghurts when any of the three penicillins (penicillin G, ampicillin and amoxicillin) were in milk at each concentration assayed.

In general, with all antibiotics at all concentrations the pH decreased significantly ($p < 0.001$) during the period of study (Tables 1 to 3). This decrease is more perceptible during the first 14 days and then became more stable. This evolution agrees with the normal evolution of pH in refrigerated yoghurts [14,15]. With regards to the effect of the concentration of antibiotics, only penicillin G showed delays on the pH evolution (Table 1).

The slow down effect of penicillins on the pH under lactic starters and yoghurts made from cow's milk have previously been reported [10-12], but the concentrations assayed were generally superior to the MRL. Suhren [12] reported imbalances of 0.25-0.5 units of pH at concentrations of penicillin G between 4.5-7 μg/kg. Similar results were pointed out by Mäyra-Mäkinen [10] at penicillin G concentrations of 0.005 IU/ml (near to 3 μg/kg). Moreover, levels of ampicillin of 5 μg/kg [11] or 6 μg/kg [12] forced delays on the pH decrease. The level of amoxicillin for delaying pH on yoghurts was established by Mourot and Loussouarn [11] in 10 μg/kg. This concentration was superior to the highest assayed in the present study, and therefore, there exists the possibility that concentrations superior to 6 μg/kg could cause this negative effect.

The evolution of the titratable acidity was logically opposite to the pH. Acidity increased during the storage time. However, this increase in the acidity was once again significantly inferior in yoghurts that contained penicillin G, where the acidity at the 0.5xMRL-yoghurts was 8% inferior to the control yoghurts (Table 1). The others penicillins did not show significant inferior acidity at allowed concentrations (≤ MRL). A negative effect of penicillin G has previously been described by several authors who reported that concentrations between 6 and 180 μg/kg could provoke significant imbalances in the acidity of several yoghurt starters [16].

A more detailed study of the acidification of yoghurts through the evolution of D(-) and L(+) lactic acid isomers showed that penicillins had more influence on L(+) lactic acid production (Tables 1 to 3). The production of L(+) lactic acid in yoghurts was significantly inferior when penicillin G or ampicillin were present at allowed concentrations of the MRL or superiors (Tables 1 and 2). Suhren [12] also observed that the amount of penicillin or ampicillin needed to inhibit the production of the isomer L(+) lactic acid was inferior than for the isomer D(-) lactic acid. In their study, the author observed a 50% reduction in the production of the isomer L(+) at 10 μg/kg of penicillin G, whereas it was necessary to double the concentration of this antibiotic to obtain the same percentage of reduction in the isomer D(-). Furthermore, when ampicillin was added at 4 μg/kg this author detected a reduction of the L(+) production around 20%, but needed 25 μg/kg for the D(-) isomer. Under our conditions 4 μg/kg of penicillin or ampicillin was enough to notice a 40% decrease in the L(+) lactic acid production (Table 1), and at these levels none of them manifested significant reductions in the D (-) isomer.

Table 1 Effect of different concentrations of penicillin G on the acidity and microbiological parameters of yoghurts made with ewe milk[1].

Antibiotic	Time (days)	Control	0.5 x MRL	MRL[2]	1.5 x MRL	*Anova*
pH	2	4.22 ± 0.02^{ax}	4.25 ± 0.02^{ab}	4.30 ± 0.01^{bcx}	4.37 ± 0.01^{cx}	***
	14	3.90 ± 0.01^{ay}	3.95 ± 0.01^{by}	3.97 ± 0.01^{by}	4.05 ± 0.01^{cy}	***
	21	3.88 ± 0.01^{ay}	3.92 ± 0.01^{by}	3.95 ± 0.01^{by}	3.99 ± 0.01^{cz}	***
	Anova	***	***	***	***	
Acidity (ºD)	2	139.3 ± 2.5^{ax}	128.5 ± 0.9^{bx}	121.4 ± 0.8^{bcx}	116.1 ± 0.4^{cx}	***
	14	179.4 ± 2.7^{ay}	159.1 ± 0.4^{by}	154.3 ± 0.5^{bcy}	145.7 ± 1.6^{cy}	***
	21	181.2 ± 2.4^{ay}	161.4 ± 1.1^{by}	158.7 ± 1.4^{bcz}	149.1 ± 0.8^{cy}	***
	Anova	***	***	***	***	
D(-)-lactic acid (g/100g)	2	0.94 ± 0.03^{x}	0.95 ± 0.05^{x}	0.98 ± 0.06^{x}	0.88 ± 0.02^{x}	NS
	14	1.20 ± 0.06^{xy}	1.16 ± 0.02^{y}	1.25 ± 0.04^{y}	1.17 ± 0.03^{y}	NS
	21	1.21 ± 0.09^{xy}	1.19 ± 0.03^{y}	1.18 ± 0.03^{y}	1.23 ± 0.02^{y}	NS
	Anova	*	***	**	***	
L(+)-lactic acid (g/100g)	2	0.32 ± 0.02^{a}	0.33 ± 0.03^{a}	0.19 ± 0.01^{b}	0.11 ± 0.03^{b}	***
	14	0.30 ± 0.02^{a}	0.26 ± 0.02^{ab}	0.19 ± 0.04^{b}	0.08 ± 0.02^{c}	***
	21	0.31 ± 0.03^{a}	0.27 ± 0.03^{ab}	0.17 ± 0.04^{bc}	0.10 ± 0.01^{c}	***
	Anova	NS	NS	NS	NS	
L. delbruekii subsp. *bulgaricus*	2	8.47 ± 0.08	8.45 ± 0.10	8.41 ± 0.06	8.09 ± 0.13	NS
	14	8.53 ± 0.07^{a}	8.40 ± 0.09^{ab}	8.40 ± 0.18^{ab}	7.99 ± 0.20^{b}	*
	21	8.28 ± 0.11^{a}	8.78 ± 0.15	8.31 ± 0.16	8.19 ± 0.23	NS
	Anova	NS	NS	NS	NS	
S. thermophilus	2	8.24 ± 0.05^{ax}	8.12 ± 0.08^{ax}	7.53 ± 0.16^{b}	7.21 ± 0.13^{b}	***
	14	7.92 ± 0.09^{ay}	7.71 ± 0.12^{ax}	7.63 ± 0.25^{ab}	7.15 ± 0.10^{b}	**
	21	7.82 ± 0.15^{ay}	7.61 ± 0.22^{ab}	7.66 ± 0.35^{ab}	6.86 ± 0.18^{b}	*
	Anova	*	*	NS	NS	

[1]Results are reported as means± SE; significance level for each factor is indicated as follows: *$P< 0.05$, ** $P< 0.01$, and *** $P< 0.001$; NS: not significant; $^{a, b, c}$ in the same row are significantly different ($P< 0.05$); ^{x,y,z} in the same column are significantly different ($P< 0.05$).

[2]MRL for penicillin G, ampicillin and amoxicillin in ewe milk is established by the EU in 4 μg/kg.

Neither *Enterobacteriaceae*, nor moulds and yeasts growth was detected during the 21 days of study (data not shown) nor control or antibiotic added yoghurts. These results confirmed a correct pasteurization treatment and not posterior contaminations by these microorganisms, probably due to the high acidity of those yoghurts which inhibit the growth of this spoilage flora.

Generally, the bacterial growth was slightly inhibited when concentration of penicillin or ampicillin increased (Tables 1 and 2). Ampicillin seemed to have a high inhibitory effect on the growth of *S. thermophilus* even at 0.5 x MRL-concentration (Table 2), but also penicillin at 1.5xMRL level (Table 1). The highest sensitivity of *S. thermophilus* to β-lactam antibiotics like penicillin G or cloxacillin has already been reported by Cogan [16] who compared the effect of penicillin on S. *thermophilus, L. delbruekii* subsp. *bulgaricus* and other lactic acid bacteria. The inhibitory effect of penicillins on lactic acid bacteria, and therefore on the fermentation processes, has also been underlined in meat product fermentation [17], but at concentrations higher than those used in this study.

Table 2 Effect of different concentrations of ampicillin on the acidity and microbiological parameters of yoghurts made with ewe milk[1].

Antibiotic	Time (days)	Control	0.5 x MRL	MRL[2]	1.5 x MRL	*Anova*
pH	2	4.22±0.02^{ax}	4.37±0.04^{bx}	4.29±0.02^{abx}	4.33±0.02^{bx}	***
	14	3.90±0.01^{y}	3.95±0.02^{y}	3.94±0.02^{y}	3.95±0.02^{y}	NS
	21	3.88±0.01^{y}	3.91±0.01^{y}	3.91±0.01^{y}	3.91±0.01^{y}	NS
	Anova	***	***	***	***	
Acidity (ºD)	2	139.3±2.5^{x}	134.7±2.2^{x}	135.1±2.7^{x}	134.6±1.7^{x}	NS
	14	179.4±2.7^{ay}	176.8±1.9^{ay}	169.6±2.1^{aby}	165.1±1.5^{by}	**
	21	181.2±2.4^{aby}	188.6±2.0^{bz}	180.3±2.3^{abz}	176.3±2.9^{az}	*
	Anova	***	***	***	***	
D(-)-lactic acid (g/100g)	2	0.94±0.03^{x}	0.91±0.02^{x}	1.03±0.01^{x}	0.96±0.03^{bx}	NS
	14	1.20±0.06^{axy}	1.38±0.01^{by}	1.35±0.01^{aby}	1.40±0.06^{by}	*
	21	1.21±0.09^{xy}	1.46±0.05^{y}	1.38±0.01^{y}	1.44±0.02^{y}	NS
	Anova	*	***	***	***	
L(+)-lactic acid (g/100g)	2	0.32±0.02^{a}	0.25±0.03^{ab}	0.20±0.02^{b}	0.17±0.03^{b}	**
	14	0.30±0.02^{a}	0.23±0.03^{ab}	0.21±0.03^{b}	0.21±0.03^{b}	*
	21	0.31±0.03^{a}	0.25±0.04^{ab}	0.20±0.03^{b}	0.22±0.03^{b}	*
	Anova	NS	NS	NS	NS	
L. delbruekii subsp. *bulgaricus*	2	8.47±0.08	8.25±0.11	8.55±0.06	8.47±0.19^{x}	NS
	14	8.53±0.07^{a}	7.65±0.25^{b}	8.23±0.03^{ac}	8.02±0.14^{bc}	*
	21	8.28±0.11^{a}	7.82±0.13^{ab}	7.63±0.03^{b}	7.73±0.10^{by}	**
	Anova	NS	NS	NS	*	
S. thermophilus	2	8.24±0.05^{ax}	7.59±0.12^{bx}	7.42±0.12^{bx}	7.29±0.11^{bx}	***
	14	7.92±0.09^{ay}	7.31±0.07^{ax}	6.65±0.19^{ax}	4.49±1.00^{bx}	***
	21	7.82±0.15^{ay}	6.72±0.08^{ay}	4.49±1.01^{by}	3.07±0.97^{by}	***
	Anova	*	***	**	**	

Table 3 Effect of different concentrations of amoxicillin on the acidity and microbiological parameters of yoghurts made with ewe milk[1].

Antibiotic	Time (days)	Control	0.5 x MRL	MRL[2]	1.5 x MRL	*Anova*
pH	2	4.22±0.02^{ax}	4.25±0.04^{x}	4.27±0.02^{x}	4.31±0.03^{x}	NS
	14	3.90±0.01^{y}	3.90±0.01^{y}	3.94±0.01^{y}	3.94±0.01^{y}	NS
	21	3.88±0.01^{y}	3.91±0.01^{y}	3.92±0.01^{y}	3.92±0.01^{y}	NS
	Anova	***	***	***	***	
Acidity (ºD)	2	139.3±2.5^{x}	130.5±16^{x}	142.1±1.5^{x}	137.9±1.5^{x}	NS
	14	179.4±2.7^{y}	165.2±21^{y}	179.1±3.9^{y}	172.9±1.9^{y}	NS
	21	181.2±2.4^{aby}	163.4±20^{y}	179.4±2.2^{y}	185.1±3.9^{z}	NS
	Anova	***	***	***	***	
D(-)-lactic acid[3]	2	0.94±0.03^{x}	1.08±0.05^{x}	0.99±0.05^{x}	1.00±0.05^{x}	NS
	14	1.20±0.06^{xy}	1.33±0.09^{y}	1.46±0.06^{y}	1.32±0.11^{y}	NS
	21	1.21±0.09^{xy}	1.28±0.05^{y}	1.32±0.03^{y}	1.42±0.12^{y}	NS
	Anova	*	*	***	*	
L(+)-lactic acid[3]	2	0.32±0.02^{a}	0.31±0.02	0.26±0.04	0.25±0.02	NS
	14	0.30±0.02	0.38±0.09	0.33±0.01	0.28±0.06	NS
	21	0.31±0.03	0.47±0.02	0.35±0.07	0.34±0.03	NS
	Anova	NS	NS	NS	NS	
L. delbruekii subsp. *bulgaricus*	2	8.47±0.08^{ab}	8.63±0.09^{a}	8.41±0.05^{abx}	8.22±0.03^{bx}	*
	14	8.53±0.07	8.55±0.08	8.66±0.08^{y}	8.66±0.08^{y}	NS
	21	8.28±0.11	8.51±0.18	8.41±0.06^{x}	8.18±0.07^{x}	NS
	Anova	NS	NS	*	***	
S. thermophilus	2	8.24±0.05^{abx}	8.29±0.05^{ax}	8.13±0.07^{abx}	8.04±0.03^{bx}	*
	14	7.92±0.09^{ay}	8.41±0.12^{bx}	8.02±0.05^{ax}	7.93±0.03^{ax}	**
	21	7.82±0.15^{y}	8.07±0.09^{y}	7.79±0.04^{y}	7.49±0.08^{y}	NS
	Anova	*	*	**	***	

On the other hand, the influence of penicillins on the growth of *S. thermophilus* is directly reflected on the production of the two isomers of lactic acid. Generally, L(+) lactic acid is the most affected isomer by penicillins. The production of this isomer is fundamentally attributed to *S. thermophilus*, while the production of the isomer D(-) mainly corresponds to *L. delbrueckii* subsp. *bulgaricus* [18]. According to Tamine and Robinson [18], the ratio of L(+):D(-) lactic acid could be used as marker of the quality of yoghurts, so commercial yoghurts ranged from 0.34 (prevalence of D(-) isomer) to 8.28 (prevalence of L(+) isomer), classifying as very acidic those yoghurts in which the isomer D(-) predominates. Two days after the elaboration of control yoghurts a ratio L(+):D(-) was observed which confirmed their acidic characteristics. The occurrence of penicillins in yoghurts made this ratio inferior and consequently yoghurts became even more acidic. Therefore, it can be concluded that some penicillins could change the equilibrium of the two isomers of lactic acid (Tables 1 to 3), and particularly damage the production of the isomer L(+) lactic, which is easier to assimilate by consumers [18].

Finally we observed that the level of penicillin G or ampicillin caused some alterations in the characteristics of yoghurt and the MRL were too close. Therefore, it is important to use screening tests which are sensitive enough not only to protect consumers from antibiotics, but also to protect dairy fermented products.

Acknowledgements This work was funded by the project PCI-03-006, from the Consejería de Ciencia y Tecnología (JJCC Castilla-La Mancha, Spain). Authors would also like to thank Instituto Técnico Agronómico Provincial (Diputación de Albacete, Spain) and DSM Food Specialties Spain (Barcelona) for their support.

References

[1] Commission Regulation (CEE) n° 2377/90 of the European Parliament. *Off. J. European Union.* 1990; L 224: 1-8.

[2] Diserens JM. *Compilation of commercial available screening methods for the detection of inhibitors/antibiotic residues in milk.* Draft of the Joint Action Team Organic Contaminants and Veterinary Residues from the FIL-IDF. In: FIL-IDF-ISO Analytical Week, Munich: International Dairy Federation; 2007.

[3] Althaus RL, Peris C, Montero A, Molina MP. Detection limits of antimicrobials in ewe milk by Delvotest. *Milchwissenschaft.* 2002; 57: 660-664.

[4] Molina MP, Althaus RL, Molina A, Fernández N. Antimicrobial agent detection in ewes' milk by the microbial inhibitor test brilliant black reduction test—BRT AiM. *Int. Dairy J.* 2003;13: 821-826.

[5] Montero A, Althaus RL, Molina A, Berruga I, Molina MP. Detection of antimicrobial agents by a specific microbiological meted (Eclipse 100®) for ewe milk. *Small Rum. Res.* 2005; 57: 229-237.

[6] Bergonier D, De Crémoux R, Rupp R, Lagriffoul G, Berthelot X. Mastitis of dairy small ruminants. *Vet. Res.* 2003; 34: 689–716.

[7] Berruga MI, Yamaki M, Althaus RL, Molina MP, Molina A. Performance of antibiotic screening tests in determining the persistence of penicillin residues in ewe's milk. *J. Food Prot.* 2003; 66: 2097-2102.

[8] Zorraquino MA, Roca M, Fernández N, Molina MP, Althaus R. Heat inactivation of beta- lactam antibiotics in milk. *Journal of Food Protection.* 2008;71(6):1193-18.

[9] Adriany A, Nadal R. *Test microbiológicos para inhibidores; tests screening o de confirmación.* In: Proceedings of the XIX Reunión G-Temcal "Nuevos retos en los sistemas de calidad de producción de leche para el primer decenio del siglo XXI", Palencia: Laboratorio Interprofesional Lácteo de Castilla y León; 2002: 40.

[10] Mäyrä-Mäkinen A. *Residues and quality impairment of fermented milk products and cheeses.* In: Workshop on residues of antibiotics and other antimicrobial inhibitors in raw and heat-treated milk: significance, detection and development of an integrated detection system. Copenhagen: International Dairy Federation; 1993.

[11] Mourot D, Loussouarn S. Sensibilité des ferments lactiques aux antibiotiques utilisés en médicine vétérinarie. *Rec. Med. Vet.* 1981; 157: 175-177.

[12] Suhren G.. Influence of residues of antimicrobials in milk on commercially applied starter cultures-model trials. *Kieler Milchwirtschaftliche Forschungsberichte.* 1996; 49: 131-149.

[13] Berruga MI, Molina MP, Novés B, Román M, Molina A."In vitro" study about the effect of several penicillins during the fermentation of yogurt made from ewe's milk. *Milchwissenschatf.* 2007; 62: 303-305.

[14] Katsiari MC, Voutsinas LP, Kondyli E. Manufacture of yoghurt from stored frozen sheep's milk. *Food Chem.* 2002; 77: 413-420.

[15] Tamine AY, Robinson RK. *Yoghurt: science and technology.* 2nd ed. Cambridge: Woodhead Publishing & CRC Press LC; 1999.

[16] Cogan TM. Susceptibility of cheese and yoghurt starter bacteria to antibiotics. *App. Microb.* 1972; 23: 960-965.

[17] Holley RA, Blaszyk M. Antibiotic challenge of meat starter cultures and effects upon fermentations. *Food Res. Int.* 1998; 30: 513-522.

[18] Robinson RK, Itsaranuwat P. *Properties of yoghurt and their appraisal* In: Tamine AY, ed. *Fermented milks.* Oxford: Blackwell Science; 2006: 76-94.

Exploring thermostable quorum quenching lactonases to counteract bacterial infections in cystic fibrosis

Mandrich L., Porzio E., Merone L., Febbraio F., Nucci R. and Manco G.

Institute of Protein Biochemistry, National Research Council, via P. Castellino 111, I-80131, Naples, Italy.

Recently a new family inside the amidohydrolase superfamily has been described and dubbed PTE-like lactonase family (PLL) due to structural similarity with phosphotriesterases (PTEs). Some members of this family are thermostable enzymes. These lactonases are characterised by high resistance to several stressing factors other than temperature. It has been proposed that these enzymes could work as "quencers" in the quorum sensing phenomenon. We report experiments made on *Sulfolobus solfataricus* to understand a) if the quorum sensing exists in *Sulfolobus solfataricus* and b) if PTE-like lactonases can be used to counteract *Pseudomonas aeruginosa* PO1 infective behaviour. Results suggest that no QS system works in *S. solfataricus* and that three thermostable PLLs are able to counteract the swarming motility in *P. aeruginosa.* In conclusion, thermostable PLLs are promising tools to counteract bacterial infections in cystic fibrosis.

Keywords: quorum quencing, thermostable lactonases

1. Introduction

Pseudomonas aeruginosa is the major cause of clinically relevant infections in cystic fibrosis (CF) due to antibiotic resistance, which demands for new strategies of microbe proliferation control. CF represents a model biofilm infection [1]. During the course of CF, *P. aeruginosa* starts to produce exo-polysaccharides and forms biofilms characterized by cellular aggregates embedded within the mucus layer present in the airways [2]. Biofilm formation in the CF lung is controlled by the bacterial quorum-sensing (QS) mechanism [3-6]. The question of whether there is a direct involvement of the QS system in biofilm formation has been a key issue for some time. However, in a mouse infection model, QS inhibitory drugs reduce biofilm formation [7]. In addition, human PON1 inhibits *P. aeruginosa* biofilm growth in an *in vitro* biofilm model as well as in an *in vivo* animal model [8,9]. QS is an interbacterial mode of communication accomplished through the coordinated production, secretion, and detection of chemical signals (QS signals) that trigger the expression of specific bacterial genes. The QS signals self-produced by *P. aeruginosa* are in the form of small molecules, or autoinducers, termed acyl-homoserine lactones (acyl-HSL) [10]. *P. aeruginosa* uses acyl-HSL quorum-sensing molecules to regulate the expression of genes implicated in virulence and biofilms formation. Promising new therapies that target biofilm formation include molecules (quenching enzymes or small molecules) that disrupt QS [11]. It has been shown that all of the human PONs can inactivate 3OC12-HSL in vitro [8,12,13] and in vivo [14]. These observations indicate that quorum-sensing molecules are important for *P. aeruginosa* virulence and biofilm formation and that lactonases, such as PONs, can protect the host from lethal *P. aeruginosa* infection.

Recently a new group of enzymes showing lactonase activity as the primary activity and additional promiscuous activities (phosphotriesterases, carboxylesterase) has been defined on the basis of conserved key amino acids residues and common active site features, and named PTE-like lactonase (PLL) [15]. The solved 3D structure of the *Sulfolobus solfataricus* enzyme (*Sso*Pox) with and without a bound lactone analogue allowed to confirm relationship with pPTE and defined *Sso*Pox as the first member of a new family [16]. Several enzymes have been ascribed to the PLL group coming out from different sources such as: *Mycobacterium avium, Mycobacterium sp., Mycobacterium tuberculosis CDC1551, Mycobacterium bovis, Mycobacterium tuberculosis (PPH), Rubrobacter xylanophils, Geobacillus kaustophilus, Rhodococcus erythropolis (AhlA), Brevibacterium linens, Deinococcus radiodurans, S. acidocaldarius* [17]. Many information on the two archaeal PLLs *Sso*Pox from *S. solfataricus* and *Sac*Pox from *S. acidocaldarius* [18-20], have been obtained. More recently, studies on two new thermostable members from *D. radiodurans* Dr0930 and *Geobacillus stearothermophilus*, have been reported. [21 and our unpublished results]

*Sso*Pox is a protein of 314 amino acids with 76% sequence identity to *Sac*Pox (315 a.a.); both proteins show about 30% sequence identity to pPTE [19]. Dr0930 is a 323 amino acids protein from *D. radiodurans* sharing about 35% sequence identity with pPTE, *Sso*Pox and *Sac*Pox. In all this enzymes the residues which in pPTE are involved in metals binding are strictly conserved, as well as the lysine corresponding to Lysine 169 in pPTE (which is carboxylated) [19,20] .

What differentiates the group from pPTE is principally the deletion of the surface loop 7 according to the amidohydrolase superfamily nomenclature and the presence of a specific sequence motif in the N-terminal region [15,16]. Moreover loop 8 in *Sso*Pox rearranges to form an hydrophobic channel in which the lateral acyl chain of the lactone analogue (C10HTL) fits nicely [16]. The structural basis of the high thermostability of *Sso*Pox has been investigated by performing a comparison between its structure and that of two mesophilic phosphotriesterases *p*PTE from *P. diminuta* and OpdA from *A. radiobacter* [22]. Results suggest that the high thermal stability of *Sso*Pox resides in the larger number of surface salt bridges, which are involved in surface networks, and an optimization of the interactions at the interface between the two monomers which stabilize the dimeric structure of *Sso*Pox. Comparative studies of thermal denaturation monitored by CD spectroscopy in the far UV demonstrate the higher stability of *Sso*Pox and *Sac*Pox with respect to *Pseudomonas* PTE [22].

We propose to exploit recently characterised phosphotriesterase-like lactonases (PLLs) from thermophilic microorganisms to counteract *Pseudomonas* infection in the CF patients.

2. Materials and Methods

2.1 Materials

Ampicillin, cloramfenicol, Isopropyl-beta-D-thiogalactopyranoside (IPTG), bacto-tryptone, agar, antifoam and substrate paraoxon were from SIGMA-Aldrich.

LB medium and Mueller Hinton medium, yeast extract and casamminoacids were from Difco.

2.2 Methods

2.2.1 Cells growth

Sulfolobus solfataricus P2 was grown as described in Merone et al. [17].

Pseudomonas P01 strain was a gift of E. Patriarca lab and was growth in Luria Bertani (LB) medium or Mueller Hinton Broth (MHB) media (broth or agar) at 30 or 37°C. *E. coli* has been grown at 37°C in LB.

Swarming motility of *P. aeruginosa* was assayed on semi-solid agar (0.3% w/v) containing 1% tryptone, 0.5% NaCl and 0.6% glucose according to Rashidi and Kornberg, 2000 [23].

2.2.2 Proteins purification

*Sso*Pox and *Sac*Pox expression in *E. coli* and purifications were performed as previously described [17,18] . Expression and purification of *D. radiodurans* lactonase will be described elsewhere.
Purified PTE from *P. diminuta* was a gift of Prof. A. Griffiths.

2.2.3 Enzyme assays

*Sso*Pox activity was followed by spectrophotometric assays as previously described [17].

After growth (1OD 600nm) cells of *Sulfolobus* were centrifuged and the supernatant was extracted with ethyl-acetate and the extracted concentrated up to tot ml. Aliquots were analysed for detection of acylhomoserinelactones (AHLs) with biosensors CV026 and *A. tumefaciens* Nxxx (gifts of Dr. Venturi) both on agar plates or after separation on TLC plates.

Table 1. Comparison of stability to different stressing agents in *Sso*Pox versus PTE

	PTE 1h	PTE 3h	PTE 16h	PTE 40h	SsoPox 1h	SsoPox 3h	SsoPox 16h	SsoPox 40h
H_2O/Buffer	100%	82%	40%	39%	100%	100%	77%	69%
10% MeOH	25%	41%	16%	15%	100%	123%	77%	80%
10% EtOH	37%	48%	46%	38%	69%	162%	92%	90%
10% Acetonitrile	90%	97%	51%	45%	185%	94%	62%	65%
10% Isopropanol	28%	30%	27%	22%	85%	107%	100%	92%
0.1%SDS	0.2%	0.4%	0.4%	0.04%	1180%	900%	623%	630%
0.5M GuHCl	35%	31%	20%	15%	192%	200%	138%	123%
2% Amuchine	46%	58%	32%	25%	115%	62%	62%	66%

Triplicate assays were performed at RT in 20mM Hepes buffer pH 8.0, Paraoxon 0.5mM. Reported values are mean of two experiments.

3. Results and Discussion

3.1 Stability of *S. solfataricus* PLL to different stressing conditions and comparison with *P. diminuta* PTE.

In order to assess the higher stability of *Sso*Pox to other stressing conditions other than temperature we incubated at 25°*C P. diminuta* PTE (pPTE) and *Sso*Pox in the presence of: 10% methanol, ethanol, isopropanol or acetonitrile; 0.1% SDS; 0.5 M guanidinum hydrochloride; 2% Amuchine for the lengths of time indicated in Table 1. After incubation the PTE enzyme activities were assayed at 25°C at their respective optimal pH conditions and reported for each enzyme as percentage of the activity in buffered solution after 1 hour incubation.

As shown, pPTE loses a fair part of its activity after 1h incubation under each condition except 10% acetonitrile. Under this condition stability is similar to *Sso*Pox up to 4h. For longer incubation times *Sso*Pox is quite more stable. In addition, *Sso*Pox is stable under any condition for up to 40 h with respect to PTE. Finally, under specific conditions such as 10% acetonitrile, 0.5M guanidinium hydrochloride and 0.1% SDS an activation phenomenon for *Sso*Pox activity is observed, which is stable over time. Particularly relevant is the SDS activation which is more than ten fold with respect to the sample incubated in buffer only. This activation has been analysed in a recent work [24] showing an effect on the Km for substrate which is also dependent from temperature in the sense that the activation is only 5-6 fold at 70°C. In the same paper it has been demonstrated the *Sso*Pox higher performance when tested against PTE for the ability to detoxificate a pesticide covering a polystirene surface [24].

3.2 Is there a AHL-based quorum sensing phenomenon acting in *S. solfataricus*?

The *Sso*Pox enzyme is a lactonase showing a specificity constant of around $4x10^5$ $M^{-1}s^{-1}$ estimated by using the synthetic lactone TBBL [15]. The phosphotriesterase activity is around 100 fold less. By using CV026 biosensor it has been demonstrated activity against natural AHL [15]. These discoveries pose the problem of the physiological role of *Sso*Pox. It has been argued that the lactonases could act as controller of quorum sensing. In order to shed some light on this point we tested *S. solfataricus* cultures for lactones production by using the biosensors CV026 and *Agrobacterium tumefaciens* NT1 and by extracting lactones with ethyl-acetate from the medium after growth in order to increase their concentration. In both cases no AHLs were detected. However we observed the presence of molecules in the growth medium able to activate biosensors (data not shown). By addition of enzymes to the *S. solfataricus* culture we observed a statistically significant increase in the biomass with respect to a control in which we added the same amount of BSA (Fig. 1). Starting from 15 and up to 50 h of growth a 30% higher OD_{600} for *S. solfataricus* with the addition of *Sso*Pox (4 μg/ml) with respect to the control was observed.

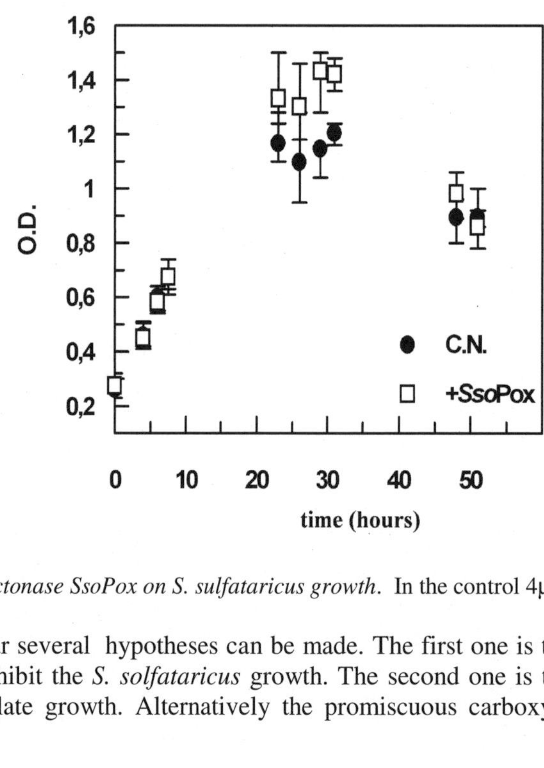

Fig. 1 *Effect of lactonase SsoPox on S. sulfataricus growth.* In the control 4μg/ml of BSA were added.

To explain this behaviour several hypotheses can be made. The first one is the presence in the medium of a lactone which is able to inhibit the *S. solfataricus* growth. The second one is the presence of a lactone that if destroyed is able to stimulate growth. Alternatively the promiscuous carboxylesterase or phosphotriesterase activities can generate compounds that stimulate growth or can destroy putative inhibitors. The fact that in the M182 medium there are compounds able to activate biosensors let to guess the presence of lactone analogues which have an effect on *S. solfataricus*. However it does not seem that *S. solfataricus* produces AHLs. This result is in agreement with the fact that in the *S. solfataricus* genome the genes involved in the QS system appears to be missing in large part. Finally, in Archaea there is only a single report of AHLs production from a species of *Natrosonococcus* [25]. Considering that AHLs are quite unstable at high temperature and that in other thermophilic microrganisms QS is present but based on the more stable short peptides it can be concluded that at least in *Sulfolobus* no AHL based QS system is present.

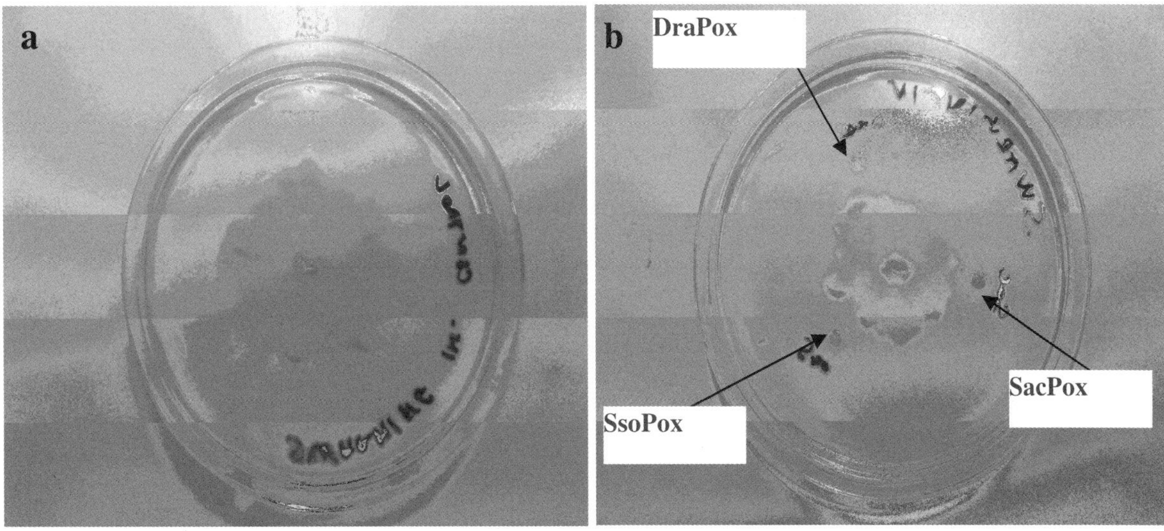

Fig. 2 Swarming assay of PA01. In panel (a) is shown the control. In (b) 30 micrograms of each enzymes were placed in the wells indicated by arrows.

The high lactonase activity and stability of PLLs invited us to exploit these properties for interfering in the QS phenomenoun in *Pseudomonas aeruginosas* (PA01). In *P. aeruginosa* several pathogenicity markers such as pyocyanin and elastase production, motility and formation of biofilm are under QS control and are the main cause of mortality in diseases like Cystic fibrosis and others. We have started a project to exploit the possibility to interfere with *Pseudomonas* QS behaviour *in vitro* and *in vivo*. We have tested for now the capability of thermostable enzymes to interfere with swarming motility in PA01. As shown in FIG. 2b all three wt enzymes were able to counteract swarming motility with respect to the control (a). It is worth to note that the experiments were performed at 37 °C where the activity of lactonases is not optimal (Topt≥70°C). We will also check in the future the effect of mutants with optimal activity at lower temperatures. In the next two years in the frame of a grant founded by the Italian Cystic Fibrosis Research Foundation we will evaluate all pathogenicity markers and will perform *in vivo* experiments in animal models by comparing results with human enzymes PON1 and 2. The hope is that these very robust enzymes could represent an additional weapon in the race for the defeat of inections in the Cystic Fibrosis disease.

Acknowledgements We thank Prof. A. G riffiths for PTE enzyme, A. Riccio and Dr. E. Patriarca for PA01 strain, Prof. Venturi V for CV026 and *A .tumefaciens* NT1 biosensors strains.

References

[1] Parsek, MR, and Singh, PK.. Bacterial biofilms: an emerging link to disease pathogenesis. Annu. Rev. Microbiol. 2003, 57: 677-701.

[2] Lam, J, Chan, R, Lam, K, and Costerton, JW. Production of mucoid microcolonies by *Pseudomonas aeruginosa* within infected lungs in cystic fibrosis. Infect. Immun. 1980, 28: 546-556.

[3] McClean K H, Winson M K, Fish L, Taylor A, Chhabra SR, Camara M, Daykin M, Lamb JH, Swift S, Bycroft BW, Stewart GS, Williams P. Quorum sensing and Chromobacterium violaceum: exploitation of violacein production and inhibition for the detection of N-acylhomoserine lactones. Microbiology, 1997, 143: 3703-3711.

[4] Reimmann C, Ginet N, Michel L, Keel C, Michaux P, Krishnapillai V, Zala M, Heurlier K, Triandafillu K, Harms H, Défago G, Haas D. Genetically programmed autoinducer destruction reduces virulence gene expression and swarming motility in Pseudomonas aeruginosa PAO1. Microbiology, 2002, 148: 923-932.
[5] Wagner VE, Bushnell D, Passador L, Brooks AI, Iglewski BH. Microarray analysis of Pseudomonas aeruginosa quorum-sensing regulons: effects of growth phase and environment. J Bacteriol, 2003, 185: 2080-2095.
[6] Wopperer J, Cardona ST, Huber B, Jacobi CA, Valvano MA, Eberl L. A quorum-quenching approach to investigate the conservation of quorum-sensing-regulated functions within the Burkholderia cepacia complex. Appl Environ Microbiol, 2006, 72: 1579-1587.
[7] Christensen LD, Moser C, Jensen PØ, Rasmussen TB, Christophersen L, Kjelleberg S, Kumar N, Høiby N, Givskov M, Bjarnsholt T. Impact of Pseudomonas aeruginosa quorum sensing on biofilm persistence in an in vivo intraperitoneal foreign-body infection model. Microbiology. 2007 153:2312-20.
[8] Ozer EA, Pezzulo A, Shih DM, Chun C, Furlong C, Lusis AJ, Greenberg EP, Zabner J. Human and murine paraoxonase 1 are host modulators of Pseudomonas aeruginosa quorum-sensing. FEMS Microbiol. Lett. 2005 253:29-37.
[9] Stoltz DA, Ozer EA, Taft PJ, Barry M, Liu L, Kiss PJ, Moninger TO, Parsek MR, Zabner J. Drosophila are protected from Pseudomonas aeruginosa lethality by transgenic expression of paraoxonase-1.J. Clin. Invest. 2008, 118:3123-3131.
[10] Dong, YH, Xu, J, Li X, and Zhang L. AiiA, an enzyme that inactivates the acylhomoserine lactone quorum-sensing signal and attenuates the virulence of Erwinia carotovora. Proc. Natl. Acad. Sci. USA 2000, 97: 3526-3531.
[11] Murray TS, Egan M, Kazmierczak BI. Pseudomonas aeruginosa chronic colonization in cystic fibrosis patients. Curr Opin Pediatr. 2007 19:83-8.
[12] Draganov DI, Teiber JF, Speelman A, Osawa Y, Sunahara R, La Du BN.Human paraoxonases (PON1, PON2, and PON3) are lactonases with overlapping and distinct substrate specificities. J. Lipid. Res. 2005, 46:1239-1247.
[13] Stoltz DA, Ozer EA, Ng CJ, Yu JM, Reddy ST, Lusis AJ, Bourquard N, Parsek MR, Zabner J, Shih DM. Paraoxonase-2 deficiency enhances Pseudomonas aeruginosa quorum sensing in murine tracheal epithelia. Am. J. Physiol. Lung Cell. Mol. Physiol. 2007, 292:L852-L860.
[14] Stoltz DA, Ozer EA, Taft PJ, Barry M, Liu L, Kiss PJ, Moninger TO, Parsek MR, Zabner J. Drosophila are protected from Pseudomonas aeruginosa lethality by transgenic expression of paraoxonase-1. J. Clin. Invest. 2008, 118:3123-3131.
[15] Afriat L, Roodveldt C, Manco G, Tawfik DS. The latent promiscuity of newly identified microbial lactonases is linked to a recently diverged phosphotriesterase. Biochemistry. 2006, 45:13677-86.
[16] Elias M, Dupuy J, Merone L, Mandrich L, Porzio E, Moniot S, Rochu D, Lecomte C, Rossi M, Masson P, Manco G, Chabriere E. Structural basis for natural lactonase and promiscuous phosphotriesterase activities. J Mol Biol. 2008 379:1017-28.
[17] Merone L., Mandrich L., Rossi M., Manco G. A thermostable phosphotriesterase from the archaeon Sulfolobus solfataricus: cloning, overexpression and properties, Extremophiles. 2005, 9:297-305.
[18] Porzio E, Merone L, Mandrich L, Rossi M, Manco G. A new phosphotriesterase from Sulfolobus acidocaldarius and its comparison with the homologue from Sulfolobus solfataricus. Biochimie. 2007, 89:625-636.
[19] Merone L, Mandrich L, Rossi M and Manco G. Curr. Chem. Bio., 2008, 2, 237-248 (September issue) review
[20] Mandrich L, Merone L, Manco G. Hyperthermophilic phosphotriesterases/lactonases for the environment and human health. Environ Technol. 2010, 31:1115-27.
[21] Hawwa R, Aikens J, Turner RJ, Santarsiero BD, Mesecar AD. Structural basis for thermostability revealed through the identification and characterization of a highly thermostable phosphotriesterase-like lactonase from Geobacillus stearothermophilus. Arch Biochem Biophys. 2009, 488:109-20.
[22] Del Vecchio P, Elias M, Merone L, Graziano G, Dupuy J, Mandrich L, Carullo P, Fournier B, Rochu D, Rossi M, Masson P, Chabriere E, Manco G. Structural determinants of the high thermal stability of SsoPox from the hyperthermophilic archaeon Sulfolobus solfataricus. Extremophiles. 2009, 13:461-70.
[23] Rashid M H, Kornberg A Inorganic polyphosphate is needed for swimming, swarming, and twitching motilities of Pseudomonas aeruginosa. Proc Natl Acad Sci USA, 2000, 97: 4885-4890.
[24] Merone L, Mandrich L, Porzio E, Rossi M, Müller S, Reiter G, Worek F, Manco G. Improving the promiscuous nerve agent hydrolase activity of a thermostable archaeal lactonase. Bioresour. Technol. 2010, 101: 9204-12.
[25] Paggi RA, Martone CB, Fuqua C, De Castro RE. Detection of quorum sensing signals in the haloalkaliphilic archaeon Natronococcus occultus. FEMS Microbiol Lett. 2003, 221:49-52.

Furanones and Thiophenones in Control of *Staphylococcus epidermidis* Biofilm Infections?

J. Lönn-Stensrud*[,1], T. Benneche[2] and A. Aa. Scheie[1]

[1] Department of Oral Biology, Faculty of Dentistry, University of Oslo, P.O. Box 1052 Blindern, N-0316 Oslo, Norway

[2] Department of Chemistry, Faculty of Mathematics and Natural Science, University of Oslo, P.O. Box 1033 Blindern, N-0315 Oslo, Norway

* Corresponding author: e-mail: jessiclo@odont.uio.no, Phone: +47 22 84 03 40

In the past decade, *Staphylococcus epidermidis* has emerged as one of the most frequently isolated pathogens from biofilm on implants and catheters. Since infections associated with biofilms are difficult to treat due to enhanced resistance to antibiotics, there is need of alternative approaches to prevent and to treat biofilm related infections. Quorum sensing is one of the regulatory mechanisms suggested to be involved in coordinating biofilm formation. Many studies have identified furanones as potential inhibitors of biofilm formation and quorum sensing. Here we aimed at investigating and comparing the effect of furanone and thiophenone on biofilm formation and quorum sensing. We show that both furanone and thiophenone are effective in reducing biofilm formation and bacterial communication. Thiophenones may be even more effective than furanones.

Keywords: Biofilm; Furanone; Quorum sensing; *S. epidermidis*; Thiophenone

1. Introduction

Biofilm is the preferred growth mode for most bacteria. Bacterial biofilms may be found everywhere, on biotic and abiotic surfaces. Biofilms may cause severe problems in various kinds of industrial and non-industrial settings resulting in system breakdowns [1]. In medicine, biofilms can be a protection or a threat to the host, depending on the composition and the balance of the biofilm. If the balance shifts, biofilms on human surfaces such as teeth, skin or urinary tract, may result in disease [2, 3]. The biofilms may also cause severe damage when formed on medical devices, causing chronic infections that are difficult to treat. The bacteria inhabiting the biofilm are protected from host defense mechanisms and antibiotics [4-6]. The antibiotics of today were developed and the efficacy assessed against planktonic bacteria in the laboratory, disregarding that most bacteria prefer to grow in a protective biofilm [7].

Biofilm formation is a result of bacterial interactions and group behavior. Quorum sensing (QS) is one of the regulatory mechanisms suggested to be involved in coordinating biofilm formation [8]. The QS system is a cell-to-cell communication system used by many bacteria to assess the cell density in the local environment. The communication is based on the production, secretion and detection of small signaling molecules. When the signal concentration reaches a defined threshold, a specific quorum of bacteria is understood to be present, resulting in a coordinated change in gene-expression amongst the bacteria in the quorum [9].

Quorum sensing inhibitors (QSI) could be a novel way to fight biofilm associated infections. Many studies have identified furanones as potential inhibitors of quorum sensing and biofilm formation [10-17]. Thiophenones are molecules structurally resembling furanones, but with the difference that thiophenones has sulphur in the ring instead of oxygen. Their synthesis pathway is also different [18]. In this study we aimed at comparing the effect of a thiophenone and a furanone to function as QSIs by assessing their ability to interfere with biofilm formation and quorum sensing.

2. Material and methods

2.1 Biofilm assay

The *S. epidermidis* cells (ATCC35984) were kept as frozen stocks and were grown on Brain Heart Infusion agar plates (BHI, Difco Laboratories, Detroit, Mich., USA) for 24 h at 37°C in an aerobic atmosphere before inoculation into Brain Heart Infusion medium (BHI, Difco Laboratories, USA) for biofilm formation and growth. *S. epidermidis* was allowed to form biofilm on furanone- or thiophenone coated polystyrene discs (Nunc, Roskilde, Denmark) immersed in the wells of a polystyrene micro-titer plate (Nunc, Roskilde,

Denmark). Coatings were prepared on discs as described earlier [15]. Briefly, 1 ml of a 100 mM solution of the furanone (*Z*)-5-(bromomethylene)furan-2(5*H*)-one (Fig. 1a) or the thiophenone (*Z*)-5-(bromomethylene)thiophen-2(5*H*)-one (Fig. 1b) in ethanol were added to respective wells of a polystyrene micro-titer plate containing discs. The solvent was allowed to evaporate for 24 hours at room temperature. Plain ethanol coating was included as negative control. Second overnight cultures of *S. epidermidis* were diluted 1:100 in BHI. One ml samples were transferred to the wells and the plates were incubated for 18 h at 37°C in aerobic atmosphere. Biofilm mass was quantified after removing the discs from the wells. Unattached cells were removed by rinsing the discs twice in 0.9% saline. The discs were then immersed into new wells and the biofilms were stained for 10 min with a 0.1% solution of safranin. The discs were again transferred to new wells after rinsing with 0.9% saline. Bound stain was released using 30% glacial acetic acid, and optical density (OD) at 530 nm was measured in a Synergy HT Multi-Detection Microplate Reader (Biotek, Vt, USA).

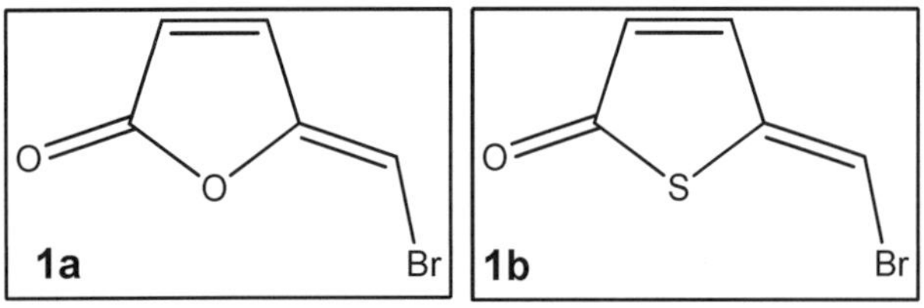

Fig. 1 Structure of (*Z*)-5-(bromomethylene)furan-2(5*H*)-one and (*Z*)-5-(bromomethylene)thiophen-2(5*H*)-one

To assess possible interference of furanone and thiophenone with *S. epidermidis* communication, the synthetic Autoinducer-2 (AI-2) molecule (*S*)-4,5-dihydroxy-2,3-pentanedione (DPD) (OMM Scientific Inc, TX, USA) was added to the growth medium at 0.25, 0.5, 1, 2 or 4 nM. Biofilm was allowed to form on furanone or thiophenone coated discs and quantified as described above.

We assessed a possible antibacterial effect of the furanone or thiophenone on *S. epidermidis* by measuring total growth, including both planktonic and biofilm cells scraped together in the wells. For this assay, biofilm was allowed to form on furanone- or thiophenone coated and uncoated wells as above, but omitting the discs. The total bacterial mass was determined by OD measurements (600 nm) after vigorous shaking.

SEM was used to visually confirm the effect of furanone and thiophenone on *S. epidermidis* biofilm formation. Biofilms were formed as described above on furanone-coated and uncoated polystyrene discs. Biofilm formed was fixed with 2.5% glutaraldehyde in 0.1 M Sørensen phosphate buffer. Samples were successively dried with 30, 50, 70 and 90% absolute ethanol for 10 min, sputtered with palladium/gold and then examined by SEM (Model XL 30 ESEM, Philips, Eindoven, Netherlands).

2.2 Bioluminescence assay

AI-2 communication was assessed as the ability of cell free culture supernatants to induce bioluminescence in *Vibrio harveyi* BB170, only sensing AI-2 (A kind gift from Prof. Bonnie Bassler). *V. harveyi* BB170 was grown as previously described [19, 20]. Incubation was with shaking, 200 rpm, in a Minitron Incubator Shaker (Infors AG, Bottningen, Schweiz) at 30°C.

Cell-free supernatants were prepared from *S. epidermidis* by centrifuging second overnight culture at 8.000 x g for 10 min and then passing the supernatants through 0.2 μm filters (Schleicher Schuell, Germany). The bioluminescence assay was performed essentially as described earlier [19, 20], with slight modifications [21]. Furanone and thiophenone were added to a final concentration of 10 μM.

The biofilm and bioluminescence assays were all performed in triplicate in three independent experiments.

3. Results

3.1 Effect on biofilm formation

Furanone and thiophenone both reduced *S. epidermidis* biofilm formation. Thiophenone seemed to be the most effective biofilm reducer (Fig. 2). The difference between thiophenone compared to control, and thiophenone compared to furanone, was statistically significant.

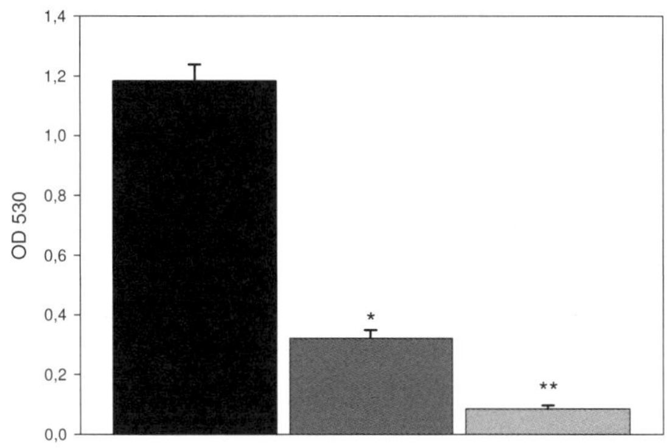

Fig. 2 *S. epidermidis* biofilm (mean and SE) formed on furanone (dark gray) or thiophenone (light gray) coated surfaces compared to control (black). *Statistically different from control. ** Statistically different from control and furanone.

At the concentrations used of the furanone and thiophenone, no effect on total bacterial growth was observed.

3.2 DPD abolished the inhibitory effect of furanone and thiophenone on biofilm formation

DPD added to the growth medium reduced the biofilm inhibitory effect of furanone and thiophenone (Fig. 3) at all concentrations used, indicating that the biofilm reduction was a result of interference with bacterial quorum sensing. The biofilm formation was completely restored by 1 and 0.5 nM DPD in case of furanone (Fig. 3a) and by 4 nM in case of thiophenone (Fig. 3b).

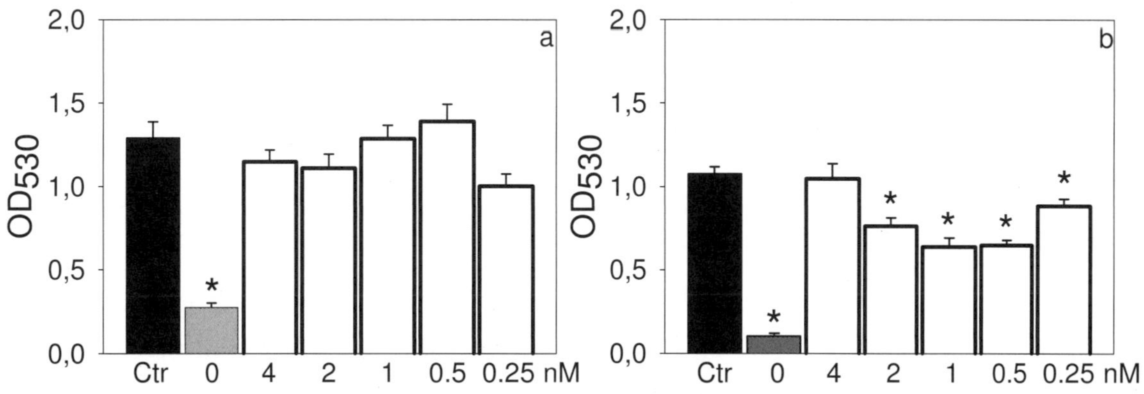

Fig. 3 *S. epidermidis* biofilm (mean and SE) formed on furanone (a) or thiophenone (b) coated surfaces (grey bars) with addition of 0.25, 0.5, 1, 2, or 4 nM of DPD to growth medium (white) compared to control (black) *Statistically different from control.

3.3 Scanning electron microscopy

SEM (Fig. 4a-d) visually confirmed the results from the biofilm assay.

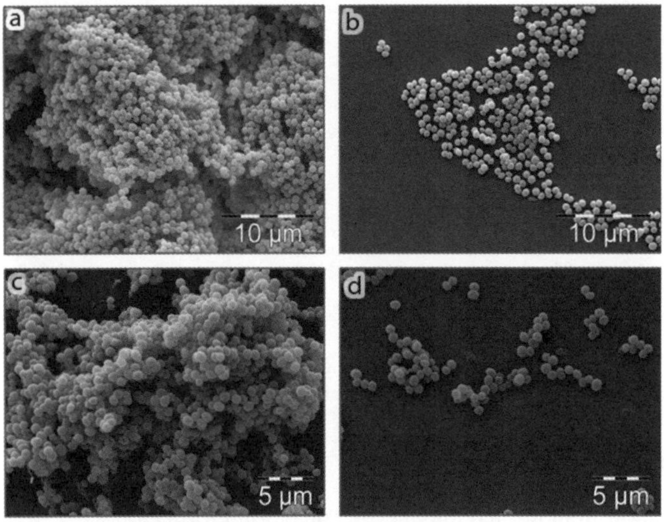

Fig. 4 *S. epidermidis* biofilm formed on furanone (b) or thiophenone (d) coated surfaces compared to control (a,c).

3.4 Furanones and thiophenones reduced bioluminescence induced by *S. epidermidis* supernatant

Bioluminescence induced by supernatant from *S. epidermidis* was reduced by both the furanone and the thiophenone with 8 and 4625 fold reduction, respectively. Thiophenone was statistically different from both control and furanone, indicating that the thiophenone is a more effective communication inhibitor than the furanone.

4. Discussion

In the present study we compared the inhibitory effect of a thiophenone to the effect of a furanone on *S. epidermidis* biofilm formation and quorum sensing.

Furanones have in several recent studies been shown to have quorum sensing inhibitory properties [13-17]. Even though furanones reduce biofilm formation to a great extent, it is of interest to find more effective compounds. We observed that both the furanone and the thiophenone reduced biofilm formation by *S. epidermidis*. From our study it seems that thiophenone reduced biofilm formation to a higher extent than furanone, indicating that thiophenones may be more suitable as a biofilm reducing agent.

Previous studies have indicated that furanones interfere with AI-2 communication in *E. coli* [22], *S. epidermidis* [15], *Streptococcus anginosus, Streptococcus intermedius* and *Streptococcus mutans* [16]. It has also been shown that externally added AI-2 regulates the expression of phenol-soluble modulins (PSMs), which are thought to be involved in biofilm development in *S. epidermidis*. Li and co-workers showed that adding external AI-2 almost completely restored the wildtype gene expression patterns of a *S. epidermidis luxS* mutant [23]. In this study we observed that the effect of both the furanone and the thiophenone could be abolished by the synthetic AI-2 precursor, indicating that thiophenone also affects biofilm formation through interference with bacterial communication.

In the bioluminescence assay, thiophenones almost extinguished the light production, with a 99,9% reduction of bioluminescence. This clearly indicates that thiophenones are able to interfere with quorum sensing, and thereby might be an even better alternative than furanone in preventing biofilm formation on surfaces.

In conclusion, this study shows that both a synthetic furanone and a thiophenone may inhibit biofilm formation by *S. epidermidis*. The effect on biofilm formation was most probably related to interference with AI-2 communication, since addition of DPD abolished the reducing effect. This would open for the use of furanones to prevent biofilm infections, for instance as coatings on implants.

Further studies on toxicity and irritability are needed to assess the *in vivo* suitability. It is also of interest to elucidate more about the mechanism of action of thiophenones on bacterial communication.

References

[1] de Carvalho CC, Biofilms: recent developments on an old battle. *Recent Pat Biotechnol.* 2007;1(1):49-57.

[2] Costerton JW, Lewandowski Z, Caldwell DE, Korber DR, Lappin-Scott HM. Microbial biofilms. *Annu Rev Microbiol.* 1995;49:711-745.

[3] Marsh PD. Are dental diseases examples of ecological catastrophes? *Microbiology.* 2003;149(Pt 2):279-294.

[4] Donlan RM. Biofilms on central venous catheters: is eradication possible? *Curr Top Microbiol Immunol.* 2008;322:133-161.

[5] Hall-Stoodley L, Costerton JW, Stoodley P. Bacterial biofilms: from the natural environment to infectious diseases. *Nat Rev Microbiol.* 2004; 2(2): 95-108.

[6] Hatt JK, Rather PN. Role of bacterial biofilms in urinary tract infections. *Curr Top Microbiol Immunol.* 2008;322:163-192.

[7] Smith AW. Biofilms and antibiotic therapy: is there a role for combating bacterial resistance by the use of novel drug delivery systems? *Adv Drug Deliv Rev.* 2005;57(10):1539-1550.

[8] Parsek MR, Greenberg EP. Sociomicrobiology: the connections between quorum sensing and biofilms. *Trends Microbiol.* 2005;13(1):27-33.

[9] Bassler BL, Losick R. Bacterially speaking. *Cell.* 2006;125(2):237-246.

[10] Benneche T, Hussain Z, Scheie, AA, Lönn-Stensrud J. Synthesis of 5-(bromomethylene)furan-2(5*H*)-ones and 3-(bromomethylene)isobenzofuran-1(3*H*)-ones as inhibitors of microbial quorum sensing. *New J. Chem.* 2008;32:1567-1572.

[11] Defoirdt T, Miyamoto CM, Wood TK, Meighen EA, Sorgeloos P, Verstraete W, Bossier P. The natural furanone (5Z)-4-bromo-5-(bromomethylene)-3-butyl-2(5H)-furanone disrupts quorum sensing-regulated gene expression in Vibrio harveyi by decreasing the DNA-binding activity of the transcriptional regulator protein luxR. *Environ Microbiol.* 2007;9(10):2486-2495.

[12] Han Y, Hou S, Simon KA, Ren D, Luk YY. Identifying the important structural elements of brominated furanones for inhibiting biofilm formation by Escherichia coli. *Bioorg Med Chem Lett.* 2008;18(3):1006-1010.

[13] Janssens JC, Steenackers H, Robijns S, Gellens E, Levin J, Zhao H, Hermans K, De Coster D, Verhoeven TL, Marchal K, Vanderleyden J, De Vos DE, De Keersmaecker SC. Brominated furanones inhibit biofilm formation by Salmonella enterica serovar Typhimurium. *Appl Environ Microbiol.* 2008;74(21):6639-6648.

[14] Kim C, Kim J, Park HY, Park HJ, Lee JH, Kim CK, Yoon J. Furanone derivatives as quorum-sensing antagonists of Pseudomonas aeruginosa. *Appl Microbiol Biotechnol.* 2008;80(1):37-47.

[15] Lönn-Stensrud J, Landin MA, Benneche T, Petersen FC, Scheie AA. Furanones, potential agents for preventing Staphylococcus epidermidis biofilm infections? *J Antimicrob Chemother.* 2009;63(2):309-316.

[16] Lönn-Stensrud J, Petersen FC, Benneche T, Scheie AA. Synthetic bromated furanone inhibits autoinducer-2-mediated communication and biofilm formation in oral streptococci. *Oral Microbiol Immunol.* 2007;22(5):340-346.

[17] Vestby LK, Lönn-Stensrud J, Møretrø T, Langsrud S, Aamdal-Scheie A, Benneche T, Nesse LL. A synthetic furanone potentiates the effect of disinfectants on Salmonella in biofilm. *J Appl Microbiol.* 2010;108(3):771-778.

[18] Universitetet i Oslo, Scheie AAa, Benneche T, Lönn-Stensrud J, Skramstad J. Patent. 2010. WO/2010/040839 (www.wipo.int/pctdb/en/wo.jsp?WO=2010040839).

[19] Schauder S, Shokat K, Surette MG, Bassler BL. The LuxS family of bacterial autoinducers: biosynthesis of a novel quorum-sensing signal molecule. *Mol Microbiol.* 2001;41(2):463-476.

[20] Surette MG, Bassler BL. Quorum sensing in *Escherichia coli* and *Salmonella typhimurium. Proc Natl Acad Sci U S A.* 1998;95(12):7046-7050.

[21] Rickard AH, Palmer RJ Jr, Blehert DS, Campagna SR, Semmelhack MF, Egland PG, Bassler BL, Kolenbrander PE. Autoinducer 2: a concentration-dependent signal for mutualistic bacterial biofilm growth. *Mol Microbiol*, 2006. 60(6): p. 1446-1456.

[22] Ren D, Bedzyk LA, Ye RW, Thomas SM, Wood TK. Differential gene expression shows natural brominated furanones interfere with the autoinducer-2 bacterial signaling system of *Escherichia coli. Biotechnol Bioeng.* 2004;88(5):630-642.

[23] Li M, Villaruz AE, Vadyvaloo V, Sturdevant DE, Otto M. AI-2-dependent gene regulation in *Staphylococcus epidermidis. BMC Microbiol.* 2008;8:4.

Innovative High Surface Area CuO Pretreated Cotton Effective in Bacterial Inactivation under Visible Light

A. Torres[1,2], C. Ruales[3], C. Pulgarin[3*], A. Aimable[2], P. Bowen[2], V. Sarria[1] and J. Kiwi[4*]

[1]UNIVERSIDAD de los Andes, Departamento de Quimica, Cra1E No 18A-10, Bogota, Colombia
[2]EPFL-STI-IMX-LTP, Powder Technology Laboratory, Materials Institute MXD 334, Station 12, CH-1015 Lausanne, Switzerland
[3]EPFL-SB-ISIC-GGEC, Station 6, CH-1015 Lausanne, Switzerland
[4]EPFL-SB-ISIC-LPI, Station 6, CH-1015 Lausanne, Switzerland

This study presents the first report on enhanced bacterial inactivation of *E. coli* by RF-plasma pretreated cotton with high surface area CuO powders compared with non-pretreated cotton textiles. The high surface area CuO (65 m/g) powder was fully characterized. The *E. coli* inactivation proceeded in the dark and was accelerated under visible and sunlight irradiation even at very low levels of visible light irradiation. The effect the RF-plasma pretreatment of the cotton on the binding of CuO, applied light dose, and initial *E. coli* concentration on the inactivation kinetics of *E. coli* is reported.

Keywords: RF-plasma; CuO high SSA, cotton, *E. coli*, visible light

1. Introduction

The manufacturing of high value added products such as bactericide textiles has increased rapidly in the last few years [1,2]. This development also opened new interest in the pre-treatment or activation of the surface of textiles by RF-plasma, Vacuum-UVC light irradiation [3,4]. Our group has recently reported the loading of Ag on different textiles like cotton, wool and polyester previously activated by RF-plasma and vacuum-UVC [3,5]. These studies indicated that the pre-treatment of textile fibers by low pressure RF-plasma and Vacuum-UVC increased the number of binding sites like: carboxylic acid -COOH, peroxides -O-O-, percarboxylic acids, epoxides, lactams and other giving raise to organic functional groups to bind Ag [3] or TiO_2 [4]. These pre-treatments roughens the textile surface also leading to: a) C-C bond scission, b) decreasing the C-H bonds due to the destruction of intermolecular H-bonding and c) introducing additional –C=O groups as observed by XPS having the effect of increasing the textile polarity and hydrophilicity which facilitates the adhesion between nanoparticles and the fabric [6].

The bactericide properties of Cu have been widely reported [7]. Recently, Borko and Gabbay [8] have reported bactericide, antiviral, algicide and fungicide textiles that sustain well washing cycles loaded with CuO 1%. This textile is effective against *E. coli*, *Staphylococcus aureus* and the *fungi Candida albicans*. These textiles are used in hospitals and laboratories using CuO against already resistant antibiotics involving nosocomial infections. Recently our laboratory has reported the synthesis of large surface area CuO effective in the *E. coli* inactivation in the dark and under visible light [9,10].

The present study addresses: a) the high surface area preparation and characterization of CuO, b) the adhesion of the nano-particulate CuO on the cotton activated by RF-plasma in comparison with non plasma activated surfaces [3,4] and c) the evaluation of *E. coli* inactivation by cotton/CuO in the dark and under light.

2. Experimental Section

2.1. Synthesis of high surface area CuO and pre-pretreatment

The preparation of high surface CuO has been reported by our laboratory [9]. The CuO (65 m^2/g) was deposited on bleached cotton received from Cilander AG, Cilanderstr. 17, CH-9101, Herisau. The cotton samples were fist activated by RF-plasma source from Harrick Corp, UK with a generator source of 13.56 MHz, 100W at a pressure of 1 torr for 30 min. Then, the samples were immersed in a sonicated solution of CuO (0.2 g/mL) and heated for 1 h at 75 °C. The loading of the samples was completed in three steps: (a) heating at 60°C for 1h and (b) followed by heating at 100 °C for 15 min and finally c) the loose CuO was sonicated out of the samples in distilled water for 5 min followed by drying at 60 °C.

2.2. Specific surface area (SSA) and particle size distribution

The (BET) specific surface areas S_{BET} (m²/g) were determined from N_2 adsorption isotherms (Micromeritics 2375) for the oxalate and CuO powders. The size of the primary particles, d_{BET} (nm) was calculated by assuming spherical monodisperse particles.

2.3. Electron Microscopy (SEM) and X-Ray fluorescence determination of Cu on textiles

The powder morphology was analyzed by scanning electron microscopy (SEM, Philips XL 30 FEG microscope). The SEM sample was prepared by dispersing the CuO powder in ethanol. The X-ray fluorescence spectrometer used was an RFX, PANalytical PW2400.

2.4. Irradiation of cotton/CuO samples during *E. coli* inactivation

Figure 4 shows in the insert the visible emission spectrum of the Osram Lumilux T8-L18W lamp (Winterthur, Switzerland) used to activate the cotton/CuO, having each lamp an output of 1.2 mW/cm^2.

2.5. Bacterial inactivation under light and in the dark

The bacterial strain *Escherichia coli* K12 was inoculated in 5 mL of Luria Bertani solution and grown during 8h at 37 ◦C with constant agitation under aerobic conditions. Aliquots of the overnight culture were inoculated into a fresh medium and incubated aerobically at 37 ◦C for 15 h. At the stationary growth phase, bacteria cells were collected by centrifugation at 500×*g* for 15 min at 4 ◦C and the bacterial pellets were washed three times and re-suspended in a NaCl/KCl solution.

The bactericide activity of cotton/CuO was tested on *E. coli* sterilizing 2 cm^2 of cotton/CuO and cotton samples by autoclaving at 121 °C for 2 h. Then, 20 μL aliquots of the culture in NaCl/KCl solution were added to the cotton/CuO and cotton samples to adsorb the bacteria before irradiation. Next, the samples were transferred into a sterile 2 mL Eppendorf tube containing 1 mL autoclaved NaCl/KCl saline solution and mixed thoroughly using a vortex for 3 min. Serial dilutions were made in NaCl/KCl solution. A 100 μL of each aliquot was spread on the nutrient agar plate. Agar plates were incubated for 37 °C for 24 h before counting the CFU. The recovery percentage by the experimental method used was 91.7 ± 2.7%.

3. Results and Discussion

3.1. Scanning electron microscopy of CuO (SEM)

From the SEM picture in Figure 1a we could observe the morphology of the CuO powder, composed of cubic particles 2 micrometers in size with a median size measured of d_{v50} =1.85 μm. Figure 1b presented the SEM of CuO powders (65 m^2/g) shows the structure of the cubic crystallites. If we take the known relation for the size of the CuO particle $d_{BET} = 6000 /(S_{BET}\ \rho)$ with BET specific surface area of CuO (65 m^2/g) and ρ the CuO density of 6.5 g/cm^3, then we can estimate the primary particle size d_{BET} is ~14 nm in the nanometer range.

Figure 1a. SEM of CuO powders having 65 m^2/g.
Figure 1b. SEM of CuO showing the detailed structure of the cubic crystallites.

3.2. *E. coli* inactivation by cotton/CuO (65 m^2/g) under visible light irradiation

The beneficial effect of the RF-plasma on the cotton surface introducing additional binding sites can be seen in Figure 2 for *E. coli* K12 bacterial inactivation. Trace 2a shows a negligible decrease of *E. coli* on cotton in the dark on samples not activated by RF-plasma because little CuO sticks to the cotton surface. Trace 2b shows an increased *E. coli* inactivation within 4 hours due to the highly oxidative species generated by CuO under light. The additional binding sites introduced by RF-plasma on the cotton binds more CuO increasing the bacterial inactivation in the dark and under light as shown in traces 2c and 2d respectively.

3.3. Effect of applied visible light intensity

Figure 3 shows the *E. coli* inactivation as a function of the applied light intensity. The bacterial inactivation time decreases as the intensity of the applied visible light was increased from 1.2 mW/cm^2 (1 lamp) to 4.7 mW/cm^2 (5 lamps). The semiconductor character of CuO has been recently described [9]. Under irradiation CuO leads to oxidative radicals by a mechanism

$$CuO + hn (< 720 \text{ nm}) \rightarrow CuO(e^-_{cb}, h^+_{vb}) \quad (1)$$

The e^-_{cb} in eq(1) is produced from the CuO (p-type) with a band-gap energy of 1.72 eV, a flat-band potential of -0.3 V SCE (pH 7) and a valence band at +1.4 V SCE [10]. The electron-hole pair is formed with photon energies exceeding the band gap of CuO. The excited electron could either react (a) directly with the O_2 forming $O_2^{\bullet-}$ (reaction 2) or by (b) reducing the Cu^{2+}-lattice to Cu^+ leading to the ensuing reactions (2-4):

$$O_2^{\bullet-} \text{ radical: } CuO(e^-_{cb}) + O_2 \rightarrow CuO + O_2^{\bullet-} \ (2);\ CuO(e^-_{cb}) \rightarrow CuO(Cu^+) \quad (3)$$

$$CuO(Cu^+) + O_2 \rightarrow CuO(Cu^{2+}) + O_2^{\bullet} \quad (4)$$

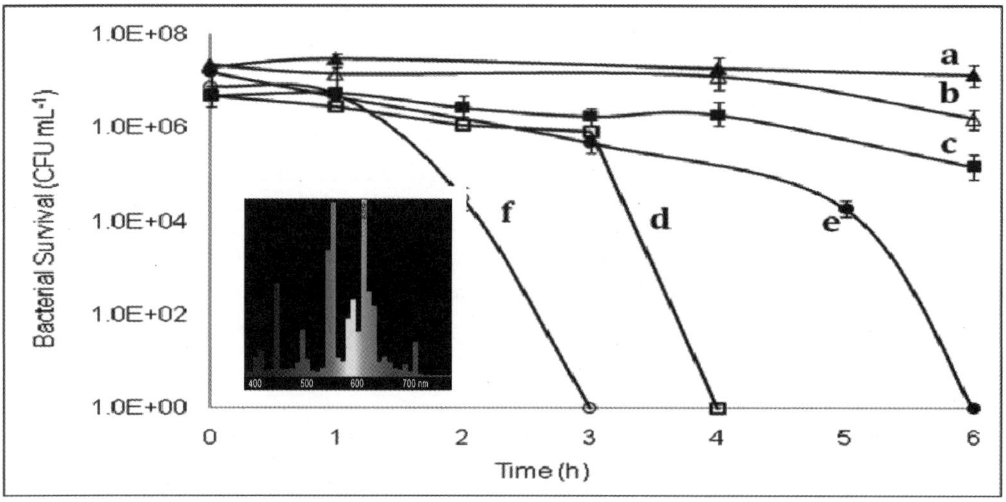

Figure 2. *E. coli* inactivation), a) cotton alone, dark b) cotton under visible light irradiation from an Osram Lumilux T8-L18W source with 5 lamps (4.7 mW/cm^2) c) cotton/CuO(65m^2/g), (0.71 %wt/wt), dark d) cotton/CuO(65m^2/g), (0.71% wt/wt), under visible light irradiation from source with 5 lamps (4.7 mW/cm^2) e) RF-plasma cotton/CuO(65m^2/g), (071% wt/wt), dark f) RF-plasma cotton/CuO(65m^2/g), (0.71% wt/wt), under visible light irradiation from 5 lamps (4.7 mW/cm^2). The insert shows the wavelength distribution of the Osram Lumilux T8-L18W lamp.

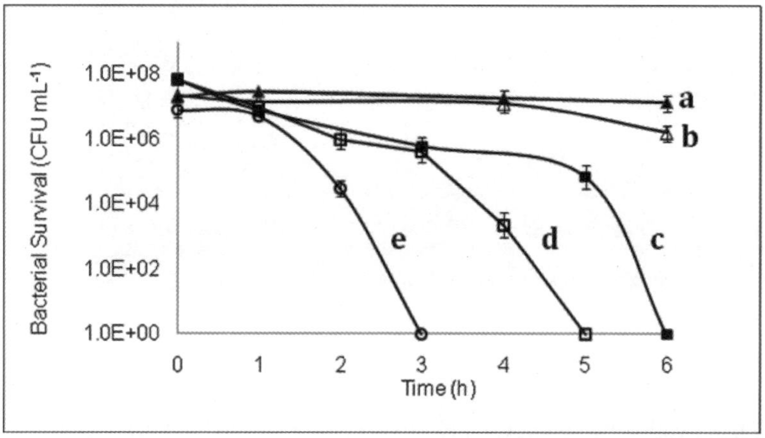

Figure 3. *E. coli* inactivation mediated by: a) cotton alone dark, b) cotton under visible light irradiation from an Osram Lumilux T8-L18W source with 5 lamps (4.7 mW/cm^2), c) RF-plasma cotton/CuO, (65 m^2/g) with a loading of 0.71% wt/wt in the dark (■) d) same as c) but under light irradiation of 1 lamp (1.2 mW/cm^2) (□), e) same as d) under light irradiation of 5 lamps (4.7 mW/cm^2) (O).

3.4. Bacterial inactivation kinetics as a function of the initial concentration of *E. coli*.

Figure 4 shows the *E. coli* inactivation kinetics for two different initial bacterial concentrations on cotton/CuO ($65m^2/g$) in the dark and under light. The lowest visible light intensity of 1.2 mW/cm^2 (1 lamp) was applied. This light intensity used was about 1% of the full sunlight intensity (AM 1) and this low light intensity was seen to have a significant effect enhancing bacterial inactivation. As noted in the captions in Figure 4, the *E. coli* inactivation kinetics increases when visible light is applied and also becomes faster as expected when the initial concentration of *E. coli* decreases.

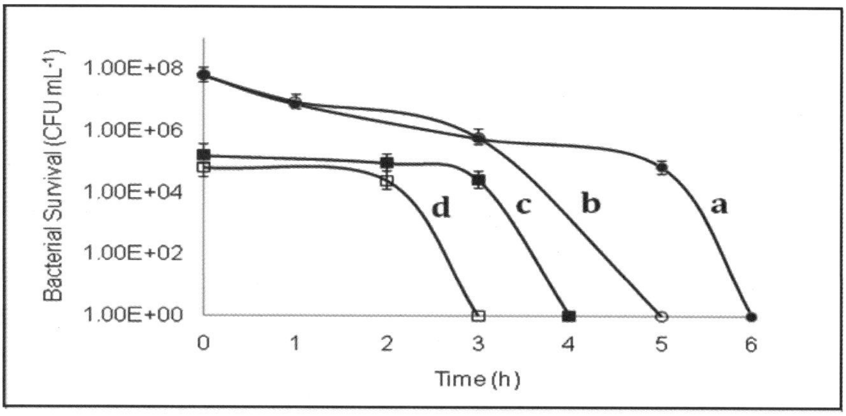

Figure 4. Bacterial inactivation, RF-plasma cotton/CuO($65m^2/g$) (0.71% wt/wt), under visible illumination as a function of initial number of *E. Coli*, 1 lamp (1.2 mW/cm^2) a) dark (●) b) light (○) c) dark (■) d) light (□).

References

[1] Yuranova, T, Rincon A, Laub D, Kiwi J. Synthesis, activity and characterization of textiles showing self-cleaning activity under daylight irradiation and bactericide action in dark processes, Catal. Today, 2007; 122: 109-117.

[2] Kiwi J, Pulgarin C. Self-cleaning and bactericide textiles, Catal. Today, 2010; 151: 2-7.

[3] Yuranova T, Rincon A, Bozzi A, Parra S, Pulgarin C, Kiwi J. Antibacterial textiles prepared by RF-plasma and vacuum-UV mediated deposition of silver, J Photochem. Photobiol. A. Chem. 2003; 161: 27-34.

[4] Bozzi A, Yuranova T, Kiwi J. Self-cleaning of Wool-polyamide and Polyester Textiles Due to Surface TiO_2-rutile Modification Under Daylight Irradiation, J. Photochem. Photobiol. A. Chem. 2005; 172: 27-34.

[5] Yuranova, T, Rincon G. A, Pulgarin C, Laub D, Xanthopoulos N, Mathieu H-J, Kiwi, J. J. Photochem. Photobiol. A. 2006; 181: 363-369.

[6] Daoud W.A, Xin, J. H, Zhang Y. H. Surf. Sci. 2005; 599: 6975- 6980.

[7] Hostynek J, Maibach I. Rev. Environ. Health, 2003; 18: 153-183.

[8] Borkow, G. and Gabbay. J. FASE, 2004: 18:1728-1730.

[9] Bandara J, Bowen P, Soare L, Jardim W, Kiwi J. Langmuir, 2005; 21: 8554-8559.

[10] Paschoalino M, Guedes N, Jardim W, Mielczarski E, Mielczarsi J, Bowen P, Kiwi, J. J. Photochem. Photobiol. A. 2008; 199: 105-111.

Is synergy of antimicrobials the effective way of management of resistance among cosmetically significant skin microflora?

BK Mohanty[1], S Gokul Shankar[2*], MS Ranjith[2] and M Prabhamanju[3]

[1]Pharmacology Unit, Faculty of Medicine, AIMST University, Semeling, Jalan Bedong, Bedong, 08100, Kedah, Malaysia [2]

[2]Microbiology Unit, Faculty of Medicine, AIMST University, Semeling, Jalan Bedong, Bedong, 08100, Kedah, Malaysia [2]

[3]R & D Centre, Cholayil Private Limited, SIDCO Industrial Estate, Ambattur, Chennai 600 098, India

*Corresponding author: e-mail: gokkavi@gmail.com

Cosmetic and personal care products (anti-acne, anti-dandruff, anti-odorant, prickly heat talc, etc) use synthetic antimicrobials like Triclosan, Farnesol, Zinc Pyrithione etc or herbal extracts for the anti-microbial 'functional' benefit. The use of single anti-microbial agent would pave way for emergence of resistance in the cosmetically significant skin micro-organisms. To combat the development of resistance and deliver the anti-microbial benefit, a combination of synergistic antimicrobials can be used. The paper reports on the use of synergistic combination of selective synthetic and natural compounds to deliver the desired anti-microbial benefit in these products.

Keywords: antimicrobial research; cosmetics; anti-dandruff; anti-acne; anti-prickly heat; herbal antimicrobials

1. Introduction

Skin and hair reflects our inner health and beauty. Cosmetics and personal care products are used by people all over the world for their skin and hair care. Several products developed to deliver anti-microbial functional benefits such as Anti-dandruff, Anti Prickly heat, Anti-acne, Anti-odorant etc need to eliminate the microorganisms that are primarily responsible for these skin conditions. The reduction in the load of these causative microorganisms results in the remission of these clinical conditions that are experienced as cure by the consumers. The causative organisms are often the normal flora of the skin/scalp whose sudden proliferation driven by several environmental and host conditions resulting in such clinical manifestations. Synthetic anti-microbial agents such as Triclosan, Farnesol, Zinc Pyrithione, Erythromycin, Chlorphenicin etc are commonly used as 'functional antimicrobials' in cosmetics and personal care products because of their efficient microbicidal activity. However, repeated use of these antimicrobials in higher concentrations pave way for the resistance among the target microbes. Further, the recommended dosage levels of use of these antimicrobials in cosmetics and personal care products are often associated with skin allergies, irritation, etc besides posing potential environmental hazards. Besides, there may be absorption of these antimicrobials through skin to produce systemic side effects. The present study reports on developing products with herbal ingredients in synergistic combination with synthetic agents thereby reducing the levels of synthetic ingredients in the formulations yet deliver the desired anti-microbial functional benefits of these products.

2. Objectives

1. To find the synergistic combination of selective synthetic and natural compounds to deliver the desired anti-microbial benefit in some cosmetic and personal care formulations viz. Antidandruff shampoo, Deodorant spray, Anti-acne cream and Prickly heat powder.
2. To compare the above formulations with popular brands available in the market (which use the same synthetic anti-microbial agents) for their efficacy against the causative organisms responsible for the respective clinical conditions.

3. Methodology [1-3]

3.1 Herbs used

Over 50 herbs with anti-microbial activity used in traditional medicine were selected through literature reserach [4-7] screened for the syngergistic microbicidal activity with synthetic antimicrobials on *as is* basis and in formulations and a selected few were short-listed (Table 1) for use in the study formulations.

3.2 Extraction

The herbal extracts were prepared using soxhlet apparatus. In short, 100 g of powdered sample was extracted using selected solvents (Table 1) separately. Then the extract was filtered and the filtrate was evaporated at low temperature. The final residue was used for evaluating the antimicrobial activity.

Table 1: Herbal extracts used in the study

Herbs shortlisted		Parts used	Solvent used for extraction
Common name	Botanical name		
Datura	*Datura metal*	Leaves	Ethanol
Albizia	*Albizia amara*	Whole plant with leaves	Ethanol
Black pepper	*Piper nigrum*	fruit	Benzyl alcohol
Turmeric	*Curcuma longa*	Rhizome	Ethanol
Wild Leadwort	*Plumbago indica*	Root	Ethyl acetate & ethanol
Tulasi	*Ocimum sanctum*	Leaves	Ethanol
Clove	Syzygium aromaticum	Flower bud	Ethanol
Grape seed	*Vitis vinifera*	Seeds	Benzyl alcohol

3.3 Formulation

Different cosmetic/personal care study formulations (anti-dandruff shampoo, anti-acne cream, prickly heat powder and alcohol based deodarant) were prepared with selective herbal and synthetic antimicrobials with the help of a cosmeticologist. For the purpose of comparative study, popular brands (names not disclosed to avoid commercial implication) of respective products were procured from the market.

3.4 Cultures used (Etiological agents of skin conditions)

Ten isolates of each of the organisms were used. Nine of the isolates were of clinical origin and one isolate of each of the organism was a standard culture

Dandruff: *Malassezia globosa* and *M.restricta*

Acne: ***Propionibacterium acnes,*** *Staphylococcus epidermidis and Micrococcus luteus*

Prickly heat and malodor: Propionibacterium acnes, Staphylococcus epidermidis and Micrococcus luteus

All the clinical isolates of the different cultures used in the study were obtained from Department of Microbiology, The New College, Chennai, India and the standard cultures were procured from Microbial Type Culture Collection IMTECH, Chandigarh, India

3.5 Methods of antimicrobial testing

MIC assay was done for test formualtions in comparison with the popular brands in the market. In addition, zone of inhibition (ZOI) determination was done for the shampoo using standard procedures [1-3]

3.5.1 MIC Determination

Agar dilution technique was employed for antidandruff/ anti acne/ anti prickly heat evaluation of the test formulations and popular market products *in vitro*. Based on the growth inhibition or presence of growth of the organisms in the media containing different concentrations of the test formulations, the activity level of the product was evaluated. In short, Sabouraud's Dextrose Agar (SDA) overlaid with olive oil was used for determining the activity of anti-dandruff shampoo and Mueller Hinton Agar (MHA) was used for determining the activity of other products. 0.1 ml of cell suspension of cultures in normal saline adjusted to 0.5 McFarland Standard was used as inoculum. The developmental products were weighed in different concentrations 10, 25, 50, 75, 100, 200, 400, 500, 750, 1000 mg quantities separately in appropriate medium. Wherever, necessary higher or lower quantities were taken for assay. One ml of 20% DMSO was added aseptically to the weighed products and was homogenized fully using a sterile glass rod. To the above-homogenized product, 10 ml of the warm agar medium was added and mixed thoroughly to achieve uniform distribution of the products in the medium. After over night air-drying of the plates at room temperature, the standardized culture suspension was inoculated onto the plates and incubated at 30 ^{0}C for 5 days for fungi and at 37 ^{0}C for bacteria for 3 days. The presence of growth represents the product is inactive and the absence of growth represents the product is active at that concentration tested. Proper control plates were maintained for testing the growth of the culture suspension. The experiments were repeated at least thrice with each of the isolates of the different test organisms.

3.5.2 ZOI Determination

ZOI test was employed to determine the efficacy of AD shampoo.. In short, 10 ml of molten medium was poured in petriplates and kept for solidification. After air-drying of the plates, a well measuring less than 10mm was cut in the center of the plates containing the medium. *Malassezia* species was swabbed over the surface of the SDA medium overlaid with olive oil so as to form a lawn culture. After drying of the culture, a well was cut in the medium at the center of the plate. One gram of the test product (shampoos) was dissolved in 1 ml DMSO and 140 µl of this suspension was filled in the well. The plates were incubated at 30 ^{0}C for the growth of yeast for 5 days and the diameter of the zones of inhibition were measured using a Zone measuring scale and the readings were tabulated. The experiments were repeated at least thrice with each of the isolates of the different test organisms.

4. Results

The test formulations were more effective in their anti-microial activity than the popular brands that used either herbal or synthetic antimicrobials individually (Tables 2-5)

Table 2: Anti-dandruff shampoo formulation - Test Vs Popular brands

S. No.	**Brand**	**Active Ingredient**	**Level in %**	Zone of Inhibition (mm)	MIC Value (mg/ml)	
					M. globosa	*M. restricta*
01	Brand 1	ZnPTO	0.92	26	10	10
02	Brand 2	Tea Tree oil	0.15	20	30	30
03	Brand 3	ZnPTO Benzyl alcohol	1.05 0.02	26	10	10
04	Brand 4	ZnPTO Benzyl alcohol	0.92 0.02	26	10	10
05	Brand 5	Climbazole	0.23	25	10	10
06	Brand 6	ZnPTO	0.45	20	10	10
07	Brand 7	Ketoconazole	0.10	26	5	5
08	Test Formulation	ZnPTO Herbal Extract 1 (*Datura metal*) Herbal Extract 2 (*Albizia amara*)	0.02 0.02 0.02	28	5	2.5
09	Test Formulation	Ketoconazole Herbal Extract 1 (*Datura metal*) Herbal Extract 2 (*Albizia amara*)	0. 02 0.02 0.02	28	5	5

Table 3: Anti-acne cream formulation - Test Vs Popular brands

S.No	Product	Ingredient	Level in %	MIC Value (mg/ml)		
				P.acnes	*S.epidermidis*	*M.luteus*
01	Brand 1	Turmeric extract Sandalwood oil	16.00 0.50	NA up to 70	NA up to 70	NA up to 70
02	Brand 2	Erythromycin	1.00	NA up to 70	50	50
03	Brand 3	Triclosan	1.00	30	30	30
04	Test Formulation	Triclosan Herbal Extract 1 (*Piper nigrum)* Herbal Extract 2 (*C*urcuma longa)	0.10 0.10 0.10	15	20	20

Table 4: Prickly Heat powder formulation - Test Vs Popular brands

S No	Product	Ingredient	Level in %	MIC Value (mg/ml)		
				P.acnes	*S.epidermidis*	*M.luteus*
01	Brand 1	Zinc oxide Chlorphenesin	16.00 1.00	15	5	5
02	Brand 2	Zinc oxide Boric acid Salicylic acid Camphor Menthol Alcohol Other herbs	10.00 4.00 1.50 0.20 0.40 0.70 17.10	15	10	10
03	Brand 3	Zinc oxide Boric acid Salicylic acid Alcohol Menthol	10.00 3.50 1.50 0.70 0.30	25	20	10
04	Brand 4	Zinc oxide Salicylic acid	5.00 0.83	15	10	10
05	Brand 5	Zinc oxide Salicylic acid Boric acid	10.00 1.50 5.00	15	10	10
06	Test Formulation	Herbal Extract 1 *Plumbago indica* Herbal Extract 2 *Ocimum sanctum* Salicylic acid Boric acid Zinc oxide	0.20 0.40 0.20 2.00 3.00	5	2.5	3

Table 5: Alcohol based deodorant spray - Test Vs Popular brands

S.No	Product	Ingredient	Level in %	MIC Value		
				P.acnes	*S.epidermidis*	*M.luteus*
01	Brand 1	Triclosan	0.50	10	10	10
02	Brand 2	Farnesol	1.00	10	10	10
03	Test Formulation 1	Triclosan Herbal Extract 1 (Grape seed) Herbal Extract 2 (Clove) Herbal Extract 3 (Black pepper)	0.01 0.10 0.10 0.20	2.5	2.5	2.5
04	Test Formulation 2	Farnesol Herbal Extract 1 (Grape seed) Herbal Extract 2 (Clove) Herbal Extract 3 (Black pepper)	0.20 0.10 0.10 0.20	5	5	5

5. Discussion

Most of the common skin conditions are associated with normal skin flora. Organisms like *Staphylococcus epidermidis, Micrococcus luteus, Propionibacterium acnes, etc* are associated with conditions like miliaria rubra (prickly heat), furuncles (boils), acne (pimples). Body odour is caused by a natural process involving sweat that occurs on the skin's surface. Sweat is odourless, but if left on the skin the bacteria that normally live there feed on it and break it down. This process releases chemicals that cause the unpleasant smell. Some areas of the skin, such as the armpits, genitals, body folds etc are more likely to produce body odour because the sweat/sebaceous glands in these areas produce proteins and oily substances. This facilitates the proliferation of normal skin bacteria like *Diptheroids* and *S. epidermidis* resulting in odorous by-product like isovaleric acid.

Malassezia species are lipid-dependent commensal flora that adapt to the narrow niche provided by sebum-rich scalp. *Malassezia globosa* and *M. restricta* initiate dandruff formation due to its high lipase activity.

Cosmetics and personal care products developed against conditions like acne, prickly heat, dandruff and body odour should have antimicrobial agents that contain the growth of these normal skin flora that are responsible for such conditions. Personal hygiene is therefore an important aspect in the management of these conditions. Functional cosmetics and personal care products with anti-microbial ingredient(s) play a vital role in proper control of these skin microorganisms. Repeated and extensive use of personal care/ cosmetic products also paves way for the emergence of resistance among microorganisms [8,9] or cause environmental pollution [10]. Of major concern is the possibility that resistance to antimicrobials used in cosmetics/ personal care products may also contribute to reduced susceptibility to clinically important antimicrobials, due to either cross-resistance or co-resistance mechanisms[9]. Hence there is a need for the development of eco-friendly and skin friendly cosmetics and personal care products by lowering the levels of synthetic anti-microbials. The use of herbal antimicrobials in synergistic combination with synthetic anti-microbial agents seems to be a better option. In our current study we have shown that the addition of one or more herbal extracts not only helps to reduce the level of synthetic anti-microbials like Triclosan, Farnesol, Zinc pyrithione, etc in the formulations but also make them more efficacious than those popular brands that use stand alone synthetic antimicrobial ingredients.

5.1 Anti-dandruff

The microbicidal activity of anti-dandruff ingredients like ZnPTO and Ketoconazole is enhanced by herbal extracts like *Datura metal* and *Albizia amara* even when used at lower concentration (0.02%) while the popular brands use higher levels of these synthetic anti-fungals (Table 2).

5.2 Anti-acne

Activity of Triclosan is enhanced even at a concentration of 0.1% by synergistic action of extracts of *Piper nigrum* and *Curcuma longa.* This combination in anti-acne formulation may serve better than the erythromycin based anti-acne creams for which there seems to be an emerging resistance among the skin microflora especially *P. acnes* (Table 3). It is also interesting to note that the extract of *Curcuma longa* (turmeric) even when used at 16% concentration in Brand 1 did not record anti-microbial activity up to 70 mg/ml, where as the test formulation using 0.1% of the *Curcuma longa* extract in combination with *Piper nigrum* (0.1%) and Triclosan (0.1%) recorded significant activity against *P. acnes* and other skin bacteria. This validates our hypothesis on synergy between the herbal and synthetic antimicrobials.

5.3 Anti prickly heat

Test formulation with the unique combination of *Plumbago indica, Ocimum sanctum,* Salicylic acid, Boric acid and Zinc oxide recorded highest antimicrobial activity against the skin microflora when compared to the popular prickly heat products in the market (Table 4)

5.4 Anti-odorant

Levels of antimicrobials like Triclosan and Farnesol which are commonly used in deodorants can be reduced to a minimum level by addition of extracts of clove, grape seed and black pepper extracts (Table 5)

6. Conclusion

In our present study, we have established that the synergistic combination of herbs in formulations helps to reduce the levels of synthetic antimicrobials and also results in increasing the efficacy of the product. Further, such unique combinations, that are arrived at after several trial and error permutation combinations, have several added advantages besides functional efficacy which includes preventing the possible development of resistance in target microbes, reducing the risk of allergy, irritation or other health related hazards in users (safe & skin-friendly) and minimizing environmental accumulation of the synthetic chemicals (eco-friendly). Further, there may also be a reduction in cost of production (consumer friendly) of these formulations because the levels of costly synthetic antimicrobials are reduced. This is a preliminary *in vitro* study and functional anti-microbial claims of the test formulations are further to be substantiated by clinical studies. However, the present study gives a different insight to the judicious use of antimicrobials in cosmetics and personal care formulations in future.

Acknowledgements The authors thank the Management, Vice Chancellor and Dean of Faculty of Medicine of AIMST University, Malaysia and the Management of Cholayil Private Limited, India for their support

References

1. Donald S Orth. *An introduction to cosmetic microbiology*. Micelle press, England;1999.
2. Dhanya N, Rajan S, Rai S, Gokul Shankar S, Babu K, Subashini S. Studies on the antimicrobial activity of Plumbago indica L on certain human skin microflora *Bull. Biol. Sci.* 2009; 8 (2): 129-134.
3. Prabhamanju M, Gokul Shankar S, Babu K, Ranjith MS. Herbal vs. chemical substances as antidandruff ingredients: which are more effective in the management of Dandruff? – An overview *Egyptian Dermatology online Journal;* 2009; 5(2)
4. Ayurvedic Pharmacopoeia of India, Vol-3, Govt. of India, Ministry of health and family welfare, Dept. of ISM & H, New Delhi; 2001.
5. Sivarajan VV, Balachandran I. *Ayurvedic Drugs and Their Sources*, Oxford & IBH Publishing Co. Pvt. Ltd., New Delhi, India; 1994.
6. Warrier PK Indian Medicinal Plants, Vol. 2, Orient Longman Ltd., Kottakkal, Kerala, India; 1994.
7. Kurup PNV, Ramadas VNK, Joshi P. Handbook of Medicinal Plants, New Delhi, India; 1979.
8. Aurélie Deveau, Amy EP, Angelyca AJ, and Deborah AH. Farnesol Induces Hydrogen Peroxide Resistance in *Candida albicans* Yeast by Inhibiting the Ras-Cyclic AMP Signaling Pathway. *Eukaryotic Cell.* 2010; 9(4):569-577.
9. Yazdankhah SP, Scheie AA, Høiby EA, Lunestad BT, Heir E, Fotland TØ, Naterstad K, Kruse H. Triclosan and antimicrobial resistance in bacteria: an overview. *Microb Drug Resist.* 2006; 12(2):83-90.
10. Rule KL, Ebbett VR, Vikesland PJ. Formation of chloroform and chlorinated organics by free-chlorine-mediated oxidation of triclosan. *Environ Sci Technol.* 2005; 39 (9): 3176–85.

Microencapsulation of *Melaleuca alternifolia* (tea tree) oil as biocide for footwear applications

M. M. Sánchez-Navarro*, N. Cuesta-Garrote, F. Arán-Aís and C. Orgilés-Barceló

INESCOP. Footwear Research Institute. Pol. Industrial Campo Alto s/n. 03600 Elda (Alicante). Spain.

* Corresponding author: msanchez@inescop.es, +34 965 395 213

Functional materials with antimicrobial properties are being investigated in order to provide shoes with new properties and added value by incorporating natural antimicrobial agents to conventional materials (leather, fabrics, foams, etc). Microencapsulation is an effective method to protect these functional natural biocides from reactions with moisture, light, and oxygen. If a footwear material is treated with microencapsulated biocides agents, higher durability of functionality is expected. This article reports the evaluation of *Melaleuca alternifolia* (tea tree) oil which is a natural essential oil, as antimicrobial agent for footwear application. Furthermore, the synthesis of microcapsules containing this essential oil by *in situ* polymerization method using a melamine-formaldehyde resin as shell material was studied in order to increase the durability of the antimicrobial effect in footwear materials. Microcapsule characterization, such as particle size, morphology and chemical properties was carried out for different oils to polymer ratios in the microcapsules. Finally, the incorporation of the microencapsulated biocides into different footwear materials was also evaluated.

Keywords melamine-formaldehyde; antimicrobial activity; Tea tree oil (TTO); O/W emulsions; *Staphylococcus aureus*, microencapsulation, *in situ* polymerization.

1. Introduction

Microencapsulation presents a new option for the shoe industry as its application can transform traditionally used materials or products into smart materials or products capable of interacting with feet. For instance, they can improve quality of life by incorporating products for foot care such as properly dosed essential oils. The microencapsulation of active substances to be incorporated in different footwear components in order to obtain an "active shoe" presents an opening up of a new way of innovation.

In recent years, interest in natural medicinal products, essential oils and other botanicals, has grown in response to the ever increasing incidence of adverse side effects associated with conventional drugs, and the emergence of resistance to antibiotics, synthetic disinfectants and germicides. Particularly, there has been resurgence of interest in Australian Tea Tree Oil (TTO) which has been employed for its germicidal activity since 1925 [1-3]. This oil is obtained by steam distillation of the leaves of *Melaleuca alternifolia* and contains almost 100 components [4], which are mostly terpene hydrocarbons, mainly monoterpenes, sesquiterpenes and related alcohols; it was used early the last century to treat various ailments with demonstrated activity against bacteria, fungi and viruses [5-11].

The *in situ* polymerization allows the formation of microcapsules containing water-immiscible dispersed phase, with improved mechanical properties and thermal stability. The properties of the membrane depend not only on its chemical structure but also on all the synthesis conditions. The polycondensation of the amino resin occurs in the continuous phase, and the phase separation is linked to the pH and the melamine-formaldehyde molar ratio [12-15].

In this study a series of melamine-formaldehyde (MF) microcapsules containing TTO was prepared by *in situ* polymerization (O/W) method to be applied to footwear materials (lining, insoles, etc…) as antimicrobial agent and their physicochemical properties have been characterized by different experimental techniques.

2. Materials

Melamine (99.9 % purity) and formaldehyde were obtained from Quimidroga S.A. (Barcelona, Spain) and used as received. Formaldehyde was a 36,5% solution containing around 4,8% of methanol. *Melaleuca alternifolia* oil was purchased from Guinama S.L.U. (Alboraya (Valencia), Spain). Sodium dodecylsulphate (SDS) from Sigma-Aldrich was used without further purification. Water was purified with Millipore automatic Sanitization Module.

3. Experimental

3.1. Oil antimicrobial activity determination.

Previously to the microencapsulation process, the antimicrobial activity of TTO against different microorganisms typically found in used footwear [16] (*Escherichia coli SG13009 (QIAGEN), Bacillus subtilis 168 ATCC 23857, Klebsiella pneumoniae CECT 141 and Staphylococcus aureus CECT 239*) was analyzed.

The antimicrobial evaluation of TTO was carried out by *in vitro* assays. On the one hand, the oil antimicrobial activity was measured in liquid media, Luria-Bertani medium (LB) [17] and nutrient broth (NB) [16]. On the other hand, in solid media the inhibition halos assay was carried out in LB and NB agar medium plates following the protocol described in ISO 20645:2004 [18].

To test the antimicrobial activity of TTO in liquid media, two tubes containing almond oil as positive control and tea tree oil as antimicrobial compound were prepared for each species and the above mentioned microorganisms were inoculated in LB (*Escherichia coli and Bacillus subtilis*) and NB (*Klebsiella pneumoniae and Staphylococcus aureus*), respectively. Moreover, an oil-free tube was inoculated as negative control. All of them have grown up at 37° C in an orbital incubator Gallenkamp INR-200. After 24 hours of growth, the absorbance at 600 nm was evaluated in an Ultraviolet-Visible Cecil CE3021 3000 series spectrophotometer[19,20]. In solid media, the evaluation of the inhibition halos assay was carried out by seeding three agar medium plates for each type of bacteria. It was placed on them a disc of cloth impregnated with the oils just mentioned: almond oil, tea tree oil and an oil-free cloth disc (negative control). The plates were left to growth at 30ºC in a Memmert INE 500 incubator with natural ventilation. After 24 hour of growth, the inhibitory halos were evaluated.

3.2. Synthesis of melamine-formaldehyde microcapsules containing tea tree oil.

First of all, a melamine-formaldehyde (MF) resin to be used as microcapsules shell was prepared. In a round bottom glass flask melamine, water and formaldehyde were added. The vessel was immersed in a water bath heated at 70ºC and the mixture was magnetically stirred for 1 hour. Finally, a clear solution was obtained.

After that, an O/W emulsion was prepared. The oil phase was composed of TTO and the aqueous one was constituted of sodium dodecyl sulphate (SDS) as surfactant and distilled water. Both phases were mixed and then emulsified by a sonifier (Branson) for 90 seconds at 50% amplitude. The result was a milky emulsion. Subsequently, the emulsion was maintained at 45 °C and magnetically stirred, while the MF resin prepolymer obtained in the previous section was added drop by drop. After the addition, the pH of the mixture was adjusted to 4 and the emulsion was left stirring for 1 hour. After this time the temperature rose to 85 °C and kept stirring for 2 hours. Finally, it was allowed to cool down to room temperature and the pH was adjusted to 10. Finally the synthesized microcapsules were freeze-dried, obtaining a fine white powder. In this study, microcapsules with three different MF/oil ratios were synthesized and characterized.

3.3. Characterization of microcapsules.

Physicochemical properties of the synthesized microcapsules containing TTO have been characterized by different experimental techniques. The size of the microcapsules was determined using a particle size analyzer Coulter LS 230; the chemical structure of the oil and the microcapsules obtained was analyzed by Fourier transform infrared spectroscopy, (FTIR). Finally, the incorporation of microcapsules to footwear materials was analyzed by scanning electron microscopy (SEM).

4. Results and Discussion

In Table 1 the behaviour of the microorganisms to the different oils studied (almond oil and tea tree oil) in order to test the antimicrobial activity are displayed. The results showed the effective antimicrobial activity for the *Melaleuca alternifolia* oil for the most microorganisms.

Table 1: Antimicrobial activity test results for different oils and for the control with no-antimicrobial substance.

Sample	*E. coli*		*B. subtilis*		*K. pneumoniae*		*S. aureus*	
	Liq	Sol	Liq	Sol	Liq	Sol	Liq	Sol
Control	-	-	-	-	-	-	-	-
Almond oil	-	-	-	-	-	-	-	-
TTO	++	-	++	++	+++	-	+++	++

(-) no antimicrobial activity was observed.
(++) moderate antimicrobial activity observed.
(+++) strong antimicrobial activity observed.

After the evaluation of the antimicrobial activity of the tea tree oil, the synthesis of microcapsules containing this essential oil by *in situ* polymerization method using a melamine-formaldehyde resin as shell material was carried out using different polymer to oil ratios.

Figure 1 shows the distribution of microcapsule diameter obtained for different MF to oil ratios during the synthesis. Mainly a bimodal distribution curve was obtained for all the microcapsules dispersion samples. The average diameters observed for the microcapsules synthesized were 0.706, 0.695, 1.092 and 2.133 μm for the MF1TTO, MF2TTO, MF3TTO and MF4TTO samples, respectively. As the resin molar rate in the microcapsules increases the amount of smaller particles were decreasing and the largest ones were increasing. The microcapsule diameter distribution shifts to larger size as the MF polymer content in the microcapsule increases. These results indicate that the thickness of the MF coating wall (shell) increases as the MF resin amount increases, according to the literature [21].

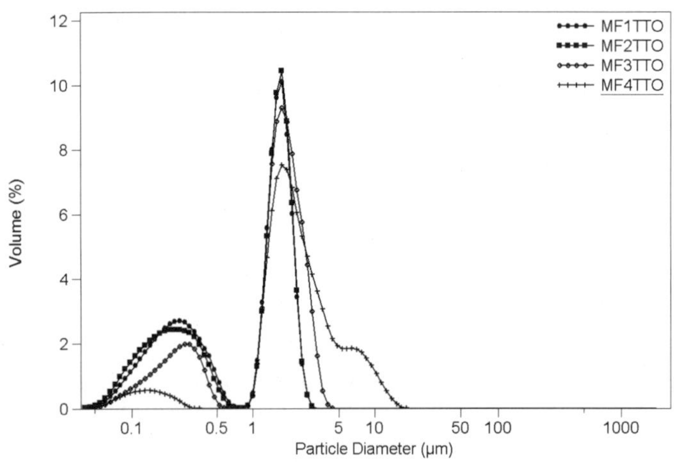

Figure 1. Particle size distribution in volume of the microcapsules containing Tea Tree oil synthesized.

Chemical properties of the microcapsules containing tea tree oil were analysed by FTIR spectroscopy. On the one hand, Figure 2.a. includes spectra corresponding to the melamine-formaldehyde resin synthesized to produce microcapsule shell and the tea tree oil composition used. MF resin shows a typical spectrum for this kind of polymers [12,13,22]. A characteristic broad band responsible for hydroxyl, imino and amino stretching was observed around 3335 cm^{-1}. Alkyl C-H stretching vibration was found around 2930 cm^{-1}. The C-N multiple stretching in the triazine ring was observed around 1537 cm^{-1}. C-H bending vibration in CH_2 was found at 1450 cm^{-1} and 1369 cm^{-1} due to methylene bridges. The characteristic absorption bands of aliphatic C-N vibration appeared between 1200 and 1170 cm^{-1}. Characteristic triazine ring bending at 812 cm^{-1} can also be observed.

The spectrum corresponding to the tea tree oil shows typical bands related to its chemical composition. A broad band responsible of hydroxyl stretching was observed around 3430 cm^{-1}. Alkyl C-H stretching vibration was found between 2960 and 2854 cm^{-1}. The band at 1744 cm^{-1} corresponding to double bond C=O vibration characteristic of oil components. Between 1464 and 1443 cm¯1 was found the C-H bending vibration. Characteristic absorption band of vibration C-O-C aromatic rings appeared at 927 cm^{-1}.

On the other hand, FTIR spectra of the microcapsules containing TTO obtained after freeze-dried proved to be the combination of the characteristic bands of MF resin and the *M. alternifolia* oil as it is shown in Figure 2.b. As the polymer to oil ratio in the microcapsule increases the intensity of the corresponding MF resin bands increases. Nevertheless, there are no so much differences on intensity between the spectra corresponding to the microcapsules with higher polymer to oil ratios.

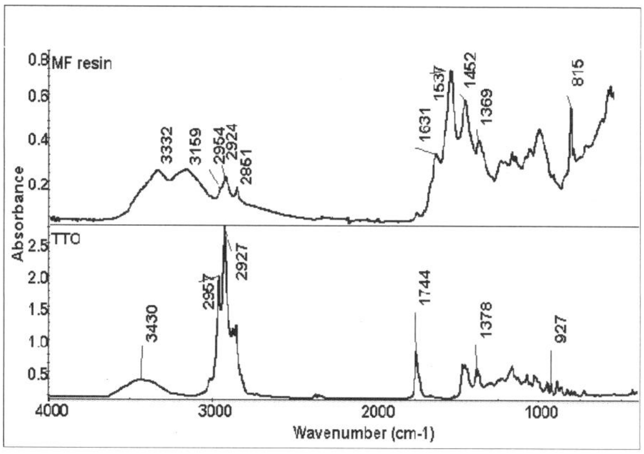

Figure 2.a: FTIR spectra of the melamine-formaldehyde resin and Tea Tree oil.

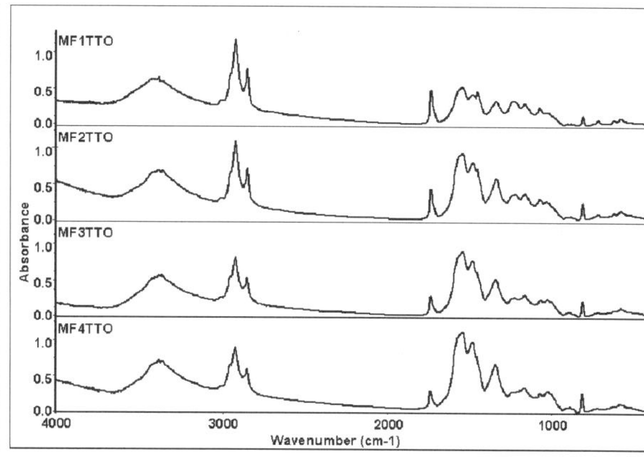

Figure 2.b: FTIR spectra of the melamine-formaldehyde microcapsules synthesized with different resin/oil ratio.

Finally, the microcapsules containing the tea tree oil with a polymer to oil ratio equal to 2 were incorporated to some materials as leather and fabrics employed in the footwear industry. In that process, a piece of each material was immersed into 50 mL of the microcapsules dispersion for 1 hour under mechanical agitation at room temperature. After that, the samples were left to dry at room temperature overnight. The anchorage of the microcapsules to the each material was analysed by scanning electron microscopy (SEM). Figures 3.a and 3.b. show SEM images of the microcapsules attached to the leather and the fabric fibres, respectively.

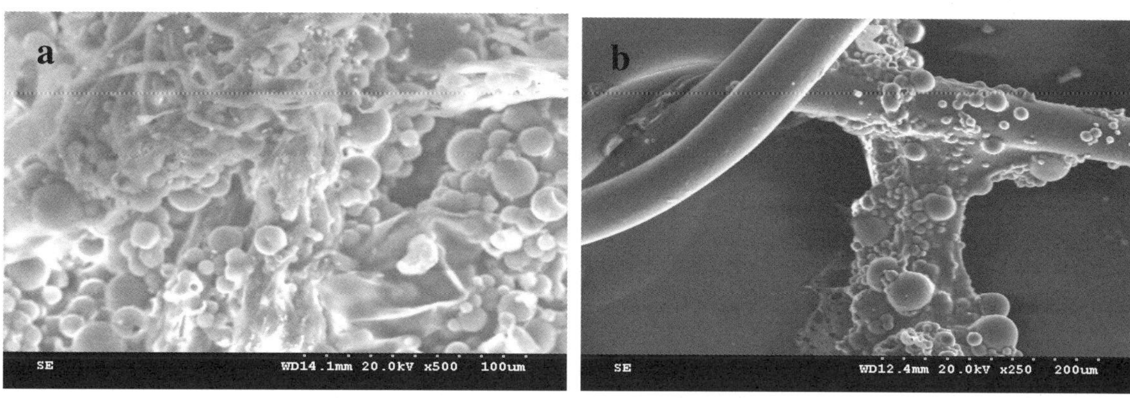

Figure 3: Microcapsules added to a) leather and b) fabrics used in the footwear industry.

5. Conclusions

Melaleuca alternifolia oil showed suitable antimicrobial activity against different microorganisms found in foot skin and worn shoes, to be used as a natural biocide for footwear applications.

Regarding the microencapsulation of this oil to be incorporated into different footwear materials, the relationship between the melamine-formaldehyde resin mass that forms the shell to TTO contained largely determines the efficiency of the microencapsulation process and the morphology of the microcapsules, as well as the particle size distribution thereof.

The results from the incorporation of microcapsules in leather and fabrics obtained have demonstrated the feasibility of this technology for use in shoes to achieve the concept of "active shoe". Furthermore, it is expected that the thickness of the microcapsule wall determines the durability of the antimicrobial effect of these functionalised materials.

On the other hand, the microencapsulation of TTO constituted an innovative solution for the control release of this active substance in footwear. Nevertheless, further optimization of the microcapsules characteristics should be needed in order to get the best properties for a long lasting antimicrobial effect in footwear applications depending on the type of shoe, user, materials.

Acknowledgements. The Research was partially funded by the Institute of Small and Medium Industry of the Generalitat Valenciana and FEDER, through Action 1 of its R&D Programme for Technological Institutes 2010 (project IMIDIC/2010/106).

References

[1] Anonymous. An Australian antiseptic oil. Br. Med.J. 1933 i: 966.
[2] Anonymous. A retrospect. Med.J. Aust. 1930 i: 85-89.
[3] Anonymous. Ti-trol oil. *Br. Med. J.* 1933 ii: 927.
[4] Brophy, J.J., Davies, N.W., Southwell, I.A., Stiff, I.A., Williams, L.R. Gas chromatographic quality control for oil of Melaleuca terpinen-4-ol type (Australian tea tree). *J. Agric. Food Chem.* 1989, 37: 1330-1335.
[5] Bassett, I.B., Pannowitz, D.L. and Barnetson, R. St.C. A comparative study of tea tree oil versus benzoylperoxide in the treatment of acne. *Med. J. Aust.* 1990, 153:455-458.
[6] Enshaied, S., Jooya A., Siadat A.H., Iraji F. The efficacy of 5% tea tree oil gel in mild to moderate acne vulgaris. A randomized, double-blind placebo-controlled study. *Indian J. Dermatol. Venereol. Leoprol.* 2007, 73:22-25
[7] Jandourek, A., Vaishampayan, J. K. and Vazquez, J.A. Efficacy of Melaleuca oral solution for the treatment of fluconazole refractory oral candidiasis in AIDS patients. *AIDS.* 1998, 12:1033-1037.
[8] Vazquez, J.A. and Zawawi, A.A. Efficacy of alcohol-based and alcohol-free Melaleuca oral solution for the treatment of fluconazole-refractory oropharyngeal candidiasis in patients with AIDS. *HIV Clin. Trials.* 2002, 3:379-385
[9] Carson, C.F., Ashton, L., Dry, L. et al. Melaleuca alternifolia (tea tree) oil gel (6%) for the treatment of recurrent herpes labialis. *J. Antimicrob. Chemother.* 2001, 48:450-451.
[10] Satchell, A.C., Saurajen, A., Bell, C. et al. Treatment of dandruff with 5% tea tree oil shampoo. *J. Am. Acad. Dermatol.* 2002, 47:852-855.
[11] Satchell, A.C., Saurajen, A., Bell, C. et al. Treatment of interdigital tinea pedis with 25% and 50% tea tree oil solution: a randomized, placebo-controlled, blinded study. *Australas. J. Dermatol.* 2002, 43:175-178.
[12] Lee, H.Y., Lee, S.J., Cheong, W., Kim, J.H. Microencapsulation of fragrant oil via in situ polymerisation: effects of pH and melamine-formaldehyde molar ratio. *J. Microencapsul.* 2002, 19:559-569.
[13] Salaün, F., Vroman, I. Influence of core materials on thermal properties of melamine–formaldehyde microcapsules. *Eur. Polym. J.* 2008, 44:849–860.
[14] Wei, L., Xing-Xang, Z., Xue-Chen, W., Jian-Jin, N. Preparation and characterization of microencapsulated phase change material with low remnant formaldehyde content. *Mater. Chem. Phys.* 2007, 106:437–442.
[15] Hwang, J.S., Kim, J.N., Wee, Y.J., Jang, H.G., Kim, S.h., Ryu, H.W. Factors affecting the characteristics of melamine resin microcapsules containing fragrant oils. *Biotechnol. Bioprocess Eng.* 2006, 11: 391-395.
[16] ISO/NP 16187. Footwear- Test methods for uppers, lining and insocks-antibacterial activity.
[17] Sambrook, J., Fritsch, E. F., Maniatis, T. Molecular Cloning, (appendix 1)(A Laboratory Manual-3) Second Edition Cold Spring Harbor Laboratory Press. 1989.
[18] ISO 20645:2004 Textile fabrics. Determination of antibacterial activity. Agar diffusion plate test.
[19] Madigan, T., Martinko, J. M., Parker, J. Brock: Biology of Microorganisms. Eighth Edition. Michael Prentice Hall International, INC. 1997.
[20] Monod, J.. The growth PF bacterial cultures. *Annu. Rev. Microbiol.* 1949, 3: 371–394.
[21] Lee, Y.H., Kim, C.A., Jang W. H., Choi, H.J., Jhon, M. S.. Synthesis and electrorheological characteristics of microencapsulated polyaniline particles with melamine-formaldehyde resins. *Polymer* . 2001, 42:8277-8283.
[22] Luo, W., Yang, W., Jiang, S., Feng, J., Yang, M. Microencapsulation of decabromodiphenyl ether by in situ polymerization: preparation and characterization. *Polym. Degrad. Stabil.* 2007, 92:1359-1364.

Plasma polymerisation and retention of antibacterial properties of terpinen-4-ol

Kateryna Bazaka[1], **Mohan V. Jacob**[1], **Russell J. Crawford**[2] **and Elena P. Ivanova**[2,*]

[1] Electronic Materials Research Lab, School of Engineering and Physical Sciences, James Cook University, Townsville, Queensland 4811, Australia
[2] Faculty of Life and Social Sciences, Swinburne University of Technology, PO Box 218, Victoria 3122, Australia

* Corresponding author: e-mail: eivanova@swin.edu.au, Phone: +61 392145137

Plasma polymerisation was used to deposit thin oligomeric films of terpinen-4-ol on a range of substrates. The coatings were examined in terms of their chemical properties and surface architecture to ascertain the changes in chemical composition as a result of exposure to the plasma field. The antifouling and antimicrobial activity of oligomeric terpinen-4-ol coatings were then examined against such human pathogens as *Staphylococcus aureus*, *Pseudomonas aeruginosa* and *Staphylococcus epidermis*. The bacterial adhesion patterns were investigated using scanning electron microscopy (SEM), atomic force microscopy (AFM) and confocal scanning laser microscopy (CSLM).

Keywords terpinen-4-ol; plasma enhanced deposition

1. Introduction

A significant body of reports supported by contemporary in vitro and clinical studies indicate tea tree oil to be effective alternative to conventional antimicrobials, particularly in the case those organisms that developed resistance against systemic antibiotics. Most bacteria tested were found to be susceptible to tea tree oil at concentrations of 1.0% or less, with minimal inhibitory concentration (MIC) of above 2% reported for commensal skin staphylococci and micrococci organisms, *Enterococcus faecalis*, and *Pseudomonas aeruginosa* [1]. In the case of methicillin-resistant *Staphylococcus aureus*, the MICs and minimal bactericidal concentrations (MBC) were identified to range between 0.25-0.31% and 0.5-0.63%, respectively, depending on the origin of the isolates. In its vaporized form, tea tree oil has been demonstrated to be effective against *Mycobacterium avium*, *Escherichia coli*, *Haemophilus influenzae*, *Streptococcus pyogenes*, and *Streptococcus pneumoniae*, reducing the incidence of nosocomial infections in hospital patients [2-4]. Despite promising in vitro and in vivo research data, many aspects of oil's bioactivity are yet to be explored. In particular, most of the in vitro studies and all of clinical research has been performed using liquid or aerosol formulation of tea tree oil.

Recent developments in plasma aided deposition technologies allow for the activation of simple organic molecules without destroying their basic chemical functionality and offer high degree of control over the chemical and physical properties of the coatings fabricated [5]. Such coatings can be deposited without compromising the inherent properties of the substrate material. Immobilisation of molecules with specific functionality are of particular interest since they produce surfaces that are highly specific in terms of their chemical reactivity, and hence can exert direct influence on the behaviour and biological function of cells in its immediate environment [6].

With biomaterial-associated infections recognised as a major hindrance to the long-term utilisation of most implanted or intravascular devices, the focus of this study was to examine the possibility of using plasma aided deposition to coat three-dimensional solid objects with ultra-thin films characterised by desirable biophysical and biochemical properties, with particular focus on antifouling and antibacterial characteristics. The monomer chosen for the study, terpinen-4-ol is the major constituent of *Melaleuca alternifolia* essential oil believed to be primarily responsible for its biocidal activity against a broad range of human pathogenic bacteria.

2. Materials and Methods

Plasma polymerisation was used to deposit thin oligomeric films of terpinen-4-ol (Australian Botanical Products, Ltd) on a range of substrata. The chemical structure of terpinen-4-ol, otherwise known as 4-methyl-1-(1-methylethyl)-3-cyclohexen-1-ol, is presented in Fig. 1(a). Glass substrata (BioLab Scientific, Ltd.) and KBr (International Crystal Laboratories) windows were cleaned and placed into the custom built deposition chamber, as depicted in Fig.1 (b) [7]. The RF energy (13.56 MHz) was delivered into the chamber by means of two capacitively coupled copper electrodes [8]. The substrata were treated with Ar plasma to activate the substrate surface and hence promote subsequent adhesion of the monomer molecules. The chamber was then evacuated to 50 mTorr, and the active monomer was gradually delivered into the deposition chamber. RF power of 7 W was found to be just sufficient enough to initiate the plasma state in the process terpinen-4-ol gas and facilitate oligomer assembly at the surface of the substrate.

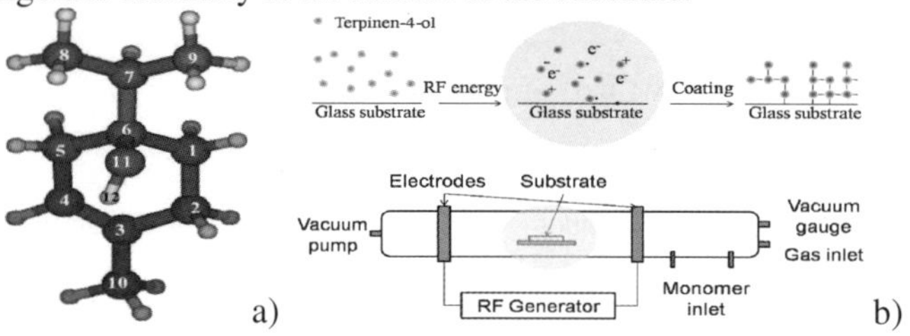

Fig. 1 a) Chemical structure of terpinen-4-ol molecule, $C_{10}H_{18}O$, b) Schematics of deposition process of thin terpinen-4-ol coatings onto glass substrates.

The coatings were examined in terms of their chemical properties and surface architecture to ascertain the changes in chemical composition as a result of exposure to the plasma field. A single beam FTIR Spectrometer (Nicolet Maxim) was used to acquire the FTIR spectrum of the films, with the spectrum for liquid terpinen-4-ol employed as a reference. Each sample was scanned 32 times in transmission mode over the region of 4000 – 400 cm^{-1} at a resolution of approximately 2 cm^{-1}. The elemental composition of the films was analysed using an Axis Ultra spectrometer (Kratos Analytical Ltd., UK), equipped with a monochromatic X-ray source (Al Kα, hν = 1486.6 eV) operating at 150 W, as outlines previously in [9]. The surfaces of the thin film samples were scanned using an atomic force microscope (AFM) (NT-MDT) in semi-contact (tapping) mode to observe the surface topography and to quantitatively estimate the extent of surface roughness. Semi-contact mode is commonly used for the characterization of soft materials such as polymer films and biological samples as the short intermittent contacts between the tip and the sample surface characteristics reduce the inelastic deformations to the surface of the sample compared to contact mode, where the high tip-to-surface forces can mechanically deform the surface under the investigation. The carbon "whisker" type silicon cantilevers (type NSC05, NT-MDT) with a spring constant of 11 N/m, tip radius of curvature of 10 nm, aspect ratio of 10:1 and resonance frequency of 150 KHz were used to obtain topographic resolution. Scanning was performed perpendicular to the axis of the cantilever at a rate of typically 1 Hz.

The antifouling and antimicrobial activity of oligomeric terpinen-4-ol coatings were then examined against such human pathogens as *Staphylococcus aureus*, *Pseudomonas aeruginosa* and *Staphylococcus epidermidis*. The bacterial adhesion patterns were investigated using scanning electron microscopy (SEM), atomic force microscopy (AFM) and confocal scanning laser microscopy (CSLM).

3. Results

3.1 Fundamental properties

Ellipsometric studies of terpinen-4-ol thin coatings deposited onto the surface of glass substrates indicated the deposition rate to be at approximately 50 nm/min. Previous investigation on thin organic coatings produced using the experimental set-up described in Fig.1 (b) demonstrated that the deposition rate can be significantly altered by changing such operational parameters as input power, monomer flow, or pressure within the reactor. For example, a notable reduction in the coating rate of several essential oils and their individual components was noted for increased levels of input power, with the decrease attributed to a higher degree of fragmentation of the monomer molecule and subsequent increase in the degree of cross-linking of the resultant polymer film [10-13]. The coatings were determined to be transparent to the visible region of the spectrum, with refractive index slightly above that of the glass substrate, as depicted in Fig. 2 (b). Surface roughness layer modelled as 50% film and 50% void, was estimated to be approximately 0.4 nm of the sample films, the values later confirmed by AFM analysis.

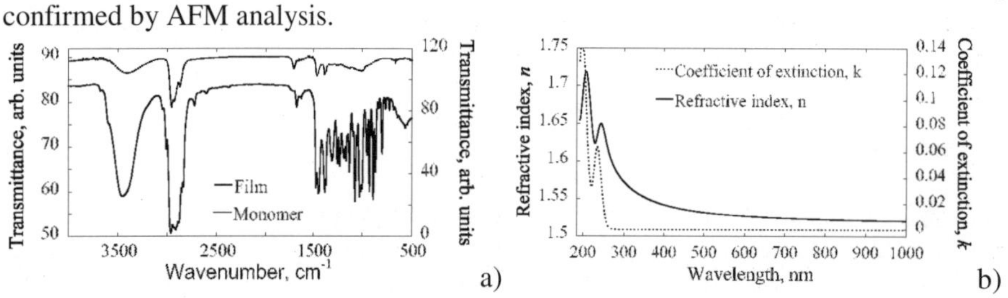

Fig. 2 Chemical (a) and optical (b) properties of terpinen-4-ol thin coatings formed as a result of exposure to RF plasma field of input energy of 7 W.

Plasma deposited terpinen-4-ol coatings were predominantly hydrocarbon in nature, with oxygen-containing groups being present within the structure in the form of alcohol and carboxylic acid groups. Under the RF plasma field conditions, the bond structures associated with vertices C(6), C(7) and C(3), such as C(6)–OH and C(6)–C(7) may dissociate to form fragments, and the C(3)=C(4) double bond may become saturated to produce C(3)–C(4). FTIR spectra illustrated in Fig. 3 indicate that the number of absorption frequencies decreased significantly upon exposure to the plasma field. Table 1 presents the respective peak assignment for the spectra depicted in Fig. 2. There was a substantial reduction in the magnitude of the H–bonded O–H stretch band and the absorptions corresponding to methyl C-H asymmetric and symmetric stretching (~ 2957, 2931, and 2873 cm^{-1}) and bending (around 1460 and 1383 cm^{-1}). Bands associated with the unsaturated C=C–H moiety between 3150 and 3000 cm^{-1} disappeared. A relatively strong experimental peak at 1706 cm^{-1} appeared, indicating the presence of saturated carboxylic acid (C=O stretch). It is likely that post-deposition ageing, particularly oxygen uptake and radical termination, would also contribute to the formation of moieties observed in the spectrum of the coating.

Table 1 Assignment of group frequencies for terpinen-4-ol monomer and coatings fabricated from terpinen-4-ol at 7 W.

Terpinen-4-ol monomer		Terpinen-4-ol coating		Assignment
Wavenumber [cm^{-1}]	Intensity	Wavenumber [cm^{-1}]	Intensity	
3450	str & brd	3407	med	(O-H), H-bonded
2963, 2959	v. str	2957, 2931	v. str-str.	$_a$(C–H) CH_3
2880	v. str	2873		(C–H) CH_3
		1706	med	(C=O)
1678, 1629	v. wk	1639	v. wk	(C=C)
1463	str	1460	wk	(C–H) CH_2
1444	str	1459	med	$_a$(C–H) CH_3
1376	str	1382	med	(C–H)
1300-700	str-med	1298, 1248	v. wk	(C–C) CH_2
1128, 1070, 1050	med	1128, 1150, 1050	v. wk	(C-O)
1026, 948, 924	str-med	969	str-med	(C-C), ring
900-670	med-v.wk	884	str	(C–H), out-of-plane

str = strong, wk = weak, med = medium, brd = broad

XPS analysis of the surface chemistry of the native and coated substrate confirmed the carbohydrate nature of the terpinen-4-ol coating. Furthermore, it showed that the surface chemical characteristics of terpinen-4-ol coating differed significantly from those of native glass substrate, the latter being composed primarily from Si and O, with C present as (C–O, C–N), (C=O) and (CO_3). Whilst Si and N were present in minute quantities, terpinen-4-ol coatings were predominantly comprised of C, O and H bonded in the form of *O--C=O, C–O, C=O, C–C and C–H. These results are consistent with the result from FTIR.

3.2 Surface architecture and surface hydrophobicity

The contact angle and surface roughness characteristics for native and terpinen-4-ol coated substrates were investigated and are shown in Table 2. The native glass surfaces were found to be moderately hydrophilic, with a contact angle of approximately 46°±7°, while the terpinen-4-ol films were found to be hydrophobic, characterised by contact angles of approximately 77°±5°). The difference in contact angles was determined to be statistically significant ($p < 0.05$). The glass substrates possessed electron donor components due to the presence of hydroxyl groups (–OH) on the glass surface.

Table 2 Comparative analysis of surface characteristics of terpinen-4-ol coatings and unmodified substrates.

	Unmodified substrate	Terpinen-4-ol coating
Roughness parameters, nm		
maximum peak height (R_{max})	11.7±0.14	6.4±0.09
average roughness (R_a)	3.5±0.04	2.4±0.03
root mean square (R_q)	0.8±0.02	0.4±0.02
surface skewness (R_{skw})	2.1±0.17	1.0±0.08
coefficient of kurtosis (R_{kur})	8.2±0.7	3.9±0.5

Fig. 3 AFM images of glass (a) and terpinen-4-ol treated (b) surfaces, scanning size 10μm × 10 μm

High resolution AFM scanning (Fig. 3) demonstrated that coating of the substrate resulted in approximately 1.5 times reduction in average surface roughness (R_a) and approximately 2 times decrease in maximum peak height roughness (R_{max}) and root mean square roughness (R_q) parameters. Average surface roughness and root-mean-square roughness describe the average absolute deviation of the surface irregularity from the mean line the standard deviation of the distribution of surface height over one sampling length, respectively, and the peak-to-peak roughness which represents the vertical distance between the highest peak and the lowest valley along the assessment length of the surface profile. Statistical analysis found the coated substrates to be statistically smoother compared to their unmodified counterparts ($p < 0.05$). Similarly, surface skewness (R_{skw}) and coefficient of kurtosis (R_{kur}) decreased 2 times upon application of the terpinen-4-ol coating. Surface skewness was more than 0 for both surfaces, suggesting the surfaces were comprised of narrower peaks and wide valleys. Since R_{skw} for unmodified glass substrate deviated further from 0, the height distribution was less symmetrical compared to that of terpienen-4-ol coated surfaces. With kurtosis value of 3.9, the surface of terpinen-4-ol was slightly more centrally distributed than a Gaussian surface which has a kurtosis value of 3. At R_{kur} of 8.2, the surface of unmodified glass was found to be centrally distributed in terms of height profile.

3.3 Antibacterial and antifouling properties

Application of terpinen-4-ol coating resulted in a statistically significant decrease in the attachment of all three pathogens tested, namely *S. aureus*, *S. epidermidis*, and *P. aeruginosa*, independent of their taxonomic designation. The attachment patterns of the bacterial species on glass and coated substrate also differed, as presented in Fig. 4, with cells attaching in small clusters to terpinen-4-ol coating and aggregating to form larger colonies when adhering to glass slides, particularly in the case of *P. aeruginosa*. The viability and metabolism of the attached bacteria were investigated by staining the cells and surrounding extracellular polymeric material (EPS) produced by these cells with appropriate dyes and visualising those with CSLM. The results of CSLM analysis showed that the amount of EPS produced by the attached cells was significantly lower in the case of terpinen-4-ol coating. Similarly, the proportion of dead to live cells was significantly higher for terpinen-4-ol coatings compared to that of bacterial cells attached to untreated glass.

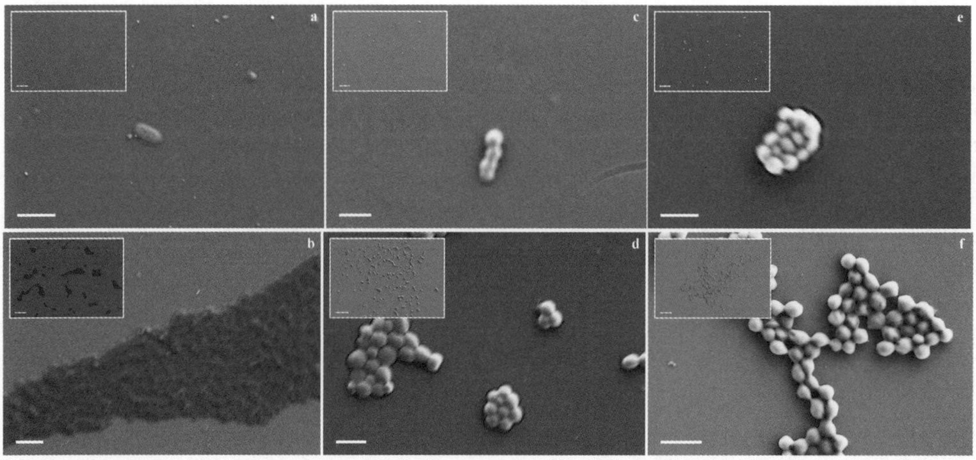

Fig. 4 SEM images of *P. aeruginosa* (a, b), *S. aureus* (c, d) and *S. epidermidis* (e, f) bacteria attached to terpinen-4-ol treated surfaces (a, c, e) and surfaces of untreated substrate (b, d, f). Scale bar 2 μm, inset scale bar 20 μm.

4. Conclusion

Application of terpinen-4-ol coating significantly altered the physico-chemical properties of the glass substrates, significantly increasing the hydrophobicity and decreasing the roughness of the substrate. Compared to unmodified substrate, the inherent differences in the chemical structure, surface architecture, and surface energy of terpinen-4-ol plasma coatings fabricated using RF plasma polymerisation can explain the significant variation in the attachment, proliferation, and EPS production of *S. aureus, P. aeruginosa,* and *S. epidermidis* cells. Favourable fabrication conditions such as low RF power just sufficient enough to initiate the ionisation of the gas and low substrate temperature encouraged condensation and incorporation of biologically active monomer into the film matrix on the surface of the substrate. Hence, plasma polymerised coatings of terpinen-4-ol can be potentially applied as antimicrobial and antifouling coating for medical and other clinically relevant devices to prevent growth of human pathogenic bacteria and to reduce bacteria-related host responses. Being optically transparent, the coating has potential for applications where transparency is essential, such as in contact lenses.

Acknowledgements This study was supported in part by Advanced Manufacturing Co-operative Research Centre (AMCRC). K. B. is a recipient of APA and AINSE PGRA; K. B. is grateful to the funding from RIRDC and DAFF.

References

[1] Carson CF, KA Hammer, TV Riley. Melaleuca alternifolia (Tea Tree) Oil: a Review of Antimicrobial and Other Medicinal Properties. *Clinical Microbiology Reviews*. 2006; 19(1): 50-62.

[2] Carson CF, BJ Mee, TV Riley. Mechanism of Action of Melaleuca alternifolia (Tea Tree) Oil on Staphylococcus aureus Determined by Time-Kill, Lysis, Leakage, and Salt Tolerance Assays and Electron Microscopy. *Antimicrobial Agents and Chemotherapy*. 2002; 46(6): 1914-1920.

[3] Loughlin R, BF Gilmore, PA McCarron, MM Tunney. Comparison of the cidal activity of tea tree oil and terpinen-4-ol against clinical bacterial skin isolates and human fibroblast cells. *Letters in Applied Microbiology*. 2008; 46(4): 428-433.

[4] Mondello F, F De Bernardis, A Girolamo, G Salvatore, A Cassone. In vitro and in vivo activity of tea tree oil against azole-susceptible and -resistant human pathogenic yeasts. *Journal of Antimicrobial Chemotherapy*. 2003; 51(5): 1223-1229.

[5] Denis L, P Marsal, Y Olivier, T Godfroid, R Lazzaroni, M Hecq, J Cornil, R Snyders. Deposition of functional organic thin films by pulsed plasma polymerization: a joint theoretical and experimental study. *Plasma Processes and Polymers*. 2010; 7(2): 172-181.

[6] Förch R, AN Chifen, A Bousquet, HL Khor, M Jungblut, LQ Chu, Z Zhang, I Osey-Mensah, EK Sinner, W Knoll. Recent and Expected Roles of Plasma-Polymerized Films for Biomedical Applications. *Chemical Vapor Deposition*. 2007; 13(6-7): 280-294.

[7] Bazaka K,MV Jacob. Synthesis of radio frequency plasma polymerized non-synthetic Terpinen-4-ol thin films. *Materials Letters*. 2009; 63(18-19): 1594-1597.

[8] Bazaka K,MV Jacob. Post-deposition ageing reactions of plasma derived polyterpenol thin films. *Polymer Degradation and Stability*. 2010; 95(6): 1123-1128.

[9] Bazaka K, MV Jacob, VK Truong, F Wang, WA Pushpamali, J Wang, A Ellis, CC Berndt, RJ Crawford, EP Ivanova. Effect of plasma-enhaced chemical vapour deposition on the retention of antibacterial activity of terpinen-4-ol. *Biomacromolecules*. 2010; 11(8): 2016-2026.

[10] Anderson LJ,MV Jacob. Effect of RF power on the optical and morphological properties of RF plasma polymerised linalyl acetate thin films. *Applied Surface Science*. 2010; 256(10): 3293-3298.

[11] Bazaka K, MV Jacob, EP Ivanova. A study of a retention of antimicrobial activity by plasma polymerized terpinen-4-ol thin films. *Materials Science Forum*. 2010; 654-656: 2261-2264.

[12] Bazaka K, MV Jacob, RA Shanks. Fabrication and Characterization of RF Plasma Polymerized Thin Films from 3,7-Dimethyl-1,6-octadien-3-ol for Electronic and Biomaterial Applications. *Advanced Materials Research*. 2010; 123-125: 323-326.

[13] Easton CD, MV Jacob, RA Shanks. Fabrication and characterisation of polymer thin-films derived from cineole using radio frequency plasma polymerisation. *Polymer*. 2009; 50(15): 3465-3469.

Preparation of silver or zinc loaded nanocapsules with core-shell architecture and their application as metal-ion release agents in plastics leading to antibacterial and fungicidal surface properties

M. Gladitz[1], S. Reinemann[*,1], J. Bauer[1] and H. J. Radusch[2]

[1] Department of Plastic Research, Thuringian Institute of Textile and Plastics Research, Breitscheidstr. 97, 07407 Rudolstadt. Germany

[2] Polymer Engineering, Center of Engineering Sciences, Martin Luther University Halle-Wittenberg, 06099 Halle/ Saale, Germany

*Corresponding author: e-mail: reinemann@titk.de, Phone: +49 3672379400, Fax: +49 3672379379

The broad spectrum of microbiocidal activity of silver and zinc ions is known as oligodynamic effect. The aim of this work was to provide a facile and robust approach to finish polymeric surfaces with microbiocidal functions, basing on hyperbranched polymers as carriers for silver or zinc ion release agents. Due to the encapsulation of the metal species a high compatibility with various finishing techniques (coating or bulk incorporation) and a controlled release of metal ions in the end use application can be realized. This ensures a high surface activity at low agent content. Both effects could be demonstrated for different polymer material applications.

Keywords hyperbranched polymers; silver; zinc; antimicrobial

1. Introduction

In the last years a great increase in the use of engineering plastics has occurred in particular in medical-technical and health care applications. However, pathogenic and dangerous germs could often more readily colonise on plastic surfaces than on metallic or ceramic ones. This repeatedly leads to the so called medical device associated infections [1, 2]. Therefore, different plastic surface finishings were developed and applied today. Plastic device surfaces impregnated with antibiotics are state of the art (e.g. BACTISEAL® Antimicrobial Catheter System). Furthermore there is a wide range of especially silver-based additives (e.g. silver-zeolite or nanosized silver-powders) for this purpose. But both ways are still problematic. The use of antibiotics causes the dramatic problem of new multi-drug resistant bacterial germs, e.g. MRSA [3]. While silver nanoparticles have a higher bioactivity as micron-sized silver, the handicap of the aggregation of nanosized silver frequently causes processing problems and leads to insufficient properties in the end use applications.

Dendrimers and hyperbranched polymers (HBPs) are a unique class of macromolecules with a tree-shaped, highly branched spherical structure and high density of terminal functional groups which could take part in numerous reactions. They are able to act as carriers for e.g. nanoparticles, metal ions or small molecules. Because of the outstanding properties of dendritic branched macromolecules, they are suitable for a wide range of applications, e.g. catalyst systems, nanoparticle templates, DNA transport or gene transfer systems.

The complex synthesis of perfect branched dendrimers is still an unsolved problem and makes them cost-expensive materials. Therefore, lesser perfect branched HBPs are more in the point of interest to modify polymeric materials by additives or some chemical treatment. Beside the class of HBPs from polyesters, HBPs from polyamines and polyesteramides were developed and are commercially available. Especially the chemical behavior of branched polyethyleneimines has been detailed studied by the groups of Mecking, Haag and Tiller [4 - 6]. The polyethyleneimines are a class of highly branched nanoscopic homopolymers synthesized from aziridine which carry terminal primary amines and secondary or tertiary amines at the branching points.

In this work, the study is focused on syntheses of silver or zinc loaded nanocapsules based on hyperbranched polyethyleneimines suitable for use as agents / additives in coatings and polymeric bulk materials.

2. Preparation of silver or zinc loaded nanocapsules with core-shell architecture

Hyperbranched polyethyleneimine (PEI, M_w 25000, degree of branching of approx. 60%) was obtained from BASF SE and modified with palmitic acid according to the procedure described in literature [6]. The modification was performed to 50% of the nominal value of primary amino groups. The amphiphilic reaction product (PEIm) has a core-shell-architecture with the aliphatic fatty acid as a hydrophobic shell as schematically illustrated in Figure 1 and could be used for further preparations without purification.

The synthesis of the silver loaded nanocapsules in chloroform ($CHCl_3$) was previously described in detail [7]. The realized molar loading ratio of silver ($AgNO_3$) to nitrogen (N atoms of the PEI-backbone) was 1 : 8. The coordinated silver ions were reduced to silver atoms by using $NaBH_4$ as reducing agent. The obtained silver dispersion could be used as received for coating purposes. Otherwise $CHCl_3$ was distilled off and the remaining precipitate was dried at room temperature in a vacuum dry chamber until constant weight. The resulting brown powder material (PEImAg(0)) was finely ground before further using.

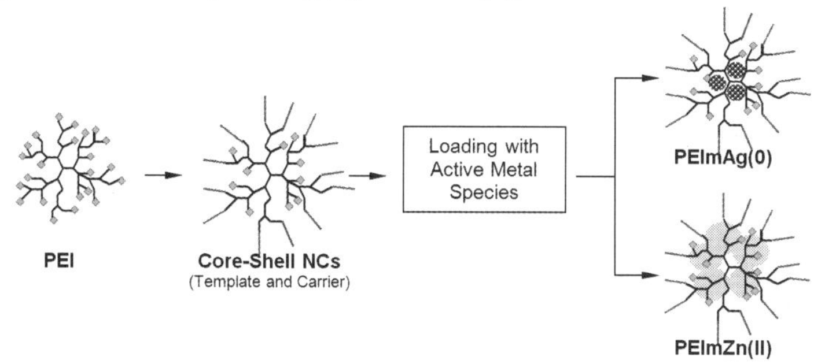

Fig. 1 Schematic illustration of the preparation steps and the architecture of the silver or zinc loaded nanocapsules (NCs)

The zinc loaded nanocapsules were synthesized by adding an appropriate amount of $Zn(CH_3COO)_2{\cdot}2H_2O$ (molar ratio of Zn to N was 1 : 6) to a solution of PEIm in $CHCl_3$. Although that the zinc salt is insoluble in $CHCl_3$ it slowly dissolves in the presence of PEIm in $CHCl_3$. Afterwards the formation of the zinc-amino-complex was completed $CHCl_3$ was distilled off from the solution and the remaining precipitate was vacuum-dried. The resulting light yellow powder material (PEImZn(II)) was finely ground before further using.

3. Results and discussion

3.1 Characterization of the silver or zinc containing nanocapsules

The nanocapsules based on the amphiphilic modified PEIm with its core-shell architecture have a high uptake capacity. They could be loaded up to a molar ratio of 1:8 for silver and 1:6 for zinc, respectively. A higher loading leads to unstable conditions (e.g. repining, gelation). Evidence of the silver nanoparticles formation was obtained from UV-Vis-spectroscopic measurements as shown in Figure 2. The silver containing nanocapsules PEImAg(0) have a yellowish brown colour in solution. This colour arises due to excitation of surface plasmon vibrations in the metal nanoparticles [8], with a surface plasmon resonance (SPR) band centered at 420 nm. The zinc containing nanocapsules PEImZn(II) remain colourless. XRD measurements support the crystallinity of the silver nuclei in PEImAg(0), see Figure 3. The broad peaks at 2 Theta 38.2°, 44.4° and 64.5° in the XRD pattern of PEImAg(0) can be assigned to the (111), (200) and (220) planes of a face centered cubic (fcc) lattice of silver (Figure 3b, Inset). Furthermore in the XRD patterns of PEIm, PEImZn(II) and PEImAg(0) a discrete peak at 2 Theta 21.6° appears. This can be assigned to the long chain crystallization of the fatty acid hydrophobic shell of the PEIm. The broad amorphous background overlapping with the peak at 2 Theta 21.6° is the result of the PEI-backbone having no long-range order. The decrease of the amorphous background feature indicates a higher state of order after intra-molecular incorporation of zinc ions or silver particles. But after heating above melting temperature (T_m) a complete X-ray-amorphous state is observed in the case of PEImZn(II).

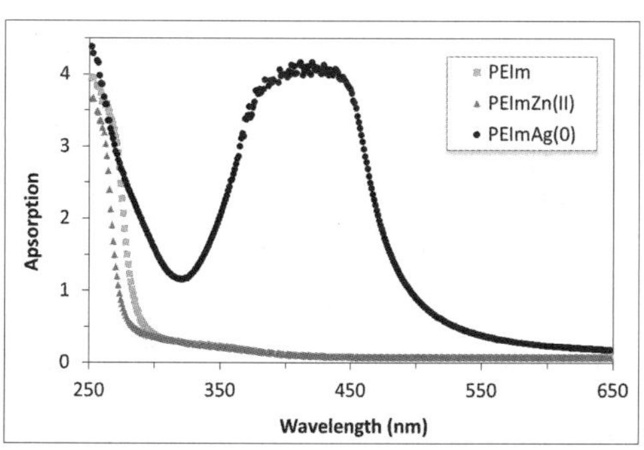

a)

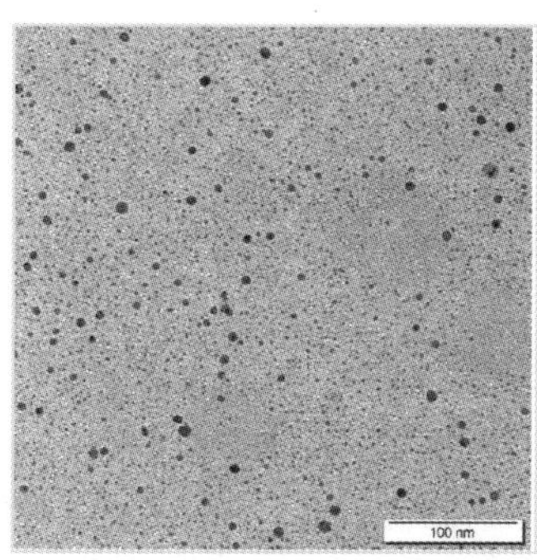

b)

Fig. 2 (a) UV-Vis spectra of solutions in $CHCl_3$ (c = 0.01g/ml); (b) TEM image of the PEImAg(0) solution

This change in crystallinity arises due to the loading of the PEIm with zinc acetate inhibiting the re-crystallization of the long alkyl-chains of the hydrophobic shell. These results are in agreement with the thermal behavior obtained by DSC measurements and the solubility properties of the amphiphilic nanocapsules as summarized in Table 1. As evident from the DTG values in Table 1 the synthesized nanocapsules containing silver or zinc are also very thermally stable. This makes them promising materials not only for coating applications but also for polymer bulk incorporation. The incorporation in the plastic bulk by melt processing techniques (e.g. extrusion) usually requires stability until temperatures of up to 300°C.

Table 1 Characteristic properties of the synthesized loaded nanocapsule and the unloaded carrier material

Sample	T_m / T_g (°C)	DTG_{max} *) (°C)	Solubility water / toluene	Metal content (%)	Moisture content (%)
PEI	– / -55	363	Y / N	n/a	≤ 1.0
PEIm	41 / –	399	N / Y	n/a	≤ 1.5
PEImAg(0)	41 / –	337	N / Y	13.1	1.9 (approx.)
PEImZn(II)	42^a / 42^b	395	Y / Y	9.3	2.7 (approx.)

*): thermal decomposition relating to the derivative of the TGA-signal with respect to time (□ = 20K/min); a: first heating, b: second heating run in DSC (□ = 10K/min); Y: soluble; N: insoluble

The average particle size of the silver nuclei in PEImAg(0) can estimated by calculation according the Scherrer equation using the half width of the intense (111) reflection of the XRD pattern in Figure 3b. The calculated value of 8.3 nm is in good agreement with the value obtained from TEM images of the regular shaped encapsulated silver particles (Figure 2b).

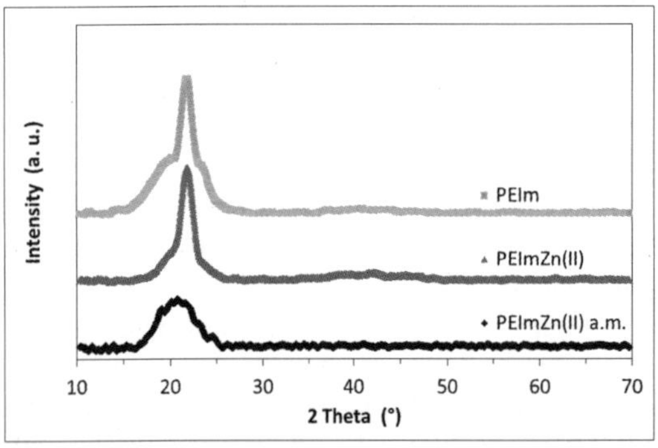

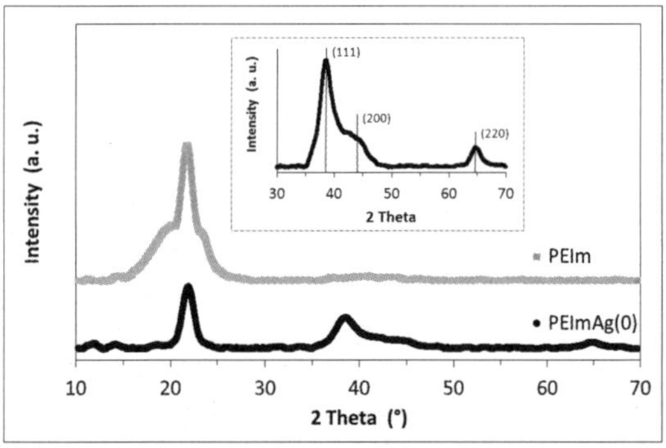

Fig. 3 (a) XRD patterns of PEIm, PEImZn(II) and PEImZn(II) after melting (a.m.); (b) XRD patterns of PEIm and PEImAg(0), Inset: Detail of pattern of PEImAg(0) with peaks assigned to the planes of a fcc-lattice of silver. (Values shifted for better visualization.)

FT-IR measurements reveals possible coordinative interactions between groups of the dendritic carrier (PEIm) and silver or zinc ions in the nanocapsule materials as shown by the results presented in Table 2.

Table 2 Major FT-IR frequencies for PEIm, PEImAg(0) and PEImZn(II)

IR bands	Frequencies (cm^{-1})		
	PEIm	PEImAg(0)	PEImZn(II)
N-H stretching	3290, w	3274, w	3242, w
C-H stretching	2919, 2850, s, s	2918, 2850, s, s	2921, 2852, s, m
N-H bending / “Amide I”	1641, m	1645, s	1641, w
“Amide II”	1551, m	1547, m	1568, s
C-H bending	1465, m	1466, m	1464, w
C-H vibration, sym.	–	–	1390, s
N-O vibration	–	1330, s	1330, w
C-N stretching	1119, 1054, w, w	1119, 1051, w, w	1052, w
	–	–	1011, w
N-H wagging	719, m	720, m	720, w

s: strong; m: medium; w: weak

The additional intensive broad absorption band at 1330 cm^{-1} for PEImAg(0) is assumed to the coordination of N atoms at the silver surface, taking into account that the particle surface is covered by AgO_x. In the IR-spectrum of PEImZn(II) the bands at 3290 and 1551 cm^{-1} are shifted to 3242 and 1568 cm^{-1} and in the intensive absorption band at 1568 cm^{-1} probably the band of the "Amide II" and the carboxylate bands of the anti-symmetric stretch for zinc actetate [9] are overlapped. The additional sharp absorption band at 1390 cm^{-1} could be addressed to the symmetrical C-H deformation vibration of the RCH_3 group. This is in good accordance with the band at 1011 cm^{-1}, which is also related to the zinc(II)-complex formation that involves in the counter ions COO^- groups. The changes in the bands of PEImZn(II) indicates that the coordination take place by N atoms of the PEI-backbone.

3.2 Modified plastics having antibacterial and fungicidal surface properties

3.2.1 Coating route

The surface of silicone catheters was modified by coating with PEImAg(0). The catheter material (NuSil 4065 med) was incubated for up to 8 hours at room temperature in a solution of PEImAg(0) in $CHCl_3$, afterwards rinsed with pure $CHCl_3$ and washed several times with bi-distilled H_2O. The depth of penetration reached approx. 540 µm and a corresponding silver content of 600 ± 90 ppm Ag was experimental proved. The evaluated inhibitor efficiency is presented in Figure 4a as a function of incubation time.

In another approach a thermoplastic clear coat was antibacterial finished. A solution of PEImAg(0) in $CHCl_3$ was pre-mixed with a solution of a suitable co-polyamide (Ultramid 1C) soluble in methanol. The coating solution was applied to a PET-foil by doctor blade technique. The antibacterial properties of the thermoplastic thin film coating containing 10 wt-% of the PEImAg(0) are presented in Figure 4b.

As demonstrated by the results presented in Figure 4 the stable, non-leaching microbial inhibitor PEImAg(0) provides a promising approach in means to effectively control the surface growth of bacteria germs in a meaningful manner.

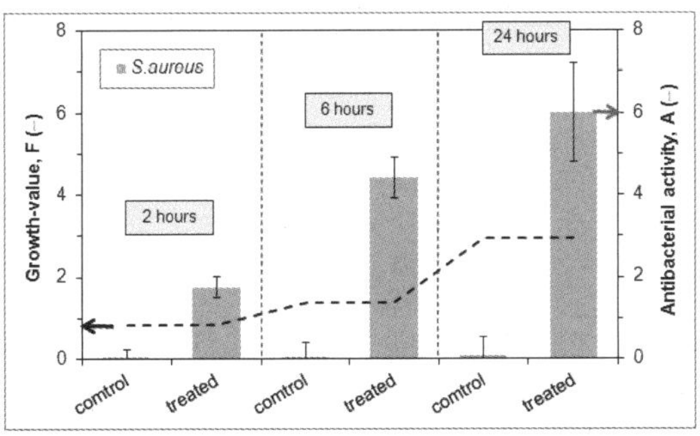

a)

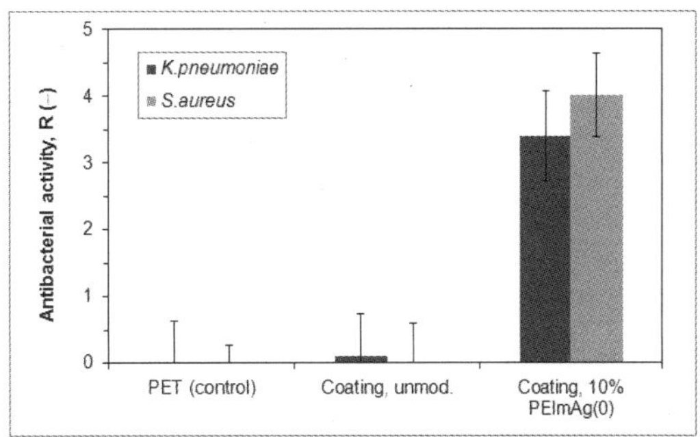

b)

Fig. 4 Inhibition of bacterial growth according JIS L 1902 of (a) PEImAg(0) surface treated silicone catheter as a function of incubation time at 37°C and with respect to the growth-value of the bacterium, initial population of approx. 10^4 CFU/ml; (b) PEImAg(0) containing thermoplastic clear coat on PET-foil after 24 incubation at 37°C (F: approx. 2.5 ± 0.6).

3.2.2 Extrusion / bulk route

The incorporation of the nanocapsules in the plastic bulk was done by melt extrusion technique using a synchronous twin-screw extruder ZSK-25 from Coperion Werner & Pfleiderer. In a first step a compound containing 10 wt-% of the silver or zinc loaded nanocapsules was produced by feed of the powder additive into the polymer material at its molten state at a peak temperature of approx. 220°C (TPS, based on SEBS), 235°C (PP) and 285°C (PA6), respectively. In a second step the compounded resin was pre-mixed with the neat polymer resin in appropriate ratios. These mixtures were molten and shaped by injection molding into sheets with a thickness of 2.0 mm.

In a comparable study of various materials of thermoplastic nature, which were modified with the silver or zinc containing nanocapsules material a high antibacterial activity was found in both cases. In the silver doped systems a fungicidal efficiency was not initiated for concentrations of up to 0.65 wt-% of silver in the bulk material. The antibacterial and fungicidal surface effect of injection-moulded sheets comprising the zinc loaded nanocapsules is shown in Figure 5a. The results demonstrate the broad application range of the PEImAg(II) as microbiocidal agent in different polymeric materials. As Figure 5b shows, the microbial activity is attended by an almost constant slow release of zinc ions in a controlled manner. The release of zinc ions from the doped PP is still below the quantification limit after 14 days of elution. This effect is related to the hydrophobicity and crystallinity of the polymer matrix as described for silver-nanocomposites [10, 11]. Furthermore, the antimicrobial effect is not only related to metal ion release but also to a surface contact activity. Therefore a high microbiocidal surface effect is still observable even at very low metal ion release as in the case of the PP-compounds. The results demonstrate that the microbial inhibitor persists for longer time periods without burst release of the metal species. This makes such modified materials promising to prevent contaminations in a wide range of applications. Such surface active material systems in addition with standard hygiene procedures could greatly improve efforts to mitigate infections, e.g. particularly in clinical applications.

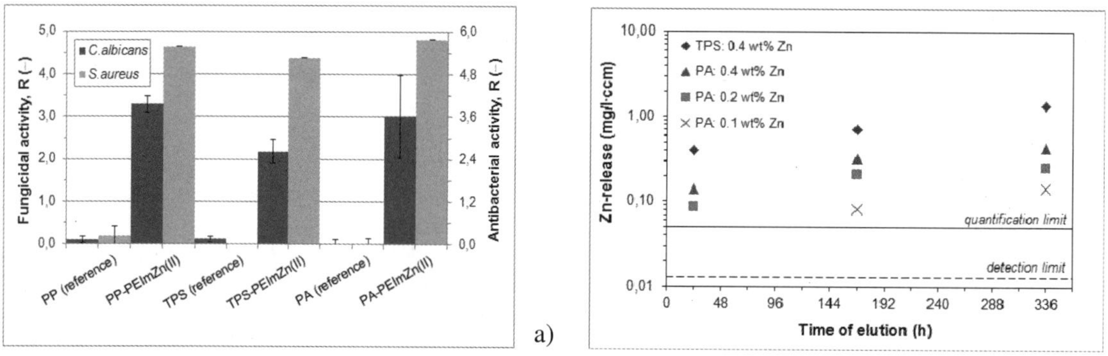

Fig. 5 (a) Microbiocidal surface activity according JIS Z 2801 of polymeric compounds doped with 0.4 wt-% zinc in the form of PEImZn(II) with regard to growth values of 1.9 ± 0.1 and 2.3 ± 0.2 for *C.albicans* and *S.aureus* after 24h incubation at 37 °C, respectively; (b) zinc ion release into bi-distilled water at 23°C, pH 4.8 (acetic acid) as a function of time and the concentration of PEImZn(II) in the material bulk.

4. Conclusions

We have demonstrated a facile method to create amphiphilic nanocapsules containing silver or zinc as antimicrobial agents. These nanocapsules could act as microbial agents to inhibit surface contaminations in selected applications, e.g. medical device surface finishes or wound dressings. The antimicrobial inhibitors can be used as solutions for surface treatment, added to coating formulations or into the material bulk. The encapsulation of the metal species led to controlled release of metal ions and to an easier dispersion by established plastic processing techniques (coating suspensions, melt extrusion or injection molding). The biocidal activity of the silver or zinc containing nanocapsules is not only a result of the controlled release of metal ions as bioavailable species in the presence of moisture but also of the surface contact with the PEImAg(0) or PEImZn(II) as our experimental results have shown. To explain the high inhibitor efficiency of the nanocapsules agents also the poly-cationic nature of the PEI-backbone must be taken into account. Especially the zinc loaded nanocapsules are able to render surfaces with antibacterial and fungicidal properties, which is of special interest particularly for many medical applications.

Acknowledgements The support of the projects ZIM, No. KF2099103AK9 and InnoWatt, No. IW091004 is gratefully acknowledged.

References

[1] Von Eiff C, Heilmann C, Peters G. New aspects in the molecular basis of polymerassociated infections due to staphylococci. *Eur J Clin Microbiol Infect Dis.* 1999; 18: 843–846.

[2] Peters KM. Exogene (unspezifische) Osteomyelitis. In: Peters KM, König DP. *Fortbildung Osteologie 1.* Darmstadt, Germany: Steinkopff; 2006: 116-119.

[3] Tacconelli E, Angelis GD, Cataldo MA, Pozzi E, Cauda R. Does antibiotic exposure increase the risk of methicillin-resistant Staphylococcus aureus (MRSA) isolation? *J Antimicrob Chemother.* 2008; 61: 26–38.

[4] Moisan S, Martinez V, Weisbecker P, Cansell F, Mecking S, Aymonier C. General Approach for the Synthesis of Organic–Inorganic Hybrid Nanoparticles Mediated by Supercritical CO_2. *J Am Chem Soc.* 2007; 129: 10602–10606.

[5] Fischer W, Calderón M, Haag R. Hyperbranched Polyamines for Transfection. *Top Curr Chem.* 2010; 296: 95-129.

[6] Aymonier C, Schlotterbeck U, Antonietti L, Zacharias P, Thomann R, Tiller JC, Mecking S. Hybrids of silver nanoparticles with amphiphilic hyperbranched macromolecules exhibiting antimicrobial properties. *Chem Commun.* 2002; 3018–3019.

[7] Gladitz M, Reinemann S, Radusch HJ. Preparation of Silver Nanoparticle Dispersions via a Dendritic-Polymer Template Approach and their Use for Antibacterial Surface Treatment. *Macromol. Mater. Eng.* 2009; 294:178–189.

[8] Mulvaney P. Surface Plasmon Spectroscopy of Nanosized Metal Particles. *Langmuir.* 1996; 12: 788–800.

[9] Ishioka T, Shibata Y, Takahashi M, Kanesaka I, Kitagawa Y, Nakamura KT. Vibrational spectra and structures of zinc carboxylates I. Zinc acetate dehydrate. *Spectrochimica Acta Part A*. 1998; 54: 1827-1835.

[10] Kumar R, Münstedt H. Polyamide/silver antimicrobials: effect of crystallinity on the silver ion release. *Polym Int*. 2005; 54: 1180-1186.

[11] Damm C, Münstedt H, Rösch A. Long-term antimicrobial polyamide 6/silver-nanocomposites. *J Mater Sci*. 2007; 42: 6067-6073.

Structure-reactivity relations for DC-magnetron sputtered Cu-layers during *E. coli* inactivation in the dark and under light

C. Castro[1,4], R. Sanjines[2], C. Pulgarin[1*], P. Osorio[1], S. A. Giraldo[4] and J. Kiwi[3*]

[1]EPFL-SB-ISIC-GGEC, Group of Electrochemical Engineering, Station 6, CH1015, Lausanne, Switzerland,
[2]EPFL-SB-IPMC-LNNME, Bât PH, Station 3, CH-105 Lausanne, Switzerland EPFL-[3]SB-ISIC-LPI, Bât Chimie, Station 6, CH-1015, Lausanne, Switzerland,
[4]UNIV Industrial Santander, Centro de Investigaciones en Catalisis, CICAT, UIS, A. A. 678, Bucaramanga, Colombia.

This study addresses unreported features for Cu DC-magnetron sputtering on cotton mediating inactivation of *Escherichia coli* K12 (from now on *E. coli*). The *E. coli* inactivation for cotton-Cu samples sputtered for 40 s was attained within 30 min under visible light (1.2 mW/cm^2) and within 120 min in the dark. For a longer sputtering time of 180s, the bacterial inactivation kinetics under light was observed within 30 min, as was the case for the 40 s sputtered sample. This suggests that Cu-ionic species play a key role *E. coli* inactivation. The 40 s sputtered samples present the highest amount of Cu-sites held in exposed positions interacting on the cotton surface (or inside the cotton) with *E. coli.* Sputtering for 40 s deposited 4.10^{16} atoms Cu/cm^2 (taking ~10^{15} atoms/cm Cu per atomic layer) and this was the threshold amount of Cu necessary for complete *E. coli* inactivation. Cu DC-magnetron sputtering lead to thin metallic semi-transparent grey-brown Cu-coating presenting hydrophobic behavior as determined by contact angle measurements.

Keywords: DC-Sputtering; cotton; *E. coli*; Cu-ionic; etching; XPS

1. Introduction

The manufacturing of high value added products such as bactericide textiles with nanoparticulate Cu has increased rapidly in the last few years [1]. Cu-particles are used for their catalytic and bactericide properties. Recently, Borko and Gabay [2] have reported Cu-textiles and Cu-composite materials presenting antiviral, algicide and fungicide properties that sustain washing cycles when loaded with CuO 1%. These textiles have been reported to be effective against *E. coli*, *Staphylococcus aureus* and the fungi *Candida albicans* and were effective against other infections being also safe to humans and not causing skin irritation when externally applied. Recently, our laboratory has shown that large surface area Cu-suspensions were effective in *E. coli* inactivation in the dark and under visible light [3].

The present study addresses: a) the relation between structure and reactivity of the Cu-clusters on cotton towards the inactivation of *E. coli,* b) the identification if the Cu-ions on the cotton textile intervening in the *E. coli* inactivation by XPS and d) the increase in hydrophobicity of the Cu-cotton as a function of Cu-loading compared to bare cotton samples.

2. Experimental Section

2.1. Cu deposition on cotton sample

In Figure 1 the positive Ar-ions near the target are accelerated towards the target surface by applying high negative voltages (0.2-2.0 kV). The 5 cm diameter Cu-disk is eroded by the Ar-ions and the ejected Cu-atoms/clusters/ions are collected on the cotton. The plasma working pressure was in the range of 0.1-1 Pa, the distance between the Cu-target and the cotton was ~10 cm and the deposition current was 30 mA. The adhesion of Cu to the cotton (ex-Cilander AG, Herisau, CH-1109 Switzerland) was strong. Friction with paper or cloth did not smear the Cu. Deposition times of 20, 40, 90, 180 and 360 seconds lead to Cu layers of 1.5, 3.0, 7.8, 16 and 28 nm thick on the cotton surface. The film thickness was determined with a profilometer (Alphastep500, TENCOR).

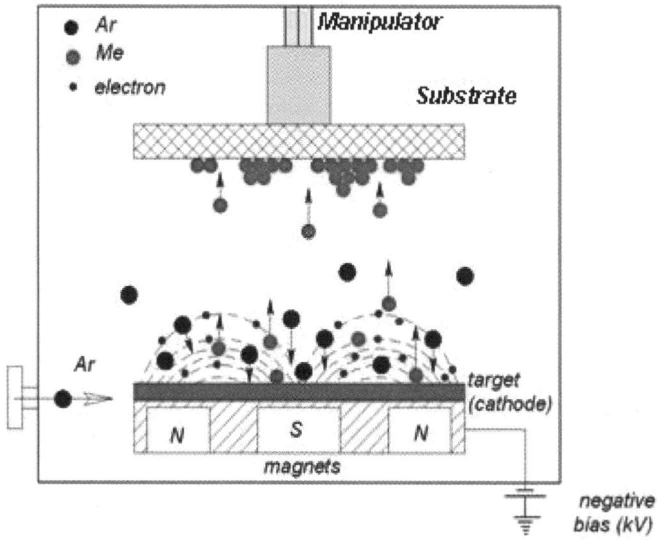

Figure 1. Schematic diagram of the DC-magnetron sputtering unit

2.2. Evaluation of the bacterial inactivation of *E. coli* K12

The bacterial strain *E. coli* K12 was inoculated in 5 ml of Luria Bertani solution and grown during 8 h at 37 ◦C with constant agitation under aerobic conditions. Aliquots of the culture were inoculated into a fresh medium and incubated aerobically at 37 ◦C for 15 h. The procedure followed the CFU count is to be found in reference [4].

2.3. High-resolution transmission electron microscopy (HRTEM*)*

A Philips HRTEM CM 300 provided with a field emission gun, 300 kV, 0.17 nm resolution microscope and a Philips EM 430 (300 kV, LaB_6, 0.23 nm resolution) were used to measure the particles sizes of Cu-clusters. The textiles were embedded in epoxy resin (Embed 812) and the fabrics were cross-sectioned with an ultra-microtome.

2.4. Irradiation of Cu-cotton samples during the inactivation of *E. coli* K12

The irradiation of the Cu cotton samples was carried out in a cavity provided with a tubular Osram Lumilux T8-L18W (ex Winterthur, Switzerland) actinic lamp having a visible emission spectrum between 400 and 700nm as shown in the insert to Figure 4 with an integral output of 1.2 mW/cm^2 (Winterthur, Switzerland).

2.5. X-ray photoelectron spectroscopy (XPS)

An AXIS NOVA photoelectron spectrometer (Kratos Analytical, Manchester, UK) equipped with monochromatic AlK_α (hν = 1486.6 eV) anode was used during the study. XPS experiments were carried out as described earlier by our laboratory [3].

3. Results and Discussion

3.1. Inactivation of *E. coli* on Cu-sputtered cotton samples

Cotton by itself did not inactivate the *E. coli*. Figure 2 shows 40s Cu DC sputtered samples in the dark lead to a complete bacterial inactivation within 120 min and 30 min under low intensity visible light (~1% of the full solar light intensityAM1). The mechanism of bacterial inactivation by CuO under light has been reported recently [3]. The sputtering for 40s, loading Cu 0.060% wt/wt on the cotton samples was the threshold amount of Cu necessary to induce complete bacterial inactivation within reasonable times. The sputtering for 180s produced cotton samples leading to a complete bacterial inactivation within 30 min minutes in the dark and 120 min in the dark in spite of having a Cu-content of 0.294% wt/wt that is about 5 times above the threshold Cu-concentration deposited with only 40s sputtering. This shows that the amount of Cu^0 is not the main species leading to bacterial deactivation since the bacterial inactivation kinetics was the same when sputtered for 40s and for 180s. The sputtering time of 40s lead to the most favorable structure-reactivity for the Cu-cotton leading to the shortest bacterial *E. coli* inactivation. This is the evidence that the 40 s sputtered samples presents the highest amount of Cu-sites held in exposed positions on the surface or inside the cotton able to interact with *E. coli*.

3.2. High-resolution transmission electron microscopy (HRTEM) of Cu-cotton

Figure 3 presents the HRTEM results for cotton samples. In this HRTEM the epoxide and cotton presented the same contrast and the particles observed in Figures 3 were observed in the first 10 cotton layers or within 2 nm. The 40 s sputtered Cu-particles on cotton present an average size of 50 nm. The Cu-clusters in the 180 s sputtered cotton show an increased Cu-particle density consistent with an increased sputtering time. As the loading increases from 0.060% Cu wt/wt for the 40s sample to 0.294% Cu wt/wt for the 180s sample, the density of the clusters increase but the catalytic activity per exposed atom decreases due to this agglomeration process.

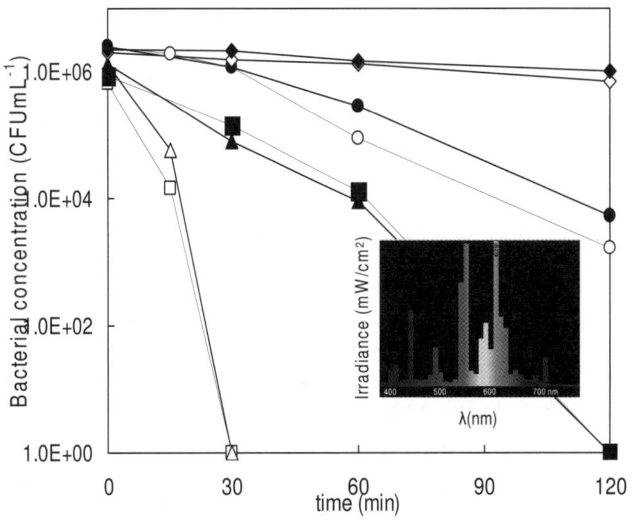

Figure 2. *E. coli* inactivation on cotton: light (◊), dark (♦) and Cu DC-magnetron sputtered cotton at: 20 s dark (●), light (○) 40 s dark (■ light (□); 180 s dark (▲ light (Δ). Insert: spectral distribution of the visible light source of 1.2 mW/cm^2

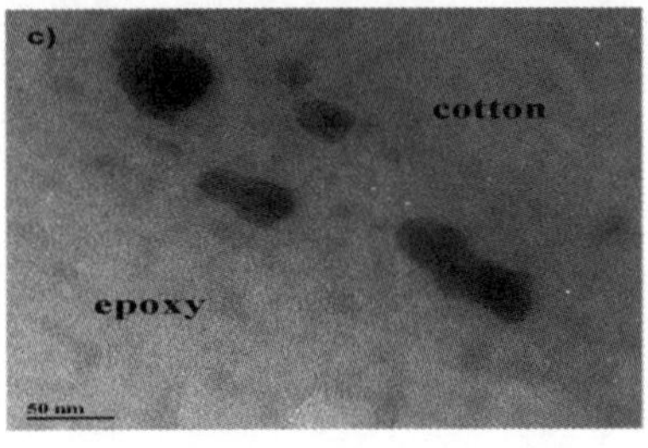

Figure 3. Electron microscopy of Cu-cotton DC-magnetron sputtered for 40 s

3.3. X-ray photoelectron spectroscopy of Cu-cotton samples (XPS)

Figure 4 shows the XPS-spectra of the deconvoluted Cu-doublet analyzed in terms of the Cu^0/Cu^{1+}, Cu^{2+} and Cu^{3+} components. The envelope Cu^0 and Cu^{1+} component at 933.1 eV could not be unambiguously deconvoluted without introducing a large approximation in the peaks for Cu^0 and Cu^{1+}. The Cu^{2+} peak at 934.2.0 eV and the Cu^{3+} peak at 937.1eV have been assigned according [5].

Table 1 shows by XPS measurements that cotton samples sputtered for 40s and 180s present various Cu-oxidation states at different depths. The amount of Cu for the 40s sample is 0.060% wt/wt and for the 180s sample 0.294% wt/wt. But the bacterial inactivation kinetics occurs within the same times in the dark and under light for both samples. Therefore, Cu^0 is not the main active responsible for *E. coli* inactivation. We conclude that the Cu ionic-species are the ones that lead to bacterial inactivation.

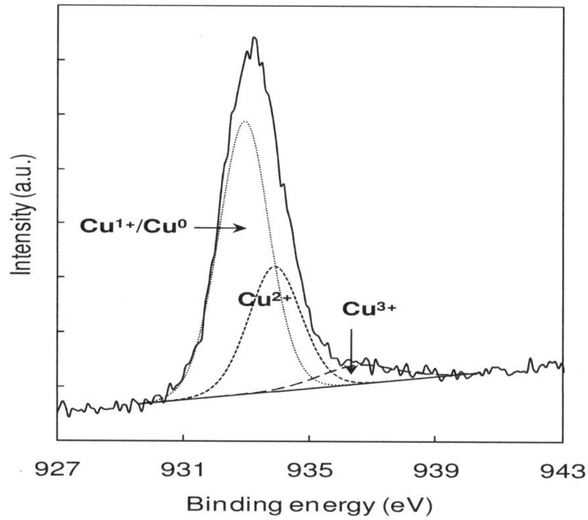

Figure 4. XPS spectra of Cu DC-magnetron sputtered cotton showing the deconvoluted peaks for the Cu-ionic species.

Table 1: Cu- species % in the sputtered cotton for 40s and 180s, etched with 5 KeV Ar^+-ions as a function of etching depth

Etched Depth (nm)	40 s			180 s		
	Cu(I)/Cu(0)	Cu(II)	Cu(III)	Cu(I)/Cu(0)	Cu(II)	Cu(III)
0.0	15.3	6.0	~	27.0	2.7	~
4.0	6.4	4.5	~	41.1	9.5	4.8
6.0	6.9	1.0	0.4	28.0	6.7	5.2
8.0	4.7	2.3	~	17.3	14.3	~

References

[1] Kiwi J, Pulgarin C. Self-cleaning and bactericide textiles, *Catal, Today*, 2010; 151: 2-7.

[2] Borkow G and Gabbay J. Biocidal textiles can help fight nosocomial infections, *Med. Hypothesis,* 2008: 70:990-994.

[3] Paschoalino M. Guedes, N, Jardim W, Mielczarski E, Mielczarsi J, Bowen P, Kiwi J. Photo-assisted Inactivation of *E coli* Mediated by High Surface Area CuO under Light Irradiation (>360 nm), *J. Photochem. Photobiol. A: Chem.* 2008; 199: 105-111.

[4] ISO Standard 20743, Sept 2007: Evaluation of the antibacterial activity of biocidal products (www.afnor.org).

[5] Wagner C. D, Riggs W. M, Davis l. E, Mullenberg G.E, (Ed), *Handbook of X-ray Photoelectron spectroscopy*, Perkin-Elmer Corporation Physical Electronics Division, Minnesota, 1979.

Study of Submerged and Biofilm Fermentation of *Bacillus subtilis* using Fish Protein for Production of Lipopeptide Antibiotic Iturin A

Umme Salma Zohora[1,3], Abdul Wahab Khan[1,3], Masahiro Okanami[2], Takashi Ano[1,2] and Mohammad Shahedur Rahman[1,3*]

[1]Chemical Resources Laboratory, Tokyo Institute of Technology, 4259 Nagatsuta, Midori-ku, Yokohama 226-8503, Japan
[2]Department of Biotechnological Science, Faculty of Biology-Oriented Science and Technology, Kinki University, 930 Nishimitani, Kinokawa-city, Wakayama, 649-6493, Japan
[3]Anise Corporation, N605 Building #C, 1-12 Minamiwatarida-cho, Kawasaki-ku, Kawasaki-shi 210-0855, Japan

*Mohammad Shahedur Rahman : e-mail: msrahman@anise.co.jp, Phone: +81 90 6340 5799

Iturin A is an environmentally safe biocontrol agent produced by *Bacillus subtilis* as a secondary metabolite. Generally iturin A is produced in conventional submerged fermentation. Recently, *B. subtilis* has received a huge interest for its nature to develop into biofilm as it shows significantly independent genetic and morphological development in biofilm compared to its planktonic culture. In this study it was attempted to compare the production of iturin A in submerged with that in biofilm fermentation using novel marine fish protein as a medium component. When fish protein was compared with commercially available peptones, it was observed that the microbial growth and iturin A productions were similar to those in the medium containing Polypepton S (originated from soybean) and higher than those in the medium containing Polypepton (originated from casein). Quicker cellular growth and secondary metabolite production was observed in submerged fermentation whereas slower but higher cellular growth and iturin A production was found in biofilm fermentation.

Keywords fish protein; *Bacillus subtilis* RB14; secondary metabolite; iturin A; biofilm

1. Introduction

Biofilm development of *B. subtilis*, a plant growth promoting rhizobacteria (PGPR), has become a hot topic in recent microbiology. In favorable conditions for growth, the bacterial cells undergo dramatic lifestyle changes; the planktonic cells settle down and establish a diverse community of biofilm [1]. *B. subtilis* cells show significantly independent genetic and morphological development during biofilm formation compared to planktonic culture [2-7]. Microbes form a biofilm in response to many factors, which may include cellular recognition of specific or non-specific attachment sites on a surface, nutritional cues or in some cases, by exposure of planktonic cells to sub-inhibitory concentrations of antibiotics [8, 9].

Generally peptones are being used as major ingredients for culture media in microbiology. The sources of peptones may be from animals or plants and in composition they are highly variable due to their diversity of origin. In recent years the researchers are trying to utilize marine resources as a source of peptone. Utilization of marine resources can open a new era in modern research. During the cultivation of microorganisms, the nitrogen source often represents the major part of the production cost in the fermentation industry [10]. Commercially available peptones, like Polypepton and Polypepton S, which are derived from soybean and milk casein respectively, are being used as standard peptones for microbial growth. Previous studies revealed that marine resources like tuna red meat hydrolysate [11], cod viscera [12], viscera from squid, yellow fin tuna, swordfish and rainbow trout [13] had been used as alternative nitrogen sources. Cod (*Gadus macrocephalus*) is an important fish resource. Generally it is used for food and food supplement such as liver oil. As a model it was tried to use the Japanese cod as a marine resource for laboratory experiments.

Iturin A is a cyclic lipopeptide [14]. Along with its wide antifungal spectrum, iturin A has low toxicity and low allergic effect on human and animals [15] thereby stands as a candidate for environmentally safe biological pesticide [16]. Iturin A has been tested to control a variety of fungi in pure cultures [15, 17, 18] and during composting [19, 20]. At present commercially available chemical agents against plant diseases are being removed from the market for their hazardous impact to the nature, creating the demand for the safer products as fungicide like iturin A. The strong efficacy of iturin A against various phytopathogenic fungi is similar to the available chemical pesticides [19].

Nowadays studies of microbial productions almost entirely are based on the submerged culture system where the microorganisms are offered with higher oxygen transfer rate by the continuous movement and or transfer of the medium. Cells remain planktonic in this culture and the entire fermentation system is based on the sum performance of the each of the individual cells. On the other hand, when the culture is incubated on a liquid medium at a static condition, wild type strain of *B. subtilis* RB14 produces thick and stable biofilm in the liquid-air surface of the culture medium and surprisingly iturin A is also found to be produced from the culture [21, 22]. In this study production of iturin A in both submerged and biofilm fermentation using FP medium was compared and described.

2. Materials and methods

2.1 Elemental analysis of Polypepton S, Polypepton and fish protein

Polypepton S, Polypepton and fish protein were dried at 105°C for 24h. After cooling in desiccators for 1 h, the samples were taken in 1500 μl polypropylene tube and elemental analysis was performed at the center for Advanced Materials Analysis (Suzukakedai), Technical Department, Tokyo Institute of Technology. The comparison of the fish protein with Polypepton and Polypepton S is shown in Table 1.

2.2 Culture medium composition and cultivation of *B. subtilis* RB14

B. subtilis RB14 were grown in LB medium containing 10 g/L of Polypepton (Nihon Pharmaceutical, Tokyo, Japan), 5 g/L of yeast extract (Oriental Co., Tokyo) and 5 g/L of NaCl, adjusted to pH 7.0 for preculture. Then the cells were inoculated in a medium which contains 30 g/l of Polypepton (Nihon Pharmaceutical), 10 g/l of glucose, 1 g/l of KH_2PO_4, and 0.5 g/l of $MgSO_4 \cdot 7H_2O$ (pH 6.8). To investigate the effect of various peptones on iturin A production, medium were prepared by replacing the Polypepton with Polypepton S and fish protein. L-agar plate containing LB medium with additional 1.5% agar (Shimizu Shokhin Co., Ltd., Japan) was used for the determination of the colony-forming units (cfu) in the culture system.

B. subtilis RB14 was transferred from culture stock (kept at −80°C in 10% glycerol solution) into 5 ml of modified LB medium and incubated at 37°C with 120 strokes per minute (spm) for 16 h for preculture. 400 μl of overnight cultured *B. subtilis* RB14 was inoculated into 40 ml of fish protein, Polypepton and Polypepton S medium in 200 ml Erlenmeyer Flasks. The flasks were then incubated for 8 days at 30°C with 120 spm for submerged fermentation and in 25°C at static condition for biofilm fermentation for maximum iturin A production.

2.3 Estimation of cellular growth and microscopic observation

At specific time intervals samples were collected from the culture flasks to measure bacterial growth by counting cfu. For the cfu count, collected culture samples were serially diluted with 0.85% NaCl and spread on L-agar plate. To detect the spore numbers, samples from the culture were heated at 80°C for 10 min in water bath, serially diluted and spreaded on agar plate. The plates were incubated at 37°C for 16 h. Time course of pH were recorded at regular intervals.

For scanning electron microscopy (SEM) [Hitachi S-5200 Nano SEM (Hitachi, Tokyo, Japan)] observation samples were rinsed with sterile distilled water then allowed to fix into 2% glutaraldehyde in 50 mM phosphate buffer saline (PBS) at 70% relative humidity. Sample was rinsed three times after 24 h in 50 mM phosphate buffer for 15 min each, followed by successive 15-min dehydrations in 50, 70 and 90% ethanol and finally in 100% ethanol three times. Samples were examined by a scanning electronic microscope operating at 1 kV.

2.4 Quantitative analysis of iturin A

Quantification of iturin A was performed using HPLC analysis and quantification was performed by using HSS-2000 and JASCO-Borwin software installed in a computer [22, 23].

3. Results

3.1 Comparative studies of carbon and nitrogen content in different peptone media

The total carbon and nitrogen contents of Polypepton, Polypepton S and fish protein were calculated in order to compare them during the production of iturin A. As the C and N contents of Polypepton S is 37.5 % and 8.7 % (Table 1) respectively, the total carbon and nitrogen contents in Polypepton S medium, which contains 30 gm/l Polypepton S and 10 gm/l glucose were calculated to be 15.25 and 2.61 gm/l. Similarly, the calculated carbon and nitrogen contents in Polypepton medium were 18.16 and 3.67 gm/l and the total carbon and nitrogen contents in fish protein medium were calculated to be 18.67 and 4.41 gm/l, respectively. These results indicate that fish protein medium contains highest amount of C and N source, thereby, supposed to be nutritionally rich for suitable microbial growth and production of desired secondary metabolites.

Table 1. Comparison of Polypepton, Polypepton S and fish protein

	Peptones used in the study		
Properties	**Polypepton**	**Polypepton S**	**Fish protein**
Origin	Milk casein	Soybean	Cod fish
Source of origin	Dairy industry	Agriculture	Sea
C	47.2%	37.5 %	48.9 %
N	13.22%	8.7 %	14.7 %
Ash	6.5 %	4.7 %	7.5%
Energy	390 Kcal	446 Kcal	395 Kcal

3.2 Comparative growth profile of *B. subtilis* in different peptone medium

The growth of *B. subtilis* was observed with in different peptone medium such as Polypepton, Polypepton S and fish protein in submerged fermentation. The cellular growth in Polypepton medium was found to be low (about 7×10^{8}) as expressed in terms of cfu determination. The spore number remained lower (6×10^{7}) than the total viable cell number. Rapid and high cellular growth and sporulation was observed in Polypepton S medium (about 5×10^{9}). Similar to Polypepton S high cellular growth was observed in fish protein medium (about 4×10^{9}), however, the sporulation rate was found relatively low (7×10^{8}).

3.3 Comparative iturin A production by *B. subtilis* in different media

In Table 2B comparative iturin A production by *B. subtilis* in submerged fermentation is shown within different medium components. In Polypepton medium the production of iturin A was found very low. The total amount of iturin A was about 140 mg/l. On the other hand, Polypepton S medium allowed high iturin A production about 550 mg/l. Fish protein medium also allowed high iturin A production. The amount of iturin A produced from fish protein medium was about 510 mg/l which was similar to Polypepton S medium but notably higher than Polypepton medium.

3.4 Biofilm formation in fish protein medium

After comparing the fish protein with other peptones, it was confirmed that it can be used as a medium component. Then fish protein was applied for biofilm formation. Generally biofilm is the natural state of microbial growth. In the static condition *B. subtilis* RB14 single-swimming cells migrate to the air–liquid interface and become enclosed in a sticky extra cellular matrix surface (EPS) where they divide and establish a diverse community of biofilm. The schematic diagram of biofilm formation in an air-liquid inter phase was shown in Fig.1A. *B. subtilis* RB14 showed biofilm development in fish protein medium. Visually a layer of biofilm was observed after overnight incubation. The thin layer of biofilm subsequently got thicker over longer incubation period. A look from the upside of biofilm showed a brain like appearance where the side views showed the uneven folded mushroom like appearance. The thickness of the mature biofilm was about 0.4 cm. The photographic image of the upper view of *B. subtilis* RB14 biofilm was shown in Fig.1B. The scanning electron microscopy image of fish protein biofilm was shown in Fig.1C.

3.5 Comparative growth profile of *B. subtilis* in biofilm and submerged fermentation

The growth profile of *B. subtilis* RB14 in fish protein medium during the comparison of submerged and biofilm fermentation is shown in Fig.1D. It was observed that cellular growth profile depend on the fermentation condition. When the cells were continuously shaken during submerged fermentation rapid cellular growth takes place. Within the second day of cultivation the cell numbers were reached to the highest level (about 4×10^{9}). Sporulation was observed earlier and at the latter stage almost all cells were found to be spore. In biofilm fermentation cellular growth was found slower compared to submerged fermentation. At the third day of incubation the total cell number reached to the upper limit (about 6×10^{9}). Sporulation was also found delayed and finally at the latter stage the spore numbers were found lower than the total cell number (2×10^{9}). The result was the average of 4 individual experiments.

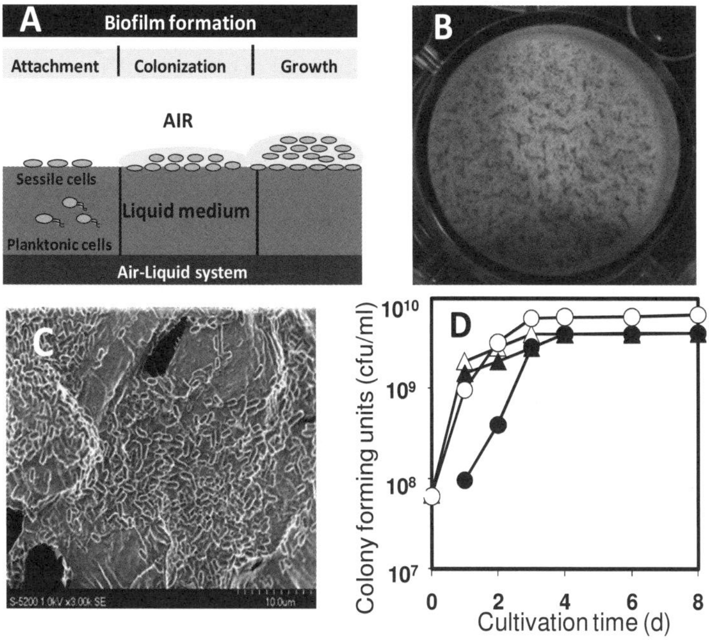

Fig. 1 Schematic diagram of biofilm formation (A); observation of biofilm formation in fish protein medium by Photographic image (B) and scanning electron microscopy (C); graphical presentation of cellular growth and spore formation in submerged (open and closed triangle) and biofilm (open and closed circle) fermentation (D) [n=4].

3.6 Comparative iturin A production by *B. subtilis* in biofilm and submerged fermentation

In Fig. 2 comparative iturin A production was shown within biofilm and submerged fermentation using fish protein medium. Production of any secondary metabolite is associated with the stationary phase of microbial growth. The amount of iturin A increased rapidly within second day when almost all viable cells remain at stationary phase but the productivity decreased along with sporulation of the cells. In submerged fermentation iturin A production was found earlier similar to cellular growth. Up to the second day of cultivation iturin A was produced more in submerged fermentation and the final production reached to 510 mg/l. In biofilm fermentation though the initial production was found lower but finally it allowed the higher production than that in submerged fermentation. The amount of iturin A produced from biofilm fermentation was about 650 mg/l. The result was the average of 4 individual experiments.

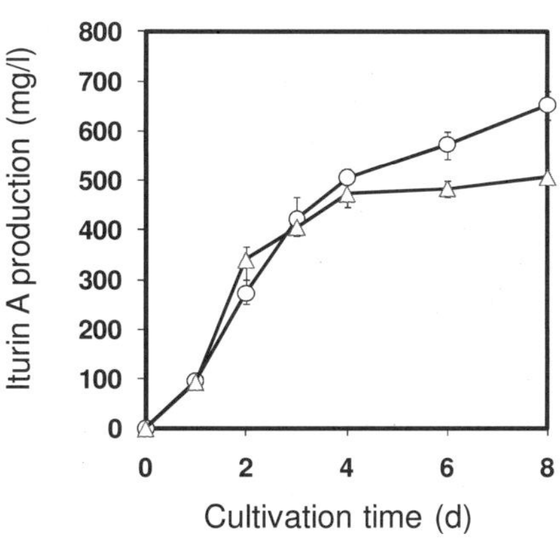

Fig. 2 Observation of iturin A production in submerged (triangle) and biofilm (circle) fermentation (n=4) in fish protein medium. To obtain the optimum iturin A production, incubation was performed at 30 °C with 120 strokes per minute for submerged fermentation and 25 °C at static condition for biofilm fermentation.

4. Discussions

Utilization of marine resources for beneficial use is a recent approach. There is a vast area of research that is still undiscovered. Fish protein can be a great candidate for microbial growth and desired secondary metabolite production as it is rich in carbon and nitrogen sources. When fish protein was compared with commercially available peptones in conventional submerged fermentation, it was observed that the cellular growth were similar to Polypepton S and higher than Polypepton medium. But the sporulation in fish protein medium was similar to Polypepton medium. In both Polypepton and fish protein medium spore numbers remained lower than viable cell numbers whereas in Polypepton S medium the total cell and spore numbers remained similar. Iturin A production in fish protein medium was similar to that in Polypepton S medium but it was significantly higher than that in Polypepton medium. In this investigation it was interestingly observed that fish protein acted like plant originated Polypepton S rather than bovine originated Polypepton, in respect of viable cell numbers and iturin A production. Cod fish is easily accessible due to its huge source and worldwide availability. This fish can be collected constantly every year by maintaining the fishing rules and regulations. Another important thing is production of peptone from bovine source can create bovine serum albumin (BSA) problem but according to present knowledge fish is a safe resource for both human and experimental use. In this study only cod fish was applied for microbial growth and secondary metabolite production, however, gradually other non edible fishes can also be used for this purpose. There are several kinds of non edible fishes, all parts of these "trash" fishes can be processed into the useful products. Therefore, production of peptone from these resources will be cheap and it will not conflict with human food in future.

When the fermentation condition was compared it was observed that the cellular growth and secondary metabolite production varied upon the fermentation condition. During submerged fermentation, *B. subtilis* planktonic cells offered rapid growth and early sporulation due to the accessibility of sufficient oxygen and nutrient as a result production of iturin A was also found earlier. In the static condition *B. subtilis* RB14 single-swimming cells migrate to the air–liquid interface and become associated to the air-liquid surface by its positioning mechanism [24] as well as begin to proliferate as long chains of non-motile cells that are bound together in parallel arrangement in standing liquid culture [4]. Successively these bundled filamentous cells give rise to the larger features of the biofilm architecture (Fig.1A, B and C). It was observed that cellular growth continues for much longer period in biofilm fermentation than in submerged fermentation (Fig. 1D). And the numbers of vegetative cells and spores were found higher in biofilm fermentation (about 6×10^9 and 2×10^9) than submerged fermentation (about 4×10^9 and 7×10^8). Lindsay et al. [25] described that spore formation of *B. subtilis* is delayed in biofilms. Using the heat sensitivity experiment, our results agreed with this phenomenon as we observed 100% sporulation on its fourth day of incubation while in submerged fermentation on its third day cells were entirely sporulated [21]. Here it is noteworthy that for cfu counting the plate count method was adopted and there was always a possibility of cell aggregation [26] that might not result with complete accuracy for the intact biofilm but could result with underestimation of the actual cell numbers.

B. subtilis produces the lipopeptide antibiotic iturin A by its exceptionally large non ribosomal peptide synthetase (NRPS synthetase) [27-30]. It is accepted that differential expression of the extra large NRPS synthetase requires higher oxygen transfer system for iturin A production [31]. Therefore, iturin A starts earlier in submerged fermentation than that in biofilm fermentation (Fig. 2). It has been reported that antibiotic production of *B. subtilis* takes place on the onset of sporulation [32]. Therefore, delayed sporulation resulted with relatively higher iturin A production in biofilm fermentation. It is considerable that the biofilm offers limited oxygen, heat and mass transfer due to the static condition and thereby, theoretically, sporulation and secondary metabolism should be impaired compared to the submerged culture. However, our findings show that biofilm is capable of producing the antibiotic. Compared to the submerged fermentation, biofilm reveals with superior productivity of lipopeptide antibiotic iturin A. Moreover, biofilm is ahead of the other fermentation processes when compared for the running production cost because of its limited energy requirement, easy recovery of cultivated cells.

References

[1] Kolter R. and Peter E. The superficial life of microbes. Nature. 2006; 441(18): 300-302.

[2] Pratt LA, Kolter R. Genetic analyses of bacterial biofilm formation. Current Opinion in Microbiology. 1999; 2: 598-603.

[3] Kuchma SL, O'Toole GA. Surface-induced and biofilm-induced changes in gene expression. Current Opinion in Biotechnology. 2000; 11: 429-433.

[4] Branda SS, Gonzalez-Pastor JE, Ben-Yehuda S, Losick R, Kolter R. Fruiting body formation by *Bacillus subtilis*. Proceedings of the National Academy of Sciences of the United States of America. 2001; 98:11621-11626.

[5] Oosthuizen MC, Steyn B, Theron J, Cosette P, Lindsay D, Von HA, Brozel VS Proteomic analysis reveals differential protein expression by *Bacillus cereus* during biofilm formation. Applied and Environmental Microbiology. 2002; 68: 2770–2780.

[6] Branda SS, Gonzalez-Pastor JE, Dervyn E, Ehrlich SD, Losick R, Kolter R. Genes involved in formation of structured multicellular communities by *Bacillus subtilis*. Journal of Bacteriology. 2004; 186: 3970-3979.

[7] Branda SS, Vik S, Friedman L, Kolter R. Biofilms: the matrix revisited. Trends in Microbiology. 2005; 13: 20-26.

[8] Karatan E, Watnick P. "Signals, regulatory networks, and materials that build and break bacterial biofilms". Microbiology and Molecular Biology Review. 2009; 73 (2): 310–347.

[9] Hoffman LR, D'Argenio DA, MacCoss MJ, Zhang Z, Jones RA, Miller SI. "Aminoglycoside antibiotics induce bacterial biofilm formation". Nature. 2005; 436 (7054): 1171–5.

[10] Broise DL, Dauer D, Gildberg A G, Guerard F. Evidence of positive effect of peptone hydrolysis rate on *Escherichia coli* culture kinetics. Journal of Marine Biotechnology. 1998; 6: 111–115.
[11] Raghunath MR. Enzymatic protein hydrolysate from tuna canning wastes: standardisation of hydrolysis parameters. Fish Technology.1993; 30: 40–45.
[12] Dufosse L, De La Broise D, Guerard F. Evaluation of nitrogenous substrates such as peptones from fish: a new method based on gompertz modeling of microbial growth. Current Microbiology. 2001; 42: 32–38.
[13] Vazquez JA, Gonzalez MP, Murado MA. Peptones from auto hydrolysed fish viscera for nisin and pediocin production. Journal of Biotechnology. 2004; 112: 299–311.
[14] Peypoux F, Guinand M, Michel G, Delcambe L, Das BC, Lederer E. Structure of iturin A, a peptidolipid antibiotic from *Bacillus subtilis*. Biochemistry. 1978; 17: 3992-3996.
[15] Delcambe L, Peypoux F, Besson F, Guinand M, Michel G. Structure of iturin and iturin-like substances. Biochemical Society Transactions. 1977; 569th Meeting, Sussex, UK.
[16] Phae CG, Shoda M, Nobuhiro K, Nakano M, Ushiyama K. Biological control of crown and root rot and bacterial wilt of tomato by *Bacillus subtilis* NB22. Annual Meeting of the Phytopathological Society. 1992; 58: 329-339.
[17] Quentin MJ, Besson F, Peypoux F, Michel G. Action of peptidolipidic antibiotics of the iturin group on erythrocytes. Effect of some lipids on hemolysis. Biochimica et Biophysica Acta. 1982; 684: 207-211.
[18] Gueldner RC, Reilly CC, Pusey PL, Costello CE, Arrendale RF, Cox RH, Himmelsbach DS, Crumley FG, Cutler HG. Isolation and identification of iturin as antifungal peptides in biological control of peach brown rot with *Bacillus subtilis*. Journal of Agriculture and Food Chemistry. 1988; 36: 366-370.
[19] Phae CG, Shoda M. Expression of suppressive effect of *Bacillus subtilis* on phytopathogens in inoculated composts. Journal of Fermentation and Bioengineering. 1990; 70: 409–414.
[20] Phae CG, Shoda M, Kubota H. Suppressive effect of *Bacillus subtilis* and it's products on phytopathogenic microorganisms. Journal of Fermentation and Bioengineering. 1990; 69:1-7.
[21] Rahman MS, Ano T, Shoda M. Biofilm fermentation of iturin A by a recombinant strain of *Bacillus subtilis* 168. Journal of Biotechnology. 2007; 127: 503–507.
[22] Zohora US, Rahman MS, Ano T. Biofilm formation and lipopeptide antibiotic iturin A production in different peptone media. Journal of Environmental Science. S. 2009; 21: S24–S27.
[23] Asaka O, Shoda M. Biocontrol of Rhizoctonia solani damping off of tomato with *Bacillus subtilis* RB14. Applied Environmental Microbiology 1996; 62: 4081–4085.
[24] Davey ME, O'Toole GA. Microbial biofilms: from ecology to molecular genetics. Microbiol. Molecular Biology Review. 2000; 64: 847-867.
[25] Lindsay D, Brozel VS, Holy A. Spore formation in *Bacillus subtilis* biofilms. Journal of Food Protection. 2005; 68: 860-865.
[26] Costerton JW, Lewandowski Z, Caldwell DE, Korber DR, Lappin-Scott HM. Microbial biofilms. Annual Review of Microbiology. 1995; 49:711-745.
[27] Marahiel MA. Protein templates for the biosynthesis of peptide antibiotics. Chemistry and Biology. 1997; 4: 561-567.
[28] Konz D, Marahiel MA. How do peptide synthetases generate structural diversity? Chemistry and Biology.1999; 6: R39-48.
[29] Keating TA, Walsh CT. Initiation, elongation, and termination strategies in polyketide and polypeptide antibiotic biosynthesis. Current Opinion in Chemistry and Biology. 1999; 3: 598-606.
[30] Tsuge K, Akiyama T, Shoda M. Cloning, sequencing, and characterization of the iturin A operon. Journal of Bacteriology.2001; 183: 6265-6273.
[31] Stanley NR, Britton RA, Grossman AD, Lazazzera BA. Identification of catabolite repression as a physiological regulator of biofilm formation by *Bacillus subtilis* by use of DNA microarrays. Journal of Bacteriology. 2003; 185: 1951-1957.
[32] Walker JE. Antibiotic production and sporulation in *Bacillus subtilis*. Biochemistry Journal. 1971; 121(3): 571–573.

The influence of DC air plasma and cellulase enzyme on the antimicrobial activity of Azadirachtin (neem leaf extract) treated cotton fabric

E. Nithya[1], S.Jayakumar[*1], K.Vaideki[1] and R.Rajendran[2]

[1]Thin film centre, Department of Physics, PSG college of Technology, Coimbatore- 641 004, Tamil Nadu, India

[2]Department of Microbiology, PSG College of Arts and Science, Coimbatore- 641 004, Tamil Nadu, India.

*Corresponding author: e-mail: s_jayakumar_99@yahoo.com, Phone: +91 0422-4344453

The need for antimicrobial textiles is increasing nowadays, due to the detrimental effects of micro-organisms on textiles as well as human hygiene. So it is essential to control or inhibit the growth of these organisms on textile fabrics which turns out to be undesirable for the wearer as well as the textile itself. The natural and ecofriendly, antimicrobial agents for textile applications are gaining interest in recent times. Eucalyptus oil, tulsi leaf, Aloe vera, neem leaf extracts, etc., can be used as a bioactive agent for finishing the textile materials. Neem is known for its environmental compatibility and is harmless against non-target materials, besides being skin friendly and non-toxic. This has significantly increased the use of bioactive agents obtained from neem products in the area of antimicrobial textiles. A large amount of biocides has to be applied on to the fabric to control effectively the microbial growth and sustain the durability of the finish. In the present study, hydrophilicity of the cotton fabrics was increased by DC air plasma and cellulase enzyme treatments to improve the uptake of azadirachtin content in the fabrics. The hydrophilicity of the treated fabrics was determined using dynamic wicking test. The physical and chemical modifications on the plasma and enzyme treated fabrics were analyzed using SEM and UATR- FTIR respectively. The fabrics treated with plasma and enzyme is found to exhibit increased hydrophilicity thereby increasing the uptake of neem leaf extract. Methanolic extract of neem leaf was prepared and applied as an antimicrobial finish for the treated cotton fabrics by pad-dry-cure method. The antibacterial and antifungal properties of neem leaf extract finished cotton fabrics were evaluated for its activity against: S.*aureus*, E.*coli*, *Penicillum, Trichoderma* using standard qualitative and quantitative test methods. The synergetic effect of the two treatments on the antimicrobial activity and wash durability of the cotton fabrics were investigated and the results are discussed..

Keywords azadirachtin; antimicrobial activity; hydrophilicity; air plasma; cellulase enzyme

1. Introduction

In recent years, human beings are often infected mainly in the hospital environment by the micro-organisms such as bacteria and fungi. These pathogens multiply rapidly in the textile medium causing ill effects to the textile as well as the wearer. The detrimental effects caused by microbes on the textiles can be effectively controlled by an antimicrobial finish or by incorporating a biocide onto the fabrics [1-3]. Hence, there is a demand for natural, efficient antimicrobial agents which are durability and cost effective. Neem is found to be one such antimicrobial agent that can be used for finishing of textile materials [3].

In imparting the anti-microbial property, an essential requirement for medical textiles in surgical zone, hydrophilicity of cotton fabrics plays an important role. Plasma and enzymatic treatments of textiles were found to be low cost, eco-friendly techniques that can be exploited for improving the antimicrobial efficacy and wash durability.

Plasma treatment of textiles offers functional, environmental and economic benefits when compared with the wet chemical processing [4, 5]. Plasma treatment modifies the surface property of textile fabrics without drastically affecting the bulk properties [6, 7]. It changes the surface chemistry and morphology by effectively confining the functional groups onto the fabric surface [8, 9]. In general, plasma treatment of cotton fabric is characterized by the formation of carboxyl groups and generation of surplus quantity of free radicals on the fabric surface, making it more reactive [10].

Enzymatic hydrolysis is a catalytic action, which induces a chemical change, thereby removing the non cellulosic impurities such as pectins, waxes, proteins, ashes and other dust particles responsible for hydrophobicity from the fabric surface. Further, removal of pectins will make cotton fabrics more hydrophilic without fiber deterioration. In this work, cotton fabrics were treated with DC air plasma or enzyme, or their combinations to enhance the hydrophilicity in turn antimicrobial activity. Pre and post plasma treatment for enzymatic treated fabrics are, hereafter, referred to as PE and EP respectively, in this paper.

The effects of pre- and post plasma treatment of the enzymatic hydrolysis of cotton fabric on the hydrophilicity, anti-microbial property and wash durability were investigated.

2. Materials and Methods

2.1 Materials

The fabric used in the present study was a plain weave, 20 s count, 100% pure cotton fabric (10cm x 10cm) with an EPI/PPI count of 72/54. Cotton fabrics were initially immersed in boiling water at 90 °C for 15 minutes and dried. Commercially available Cellulase Bio-Lish 2.0 (S &A Impex, India) was used for enzyme treatment.

2.2 Plasma and Enzyme Pre-treatments

A 12'' DC plasma chamber was used for plasma pre-treatment as discussed elsewhere [11].The cellulase enzyme pre-treatment was carried out in an incubator shaker with a material to liquid ratio (MLR) of 1:20 as discussed elsewhere.

2.3 Assessment of Hydrophilicity

Hydrophilicity of the fabrics play an important role in the uptake of antimicrobial agents, hence the treated and untreated fabrics were analyzed for its hydrophilicity using dynamic wicking test (BS 4554) [11].

2.4 UATR- FTIR and SEM analysis

The chemical changes that occurred during the DC air plasma and Cellulase enzyme treatments were analyzed by UATR-FTIR (Universal Attenuated Total Internal Reflection Fourier Transformation Infrared) spectra, recorded using a Perkin Elmer (Spectrum100) FTIR spectrometer in the range of 4000-650 cm^{-1} with a resolution of 4 cm^{-1}. The surface morphology of untreated and treated cotton samples was studied using SEM (JEOL-JSAL 6360). Instead of conventional metal coating, carbon tape was used as a conducting material to analyze the sample.

2.5 Preparation of neem leaf extract and Antimicrobial Finish Treatment

The neem leaves (*Azadirachtin indica*) was collected in and around Coimbatore and were shadow dried at room temperature to reduce the moisture content. The leaves were then powdered and sieved. 20 grams of the ground neem powder was suspended in 100ml methanol and incubated overnight. The supernatant was filtered twice using Whatman No.1 filterpaper and the filtrate was further used for the finishing process. The methanolic extract of bioactive substance in neem leaf was prepared as an anti-microbial finish for the cotton fabric using exhaustion method. The fabric samples were immersed in the extract at a MLR of 1:20 for 30 mins and were padded between the rollers. It was then cured at a temperature of 50°C for 30 minutes and then dried. Eight percent of citric acid was used as a cross linking agent to fix the antimicrobial agent and increase the durability of the finish.

2.6 Assessment of Antimicrobial Activity

The anti-microbial efficacy of treated fabrics was assessed qualitatively using Agar diffusion test (SN 195920) and Parallel streak test (AATCC test method 147-1988), quantitatively by AATCC 100 test [12]. In the above methods, test organisms used as representative gram positive and gram negative organisms were *Escherichia coli and Staphylococcus aureus* respectively.AATCC test method 30 procedure was used for evaluating the antifungal activity of the neem leaf extract treated fabric against *Trichoderma* and *Penicillium citrinum.*

2.7 Wash durability test (IS: 687-1979)

Washing test was carried out as per IS: 687-1979 by using a neutral soap (5gpl) at 40°C 2°C for 30 minutes, keeping the material to liquid ratio at 1:50, followed by rinsing, washing and drying. After drying, the test samples were assessed for antimicrobial activity using AATCC 100 procedure up to 30 washing cycles.

3. Results and Discussion

3.1 Hydrophilicity of DC air plasma and cellulase enzyme pre- treated cotton fabrics- UATR-FTIR

Hydrophilicity of the untreated and treated fabric samples was assessed using dynamic wicking test and was measured in terms of wicking height as discussed elsewhere[11] . Figure 1 shows the UATR-FTIR spectra of the DC air plasma (P), cellulase enzyme (E), PE and EP treated cotton fabrics. The spectra reveals the presence of all the cellulosic peaks corresponding to various functional groups both in the untreated (control) and treated fabrics.

It is evident from the figure that, the intensity of 1734 cm^{-1} peak corresponding to C=O stretching in COOH group is increasing for the samples in the order: control< Plasma< Enzyme< PE < EP. This increase in the carbonyl group on the fabric surface has enhanced the hydrophilic property of the fabric in addition to the physical changes.

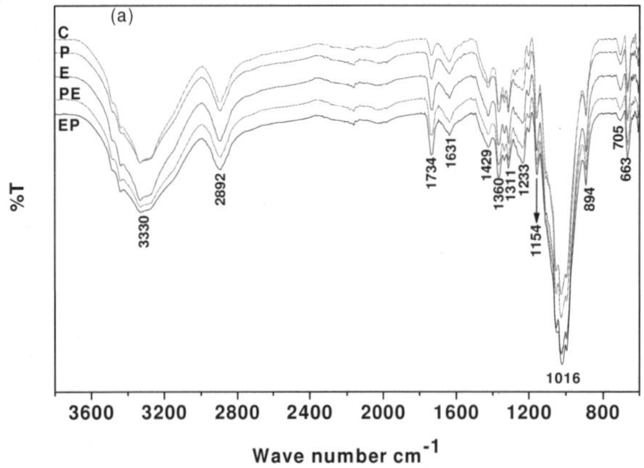

Figure 1. UATR-FTIR Spectra of Control (C), DC air plasma (P), Cellulase Enzyme (E), PE, EP treated cotton fabrics.

In plasma treated cotton fabric, the increase in concentration of carbonyl group (1734cm^{-1}) may be attributed to the dehydrogenation at C_6 carbon and oxidation of primary alcohol leading to the formation of -COOH on the fabric surface.The slight decrease in the concentration of C_1-O-C_4 group is due to the cleavage at ether linkage followed by oxidation which would have contributed to the increase in carbonyl group. In the enzyme treated fabrics, there is a decrease in the intensity of C_1-O-C_4 group, due to cellulase enzyme hydrolysis of β (1-4) glucosidic linkages in cellulose. The increase in intensity of C=O group (1734 cm^{-1}) in enzyme treated fabric may be attributed to the oxidation of C_1 or C_4 carbon after enzyme hydrolysis on the fabric surface due to the presence of acidic pH during the treatment. The presence of 894 cm^{-1} peak in enzyme treated fabric indicates that the enzyme has not hydrolyzed the cotton fabric (cellulose) completely to glucose. Thus enzyme and plasma treatments increase the polar groups on the fabric surface thereby increasing the hydrophilicity of the fabric. In PE and EP treated fabrics, the synergetic effect of plasma and enzyme treatments has increased the concentration of the carbonyl group further which is evident from the spectra. It is clear from the FTIR and dynamic wicking test results that, increase in hydrophilicity may be attributed to the chemical modification of the fabric by enzyme and plasma treatments.

3.2 SEM analysis

Figures 2 and 3 show the SEM micrographs of untreated and EP treated cotton fabric respectively. SEM micrographs confirm that the DC air plasma treatment has etched the hair like protrudants present on the surface of the fabric which resulted in the increase of interstitial pore size, thus increasing the hydrophilicity of the fabric. When treated with enzymes, it was found to be smoother, which revealed that the surface fibrils and other impurities had been removed, thereby enhancing the hydrophilicity and in turn the antimicrobial activity of the fabrics. EP treated samples was found to be smoother when compared to control fabric, which may be due to the smoothening and etching on the surface due to the effect of enzyme and plasma (Fig 1) respectively.

Figure 2. SEM images of untreated cotton fabric (C)

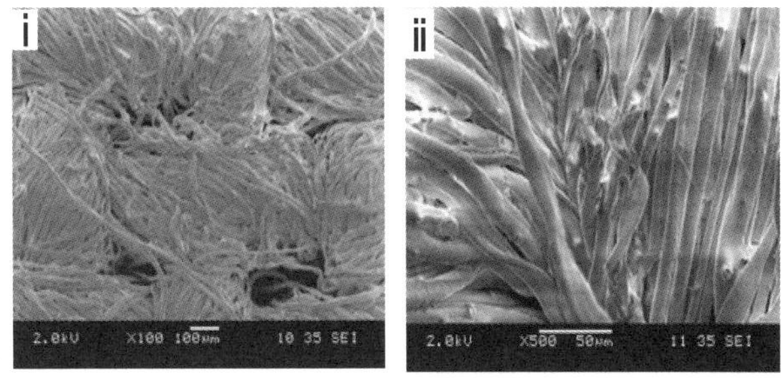

Figure 3. SEM images of EP treated cotton fabric

3.3 Antimicrobial activity

Plasma, Enzyme, PE and EP pre-treated fabric samples were finished with neem leaf extract and the extent of their antimicrobial activity (zone of inhibition in mm) was found using Agar diffusion (figure 4a) and Parallel streak methods and the results are shown in figure 4b.

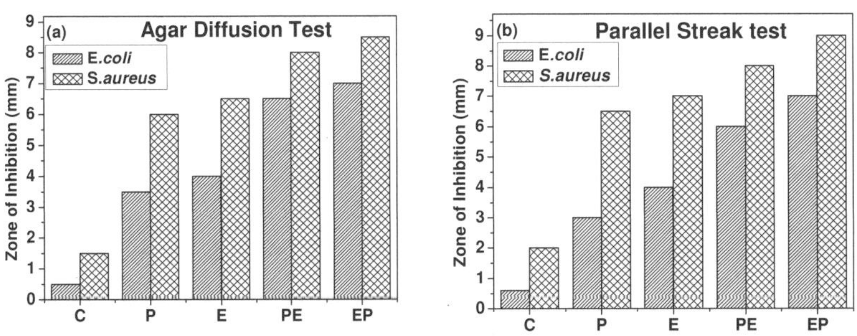

Figure 4. Antimicrobial activity of the treated (P, E, PE, EP) and untreated(C) cotton fabricsusing (a) Agar diffusion test (b) Parallel streak test for *S.aureus* and *E.coli* test organisms

Plasma, enzyme, PE and EP treated fabrics show an increase in antimicrobial activity for both the organisms which is attributed to the increase in hydrophilicity. The increase in hydrophilicity due to plasma and enzyme treatments has enhanced the uptake of antimicrobial finish, thereby increasing the antimicrobial efficacy of fabrics. Maximum zone of inhibition was found to be 8.5 mm for *S.aureus a*nd 7mm for *E.coli* in Agar diffusion test (figure 4a) for EP treated fabrics. In parallel streak test, the maximum zone of inhibition of 9 mm (*S.aureus*) and 7 mm (*E.coli*) was also observed for those samples treated with EP (Fig 4b). This is due to synergic effect of enzyme and plasma on the cotton fabrics. The enzyme pre-treatment has facilitated effective impingement of the plasma particles on the fabric surface, thereby increasing reactive sites on it.

3.4 Qualitative Assessment of Antimicrobial activity (AATCC 100)

The antimicrobial activity of the treated fabrics was estimated quantitatively using AATCC 100 method and the results are tabulated in table 1.The results reveal that all the treated fabrics show 100% bacterial reduction for S.*aureus* organism. EP treated cotton fabrics show 100% bacterial reduction for both the organisms. For Plasma, enzyme, PE treated cotton fabrics the there is approximately 98% bacterial reduction. These results are in good agreement with the agar diffusion and parallel streak test.

3.5 Antifungal activity (AATCC 30)

The activity against fungi was also estimated using AATCC 30 test method against Trichoderma and Penicillium and the results are shown in table 1. There was no zone of inhibition for the control fabric, but there was activity only under the fabric as indicated as 3.8 cm. The highest zone of inhibition (5cm) was observed for EP treated fabrics against Trichoderma as well as Penicillium. The activity against fungi also follows the same trend as that of bacteria. The anti fungal activity for the treated fabrics are in the order as follows: C < P < E < PE < EP.

Table 1. Quantitative Bacterial Reduction (AATCC 100) and Antifungal activity of untreated and treated fabrics (AATCC 30)

Treatments	Zone of Inhibition (cm)		Percentage Bacterial Reduction (%)	
	TrichodermaSp.,	*Penicillium Sp.,*	*S.aureus*	*E.coli*
Control	3.8	**3.8**	-	-
Plasma	4.6	4.5	100	97.6
Enzyme	4.8	4.6	100	98.5
PE	4.9	4.7	100	98.1
EP	5	4.8	100	100

3.6 Wash Durability Test

The durability of the antimicrobial finish was assessed by wash durability test. The percentage reduction in the antibacterial activity up to 30 washing cycles was determined using AATCC 100 test method. The fabric samples treated with plasma, enzyme, PE and EP was given finish with the neem leaf extract. The binding agent, citric acid was added to the extract to increase the durability of the finish. The fabrics were then tested for the percentage reduction of the micro-organisms (*S.aureus* and *E.coli)* using AATCC 100.

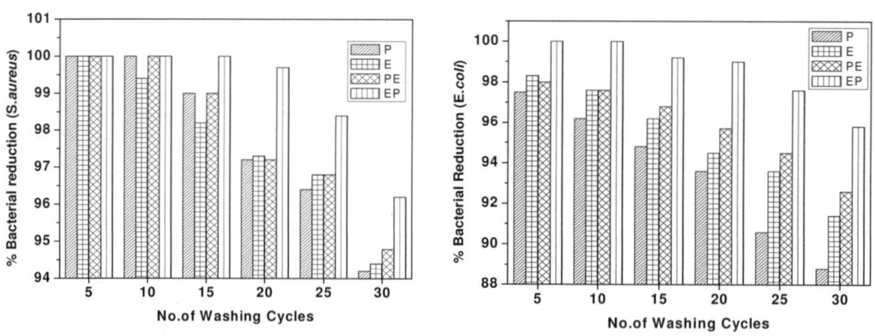

Figure 5. Percentage Bacterial reduction (AATCC 100) (a) S.*aureus* & (b) E.*coli* of the treated fabrics for 30 washing cycles

The results are presented in Figs 5a & 5b.It is evident from the figures that EP treated samples have 100% bacterial reduction up to 10 washing cycles, and thereafter there was a decrease in the percentage reduction for both the organisms. Plasma, Enzyme and PE treated fabrics showed a decrease in the activity from the fifth wash. The durability of antimicrobial finish was found to be highest for EP treated fabrics. The increase in wash durability of the fabric is attributed to the cross linking of cellulose and antimicrobial finish by citric acid (binding agent) and its increase in uptake capacity of antimicrobial finish due to hydrophilicity.

4. Conclusions

Natural antimicrobial agent, neem leaf extract, has been used for imparting antimicrobial property to the cotton fabric. Plasma and enzyme pretreatments were given to the cotton fabrics under optimized conditions to achieve maximum hydrophilicity. The increase in hydrophilicity is attributed to the physical and chemical modifications on the fabric surface which is evident from FTIR and SEM studies. The FTIR results reveal that the increase in polar groups on the fabric surface has enhanced the hydrophilicity. SEM micrographs reveal etching of the fabric due to plasma treatment and smoothening due to enzyme treatment improving the hydrophilicity. It was found that the anti bacterial and antifungal activity was highest for neem leaf extract treatedvEP treated fabric in all the test methods. The increase in hydrophilicity is responsible for the increase in anti-microbial efficacy of the treated fabrics. The investigation concludes that, EP treated fabric samples are found to exhibit higher antimicrobial efficacy when compared to plasma, enzyme and PE treatments. The wash durability of antimicrobial finish is also found to be higher for EP treated fabric than the other treated samples.

Acknowledgements: The authors would like to thank the Management and Principal, PSG college of Technology for providing the necessary infra structure.

References

[1] Yuan Gao, Robin Cranston,Recent Advances in Antimicrobial treatment of Textiles, *Textile Research Journal,* 2008;78: 60-72.

[2] Francois N.R.Renaud, Jeanne Dore, H.Jean Freney, Bernard Coronel and Jean-Yves Cusseau,Evaluation of Antibacterial properties of a textile product with antimicrobial finish in a hospital environment, *Journal of Industrial Textiles*, 36 (2006) 89-94.

[3] M.Joshi, S.Wazed Ali, R.Purwar, S.Rajendran, Ecofriendly antimicrobialfinishing of textiles using bioactive agents based on natural products, *Indian Journal of Fiber and Textile Research*, 34 (2009) 295-304.

[4] M.Radetic, P.Jovancic, N.Puac, Z.Lj.Petrovic,Environmental impact of plasma application to textiles, *Journal of Physics*, 71 (2007) 012017.

[5] H.U Poll, U. Schladitz, S.Schreiter, Penetration of plasma effects into textile structures, *Surf.Coat.Tech.*, 2001;142-144:489-493.

[6] C.X.Wang, Y.Liu, H.L.Xu, Y.Ren, Y.P.Qiu, Influence of atmospheric pressure plasma treatment time on penetration depth of surface modification into fabric, *Appl.Surf.Sci.*,2008;254: 2499-2505.

[7] S.Pane, R.Tedesco, R.Greger, Acrylic fabrics treated with plasma for outdoor applications, *J.Industrial Textiles*, 2001;31:135-145.

[8] Hartwig Hocker, Plasma treatment of textile fibres, *Pure Appl.Chem*, 2002;74:423-427.

[9] F.Ferrero,Wettability measurements on plasma treated synthetic fabrics by capillary rise method, *Polymer testing,* 2003; 22:571-578.

[10] Jie Rong Chen, Free radicals of fibers treated with low temperature plasma, *J.Appl.Poly.Science*, 1991; 42:2035-2037

[11] Nithya E, Radhai R, Rajedran R, Shalini S, jayakumar S, Synergetic effect of DC air plasma and cellulase enzyme on the hydrophilicity of cotton fabrics,*Carbohydrate polymers*, In press, 2010

[12] ATCC Technical Manual , (2001) ,Vol 76.

Urolithins, Metabolites Produced by Human Colonic Microflora, Act as *Quorum Sensing* Inhibitors of *Yersinia enterocolitica* Affecting its Gene Expression

J.A. Giménez, P. Truchado, M. Larrosa, J.C. Espín, F. Tomás-Barberán, M.T. García-Conesa and A. Allende*

[1] Research Group on Quality, Safety and Bioactivity of Plant Foods. CEBAS-CSIC. Espinardo, Murcia, Spain

* Corresponding author: e-mail: aallende@cebas.csic.es, Phone: +34 968396377

Yersinia enterocolitica is a mammalian enteropathogen which may cause gastrointestinal syndromes. *Yersinia* spp. produces N-acyl homoserine lactone (AHL), *Quorum Sensing* (QS) signal molecules which are involved in the cell population density dependent regulation of virulence, secondary metabolite production, and biofilm maturation. Urolithins (Uro-A and Uro-B), metabolites produced after the consumption of ellagitannins, have anti-inflammatory action in colitis and modulate microbiota. Our objective was to evaluate the effect of Uro-A and B, as QS inhibitors in *Y. enterocolitica* and to determine whether or not they affect the expression of specific genes involved in virulence processes. It was found that Uro-A and B were effective inhibiting QS of *Y. enterocolitica* when applied at 5 and 3 μM, respectively. The data obtained by Real-time PCR showed that urolithins increased the transcription of QS regulated genes when low concentrations were used. Thus, urolithins have an antipathogenic effect against *Y. enterocolitica* and can be used as QS inhibitors increasing the expression of virulence related genes.

Keywords Quorum quenching; N-acylhomoserine lactones; *yenI*; *yenR*; ellagitannins.

1. Introduction

Most of the health benefits of pomegranate, black raspberries, raspberries, strawberries and walnuts, among other fruits, are attributed to the elevated content of a diverse range of phytochemicals of which polyphenols, such as ellagitannins (ETs), highly contribute due to its potent antioxidant action [1, 2]. ETs are not absorbed as such but are metabolized by the intestinal flora to yield urolithins, mostly urolithin-A (Uro-A, 3,8-dihydroxy-6H-dibenzo(b,d)pyran-6-one) and urolithin-B (Uro-B, 3-hydroxy-6H-dibenzo(b,d)pyran-6-one) [3]. In previous studies, it was observed that urolithins exhibited some anti-inflammatory effects and modulated favourably the gut microbiota [4, 5]. These results led us to hypothesise that urolithins might have an additional beneficial effect by modulating the growth and pathogenicity of mammalian enteropathogens.

Yersinia enterocolitica is a mammalian enteropathogen which may cause gastrointestinal syndromes after the consumption of contaminated food or through direct inoculation following a blood transfusion in humans [6]. The three human pathogenic *Yersinia* spp. produce N-acyl homoserine lactone (AHL), *Quorum Sensing* (QS) signal molecules, which are involved in the cell population density dependent regulation of virulence, secondary metabolite production, and biofilm maturation [7]. Recent studies have associated the pathogenic activity of *Y. enterocolitica* to specific QS systems.

Y. enterocolitica synthesise two AHLs, *N*-hexanoyl-L-homoserine lactone (C6-HSL) and *N*-(3-oxohexanoyl)-L-homoserine lactone (3-oxo-C6-HSL), which are capable of activating the *Photobacterium fischeri* bioluminescence (lux) operon. A gene, *yenI*, was isolated from *Y. enterocolitica* and demonstrated to direct the synthesis of both C6-HSL and 3-oxo-C6-HSL. This gene encodes a protein with approximately 20% identity to the LuxI family of proteins [8]. Inactivation of *yenI* abolished the production of AHLs and confirmed the function of the gene as an AHL synthase. Analysis of the DNA sequence downstream of *yenI* identified a second gene named *yenR* with significant analogy to the LuxR family of quorum-sensing transcriptional activators [8]. Unlike the LuxI/LuxR system, the expression of *yenI* is not part of an auto-inducible system and is independent of the growth phase [9].

The objective of this study is to determine the effect of the main microbiota-derived metabolites of ETs, Uro-A and Uro-B, as modulators of *Y. enterocolitica* growth and/or pathogenesis as a result of putative antimicrobial and/or anti-QS activities. Additionally, we will determine whether exposure of *Y. enterocolítica* to the ETs-derived metabolites, at concentrations that can be achieved in the gut through the diet, affects the production of AHLs by modulating the expression of specific genes, *yenI* and *yenR*, involved in the synthesis of AHLs and virulence processes, as well as the expression of genes that might be regulated by *yenI* and *yenR*, such as those involved in cell motility.

2. Materials and Methods

2.1 Urolithins and Acyl-Homoserine Lactones

Urolithins, Uro-A (3,8-dihydroxy-6H-dibenzo(b,d)pyran-6-one) and Uro-B (3-hydroxy-6H-dibenzo(b,d)pyran-6-one) were chemically synthesised. Urolithins were initially dissolved in DMSO (maximum concentration of 0.1 %) and filtered through a sterile Millex-GP 0.22 μm filter (Millipore Corp., USA). A control solution containing only DMSO was made in MilliQ water. Different concentrations of urolithins were added to the culture media to obtain final concentrations of 4, 5, 6, 8 and 9 μM for Uro-A and 2, 3, 4, 5, 6 μM for Uro-B. The AHL standards used in this study, 3-oxo-C6-HSL and C6-HSL, were supplied from Sigma (St. Louis, MO).

2.2 Strains and culture conditions

Chromobacterium violaceum wild-type strain CECT 494 and *Yersinia enterocolitica* strain CECT 4315 were obtained from the Spanish Type Culture Collection (Valencia, Spain). Bacteria were routinely grown aerobically with shaking in Luria-Bertani broth (02-385, LB broth acc. to MILLER, Scharlau Chemie, S.A. Barcelona, Spain) and incubated at 30 °C for 24 h.

2.3 Antimicrobial activity

The antimicrobial activity of urolithins against *Y. enterocolitica* was evaluated. Ten mL of LB broth were inoculated with 10 μL of a *Y. enterocolitica* working culture and urolithins at a final concentration of 5 μM and 3 μM for Uro-A and Uro-B respectively. Inoculated urolithin solutions were mixed followed by incubation at 30 °C for 20 h. Inoculated and incubated urolithin samples (1 mL) were diluted in 1% sterile buffered peptone water (BPW) (AES Laboratoire, Combourg, France) (1:10 dilution). Appropriate dilutions were then spread onto LB agar (1%) (Scharlau Chemie, S.A.) and incubated at 30 °C for 24 h. Microbial counts were expressed as log cfu/mL.

2.4 Detection of the QS inhibitory capacity using a bacteriological monitoring system

Flask-incubation assays were carried out to quantify the inhibitory activity of urolithins. The bacteriological monitoring system used to determine the anti-QS activity was *C. violaceum.* This wild type of strain produces and responds to the cognate autoinducer molecules, AHLs such as C6-AHL and C4-AHL. This bacteriological monitor system generates a phenotypic response by the production of a purple pigment (violacein) when induced by the presence of AHLs. The bacterium was incubated for 18 h and inoculated to $OD_{600nm} = 0.1$ in Erlenmeyer flasks containing LB broth (Scharlau Chemie, S.A.) and LB supplemented at different concentrations with Uro-A (4, 5, 6, 8 and 9 μM) and Uro-B (2, 3, 4, 5 and 6 μM). The flasks were incubated at 30 °C in a shaking incubator. The quantification of the violacein production was carried out following the protocol described by Choo et al. [10], where 1 mL culture from each flask was centrifuged at 18,000 *g* for 10 min to precipitate the insoluble violacein. Then, the culture supernatant was discarded and the pellet was solubilized in 1 mL of DMSO, vortexed until the violacein was homogenized, and centrifuged at 18,000 *g* for 10 min to remove the cells. Absorbance of each violacein-containing supernatant was measured at 585 nm in a UV–vis spectrophotometer (Hewlet Packard 8453).

2.5 RNA isolation and quantitative reverse transcription polymerase chain reaction analysis

The RNA was isolated using the MICROBexpress kit (Ambion, Applied Biosystem). The purity and concentration were checked measuring the absorbance at 260 and 280 nm using the Nanoquant plateTM in the plate reader (Infinite M200, Tecan, Grodig, Austria). Only RNA with an Abs_{260}/Abs_{280} ratio between 2 and 2.2 was used for the experiments. Gene expression was assessed by one-step quantitative RT-PCR (Taqman system, Applied Biosystems, ABI, Madrid, Spain). Primers and probes were designed for five genes of *Y. enterocolitica* which are related to AHLs synthesis and motility: *yenI*, *yenR*, *fliA*, *fleB* and *flhDC*. 16S rRNA was used as housekeeping gene. Relative quantification was calculated using the $2^{-\Delta\Delta Ct}$ method.

2.6 Statistical analysis

At least three replications of each experiment, separately in time, were carried out. ANOVA and Brown-Forsythe tests were used depending on the homogeneity of the variances. When significant differences were observed, Tukey or Dunnett T3 test were performed using SPSS 17.0 for Windows. A significant level of $P \leq 0.05$ was selected to determine differences between samples.

3. Results and Discussion

3.1 Antimicrobial and anti-*Quorum Sensing* activities

Previous studies had found that the ETs-derived metabolites, urolithins, modulated favourably the gut microbiota [5]. The observed microbiota modulation could be due to antimicrobial and/or anti-QS activities of the urolithins. In this study we first evaluated the effect of Uro-A and Uro-B, at concentrations than can be achieved in the gut through the diet, on the growth of *Y. enterocolitica*. It was found that none of these two urolithins reduced the growth of *Y. enterocolitica* at the tested concentrations after 24 h of incubation (data not shown).

Fig. 1 shows the capacity of the different concentrations of Uro-A and Uro-B tested to inhibit violacein production by the biosensor strain *C. Violaceum* and thereby to inhibit AHL production. Compared to the control sample (CT), Uro-A and Uro-B treated samples showed a significant drop in the production of violacein, even at the low concentrations tested. The addition of 9 μM of Uro-A inhibited the production of violacein by 60% or more. For further experiments, the selected concentrations were 5 μM for Uro-A and 3 μM for Uro-B.

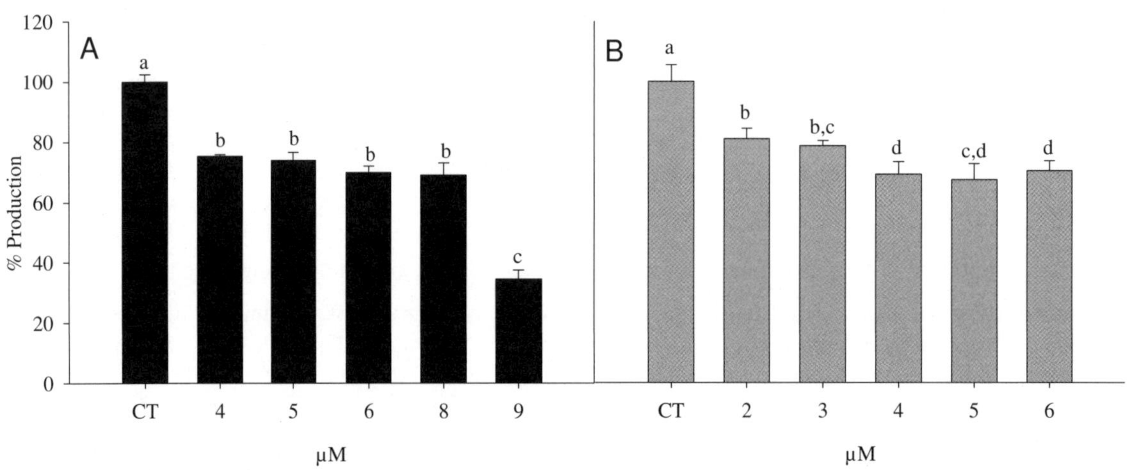

Fig. 1. Production of violacein by the biosensor strain *Chromobacterium violaceum* supplemented with different concentrations of Uro-A. (A) and Uro-B (B). Bars labeled with different letters indicate significant difference at $P < 0.05$.

These results indicated that Uro-A and Uro-B exert some anti-virulent and anti-pathogenic activities which are not related to the growth inhibition of the microorganism as found with different phytochemical extracts from dietary sources [11].

3.2 Quantitative reverse transcription polymerase chain reaction analysis

We next determined and compared the mRNA levels of *yenI* and *yenR* between *Y. enterocolitica* cells grown in control medium (C) and cells grown in medium supplemented with Uro-A (5 μM) or Uro-B (3 μM) for 14 h and 24 h (T). RT-PCR analysis (**Fig. 2**) revealed that both transcripts were expressed at 14 h and at 24 h. In cells treated with Uro-A, the levels of expression of *yenR* were slightly (although not significant) higher than in their control counterparts whereas *yenI* was transcribed similarly in both the control and the treated cells. Induction of *yenR* and *yenI* was more noticeable in cells treated with Uro-B with up to ~2.5-fold increase for *yenR* and ~2.0-fold increase for *yenI* at 24 h.

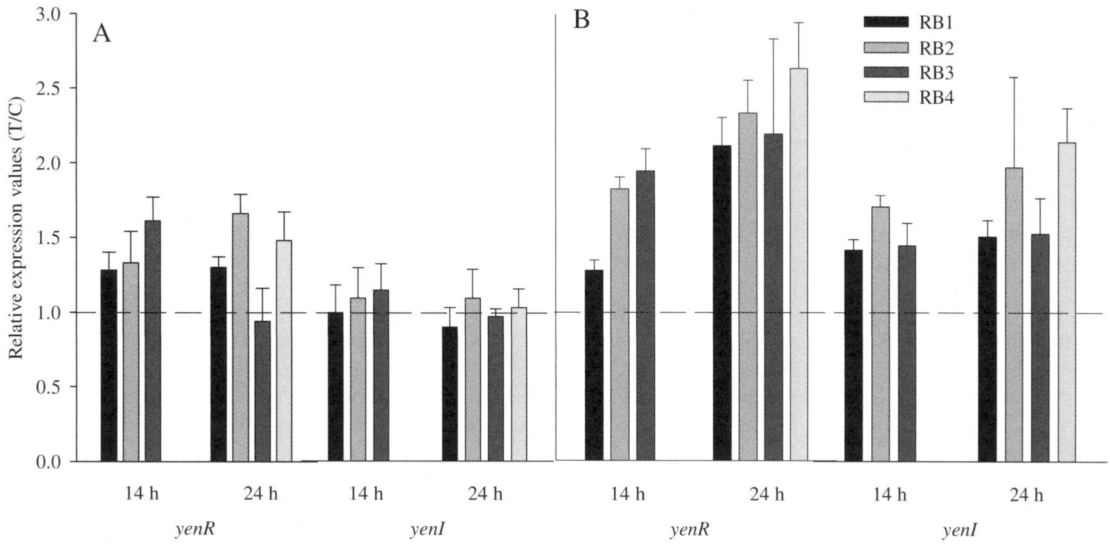

Fig. 2 Relative expression values (T/C) of genes *yenR* and *yenI* in *Y. enterocolitica* untreated (C) and treated (T) with urolithinA (A) and urolithin-B (B).

We next sought to determine whether treatment with the urolithins and associated changes in the expression of *yenI* and *yenR* influenced the expression of some key regulatory members of the flagellum cascade involved in cell motility: *flhDC*, *fleB* and *fliA*. Our analysis detected the expression of the three genes and the results (**Fig. 3**) showed that the transcripts levels of *flhDC* and *fleB* were only slightly up-regulated (not significant) after treatment of *Y. enterocolitica* with Uro-A at 24 h whereas the expression of *fliA* did not change or was even reduced under the same conditions. However, treatment of *Y. enterocolítica* with Uro-B induced the expression of the three genes at 24 h, more clearly for *flhDC*.

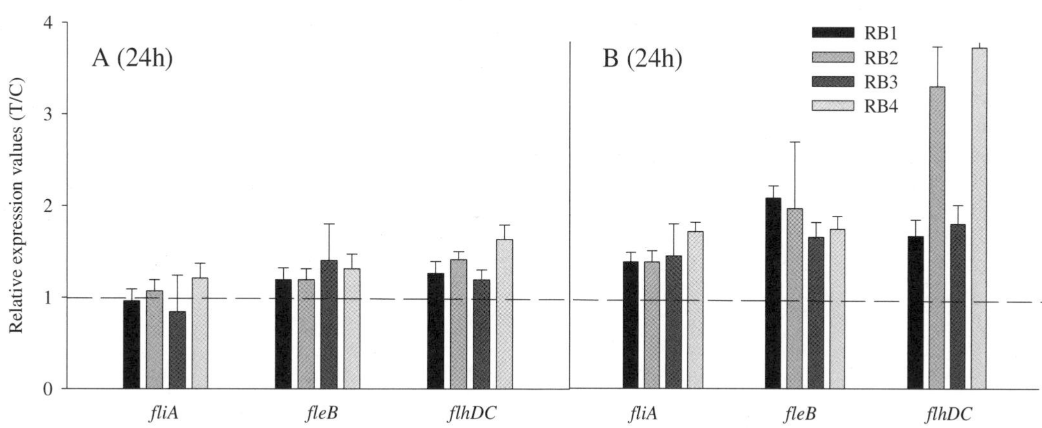

Fig. 3 Relative expression values (T/C) of genes *fliA*, *fleB* and *flhDC* in *Y. enterocolitica* untreated (C) and treated (T) with urolithinA (A) and urolithin-B (B).

Our results have shown that Uro-A and Uro-B, at concentrations that may be accumulated in the gut through the diet, decreased the production of AHLs at 24 h. In association with this phenotypic response, we detected an induction of the expression levels of *yenI* and *yenR* at 14 and 24 h after treatment of the cells, mainly with Uro-B. Based on these results, we hypothesise that, if the observed inhibition of AHLs production has been caused through transcription regulation and the down-regulation of *yenR* and (or) *yenI* expression levels, this might have occurred at earlier stages of the cell growth. The up-regulation observed at 14 h and 24 h may have been induced in the cells by other mechanisms to counterbalance the AHL reduction. We can not discard, however, other alternative mechanisms for the quenching of AHLs such as enzymatic disruption of these signal molecules by lactonases, acylases or oxidorreductases [12]. Flagellar gene expression in *Y. enterocolitica* is likely to be similar to that in other members of *Enterobacteriaceae* and to follow a hierarchical cascade consisting of three major flagellar gene classes, I, II and III. Genome analysis of *Y. enterocolitica* suggests that all the key regulatory elements of the flagellum cascade are present including: 1) class I *flhDC*, at the top of the hierarchy and required for the expression of class II genes, 2) class II *fliA*, which encodes the sigma factor σ^{28} FlgM protein and 3) class II *fleB* flagellin structural gene [13]. It has been reported that a *Y. enterocolitica yenI* mutant strain lacking AHL synthase activity is also unable to swim or swarm and lacks the flagellin protein FleB indicating that QS genes can regulate, at least in part, motility [6]. In agreement with this statement, we found that the induction observed for *yenI* and *yenR* after the treatment with the urolithins, in particular with Uro-B, was associated to an induction of the motility genes studied, more clearly in the cells treated with Uro-B, and confirm a link in the transcriptional regulation of the QS and motility genes.

Acknowledgements This work has been partially supported by the Project Consolider Ingenio 2010, CSD2007-00063 (Fun-*C*-Food) and the EU-Funded Project ''Bees in Europe and the Decline of Honeybee Colonies" (''BEEDOC", Ref: 244956 CP-FP). J.A.Giménez and M. Larrosa are holder of a JAE-Predoc and JAE-DOC fellowships respectively from the Spanish "Consejo Superior de Investigaciones Científicas" (CSIC). P. Truchado holds a PhD grant from the Seneca Fundation (Murcia, Spain).

References

[1] Seeram NP, Henning SM, Zhang Y, Suchard M, Li Z, Heber D. Pomegranate Juice Ellagitannin Metabolites Are Present in Human Plasma and Some Persist in Urine for Up to 48 Hours. *Nutr.* 2006; 136:2481-2485.

[2] Bialonska D, Kasimsetty SG, Khan SI, Ferreira D. Urolithins, intestinal microbial metabolites of Pomegranate ellagitannins, exhibit potent antioxidant activity in a cell-based assay. *J Agric Food Chem.* 2009; 11;57:10181-6.

[3] Espín JC, González-Barrio R, Cerdá B, López-Bote C, Rey AI, Tomás-Barberán FA. Iberian Pig as a Model To Clarify Obscure Points in the Bioavailability and Metabolism of Ellagitannins in Humans. *J. Agric. Food Chem.* 2007; *55*:10476–10485.

[4] González-Sarrías A, Larrosa M, Tomás-Barberán FA, Dolara P, Espín JC. NF-kappaB-dependent anti-inflammatory activity bof urolithins, gut microbiota ellagic acid-derived metabolites, in human colonic fibroblasts. *Br J Nutr.* 2010;104:503-12.

[5] Larrosa M, González-Sarrías A, Yánez-Gascón MJ, Selma MV, Azorín-Ortuño M, Totia S, Tomás-Barberán F, Dolara P, Espín JC. Anti-inflammatory properties of a pomegranate extract and its metabolite urolithin-A in a colitis rat model and the effect of colon inflammation on phenolic metabolism. *J. Nutr. Bioch.* 2010; 21:717-725.

[6] Atkinson S, Chang C-Y, Sockett, RE, Cámara M, Williams P. Quorum Sensing in *Yersinia enterocolitica* Controls Swimming and Swarming Motility. *J Bacteriol.* 2006; 188:1451–1461.

[7] Swift SJ, Downie A, Whitehead NA, Barnard AML, Salmond GPC, Williams, P. Quorum sensing as a population density-dependent determinant of bacterial physiology. *Adv. Microb. Physiol.* 2001; 45:199–270.

[8] Throup, J.P., Cámara, M., Briggs, G.S., Winson, M.K., Chhabra, S.R., Bycroft, B.W., Williams, P., Stewart, G.S. Characterisation of the yenI/yenR locus from *Yersinia enterocolitica* mediating the synthesis of two N-acylhomoserine lactone signal molecules. *Mol. Microbiol.* 1995; 17:345-356.

[9] Atkinson S, Sockett R.E, Cámara M, Williams P. Quorum sensing and the lifestyle of *Yersinia*. *Curr. Issues Mol. Biol.* 2006; 8:1-10.

[10] Choo JH, Rukayadi Y, Hwang J-K. Inhibition of bacterial quorum sensing by vanilla extract. *Lett. Appl. Microb.* 2006; 42:637–641.

[11] Vattem DA, Mihalik K, Crixell SH, McLean RJC. Dietary phytochemicals as quorum sensing inhibitors. *Fitoterapia.* 2007; 78:302–310.

[12] Czajkowski R., Jafra S. Quenching of acyl-homoserine lactone-dependent quorum sensing by enzymatic disruption of signal molecules. *Acta Biochim. Pol.* 2009; 56:1-16.

[13] Young GM. Flagella: organelles for motility and protein secretion, *In* E. Carniel and B.J. Hinnebusch (ed.), Yersinia: molecular and cellular biology. Horizon bioscience, Wymondham, United kingdom. 2004, 243-256.

Use of lactic acid bacteria biofilms as biocontrol agents

Emel Ünal*[,1], Selin Kalkan[2] and Zerrin Erginkaya[1]

[1]Çukurova University, Faculty of Agriculture, Department of Food Engineering, Adana, Turkey
[2]Osmaniye Korkut Ata University, Bahce Vocational School of Higher Education, Department of Food Technology, Osmaniye, Turkey

* Corresponding author: e-mail: unale@cu.edu.tr, Phone: +34 905365785022

A biofilm is an assemblage of microbial cells that is irreversibly associated (not removed by gentle rinsing) with a surface and enclosed in a matrix of primarily polysaccharide material (Extracellular polymeric substances-EPS). Also biofilm-associated organisms significantly differ from their planktonic (freely suspended) counterparts. Biofilms have been of considerable directly interest in the context of food hygiene. Biofilms are more resistant to antimicrobial, temperature, oxygen, protease etc. Biofilms are formed by many microorganisms. Because biofilms formed by pathogen and spoilage microorganisms are resistant to environmental stress, they increase the risk for microbial contamination in food plants. Moreover in recent days, it was discovered that some biofilms have positive properties and it was noticed that these biofilms will be available as biocontrol agents. Use of lactic acid bacteria and their metabolites is the most common and popular in methods of natural protection. Also biofilms are yet another protective agent formed by lactic acid bacteria. On the other hand, there aren't enough studies about use of biofilms as biocontrol agent. In previous surveys, it was mostly focused on antilisterial activity of lactic acid bacteria biofilms. In the other study, it was reported that lactic acid bacteria biofilms can prevent fungal attack. These studies demonstrate the significance of biofilms of LAB in the food industry, so it should be focused on use of lactic acid bacteria biofilms as biocontrol agents with new studies

Keywords: Biofilm, biopreservation, lactic acid bacteria

1. Introduction

1.1. The importance of biofilms with regard to food safety

Biofilms are currently identified as an assemblage of surface-associated microbial cells that are enclosed in hydrated extracellular polymeric substances (EPS). Polysaccharides, proteins, phospholipids, teichoic and even nucleic acids are the main components of EPS. Also biofilms call as exopolysaccharides (EPS). Biofilms have growth and survival advantages over planktonic cells, such as a documented increased resistance to antimicrobial compounds and thermal stress. Almost microorganisms especially pathogen bacteria form biofilm [1, 2, 3, 4, 5].

Inhibition of biofilm forming pathogen or spoilage microorganisms is very difficult because biofilm give resistance to environmental stress. Biofilm forming pathogen or spoilage microorganisms are food safety problems. Biofilms containing pathogens increase the risk for microbial contamination in food plants. The microbial compositions of biofilms formed on the surfaces of the processing equipment in the various food industries. These investigations revealed that biofilms were often established by various microorganisms on the equipment surfaces of the production line. They also indicated that biofilms containing pathogens became one of the major causes of contamination of food products or transmission of diseases. Therefore, it is very important to develop cleaning and disinfection methods and control systems in food-processing plants and environments. Biofilms have been of considerable directly interest in the context of food hygiene [6, 7, 8, 9].

Biofilms formed by nonstarter lactic acid bacteria, Bacillus cereus, Escherichia coli, Shigella sp., Staphylococcus aureus in dairy processing plant, Listeria monocytogenes, Shigella in ice cream plant and Enterobacteriaceae, Serratia liquefaciens, Neisseriaceae, Pseudomonas, Vibrio, Listeria in fish industry lead to quality defects in food product [10, 11].

Up to now, it has been mentioned hazard of biofilms about food safety namely disadvantages of biofilms. In addition to these, there are useful biofilms formed by lactic acid bacteria. It was reported that biofilms formed by lactic acid bacteria had antimicrobial effect on pathogens [7, 12].

Furthermore, EPS have been found so many important biological roles. EPS protect microorganism against phagocytose, bacteriophage attack, antibiotic, dehydration and toxic components. For example, cells of *L. lactis* in EPS are resistant to bacteriophage attack [13, 14].

1.2. Lactic acid bacteria biofilms as biocontrol agents

As an increasing number of consumers prefer natural foods without chemical preservatives. Nowadays, new preservation methods are focused on the use of lactic acid bacteria (LAB) or of their metabolites. Lactic acid bacteria (LAB) are well known as beneficial bacteria and include probiotic bacteria that help maintain a good enteral environment and producers of biopreservatives in dairy products and fermented foods. Metabolites of lactic acid bacteria such as organic acid, diacetil, acetoin, hydrogen peroxide, reuterine, antifungal peptide and bacteriocin used commonly as natural preservative. But there are newer strategies like use of lactic acid bacteria biofilms for the biocontrol of pathogen. Also lactic acid bacteria biofilms are approved as effective antimicrobial and biofilms provide more effective protection than planktonic cells [15, 16].
In recent days, it was discovered that some biofilms have positive properties and it was noticed that these biofilms will be available as biocontrol agents. Use of lactic acid bacteria and their metabolites is the most common and popular in methods of natural protection. Also biofilms are yet another protective agent formed by lactic acid bacteria [10, 12].

It was benefit from lactic acid bacteria biofilms for inhibition of pathogen especially *Listeria monocytogenes*. In some studies, it was demonstrated that species of *E. casseliflavus* and *Lactobacillus plantarum* inhibited growth of *Listeria monocytogenes.* In another study, it was reported that nonstarter lactic acid bacteria biofilms controlled growth of *Listeria monocytogens* in soft cheese [7, 12].

Not only lactic acid bacteria biofilm with antimicrobial properties have used preservative against pathogen and spoilage microorganisms, but also they have used preservative against biofilms formed by these microorganisms. For example, released-exopolysaccharides (r-EPS) could prevent biofilm formation by a wide range of Gram-negative and -positive pathogens. With 1.0 mg/ml of r-EPS, enterohemorrhagic *Escherichia coli* (EHEC) biofilms were dramatically decreased. r-EPS isolated from *L. acidophilus* may inhibit the initial attachment and autoaggregation of EHEC O157:H7 during biofilm development by partially affecting bacterial surface properties. Therefore, r-EPS isolated from *L. acidophilus* in this study may be applied to various food industries, and used as food-grade adjuncts for microbial biofilm control. In addition to this, it was determined in another study, EPS from *Bifidobacterium longum* prevented growth of pathogen or spoilage microorganisms [17, 18].

2. Conclusion

Because lactic acid bacteria biofilms provide effective protection against pathogen or spoilage microorganisms and show more effective inhibition than planktonic cells, it was increased interest in biofilms. These studies demonstrate the significance of biofilms of LAB in the food. So it should be studied more about biofilms and effects of biofilms should be discussed with advantages and disadvantages.

References

[1] Wu, MH, Pan, TM, Wu, YJ, Chang, SJ, Chang, MS, Hu, CY. Exopolysaccharide activities from probiotic Bifidobacterium: Immunomodulatory effects (on J774A.1 Macrophages) and antimicrobial properties, *International Journal of Food Microbiology,* 2010:*33p.*

[2] Gün İ, Ekinci FY. Biyofilmler: Yüzeylerdeki Mikrobiyal Yaşam. *Gıda*, 2009:34(3),165-173.

[3] Kubota, H, Senda, S, Nomura, N, Tokuda, H. Biofilm Formation by Lactic Acid Bacteria and Resistance to Environmental Stres, J. Biosci. Bioeng., 2008:106, 381-386.

[4] Kumar CG, Anand SK. Significance of Microbial Biofilms in Food Industry: A Review.*International Journal of Food Microbiology*, 1998:42, 9-27.

[5] Laws A, Gu Y, Marshall V. Biosynthesis, Characterisation, and Design of Bacterial Exopolysaccharides From Lactic Acid Bacteria. *Biotechnology Advances,* 2001:19, 597-625.

[6] Sharma, M, Anand, SK. Biofilms Evaluation as an Essential Component of HACCP for Food/Dairy Processing Industy-a Case. *Food Contr.* 2002:13, 469-477.

[7] Speranza B, Sinigaglia M, Corbo MR. Non Starter Lactic Acid Bacteria Biofilms:A Means to Control the Growth of *Listeria monocytogenes* in Soft Cheese. *Food Control,* 2009:20,1063-1067.

[8] Simoes, M, Simoes, LC, Vieira, MJ. A review of current and emergent biofilm control strategies, *LWT - Food Science and Technology*, 2010:43, 573–583.

[9] Mah, TFC, O'Toole, GA, 2001. Mechanisms of biofilm resistance to antimicrobial agents, *Trends in Microbiology,* 2001:9(1).

[10] Shia, X, Zhu, X. Biofilm formation and food safety in food industries, *Trends in Food Science & Technology,* 2009:20, 407-413.

[11] Somers, EB, Johnson, ME, Wong, ACL. Biofilm Formation and Contamination of Cheese by Nonstarter Lactic Acid Bacteria in the Dairy Environment, *Journal of Dairy Science*, 2001:84/9,1926-1936.

[12] Guerrieri, E, Niederhäusern, SD, Messi, P, Sabia, C, Iseppi, R, Anacarso, I, Bondi, M. Use of lactic acid bacteria (LAB) biofilms for the control of *Listeria monocytogenes* in a small-scale model, *Food Control,* 2009:20, 861–865.

[13] Badel, S, Bernardi, T, Michaud, P. New perspectives for Lactobacilli exopolysaccharides, *Biotechnology Advances*, article in press, 2010.
[14] Yılmaz, M, Çelik, GY. Bakteriyel Ekstraselüler Polisakkaritler (EPS), *Orlab On-Line Mikrobiyoloji Dergisi*, 2007:5(2),7-13.
[15] Hayaloğlu, AA, Erginkaya, Z. Gıda Endüstrisinde Kullanılan Laktik Asit Bakterileri, *Gıda Teknolojisi Derneği Yayın*, 2001, 23,1-26.
[16] Kubota, H, Senda, S, Tokuda, H, Uchiyama, H, Nomura, N. Stress resistance of biofilm and planktonic *Lactobacillus plantarum* subsp. *plantarum* JCM 1149, *Food Microbiology*, 2009:26, 592–597.
[17] Kim, Y, Oh, S, Kim, SH. Released exopolysaccharide (r-EPS) produced from probiotic bacteria reduce biofilm formation of enterohemorrhagic Escherichia coli O157:H7, *Biochemical and Biophysical Research Communications*, 2009:379, 324–329.
[18] Trias, R, Bañeras, L, Montesinos, E, Badosa, E. Lactic acid bacteria from fresh fruit and vegetables as biocontrol agents of phytopathogenic bacteria and fungi, *International Microbiology*, 2008,11:231-236.

Methods and Techniques.

Mechanisms of Action.

Physics.

A new approach for detection of bacterial contamination in cooling lubricants

D. Oberschmidt*,[1,2], A. Spielvogel[3], C. Hein[2], J.E. Langbein[2], D. Lorenz[2], U. Stahl[3] and E. Uhlmann[1,2]

[1] Fraunhofer Institut für Produktionsanlagen und Konstruktionstechnik, Pascalstr. 8-9, 10587 Berlin, Germany
[2] Institut für Werkzeugmaschinen und Fabrikbetrieb, Technische Universität Berlin, Pascalstr. 8-9 10587 Berlin, Germany
[3] Fachgebiet Mikrobiologie und Genetik, Institut für Biotechnologie, Technische Universität Berlin, Seestr. 13 13353 Berlin, Germany

* Corresponding author: e-mail: Dirk.Oberschmidt@ipk.fraunhofer.de, Phone +49 3063925105

This article presents first results of a new strategy for the detection of organic/biological contamination using the surface plasmon resonance sensor (SPR) technology. SPR sensors have been commonly applied for the detection of proteins, bacteria and chemical substances. However, only specific interactions can be detected so far that rely on specific and sensor immobilized receptor molecules. It is of importance to overcome this limitation. The here presented new approach will focus on a distinguished mode of action. The main principle of the presented sensor operates without a flow cell and allows non-specific detection of biological content in a fluid domain, because the reaction kinetics are not being taken into account. Furthermore, there is the probability to disclaim the use of an organic matrix, such as the common dextran or fibrin matrices. Besides, one of the main advantages of the sensor is the usability in a multitude of liquids such as emulsions or deeply contaminated fluids, providing a broad applicability of the sensor system.

Keywords Microbial detection, cooling lubricant, surface plasmon resonance (SPR); bacterial contamination

1. Introduction

Water-based cooling lubricants (CL) are applied as an adjuvant for several applications in manufacturing technology, above all for cutting operations. The reason for using CL is chilling and lubrication of the processing area. Moreover, it is used for chip removal and protection against corrosion.

Cooling lubricants are a highly specialized blend of a multitude of substances, containing water, oil, biocides, defoamer, disinfectants, emulsifiers and stabilizers; the exact composition being manufacturer dependent while a trade secret. They are usually delivered as a concentrate that is mixed with tap water in a fixed ratio. Cooling lubricants are typically used for a period of six months up to one year. Maintenance of the CL includes filtration of chips and other residuals, refilling of water or concentrate and testing for microbial contamination. Still cooling lubricants change considerably with time. The composition changes of CL are caused by various factors that are poorly understood yet. In the processing area, polymeric compounds are likely to be hydrolyzed due to temperature and pressure conditions. Additionally, the volatile compounds evaporate; the pH value decreases and conditions are formed that promote the proliferation of microorganisms. The disappearance of the emulsifiers and formaldehyde improves the environmental conditions for bacteria even further. Certain bacterial strains accelerate the decomposition of the CL by metabolic degradation. This leads to phase separation and functional loss of the lubricant. As a consequence of CL breakdown, tool wear rises, the accuracy diminishes and a reduced surface finish occurs. Therefore, the reduction of the cooling lubricant must be prevented. During the machining process, respireable aerosols are formed that can lead to occupational respiratory diseases and skin disorders [1, 2]. In addition, and with regard to the health impairment of included formaldehydes and biocides, a European regulation [3] will come in effect in 2013 leading to a revision of the licensed ingredients of CL. The prohibition of biocides will entail a faster contamination with microorganisms and a shortened lifetime of the lubricant. As a prerequisite to counteract this, a fast and reliable detection system for bacterial contamination in cooling lubricants is needed. The methods currently used to detect bacteria have several disadvantages. They are time-consuming, expensive, in the majority of applications strain specific, and can only be performed by trained personal. Examples are enzyme-linked immunosorbant assays (ELISA) [4], polymerase chain reaction (PCR) [5], plating and culturing or luminescence based systems. Faster detection methods such as immunosensors are available, but they have failed to gain wide acceptance because of costly labeling reagents and the high sensitivity of antibody. With regard to the working conditions, various requirements must be met in an improved sensing system. Primarily, it has to provide an unspecified detection of bacterial contamination out of cooling lubricants. Furthermore, the system must be suitable for mobile application and it must allow a quick and accurate concentration measurement of bacterial contamination with a low detection limit as well. Besides, it should provide a fully automated measurement. These requirements meet in the innovative surface plasmon resonance based sensor introduced here. The SPR-sensor allows the detection of bacterial contamination within a few minutes [6]. An antibody and DNA free biological interface reduces cost and handling problems to a satisfying amount. The usage of SPR-sensors as a biosensor system for technical fluids is a complete new field of application. The application spectrum in the industrial sector includes the detection of bacterial and chemical contamination in air, water and cooling lubricants and other working fluids.

2. Experimental Section

Chemicals: Ethanol, pentane, diethylether, 1-octanol, Na_2SO_4 (anhydrous), NH_4OH, H_2O_2, NaCl, Luria Bertani (LB) broth and agar, trypticase soy broth (TSB), yeast extract - tryptone agar (YTA) and glass slides were obtained from Carl Roth (Karlsruhe, Germany). Gold was obtained from m&k – Edelmetallrecycling (Kahla, Germany) and titanium from Leybold Oerlikon (Köln, Germany), both with a purity of 99.99 %. All chemicals, unless mentioned otherwise, were of analytical grade and were used as received.

Tested cooling lubricants: Only water soluble cooling lubricants were tested. Samples were collected from two machines, machine 1 being a milling machine tool and machine 2 a grinding machine tool. Contaminated fluids from machine 1 were taken from the CL reservoir and the CL jet. The CL tank of machine 2 was chemically and mechanically cleaned, after drying, a swab of the inner tank walls was taken, to obtain biofilm building bacteria.

Growth and identification of microorganisms in cooling lubricants: The liquid samples from machine 1 were analyzed by plating the liquid on TSB and LB agar. After colony isolation, Gram`s staining was carried out in order to confirm single species colonies. Gram negative bacteria were identified by API 20NE kit from bioMérieux (Nürtingen, Germany), additionally oxidase and KOH tests were carried out. For comparison of bacterial strains a subsequent RAPD PCR with GTG5 (5'-GTGGTGGTGGTGGTG-3') primer was conducted.

Laboratory simulation of biofilm generation and compound degradation in cooling lubricant tanks: Two 10 l tanks (tank 1 and 3) were filled with freshly prepared cooling lubricant (4 % in deionized water, pH 9.5) and one tank (tank 2) was filled with LB medium. Tank 1 was left uncontaminated, in order to examine contamination via air. Tank 2 was used to determine biofilm influence and was inoculated with a 2xYTA overnight preparatory culture of three different bacteria, which were isolated from the swab of machine 2, namely *Bacillus spp., Pseudomonas stutzeri* and *Acinetobacter spp.*, and incubated at room temperature. After two weeks the medium was completely replaced by 4 % cooling lubricant in water. Tank 3 was inoculated after two weeks with 200 ml of the bacteria-LB suspension of tank 2. The cooling lubricant and LB medium in the system circulated using an aquarium pump containing filtering material. Aerobic conditions were generated by placing the outtake hose above fluid level.

Extraction of organic compounds from cooling lubricants and analysis via GC-MS: 50 ml samples of aqueous cooling lubricant solution from tanks 1-3 were taken after 58 days incubation at room temperature. Prior to extraction, 500 µg of 1-octanol, solved in diethylether (1 mg/ml), was added as internal standard to each sample. The samples were extracted three times with 40 ml freshly distilled pentane; the organic phase was collected, dried with Na_2SO_4 and evaporated using a rotary evaporator. The residue was resolved in diethylether for GC-MS analysis. For analyzing the extracts, a Shimadzu QP2010 Plus gas chromatograph (Duisburg, Germany) with a mass spectrometer as detector was utilized. The compounds were separated on a Rxi-5ms capillary column (30 m x 0.25 mm x 0.25 µm, Restek Bellefonte, PA, USA), helium was used as carrier gas with a flow rate of 1.23 ml/min. The injection volume was 1 µl, split ratio 20; the solvent delay was 2.5 minutes. The oven temperature was programmed to hold 40 °C for 3 minutes, then ramped up to 300 °C with 10 °C/min, with a final holding time of 5 minutes at 300 °C. Injection temperature was 240 °C and detector temperature 250 °C. Mass spectra were recorded in the range of m/z 29 to m/z 600 in total ion current.

Preparation of gold coated glass surfaces: Glass slides were cleaned with $NH_4OH : H_2O_2 : H_2O$ (1:1:5) for 10 min at 80 °C, rinsed with deionized water and dried prior to coating. The glass slides are coated with a 2 nm adhesion-promoting titanium film and a 42 nm surface plasmon-active gold film by chemical vapor deposition using a metallization cluster C400 from Von Ardenne Anlagentechnik (Dresden, Germany) with an electron beam evaporator from Ferrotec (Unterensingen, Germany).

Surface characterization: Grain size and distribution of the gold coating was determined directly after coating by scanning electron microscopy with a nominal magnification of up to 100000, using a Hitachi S4800 scanning electron microscope (Tokyo, Japan) with a beam accelerating voltage of 5 kV.

Surface roughness was measured with a ZYGO scanning white light interferometer (Darmstadt, Germany), using a Mirau 10x lens and a 2 nm resolution. Measurement area was a 700x500 µm rectangle. The crystallographic structure of the Au film was measured with a D8 Advanced X-ray Diffraction from Bruker (Mannheim, Germany) with Cu-K_α-radiation ($\lambda = 1.5406$ Å, 40 mA, 40 kV) in Bragg Brentano geometry (θ-2θ Mode). Solid-water contact angles on gold coated slides were measured with a DSA 100 contact angle measure system from A. KRÜSS Optronic (Hamburg, Germany). One drop of distilled water with a volume of 1 µl was placed on the relevant surface. From the average of 10 measurements recorded within 5 s, the contact angle was calculated.

Bacterial adsorption on coated glass slides and surface plasmon resonance measurements: Cultures of gram negative (*Escherichia coli*) and gram positive (*Bacillus spp.*) bacteria were grown over night to 10^7 - 10^8 CFU/ml at 37 °C, centrifuged (6000 rpm, 10 min) and resuspended in 0.85 % NaCl in sterile water to a final concentration of 10^6 CFU/ml. 50 µl of the bacterial solution were pipeted on the gold coated glass slides and incubated for 20 min at room temperature. The slides were rinsed with 96 % ethanol or distilled water, dried and examined by microscopy, using a Zeiss confocal LSM 5 Pascal microscope (Jena, Germany).

The SPR measurements were conducted with a laboratory setup that is based on the Kretschmann setup [7]. A 635 nm and a 680 nm diode laser with 0.9 mW light output was used as light source. The majority of the materials, including the lasers, the 30 mm cage system, the optical bench and the utilized optical components as lenses, mirrors and polarizer were obtained from Thorlabs inc. (Dachau, Germany). A µEye UI-1540-M CMOS camera from IDS Imaging Development (Obersulm, Germany), was used as photodetector. Two DRTM65-DM4 rotary stages in teta-teta layout were used for variation of the angular position and system calibration. These were obtained through OWIS (Staufen, Germany). A defined volume of a bacterial culture in sodium chloride solution, with an initial concentration of 10^6 CFU/ml, was applied on the gold coated slides, as shown in table 2. The moistened slides were attached on the prism, using optical matching oil for the coupling of the light beam into the gold surface.

3. Results and Discussion

Detected microorganisms and contamination rate: The prevailing isolated microorganisms were gram negative bacteria of different species. Also a spore forming *Bacillus* and the filamentous fungus *Cephalosporium* were identified, as it is shown in table 1. This is in accordance to earlier published articles by other researchers [8, 9], who also predominantly found gram negative bacteria, especially *Pseudomonas spp.*.

Table 1 Bacteria and Fungi found in cooling lubricants.

Gram (-) Bacteria	Gram (+) Bacteria	Filamentous Fungi
Pseudomonas putida	*Bacillus spp.*	*Cephalosporium*
Pseudomonas alcaligenes		
Pseudomonas stutzeri		
Brevundimonas vesicularis		
Moraxella		
Burkholderia cepacia		
Acinetobacter spp.		

Simulation of cooling lubricant in use and GC-MS Analysis of contaminated and uncontaminated cooling lubricant: In order to analyze the proliferation of microorganism in CL and the impact of growing or persisting microorganism on the physicochemical properties of CL, three parallel running tanks differently inoculated with microorganism (see material and methods) were investigated in detail. Tank 1 showed no contamination with airborne microbes after 58 days of incubation. After initial growth of all inoculated microorganisms in tank 2, the cell number of microorganisms upon change to CL did not increase further and after 5d, only *Bacillus spp.* could be found in the bulk of the cooling lubricant in low concentration (~10^2 CFU/ml). However, it is possible that the other two species built up a biofilm in the pump filter and the glass walls of the tanks [10]. *Bacillus spp.* was the only organism found in tank 3. As *Bacillus spp.* can form endospores, it was tested, whether the samples withdrawn from the tank contain spores or vivid cells of *Bacillus*. Spore staining and steady state numbers reveal that only spores of *Bacillus spp.* are planctonically present in tank 2. In addition, the biofilm available from tank surfaces contained only *Bacillus*. The pH value did not change, in disagreement to the situation of CL in machine tools, where the pH decreases significantly. The pH reduction found in reality is not necessarily a consequence of bacterial degradation of the CL, but could also result from the harsh physico-chemical conditions at the intersection area of the workpiece, high pressure and temperature, which could lead for instance to chemical cleavage of ester bounds. In order to get an inside in the biochemical changes caused by microorganisms, samples from the tanks were analyzed by GC-MS. The resulting chromatogram is shown in figure 1, the chromatogram is normalized using the internal standard. As expected, large amounts of organic compounds were extracted; the precise structures could not be derived from mass spectra, due to the large background. What can be stated though are the numbers of carbon atoms; these are written in the chromatogram. Focus should be on the part of the chromatogram between 20 and 30 minutes, here the separation of the analytes is insufficient. The largest amounts of vaporizable compounds analyzed were found in the samples of tanks 2 and 3, while the lowest amount originated from sample 1, the uncontaminated cooling lubricant in tank 1. This finding argues for a bacterial degradation of polymeric compounds in cooling lubricants, which cannot be vaporized in their initial state due to large molecular weight. After decomposition, for instance hydrolysis of ester linkages, vaporization is possible. Also the peak labeled with 2-ethyl-hexanol at approximately 9 minutes supports this hypothesis. In the cooling lubricant concentrate used for analysis polymeric fatty acids esterified with 2-ethyl-hexanol are added. A blow up of the chromatogram, displayed in figure 1, shows that the largest relative amount of 2-ethyl-hexanol can be found in the sample of tank 2, a possible explanation is the bacterial degradation of the ester linkage of the polymeric fatty acids by *Pseudomonas spp.*, which was able to grow in tank 2 for a certain amount of time, while not in tank 3. The amount of 2-ethyl-hexanol extracted from samples of tank 3 is in the same range as in the uncontaminated tank 1. *Pseudomonas spp.* was found to be able to grow on the organic compounds (e.g. oil, petroleum sulfonates and fatty acids) of cooling lubricants as the sole carbon source [11, 12]. So in this particular cooling lubricant, 2-ethyl-hexanol could function as an indicator for bacterial growth of gram negative bacteria.

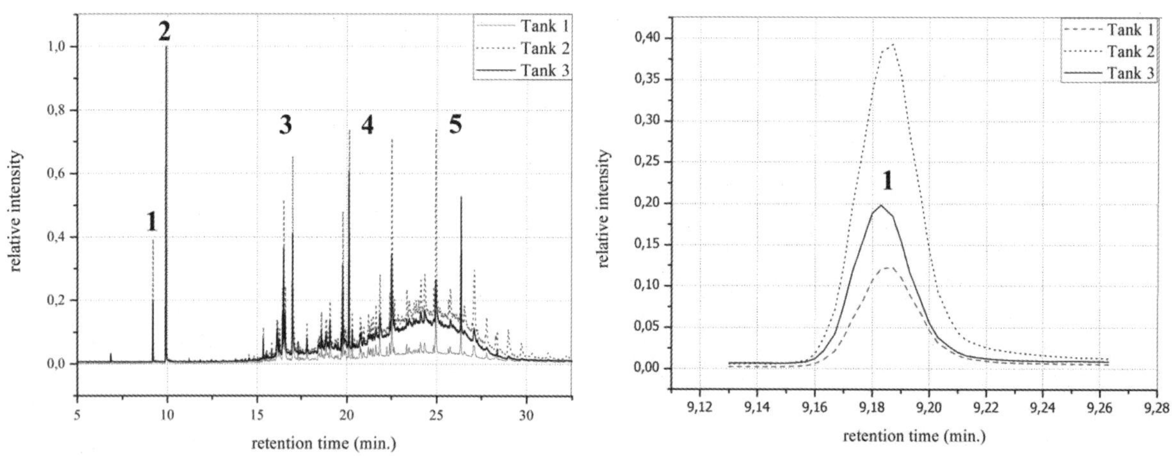

Figure 1: Normalized gas-chromatogram extracts of contaminated and uncontaminated cooling lubricants; *1: 2-Ethyl-Hexanol, 2: internal standard, 3:* $C_{14} - C_{16}$*, 4:* $C_{17} - C_{20}$*, 5:* $C_{21} - C_{24}$

Surface Characterization: A homogenous surface coating is of particular relevance for surface plasmon resonance measurement and uniform bacterial adhesion. To characterize the surface, a number of measurements were carried out. X-ray diffraction revealed that the coatings on the glass substrates are highly textured with a full Au-{111} orientation. The grain shape of the gold coating, determined by SEM, is globular with a size of approximately 40-60 nm. The surface roughness R_a, measured with a scanning white light interferometer, was below one nanometer. The surface properties of the gold coating meet the requirements for SPR measurements excellently. The wettability of the covered slides was determined by contact angle measurements. The water on gold contact angle of 83.3° speaks for a hydrophobic surface, which would favor the adsorption of the hydrophobic cell wall parts of bacteria, as found by Cunliff et al [13].

Bacterial adsorption on coated glass slides: Microscopic studies were carried out to determine the adsorption level and effectiveness of different bacteria onto the utilized coatings. *Escherichia coli* was found to adsorb in large quantities to the gold coating, while *Bacillus spp.* did not and was rinsed off the slides. This might be due to the different surface composition of gram positive and gram negative bacteria [14]. Fig 2 shows an example for *E. coli* adsorption on gold; this proves that an adsorption time of 20 minutes is sufficient.

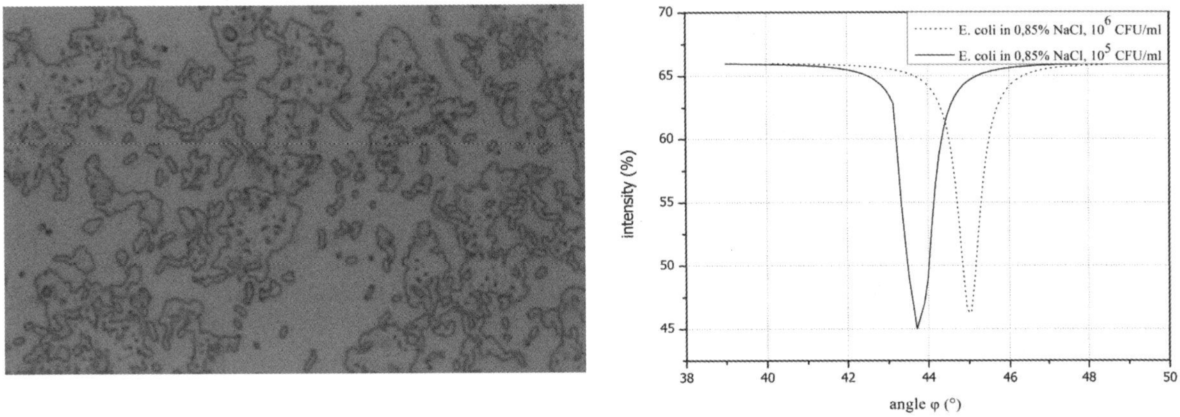

Figure 2: *E.coli* (10^6 CFU/ml) adsorbed on a gold coated surface and change of the SPR resonance angle during attachment process

SPR measurements: The experimental SPR-sensor setup provides verifiable results. The detection of gram negative *E. coli* is possible, using the Au coating with 42 nm thickness. As shown in Fig. 2, a considerable change of the dip-angle is measurable, while changing the bacterial concentration in a sodium-chloride solution. The analysis of the samples comprising *Bacillus spp.* did not provide resilient results. This is mainly attributable to the hydrophobic Au-surface, impeding the attachment of gram positive *Bacillus spp.* Therefore no significant results can be obtained. However, to gain results with a sensor system based on surface plasmon resonance principle, adhesion of the investigated subjects is necessary.

4. Summary and Outlook

The importance of monitoring cooling lubricants, with special regard to biological contamination, will increase among manufacturers and operators. The change of legislation on the one hand and the desire to increase efficiency on the other hand are the principle reasons for a rising demand on analytical capabilities for CL. The causal relation between propagation of bacteria and decomposition of cooling lubricants indicates the extraordinarily large spectrum of possible applications. The approach of using a SPR based sensor system delivers a number of advantages. A minimal detection limit, a very short measurement time and a comparatively unspecific design are indicators for the above average suitability. The sensor system complies with the specified requirements and is fundamental for future advancement. Because of the aqueous nature of the investigated cooling lubricants, the presented results, showing the adhesion of bacteria from a sodium-chloride solution, are comparable to practical application. Concerning the development of a surface coating, suitable for the adhesion of gram positive bacteria such as *Bacillus spp.* and a coating for an unspecific bacteria attachment, extensive preparations have already been made.

Acknowledgements: This work was funded by the BMBF grant 0312037. The authors would like to thank Stefanie Weber for technical assistance, Prof. Dr. Leif-A. Garbe and Konrad Neumann for access to the GC-MS and help with the analysis, Prof. Dr. Thomas Friedrich for laboratory access, and the Center for Micro- and Nanotechnology, Ilmenau, for preparation of gold coatings and SEM measurement.

References

[1] Fishwick D, Tate P, Elms J, Robinson E, Crook B, Gallagher F, Lennox R, Curran A. Respiratory symptoms, immunology and organism identification in contaminated metalworking fluid workers. What you see is not what you get. *Occupational Medicine*. 2005; 55:238-241.

[2] Gordon T. Metalworking Fluid-the toxicity of a complex mixture. *Journal of Toxicology and Environmental Health, Part A*. 2004; 67:209-219.

[3] Directive 98/8/EC of the European parliament and of the council concerning the placing of biocidal products on the market

[4] Kurupati P, Kumarasinghe G, Laa Poh C. Direct identification of *Pseudomonas aeruginosa* from blood culture bottles by PCR-enzyme linked immunosorbent assay using oprI gene specific primers. *Molecular Cell Probes*.2005, 19: 417-421.

[5] Khan I U H, Yadav J S. Real-time PCR assays for genus-specific detection and quantification of culturable and non-culturable mycobacteria ans pseudomonads in metalworking fluids. *Molecular and Cellular Probes*. 2004; 18:67-73.

[6] Homola J, Yee S S, Gauglitz G. Surface plasmon resonance sensors: review. *Sensors and Actuators B: Chemical*. 1999; 1-2:3-15.

[7] Kretschmann E. Die Bestimmung optischer Konstanten von Metallen durch Anregung von Oberflächenplasmaschwingungen. *Z. Phys.* 1971; 241:313-324

[8] Dilger S, Fluri A, Sonntag H-G. Bacterial contamination of preserved and non-preserved metal working fluids. *International Journal of Hygiene and Environmental Health.* 2005; 208:467-476.

[9] Laitinen S, Linnainmaa M, Laitinen J, Kiviranta H, Reiman M, Liesivouri J. Endotoxins and IgG antibodies as indicators of occupational exposure to the microbial contaminants of metal-working fluids. *International Archives of Occupational and Environmental Health.* 1999; 72:443-450.

[10] Huigens R W 3rd, Rogers S A, Steinhauer A T, Melander C. Inhibition of *Acinetobacter baumannii, Staphylococcus aureus* and *Pseudomonas aeruginosa* biofilm formation with a class of TAGE-triazole conjugates. *Org Biomol Chem* 2009, 7:794-802.

[11] Foxall-vanAken S, Brown J A Jr, Young W, Salmeen I, McClure T, Napier S Jr, Olsen R H. Common components of industrial metal-working fluids as sources of carbon for bacterial growth. *Applied and Environmental Microbiology.* 1986; 51:1165-1169.

[12] Van Hamme J D, Singh A, Ward O P. (2003) Recent advances in petroleum microbiology. *Microbiol Mol Biol Rev* 2003, 67:503-549.

[13] Cunliffe D, de las Heras Alarcón C, Peters V, Smith J R, Alexander C. Thermoresponsive Surface-Crafted Poly(N-isopropylacrylamide) Copolymers: Effect of Phase Transitions on Protein and Bacterial Attachment. *Langmuir.* 2003; 19:2888-2899.

[14] Silhavy T J, Kahne D, Walker S. The bacterial cell envelope. *Cold Spring Harb Perspect Biol 2*, a000414

Development of a liquid-medium assay for screening antimicrobial natural products against marine bacteria

M. Geiger[1-2], J. Dupont[3], O. Grovel[2], Y-F Pouchus[2] and P. Hess[1]

[1]Laboratoire Phycotoxines, Centre IFREMER de Nantes, rue de l'Ile d'Yeu, 44311 Nantes, France
[2]MMS group, Faculté de Pharmacie, Université de Nantes, BP 53508, 44035 Nantes Cedex 01, France
[3]LNR Microbiologie des coquillages, IFREMER Centre de Nantes, rue de l'Ile d'Yeu, 44311 Nantes, France

The marine environment is known to be a promising source of new compounds with biological activities. Many papers report on the discovery of such compounds in marine micro-organisms, especially marine eukaryotes such as micro-algae or fungi. In this context, we describe the development of antibacterial assays in a liquid-medium, to detect an activity within the lipophilic fraction of micro-organisms. Bacteria used for this assay were *Vibrio* species. Bacterial growth was carried out in a Bioscreen C, and the first step in this development was to determine the tolerance of the *Vibrio* species to solvents miscible with aqueous culture media. Methanol, ethanol and DMSO were tested and the best response was obtained with methanol, at a concentration of 1.5%. Commonly known antibiotics were then used to calibrate the assay. A preliminary antibacterial assay was carried out with extracts of *Azadinium spinosum*, a small toxic dinoflagellate, and a marine-derived strain of the fungus *Penicillium expansum*. *A. spinosum* showed no activity on *V. crassostreae* at a concentration of 26.7 mg/mL, whereas a crude extract at 5 mg/mL of *P. expansum* resulted in an inhibition of *V. crassostreae* growth. Therefore this assay seems to be able to detect an activity in crude extracts. However, as some crude extracts did not dissolve completely, emulsifying agents may improve the scope of the assay to include more lipophilic compounds.

Keywords: antibacterial activity; lipophilic compounds; microplate bioassay

1.Introduction

In the field of searching for new natural compounds with antibacterial activities for therapeutics, biofilm control or biological control of terrestrial and marine pathogens, marine organisms have been shown to be a promising source of originality in terms of chemical structure or mechanism of action (Mayer et al., 2009). Particularly, micro-organisms including phytoplankton, cyanobacteria and fungi represent a noticeable source of bioactive primary and secondary metabolites with an increasing number of original compounds isolated over the years (Blunt et al., 2009). On the other hand, some micro-algae, like dinoflagellates, are known to produce toxic compounds, which represent a threat for human health (Daranas et al., 2001). Most often, these toxins have lipophilic properties. Thus, we hypothesized that the lipophilic fraction could contain compounds with biological activities, such as antibacterial activity.

To assess an antibacterial activity, the disk-diffusion assay is widely used (Wiegand et al, 2008). However, this assay is time-consuming, arduous and need a manual reading of the results. It could be used only for preliminary antibacterial screening (Turcotte et al., 2004). So we aimed at developing an assay in a liquid-medium, using microplate for the bacterial growth. Bacterial species studied belong to the *Vibrio* genus: *V. crassostreae*, *V. gigantis*, *V. mytili* and *V. rotiferianus*. They all are shellfish pathogens. This method such developed permits to work with several strains in several conditions on the same microplate, enhancing the throughput of the experimentation. However, because of the focus of the study on lipophilic compounds, the protocols have to be adapted. First, we determined the relationship between the optical density (OD) and the number of colony forming units (CFU). Then we evaluated the tolerance of the bacterial strains to solvents: methanol, ethanol and dimethyl-sulfoxide (DMSO), to insure that they did not interfere with the assay. Following that, antibiotics were used to calibrate the assay. A preliminary antibacterial assay was finally carried out on crude extracts of *Azadinium spinosum*, a small dinoflagellate, and a marine-derived strain of *Penicillium expansum*.

2.Material and methods

2.1.Bacterial strains and growth conditions

V. crassostreae (CIP108327), *V. gigantis* (CIP108656), *V. mytili* (CIP 103929) and *V. rotiferianus* (CIP 107930) bacterial strains were grown on Brain Heart Infusion (BHI) agar or broth, adjusted at 2% of NaCl and incubated at 20°C. Strains were stored on beads at –80°C. An agar plate was inoculated by rubbing a bead and incubated for 24 h. Several isolated colonies were suspended in physiological water until turbidity is equal to a 0.5 McFarland Standard on a densitometer (ATB 1550, Biomérieux, France). Cultures were then diluted from 10^8 cfu/mL to 10^6 cfu/mL in broth medium, and 100-well microplates were inoculated at 5.10^5 cfu/mL with a final volume of 200 µl. Microplates were placed in a BIOSCREEN C (Labsystem, Helsinki, Finland), an automated microbiological growth analysis system, and the OD (600 nm) was measured every 15 min during 24 h.

2.2.Determination of the relationship between the OD and the number of CFU

A twofold serial dilution in physiological water of a 24 h-old culture was carried out in wells. Aliquots (100 µl) of the highest dilutions (10^{-6} to 10^{-8}) were used to inoculate agar plates and incubated overnight. Counts were extrapolated to the lowest dilution (1 to 10^{-4}), which were read at 600 nm. Assays were carried out in triplicate and repeated twice. Equations for the correlation were calculated using a second degree polynomial model.

2.3.Determination of the solvent tolerance

Three solvents were evaluated for their impact on the culture growth: methanol, ethanol and DMSO. A first assay investigated a large range, up to 20%. Solvent (2 to 40 µl) was added to decreasing volumes of bacterial cultures (198 to 160 µl). A second assay was carried out on a restricted range of methanol only: 1% to 2%. The protocol was slightly modified for this second assay: instead of adding directly pure methanol, a solution of methanol in medium in larger quantity is added to each well, which should enhance the repeatability. Assays were carried out in triplicate.

2.4.Calibration of the bioassay with commonly known antibiotics

Calibration of the assay was performed using antibiotics: Nafcillin, Erythromycin and Carbenicillin (Sigma-Aldrich, St. Louis, Missouri). A preliminary assay applied a concentration of 20 µM and antibiotics leading to growth inhibition were used to calibrate the assay through a range of concentration, of 4 nM up to 200 µM. Assays were carried out in triplicate.

2.5.Preliminary antibacterial assay

Crude extracts of *A. spinosum* and *P. expansum* were obtained using a mixture of dichloromethane/ethyl acetate (1:1 v/v). Solvents were evaporated using a rotary evaporator, and the extracts were re-dissolved in the lowest possible quantity of methanol (ca. 100 to 150 µL). Extracts in methanol were aliquoted into wells, and the solvent was evaporated. Culture-medium containing the maximum permitted concentration of methanol and cultures were added to the wells. The final concentration was in a range from 5 mg/mL to 26.7 mg/mL.

2.6.Growth curve parameters and statistical analysis

Parameters of the growth curves, as the specific growth rate (μ_{max}) were determined applying the Gompertz equation using the MATLAB 6.5 software. The statistical significance of the growth inhibition in the different assays was determined using ANOVA, also carried out using MATLAB 6.5.

3.Results and discussion

3.1.Growth conditions and parameters

The use of the BIOSCREEN C appeared to be well adapted to the growth of these four *Vibrio* strains. The culture presented a classical sigmoidal growth curve, adapted to the Gompertz model (Fig.1). The application of the Gompertz equation permitted the determination of the specific growth rate, the plateau and the latency period (Table inserted in Fig.1). In order to make the results comparable to other studies, the OD_{600nm} values were transformed in CFU per mL. As previously shown by Casey et al. (2004), this relationship is not linear but a second-degree polynomial-type. Equations of this polynomial are presented in Fig. 2.

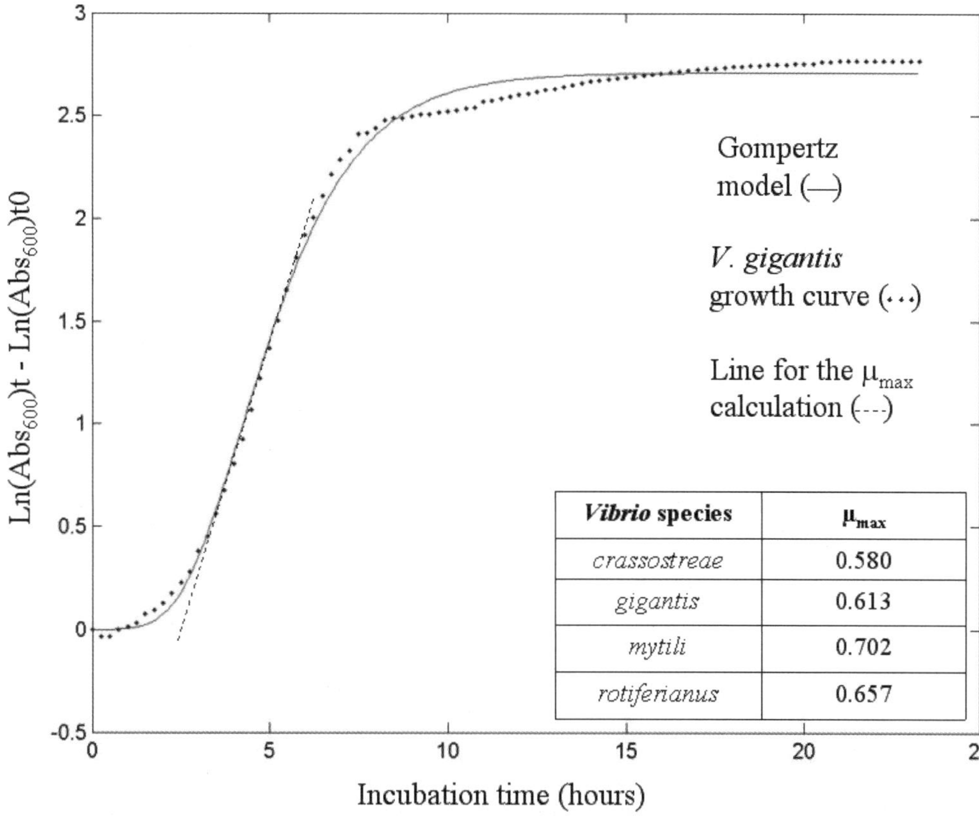

Vibrio species	μ_{max}
crassostreae	0.580
gigantis	0.613
mytili	0.702
rotiferianus	0.657

Figure 1: *V. gigantis* growth curve, corresponding to the Gompertz model.
Inserted table: average specific growth rates (μ_{max}) of the four *Vibrio* strains determined using the Gompertz equation.

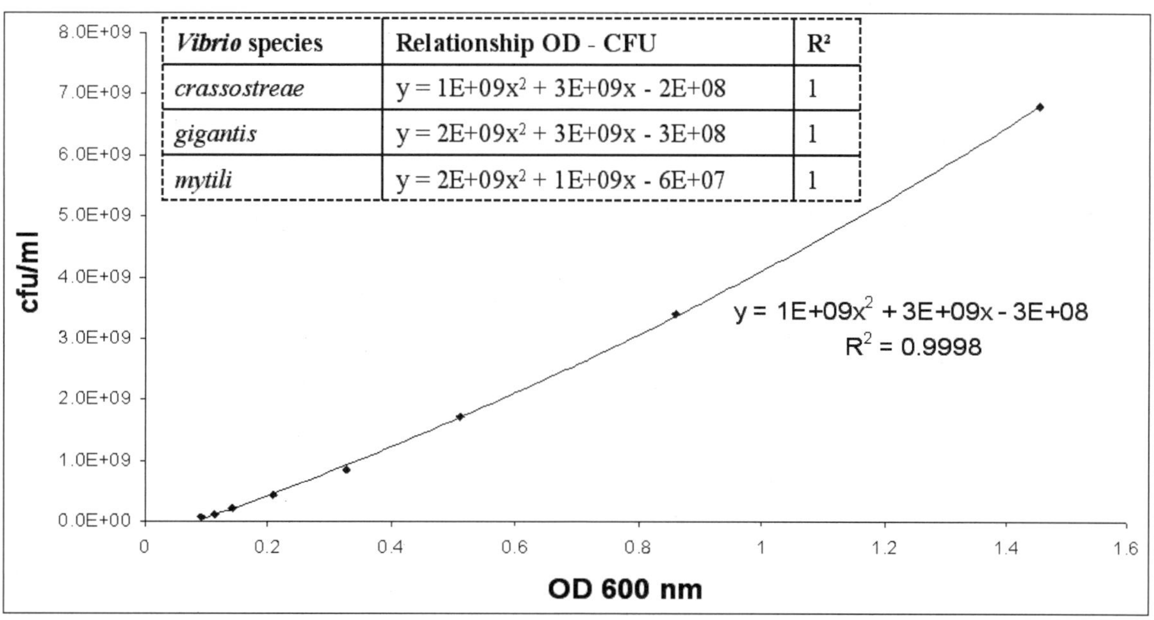

Vibrio species	Relationship OD - CFU	R^2
crassostreae	y = 1E+09x^2 + 3E+09x - 2E+08	1
gigantis	y = 2E+09x^2 + 3E+09x - 3E+08	1
mytili	y = 2E+09x^2 + 1E+09x - 6E+07	1

Figure 2: Relationship between the OD and the number of CFU calculated for *V. rotiferianus*.
Inserted table: Equations calculated for the other *Vibrio*

3.2.Solvent tolerance and calibration of the assay

The monitoring of the solvent tolerance (Fig.3 and Tab.1) showed a tolerance of methanol 1.5% for each strain, which was more interesting than the classical value of 1% of solvent for antibacterial assay. However, in the calibration tests with antibiotics, the methanol-control presented a significant difference to the medium-control, which did not seem to affect the test result. Indeed, the lowest concentration of Nafcillin, inducing no inhibition, presented the same growth curve as the methanol control. Fig.4 showed the interest of a kinetically following-up of the assay, because atypical growth curve can be observed, as the concentration of Nafcillin 10 μM.

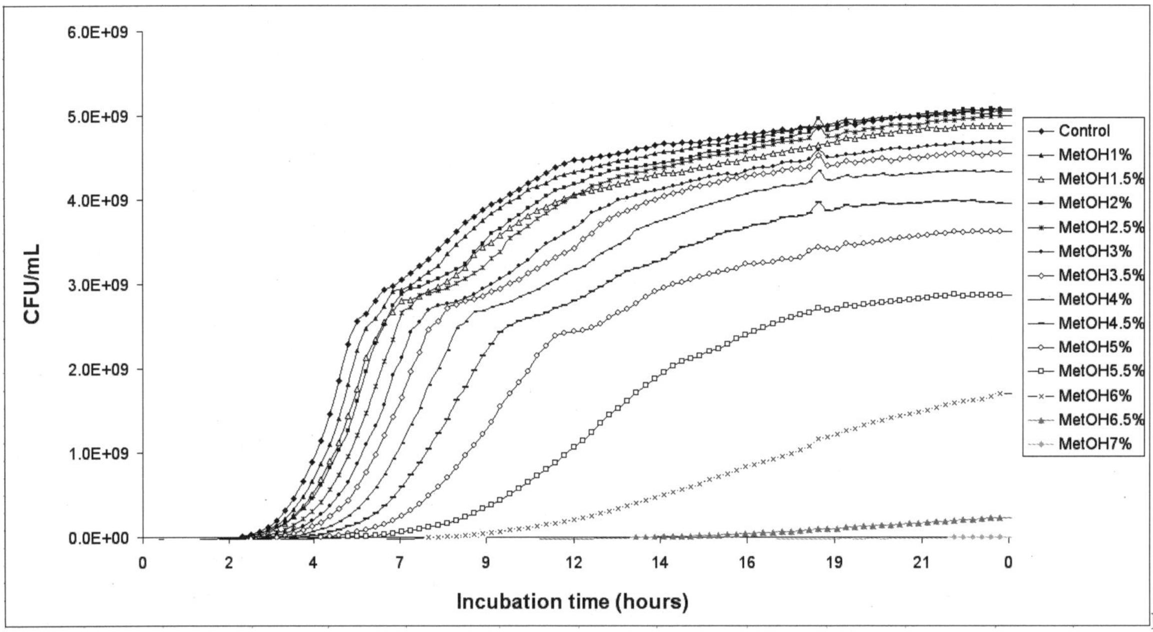

Figure 3: Evaluation of the methanol tolerance of *V. mytili*. Methanol concentration ranges to 0% up to 7%.

Table 1: Range of the solvent tolerance of the four *Vibrio*

Vibrio species	MeOH	EtOH	DMSO
crassostreae	**1.5%**	0%	0%
gigantis	1% - **1.5%** - 2%	1%	1% - 1.5%
mytili	1% - **1.5%**	1%	1% - 1.5% - 2%
rotiferianus	1% - **1.5%** - 2% - 2.5%	1% - 1.5%	1% - 1.5% - 2% - 2.5%

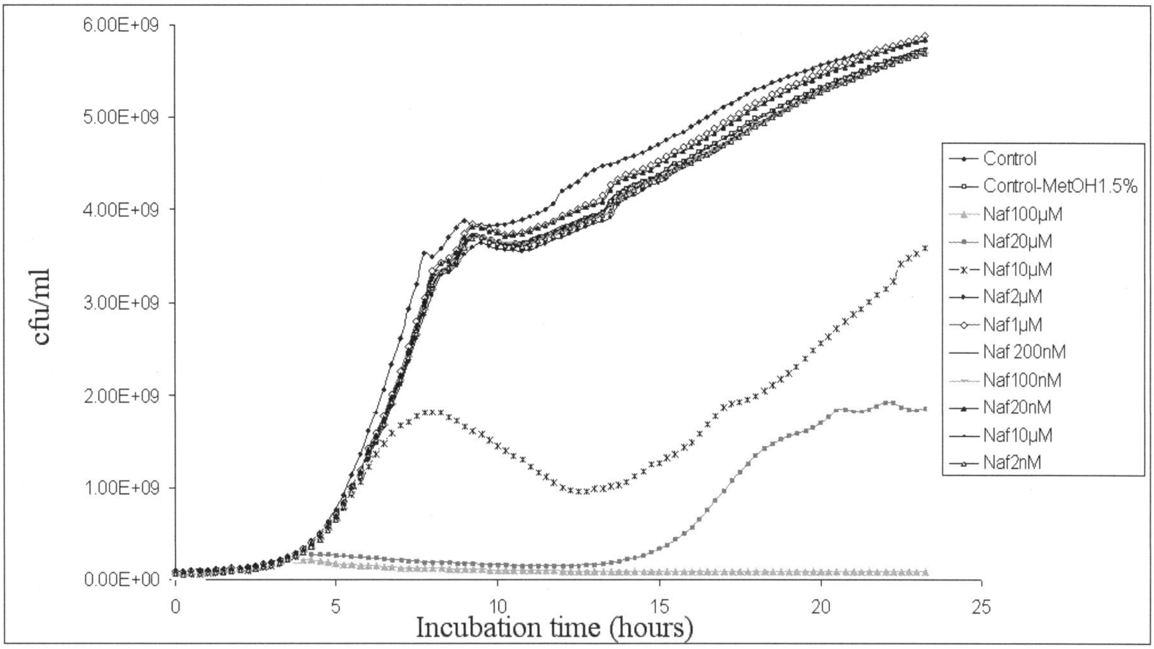

Figure 4: Calibration of the assay with a range of Nafcillin, from 100 µM to 2 nM for *V. crassostreae*.

3.3.Preliminary antibacterial assay

The evaporation of the crude extracts resulted in the formation of an insoluble film on the bottom of the wells, leading to high starting OD values. In spite of this film, normal growth of *V. crassostreae* was detected for wells containing *A. spinosum* extracts (Fig.5, concentrations of 5.4 to 26.7 mg/mL). The OD increased to result in similar shapes of growth curve as for the control.

In contrast, extracts of *P. expansum* effectively inhibited the growth of this species at 5 mg/mL.

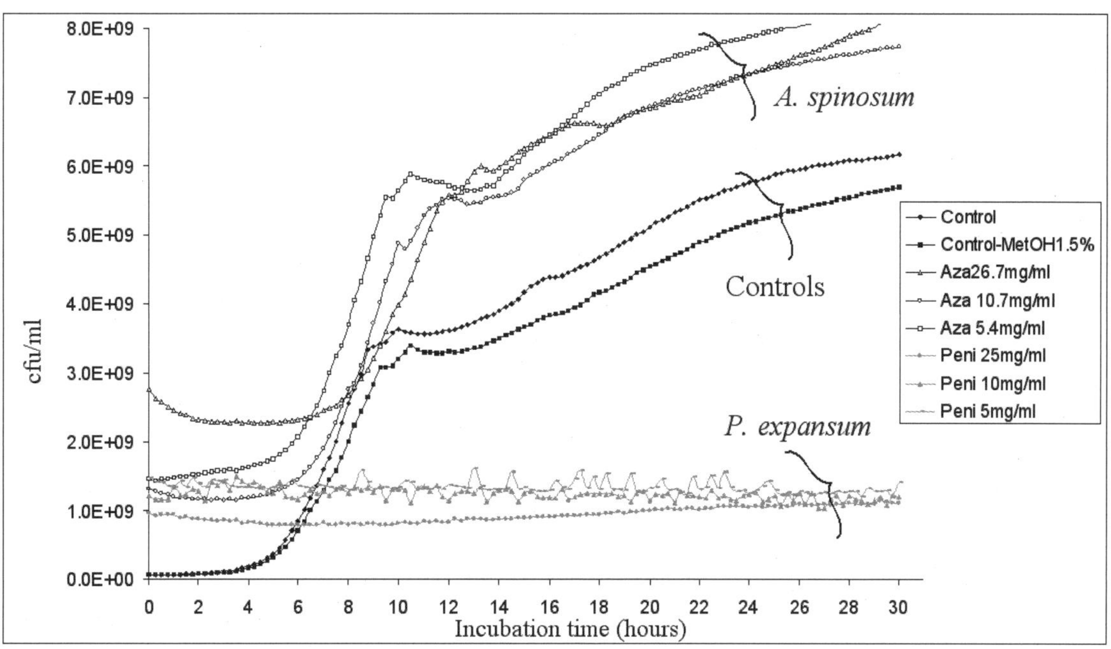

Figure 5: Preliminary antibacterial assay with crude extracts of *A. spinosum* and *P. expansum* tested on *V. crassostreae* (concentrations are given in mg dry residue crude extract per mL of final well volume).

4.Conclusion

The present study illustrates that a screening of crude extracts is feasible with this protocol, i.e. using a liquid medium, for the following *Vibrio* species: *V. crassostreae*, *gigantis*, *mytili* and *rotiferianus*. The assay has been successfully calibrated with known antibiotics. However, it may be necessary to increase the solubility of crude extracts: using emulsifying agents would enlargen the scope of the assay to include highly lipophilic compounds. A preliminary assay appeared promising as it revealed an antibacterial activity of a marine-derives strain of *P. expansum*.

Acknowledgments
The bacteria were supplied by Denis SAULNIER, of the "Laboratoire de Génétique et Pathologie", Ifremer de La Tremblade (France).
Valuable advice was obtained from the "Agence Nationale de Sécurité Sanitaire", Fougères (France), the "Laboratoire Sciences et Technologie de la Biomasse Marine", Ifremer de Nantes (France), while data processing and statistics were carried out with the help of Raymond KAAS, "Laboratoire Physiologie et Biotechnologie des Algues", Ifremer de Nantes (France).
This study has been co-financed by the "Région des Pays de la Loire" (regional funding), and by Ifremer, through the COLNACOQ-project (2009-2012).

References

- **Blunt J.W.**, Copp B.R., Hu W.-P., Munro M.H.G., Northcotec P.T. and Prinsep M.R. **(2009)** Marine natural products. *Natural Product Reports*, 2009, 26, 170–244.
- **Casey J. T.**, O'Cleirigh C., Walsh P. K. and O'Shea, D. G. **(2004)** Development of a robust microliter plate-based assay method for assessment of bioactivity. *Journal of Microbiological Methods* 58(**3**): 327-334.
- **Daranas A.H.**, Norte M. and Fernandez J.J. **(2001)** Toxic marine microalgae. *Toxicon* 39: 1101-1132.
- **Mayer A.M.S.**, Rodriguez A.D., Berlinck R.G.S. and Hamann M.T. **(2009)** Marine Pharmacology in 2005-6: Marine compounds with anthelmintic, antibacterial, anticoagulant, antifungal, anti-inflammatory, antimalarial, antiprotozoal, antituberculosis, and antivirial activities; affecting the cardiovascular, immune and nervous systems, and others miscellaneaous mechanisms of action. *Biochimica and Biophysica Acta* 1790: 283-308.
- **Turcotte C.**, Lacroix C., et al. **(2004).** A rapid turbidometric microplate bioassay for accurate quantification of lactic acid bacteria bacteriocins. *International Journal of Food Microbiology* **90**(3): 283-293.
- **Wiegand I.,** Hilpert K. and Hancock R. E. **(2008)** Agar and broth dilution methods to determine the minimal inhibitory concentration (MIC) of antimicrobial substances. *Nature Protocol* **30**(2): 163-175.

Experimental planning can help to optimize the selective photoinactivation of microorganisms

J. R. Perussi[1,2], P. L. Fernandes[1], C. Bernal[2] and H. Imasato[1,2]

[1]Interunidades Bioengenharia, Universidade de São Paulo, Av. Trabalhador sãocarlense, 400, 13560970 São Carlos (SP), Brazil
[2]Instituto de Química de São Carlos, Universidade de São Paulo, Av. Trabalhador sãocarlense, 400, 13560970 São Carlos (SP), Brazil

* Corresponding author: e-mail: janice@iqsc.usp.br, phone: +55 1633739956

The increasing resistance against antimicrobial drugs has led to the search for alternative treatment. The antimicrobial photodynamic inactivation (PDI) which is based on the oxidative destruction of biological molecules by active oxygen species generated by photo-excited molecules (photosensitizers) is a promising option. The presence of several sources of variability in PDI makes that a non trivial issue. Design of experiments is a powerful statistical tool able to eliminate several sources of variability efficiently, allowing performing fewer trials and ensuring accurate information on the responses of interest. Experimental design was used in order to optimize the experimental conditions to achieve selective photoinactivation of the fungi *C. albicans* with minimal damage to the host cells using an epithelial cell line as a model and Hypericin as photosensitizer. Central composite rotatable design (CCRD) and response surface methodology (RSM) were used for this purpose.

Keywords: selective photoinactivation; experimental design; fungi; Hypericin

1. Introduction

The increasing resistance against anti-microbial drugs has impelled the search for alternative treatments and antimicrobial photodynamic inactivation (PDI) is a propitious candidate. This procedure is based on the oxidative destruction of biological molecules by active oxygen species generated by photo-excited molecules (photosensitizers) in the presence of oxygen. Due to the large number of potential targets which produces cell damages that inactivate the microorganisms, there is no cell resistance developed using this method [1, 2].

The use of experimental design methodology has become indispensable to analyze the behavior of multi-variables process or system often affected by several sources of variability, making it a huge problem. The design of experiments proposed by R. A. Fisher around 1920 [3] has many advantages comparing with the traditional technique of varying one factor at a time, because it change the factors simultaneously allowing performing fewer but very relevant trials. In addition, the experimental design ensures accurate information on the responses of interest since the traditional method could not measure the interactions between factors. Since then different techniques of experimental design have been added to the methodology as full factorial, fractional factorial, central composite rotatable, Plackett-Burman, Box-Behnken, cubic face centered, just to mention a few [4], making it a powerful statistical tool. The choice of which design of experiments should be performed depends on the goal of the research and on the amount of a priori knowledge of the process.

The aim of this study was to employ the central composite rotatable design (CCRD) and response surface methodology (RSM) in order to optimize the experimental conditions to selective photoinactivation of the fungi *C. albicans* with the photosensitizer Hypericin causing a minimal damage to the host cells, what means to kill the microorganisms without significant level of damage to the host cells. The CCRD has been applied to various engineering processes for data interpretation and optimization [5]. It is an important technique of experimental design and allows building a model of second degree to extract the largest possible amount of information about a process with smallest number of trials through a smart placement of points in space, eliminating redundant observations and reducing the time and resources. RSM is a collection of statistical design and numerical optimization techniques used to optimize processes [6].

2. Materials and Methods

2.1 Experimental planning

The PDI process is the result of the action of individual effects and/or combination of the following factors: concentration of the photosensitizer, incubation time of cells with the photosensitizer and light dose administered to the cells containing the photosensitizer. The result of its action is given by the survival index of the cells under study. Figure 1 shows the schematic representation of PDI.

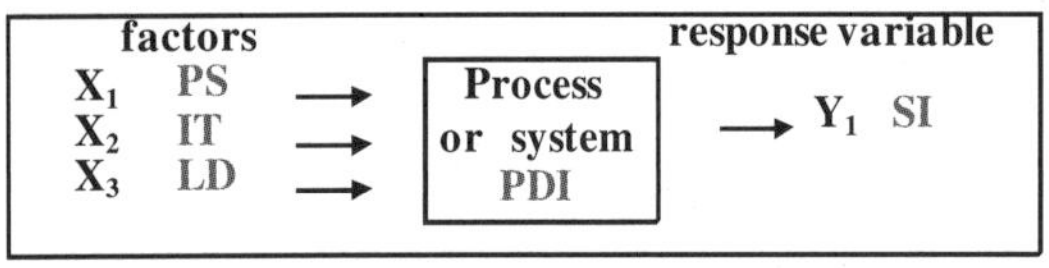

Fig. 1 Schematic representation of the PDI. The factors X_1, X_2, and X_3 are respectively the sensitizer concentration (PS) in μg mL^{-1}, incubation time (IT) in seconds and light dose (LD) in J cm^{-2}. The response Y_1 is the survival index (SI) in percentage

The factors or input variables are the controlled or independent variables and they were selected for the design of experiments because they could influence the response or dependent variable of the system. RSM is a collection of statistical and mathematical methods that involves experimental designs to achieve a basic approach for the functional relationship between factors (X_1, X_2 and X_3) and response variable (Y_1). The response surface can be expressed in its general form, as in Eq. (1)

$$Y = \text{function}\ (X_1, X_2, \ldots, X_n) \tag{1}$$

where Y is the response and X_i's are the factors of the process. As stated earlier, there are different experimental designs and the most basic one is a full factorial design. The number of design points defined by a full factorial design is calculated by Eq. (2)

$$\text{Design points} = (L)^k \tag{2}$$

where L is the levels number and k is the number of factors. The most common full factorial designs are the 2^k shown in Fig. 2a at the left. For three factors the 2^3 full factorial design allows to calculate the following effects: all main, all two factor interactions and the three factor interaction. For the fractional factorial indicated by the solid dots in Fig. 2a at the right, the main effects are mixed up with the two factor interactions. The mixed up effects cannot be estimated independently unless all, except one, are known or assumed not to exist. A first order model can be predicted by the correlation of factors and responses based on full factorial design, however in fractional planning this prediction is not allowed. Full and fractional factorial designs are used for identifying or screening important factors when the number of factors is too large to evaluate higher order effects [7].

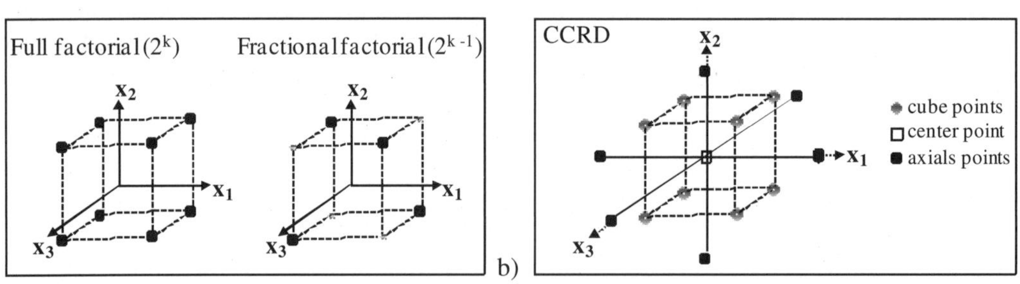

Fig. 2 a) Full (8 trials) and fractional (4 trials) factorial design with three factors; b) CCRD with its placement of points in space. Each point represents one trial with the factors set on certain levels.

An experimental design is formally a sequence of trials to be performed, expressed in terms of factors set at specified levels with predefined values. From this point on, for didactic purposes, we will fix k = 3 for equations and calculations of the numbers of trials, since PDI has three factors influencing the process response (see Fig. 1). The number of trials of a CCRD, shown in Fig. 2b, is equal to a full factorial augmented of 2k axial points, and six replicated experiments of the central point [8] to provide an independent estimate of experimental errors. The axial points are located at a certain distance from the center and they allow rotatability, which ensures that the variance of the model prediction is constant at all points equidistant from the center [9]. So the number of trials is equal to 2^3 (full factorial) + 6 (axial points) + 6 (central point) = 20.

The wide use of CCRD is due to the possibility to predict a second order model, since the quadratic terms are generated by the axial points. The vast majority of biological processes are composed of reactions in which the ratio of reactants and products are not linear and therefore a first order model cannot be used.

The correlation of the factors and the responses based on CCRD is fitted to a quadratic polynomial equation, shown in Eq. (3)

$$SI = b_0 + b_1(PS) + b_2(IT) + b_3(LD) + b_4(PS)^2 + b_5(IT)^2 + b_6(LD)^2 + \\ + b_7(PS)(IT) + b_8(PS)(LD) + b_9(IT)(LD) + b_{10}(PS)(IT)(LD) + \varepsilon \quad (3)$$

where SI (survival index) is the response of process; b_0 is the general average of responses; b_1,b_2 and b_3 are all the main effects (linear) of the factors; b_4,b_5 and b_6 are all the quadratic effects of the factors; b_7,b_8 and b_9 are all of two factors interactions; b_{10} is the three factor interaction and ε is the residual error. The value of ε is assumed to be independent and identically distributed and it is always present due to presence measurement error, such as fluctuations or turbulences inherent in a measure, and/or other sources [10]. The terms b_0 to b_{10} are called model parameters and they are estimated by the method of least squares. It was performed just one experimental planning, choosing the same values for the factors and of same response for the mammalian cells and microorganisms. The CCRD was based on previous studies performed in the group [11] and allowed to obtain the levels factor values. The assays were performed in different days and in a random way and the results of the second order models and the RSM for both cells were analyzed and compared. The results of the twenty assays for each, the mammalian cell line HEp-2 and the fungi *C. albicans* were treated using the software Statistica 8.0 with the tools that perform the statistics analysis of the planning by CCRD.

2.2 Photosensitizer and Light Source

Hypericin (HY) was synthesized from emodin at Universidade Federal ABC, Santo André (SP), Brazil. Stock solution of HY in DMSO was sterilized by filtration and kept in the dark at -20°C. The light source used was a BioTable (Instituto de Física de São Carlos, São Carlos (SP), Brazil) composed by yellow LEDs (590 nm).

2.3 Cells and microorganisms culture

An epithelial cell line (HEp-2) derived from human larynx carcinoma grown in ISCOVE'S medium containing 10% fetal bovine serum, penicillin/streptomycin at 37°C, 5% CO_2 was used [11]. *Candida albicans* strain ATCC 18804 was grown aerobically in Yeast extract, Peptone, and Dextrose medium (YPD) with 2% agar for 24 h at 37°C. The cells at logarithm phase were settled by centrifugation (10.000 rpm, 5 min), washed twice with phosphate buffered saline (PBS) pH 7.2 and ressuspended in PBS to OD of 0.1 at 625 nm, corresponding to $1x10^6$ UFC mL^{-1}.

2.4 Photodynamic treatment of mammalian cells

$1x10^5$ mammalian cells mL^{-1} were seeded in 96-well plates and incubated overnight, the medium was removed and the cells were incubated for 2-10 min in the dark at 37°C with 0.0125-6.4 ug.mL^{-1} HY in medium. The cells were PBS washed and fresh medium was added. For phototoxicity measurements, cells were irradiated at 590 nm at a light dose 2.5-6 J cm^{-2}. After 48h cell viability was evaluated by the 3-(4, 5-dimethylthiazol-2-yl)-2, 5-diphenyl tetrazolium bromide (MTT) method [12] at 570 nm. The index of cell survival (SI) was calculated for each HY concentration.

2.5 Photodynamic treatment of microbial cells

Suspensions of 100 μL of *C. albicans* (10^6 UFC mL^{-1}) were transferred into wells of a microtitre plate. Identical volume of HY in different concentrations was added to each well to a final concentration ranging from 0.0125-6.4 μg mL^{-1}. After the incubation time of 2-10 min at 37°C in the dark, the samples were irradiated at 590 nm (2.5-6 J cm^{-2}). Serial dilutions were performed of each well to determine the cell viability. The colonies counting were performed after 48 h of incubation at 37°C in order to calculate the survival index (SI).

3. Results and Discussion

Following the experimental planning, the photodynamic inactivation of *C. albicans* was performed at the same conditions as the epithelial cells with the following parameters: incubation time 2-10min; 0.125-6.4 μg mL^{-1} Hypericin, 10 mW cm^{-2} power density at 590 nm, light dose 2.5 to 6 J cm^{-2}. HEp-2 consists of epithelial cells originated from an epidermoid carcinoma which is formed with cells similar to the skin cells, containing queratine. So this cell line can be used as model for skin cells, and in the present study as model for host cell for *C. albicans*.

Through the Pareto diagram [5], a very used tool in experimental design that determines, among the effect of the factors, which are significant for the process (p-value > 0.05). The results of this study (not shown) infer that, for minimal damage in mammalian cells to selectively inactivativate this fungi, the factor photosensitizer concentration in its linear as well as in its quadratic parameters is the more significant followed by the incubation time and finally the interaction between the photosensitizer concentration and the incubation time. The other effects of the factor and their interactions have no significance and for this reason were not considered in the second grade codified model fitted and valid for the studied region. The coefficient of determination (R^2), parameter that measures the quality of the fit of the regression curve to the experimental points [10], was equal 90%. The RSM for HEp-2 line is shown in Eq. (4) and the terms with no significance in the Pareto diagram were omitted:

$$SI = 6.94 - 19.70(PS) + 15.13(PS)^2 - 8.37(IT) + 10.79\,(PS)(IT) \quad (4)$$

where PS is the photosensitizer concentration (the linear parameter) in μg mL^{-1}, PS^2 is the photosensitizer concentration (the quadratic parameter) in μg mL^{-1}, IT is the incubation time (in seconds), and PS IT is the effect of the interaction between the PS and the IT.

It is important to note that the fact of the incubation time be the second factor more significant may be due to the rate of internalization of the photosensitizer in the cells. The results showed that the light dose has no importance for human cells, a consequence of the slow accumulation of the photosensitizer.

In the same way, the Pareto diagram for *C. albicans* (not shown) indicated the more significant factors for the process being the linear and the quadratic parameters of the photosensitizer concentration the first ones, followed by the linear parameter of the light dose and then the interaction of the photosensitizer and the light dose, originating the second grade codified model fitted valid for the studied region with the coefficient of determination (R^2) equals 88%. The RSM for *C. albicans* is shown in Eq. (5) without the terms that had no significance in the Pareto diagram:

$$SI = 4.45 - 16.11(PS) + 10.81(PS)^2 - 5.82(LD) + 7.43(PS)(LD) \quad (5)$$

where PS is the photosensitizer concentration (the linear parameter) in μg mL^{-1}, PS^2 is the photosensitizer concentration (the quadratic parameter) in μg mL^{-1}, LD is the light dose (in J cm^{-2}), and PS LD is the effect of the interaction between the PS and the LD.

The model suggests that the light dose plays a considerable role in the photodinactivation of this microorganism, what may be explained by the very fast accumulation of the photosensitive molecule in this kind of cells [11]. It may suggest that ratio surface area/cell volume is relevant in this process. Employing the above equations it is possible to determine the survival index for HEp-2 and *C. albicans* in any point of the studied region.

Figure 3 shows the response surfaces obtained by the RSM for HEp-2 cells and *C. albicans* using the above models. For HEp-2 (Fig. 3a), it is shown just the effect of Hypericin concentration and the incubation time since the light dose had no significance. For *C. albicans* (Fig. 3b) it is shown just the effect of Hypericin concentration and the light dose since the incubation time was not significant in this case. It can be seen that the model is not linear and the region of interest, for selectivity, i.e., the lower effect on mammalian cells or the higher survival rate and the higher effect on *C. albicans* or the lower survival rate for the studied region occurs for low HY concentration, incubation time and light dose.

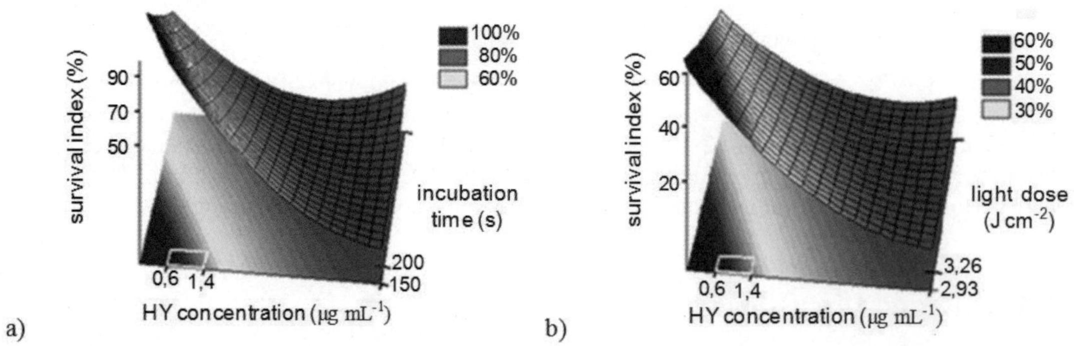

Fig. 3 Response surfaces obtained by RSM for: a) HEp-2 cells; b) *C. albicans*, both for the complete interval studied [14].

4. Conclusions

The results of this study employing experimental design suggest that a careful control of the experimental parameters permits achieving a privileged phototoxicity of Hypericin against *Candida albicans*. It was shown that an inactivation of *C. albicans* cells occurs at inferior Hypericin concentrations, incubation time and light doses than those necessary to injure important constituents of host tissues, such as epithelial cells used in this study. Hypericin accumulates very slowly in HEp-2 cells [11] so that in the low incubation time used in this study, no significant photosensitizer internalization and consequent cellular death occur in the mammalian cells. A comparison of the two RSM for mammalian and microorganism cells indicated that the region of highest selectivity was around 90% and 44% survival rate, respectively. The results showed that the best values of incubation time and light dose are very close to the inferior limits of the chosen region for these factors, suggesting that the study region could be shifted to lower values in order to refine the results. The results were considered good by being the first CCRD for this purpose, since the choice of the limits or levels of the factors was based on preliminary information and were not tested. It should be noted that the RSM indicated the direction in which these limits should be altered to achieve the optimal region, enabling it to get closer to the goal set for each CCRD performed. Finally, this approach can be very useful to search for the best conditions for PDI of *C. albicans* causing minimum damage to important constituents of host tissues. So, the experimental planning is a potent tool to optimize the PDI process.

Acknowledgements We are thankful to Prof. V. S. Bagnato of IFSC-USP for the Biotable development and the financial support of FAPESP.

References

[1] Wainwright M. Photoantimicrobials-so what's stopping us? *Photodiagnosis Photodyn Ther.* 2009; 6(3-4):167-169.
[2] Dai T, Huang YY, Hamblin MR. Photodynamic therapy for localized infections-state of the art. *Photodiagnosis Photodyn Ther.* 2009; 6(3-4):170-188.
[3] Yates F. Sir Ronald Fisher and the Design of Experiments. *Biometrics.* 1964; 20(2):307–321.
[4] EnginSoft. Virtual prototyping technologies and methods page. Available at: http://www.enginsoft.com/software/modefrontier/documentation/doe.html. Accessed November 20, 2010.
[5] Montgomery DC. *Design Analysis of Experiments*, 3rd ed., New York, John Wiley and Sons Inc; 1991.
[6] Myers RH, Montgomery DC, Vining GG, Borror CM, Kowalski SM. Response surface methodology: A retrospective and literature survey. *Journal of quality technology.* 2004; 36(1):53-78.
[7] Hyder MN, Huang RYM, Chen P. Pervaporation dehydration of alcohol-water mixtures: Optimization for permeate flux and selectivity by central composite rotable design. *Journal of Membrane Science.* 2009; 326(2):343-353.
[8] Myers RH, Montgomery DC. Response Surface Methodology: Process and Product Optimization Using Designed Experiments, 2nd ed., New York, John Wiley and Sons Inc; 2002.
[9] Kiran KR, Manohar B, Divakar S. A central composite rotatable design analysis of lipase catalyzed synthesis of lauroyl lactic acid at bench-scale level. *Enzyme Microbial Technology.* 2001; 29:122–128.
[10] Rodrigues MI, Iemma AF. *Planejamento de experimentos e otimização de processos: Uma estratégia sequencial de planejamentos*, Campinas, Casa do Pão Editora; 2005.
[11] Bernal C, Rodrigues JAO, Guimarães APP, Ribeiro AO, Oliveira KT, Imasato H, Perussi JR Selective photoinactivation of *C. albicans* and *C. dubliniensis* with hypericin. *Laser Physics.* 2010, *in press.*
[12] Menezes PFC, Bagnato VS, Johnke RM, Bonnerup C, Sibata CH, Allison RR, Perussi JR. Photodynamic therapy for Photogem and Photofrin using different light wavelengths in 375 human melanoma cells. *Laser Phys.* 2007; 4(7):546-551.
[13] Denizot F, Lang R Rapid colorimetric assay for cell growth and survival: Modifications to the tetrazolium dye procedure giving improved sensitivity and reliability. *J. Immunology.* 1986; 89(2): 271-277.
[14] Fernandes, P.L. et al, results not published.

Fungal evaluation on oolong tea irradiated with different water activities

G. B Fanaro[*1], R. C. Duarte[1], F. T. Rodrigues[1], B. Corrêa[2] and A.L.C.H.Villavicencio[1]

[1]Instituto de Pesquisas Energéticas e Nucleares (IPEN/CNEN). Centro de Tecnologia das Radiações. Av. Professor Lineu Prestes, 2242. Zip code 05508-000 São Paulo, SP, Brazil.
[2]Universidade de São Paulo. Instituto de Ciências Biológicas - Departamento de Micologia. Av. Professor Lineu Prestes 1374. Zip code 05508-900 São Paulo, SP, Brazil.

* Corresponding author: e-mail: gbfanaro@ipen.br, Phone: + 55 11 3133 9803

The aim of this study is evaluate the fungal contamination in oolong tea irradiated with different radiation doses and water activities (Aw). Samples were irradiated in 60Co irradiator at doses of 0, 2.5, 5.0, 7.5 and 10.0kGy with three different water activities. In the sample with decreased water activity, the count of fungi was lower than others samples followed by original Aw and the samples with the higher Aw, however there is no difference between the increased and decreased water activities samples after the irradiation on fungi contamination at dose of 2.5kGy.

Keywords oolong tea; *Camellia sinensis*; irradiation; fungi contamination; water activity

1. Introduction

Tea is a widely popular beverage consumed in the world for over several thousand years after water [1]. Reasons for its great popularity range from cultural traditions to purported health benefits. Results from epidemiological studies as well as laboratory experiments suggest that consumption of tea confers protection against the development of chronic diseases, such as cardiovascular and cancer [2].

Tea is processed from tender shoots of *Camellia sinensis* (L.) O. Kuntze and is mainly classified into green tea (unfermented), oolong tea (partially fermented), and black tea (fully fermented) according to the degree of fermentation during their preparations, where the term “fermentation” refers to natural browning reactions induced by oxidative enzymes in the cells of tea leaves. Oolong tea, possessing a taste and color somewhere between green and black teas, is manufactured predominantly in China and Taiwan [3].

Among the benefits associated to oolong tea consumption are the increasing of energy expenditure (acting on weight loss), the prevention of cardiovascular disease [4], reduce the LDL oxidation, total cholesterol and triglyceride and may be beneficial in the prevention of coronary heart disease [5], anti-cancer, anti-oxidant, anti-inflammatory and anti-bacterial activity [6] and suppressing the development of hypertension in rats [7].
Plant materials are highly susceptible to microbial contaminate due to the medium (water and soil) in which they grow. The current practices of harvesting, handling, storage and processing may cause additional contamination and microbial growth [8].

Irradiation with ionizing radiation is one of the most effective means to disinfecting dry food ingredients. The treatment by irradiation can inhibit cellular life division, like microorganisms and promoting a molecular structural modification [9]. The potential application of ionizing radiation in food processing is based that ionizing radiations inactivated microorganisms, insect gametes and plant meristems, prevented than from reproducing, resulting in various preservative effects as a function of the absorbed radiation dose [10]. Safety and efficiency of food irradiation has been approved by several authorities (FDA, USDA, WHO, FAO) and scientific societies based on extensive research [11].

Once absorbed by a biological material, ionizing radiation may have direct or indirect action. The indirect effect is caused by the interaction of radiation with a molecule of water (radiolysis), generating several kinds of free radicals that will interact with cells components, including DNA, and the largest damage in irradiated cell occurs by this action [12].

The radiolysis is the most important effect to be studied, once all food, even the dry foodstuffs, has water in its composition, then, the greater the amount of water (water activity) on food, greater should be the gamma radiation effect. The aim of this study is evaluate the fungal contamination in oolong tea irradiated with different radiation doses and water activities.

2. Material and methods

2.1 Samples

The oolong tea was purchased from a market in a traditional oriental neighborhood of São Paulo, Brazil.

2.2 Water activity (Aw)

The water activities were measured in Aqualab (3TE model) equipment.

To increase the water activity, in a desiccator, distilled water was added until the perforated mesh level for 48h. To decrease the water activity the samples were placed in an oven at 37°C/48h. In both case, the samples were placed in petri dish.

2.3 Irradiation

The samples were packed in plastic (polyethylene) bags, sealed and identified with their respective radiation doses. They were irradiated at room temperature in a ^{60}Co source Gammacell 220 (Nordion Ltd., Canada) with a dose rate of 1.83kGy/h, at doses of 0, 2.5, 5.0, 7.5 and 10.0kGy. Harwell Amber 3042 dosimeters were used to measure the radiation dose.

The samples were irradiated less than 12h of Aw measurement.

2.4 Fungal contamination

10g of sample were stirred with 90mL of 0.1% peptone buffered water for 20min/25°C in stomacher bags. Each suspension was diluted (10^{-1} to 10^{-4}) with the same diluent and then 0.1mL aliquot was spread on the surface of agar plate (dichloran glycerol agar - DG 18) and incubated at 25°C for 7 days. Counts were recorded in colony-forming units per gram (cfu/g) [13]. It was perform in triplicate.

2.5 Statistical analysis

The data were analyzed using one-way ANOVA and Tukey test with significance level of 5%.

3. Results and discussion

The fungal profile and the water activities values are shown in table 1.

Table 1. Water activity and effects of □-radiation on fungal contamination (cfu/g) in oolong tea.

Water activity		Doses (kGy)				
		0	**2.5**	**5.0**	**7.5**	**10.0**
Increased	0.806	$11.3x10^{3a}$	ND	ND	ND	ND
Original	0.517	$11.0x10^{3a}$	$1.7x10^{2}$	ND	ND	ND
Decreased	0.380	$2.3x10^{3b}$	ND	ND	ND	ND

Different lowercase letters in the same column means statistical difference ($p<0.05$).
ND: Not detected

In the sample with decreased water activity, the count of fungi was lower, which is explained by the value of Aw itself, because is low values limits the development of microorganisms [13]. The count of fungi in the higher Aw was slightly higher than the original Aw sample, however there was no statistical difference between them.

Despite has an initial contamination greater than the other samples, the oolong tea with Aw increased the level of contamination was "not detected" when the dose of 2.5kGy was applied, being the same level of the lowest water activity sample, while the original Aw sample needed 5kGy to achieve the same level.

This shows that radiolysis is an important effect and should be considered when using the food irradiation process manly when the sample has a contamination higher than recommended by sanitary agencies or health organizations, on the other hand, samples with lower Aw has lower contamination; however the use of gamma radiation never can be used as a substitute of good manufacture practices and good agricultural practices.

Thomas et al. (2008) showed that irradiation dose of 7kGy can be effective to control microbial growth in black tea and in extending their shelf life without any significant deterioration of quality constituents. Kumar et al. (2010) demonstrated that the radiation at dose up to 10kGy was sufficient to ensure microbiological safety of several herbals and formulations products without affecting their biochemical characteristics.

Furgeri et al. (2009) demonstrated that the irradiation at doses up to 10kGy had no effect on phenolic compounds in maté (*Ilex paraguariensis*). Fanaro et al. (2011) showed that there is no difference on volatile organic compounds with odors in green tea irradiated at doses up 20kGy.

4. Conclusions

There is no difference between the increased and decreased water activities samples after the irradiation on fungi contamination at dose of 2.5kGy, however to irradiate the oolong tea, one of these treatment must be chosen once they showed be more effective to fungal decontamination than when no pretreatment was applied.

Acknowledgment We are thankful for IPEN, CNEN, FAPESP and CNPq for financial support.

References

[1] Dou J, Lee VSY, Tzen JTC, Lee, MR. Identification and comparison of phenolic compounds in the preparation of oolong tea manufactured by semi fermentation and drying processes. *J. Agric. Food Chem.* 2007;55:7462-7468.

[2] Santana-Rios G, Orner GA, Amantana A, Provost C, Wu SY, Dashwood RH. Potent antimutagenic activity of white tea in comparison with green tea in the *Salmonella* assay. *Mutat. Res.* 2001;495:61-74.

[3] Lee VSY, Dou J, Chen RJY, Lin RS, Lee MR, Tzen JTC. Massive accumulation of gallic acid and unique occurrence of myricetin, quercetin, and kaempferol in preparing old oolong tea. *J. Agric. Food Chem.* 2008;56:7950-7956.

[4] Yamamoto S, Komatsu T, Matsushita H, Nakamori M, Dodo K, Wakabayashi K, Mizoguchi J, Komatsu K, Okamura M, Hosoda K. Oolong tea increases energy expenditure. In: *XIIth International Symposium on Atherosclerosis*, Stockholm, Sweden; 2000.

[5] Hosoda K, Yamamoto S, Toyoda Y, Ling C, Yi-Ping L, Itakura H. The effect of oolong tea on LDL oxidation in hyperlipidemia. In: *XIIth International Symposium on Atherosclerosis*, Stockholm, Sweden; 2000.

[6] Chen Y, Jiang Y, Duan J, Shi J, Xue S, Kakuda Y. Variation in catechin contents in relation to quality of 'Huang Zhi Xiang' oolong tea (*Camellia sinensis*) at various growing altitudes and seasons. *Food Chem.* 2010;119:648-652.

[7] Tanida M, Tsuruoka N, Shen J, Kiso Y, Nagai K. Effects of oolong tea on renal sympathetic nerve activity and spontaneous hypertension in rats. *Metab. Clin. Exp.* 2008;57:526-534.

[8] Khattak KF, Simpson TJ, Ihasnullah. Effect of gamma irradiation on the microbial load, nutrient composition and free radical scavenging activity of *Nelumbonucifera rhizome*. *Radiat. Phys. Chem.* 2009;78:206-212.

[9] Villavicencio ALCH, Fanaro GB, Araújo MM, Aquino S, Silva PV, Mancini-Filho J. Detection of *Phakopsora pachyrhizi* by polymerase chain reaction (PCR) and use of germination test and DNA comet assay after e-beam processing in soybean. *Radiat. Phys. Chem.* 2007;76:1878-1881.

[10] Farkas J. Irradiation for better foods. *Trends Food Sci. Technol.* 2006;17:148– 152.

[11] Villavicencio ALCH, Araújo MM, Fanaro GB, Rela PR, Mancini-Filho, J. Sensorial analysis evaluation in cereal bars preserved by ionizing radiation processing. *Radiat. Phys. Chem.* 2007;76:1875–1877.

[12] World Health Organization. High-dose irradiation: wholesomeness of food irradiated with doses above 10 kGy. *WHO Technical Report Series 890*; 1999.

[13] Pitt, J, Hocking, AD. *Fungi and food spoilage*. 3th ed. New York: Springer; 2009.

[14] Thomas J, Senthilkumar RS, Raj Kumar R, Mandal AKA, Muraleedharan N. Induction of c irradiation for decontamination and to increase the storage stability of black teas. *Food Chem.* 2008;106:180-184.

[15] Kumar S, Gautam S, Powar S, Sharma A. Microbial decontamination of medicinally important herbals using gamma radiation and their biochemical characterisation. *Food Chem.* 2010;119:328-335.

[16] Furgeri C, Nunes TCF, Fanaro GB, Souza MFF, Bastos DHM, Villavicencio ALCH. Evaluation of phenolic compounds in maté (*Ilex paraguariensis*) processed by gamma radiation. *Radiat. Phys. Chem.* 2009;78:639-641.

[17] Fanaro GB, Duarte RC, Araújo MM, Purgatto E, Villavicencio ALCH. Evaluation of □-radiation on green tea odor volatiles. *Radiat. Phys. Chem.* 2011;80:85-88.

Gamma radiation effects on bacteria and fungi in coffee (*Coffea arabica* L.)

F.T. Rodrigues*[,1], R.C. Duarte[1], G.B. Fanaro[1] and A.L.C.H. Villavicencio[1]

[1]Instituto de Pesquisas Energéticas e Nucleares (IPEN/CNEN). Av. Professor Lineu Prestes, 2242. Cidade Universitária, 05508-000 São Paulo, SP. Brazil.

* Corresponding author: e-mail: fthihara@ipen.br Phone: +55 1131339803

Microbiological contamination of coffee (*Coffea arabica L.*) is a problem in the industry, as coffee is especially sensitive to contamination. The occurrence of fungi and bacteria causes modification in the color and flavor, also loss of weight and the development of secondary metabolites. Irradiation is a decontamination method used for a variety of foodstuffs, being very feasible, effective and environment friendly without any significant change of sensory properties of the coffee. The aim of this study was to verify the effect of different gamma irradiation doses (0; 1, 5 and 3kGy to control bacteria and 0; 3; 6 and 9kGy to control fungi) on green coffee grains to determine if the microbiological load which is often present in coffee was destroyed. The results shows that microbiological contamination of coffee decrease when increase radiation doses.

Keywords coffee; irradiation; fungi; Salmonella ssp; coliforms

1. Introduction

Coffee is an important agricultural commodity produced in more than 60 countries. It generates approximately US$10-12 billion per year and ranks second on international trade exchanges, representing a significant source of income to several developing countries in Latin America, Asia and Africa. Brazil is responsible for more than one third of the global coffee production and exports. This fact ranks coffee amongst the most important commodities in the Brazilian trade balance [1].

Coffee quality is evaluated by key factors including the selection of the Coffea arabica variety, climatic conditions during different stages of growing, harvesting, processing, transport and storage conditions. Green coffee presents a real problem related to microbiological load due to the fact that it is extremely sensitive to contamination [2].

No significant damage to coffee by bacteria is known, since they require a relative humidity higher than 90% to survive and a medium with a neutral or basic pH. Damage is occurs incurred in cases where the coffee beans are already considerably deteriorated [3].

Fungi are another very important concern for storage [4-5]. Fungi which chiefly attack coffee are *Aspergillus ssp.* and *Penicillium verrucosum*. The development of these fungi is favored when the moisture content of beans is higher than 15 % and the relative humidity above 75 % [3].

In addition to their impact on the appearance, the aroma and flavor of coffee, fungi also produce toxic substances with can be harmful when consumed [3].

The fungal contamination found in Brazilian coffee predominate at the end of the processing and during storage, this may affect the quality and safety of the final product due to production of mycotoxins [4-5-6].

As a tropical country, it is likely that environmental conditions in Brazil are frequently conducive to the production of mycotoxins [7]. Mycotoxins have been detected infrequently and at low levels in green coffee beans [8-9].

The most important of these substances are mycotoxins of which there are various types suck as aflatoxins and ochratoxins for witch maximum tolerance levels have been ascertained [3].

Ionizing radiation can penetrate protected areas of produce and destroy the bacteria. Radiation processing has been employed in several countries as a mean of treatment to assure microbiological safety food [10].

Gamma irradiation is effective in reducing microorganisms and is known as a good method protect against pathogens agents in food materials [11]. One of the biggest advantages of gamma irradiation is that it can be easily applied to productsthat have limited storage temperatures and cannot be sterilized through an existing method such as heating. Gamma irradiation is able to perform sterilization without changing the temperature of commodity [12].

Efficiency and safety of food irradiation has been approved by several authorities such as Food and Agriculture Organization (FAO), World Heath Organization (WHO) and International Atomic Energy Agency (IAEA).

The aim of this study was to evaluate the effect of different gamma irradiation doses (0; 1, 5 and 3kGy to control bacteria and 0; 3; 6 and 9kGy to control fungi) on green coffee grains to determine if the microbiological load which is often present in coffee was destroyed.

2. Materials and methods

2.1. Samples

Samples of green coffee grains were purchased from local retail market in São Paulo, Brazil. The samples of 25g green coffee grains were packed up in polyethylene bags.

2.2. Irradiation

The samples were irradiated at Instituto de Pesquisa Energéticas e Nucleares - IPEN/CNEN (São Paulo, Brazil) using a Gammacell 220 cobalt-60 (MDS Nordion Ottawa, Canadá). The applied doses were 0; 1, 5 and 3kGy to control bacteria and 0; 3; 6 and 9kGy to control fungi.

2.3. Microbiological analysis

Samples were test immediately after irradiation. The analysis were performed to determine Salmonella ssp and coliforms presence in green coffee samples irradiated and unirradiated were measured according to the method described by Andrews, et al. [13].

The analysis of fungi were performed by counting the colony forming units (CFU) by plating on the surface, using the dilution of 25g of product in 225ml of 0,1% peptone water (1:10) up to 105 dilution. The dilutions were removed and transferred to 0,1 aliquots to Petri dishes with medium Potato Dextrose Agar (PDA), in triplicate. The plates were kept in incubator for 5 days at 25 ° C ± 1, and subsequently counted within existing settlements [14].

3. Results and discussion

3.1 Microbiological analysis

Microbial action detrimental to the quality and safety of the final product will depend on environmental conditions as well as crop and product management.

Microbiological contamination of coffee (*Coffea arabica* L.) was determited both for bacteria and fungi. Its was observed from that green coffee samples that the bacterial count decreased after irradiation with 1,5 kGy with no microorganisms present after 3 kGy irradiation (see Table 1).

Table 1 *Salmonella ssp* and coliforms presence in green coffee samples irradiated and unirradiated

Dose (kGy)	0	1,5	3
Coliforms (CFU/g)	<100	<50	absence
Salmonella ssp (CFU/g)	absence	absence	absence

There was a reduction of microbial load in the samples irradiated at 3 kGy dose while those irradiated at 6 and 9 kGy dose showed complete absence of microorganisms (see Table 2).

Nemtanu et al. (2005), irradiated in electron beams of 6 MeV green coffee with doses up to 40 kGy and showed that there was no microbial contamination with doses de 10 kGy [15].

Ahmad et al. (2003) demonstrated that coffee beans (both Arabica and Robusta variety) were subjected to irradiation treatments of 5 kGy (kilo Gray) and 10 kGy doses in order to achieve the above objective. Irradiation led to reduction of natural mircoflora with the 10 kGy dose being more efficient than the 5 kGy dose [16].

Table 2 Fungi contamination presence in green coffee samples irradiated and unirradiated

Dose (kGy)	0	3	6	9
Fungal Count (CFU/g)	3×10^3	1×10^3	absence	Absence

4. Conclusions

The results shows that microbiological contamination of coffee decrease when increase radiation doses. A dose of 3 kGy decreases the bacterial count and 6 kGy destroyed the microbial fungi. Considering that only a reduction of microbial count is necessary in food applications, irradiation is efficient to decontaminate green coffee even at low doses.

Acknowledgements The authors gratefully acknowledge the financial support of IPEN , CNEN, CNPQ, CAPES and FAPESP

References

[1] Vieira LGE et al. Brazilian coffee genome project: an EST-based genomic resource. *Brazilian Journal of Plant Physiology.* 2006;18:95-108.
[2] Silva CF, Schwan, RF, Dias, ES, Wheals, AE. Microbial diversity during maturation and natural processing of coffee cherries of (Coffea arabica L.) in Brazil. *International Journal of Food Microbiology. 2000;60:251-260.*
[3] Wintgens JN. *Coffee: Growing, Processing, Sustainable Production: A Guidebook for Growers, Processors, Traders, and Researchers.* 2nd revised ed. Weinheim, wiley vch verlag GmbH & Co. KgaA; 2009.
[4] Batista LR, Chalfoun SM, Prado G, Schwan, RF, Wheals, AE. Toxigenic fungi associated with processed (green) coffee beans (*Coffea arabica* L.). *International Journal of Food Microbiology. 2003;85:293-300.*
[5] Taniwaki MH, Pitt JI, Teixeira AA, Iamanaka, BT. The source of ochratoxin A in Brazilian coffee and its formation in relation to processing methods. *International Journal of Food Microbiology. 2003;82:173-179.*
[6] Taniwaki MH. An update on ochratoxigenic fungi and ochratoxin A in coffee. *Advances in Food Mycology. Advances in Experimental Medicine and Biology.* 2006;571:189–202.
[7] Urbano GR, Taniwaki MH, Leitao MF, Vicentini MC. Occurrence of ochratoxin A-producing fungi in raw Brazilian coffee. *Journal of Food Protection.* 2001; 64:1226–1230.
[8] Solim KM. Incidence, level and behaviour of aflatoxins during coffee bean roasting and decaffeination. Journal of Agricultural and Food Chemistry. 2002; 50:7477–7481.
[9] Viani R. Effect of processing on ochratoxin A (OTA) content of coffee. Advances in experimental medicine and biology. 2002; 504:189-193.
[10] Sommers, CH, BOYD G. Radiation sensation and postirradiation growth of foodborne pathogens on ready to eat frankufurter on a roll product in the presence of modified atmosphere and antimicrobials. *Journal of Food Protection.* 2006; 69:2436-2440.
[11] Villavicencio ALCH, Araújo M M, Fanaro GB, Rela PR, Mancini-Filho J. Sensorial analysis evaluation in cereal bars preserved by ionizing radiation processing. Radiation Physics and Chemistry. 2007; 76:1975-1877.
[12] Morehouse KM. Food irradiation-US regulatory considerations. Radiation Physics and Chemistry. 2002; 63:281-284.
[13] Andrews WH, Flowers RS, Silliker J, Bailey JS. Salmonella. In: Compendium of Methods for the Microbiological Examination of Foods, 4th ed. Washington, Md: American Public Health Association. Frances Pouch Downes & Keith Ito (Eds.); 2001.
[14] Pitt JI, Hocking AD Fungi and Food Spoilage, 2nd eds. *London Blackie*; 1997.
[15] Nemtanu MR, Brasoveanu M, Grecu MN, Minea R. Green coffee decontamination by electron beam irradiation. *Nuclear Instruments and Methods in Physics Research Section B: Beam Interactions with Materials and Atoms.* 2005; 240:83-86.
[16] Ahmad R, Tharappan B, Bongirwar DR. Impact of gamma irradiation on the monsooning of coffee beans. *Journal of Stored Products Research.* 2003; 39:149-157.

Irradiation effect on antifungal potential of clove essential oil

R. C. Duarte*[1]; G. B. Fanaro[1]; F. T. Rodrigues[1] and A. L. C. H. Villavicencio[1]

[1]Instituto de Pesquisas Energéticas e Nucleares (IPEN/CNEN). CTR. Av. Professor Lineu Prestes, 2242. Cidade Universitária, Zip code 05508-000 São Paulo, SP. Brazil. e-mail: renatocduarte@yahoo.com.br

The clove (*Syzygium aromaticum*) from the "Spice Islands" or Moluscas, Maluku Indonesian province. The clove portion used in the industry is the dried bud flower, leaf and stem. The clove has properties anti-irritant, soothing, relief of symptoms associated with motion sickness, nausea and vomiting. The clove essential oil is a phenolic compound from leaves and flowers (including stems) obtained from distillation. When extracted by steam distillation, high Eugenol rates, an aromatic compound is founded. The high eugenol content in essential oil can provide antimicrobial, antiseptic, bactericides, fungicides and fungicidal. The clove also acts as a food preservative due the high eugenol and tannins content. The oil is also used in flatulent colic treatment and some dental procedures. In food alimentation is used as a natural flavoring or added in small quantities and still has important applications in perfumery and cosmetics. Spices irradiation is a worldwide process used and this technique is an effective pathogenic microorganisms control providing consumers food security. By the fact cloves not only be used in food, but also as essential oil, this study intents to investigated the different irradiation doses influence on the possible antimicrobial potential of cloves oil and to evaluate the antifungal potential oil from unirradiated and irradiated clove in the fungus *Guignardia citricarpa* that causes serious damage in orange plantations. The clove samples were irradiated in ^{60}Co irradiator (Gammacell 220, Nordion Ltd., Canada) at doses of 0; 2.5; 5.0; 7.5 and 10.0kGy. The essential oils can demonstrate antifungal properties. The literature has recorded the extracts efficiency from a wide range of plant species to promote inhibition of several fungal phytopathogens. The clove essential oil showed a possible antifungal effect against the fungus *Guignardia citricarpa*

Keywords: cloves; irradiation; essential oil; *Guinardia citricarpa*; antifungal action

1. INTRODUCTION

Citrus black spot (CBS) caused by *Guignardia citricarpa* has become a serious widespread problem for citrus production [1, 2]. Develop on mature fruit and are characterized as hard spot, false melanose, freckle spot, virulent spot and in Brazil, the recently described cracked spot type [3]. Symptoms on fruit usually develop late in the season, and fruit that are symptomless at harvest may develop CBS during transportation or storage. If symptoms develop before maturity, fruit often drop, resulting in yield loss. Leaf spots may occur as well, especially on those citrus species most susceptible, such as lemon. However, fruit may be severely infected with relatively minor symptoms on foliage. *G. citricarpa* produces ascospores in the decomposing leaf litter on the grove floor [4, 5]. Ascospores are produced primarily in the summer and are dispersed by air currents and deposited on leaf and fruit surfaces, where they germinate and form quiescent infections on leaves and fruits [6]. Due to the presence of CBS in many countries in the Southern Hemisphere, the European Community (EC) and the United States (U.S.) severely limit importation of fresh citrus fruit from those countries where the disease is present [7]. Thus, fruit from countries with CBS can not be exported. Exposure of fruit to intense light and high temperatures speeds development of CBS symptoms [8]. In the protocol that allowed shipment of lemons from Argentina to the United States, samples from fruit lots for export were exposed to these conducive conditions prior to shipment to be sure they were free of CBS. There has been an increasing consumer demand for foods free or with low added synthetic preservatives because synthetic preservatives could be toxic to humans [9]. Concomitantly, consumers have also demanded for wholesome and safe food with long shelf lives. These requirements are often contradictory and press on the food industry for progressive removal chemical preservatives and adoption of natural alternatives to obtain safe food with long shelf lives. Some new technologies, which include modified atmosphere packaging and natural antimicrobial compounds has been used for food preservation [10]. There are many plants that demonstrate antimicrobial activity [11] and these plants have found application in the food industry as antibacterial and antifungal agents [12]. Being plants, spices appeal to consumers who question the safety of synthetic preservatives [13]. Although, the primary purpose of spices is to impart food flavor, the spices medicinal, antimicrobial and antioxidant properties have also been exploited [14]. The spices antimicrobial activity was documented and interest continues to the present [15 – 22]. The increase of fungal resistance to classical drugs, the treatment costs, and the fact that most available antifungal drugs have only fungistatic activity, justify the search for new strategies. In spite of the introduction of new antifungal drugs, they are limited in number, so aromatic plants have been widely used in folk medicine. It is known that most of their properties are due to their volatile oils. Essential oils from many plants are known to possess antifungal activity [23]. Clove (*Syzygium aromaticum*) bud oils have biological activities, such as antibacterial, antifungal, insecticidal and antioxidant properties, and are used traditionally as flavoring agent and antimicrobial material in food [24-26]. For example, clove oil was effective against L. monocytogenes and S. Enteritidis in tryptone soya broth (TSB) and cheese [27, 28]. The high levels of eugenol contained in clove essential oil give it strong biological activity and antimicrobial activity. This phenolic compound can denature proteins and reacts with cell membrane phospholipids changing their permeability [29, 30]. Clove oil also has several therapeutic effects, including antiphlogistic, antivomiting, analgesic, antispasmodic, anticarminative, kidney reinforcement, antiseptic, HCMV extracorporeal restraining effect [31, 32]. In Korea, clove oil has been successfully used for asthma and various allergic disorders by oral administration [33]. The essential oil is traditionally obtained by hydrodistillation, steam distillation or solvent extraction [34, 35]. The potential application of ionizing radiation in food is based mainly on the fact that ionizing radiation damage very effectively the DNA, living cells become inactivated, therefore microorganisms, insect gametes and plant meristems are prevented from reproducing, resulting in preservative effects. At the same time other chemical changes in food are minimal [36]. By the fact the ionizing radiation is widely used in spices decontamination and the spices provide the essential oils, the aim of this study is to investigate the irradiation influence in the cloves essential oil potential fungicidal.

2. MATERIALS AND METHODS

2.1. Samples

The cloves (100g per sample) were purchased from HiKari Industria e Comércio LTDA – BRAZIL.

2.2. Irradiation

The cloves samples were irradiated in polyethylene package at IPEN-CNEN/SP, at dose levels of 0; 2.5; 5.0; 7.5 and 10.0kGy using a ^{60}Co gamma facility (Gammacell 220, A.E.C.L., dose rate: 1.83 kGy/h) Harwell Amber 3042 Dosimeters were used for the measurement of radiation dose.

2.3. Hydrodistillation

The essential oils (EO) were extracted by the hydrodistillation method [37].

2.4. Fungal Isolation and Identification

The fungal (*Guignardia citricarpa*) isolation and identification [38].

2.5. Antifungal test preparation

Four small fungus (*Guignardia citricarpa*) portions were placed in each petri dish (triplicate) containing agar (potato dextrose Agar) and a drop of essential oil was placed on each fungus piece

3. RESULTS AND DISCUSSION

The essential oils demonstrate antifungal properties as reported by some authors [39 - 42]. The literature has recorded the extracts efficiency from a wide range of plant species to promote inhibition of several fungal phytopathogens [43]. The clove essential oil showed a possible antifungal effect against the fungus *Guignardia citricarpa* (Figure 2; Figure 3; Figure 4; Figure 5 and Figure 6).

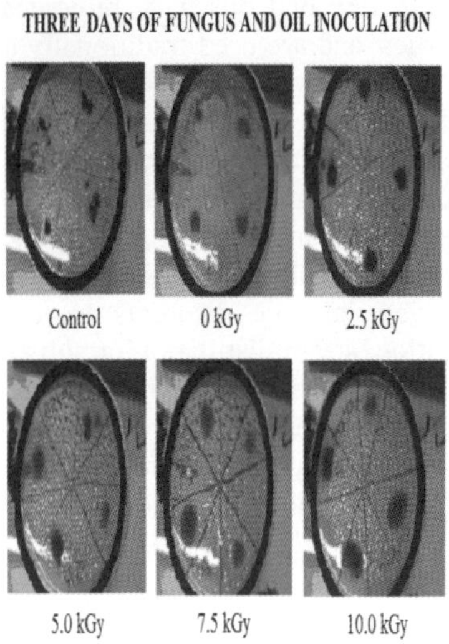

Figure 1. Three days of inoculation

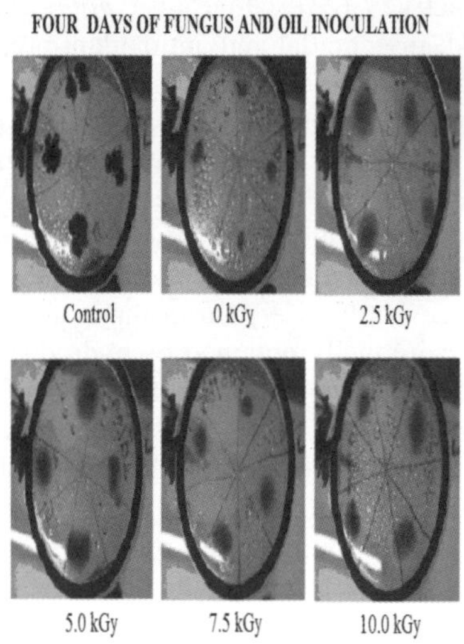

Figure 2. Four days of inoculation

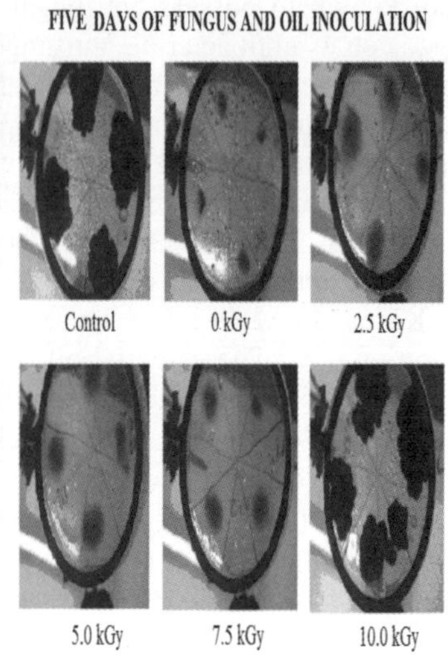

Figure 3. Five days of inoculation

In the figures 1 and 7 was possible to observe in the control samples (without essential oil) 100% of fungus growth and 33.33% in samples which the essential oils was obtained by the cloves that was irradiated with 10 kGy. This fungal growth frequency was maintained until the tenth day of analysis (Figures 2, 3, 4, 5 and 6).

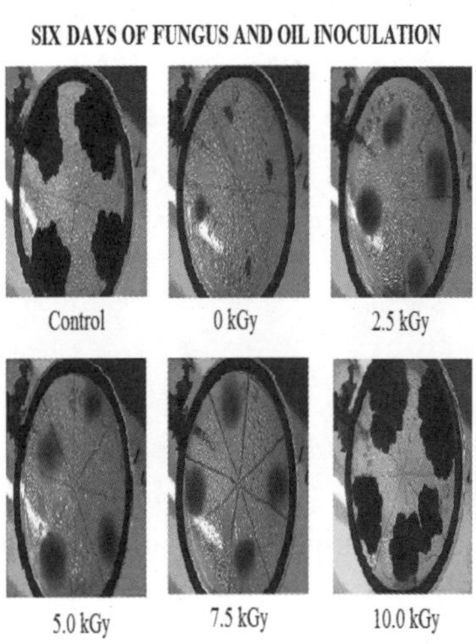

Figure 4. Six days of inoculation

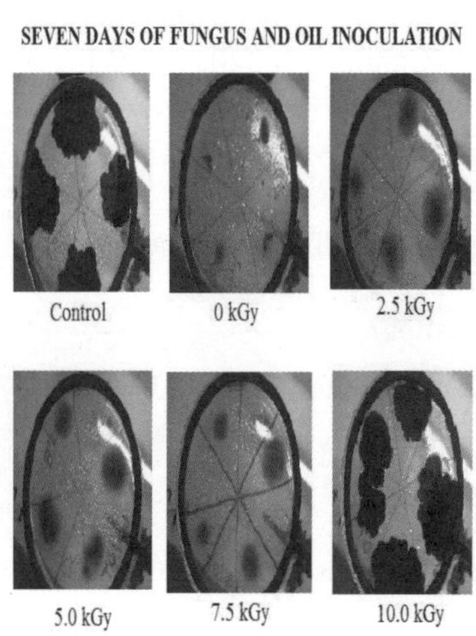

Figure 5. Seven days of inoculation

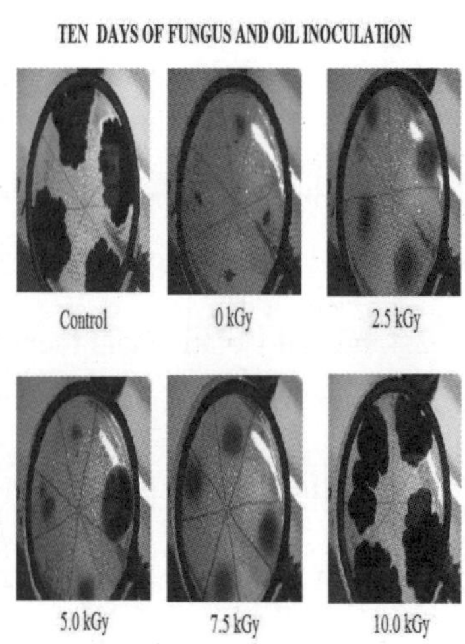

Figure 6. Ten days of inoculation

The samples irradiated until 7.5 kGy had the same antifungal results verified in the samples were not submitted by radiation processing. The fungus growth found in samples irradiated with 10 kGy may be related to the irradiation increasing dose, but it is not possible only with this study to conclude that the clove irradiation process decrease the essential oil fungicidal potential. The small difference was checked after 10 days of analysis (Figures 6 and 7). The growth of a portion of the fungus (*Guignardia citricarpa*) in only one of the triplicate samples (8.33%) may occurs by a fungus growth delay in the essential oil antifungal effect function.

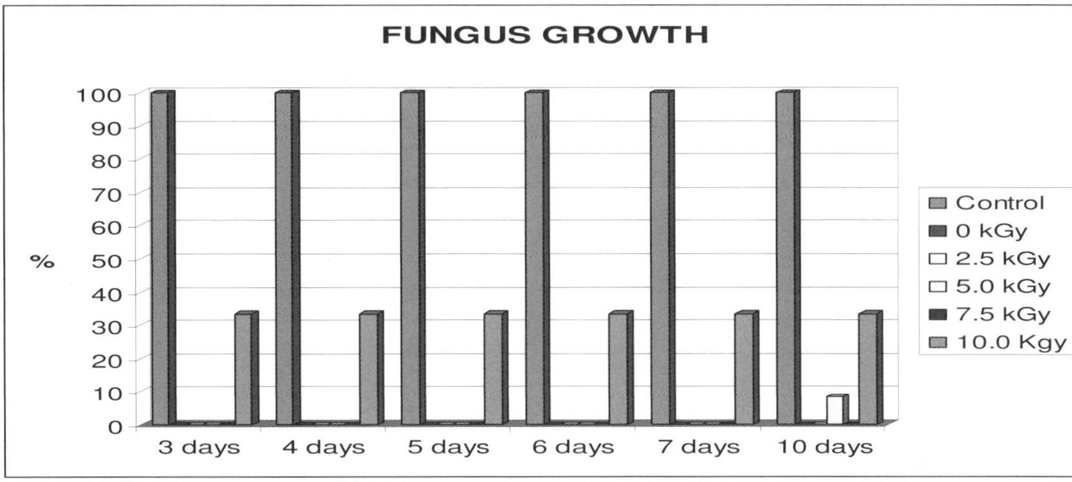

Figure 7. *Guignardia citricarpa* fungus growth frequency

4. CONCLUSIONS

It is necessary others studies to verify that the ionizing radiation influence in the clove essential oil antifungal activity decrease and the oil (*Syzygium aromaticum*) application efficiency in the field against the fungus (*Guignardia citricarpa*), the humans, environment toxic effects and negative interactions, so that we recommend its use for commercial-scale farmers, furthermore, the essential oil biological activity crude and fractions against other fungi and pests, since few studies report such activities. Thus we could know that the *Syzygium aromaticum* oil use or any of its components is a viable alternative to pests control.

Acknowledgements The authors are grateful for the financial support received from IPEN-CNEN/SP, FAPESP, CNPq and CAPES for the collaboration and participation in this study.

References

[1] Garrán SM. "Citrus black spot in the Northeast of Entre Rios: Etiology, epidemiology and control", *Proc. Int. Soc. Citricult.* 1996; 1: 466-470.

[2] Goes A. "Efeito da combinação de fungicidas sistêmicos e protetores no controle da mancha preta dos frutos cítricos causada por *Guignardia citricarpa*". *Summa Phytopathol.* 2000; 228: 9-13,.

[3] Kotzé JM. "Black spot", In: "Compendium of Citrus Diseases", L. W. Timmer, S. M. Garnsey, J. H. Graham, *American Phytopathological Society*, St. Paul, MN. 2000: 23-25.

[4] Kiely TB. "Preliminary studies on *Guignardia citricarpa* n. sp., the ascigerous stage of Phoma citricarpa McAlp., and its relation to blackspot of citrus", *Proc. Soc. N.S.W.* 1948; 73: 249-292.

[5] Kotzé JM. Epidemiology and control of citrus black spot in South Africa", Plant. 1981; 65: 945-950.

[6] McOnie KC. "Germination and infection of citrus by ascospores of *Guignardia citricarpa* in relation to control of black spot", *Phytopathology.* 1967; 57: 743-746.

[7] Baayen RP, Bonants PJ, Verkley MG, Carroll GC, Van Der HA, DeWeerdt M. "Nonpathogenic isolates of citrus black spot fungus, *Guignardia citricarpa* identified as a cosmopolitan endophyte of woody plants, G. mangiferae (Phyllosticta capitalensis)", *Phytopathology.* 2002; 92: 464–477.

[8] Timossi AJ, Goes A, Kupper KC, Baldassari RB, Reis RF. "Influência da temperatura e da luminosidade no desenvolvimento de *Guignardia citricarpa*, agente causal da mancha preta dos frutos cítricos", *Fitopatol. Bras.* 2003; 28: 489-494.

[9] Bedin C, Gutkoski SB. Wiest, J. M. Atividade antimicrobiana das especiarias. *Higiene alimentar.* 1999; 13: 26 –29.

[10] Brull S, Coote P. Preservative agents in foods: mode of action and microbial resistance mechanisms. *International Journal of Food Microbiology.* 1999; 50: 1 –17.

[11] Beales N. Food Ingredients as Natural Antimicrobials, CCFRA Review No 31, *Campden & Chorleywood Food Research Association*, United Kingdom. 2002.

[12] Lanciotti R, Gianotti A, Patrignani N, Belleti N, Guerzoni NE, Gardini F. Use of natural aroma compounds to improve shelf-life of minimally processed fruits. *Trends in Food Science and Technology.* 15, 201 –208, 2004.

[13] Sagdic O, Kuscu A, Ozcan M, Ozcelik S. Effect of Turkish spice extracts at various concentrations on growth of E. coli: 0157:H7. *Food Protection.* 2003; 19; 473 –480.

[14] Souza EL, Stamford TLM, Lima EO, Trajano VN, Filho JB. Antimicrobial effectiveness of spices: an approach for use in food conservation systems. *Brazilian Archives of Biology and Technology*. 2005; 48: 549 –558.
[15] Nychas GJE. Natural antimicrobials from plants. In New Methods of Food Preservation (ed. Gould, G. W.), *Blackie Academic and Professional*, Glasgow. 1995: 58 –89.
[16] CAST. Naturally occurring antimicrobials in foods. Council for agricultural science and technology, Task force Report. 1998: 132.
[17] Cosentino S, Tuberoso CIG, Pisano B, Satta M, Mascia V, Arzedi E, Palmas F. (1999), In vitro antimicrobial activity and chemical composition of Sardinian Thymus essential oils. *Letters in Applied Microbiology*. 1999; 29: 130 –135.
[18] Domans HJD, Deans SG. Antimicrobial agents from plants: antibacterial activity of plant volatiles oils. *Journal of Applied Microbiology*. 2000; 88: 308 – 316.
[19] Ristori CA, Pereira MS, Gelli DS. O efeito da pimento do reino moida frente a contaminacao in vitro com Salmonella rubislaw. *Revista do Instituto Adolfo Lutz*. 2002; 61: 131 –133.
[20] Radhakrishanan SS, Velusamy RR. Antifungal activity of some essential oils. *Journal of Agricultural and Food Chemistry*. 2003; 51: 7596 –7599.
[21] International Trade Centre - U N C T A D / W T O. Market Brief. The Market of Cloves in European Union p. 1. 2006. http://www.epices-comores.com/pdf_doc_gie/CCI_Girofle_Cloves.pdf
[22] Reverchon E. Supercritical fluid extraction and fractionation of essential oilsandrelated products. *Journal of Supercritical fluids*. 1997; 10: 1–37.
[23] Kalemba D, Kunicka A. Antibacterial and antifungal properties of essential oils. *Curr Med Chem*.2003; 10: 813–829.
[24] Huang Y, Ho SH, Lee HC, Yap YL. Insecticidal properties of eugenol, isoeugenol and methyleugenol and their effects on nutrition of Sitophilus zeamais Motsch. (Coleoptera: Curculionidae) and Tribolium castaneum (Herbst) (Coleoptera: Tenebrionidae). *Journal of Stored Products Research*. 2002; 38: 403–412.
[25] Lee KG, Shibamoto T. Antioxidant property of aroma extract isolated from clove buds [*Syzygium aromaticum* (L.) Merr. et Perry]. *Food Chemistry*. 2001; 74: 443–448.
[26] Velluti A, Sanchis V, Ramos AJ, Marı´n S. Inhibitory effect of cinnamon, clove, lemongrass, oregano and palmarose essential oils on growth and fumonisin B1 production by Fusarium proliferatum in maize grain. *International Journal of Food Microbiology*. 2003; 89: 145–154.
[27] Smith PA, Stewart J, Fyfe L. Antimicrobial properties of plant essential oils and essences against five important foodborne pathogens. *Letters in Applied Microbiology*. 1998; 26: 118–122.
[28] Smith PA, Stewart J, Fyfe L. The potential application of plant essential oils as natural food preservatives in soft cheese. *Food Microbiology*. 2001; 18: 463–470.
[29] Briozzo J. Antimicrobial activity of clove oil dispersed in a concentrated sugar solution. *Journal of Applied Bacteriology*. 66, 69–75, 1989.
[30] Deans SG, Ritchie G. Antibacterial properties of plant essential oils. *International Journal of Food Microbiology*. 1987; 5: 165–180.
[31] Liu H, Mao P, Hong S. Study on the virus function of extracorporeal restraining HCMV by Chinese traditional medicine Eugenia caryophyllata Thunb. Medical Journal of Chinese People's Liberation Army. 1, 73–75, 1997).
[32] National Pharmacopoeia Committee-China Pharmacopoeia. Beijing: Chemical Industry Press. 2000.
[33] Kim HM, Lee EH, Hong SH. Effect of *syzygium aromaticum* extract on immediate hypersensitivity in rats. *Journal of Ethnopharmacology*. 1998; 60: 125–131.
[34] Mostafa K, Yadollah Y, Fatemeh S, Naader B. Comparison of essential oil composition of Carum copticum obtained by supercritical carbon dioxide extraction and hydrodistillation methods. *Food Chemistry*. 2004; 86: 587–591.
[35] Reverchon E, Marrone C. Supercritical extraction of clove bud essential oil: isolation and mathematical modeling. *Chemical Engineering Science*. 1997; 20: 3421–3428.
[36] Thayer DW. Food irradiation: Benefits and concerns, *Journal of Food Quality*. 13, 147–169, 1990.
[37] Oliveira PR. Destilação por arraste de vapor. Ministério da Educação. Universidade Tecnológica do Paraná – Curitiba, 2009. Disponivel em: http://pessoal.utfpr.edu.br/poliveira/arquivos/aulapratica04destilacaoporarraste devapor.pdf. Acesso em: 15 de setembro de 2010.
[38] Michel MA, Gustavo BF, Ana PMS, Vladimir DR, Benedito C, Anna LCHV. Preliminary study of e-beam processing as a phytosanitary treatment against *Guignardia citricarpa*. In: International Nuclear Atlantic Conference – (INAC) – IX ENAN, 2009, Rio de Janeiro – RJ. *Nuclear Atlantic Conference* – (INAC).
[39] Ranasinghe L, Jayawardena B, Abeywickrama K. Fungicidal activity of essential oils of Cinnamomum zeylanicum (L.) and *Syzygium aromaticum* (L.) L.M.Perry against crown rot and anthracnose pathogens isolated from banana. Letters in Applied Microbiology. 2002; 3: 208–211.
[40] Chebli B, Mohamed A, Idrissi HLM, Mohamed H. Chemical composition and antifungal activity of essential oils of seven Moroccan Labiatae against Botrytis cinerea. *Journal of Ethnopharmacology*. 2003; 1: 165-169.
[41] Eugénia P, Cidália PV, Lígia S, Maria JG, Sofia CO, Carlos C, Ana P, Acácio R, José MO. Antifungal activity of the essential oil of Thymus pulegioides on Candida, Aspergillus and dermatophyte species. *J Med Microbiol*. 2006; 55: 1367-1373.
[42] Javad SG, Atefeh AAAntimicrobial and antifungal properties of the essential oil and methanol extracts of *Eucalyptus largiflorens* and *Eucalyptus intertexta*. Mag. 2010; 23: 172-175.
[43] Duarte, M. C. T.; Figueira, G. M.; Sartoratto, A.; Rehder, V. L. G.; Delarmelina, C. Anti-Candida activity of Brazilian medicinal plants, *Journal of Ethnopharmacology*. 2005; 97: 305–311.

Microarray for rapid detection of microbial resistance genotypes

Tomasz A. Leski*[1], Gary J. Vora[1], Brian Barrows[1], Guillermo Pimentel[2] and Chris R. Taitt[1]

[1]CBMSE, Naval Research Laboratory, 4555 Overlook Ave. S.W. Washington DC 20375, USA.
[2]Naval Medical Research Unit No.3, 3A Imtidad Ramses St, Abbassia, Cairo, Egypt 11517.

*Corresponding author: email: tomasz.leski@nrl.navy.mil, Phone: +1(202)767-0140, Fax: +1(202)767-9594

Antibiotic resistance in microbial pathogens causing hospital infections is a growing concern worldwide. The rapid spread of resistance genes can lead to the emergence of multiple drug resistant strains that in some cases are not treatable with any clinically available antibiotics. In this paper we present a new technology, the Antimicrobial Resistance Determinant Microarray (ARDM) that permits the rapid and simultaneous identification of a broad-range of antibiotic resistance genes from environmental or clinical samples. The ARDM is based on the CustomArray ElectraSense platform and is capable of detecting ~300 resistance determinants in 24 hours using a single high-density microarray. The validation results of the ARDM presented here show that this assay may be a useful tool for clinicians for the selection of appropriate antibiotic therapies before resistance phenotype information is available.

Keywords antibiotics; multidrug resistance; microarray; nosocomial infections

1. Background

Increasing resistance to antibiotics among pathogenic bacteria poses a serious threat for public health around the world. The evolution of antibiotic resistance that has been driven by antimicrobial drug use in human medicine and agriculture for more than 70 years has resulted in the selection of a very large number of antibiotic resistance genes. This process, facilitated by the spread of resistance genes on mobile genetic elements, appears to have accelerated during the last 20 years leading to the recent emergence of many strains that are resistant to the most clinically useful groups of antimicrobial agents, in some cases, leaving clinicians without therapeutic options [1].

Strains resistant to multiple classes of antibiotics (multidrug resistant – MDR strains) have emerged within important Gram-positive pathogens such as *Staphylococcus aureus*, *Streptococcus pneumoniae* and *Enterococcus faecalis*. Methicillin-resistant *S. aureus* (MRSA), which can be resistant to all β-lactam antibiotics and also frequently to all other clinically useful antibiotic groups with the exception of glycopeptides, is widespread in hospitals globally. However, the situation in this group of pathogens appears to have improved slightly in recent years due to the implementation of more efficient infection-control procedures in hospitals, introduction of a few novel antibacterial compounds (daptomycin, tigecyclin) and vaccination campaigns in order to prevent *S. pneumoniae* infections [2]. Unfortunately, the current situation seems much worse in case of Gram-negative pathogens. One of the most important problems in this group is the rapid spread of enzymes able to hydrolyze the wide spectrum of β-lactam antibiotics. These enzymes, called extended-spectrum β-lactamases (ESBLs), are currently widespread among *Enterobacteriaceae* and are responsible for resistance to third-generation cephalosporins. In practice, this leaves carbapenems (the most powerful β-lactams in use) as the only viable option to treat infections caused by ESBL carrying organisms [3]. The increased use of carbapenems has already led to the emergence of a number of enzymes (KPC, VIM and others) that are able to hydrolyze these compounds [4]. Infections with the strains carrying both an ESBL enzyme and a carbapenemase are practically untreatable [5]. Non-fermenting Gram-negative bacteria such as *Acinetobacter baumainii* and *Pseudomonas aeruginosa*, other major nosocomial pathogens that are naturally and intrinsically highly resistant to many antibiotics, are becoming MDR due to the acquisition of novel β-lactamases [6] combined with other resistance mechanisms targeting multiple antibiotic groups such as multidrug efflux systems and membrane permeability modifications [7].

Although the issue of MDR pathogens is equally relevant for both civilians and military personnel, the US Department of Defense has become especially concerned with this problem in the wake of a number of recent studies that have revealed extremely high rates (>80%) of infections with MDR pathogens in multiple military treatment facilities (MTFs) [8-10]. Bacterial infections in the last few years have become increasingly common among military personnel wounded in operations conducted in Iraq and Afghanistan [10, 11]. This is a result of dramatic increase of numbers of wounded soldiers surviving combat injuries. This increase is attributed to improvements in front-line trauma care, increasingly efficient medical evacuation and ubiquitous use of body armor. The high incidence of MDR strains that are resistant to three or more classes of antimicrobial agents in

MTFs may result from a number of factors. Close proximity of American soldiers and members of local population working and being treated together facilitates genetic exchange between formerly geographically distant bacterial strains. Another culprit is the widespread prophylactic use of broad-spectrum antibiotics for in-theater treatment immediately after injury [12]. Although this approach provides significant protection against infection, it also promotes genetic exchange, selection of antimicrobial resistance genes and emergence of MDR strains. The more targeted approach of treating wound infections only after the infection develops requires efficient methods of testing for antimicrobial resistance. This kind of information is needed in order to use the appropriate narrow spectrum antibiotic. Unfortunately, in the current system, antimicrobial resistance testing requires sample shipment to more specialized (echelon 4) MTF for obtaining pure bacterial cultures and subsequent analyses of antimicrobial resistance phenotypes (antibiogram). This process can take up to two weeks [13]. As such, an obvious technological gap exists indicating that a rapid system for antimicrobial resistance evaluation for in-theater use is needed. This kind of system would also potentially be useful in any clinical facility dealing with nosocomial infections in order to optimize antibiotic use and prevent the selection of MDR organisms.

2. Antimicrobial Resistance Determinant Microarray Design

To address the issue of rapid detection of antibiotic resistance our group has developed the Antimicrobial Resistance Determinant Microarray (ARDM). The ARDM is capable of rapid detection of a broad-range of antibiotic resistance genes present in the analyzed sample. CustomArray ElectraSense technology was chosen as a basis for this assay. The distinctive feature of ElectraSense microarrays is that the detection of hybridization of labeled DNA to microarray probes is based on electrochemical principles (Figure 1), in contrast to most currently used microarray technologies that employ optical detection of fluorescent labels. The ElectraSense reader required for acquisition of the hybridization data is small and portable and does not use precision laser optics, which is more expensive and generally more sensitive to damage and loss of alignment in portable devices.

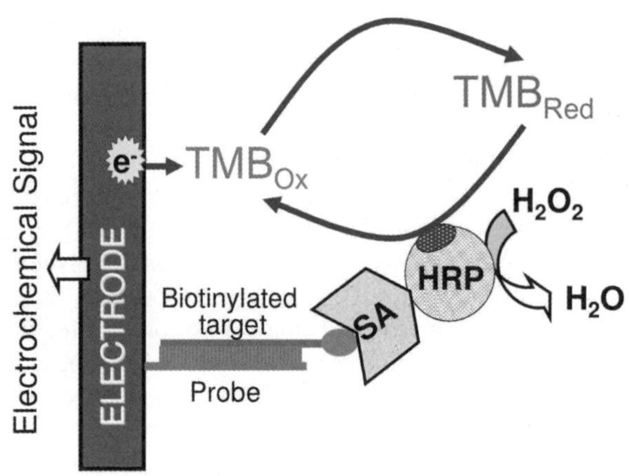

Fig. 1 *The principle of electrochemical signal generation on ElectraSense microarrays.*
Target DNA is labeled with biotin and hybridized to the array and subquently incubated with a streptavidin-HRP conjugate. HRP catalyzes the oxidation of a substrate (TMB). Electroreduction of the oxidized TMB generates a current flux. The ElectraSense reader performs amperometric detection of this current flux. Source: CombiMatrix Corp.

For our custom ARDM, the 4 x 2K format microarray was selected. Each microarray (the size of a microscope slide) contains four sub-arrays and each sub-array contains 2240 oligonucleotide probes. The oligonucleotide probes on the ElectraSense ARDMs are synthesized *in situ* using an electrochemical process. The advantage of this technology is that modification of the microarray content is simple and relatively inexpensive.

The sample to be analyzed on the ARDM needs to be amplified, fragmented and labeled before hybridization to the microarray. This process utilizes the GenomiPhi V2 amplification kit (GE Healthcare) for unbiased whole genome amplification of all of the DNA present in a sample. The resulting DNA is subsequently fragmented using DNase I and labeled with biotin using ULS Platinium Bright labeling kit (Kreatech). The amplified, fragmented and labeled DNA is then hybridized to the microarray and read using ElectraSense reader. The total processing time is relatively short and ranges from 11 to 24 hours depending on the length of hybridization.

The ARDM was designed to contain 1214 *unique* probes with probe lengths between 29 to 34nt in each of the sub-arrays. The probes are complementary to 278 antimicrobial resistance determinants responsible for resistance to antibiotics belonging to 12 different classes (Table 1) and derived from both Gram-positive and Gram-negative pathogens.

Table 1 Content overview of the ARDM assay.

Antibiotic class	Example of a compound from this class	Examples of resistance genes included on ARDM	No. of genes on ARDM*
ansamycins	rifamycin	*arr-3*	1
aminoglycosides	gentamycin	*aac(3), aac(6'), aad(A1)*	35
β-lactams	penicillin	TEM, CTX-M, KPC	64
chloramphenicol	chloraphenicol	*cat, cfr*	22
diaminopyrimidines	thrimethoprim	*dfrA1*	30
glycopeptides	vancomycin	*van(A), van(B)*	16
lincosamides	clindamycin	*lmr(A), lnu(C)*	14
macrolides	erythromycin	*car(A), ere(A), mac(A)*	58
quinolones	ciprofloxacin	*norA, qnr*	4
streptogramins	virginamycin	*lsa(A), mdt(A), vat(A)*	23
sulfonamides	sulfamethoxazole	*sul(I), sul(II), sul(III)*	3
tetracyclines	tetracycline	*tet(A), otr(A), vcaM*	40

**Some of the genes encode resistance to multiple classes of antibiotics, thus the numbers in this column total more than 278, which is the number of all genes detected.*

3. ARDM – Validation Results and Discussion

The validation of the ARDM was conducted using a collection of 29 MDR strains that were well characterized regarding their antibiotic resistance profiles and presence of antibiotic resistance genes using standard phenotypic (disk diffusion, E-test) and molecular (PCR, sequencing) methods. Since the detection of antimicrobial resistance genotypes of Gram-negative pathogens is currently our most important priority, this first validation set contained only Gram-negative bacteria. The set included seven strains of *Acinetobacter baumainii*, nine strains of *Escherichia coli* and 13 strains of *Klebsiella pneumoniae*. Almost all of the strains were resistant to multiple antibiotics. The resistance phenotypes included resistance to third-generation cephalosporins indicating the presence of ESBL enzymes. All of the *A. baumainii* strains were additionally resistant to carbapenems. Most of the strains were resistant to chloramphenicol and tetracycline and approximately half of them resistant to aminoglycosides (amikacin and/or gentamycin). Similarly, approximately half of the strains were resistant to ciprofloxacin (a fluoroquinolone).

Here we present the results obtained from 10 representative strains from the validation group. Table 2 contains a comparison of the ARDM-based detection compared with detection conducted using PCR amplification. Due to space constraints, we only present the results of the detection of genes coding for resistance to β-lactams. The table also includes the most important traits of antimicrobial resistance phenotype of each strain.

Table 2 Validation results – comparison of PCR and ARDM for the detection of β-lacamases.

#	Pathogen*	Strain	Assay	TEM	SHV	OXA	CTX-M	other	Phenotype#
1	*A.b.*	2	ARDM	1,10	-	-	-	LAT-1, CcrA	CAZ, CTX, IMP
			PCR	+	-	+	-		
2	*A.b.*	6	ARDM	-	-	-	8,14	KPC-3, VIM-2, LAT-1	CAZ, CTX, IMP
			PCR	+	-	+	-		
3	*E.c.*	7	ARDM	-	LEN-1	1	FEC1,1,3,12,15,32		ESBL
			PCR	+	-	+	+		
4	*A.b.*	9	ARDM	1,10	-	-	-	KPC-3	MBL, CAZ, CTX, IMP
			PCR	+	-	-	-		
5	*E.c.*	16	ARDM	1,10	-	1	FEC1,1,3,12,15,32	KPC-3, penA	ESBL, CAZ, CTX
			PCR	+	+	+	+		

6	*K.p.*	17	ARDM	1,10	1,5,7	9	9,13		ESBL, CAZ, CTX
			PCR	+	+	-	+	penA	
7	*E.c.*	24	ARDM	1,10	-	1	FEC1,1,3,12,15,32,9,13,14,24		ESBL, CAZ, CTX
			PCR	+	-	+	+		
8	*K.p.*	25	ARDM	1,10	1,5,7, LEN-1	1,9	FEC1,1,3,12,15,32		ESBL, CAZ, CTX
			PCR	+	+	+	+		
9	*K.p.*	26	ARDM	1,10	1,5	9	1		ESBL, CAZ, CTX
			PCR	+	+	-	+		
10	*K.p.*	29	ARDM	1,10	-	1,9	1,3,15,32		ESBL, CAZ, CTX
			PCR	+	+	+	+		

* *A.b. = Acinetobacter baumainii, E.c. = Escherichia coli, K.p=Klebsiella pneumoniae.*

\# Resistance phenotypes: ESBL – extended spectrum β-lactamase detected, CAZ – ceftazidime resistance, CTX – cefotaxime resistance, IMP – imipenem (carbapenem) resistance.

The results show that in most cases the results of the ARDM analyses agree with the PCR results. In many cases, the ARDM assay is able to detect not only presence of a given β-lactamase but also identify the specific allele present in the analyzed strain. This feature is important in the case of β-lactamase families since the particular enzymes may differ in their substrate preference resulting in different phenotypes. In a few cases the PCR and ARDM results disagreed. In some situations the cause for this can be explained by the properties of PCR primer design. For example, we found that the PCR primers for the detection of the OXA enzymes were unable to detect OXA-9 allele, which was readily detectable by ARDM (see sample (strain) 6(17) and 9 (26) in the table above). In other cases (e.g. TEM β-lactamases in samples 1(2) and 2 (6)) the ARDM was unable to confirm positive PCR amplification. This is most likely the result of the absence of some TEM specific probes on the microarray. These 'false negative' results will be taken into account when designing the next generation ARDM (v2.0).

The validation results (as exemplified here by the β-lactamase detection results) confirm that the ARDM is capable of rapidly determining the antimicrobial resistance genotype of a pathogen. We estimate that the approximate cost of a single ARDM test should be relatively low and not exceed $40 per sample. The results obtained from the ARDM are equivalent to nearly 300 PCR assays and give an insight in the antibiotic resistance potential of the analyzed pathogen long before phenotypic antibiotic testing results are available. Therefore we believe that the ARDM can become an important tool to assist clinicians in making informed and more appropriate decisions regarding antibiotic therapy.

The ARDM is currently in its first version (1.0). Assay optimization is still under development and we are planning to introduce a number of improvements in future versions of the ARDM. The improvements include modifications of the probe-set to replace poorly performing probes, eliminating overlapping and redundant probes, adding bacterial species-specific probes and adding probes that reflect the newest developments in the field (e.g. probes for detecting NDM-1 β-lactamase). Software for the automated analysis of the microarray hybridization results is also currently under development.

Acknowledgements The authors would like to gratefully acknowledge the support of the Defense Medical Research and Development Program/Military Infectious Disease Research Program (DMRDP/MIDRP) who funded this study as a part of the project #D10_I_AR_J2_628.

References

[1] Hawkey PM and Jones AM. The changing epidemiology of resistance. *J Antimicrob Chemother.* 2009; 64 Suppl 1: i3-10.
[2] Livermore DM. Has the era of untreatable infections arrived? *J Antimicrob Chemother.* 2009; 64 Suppl 1: i29-36.
[3] Paterson DL and Bonomo RA. Extended-spectrum beta-lactamases: a clinical update. *Clin Microbiol Rev.* 2005; 18: 657-86.
[4] Walsh TR. Clinically significant carbapenemases: an update. *Curr Opin Infect Dis.* 2008; 21: 367-71.
[5] Bush K. Alarming beta-lactamase-mediated resistance in multidrug-resistant Enterobacteriaceae. *Curr Opin Microbiol.* 13: 558-64.
[6] Livermore DM and Woodford N. The beta-lactamase threat in Enterobacteriaceae, Pseudomonas and Acinetobacter. *Trends Microbiol.* 2006; 14: 413-20.

[7] Zavascki AP, et al. Multidrug-resistant Pseudomonas aeruginosa and Acinetobacter baumannii: resistance mechanisms and implications for therapy. *Expert Rev Anti Infect Ther*. 2010; 8: 71-93.
[8] Zapor MJ, et al. Emergence of multidrug resistance in bacteria and impact on antibiotic expenditure at a major army medical center caring for soldiers wounded in Iraq and Afghanistan. *Infect Control Hosp Epidemiol*. 2008; 29: 661-3.
[9] Hujer KM, et al. Analysis of antibiotic resistance genes in multidrug-resistant Acinetobacter sp. isolates from military and civilian patients treated at the Walter Reed Army Medical Center. *Antimicrob Agents Chemother*. 2006; 50: 4114-23.
[10] Petersen K, et al. Trauma-related infections in battlefield casualties from Iraq. *Ann Surg*. 2007; 245: 803-11.
[11] Scott P, et al. An outbreak of multidrug-resistant Acinetobacter baumannii-calcoaceticus complex infection in the US military health care system associated with military operations in Iraq. *Clin Infect Dis*. 2007; 44: 1577-84.
[12] Mellor SG, Cooper GJ, and Bowyer GW. Efficacy of delayed administration of benzylpenicillin in the control of infection in penetrating soft tissue injuries in war. *J Trauma*. 1996; 40: S128-34.
[13] Van Syckle FJ. Infectious disease: challenges in a combat environment (Iraq). *Presented at CENTCOM HQ*. 2008.

Photosensitization as human and environmentally friendly antimicrobial tool

Zivile Luksienė

Institute of Applied Research, Vilnius University, Sauletekio 9, LT-10223, Vilnius, Lithuania
Corresponding author: Zivile Luksienė Zivile.Luksiene@tmi.vu.lt

Keywords: photosensitization, inactivation, food pathogens

1. Summary

Photosensitization is a treatment involving the interaction of the two non-toxic factors, photoactive compound and visible light, which in the presence of oxygen results in the selective destruction of the target cell. Different microorganisms, such as multidrug-resistant bacteria, yeasts, microfungi and viruses are susceptible to this treatment. Therefore, a photosensitization phenomenon might open a new avenue for the development of non-thermal, effective and ecologically friendly antimicrobial technology, which might be applied for food safety.

2. Introduction

Despite tremendous progress in food microbiology, the number of reported food-borne diseases continues to rise. Health experts estimate that every year food-borne illnesses in USA cost 5-6 billion US dollars in direct medical expenses and lost productivity [1].

Traditional thermal technologies are effective antimicrobials, but usually they induce a plenty of uncontrolled chemical reactions in the food matrix and subsequently reduce the nutrition quality [2].
Inevitably, a new approach to inactivate pathogenic and harmful microorganisms in food manufacturing and safety in cost-effective and environmentally friendly way is highly needed. To this end, modern biophotonic technology based on photosensitization might serve as a promising tool to decontaminate food and food-related surfaces from different pathogens.

In general, photosensitization is a treatment involving the administration of a photoactive compound that selectively accumulates in the target cells and the following illumination. The interaction of two non-toxic elements, photoactive compound and visible light, in the presence of oxygen results in a plethora of cytotoxic reactions and consequently induces selective destruction of target microorganism [3].

As a rule, the photosensitizers are based usually on tetrapyrrole nucleus including porphyrins, chlorines. Some bacteria are known to produce significant amounts of endogenous porphyrins from exogenously applied aminolevulinic acid (ALA) [4].

Common porphyrin and chlorine derivatives have a characteristic absorption band between 400-430 nm and 600 nm [3].

Photosensitization was performed by using conventional gas discharge and incandescent lamps, metal halide lamps, short-arc xenon lamps UV lamp, light emitting diodes [3].

As depicted in Fig. 1, accumulation of the photosensitizer in the bacteria is the main prerequisite for its photoinactivation. As a rule, for Gram (+) bacteria and yeasts the photosensitizer accumulates in the cell wall. After irradiation by visible light, reactive oxygen species induce rapid disruption of the cell wall (Fig. 1).

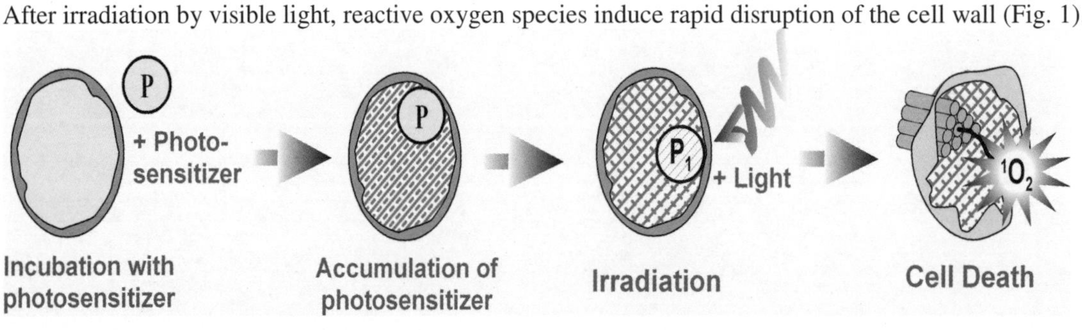

Fig.1. Mechanism of destructive action of photosensitization in the cell: P -photosensitizer, P_1 - excited state of photosensitizer after absorption of light, $^1O_2$ – singlet reactive oxygen.

Contamination of organic wheat sprouts by microfungi in control (A) and treated by photosensitization (B) samples.

Reactive ROS interacts with unsaturated fatty acids, amino acid residues, nucleic acid bases of DNA [3].

3. Antimicrobial efficiency of photosensitization

ALA-based photosensitization was used to inactivate main food pathogens. According to the obtained data, fast and significant inactivation (6.5 log) of *Bacillus cereus, Listeria monocytogenes* (4 log*), Salmonella typhimurium* (6 log) were obtained after this treatment.Moreover, spores are susceptible to photosensitization. Other physiological condition- *L. monocytogenes* biofilms can be inactivated by 3.1 log after ALA-based photosensitization [5].

It seems that plethora of harmful micromycetes that destroys food might be inactivated by photosensitization, a method that is completely safe, reproducible, nonmutagenic, noncarcinogenic, environmentally and human friendly [6].

Photosensitization being environmentally friendly, effective and free from chemical substances can be used for decontamination of seeds. Mostly prevailing in wheats *Alternari, Penicillium, Mucor* after the photosensitzation treatment was reduced by 75 %. [6,7].

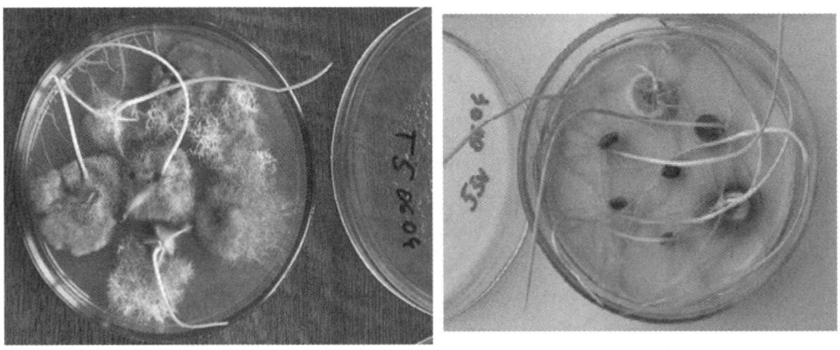

A B

Fig.2. Effects of photosensitization in decontamination of wheat sprouts: a) before treatment, control, b) after treatment.

Due to very high resistance of bacterial spores to UV, germicidal lamps are insufficient to ensure sterilization of surfaces. More than 90% of packaging contamination is composed of aerobic, spore-forming bacteria [9]. Decontamination of packaging material from *B. cereus* reduced it by 4 log [8].

In addition, the inactivation of Listeria on the surface of packaging after ALA-based photosensitization can reach 2.3 - 3.7 log. The effective inactivation of Listeria biofilms indicates that this treatment has potential to combat biofilms as well [5].

4. Conclusions

Multi-antibiotic resistance of pathogens is a rapidly growing and alarming phenomenon. Hence, the discovery of novel, cost–effective and human friendly technologies to inactivate harmful and pathogenic microorganisms seems an imperative. Photosensitization of bacteria has repetitively been shown to be independent of the antibiotic resistance-spectrum, is effective against bacterial spores and biofilms, is not-mutagenic or genotoxic. In our opinion, this phenomenon opens a new and prospective avenue for the development of effective, human and environmentally friendly antimicrobial treatment. Its proper application for the treatment of food, packaging and processing equipments might be really useful to increase microbial food control and subsequently decrease foodborne diseases.

Acknowledgements This work was partially supported by European Commission (FP6, STREP project HighQRTE).

References

[1] Center for Disease Control and Prevention (CDC), Division of Bacterial and Mycotic Diseases (2003). Salmonellosis. Available at: http://www.cdc.gov/ncidod/dbmd/diseaseinfo/salmonellosis_t.htm (accessed 22 December 2008).

[2] Manas P and Pagan R. Microbial inactivation by new technologies of food preservation. *Journal of Applied Microbiology*. 2005; 98: 1387-1399.

[3] Luksiene Z and Zukauskas A. Prospects of photosensitization in control of harmful and pathogenic microorganisms, *Journal of Applied Microbiology.*, 2009; 107: 1415-1424.

[4] Meffert H, Gaunitz K, Gutewort T and Amlong UJ Therapy of acne with visible light. *Dermatol Monatsschr.* 1990; 176: 597-600.

[5] Buchovec I, Vaitonis Z and Z Luksiene Novel approach to control *Salmonella enterica* by modern biophotonic technology: photosensitization, *Journal of Applied Microbiology.* 2009; 106: 748-754.

[6] Luksiene Z. New approach to inactivate harmful and pathogenic microorganisms : photosensitization. *Food Technology and Biotechnology.* 2005; 43: 1-8.

[7] Luksiene Z, Danilcenko H, Taraseviciene Z, Anusevicius Z, Maroziene A. and Nivinskas H. New approach to the decontamination of germinated wheat from microfungi: effects of aminolevulinic acid. *International Journal of Food Microbiology.* 2007; 116:153-158.

[8] Luksiene Z, Buchovec I and Paskeviciute E Inactivation of food pathogen *Bacillus cereus* by photosensitization *in vitro* and on the surface of packaging material. *Journal of Applied Microbiology.* 2009; 107: 2037-2046.

[9] Pirttijarvi TSM, Graeffe TH, Salkinoja-Salonen MS. Bacterial contaminants in liquid packaging boards: assessment of potential for food spoilage. *Journal of Applied Bacteriology.* 1996; 81: 445-458.

Rapid Diagnosis of Neonatal Bacteremia Using Polymerase Chain Reaction

Amany El-Sharif [1,*] and Raghdaa Ali [2]

[1] Departments of Microbiology & Immunology Faculty of Pharmacy (girls) Al Azhar Univ., [2] Pediatrics and Neonatology, Ahmad Maher Teaching Hosp., - Cairo, Egypt.

[*] Corresponding author: Department of Microbiology and Immunology, Faculty of Pharmacy(Girls), Al-Azhar University, Yousef Abbas Street, Cairo, Egypt. e-mail: amanyelsharif@yahoo.com, Phone: +20169949155

Infections in the neonate are the most important causes of mortality and hospitalizations in the neonatal practice. In the present study, 180 neonates with criteria for probable sepsis admitted to neonatal intensive care unit in Ahmad Maher Teaching hospital, Cairo Egypt were investigated for evidence of sepsis. The investigation protocol included sepsis screen, blood culture and molecular analysis by polymerase chain reaction (PCR) for bacterial DNA component encoding 16S rRNA in all cases. We compared the results of PCR with blood culture and other markers of sepsis screen. Blood cultures were positive in 80 (44.4%) cases with sensitivity of 91.9%. PCR was positive in 87(48.3%) cases (49 of them are with positive blood culture, CRP and buffy coat test). The sensitivity of PCR was 100% and specificity was 93%. The negative predictive value was 100%. The study concluded that although blood culture is a reliable method for diagnosis of neonatal sepsis, the universal bacterial primer PCR is a useful test and superior to blood culture for early diagnosis of sepsis in neonates. Consequently unnecessary treatment with antibiotics could be avoided.

Keywords: neonatal sepsis; bacteria; PCR

Introduction

Neonatal sepsis is a systemic disease caused by the spread of microbes or their toxins via the blood stream in the first month of life [1]. Infections due to primary pathogen or complication of other illness is a major cause of neonatal mortality and morbidity all over the world[2]. Clinical diagnosis of sepsis in new born infants is not easy as symptoms and signs are not specific[3]. There is no laboratory test confere100% specificity and sensitivity [4]. Neonates have risk factors for early sepsis, and /or clinical picture suggestive of sepsis or alterations in other laboratory tests are thus treated with broad spectrum antibiotics which could be not required [5-6]. However Blood culture has been the gold standard for confirmation of diagnosis, molecular detection offers a great advantage to culture techniques in circumstances where there may be minute amount of a specific microorganisms or the patient may have received antibiotics prior to specimen collection in addition to the fact that the results are available only after 48-72 hours [7].

The goal of this study was to evaluate the molecular diagnosis of sepsis using PCR amplification of 16S rDNA in Egyptian neonates with risk factors for sepsis or those who have clinical evidence of sepsis. We also compared the results of PCR with blood culture and other markers of sepsis screen.

Patients and methods

This study was carried on 180 infants selected from the neonatal intensive care unit in Ahmad Maher Teaching hospital, Cairo, Egypt in the period of 2008-2009. The selected group was neonates with their age ranged from one day to 4 weeks. Neonates with a new episode of clinically suspected sepsis were enrolled in this study. Baseline assessments included Full history, type of delivery, complete physical examination, complete blood count, qualitative detection of C-reactive protein (CRP) and microscopical examination of buffy coat layer.

- Venous blood were cultured on Signal oxoid blood culture media.
- Polymerase Chain Reaction (PCR) was carried out for detection of bacterial DNA using ***universal primers***. The DNA sequences present in all bacteria, such as portions of the DNA encoding the 16S ribosomal RNA (861 bp fragment) was used to define bacterial antigen using the following primers F-5'-AGAGTTTGATCCTGGCT-CAG-3' and R-5'- GGACTACCAGGGTATC TAAT-3', those sequences were amplified by PCR. Following the protocol of Laforgia *et al.*, [4] with a modification of the test by DNA extraction from whole blood using QIAamp DNA blood kit (Quiagen Inc.).

QIAamp DNA blood kits are designed for rapid purification of an average of 6 µg of total DNA from 200 ul of whole human blood. A small volume of the sample (2ul) was introduced in the PCR reaction mix., (briefly 10 mmol/L Tris HCl (pH 8.8), 50 mmol/L KCl, 1.5 mmol/L $MgCl_2$, 200 µL of each dNTP, < 0.2 µmol/L of each primer, 2.5 units of Taq polymerase in a final volume of 100 μL. The computerized thermocycler was programmed for the following conditions: an initial precycling hold at 94^{0}C for 5 min. then 35 cycles of denaturation at 94^{0}C, one min. annealing at 55^{0}C for one min. and extension at 72^{0}C for one min. In the last cycle extension was prolonged for 7 min. The amplification fragments were separated in 2% agarose gel stained with ethidium bromide.

Blood samples from twenty healthy adult volunteers were used as negative controls. One ml of each sample was infected with different clinical isolates as *Escherichia coli, Klebsiella pneumoniae* and *Staphylococcus epidermidis* at a concentration of 200 CFU/ml were used as positive controls.

Results

Of the 200 patients who were evaluated, 180 were included. Exclusion criteria were major malformations, life expectancy less than 12 hours, or contaminated blood cultures. Mean gestation was 33.1 ± 4.1 weeks, and mean birth weight was 1730.3 ± 597.3 g. Different methods of detection of bacterial infections were compared. Parameters of sepsis screen were studied. Of the 180 neonates in the study group, positive CRP qualitative assay yielded positive result in 64 cases (35.5%). Seventy two cases (40%) had leucocytosis, 116 cases (64.4%) had high MCH and 83 cases (46.1%) had high MCV.

Blood culture yielded positive results in 80 cases (44.4%) of the study group and had a sensitivity of 91.9% for diagnosis of neonatal sepsis in relation to PCR, while eighty seven (48.3%) had positive PCR. In comparison with blood culture, the Buffy coat method had sensitivity of 78.8%, specificity of 100%, and positive predictive value of 100% and negative predictive value of 85.5% (table 1).

Table (1): Utility of Buffy coat method for diagnosis of neonatal sepsis.

Blood culture	Positive buffy coat	Negative buffy coat	Total
Positive culture	63	17	80
Negative culture	0	100	100
Total	63	117	180

Sensitivity= 78.8%, Specificity= 100%
Positive predictive value= 100%,
Negative predictive value= 85.5%

CRP test had sensitivity of 61.3%, specificity of 85%, positive predictive value of 76.5%, and negative predictive value of 73.3 % (Table 2).

PCR amplification for 861 bp fragment was positive in 87 cases (48.3%) (Fig. 1 a-b). All patients had positive blood culture were positive PCR, whereas 7 patients had positive PCR but negative blood culture. Those were also included all cases with positive Buffy coat. Thus PCR is proved to have sensitivity of 100%, specificity of 93%, positive predictive value of 91.9 %, and negative predictive value of 100%. The lower specificity of PCR is because of all the 7 cases with positive PCR but sterile cultures being true positives and not false positives and can be explained by the lower sensitivity of the gold standard (91.9%) (Table 3).

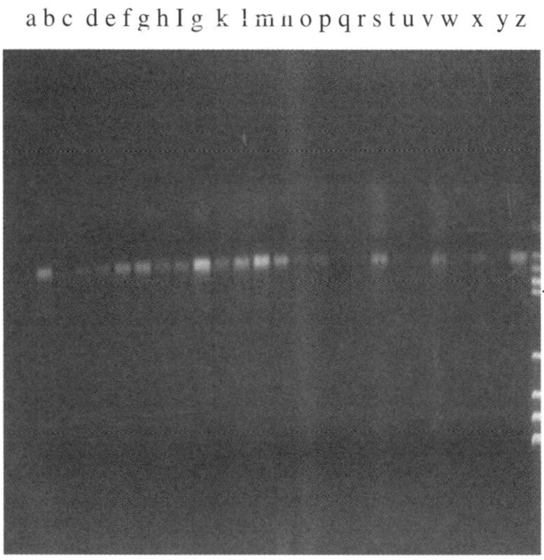

Fig.1.a. PCR for DNA region encoding the16S rDNA at 861bp (universal bacterial DNA) in 25 neonatal blood samples with and without sepsis.

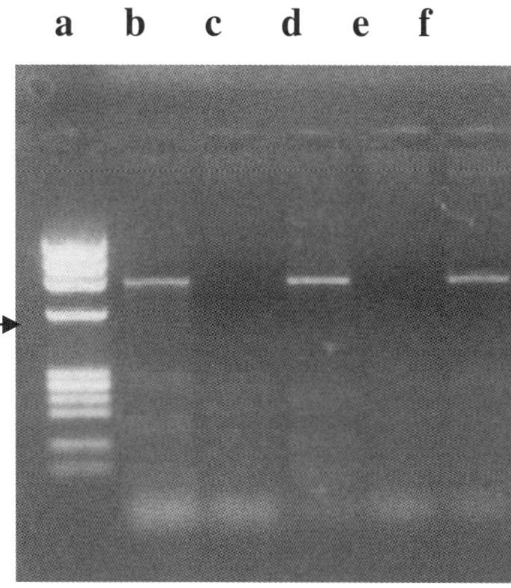

Fig.1.b. for DNA region encoding the16s rDNA at 861bp (universal bacterial DNA) in six blood samples from healthy adult volunteers, before (lanes c, e) and after artificial infection with *Staphylococcus epidermidis, P. aeruginosa* and *E.coli* (lanes b, d & f respectively).

Table (2): Utility of CRP for diagnosis of neonatal sepsis.

Blood culture	Positive CRP	Negative CRP	Total
Positive culture	49	31	80
Negative culture	15	85	100
Total	64	116	180

Sensitivity= 61.3%, Specificity= 85%
Positive predictive value= 76.5%,
Negative predictive value= 73.3%

Table (3): Utility of PCR in diagnosis of neonatal sepsis.

Blood culture	Positive PCR	Negative PCR	Total
Positive culture	80	0	80
Negative culture	7	93	100
Total	87	93	180

Sensitivity= 100%, Specificity= 93%
Positive predictive value= 91.9 %,
Negative predictive value= 100%

Table (4): Microorganisms isolated from the eighty positive blood cultures.

Organism	No.	%
Eschericia coli	55	68.7
Staphylococcus epidermidis	14	17.5
Klebsialla pneumoniae	5	6.2
Pseudomonas aeruginosa	3	3.7
Streptococcus agalactiae	3	3.7
Total	80	100

The identification of the isolated bacteria of positive blood culture revealed that, *E. coli* was the predominant bacterial species isolated from blood samples of the studied group, represent 68.7% of positive blood culture samples, followed by *S. epidermidis*, represent 17.5 %. *K. pneumoniae*, represent 6.2%, while the least common pathogens were identified as *P. aeruginosa* & *S. agalactiae*, representing 3.7 % for each (Table 4).

Discussion

Infections are the most common causes of death in neonatal period [8]. Early diagnosis and timely treatment of neonatal infections is essential [3].

In our study sepsis detection by CRP and Buffy coat method had been used but their sensitivity and negative predictive values are far from 100%. In comparison with blood culture, sepsis screen couldn't be used as a single tool for evaluation of sepsis [8-9]. Our results proved that sepsis episodes had a normal initial CRP. This can be explained by the fact that there is generally a delay of up to 24 hours between the onset of symptoms of infection and a rise in serum CRP [10]. Thus CRP should not be checked as a routine part of a septic screen in neonates.

PCR with sensitivity of 100%, specificity of 93% and 100% negative predictive value is a useful tool because of the short time required to reach confirmation. The PCR has shown 100% correlation with microbiological methods in all infected cases and was positive in seven additional cases with sterile blood cultures. The present results are in accordance with that of Laforgia *et al.* [4]. Another study illustrated the close agreement between PCR and bacterial culture in 15 of 16 culture-positive amniotic fluid samples obtained from women in premature labor [6]. With the sensitivity of 91.9%, blood culture alone would have missed another 7 cases from our studied patients. Yadav *et al.* [5] explained the low sensitivity of blood culture in their study to be related to the use of antibiotics in large number of mothers.

Concerning bacterial identification in our study, five bacterial species were identified from the eighty neonates had positive blood cultures, *E. coli* was the most predominant species, followed by *S. epidermidis, K. pneumoniae, P. eruginosa* and *S. agalactea.* In another study by Kaushik *et al.* [11], the most common organisms causing neonatal sepsis were *S. pyogenes, E.coli, Klebsiella, S. epidermidis* and *Enterobacter species. S. epidermidis* have emerged in recent years as pathogens in a growing number of serious nosocomial infections in neonatal intensive care units [12]. Gram negative organisms were more common as a whole, in a Bangladeshi study *E coli* was shown as the leading organism[13]. However, *klebsiella spp.* was the dominate organism in an Indian study [8]. These results show that every neonatal unit has their own microbial flora and require separate antimicrobial combinations, which should be modified according to culture results.

Blood culture couldn't be completely replaced by a nucleic acid amplification technique anytime nearby, as pure isolates remain essential for antimicrobial drug susceptibility testing. But PCR seems to be an excellent diagnostic alternative for a rapid mean of excluding bacterial sepsis in certain patient populations [14].

There are many protocols of PCR-based assays for detecting bacterial infection [14-15]. However, most of those approaches use primers that target specific bacterial species, so suspected bacteria causing the infection should be known in advance.

On contrary, we chose a set of universal primers that recognize a highly conserved region, within the 16S rDNA target that successfully detects the bacterial pathogens relevant to neonatal sepsis. Also the present study revealed the successful amplification of the 16 S rDNA gene from five different species of bacteria, including gram-negative and gram-positive organisms. Due to our results, unnecessary treatment with antibiotics could be avoided in 93 cases out of 180 newborns with suspected sepsis but show sterile blood culture and negative PCR. As we are keen to improve child health, it is vital to address the importance of neonatal deaths from infections. Recently, the rapid result of PCR technique in molecular testing has become more appreciated and used successfully to detect a variety of bacterial pathogens from different body sites [16-18].

We conclude that although blood culture is the gold standard for diagnosis of neonatal sepsis, PCR is superior to blood culture for early diagnosis of neonatal sepsis. Consequently unnecessary treatment with antibiotics and hospitalization could be avoided. The universal bacterial primer PCR is a useful and optimum tool for sepsis diagnosing especially where the enhanced sensitivity and rapid results outcome are overmuch the increased cost.

References

[1] Gotoff, S.P., *Chemoprophylaxis of early-onset group B streptococcal disease in 1999.* Curr Opin Pediatr, 2000. **12**(2): p. 105-10.

[2] Orogade, A.A. and R.M. Akuse, Changing patterns in sensitivity of causative organisms of septicaemia in children: the need for quinolones. Afr J Med Med Sci, 2004. **33**(1): p. 69-72.

[3] Decamp, L.R.A.F. Dempsey, and B.A. Tarini, Neonatal sepsis: looking beyond the blood culture: evaluation of a study of universal primer polymerase chain reaction for identification of neonatal sepsis. Arch Pediatr Adolesc Med, 2009. **163**(1): p. 12-4.

[4] Laforgia, N.B. CoppolaR. CarboneA. GrassiA. Mautone, and A. Iolascon, *Rapid detection of neonatal sepsis using polymerase chain reaction.* Acta Paediatr, 1997. **86**(10): p. 1097-9.

[5] Yadav, A.K.C.G. WilsonP.L. Prasad, and P.K. Menon, *Polymerase chain reaction in rapid diagnosis of neonatal sepsis.* Indian Pediatr, 2005. **42**(7): p. 681-5.

[6] Jordan, J.A.M.B. DursoA.R. ButchkoJ.G. Jones, and B.S. Brozanski, Evaluating the near-term infant for early onset sepsis: progress and challenges to consider with 16S rDNA polymerase chain reaction testing. J Mol Diagn, 2006. **8**(3): p. 357-63.

[7] Miller, M.B., *Molecular diagnosis of infectious diseases.* N C Med J, 2007. **68**(2): p. 115-8.

[8] Choudhury, S.R.R. ChadhaA. MishraV. KumarV. Singh, and N.K. Dubey, *Lung resections in children for congenital and acquired lesions.* Pediatr Surg Int, 2007. **23**(9): p. 851-9.

[9] Al-Zwaini, E.J., C-reactive protein: a useful marker for guiding duration of antibiotic therapy in suspected neonatal septicaemia? East Mediterr Health J, 2009. **15**(2): p. 269-75.

[10] Franz, A.R.G. SteinbachM. Kron, and F. Pohlandt, Reduction of unnecessary antibiotic therapy in newborn infants using interleukin-8 and C-reactive protein as markers of bacterial infections. Pediatrics, 1999. **104**(3 Pt 1): p. 447-53.

[11] Kaushik, S.L.V.R. ParmarN. GroverP.S. Grover, and R. Kaushik, *Neonatal sepsis in hospital born babies.* J Commun Dis, 1998. **30**(3): p. 147-52.

[12] Villari, P.C. Sarnataro, and L. Iacuzio, Molecular epidemiology of Staphylococcus epidermidis in a neonatal intensive care unit over a three-year period. J Clin Microbiol, 2000. **38**(5): p. 1740-6.

[13] Kumhar, G.D.V.G. Ramachandran, and P. Gupta, Bacteriological analysis of blood culture isolates from neonates in a tertiary care hospital in India. J Health Popul Nutr, 2002. **20**(4): p. 343-7.

[14] DeCamp, L.R., Universal primer polymerase chain reaction can diagnose neonatal sepsis when performed before starting antibiotics. J Pediatr, 2009. **155**(1): p. 146.7-

[15] Barclay, A.R.C.E. PaxtonP. GillettD. HooleJ. LivingstoneD. YoungG. MenonF. Munro, and D.C. Wilson, *Regionally acquired intestinal failure data suggest an underestimate in national service requirements.* Arch Dis Child, 2009. **94**(12): p. 938-43.

[16] van Haeften, R.S. PalladinoI. KayT. KeilC. Heath, and G.W. Waterer, *A quantitative LightCycler PCR to detect Streptococcus pneumoniae in blood and CSF.* Diagn Microbiol Infect Dis, 2003. **47**(2): p. 407-14.

[17] Golden, S.M.D.M. StamilioB.M. FauxW.P. dela CruzC.T. ShoemakerC.L. BlackmonS.D. StassenV.M. ClarkJ.W. Smith, and O.L. Johnson, *Evaluation of a real-time fluorescent PCR assay for rapid detection of Group B Streptococci in neonatal blood.* Diagn Microbiol Infect Dis, 2004. **50**(1): p. 7-13.

[18] Makhoul, I.R.T. SmolkinP. SujovI. KassisA. TamirR. Shalginov, and H. Sprecher, *PCR-based diagnosis of neonatal staphylococcal bacteremias.* J Clin Microbiol, 2005. **43**(9): p. 4823-5.

Shelf life extension of semi-dried fish Nile Tilapia (*Oreochromis niloticus*) by ultraviolet and infrared irradiations

Narumol Matan*

Food Technology, School of Agricularal Technology, Walailak University, Nakon Si Thammarat 80160, Thailand

*Corresponding author: e-mail: nnarumol@wu.ac.th, Phone: +66 75672359

Shelf life extension of semi-dried fish, Nile Tilapia (*Oreochromis niloticus*), was investigated by means of ultraviolet (UV-C) irradiation, infrared (IR) spectrum or a combined UV and IR irradiations for 5, 10 and 15 min. Microbial qualities (i.e. total bacteria, total yeast and mold, *Escherichia coli*, *Staphylococcus aureus*) on the semi-dried fish were examined during incubation at 4 °C for up to 21 days. Quantitative descriptive analysis (QDA) was also employed to quantify various sensorial attributes (i.e. color, appearance, odor, hardness, taste and texture) of the semi-dried fish product during storage. It was found that the combined irradiation of UV for 7.5 min followed by IR for 7.5 min successfully extended the shelf life of the semi-dried fish up to 21 days. Pathogens on the semi-dried fish were reduced to an acceptable level. All sensorial attributes of the semi-dried fish after 21 days storage was decreaed at the levels obtained just after irradiation except appearance. These findings illustrate a possibility of using the combined irradiation of UV-C and IR to improve the quality of the semi-dried fish to the required shelf life.

Keywords semi-dried fish; Nile Tilapia (*Oreochromis niloticus*); ultraviolet; infrared; shelf life; sensory; microbial qualities

1. Introduction

Semi-dried fish is one of the most serious sources of *Escherichia coli*, *Staphylococcus aureus* induced disease in humans [1-2]. Mishandled or undercooked fish with pathogens can cause fatal illnesses such as diarrheal [2]. Therefore, semi dried fish processing procedures must consider means of decreasing contamination by foodborne pathogens [3]. Since dried fish is normally dried by sun drying before packaging, microbial cross-contamination is therefore inevitable. Microbiological performance standards of the dried fish product should include the reduction of external infection with pathogens by surface decontamination techniques [4]. In order to improve the safety of dried fish products, various processing methods such as sorbic acid have been applied for the reduction of microbial counts [5]. However, these processes may change the texture and nutritional components of the dried fish, and the chemical elements can also remain on the surface and cause health problems [6].

Ultra-violet (UV) and infrared (IR) radiations have numerous advantages over several existing sanitation methods. They do not require chemicals or heat, and are not expensive. Presently, UV radiation technology is used as an alternative to chemical sterilization in food products [7]. UV-C (wavelength of 220–300 nm with 90% of emission at 253.7 nm) in particular has been approved by the FDA for use on food products to control surface microorganisms [8]. Inactivation study using the UV-C with wavelength of 253.7 nm has also been shown to be effective in reducing pathogenic bacteria in vegetables and fruits [8]. IR radiation has been widely used as an heat source to dry various kinds of grains and cereal products in food industries. During irradiation, thermal energy of IR is absorbed on the food surfaces with limited penetration capability because thermal conductivity of most solid foods are generally low. Heating is therefore confined on the food surfaces. This local heating should reduce foodborne pathogens contaminating on the food surfaces through the pasteurization process. This study intended to examine the effectiveness of UV-C and IR irradiations on the inactivation of foodborne pathogens in the semi-dried fish of Nile Tilapia (*Oreochromis niloticus*) and their qualities upon storage.

2. Material and Methods

2.1 Preparation of samples

Nile Tilapia (*Oreochromis niloticus*) were obtained from a local fish market in Thasala district, Nakhon Si Thammarat, Thailand. Fish were headed and gutted, then cut into two pieces. The fish were pickle salted for 30 minutes in a clean, lid-covered plastic box previously washed with 70% ethanol. Microwave drying was performed in a laboratory microwave oven (ThEP center, Walailak University) at 200 W for 10 hrs. The final water activity of dried samples was controlled at 0.84.

2.2 Ultraviolet (UV) and infrared (IR) irradiation

UV-C irradiation was performed using the Phillips UV-C 30W lamp (Phillips, Thailand). The dried fish were placed on a stainless-steel tray and irradiated with the UV-C lamp on the upper surfaces at a distance of 30 cm at different exposure times of 5, 10, and 15 min. The IR heating apparatus was designed for surface pasteurization using two ceramic emitters (60×245 mm) 425 watt The dried fish sample were put on a stainless-steel tray before irradiating with IR at a distance of 30 cm for 5, 10 and 15 min. Combined effect of UV-C and IR were also performed by irradiation of UV-C followed by IR at 5min (2.5min UV and 2.5min IR), 10min (5.0min UV and 5min IR), and 15min (7.5min UV and 7.5min IR). After irradiation, the samples were individually packed using polyethylene terephthalate containers and stored at 4 ± 1 °C for 21 days.

2.3 Microbiological evaluation

The dried fish (25 g) was removed every 7 days until 21 days using a sterile scalpel and mixed with 225 mL of peptone water in a sterile stomacher bag. The samples were then homogenized using a stomacher (MIX 2, AES Laboratoire; Combourg, France) for 3 min, dried fish through sterile cheesecloth, and diluted with peptone water to determine the microbial count. Serial dilutions were performed on the agar plates. Total bacteria, total yeast and mold, *Escherichia coli*, *Staphylococcus aureus* counts were determined by using Compact Dry TC, YM, EC and SA (Nissui Co. Ltd, Japan), respectively. Microbial count determinations were performed in three replications and expressed as log CFU/g.

2.4 Sensory evaluation

The dried fish samples treated with UV for 7.5 min followed by IR for 7.5 min were subjected to sensory evaluation test. After irradiation, the samples were individually packed and stored at 4 ± 1 °C for 21 days. Sensory test was carried out every week until 21 days. The semi dried fish were fired in vegetable oil for 1-2 minutes and served to panellists immediately. Trained quantitative descriptive analysis (QDA) panels of 10 were used. Panellists were selected from students and staff at Walailak University, Nakon Si Tammarat, Thailand. QDA screening consisted of 5 triangle and 5 duo trio tests carried out in duplicate to assess individual ability to perceive product differences. These screening procedures were specific to the product category under investigation. Assessors who discriminated correctly in at least 65% of the tests were included on the QDA panels. The training programme aimed to develop and ensure competent usage of sensory descriptors by the panels. A total of 6 attributes (color, appearance, odor, hardness, taste and texture) were developed and agreed by the panels. Each agreed sensory descriptor was rated on a 15 cm continuous line scale with anchor points at each end.

3. Results and Discussion

3.1 Microbial growth on the semi-dried fish

Growth of bacteria, yeast and mold on the dried fish treated with UV and/or IR at various conditions during incubation at 4°C for 21 days are shown in Figures 1a and 1b. According to the Thai Industrial Standard for semi-dried fish [11], the total bacteria and total yeast and mold on the dried fish surface must be $\leq 10^5$ and $\leq 10^2$, respectively. The combined effect of UV-C and IR at 15min (7.5min UV-C and 7.5min IR) was effective to preserve the semi-dried fish from bacteria, yeast and mold for up to 21 days. Semi-dried fish preserved under other conditions examined including the controls were found to be spoilage within 3 days.

Both *E. coli* and *S. aureus* on the semi-dried fish treated with UV-C and IR at all conditions were not detected upon storage for up to 21 days (Table 1). It is possible that *E. coli* and *S. aureus* would not be contaminated in this products before testing because of the sanitaly technique performed in the laboratory such as fish washing. In addition, drying of the fish used within this work was carried out in a closed system within the microwave oven. As a result, contanmination of both bacteria was then avioded.

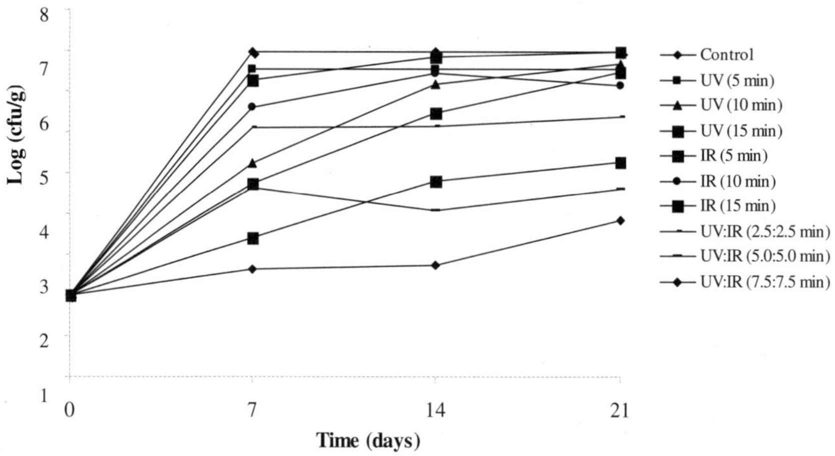

(a)

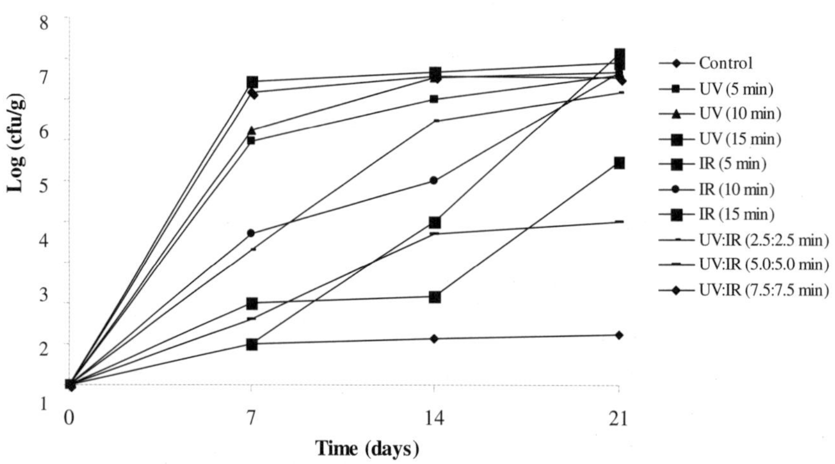

(b)

Fig.1 Effect of ultraviolet (UV) irradiation, infrared (IR) spectrum on (a) the bacterial population (CFU/g) and (b) the yeast and mold population (CFU/g) of semi-dried fish during storage at 4°C.

From the results, it is clear that the combination of UV-C and IR radiation for 15 min was sufficient to inactivate growth of bacteria, yeast and mold on semi-dried fish for at least 21 days. Inactivation of bacteria, yeast and mold by UV radiation were also reported by various authors [12-14]. UV-C has a shorter wavelength, therefore has higher energy and greater penetration depth into cell than UV-A and UV-B. The most severe cell damage has been reported to occur when nucleic acids absorb UV-C light, which can result in crosslinked DNA [15]. However, the UV raditaiton has been known to be lethal and mutagenic in a variety of organisms, inculding bacteria [16]. Therefore, bacteria muation on dried fish products after UV treatment should be examined in the future. Local surface heating as a result of IR irradiation would also raise the surface temperature to a lethal degree to microorganisms. This can be achieved without causing the interior to substantially increase in temperature [10]. The ongoing research is focused on investigating the mode of action of the combined UV-C and IR irradiations on cell of bacteria.

3.2 Sensory analysis

Significant differences were observed in the sensory profiles of the semi-dried fish during storage in comparison with those at the beginning of storage (Figure 2). The semi-dried fish just after the treatment had strong taste and odor of fresh fish with soft texture. It appears that all sensorial attributes decreased duirng 21 days storage except the appearance. Further work on scavenging effects of odor (off-flavor) by UV and IR irradiations is required.

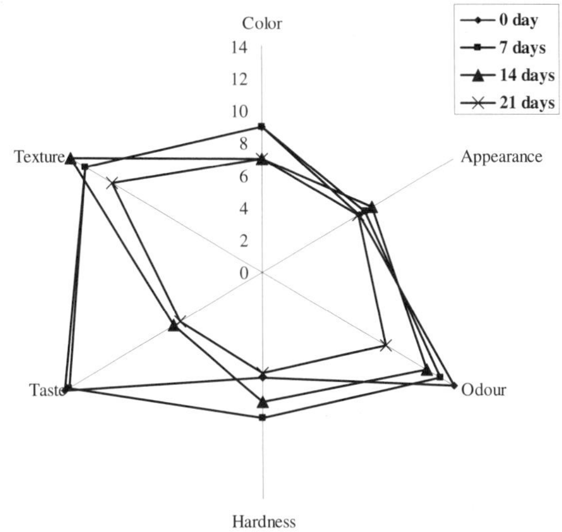

Fig.2 Sensory profiles for semi-dried fish at 0, 7, 14 and 21 days presented as a spider diagram.

4. Conclusion

The pasteurization and quality improvement of semi-dried fish (Nile Tilapia; *Oreochromis niloticus*) by the combination of UV-C and IR irradiations for 15 min (7.5 min UV-C and 7.5 min IR) was capable of reducing total bacteria and total yeast and mold contaminated on the semi-dried fish within the accepted standard for at least 21 days. The results indicated that combination between UV and IR irradiation can be used for ensuring hygienic quality of semi-dried fish.

Acknowledgements This study was supported by the Research and Innovation Network for Technology Transfer to Community, Commission on Higher Education Upper Southern, Walailak University, THAILAND.

References

[1] Olsvik Ø, Wasteson Y, Lund A, Hornes E. Pathogenic *Escherichia coli* found in food. *International Journal of Food Microbiology*. 1991;12(1):103-113.
[2] Huong BTM, Mahmud ZH, Neogi SB, Kassu A, Nhien NV, Mohammad A, Yamato M, Ota F, Lam NT, Dao HTA, Khan NC. Toxigenicity and genetic diversity of *Staphylococcus aureus* isolated from Vietnamese ready-to-eat. *Food Control*. 2010;21(2):166-171.
[3] Jackson V, Blair IS, McDowell DA, Kennedy J, Bolton DJ. The incidence of significant foodborne pathogens in domestic refrigerators. *Food Control*. 2007;18(4):346-351.
[4] Jo C, Lee NY, Kang HJ, Shin DH, Byun MW. Inactivation of foodborne pathogens in marinated beef rib by ionizing radiation. *Food Microbiology*. 2004;21(5):543-548.
[5] Guillard V, Issoupov V, Redl A, Gontard N. Food preservative content reduction by controlling sorbic acid release from a superficial coating. *Innovative Food Science & Emerging Technologies*. 2009; 10(1):108-115.
[6] Shtenberg AJ, Ignat'ev AD. Toxicological evaluation of some combinations of food preservatives. *Food and Cosmetics Toxicology*. 1970;8(4):369-380.
[7] Chun HH, Kim JY, Lee BD, Yu DJ, Song KB. Effect of UV-C irradiation on the inactivation of inoculated pathogens and quality of chicken breasts during storage. *Food Control*. 2010;21(3):276-280.
[8] Gil MI, Selma MV, López-Gálvez F, Allende A. Fresh-cut product sanitation and wash water disinfection: Problems and solutions. *International Journal of Food Mcirobiology*. 2009;134(1-2): 37-45.
[9] Nowak D, Lewicki PP. Infrared drying of apple slices. *Innovative Food Science & Emerging Technologies*. 2004;5(3):353-360.
[10] Huang L. Infrared surface pasteurization of Turkey frankfurters. *Innovative Food Science & Emerging Technologies*. 2004;5(3):345-351.
[11] Thai Industrial Standard. Dried fish 1204-2536. 2536.

[12] Chun H, Kim J, Chung K, Won M, Song KM. Inactivation kinetics of *Listeria monocytogenes, Salmonella enteric* serovar Typhimurium, and *Campylobacter jejuni* in ready-to-eat sliced ham using UV-C irradiation. *Meat Science.* 2009;83(4):599-603.

[13] Cia P, Pascholati SF, Benato EA, Camili EC, Santos CA. Effects of gamma and UV-C irradiation on the postharvest control of papaya anthracnose. *Postharvest Biology and Technology.* 2007;43(3): 366-373.

[14] Stevens C, Khan VA, Lu JY, Wilson CL, Pusey PL, Kabwe MK, Igwegbe ECK, Chalutz E, Droby S. The germicidal and hermetic effects of UV-C light on reducing brown rot disease and yeast microflora of peaches. *Crop Protection.* 1998;17(1): 75-84.

[15] Schenk M, Raffellini S, Guerrero S, Blanco GA, Alzamora SM. Inactivation of *Escherichia coli, Listeria innocua* and *Saccharomyces cerevisiae* by UV-C light: Study of cell injury by flow cytometry. *Food Science and Technology.* 2011;44, (1):191-198.

[16] Wehner J, Horneck G. Effects of vacuum UV and UVC radiation on dry Escherichia coli plasmid pUC19 II. Mutational specificity at the *lac*Zgene. *Journal of Photochemistry and Photobiology B: Biology*, 1995;30(2-3):171-177.

Study on the Effect of RF Plasma Pretreatment on the Antimicrobial Efficacy of Neem Leaf Extract Processed Cotton Fabric

K. Vaideki[1*], S. Jayakumar[1], E. Nithya[1] and R. Rajendran[2]

[1] Department of Physics, PSG College of Technology, Coimbatore, India
[2] Department of Microbiology, PSG College of Arts and Science, Coimbatore, India

*Corresponding author: vaidhu1@yahoo.com; Phone: +91422-4344453

In this paper, a detailed investigation on the effect of RF plasma treatment in improving the hydrophilicity of cotton fabric has been carried out. The hydrophilicity of plasma treated samples has been assessed and the process parameters have been optimized for maximum hydrophilicity. The fabric samples that exhibited maximum hydrophilicity have been subjected to chemical and ATR-FTIR analysis to study the reaction mechanism that has occurred during plasma treatment. SEM analysis has been carried out to understand the morphological changes that occurred during plasma treatment. The effect of plasma treatment on the mean pore diameter of the fabric matrix has been analyzed using dynamic wicking test and air permeability test. An attempt has been made to understand the role of plasma treatment in improving the antimicrobial efficacy of this fabric when treated with methanolic extract of neem leaves. FTIR analysis of the antimicrobial finish treated samples has been carried out to understand the improvement in efficacy and durability.

Keywords: Cotton fabric, Plasma pretreatment, Hydrophilicity, Antimicrobial finish, Methanolic extract of neem leaves, Chemical analysis, ATR-FTIR spectrum, , SEM micrographs

1. Introduction

Research on neem tree by various groups has indicated that all the parts of neem tree (bark, leaves, seed etc.,) contains a plethora of triterpeniods like azadirachtin, nimbin, nimbidin, nimbolide, gedunin etc., which are bioactive [1]. Extracts of parts of neem trees are used as antimicrobial finishes for textiles [2,3,4]. To improve the antimicrobial efficacy and durability of these finishes, using suitable pretreatment techniques viz., chemical, physicochemical and biochemical, certain surface properties of the fabric have to be altered [5]. Plasma treatment of textile materials is one of physiochemical technique which is highly effective in altering the surface properties without affecting the bulk properties and most importantly it is an ecofriendly process as toxic effluents are not let out during the process [6,7]. In the present study, 20 count 100% pure finished cotton fabric used in the surgical zone has been treated with RF plasma using air, oxygen and argon as process gas to improve its hydrophilicity prior to the application of methanolic extract of neem leaves. The hydrophilicity and antimicrobial efficacy are assessed using standard tests.

2. Experimental Work

2.1 Plasma treatment

The fabric material considered for the present study is a 20 count finished cotton fabric (100% pure) with an EPI/PPI (Ends per inch / Picks per inch) ratio equal to 130/140 of size 10cm x 10cm. The fabric sample has been treated with RF plasma using RF plasma chamber of frequency 13.56 MHz. The process gases used to produce plasma are commercial grade air with a relative humidity of 48%, 99.99% pure oxygen and argon. It is to be noted that for all the plasma treatments the base pressure was maintained at 5×10^{-5} mbar. As the plasma is distributed uniformly within the space between the two electrodes in RF plasma, the sample was placed on the lower electrode. The process parameters such as process gas pressure, electrode gap, RF power and exposure time has been varied.

2.2 Assessment of hydrophilicity

The standard static immersion test (BS 3449) [8] and wettability test [8] were used to assess the hydrophilicity of the sample.

2.3 Estimation of carboxyl group concentration in cotton fibres

Sodium Bicarbonate-Sodium Chloride test is a TAPPI (Technical Association of Pulp and Paper Industry) standard test [9, 10], to determine the concentration of carboxyl group in the cotton fibers.

2.4 Carbonyl group identification

Colorimetric test was carried out to identify the carbonyl group [10,11].

2.5 Estimation of mean pore diameter of the fabric

Dynamic wicking test has been carried out to determine the mean pore diameter of the cotton fabric [12,13].

2.6 Antimicrobial finish treatment

The methanolic extract of the neem leaves was prepared at a concentration of 50 g/l (grams per liter) with M:L (material : liquid in g:ml) of 1:20. The fabric sample was immersed in the solution for 30 minutes and was padded between the rollers. It was cured at a temperature of 500C for 30 minutes and then dried. The fabric was given a post treatment with citric acid of 8% OWF (on weight of the fabric), at 500C for 50 minutes. Citric acid is used as a cross linking agent to increase the durability of the antimicrobial finish. The sample was then dried.

2.7 Assessment of antimicrobial activity

The antimicrobial activity of the sample was studied by performing Agar Diffusion test (SN 195920) [14] and modified Hohenstein test (JIS L 1902) [14].

3. Results and Discussion

3.1 Effect of RF plasma treatment on the hydrophilicity of cotton fabric

During oxygen plasma treatment, the RF power is fixed as 40 W (power density: 2.04 W/cm^2). A low power density is preferred so as to avoid the possibility of sputtering of the stainless steel electrode. While treating the fabric sample with RF air plasma, the gas pressure is fixed as 0.5 mbar, which is the highest possible pressure for which we are able to strike and sustain the plasma. Beyond this pressure the discharge could not be sustained. In both these treatments, the other three process parameters have been varied and optimized to yield highest hydrophilicity based on the results of static immersion test. The RF argon plasma treatment involved the variation in all the four parameters and has been optimized to give maximum hydrophilicity based on the results of wetting time test. The parameters thus optimized are presented in Table 1.

Table 1 Optimized process parameters

RF Plasma Treatment	**Optimized process parameter**			
	RF power (W)	**Process gas pressure(mbar)**	**Electrode gap(cm)**	**Plasma exposure time (minutes)**
Oxygen plasma	40	0.06	3	10
Air plasma	50	0.5	3	10
Argon plasma	50	0.5	2	5

3.2 Reaction Mechanism Due to Plasma Treatment - ATR-FTIR Analysis

The ATR-FTIR spectra of the untreated and plasma treated samples are recorded to analyze the nature of chemical reaction that has taken place on the fabric surface. The spectra of untreated and plasma treated cotton fabric are presented in Figure 3.1 and Figure 3.2.

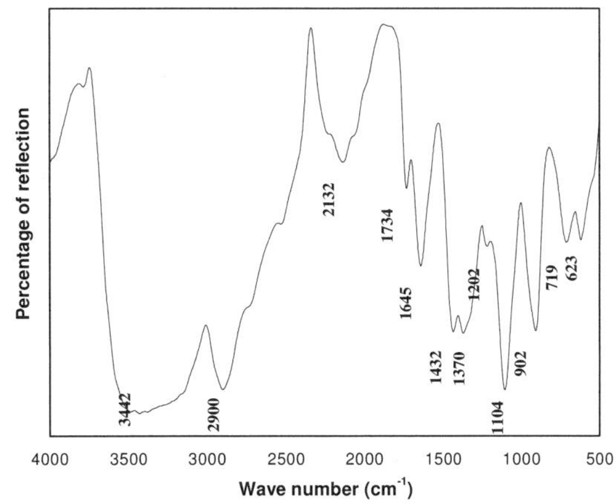

Figure 3.1 ATR-FTIR spectrum of cotton fabric

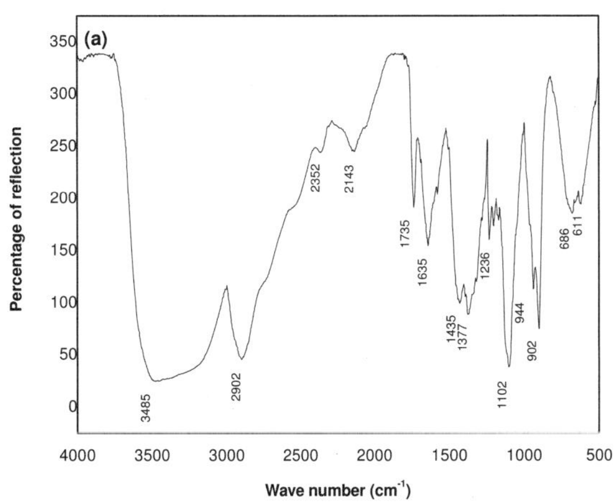

(a) Spectrum of RF oxygen plasma treated fabric

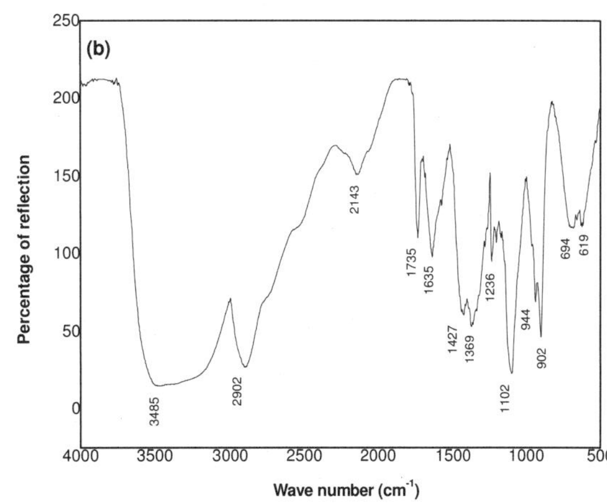

(b) Spectrum of RF air plasma treated fabric

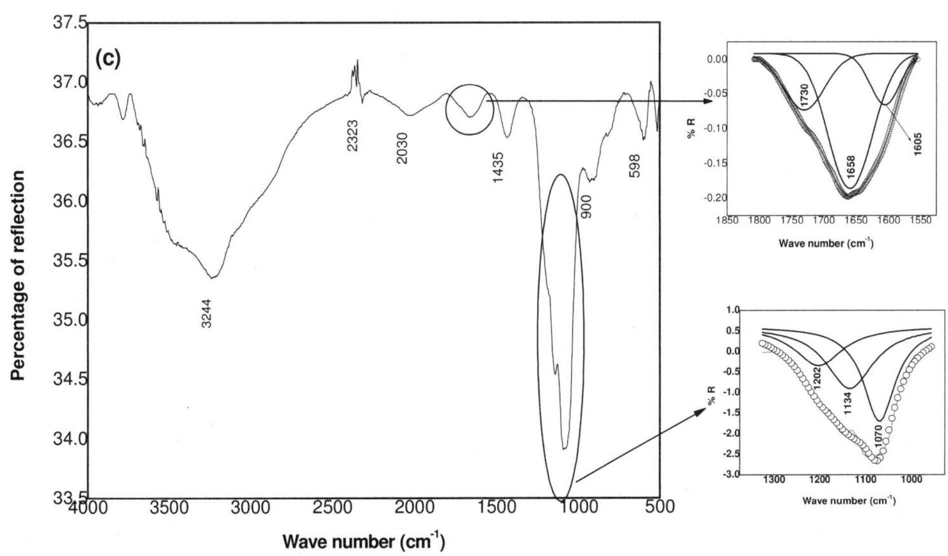

(c) Spectrum of RF Ar plasma treated fabric

Figure 3.2 ATR-FTIR spectra of RF plasma treated cotton fabric

The peaks present in the spectra (Figure 3.1 and Figure 3.2) confirm the fact that the constituents present in the cotton fabric before plasma treatment are also present in the fabric after the plasma treatment. In Figure 3.12 (c), a broad peak has been observed at 1650 cm^{-1} and 1070 cm^{-1}. In order to under stand the reaction mechanism, these peaks have been deconvoluted using Lorentizian fit. Three peaks have been identified at 1730 cm^{-1}, 1658

cm^{-1} and 1605 cm^{-1} for the broad peak at 1650 cm^{-1} which is attributed to C=O stretching and OH stretching. Similarly, three peaks have been identified at 1202 cm^{-1}, 1134 cm^{-1} and 1070 cm^{-1} for the broad peak at 1070 cm^{-1} which is attributed to OH bending and asymmetric ring stretch. The increase in hydrophilicity indicates that polar functional groups must have been introduced on the fabric surface during the plasma treatment. The hydroxyl (-OH) is the polar group present in the structure of cellulose (figure 3.3). But few hydroxyl group are converted to carboxyl (HO-C=O) group during the growth of the plant(C=O stretching at 1734cm^{-1} in Figure 3.1)

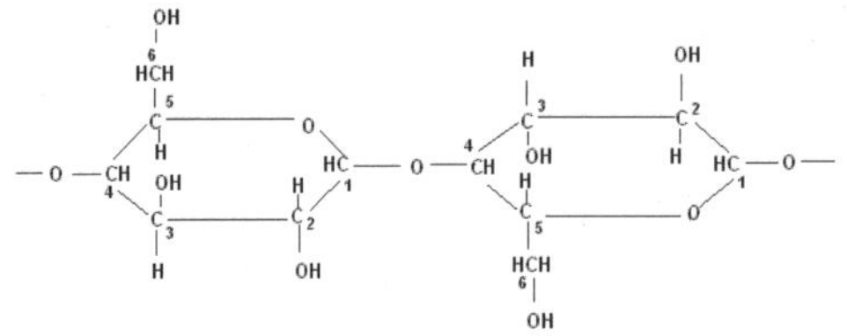

Figure 3.3 Anhydroglucose units linked by glycosidic bonds as in cellulose

It has been reported by Ward et al [15] that different cellulosic radicals are formed on the fabric surface based on the five possible reaction mechanisms that can occur during plasma treatment. The tests for hydrophilicity substantiate that the hydroxyl groups must have been converted to carboxyl group during the plasma treatment. This can be understood by analyzing the reaction mechanism that has occurred during the plasma treatment. The presence of peaks near 898 cm^{-1} and 1178 cm^{-1} confirms that plasma treatment has not affected the β glucosidic bond (C_1-O-C_4) and the cyclic ether bond (C_5-O-C_1). This establishes the fact that the reaction mechanisms resulting in the bond breakage between C_1 and glucosidic bond oxygen, bond breakage between C_1 and ring oxygen and dehydrogenation and dehydroxylation between C_2 and C_3 after ring opening of anhydroglucose has not taken place during the plasma treatment. Moreover the mechanism resulting in dehydroxylation at C_6 cannot result in the formation of –COOH. As a result plasma species that has impinged on the surface of the cotton fabric has resulted in the formation of cellulosic radical by dehydrogenation at C_6.

These cellulosic radicals on further reaction with the oxygen ions present in the plasma get oxidized to carboxylic acid. The oxidation of hydroxyl to carboxylic acid can be confirmed by the colorimetric test and sodium bicarbonate - sodium chloride test the results of which are presented in Table 2.

Table 2 Identification of aldehyde group and estimation of carboxyl content in the untreated and RF oxygen/air/argon plasma treated cotton fabric

S.No	Sample	Inference of colorimetric test (carbonyl content)	Carboxyl content (meq / 100 g)
1.	Untreated cotton fabric	Traces of red precipitate	2.49
2.	Cotton fabric treated with RF oxygen plasma using optimized process parameters	Traces of red precipitate, but lesser than that of untreated fabric	3.72
3.	Cotton fabric treated with RF air plasma using optimized process parameters	Traces of red precipitate, but greater than that of untreated fabric	3.6
4.	Cotton fabric treated with RF argon plasma using optimized process parameters	Traces of red precipitate are not observed.	2.8

Table 2 confirms the fact that the hydroxyl group present in the cellulose has been oxidized to carboxyl group via the formation of cellulosic radical by dehydrogenation at C_6. The data presented in Table 2 proves that the aldehyde formation is much pronounced when the fabric is treated with RF air plasma than oxygen plasma. This is because effective oxidation (conversion of carbonyl to carboxyl group) has taken place in the case of RF oxygen plasma treatment than air plasma treatment. The fact that the concentration of carboxyl content is more in the RF oxygen treated cotton fabric is validated from the data presented in table 2. This is attributed to the higher concentration of oxygen ions present in the oxygen (O_2^+, O^+ and O^- [16] plasma rather than in air (O_2^+, O^+, O^-, N_2^+ and NO^+ [17] and argon (Ar^+, Ar^{++}, Ar_2^+, ArH^+, H_3O^+ [18,19] plasma. This justifies that the hydrophilicity is improved to a greater extent in the RF oxygen plasma treated cotton fabric rather than the RF air and argon plasma treated cotton fabric.

3.3 Surface Morphology- SEM Analysis

The hydrophilicity of a fabric can also be measured in terms of its wicking ability. The micropores are responsible for the wickability of a fabric. To analyze whether the plasma treatment has any effect on the micropores present in the fabric matrix, the surface morphology of the untreated and plasma treated fabric samples are studied using scanning electron microscope. Figures 3.4 and 3.5 shows the SEM micrographs of untreated and RF oxygen, air and argon plasma treated cotton fabric with a magnification of 100 x and 500 x

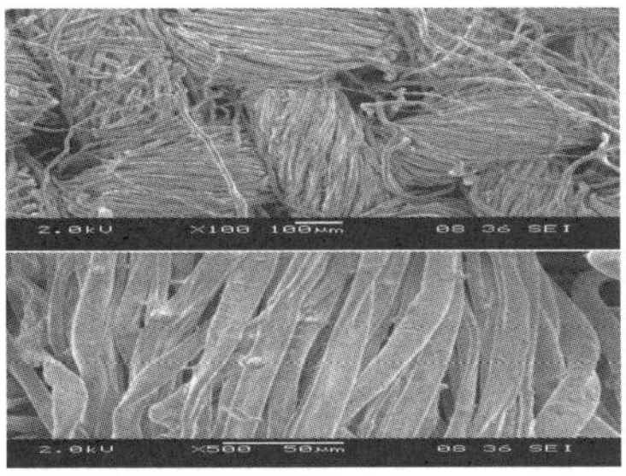

Figure 3.4 SEM micrographs of cotton fabric

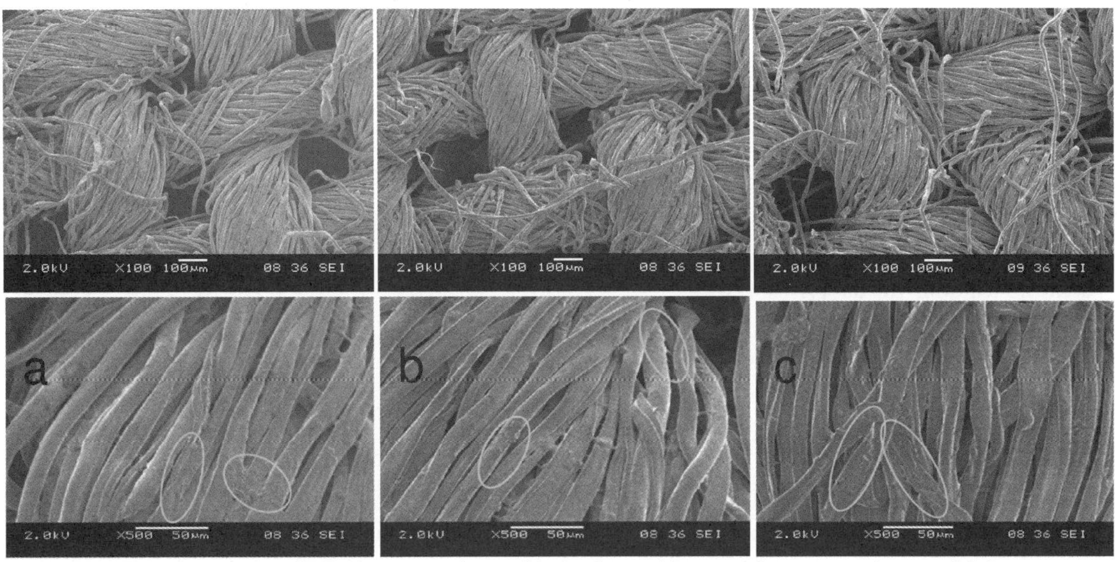

Figure 3.5 SEM micrographs of RF (a) oxygen plasma (b) air plasma (c) argon plasma treated cotton fabric

SEM micrographs reveal that the void between the warp and the weft has increased by the etching of hair like protruding structures on the surface. Figure 3.4 which shows traces of starch even after the RF argon plasma treatment. The effect of plasma treatment on the morphology of the fabric has also been justified by performing the dynamic wicking test and air permeability test using Kawabata air permeability tester and the results are tabulated in Table 3.

Table 3 Mean pore radius and air permeability of untreated and RF oxygen/air/argon plasma treated cotton fabric

S.No	Sample	Mean pore radius (μm)	Air permeability (KPa-s/m)
1.	Untreated cotton fabric	0.31	0.212
2.	Cotton fabric treated with RF oxygen plasma using optimized process parameters.	0.52	0.132
3.	Cotton fabric treated with RF air plasma using optimized process parameters.	0.51	0.134
4.	Cotton fabric treated with RF Ar plasma using optimized process parameters.	0.36	0.177

The results confirm the fact that the bombardment of plasma particles on the surface of the fabric removes the hair like protruding structure and increases the mean diameter of the pores which in turn improves the wickability of the fabric. From the data presented in the Table 3 the mean diameter of the pores is almost the same for RF oxygen and air plasma treated cotton fabric and is more when compared with the RF argon plasma treated cotton fabric which is ascribed to the extent of etching. On comparing the RF oxygen and air plasma treatment, as the extent of etching is the same, the contribution to the increment in hydrophilicity is more due to chemical reaction in the case of RF oxygen plasma treatment. This is attributed to the higher concentration of oxygen ions in the oxygen plasma. The above investigations prove that the hydrophilicity of the fabric has been improved by RF plasma treatment by chemical reactions as well as by etching process and the maximum hydrophilicity of the fabric has been exhibited by the RF oxygen plasma treated cotton fabric.

3.4 Antimicrobial Activity of the RF Plasma and Neem Leaf Extract Treated Cotton Fabric

The RF oxygen, air and argon plasma treated fabric sample (using the optimized system process parameters) are treated with the antimicrobial finish and the efficacy is assessed using agar diffusion test. The results are tabulated in Table 4.

Table 4 Antimicrobial activity-Agar diffusion test & Modified Hohenstein test

S. No	Sample	Agar diffusion test- Zone of inhibition (mm)				Modified Hohenstein test- Percentage reduction(%)	
		Test Bacteria		Test Fungi			
		S.aureus	*E.coli*	*Penicilium funiculosum*	*Trichoderma riridea*	*S.aureus*	*E.coli*
1.	Fabric	0	0	0	0	0.0	0.0
2.	Fabric treated with the antimicrobial finish	5.4	5.1	5	3.9	97.5	96.0
3.	Fabric treated with RF oxygen plasma using optimized process parameters and antimicrobial finish	7.2	7.0	6.1	4.4	100.0	100.0
4	Fabric treated with RF air plasma using optimized process parameters and antimicrobial finish	6.9	6.5	5.9	4.2	100.0	99.0
5.	Fabric treated with RF argon plasma using optimized process parameters and antimicrobial finish	6.0	5.9	5.5	4.0	98.6	98.0

Table 4 reveals that the RF oxygen plasma treated cotton fabric because of better hydrophilicity has absorbed more neem leaf extract and hence is more active against the microbes than the RF air and argon plasma pre treated fabric.

3.5 Wash Durability

It is important to assess the durability of a functional finish from the commercial perspective. So, the sample that yielded the maximum antimicrobial activity is subjected to wash test. The antimicrobial activity up to 50 washes has been assessed using challenge test. The wash test result is provided in Table 5.

Table 5 Wash test result

No.of laundering cycles	% Bacterial Reduction			
	Cotton fabric treated with neem leaf extract and citric acid		Cotton fabric treated with RF oxygen plasma, neem leaf extract and citric acid	
	S. aureus	*E. coli*	*S. aureus*	*E. coli*
0	97.5	96.0	100.0	100.0
10	96.9	91.4	100.0	98.6
20	94.3	84.2	97.2	93.6
30	74.3	71.6	94.3	86.3
40	54.0	49.0	74.6	66.2
50	40.8	32.0	63.3	49.1

An attempt has been made to understand the reaction mechanism between the plasma treated cotton fabric, neem leaf extract and citric acid. The new functional groups formed as a result of these reactions have been identified using FTIR spectra of neem leaf extract, citric acid, untreated cotton fabric and RF oxygen plasma, neem leaf extract and citric acid treated cotton fabric recorded in the transmittance mode (Figure 3.6, Figure 3.7, Figure 3.8 and Figure 3.9).

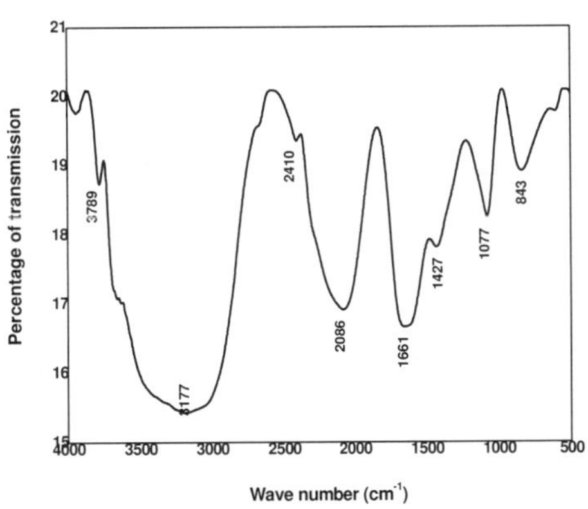

Figure 3.6 FTIR spectrum of neem leaf extract

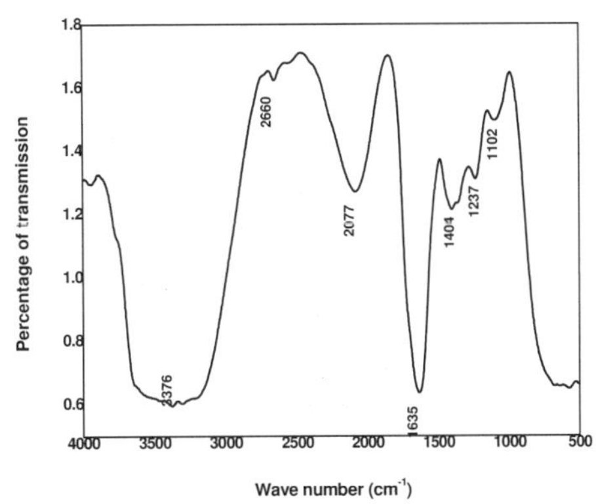

Figure 3.7 FTIR Spectrum of neem leaf extract with 8% citric acid

The FTIR spectrum of neem leaf extract is shown in Figure 3.6. As it has been already been proved using HPLC chromatograph [4], that the methanolic extract of the neem leaves contain azadirachtin, the peaks present in FTIR spectrum of the extract correspond to the functional groups present in the azadirachtin. The reaction between neem leaf extract and the citric acid can be understood by comparing the FTIR spectra of neem leaf extract (azadirachtin) and azadirachtin with citric acid (Figure 3.7). The hydroxyl group in citric acid reacts with hydroxyl group at any one of C_7, C_{11} or C_{20} sites of azadirachtin [20] present in the neem leaf extract to form an ether which is evident from the peaks at 1237 cm^{-1} and 1102 cm^{-1} (Figure 3.7) that correspond to C-O stretching in alkyl aryl ethers. These peaks are absent in the FTIR spectrum of neem leaf extract (Figure 3.6).The reaction between the oxidized cellulose (plasma treated fabric) and the antimicrobial finish can be understood by comparing the FTIR spectrum of untreated fabric (Figure 3.8) and fabric treated with RF oxygen plasma using optimized process parameters and the antimicrobial finish (Figure 3.9).

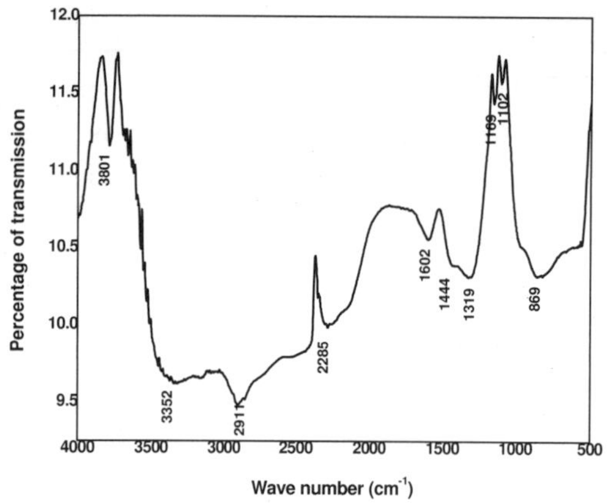

Figure 3.8 FTIR spectrum of untreated fabric RF O_2

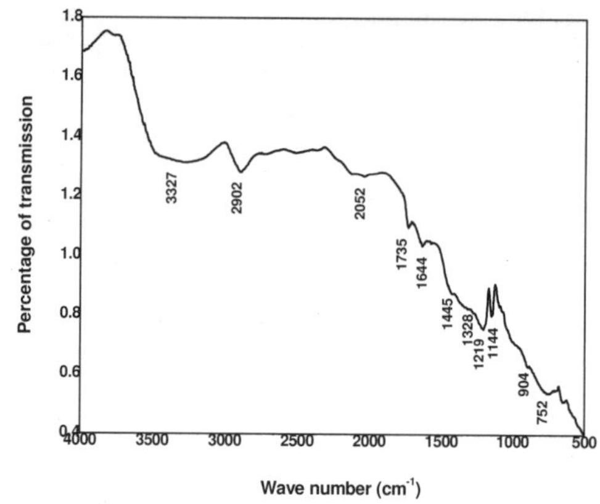

Figure 3.9 FTIR spectrum of the fabric treated with plasma, neem leaf extract and citric acid

On comparing the two spectrum (Figure 3.8 and Figure 3.9) it is observed that the peak at 1735 cm^{-1} corresponding to C=O stretching and at 1219 cm^{-1} corresponding to C-O stretching are the new peaks formed after the RF oxygen plasma and antimicrobial finish treatment. This may be attributed to the formation of either ester or acid anhydride due to the reaction between the carboxyl group of citric acid and oxidized cellulose. The bonds thus formed are less water soluble, thereby increasing the durability of the antimicrobial activity of the fabric.

4. Conclusion

The cotton fabric when treated with RF oxygen, air and argon plasma showed improved hydrophilicity. The reaction mechanism leading to this has been analyzed using colorimetric test, sodium bicarbonate- sodium chloride test, ATR-FTIR analysis. The results of the tests show that cellulosic radical has been formed during the plasma treatment due to dehydrogenation at C_6. This reactive radical on further reaction with plasma particles gets converted to carboxyl group which is more polar than hydroxyl group. The SEM analysis, dynamic wicking and air permeability tests confirm the increase in mean diameter of the pores due to etching which enhances the wick ability of the fabric. The synergy of increased carboxyl group and increase in mean pore diameter enhances the hydrophilicity which in turn enhances the neem leaf extract uptake and so the antimicrobial efficacy is increased. The entire investigations reveal that the fabric treated with RF oxygen plasma prior to the processing with neem leaf extract showed the highest efficacy than the other samples. This sample when subject to wash test showed that it is able to withstand more than 50% of the activity for about 50 washes.

AcknowledgementsThe authors wish to thank the Management and Principal of PSG College of Technology and PSG College of Arts and Science, Coimbatore for providing the necessary infrastructure.

References

[1] Butterworth J.H. and Morgan E.D. Investigation of the Locust Feeding Inhibition of the Seeds of the Neem Tree, Azadirachta Indica, *Journal of Insect Physiology*, 1971; 17:969-977.

[2] Vaideki K, Jayakumar S, Rajendran R, Thilagavathi G. A study on the antimicrobial efficacy of RF oxygen plasma and neem extract treated cotton fabrics. *Applied Surface Science*,2007; 253:7323-7329.

[3] Vaideki K, Jayakumar S, Rajendran R, Thilagavathi G. Investigation on the effect of RF air plasma and neem leaf extract treatment on the surface modification and antimicrobial activity of cotton fabric. *Applied Surface Science*, 2008; 254:2472-2478.

[4] Vaideki K, Jayakumar S. and Rajendran R. Investigations on the enhancement of antimicrobial activity of neem leaf extract treated cotton fabric using DC air and oxygen plasma. Plasma Chemistry and Plasma Processing, 2009;29: 515-534.

[5] Wei Q. *Surface Modification of Textiles*. England, Woodhead Publishing Ltd, 2009.

[6] Wong KK, Tao XM, Yeung KW. Topographical Study of Low Temperature Plasma Treated Flax Fibers.*Textile Res. J*; 2000:70:886-893.

[7] Wong KK, Tao X.M, Yeung KW. Low Temperature Plasma Treatment of Linen.*Textile Res. J*; 1999: 69:846-855.

[8] Savile BP. *Physical testing of textiles*, Woodhead Publications Ltd., 1999.

[9] TAPPI Standard Test Method, Carbonyl content of pulp, T237 os-77, 1-7,1977.
[10] Reza MA Malek, Ian Holme. The effect of plasma treatment on some properties of cotton. *Iranian Polymer Journal*, 2003; 12(4) :271-280
[11] Garner W. *Textile laboratory manul.* National trade, London, 42-43,1949.
[12] Kalogianni EP, Savopoulos T, Karapantsios TD, Raphallides SN. A dynamic wicking technique for determining the effective pore radius of pregelatinized starch sheets. *J. Colloid Interf. Sci,*2004;35 :159-167.
[13] Navaneetha Pandiyaraj K , Selvarajan V. Non-thermal plasma treatment for hydrophilicity improvement of grey cotton fabrics. *Journal of Materials Processing and Technology*, 2008;199 (1-3) : 130-139.
[14] *AATCC Technical Manual.* Published by American Association of Textile Chemists and Colorists, USA, Vol 76, 2001.
[15] Ward TL, Jung HZ, Hinojosa O, Benerito RR. Characterization and use of radio frequency plasma activated natural polymers. *Journal of Applied Polymer Science.* 1979; 23:1987-2003.
[16] Kim J.S., Kim G.H., Chung T.H., Yeom G.Y. and Kwon K.H. Characterization of oxygen plasma by using a Langmuir probe in an inductively coupled plasma. *Journal of Korean Physical Society*,2001;38(3);259-263.
[17] Becker K.H., Kogelschatz K., Schoenbach K.H. and Barker R.J. (2005), First Edition, IOP Publishing Ltd, Bristol UK.
[18] Olthoff J.K., Van Brunt R.F. and Radovanov S.B. Ion kinetic energy distribution in argon RF glow discharges. *Journal of Applied Physics*, 1992; 72(10):4566-4574.
[19] Rees J.A., Greenwood C.L., Carew A.J., Neale I.D., Aranda Gonzalvo Y. and Buckly M.E. Ion energy distributions for a DC plasma . Proceedings of *International Conference on Phenomena in Ionized Gases (ICPIG-2003),2003.*
[20] Baldoni H.A., Enriz R.D., Jauregui E.A. and Csizmadia I.G. A Theoretical Study on the Conformations of Azadirachtin. *Journal of Molecular Structure (Theochem)*, 1996; 363: 167-178.

Resistance and Susceptibility

A 3-year review on the profile of multidrug-resistant Gram-negative in a tertiary teaching hospital in Malaysia

H. Habsah*, **Z.Z. Deris, M. Zeehaida, A.R. Zaidah, H. Siti Asma' and I. Nabilah**

Corresponding author: Habsah Hasan , Department of Medical Microbiology and Parasitology, School of Medical Sciences, Universiti Sains Malaysia, 16150 Kubang Kerian, Kelantan, MALAYSIA

*E mail: habsah@kb.usm.my

The objective of this study is to describe the profile of multidrug-resistant (MDR) Gram-negative bacteria in a tertiary teaching hospital at North-Eastern State, Malaysia over a 3-year period (2007-2009). The organisms were identified by conventional method or/and API E or/and API NE. The antimicrobial sensitivity was determined by modified Kirby Bauer method and the sensitivity breakpoints were interpreted according to the Clinical Laboratory Standard Institute. A total of 6978 Gram negative were isolated during that period. 840 (12%) were resistant to 3rd and 4th generation cephalosporin, ciprofloxacin and piperacillin tazobactam. They were *Acinetobacter* spp. (69%), *Klebsiella pneumoniae* (14 %) and *Pseudomonas aeruginosa* (6%). Other organisms found to be resistant were *Enterobacter* spp (2%), *Escherichia coli* (2%), *Klebsiella* spp (3%) and *Stenotrophomonas maltophilia* (1%). Most of the MDR Gram-negative bacteria were from tracheal aspirates (35%), wound specimens (21%), blood (16%) and urine (13%). They were mostly from intensive care unit (38%) and hospitalized patient (59%). The percentage of organisms sensitive to other antibiotics were to amikacin (42%), gentamicin (14.8%), netilmicin (49.5%), cefoperazone/sulbactam (74.8%), imipenem (25.7%) and meropenem (24.6%) In conclusion, MDR Gram-negative bacteria were high in hospitalized patients, especially from surgical based department. They were mostly isolated from tracheal aspirates and wounds. Cefoperazone/sulbactam and amikacin exhibit the highest percentage of susceptibility.

Keywords: multidrug-resistant; Gram negative infections; profile

1. Introduction

Multidrug-resistant (MDR) Gram-negative bacteria are a threat to hospitalized patient. They are important causes of urinary tract infections, bloodstream infections, health-care associated pneumonia and intra-abdominal infections. They are also associated with significant morbidity and the cause of high mortality in patients with nosocomial infections. Worldwide, an increasing prevalence of MDR Gram-negative bacteria is now recognized [1,2,3]. Infection control personnel should analyze the profile of these organisms in the hospitals to implement appropriate measures for prevention. Therefore, the study aims to describe the profile of MDR Gram-negative bacteria in Hospital Universiti Sains Malaysia, a tertiary teaching hospital located in North-Eastern State in Malaysia.

2. Methodology

All significant Gram-negative isolates and their antimicrobial profiles from 2007 to 2009 were analyzed
The organisms were identified by conventional method and/or API E® and/or API NE® (bioMérieux, France). The antimicrobial sensitivity was determined by modified Kirby Bauer method and their sensitivity breakpoint were interpreted according to the Clinical Laboratory Standard Institute (CLSI)
Multidrug-resistant is defined as bacteria resistant to three or more classes of antibiotics. In this study, all Gram-negative bacteria resistant to ceftriaxone/cefotaxime/ceftazidime/cefipeme, ciprofloxacin and piperacillin/tazobactam were included and analyzed.

3. Results

6978 Gram-negative bacteria were isolated during that period. 840 (12%) were resistant to 3rd or 4th generation cephalosporin, ciprofloxacin and piperacillin/tazobactam and their profiles were further analyzed. The three commonest isolates were *Acinetobacter* spp. (69%), *Klebsiella pneumoniae* (14%) and *Pseudomonas aeruginosa* (6%). Other organisms were *Enterobacter* spp (2%), *Escherichia coli* (2%), *Klebsiella* spp (3%) and *Stenotrophomonas maltophilia* (1%). Most of the MDR Gram-negative bacteria were from in-patient (hospitalized patient) (59%) and intensive care unit (ICU) (38%) as shown in Figure 1. By specialties, MDR Gram-negative bacteria were mostly isolated from Surgical and Neurosurgical Department (40%) followed by

Medical and Orthopaedic Department, 27% and 13% respectively as shown in Figure 2. Most of the MDR Gram-negative bacteria were from tracheal aspirates (35%), wound specimens (21%), blood (16%) and urine (13%) as shown in Figure 3. The percentage of organisms sensitive to other antibiotics were to amikacin (42%), gentamicin (14.8%), netilmicin (49.5%), cefoperazone/sulbactam (74.8%), imipenem (25.7%) and meropenem (24.6%) (Table 1)

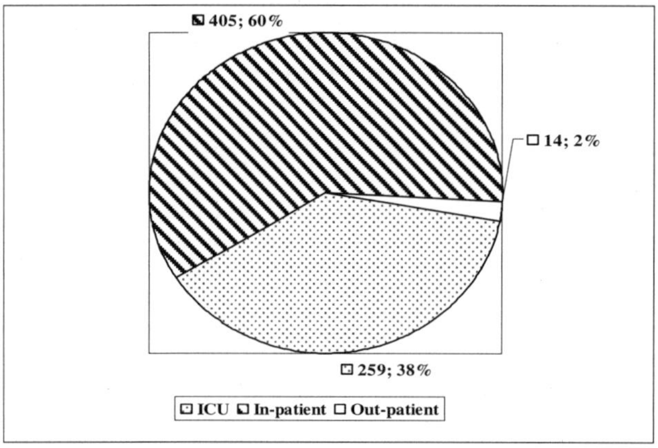

Fig 1: Source of MDR Gram negative bacteria in the hospital by location type

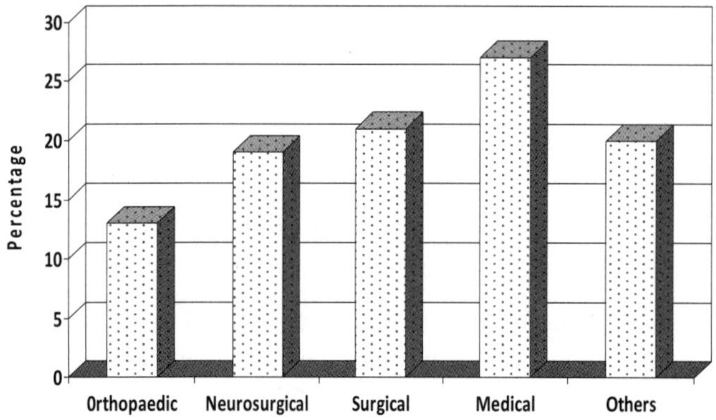

Fig 2: Distribution of MDR Gram-negative bacteria by specialties

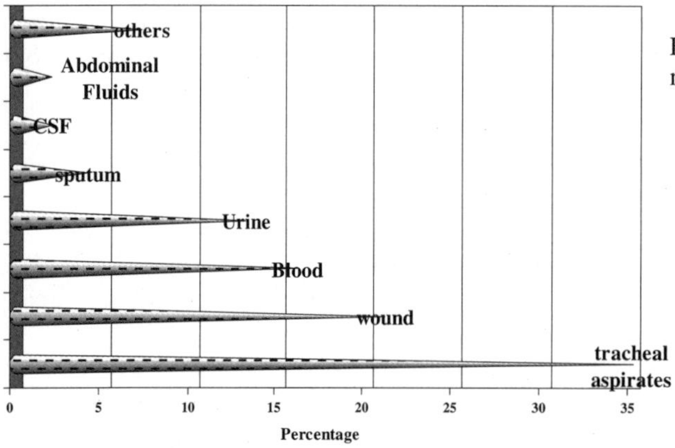

Fig 3: Types of specimens where MDR Gram-negative bacteria were isolated

Table 1: Antimicrobial profiles of MDR Gram-negative bacteria

Code	Antibiotic name	Number of isolates tested	Sensitive(%)
AMK_ND30	Amikacin	839	42
CFP_ND75	Cefoperazone	592	0
CTX_ND30	Cefotaxime	623	0.2
CAZ_ND30	Ceftazidime	840	0
CRO_ND30	Ceftriaxone	558	0
CXM_ND30	Cefuroxime sodium	621	0
CHL_ND30	Chloramphenicol	523	13.2
CIP_ND5	Ciprofloxacin	840	0
SXT_ND1.2	Trimethoprim/Sulfamethoxazole	620	13.7
GEN_ND10	Gentamicin	790	14.8
IPM_ND10	Imipenem	840	25.7
NET_ND30	Netilmicin	792	49.5
POL_OD50	Polymixin B	178	99.4
AMC_ND20	Amoxicillin/Clavulanic acid	624	0.5
X_1_ND75	cefoperazone/sulbactam	250	74.8
TZP_ND100	Piperacillin/Tazobactam	840	0
FEP_ND30	Cefepime	840	0
MEM_ND10	Meropenem	838	24.6
SAM_ND20	Ampicillin/Sulbactam	732	13.8

4. Discussion

In our study, majority of the MDR organisms (98%) were isolated from in-patients and ICU patients. Only 2% were from outpatient. This finding was in agreement with many other studies [4,5,6]. Hospitalized patient has higher risk of being colonized by MDR Gram negative bacteria. Globally, patients in the ICU have encountered an increasing emergence and spread of antibiotic-resistant pathogens. The worldwide median rate is 9 infections per 100 discharged ICU patients or 23.7 infections per 1000 patient days [7]. Interestingly, we also found that patients from surgical based disciplines are more frequent to get infection by MDR Gram-negative bacteria.

Acinetobacter spp is the commonest MDR Gram-negative in our study which is not different from other recent study. In a military medical center, Texas, the percentage of MDR *A. baumannii* and MDR *P. aeruginosa* isolates were 55% and 8% respectively [8].

We found that MDR organisms were mostly isolated from tracheal aspirates, wound and blood which is comparable to other study. Keen *et al*, in other study shows that respiratory tract specimens had a higher percentage of MDR *A. baumannii* isolates (49%), followed by blood (30%), wound sites (24%), or urine (19%). No difference in the percentages of MDR *P. aeruginosa* isolates was observed with regards to source of specimen [8].

Review of the National Nosocomial Infections Surveillance (NNIS) data from 1992 to 2004 showed that the overall antimicrobial resistance rate has remained elevated. The third-generation cephalosporin resistance rate for *K. pneumoniae* was reported at 20%, while *E. coli* and *Enterobacter* species were reported as 5.6% and 31%, respectively [9]. In a Malaysian study, although 27 of the 51 *K. pneumoniae* strains were MDR, the majority of the strains (98%) were sensitive to imipenem [10].

In this study, the bacteria that are resistant to 3rd and/or 4th generation cephalosporin, ciprofloxacin and piperacillin/tazobactam still shows good susceptibility to polymyxin and cefoperazone/sulbactam, moderately susceptible to netilmicin and amikacin, but markedly reduced susceptibility against carbapenem. This information is useful for physician to start empirical treatment for cases suspected to have infection by MDR Gram-negative bacteria.

5. Conclusion

MDR Gram-negative bacteria were high in hospitalized patients, especially from Surgical Department. They were mostly isolated from tracheal aspirates and wounds. Polymyxin, cefoperazone/sulbactam and amikacin exhibit the highest percentage of susceptibility against MDR Gram negative bacteria.

Acknowledgment I would like to thank Universiti Sains Malaysia, Malaysia for the financial support under the Research University Grant 1001/PPSP/812018. This study is a part of a bigger project entitled 'Multifaceted Approach of Hospital Associated Infection Control Program; Towards Zero Tolerance'.

References

[1] Bell JM, Turnidge JD, Gales AC, Pfaller MA, Jones RN. Prevalence of extended spectrum beta-lactamase (ESBL)-producing clinical isolates in the Asia-Pacific region and South Africa: regional results from SENTRY Antimicrobial Surveillance Program (1998-99). Diagn Microbiol Infect Dis. 2002;42:193-8.

[2] National Nosocomial Infections Surveillance (NNIS) System Report, data summary from January 1992 through June 2004. Am J Infect Control. 2004; 32:470-85.

[3] Falagas ME, Bliziotis IA. Pandrug-resistant gram-negative bacteria: the dawn of the post-antibiotic era? Int J Antimicrob Agents. 2007;29: 630-6.

[4] Al Johani SM, Akhter J, Balkhy H, El-Saed A, Younan M, Memish Z. Prevalence of antimicrobial resistance among gram-negative isolates in an adult intensive care unit at a tertiary care center in Saudi Arabia. Ann Saudi Med. 2010; 30(5): 364–369

[5] Eagye KJ, Kuti JL, Sutherland CA, Christensen H, Nicolau DP .In vitro activity and pharmacodynamics of commonly used antibiotics against adult systemic isolates of *Escherichia coli and Pseudomonas aeruginosa* at Forty US Hospitals. Clin Ther. 2009; 31(11):2678-88

[6] Clark NM, Patterson J, Lynch JP 3rd . Antimicrobial resistance among gram-negative organisms in the intensive care unit. Curr Opin Crit Care. 2003; 9(5):413-23.

[7] Fridkin SK. Increasing prevalence of antimicrobial resistance in intensive care units. Crit Care Med. 2001;29:64–8.

[8] Keen EF 3rd, Murray CK, Robinson BJ, Hospenthal DR, Co EM, Aldous WK. Changes in the incidences of multidrug-resistant and extensively drug-resistant organisms isolated in a military medical center. Infect Control Hosp Epidemiol. 2010;31(7):728-32.

[9] Anjali N. Kunz & Itzhak Brook. Emerging Resistant Gram-Negative Aerobic Bacilli in Hospital-Acquired Infections. Chemotherapy.2010;56:492–500.

[10] Lim KT, Yeo CC, Rohani MY, Balan G and Thong KL. Characterization of multidrug-resistant and extended-spectrum b-lactamase-producing *Klebsiella pneumoniae* strains from Malaysian Hospitals. Journal of Medical Microbiology.2009; 58, 1463–1469

Antimicrobial susceptibility in clinical isolates of *Staphylococcus aureus* harbouring of *mec*A and *luk*FS-PV genes in Northern Portugal

N. Silva[1,2,3], C. Prudêncio[1,2], C. Tomaz[3] and R. Fernandes*[1,2]

[1]Ciências Químicas e das Biomoléculas, Escola Superior de Tecnologia da Saúde do Porto, Instituto Politécnico do Porto, 4400-330, Vila Nova de Gaia, Portugal
[2]Centro de Farmacologia e Biopatologia Química (U38-FCT), Laboratório de Bioquímica, Faculdade de Medicina, Universidade do Porto, 4200 - 319 Porto, Portugal
[3]Departamento de Química, Universidade da Beira Interior, 6201-001 Covilhã, Portugal

* Corresponding author: e-mail: rfernandes@eu.ipp.pt, Phone: +351 222061004

Methicillin-resistant *Staphylococcus aureus* (MRSA) is a highly pathogenic multiple-drug resistant (MDR) microorganism that is most prevalent in the community. It has been found that MRSA strains can also contain genes that encode the panton valentine leukocidin toxin (PVL) and several genes that confer resistance to a various number of antimicrobial agents. The aim of this study was to determine the presence of MRSA (*mec*A gene) and PVL toxin genes (*luk*S-PV and *luk*F-PV) in strains isolated from two hospitals. The susceptibility to a various number of antimicrobial agents was tested and its clonality also studied. The results indicated a frequency of approximately 50% of MRSA and a frequency of 9% of isolates carrying the *luk*FS-PVL gene. Regarding the susceptibility patterns of isolates under study it was concluded that for most antimicrobial agents used MRSA strains reveal to be resistant to more antimicrobials than the isolates that lack *mec*A gene. Regarding fingerprinting method it was possible to observe a quite fixed number of strains among pathogen causing hospital-acquiring infection. The explanation for these findings is complex and multi-factorial. The high diversity of clones suggests diverse genetic backgrounds rather than the global spread of a single clone.

Keywords MRSA; PVL; MDR; mPCR; RAPD

1. Introduction

S. aureus is a type of bacteria which is carried on the skin and nose (mucous membranes) of healthy people and animals, where the constant humidity, temperature, and various nutrients are supplied by the host. About one third of the humans are colonized with *S. aureus* in the nose, however, this does not cause infection. Asymptomatic colonization by *S. aureus* is much more common than infection. However, if it enters inside the human body it can trigger a mechanism of infection, in which local colonization serves as a reservoir for the spread of this agent [1, 2]. *S. aureus* has three characteristics that make it different from the most of microorganisms. It can express a variety of virulence factors, has the ability to develop and expand resistance to a broad spectrum of antimicrobial drugs, and is prominently found in various environments such as hospitals and community [3].

Currently, the emergence of resistance to antibiotics used in clinical practice is something that concerns scientific community. Every day is documented new resistance and new strains are discovered possessing resistance genes previously unknown. One of the most worrying issues is the case of MRSA. The MRSA is worldwide distributed and is, characteristically, highly resistant to various antibiotics. Additionally, this infectious agent has the ability to bear several virulence genes. Thus, the decrease in immunological defence of the host caused by the virulence factors, together with reduced susceptibility to chemotherapeutic agents, makes the virulence genes carrier MRSA an infectious agent much feared [1, 4-7]. MRSA has, by definition, the *mec*A gene responsible for methicillin resistance [6]. The *mecA* gene encodes a modified penicillin-binding protein (PBP), of 78-kDa, called PBP2a or PBP2', which is a peptidoglycan transpeptidase, and that differs from endogenous *S. aureus* PBP. This PBP2a maintains its normal enzyme activity, but differs from PBP2 because it has the recognition site for β-lactam modified. Thus, when the peptidoglycan synthetases are linked to β-lactam they become inactive with the exception of PBP2a due to its insensibility to these antibiotics, including methicillin. Therefore, despite being linked to methicillin, the PBP2a can also promote cell wall synthesis [6, 8, 9].

For the treatment of infections caused by MRSA, vancomycin is the antibiotic of choice and often of last resort. The vancomycin is a glycopeptide antimicrobial that does not act through direct inhibition of transglycosylases or transpeptidases, but through complexation with the structural units of peptidoglycan with terminations D-Ala-D-Ala [4]. Nevertheless it was recently reported the development of resistance to this new antimicrobial agent. In 2002, it was discovered in the United States, *S. aureus* resistant to vancomycin (VRSA), and since then there is not known another antimicrobial that could be universally effective against *S. aureus*. Therefore, future therapies for infection by *S. aureus* have to rely less on the effectiveness of only one drug and more on the combination of several [6].

The leukotoxins, which are produced by *S. aureus* belong to a family of pore-forming toxins, which are composed of two distinct components. The toxic effect depends on the synergistic action of both proteins: the class S (slow elution) and the class F (fast elution) on polymorphonuclear cells (PMN) and monocytes. In this family is included the Panton-Valentine Leukocidin (PVL) which consists of the subunits LukS-PV and LukF-PV [9, 10]. It is a toxin that besides destroying certain white blood cells, also mediates tissue necrosis. It has been associated with infections of skin and soft tissue, and severe necrotizing pneumonia among healthy children and young adults [11].

Infections caused by MSSA in comparison with others caused by MRSA are associated with a significant increase in morbidity, mortality, length of hospital stay and costs [12]. Understanding the prevalence of MRSA, their resistance patterns, virulence and the risk of infection is essential to help physicians to choose a suitable treatment.

2. Materials and Methods

A total of 113 strains collected from two different hospitals in the Northern region of Portugal were studied by means of multiplex-PCR (mPCR) that simultaneous detects the gene that confers resistance to methicillin, the *mec*A gene and the *PVL* genes. Clonality of the strains in study was evaluated by random amplification of polymorphic DNA (RAPD) discriminated by three primers and the dendrograms were generated by unweighted pair group method with arithmetic mean (UPGMA) analysis of the agarose gels using computational tools. The antimicrobial susceptibility was tested by the Kirby-Bauer disk-diffusion method using several antimicrobial agents. As controls were used 2 strains, JCSC 2958 and 81/108, both encoding the *mec*A and PVL genes.

2.1 mPCR assays

A multiplex PCR (mPCR) assay targeting 16S rRNA (*Staphylococcus* genus specific), *mec*A and *luk-PV* was used in our study. mPCR assays were regularly performed directly with bacterial suspensions obtained after DNA extraction according to the method phenol-chloroform. An aliquot of 2 µl of bacterial suspension was added to a 23-µl PCR mixture containing 50 mM KCl, 20 mM Tris-HCl (pH 8.4), 2.5 mM $MgCl_2$, 0.2 mM each deoxynucleoside triphosphate (dATP, dUTP, dGTP, and dCTP) (Invitrogen, Inc., Carlsbad, Calif.), 0.12 µM each 16S rRNA, mecA and *luk-PV* primers and 1.0 U of Platinum *Taq* DNA polymerase (Invitrogen). Amplification was performed by using a GeneAmp PCR system 9700 or 9600 thermal cycler (Applied Biosystems, Foster City, Calif.). Amplification conditions and the primers used are presented in Table 1. The mPCR amplifications were visualized by using a UV light box after electrophoresis on a 1 % agarose gel containing 0.5 µg of ethidium bromide/ml [13]

Table 1 Primers and PCR conditions form PCR and RAPD assays.

Primer	Oligonucleotide sequence (5'→ 3')	Size (bp)	Time (seconds) / Temperature (°C)		
			Denaturation	Annealing	Extension
$Staph756_{FW}$	aac tct gtt att agg gaa gaa ca	756	45 s	45 s	75 s
$Staph756_{RV}$	cca cct tcc tcc ggt ttg tca cc		94 °C	55 °C	72 °C
$MecA_{FW}$	gta gaa atg act gaa cgt ccg ata a	310		(10 cicles)	
$MecA_{RV}$	cca att cca cat tgt ttc ggt cta		45 s	45 s	75 s
$Luk\text{-}PV_{Fw}$	atc att agg taa aat gtc tgg aca tga tcc a	433	94 °C	50 °C	72 °C
$Luk\text{-}PV_{RV}$	gca tca agt gta ttg gat agc aaa agc			(30 cicles)	
RAPD 2	gtt tcg ctc c	-	60 s	60 s	120 s
RAPD 4	aag agc ccg t	-	95 °C	36 °C	72 °C
RAPD 6	ccc gtc agc a	-		(45 cicles)	

2.2 RAPD assays

For this assays were used the RAPD primers 2, 4 and 6 from the Ready-To-Go RAPD Analysis Beads Kit (GE Healthcare, 2006). The reaction tube contained the same master mix as the pcr assays and the same amount of DNA was added to mix. The RAPD reaction was performed using a GeneAmp PCR system 9700 or 9600 thermal cycler (Applied Biosystems, Foster City, Calif.). The RAPD reaction conditions and the primers used are also presented in Table 1. The RAPD reaction products were visualized by using a UV light box after electrophoresis on a 1 % agarose gel containing 0.5 µg of SYBR green/ml. The results were analised in the software FPQuest 5.1.

2.3 Antimicrobial susceptibility assays

A large number of antimicrobial agents from different classes such as ß-lactams, macrolides, tetracyclines, among others were used to determine the susceptibility of the isolates. With this purpose was used the Kirby-Bauer disk-diffusion method. Antimicrobial concentrations are predetermined and available commercially. For each assay a petri dish was incubated with the microorganism and the antimicrobial agents at 35-37 °C for 18-24 hours [14].

3. Results

From the 113 strains studied, 54 shown to harbor the *mecA* gene, whereas only 7 possessed the PVL genes. Figure 1 shows an electrophoretic gel with some samples and the identification of the 16S rNA gene (an internal control), the *mecA* gene and the l*uFS-PV* genes. Figure 2 shows two Petri dishes with a strain and several antimicrobial agents where it can be observed resistance to the antimicrobial agent (normal growing of the bacteria) or susceptibility (formation of a halo). As expected MRSA strains reveal to be resistant to more antimicrobials than the isolates that lack *mec*A gene. The 54 MRSA strains were then subjected to fingerprinting methods by a PCR reaction using different primers. The number of clusters obtained is presented in Table 2 and an example of the phylogenetic tree using RAPD4 primer and the UPGMA method is presented in Figure 3. Regarding the results themselves to a degree of similarity of 100%, except for a cluster, all others are formed by strains of the same hospital, while 99% from the frequency between the strains in two hospitals within the same cluster becomes higher. For the reactions with the three primers, although there are some minor differences, most clusters are the same.

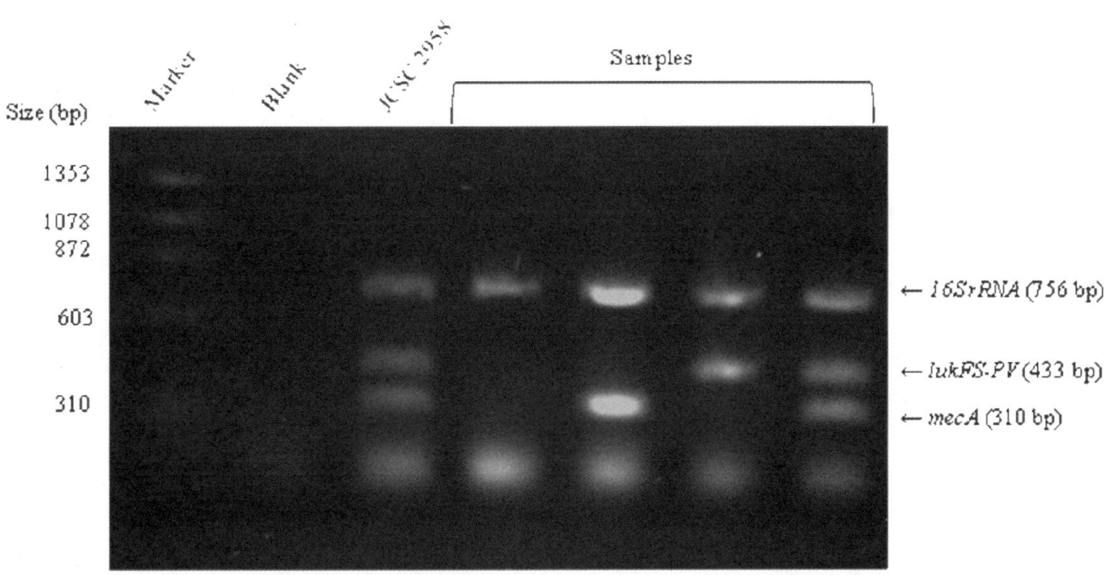

Figure 1: Agarose gel at 0,7%% revealed with ethidium bromide. Marker: DNA Lambda Hind III/phiX174 Hae III.

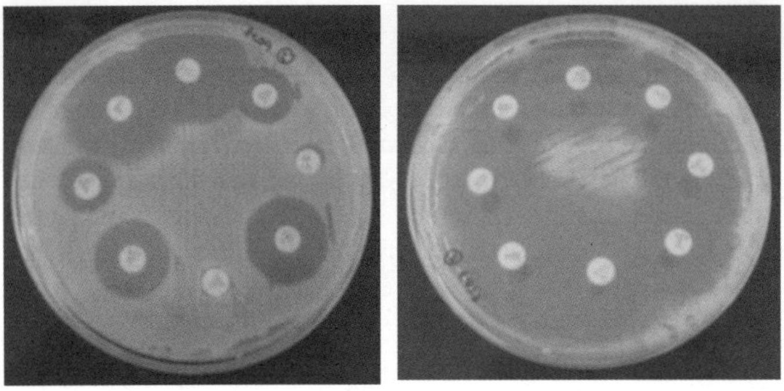

Figure 2: Petri disk used for the antimicrobial susceptibility test. Leftmost figure shows a strain where there is both resistance and susceptibility to antimicrobial agents while in the rightmost figure shows a strain with total susceptibility to all antimicrobial agents used.

Table 2 Clonality of MRSA strains after fingerprinting by a RAPD reaction with different primers.

Primer	Cluster		
	100%	99%	97%
RAPD2	4	10	13
RAPD4	4	6	13
RAPD6	6	11	8

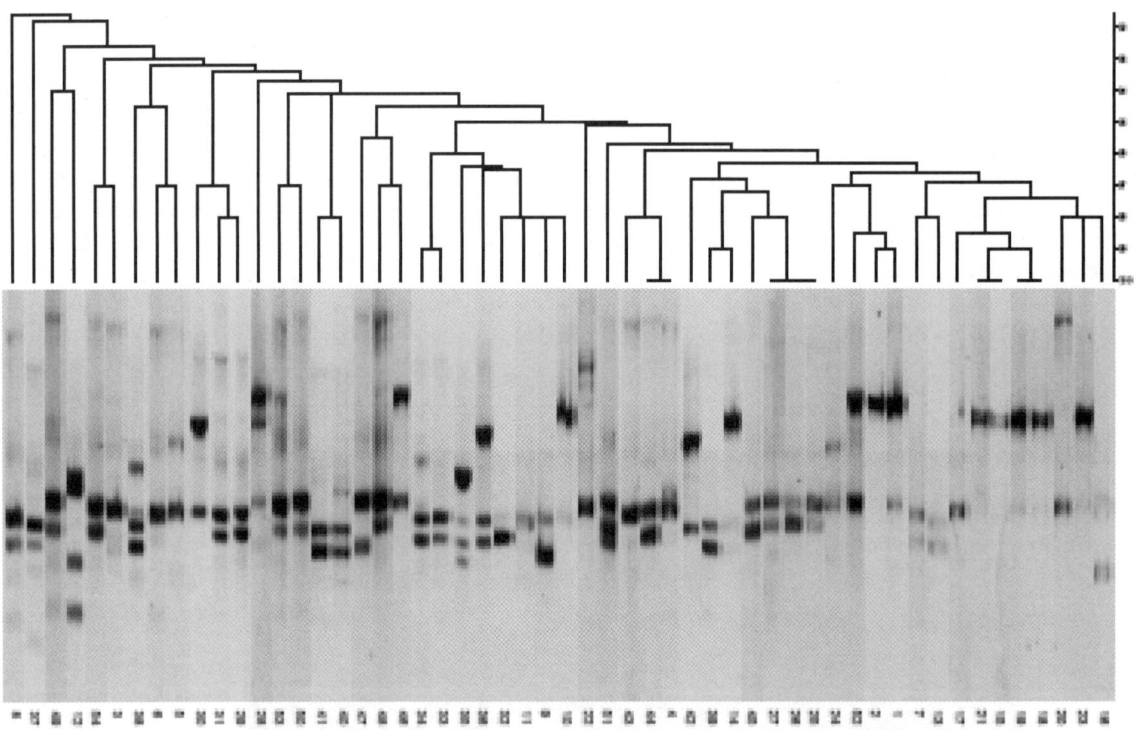

Figure 3: Comparison of MRSA strains by UPGMA method after the RAPD reaction with primer RAPD4.

4. Discussion

The frequency of isolates carrying the *mec*A gene shows a worrying raising trend through the world [15]. In this study, was found that about 50% of the isolates carried the *mec*A gene. These data are consistent with the study published in 2007 by the European Antimicrobial Resistance Surveillance System [16]. Regarding the isolates carriers of the *lukFS-PV* genes, there was not a frequency as high as the one observed for the *mec*A gene. Thus, it was found that only 6% of isolates are carriers of this gene. This finding is consistent with the data documented in the literature that describe an increasing *S. aureus* strains carrying the gene *lukFS-PV*, although these are not as frequent as the *S. aureus* carrying the *mec*A gene [6, 11]. One factor that may have contributed to the low frequency of isolates carrying the gene that encodes the protein PVL in this study is that the PVL gene is more frequent in isolates that cause infection in the community. However, the isolates used came only from infections in hospital environment [6, 11, 17].

The highly resistance to most antimicrobial agents presented by MRSA strains may be explained by the fact that several molecular mechanisms may be involved in the gene exchange between species or horizontal transfer, and by the acquisition of new resistance genes from the environment or transformation [3, 4, 18]. This ability may be justified by the fact that antimicrobial resistance genes lie within molecular cassette flanked by mobile genetic sequences (transposons, insertion sequences) that have high rates of recombination. These mechanisms are triggered mainly by selective pressure caused by antimicrobial therapy [19, 20].

The appearance of the same strain in different hospitals during the same period reveals the possibility of transmission of that infectious agent within several healthcare institutions. These results illustrate the potential for spread of MRSA and the need to alert their control in hospitals.

Resistance to antimicrobial agents has several implications. People infected with resistant microorganisms such as MRSA, face higher mortality rates. This is partly due to the fact that an infection of a resistant organism is associated with risk of initial antimicrobial therapy is inappropriate. In addition, the increased risk of death may reflect the fact that the organism causing the infection has more virulence factors. Besides mortality, the resistant microorganisms contribute to high hospital costs. Infections with MRSA, for example, lead to a prolonged hospitalization. In other words, MRSA is associated with adverse outcomes of mortality and morbidity [21]. So, further efforts should be made to develop and implement strategies for public health and education for health in order to slow the rise and spread of disturbing and threatening resistance to antibiotics.

Acknowledgments: The study of methicillin-resistant *S. aureus* was possible due to a collaboration between the Universidade da Beira Interior and the Escola Superior de Tecnologia da Saúde. We are grateful to Professor Hermínia de Lencastre, Universidade Nova de Lisboa and ITQB (Instituto de Tecnologia Química e Biológica) for the control strains used in this study.

References

[1] Persing D. *Molecular microbiology: diagnostic principles and pratice.* Washington, DC: ASM Press, 2004.

[2] Weber C. Update on methicillin-resistant Staphylococcus aureus (MRSA). *Urologic Nursing.* 2008, 28 (2): 143-145.

[3] Decker C. Pathogenesis of MRSA infections. *Disease a Month.* 2008, 54 (12): 774-779.

[4] Walsh C. *Antibiotics: actions, origins, resistance.* Washington, DC: ASM Press, 2003: 92-93.

[5] Enright M. The evolution of a resistant pathogen – the case of MRSA. Current Opinion in Pharmacology. 2003, 3: 474–479.

[6] Boyle-Vavra S. and Daum R. Community-acquired methicillin-resistant Staphylococcus aureus: the role of Panton–Valentine leukocidin. *Laboratory Investigation.* 2007, 87: 3–9.

[7] Corriere Mark D. and Decker CF. MRSA: An evolving pathogen. *Disease a Month.* 2008, 54 (12): 751-755.

[8] Gordon R. and Lowy F. Pathogenesis of methicillin-resistant Staphylococcus aureus infection. *Clinical Infection Diseases.* 2008, 46 (5): 350–359.

[9] Shukla S. Community-associated methicillin-resistant Staphylococcus aureus and its emerging virulence. *Clinical Medicine & Research.* 2005, 3 (2): 57-60.

[10] Yamada T., et al. Leukotoxin family genes in Staphylococcus aureus isolated from domestic animals and prevalence of lukM–lukF-PV genes by bacteriophages in bovine isolates. *Veterinary Microbiology.* 2005, 110: 97-103.

[11] Maltezou H. and Giamarellou H. Community-acquired methicillin-resistant Staphylococcus aureus infections. *International Journal of Antimicrobial Agents.* 2006, 27: 87–96.

[12] Livermore D. and Pearson A. Antibiotic resistance: location, location, location. *Clinical Microbiology and Infectious Diseases.* 2007, 13: 7–16.

[13] Zhang K. et al. New quadriplex PCR assay for detection of Methicillin and Mupirocin resistance and simultaneous discrimination of Staphylococcus aureus from Coagulase-negative Staphylococci. *Journal of Clinical Microbiology.* 2004, 42 (11): 4947-4955.

[14] CLSI, Clinical and Laboratory Standards Institute. Performance standards for antimicrobial susceptibility testing; Fifteenth informational upplement. *CLSI document M100-S15 (ISBN 1-56238-556-9).* 2005, 25(1).

[15] Appelbaum P. MRSA - The tip of the iceberg. *Clinical Microbioly Infections.* 2006, 12 (2): 3-10.

[16] EARSS, European Antimicrobial Resistance Survailance System. Available at http://www.ricm.nl/earss. Accessed June 26, 2009.

[17] Levenhagen K. Community-acquired methicillin-resistant Staphylococcus aureus: an emerging concern for physical therapists: Discussion. *Physiotherapy Research International.* 2008, 13(1): 9–17.

[18] Bryskier A. *Antimicrobial agents: Antibacterials and antifungals.* Washington, DC: ASM Press, 2005.

[19] Lencastre H, Oliveira D and Tomasz A. Antibiotic resistant Staphylococcus aureus: a paradigm of adaptive power. *Current Opinion in Microbiology.* 2007, 10: 428–435.
[20] Kuroda M. et al. Whole genome sequencing of meticillin-resistant Staphylococcus aureus. *Lancet.* 2001, 357: 1225-1240.
[21] Shorr A. Epidemiology of Staphylococcal resistance. *Clinical Infectious Diseases.* 2007, 45: 171–176.

Antimicrobial susceptibility profile and effect of stem bark extracts of *Curtisia dentata* on multi-drug resistant verotoxic *Escherichia coli* and *Acinetobacter* spp. isolates obtained from water and wastewater samples

[*]Hamuel James Doughari, Patrick Alois Ndakidemi, Izanne Susan Human and Spinney Benade

Applied Sciences Faculty, Cape Peninsula University of Technology, P.O. Box 652 Cape Town 8000, South Africa.

*Corresponding author's email address: jameshamuel@yahoo.com Phone: +27 (0) 73355 0274; +234 (0) 808 381 9054

Twenty one (21) *Escherichia coli* (including serotypes 026:H11, 055, 0111:NM, 0126, 044, 0124, O96:H9, O103:H2, O145:NM and O145:H2), 10 *Acinetobacter lwoffii* (A1-10) and 4 *A. haemolyticus* (B1-4) strains were isolated from 32 water (18 wastewater and 14 river water) samples collected from different sources in Cape Town, South Africa. Samples were first subjected to membrane filtration and the membrane filters inoculated into MacConkey broth (MB) and Baumann's enrichment medium (BEM) and incubated at 37ºC for 24-144 h, after which a loop full of the MB and BEM were then surface inoculated onto Eosin Methylene Blue (EMB) and Leeds Acinetobacter Medium (LAM) and further incubated for 24-48 h at 37ºC in order to isolate *E. coli* and *Acinetobacter* spp. respectively. The isolates obtained were presumptively identified biochemically using oxidase test strips, sulfide indole motility (SIM) medium and Erchlich's reagent. *E. coli* isolates were then confirmed serologically using *E. coli* polyvalent (Bioweb, SA) and *Acinetobacter* spp. biochemically using RapID™ NF plus test kit (Bioweb, SA). Screening for verotoxin production using Glisa Duopath Verotoxins® (Merck, Germany) test kit showed that 71% (15) of all the *E. coli* and 75% (3) of *A. haemolyticus* isolates produced verotoxin, while none of the *A. lwoffii* produced verotoxin. Results of antimicrobial susceptibility testing using the disc diffusion method showed that all the isolates were resistant to more than four antibiotics indicating multidrug resistance (MDR), with 61% of the 21 *Escherichia coli* isolates resistant to ampicillin (10 µg), cefuroxime, cephalexin, ceftazidime and tetracycline (30 µg in each case), 44% of the 10 *Acinetobacter lwoffii* isolates resistant to ampicillin (10 µg), ofloxacin (5 µg), cefuroxime and ceftazidime (30 µg in each case), while 50% of the 4 *A. haemolyticus* strains obtained were resistant to ceftriaxone, cefuroxime, nalidixic acid, amikacin and tetracycline (30 µg in each case). All the isolates were susceptible to aztreonam (5 µg), gentamicin (10 µg), cefotaxime (30 µg) and ciprofloxacin (5 µg). Results also showed that 50-68% of the MDR isolates were susceptible when tested (disc diffusion method) against ethanol stem bark extracts of *Curtisia dentata* with MIC values ranging between 2.5-20 mg/ml. The study has revealed that verotoxic multidrug resistant *E. coli* and *Acinetobacter* prevail in the environment and that *C. dentata* stem bark extracts has the potential to provide alternative source of antimicrobials that can be used in controlling these multi-drug resistant pathogenic bacteria.

Keywords: *Acinetobacter* spp.; Baumann's enrichment medium; *Curtisia dentate; Escherichia coli;* multi-drug resistance; plant extracts; verotoxins.

1.0 Introduction

Escherichia coli (Gram negative bacilli, Enterobacteriaceae family) has been the causative agent of gastrointestinal infections especially diarrhea in humans. A natural inhabitant of the human and animal intestines, the organism has been implicated in contamination of food and water through faecal pollution. The diarrhea is often contracted via consumption of water and food contaminated by the bacteria. Consequently, diarrhea accounts for major cause of morbidity and mortality worldwide with 2.0 million deaths per annum, particularly in developing countries [1]. The emergence of non O157:H7 VTEC serotypes including O111:H, 026:H11, 0103:H2 and 0145 in addition to the well known agent of bloody diarrhea, the verotoxin producing *E. coli* O157:H7 (VTEC) [2, 3] is a call for concern. *Acinetobacter* spp. (Gram-negative cocco-bacilli, Moraxellaceae family) have also emerged as significant nosocomial pathogens in the hospital settings ranked 9th after *S. aureus, E. coli, Klebsiella* spp., *P. aerugenosa, C. albicans, Enterococci*, *Serratia* and *Enterobacter* as agents of nosocomial blood stream infections (BIs). The organisms also account for 34% mortality and 43% of deaths due to hospital acquired infections [4]. *A. baumannii* (predominantly clinical isolates), *A. johnsonnii, A. haemolyticus, A. junii,* and *Acinetobacter* genomospecies 3, and 13 (all mainly environmental isolates) are the most medically important, with *A. haemolyticus* associated with endocarditis and verotoxin production [5], hence bloody diarrhea [6]. Increasing incidences of multidrug resistant (MDR) infections and food and water contamination [7] amongst these two bacterial genera is a call for concern. The threat posed by the global increase of antibiotic resistance is not only limited to the hospitals and other medical institutions, but has resulted in the spread of resistant pathogens to the community at large. The potential for co-existing *E. coli* and *Acinetobacter* spp. in environmental samples to undergo rapid transformation is a further means of spreading resistant strains and their pathogenic effects [8]. Complexities arising from antibiotic treatment of verotoxin-related bacterial infections [9] such as bacterial cell wall rapture resulting in further release of nephrotoxins into the system underlie the need to search for alternative treatment strategies. Although few reports are available on the incidences of *E. coli* O157:H7 in Africa [10-12], there is paucity of information on other verotoxin producing *E. coli* pathotypes or 'non-verotoxic *E. coli* strains; and none at all on *Acinetobacter* spp. [12]. The inability of commonly prescribed antibiotics to treat some common infections has made the use of traditional medicinal plants popular in Africa even among urban dwellers. Complications arising from the antibiotic treatment of verotoxic bacteria should be a further inducer to investigate alternative treatment sources especially from plants.

Curtisia dentata (Cornaceae or dogwood family) commonly called in South Africa as assegaai (Afrikaans), uSirayi, umGxina (Xhosa), umLahleni (Xhosa, Zulu) and assegai (English), is a traditional medicinal plant employed in the treatment of diarrhea and related stomach ailments in South Africa [13]. It is therefore significant to investigate its efficacy against verocytotoxin and other toxin producing bacterial pathogens with a view of discovering more effective drugs for the control of MDR verotoxic bacteria. In this paper therefore, *Escherichia coli* other than *E. coli* O157:H7 and *Acinetobacter* spp. isolates from some river and wastewater samples, were investigated for verotoxin production and antibiotic resistance. The effect of stem bark extracts of *Curtisia dentata* were also determined against the verotoxin-producing multidrug resistant isolates.

2.0 Materials and Methods

2.1 Sample collection and preparation of plant material

Thirty two (32) water (18 wastewater samples - from a wastewater treatment plant and an Abattoir, and 14 river water samples - from River Berg, River Plankenburg) all in Cape Town South Africa, were collected using the shoreline method [14]. The collected water samples were placed in a cooler box with temperature maintained between 4-10^{0}C using ice packs and then immediately transported to the Microbiology Laboratory of the Biotechnology Department of Cape Peninsula University of Technology, Cape Town South Africa where they were analyzed within 3-6 h. Fresh stem backs of *Curtisia dentata* (authenticated as well as provided by Dr. Charles Laubscher from his plant collections in the Glass House of the Horticulture Department, Cape Peninsula University of Technology, Cape Town South Africa.) were dried to constant weight in an oven for 6-12 h at 45ºC. The dried stem barks were coarsely grated in a pistil and mortar and then reduced to powdered form using an electric grinder. The powdered plant materials were transferred to brown bottles and stored at ambient temperature until use.

2.2 Cultivation, isolation and identification of bacteria

Water samples (1 ml) were first inoculated into 9 ml of MacConkey Broth (MB, Oxoid) (for isolation of *E. coli*) or Bauman's Enrichment medium (for isolation of *Acinetobacter* spp.) and incubated at 37ºC for 24 h. The MB was inoculated onto plates of Eosin Methylene Blue (EMB, Oxoid) and further incubated at 37°C for 24 h (for isolation of *E. coli*) while the BEM was incubated onto plates of Leeds Acinetobacter Medium (LAM, Hardy Diagnostics USA) and further incubated for 24-144 h (1-6 days). Isolates were identified by Gram staining then biochemically using Sulfide-indole-motility medium (SIM), Erchlich's reagent and REMEL RapID™ NF plus kit (Bioweb, South Africa). *E. coli* were further identified serologically using *E. coli* polyvalent antisera 2, 3 & 4. *Escherichia coli* ATCC 25922 and *Acinetobacter haemolyticus* ATCC 19002 were used as control. Identified isolates were inoculated onto Nutrient agar (NA, Oxoid) slants (*E. coli*) and Trypton soy agar (TSA, Oxoid) slants and incubated for 24 h at 37 °C, and then stored at 4 °C until use [15].

2.3 Differentiation of verotoxic from non-verotoxic bacteria

Isolates inoculated into TSB were incubated for 18-20 h at 37°C with shaking at 120-150 rpm. Bacterial suspension was centrifuged for 20 min at 4,000 rpm. Supernatant was transferred to new tubes and then used for verotoxin screening. Verotoxin production from all the isolates was screened using antibody-based rapid slide agglutination assays with the Glisa Duopath Verotoxins® (Merck, Germany) test kit according to the manufacturer's instructions. Isolates were then separated into verotoxic and non-verotoxic strains and stored accordingly as earlier described and verotoxic isolates were used for antimicrobial susceptibility testing.

2.4 Antimicrobial susceptibility testing and determination of multidrug resistant (MDR) index

This was carried out on molten Mueller-Hinton agar (MHA) plates using the disc diffusion method [16]. Standard antibiotics commonly used in chemotherapy of Gram-negative infections as set out elsewhere [17] were used for the susceptibility tests. Resistance to four or more antibiotics was taken as multidrug resistance (MDR). MDR index (MDRI) of individual isolates was calculated by dividing the number of antibiotics to which the isolate was resistant by the total number of antibiotics to which the isolate was exposed. While isolates with MDR index of more than 0.2 or 20% were considered as originating from a high source of contamination [18].

$$\% \text{ MDRI} = \frac{\text{Number of antibiotics resisted}}{\text{Total number of antibiotics used}} \times 100$$

2.5 Extraction and determination of phytoconstituents from stem bark extracts of *Curtisia dentata*

To extract phytoconstituents from the plant material, 5 g ground plant stem barks was soaked in 200 ml of solvent (ethanol and water) for 2 h followed by filtration; the procedure was repeated three times. The filtered extracts obtained from extraction with any one solvent was combined, and dried by evaporation under laminar flow at ambient conditions. The extract obtained was then used to screen for the presence of phytoconstituents [19].

2.6 Determination of antibacterial effects and minimum inhibitory concentration (MIC) of the stem bark extracts of *Curtisia dentata* against *E. coli* and *Acinetobacter* spp.

Antibacterial activity determination was carried out using the filter paper disc diffusion method [19]. The antimicrobial activity (expressed as percentage relative inhibition zone diameter) was calculated by applying the expression:

$$\% \text{ RIZD} = \frac{\text{IZD sample - IZD negative control}}{\text{IZD antibiotic standard}} \times 100$$

Where RIZD is the percentage of relative inhibition zone diameter and IZD is the inhibition zone diameter (mm). The equation compensates the possible effect of the solvent (blank) other than water on the IZD. The test was considered negative (-) when the IZD of the sample equaled the IZD of the blank [20]. Filter paper discs soaked with distilled water or 30 μg/ml ampicillin were used as negative and positive controls respectively.

To determine the MIC of the plant extracts against the test bacteria, the micro-broth dilution method was used. Organisms were inoculated into test tubes containing varying concentrations of 20.0, 10.0, 5.0, 2.5 1.0 and 0.5 mg/ml of plant extract in triplicates. To obtain these concentrations, varying concentrations (1 ml) of the extracts containing double strength of the concentrations (40.0, 20.0, 10.0, 5.0 2.0 and 1.0 mg/ml) were constituted in different test tubes, and 1 ml of nutrient broth (NB) was added. To determine the MIC, a loopful of the test bacteria previously diluted to 0.5 McFarland turbidity standard, was introduced into each broth sample. The procedure was repeated on the test organisms in test tubes containing NB and the standard antibiotic ampicillin (as negative control), or NB only (as positive control). All the culture tubes were then incubated at 37^0C for 24 h. After incubation, they were examined for bacterial growth by observing/measuring of turbidity.

3.0 Results and Discussion

The number of isolates obtained from each of the samples investigated is shown in Table 1. A total of 21 *E. coli* isolates including the serotypes O103:H2, O145:NM, O145:H2, O96:H9, 0126, 026:H11, 055, 0111:NM, O96:H9, 044 and 0124 were isolated from 32 wastewater and water samples (9 each from a wastewater treatment plant and an abattoir and 14 from Rivers Plankenburg and River Berg). Results also showed that a total of 14 *Acinetobacter* spp. were isolated comprising of 10 *A. lwoffii* (A1-10) and 4 *A. haemolyticus* (B1-4) isolates. The wastewater samples yielded the highest (13) *Acinetobacter* isolates while the river yielded only 1 isolate. The *Acinetobacter* spp. isolated has been reported to be predominantly environmental bacteria [21, 22].

Table 1. Number of isolates obtained from water and wastewater samples investigated.

Sampling station and number of samples	Number of samples	Number of isolates		
		Escherichia coli /serotypes	**Acinetobacter* spp. (A1-10)	**Acinetobacter* spp. (B1-4)
Wastewater (Plant)	9	10 (O103:H2, O145:NM, O145:H2, O96:H9, 0126)	5	3
Abattoir wastewater	9	8 (026:H11, 055, 0111:NM, O96:H9)	4	1
River water	14	3 (044, 0124)	1	0
Total	32	21	10	4

**Acinetobacter* spp. (A1-10) = *A. lwoffii; Acinetobacter* spp. (B1-4) = *A. Haemolyticus*

The organisms have recently gained increasingly significant attention due to their ability to cause severe nosocomial infections and to develop multidrug resistance [7, 21]. *A. baumannii* the most clinically significant, is responsible for outbreaks, and is currently the third commonest isolate from Gram-negative sepsis in immunocompromised patients, posing risk for high mortality [23]. When the isolates were screened for verotoxin production, 71 and 75% of the *E. coli* and *A. haemolyticus* isolates respectively were positive (Figure 1).

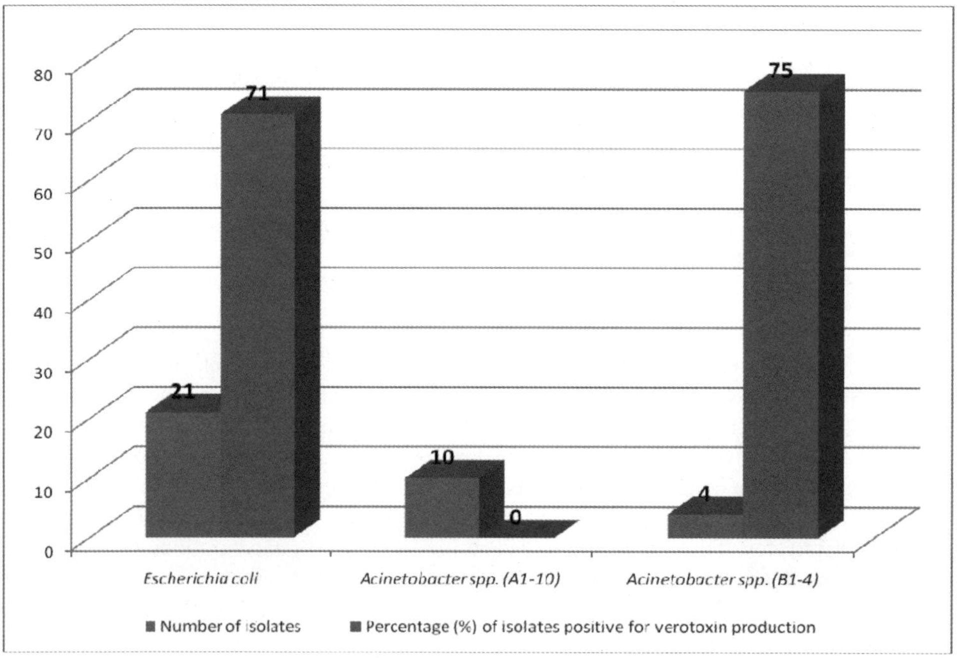

Figure 1. Number and percentage (%) of bacterial isolates obtained from water and waste water samples that were positive for verotoxin production (*Acinetobacter* spp. (A1-10) = *A. lwoffii; Acinetobacter* spp. (B1-4) = *A. haemolyticus*).

Verocytotoxin or shiga-like toxin-producing bacterial strains are associated with a broad spectrum of human illnesses throughout the world, ranging from mild diarrhea to hemorrhagic colitis (HC), hemolytic uremic syndrome (HUS) and thrombotic thrombocytopenic purpura (TTP) [12]. In this study several of these verotoxin producing non VTEC *E. coli* were isolated, thus revealing their presence in the environment. The significance of non-O157:H7 VTEC strains in human disease is less well understood than that of VTEC O157:H7 and is probably underestimated [24].

Table 2. Percentage of isolates resistant to four or more antibiotics, their multidrug resistance index (MDRIs) and *Curtisia dentata* mean MIC (mg/ml) values.

Isolate	Number of isolates	% of isolates resistant to four or more antibiotics	MDR index of isolates	Mean MIC (mg/ml) for *C. dentata*
Escherichia coli	21	61 (13)	38.5%	20
Acinetobacter spp. (A1-10)	10	44 (4)	30.8%	12.5
Acinetobacter (B1-4)	4	50 (2)	46.2%	2.5

Acinetobacter spp. (A1-10) = *A. lwoffii; Acinetobacter* spp. (B1-4) = *A. haemolyticus*

The antimicrobial susceptibility profiles and multidrug resistant index (MDRI) of all the bacteria (*E. coli* 61%, *A. lwoffii* 44% and *A. haemolyticus* 50%) isolates showed that all were resistant to more than four antibiotics indicating multidrug resistance (MDR) with multidrug resistance index above 46.2%. (Figures 2 and Table 2.), Antibiotics resisted include ampicillin, cefuroxime, cephalexin, ceftazidime, tetracycline, ofloxacin, ceftriaxone, nalidixic acid and amikacin. Multiple drug resistance among *Acinetobacter* spp. and *E. coli* has generated a lot of concern among hospital and veterinary workers because of difficulties in chemotherapy [5, 7, 17, 25]. Effect of stem bark extracts of *Curtisia dentata* on the verotoxic MDR isolates showed that at 12.5 mg/ml/disc, the extracts exhibited antibacterial activity against all the bacterial isolates (MIC values, 2.5-20 mg/ml) (Table 2).

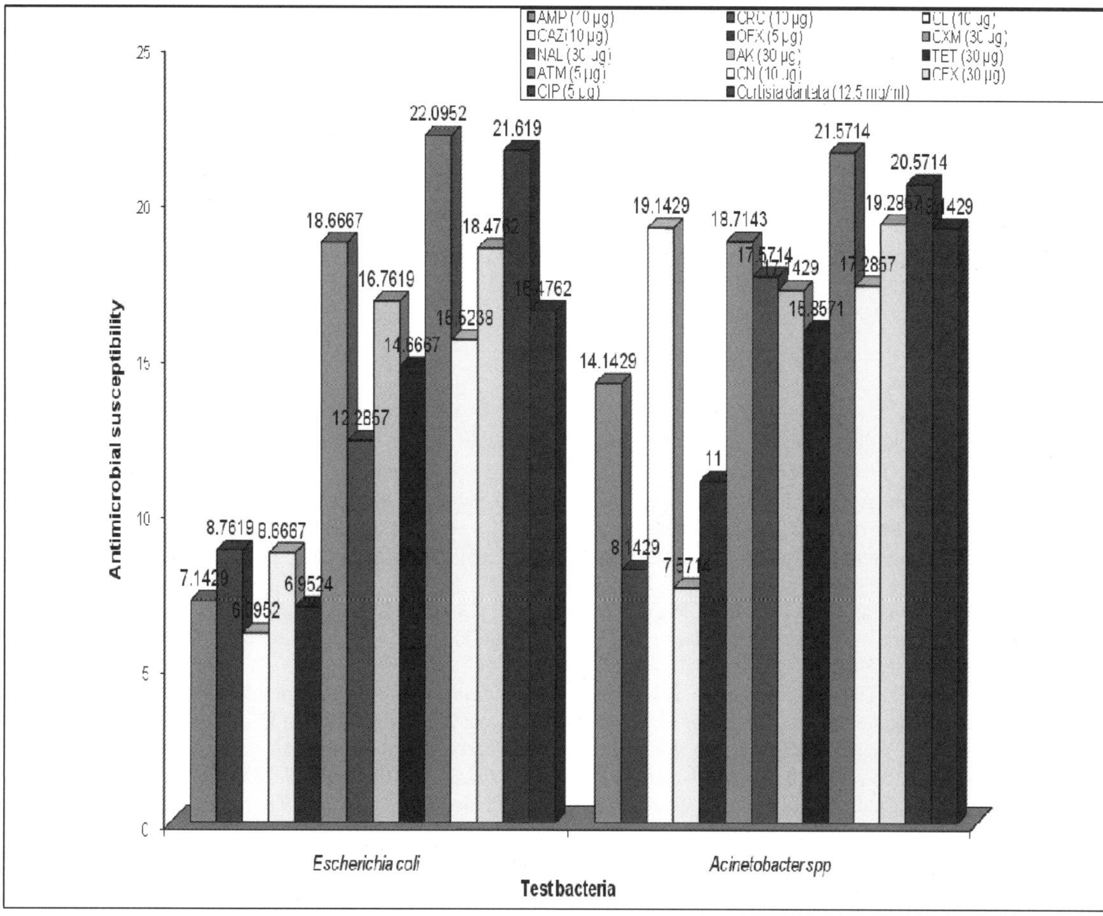

Figure 2. Mean Antimicrobial susceptibility profiles of *Escherichia coli* and *Acinetobacter* spp. isolates obtained from water and wastewater samples against some selected antibiotics and stem bark extracts of *Curtisia dentata.*

The plant extracts also possessed phytoconstituents including saponins, tannins, glycosides and anthraquinones (Table 3). The presence of these phytoconstituents accounts for the demonstration of antibacterial activity against the verotoxic MDR *E. coli* and *A. haemolyticus.* The effect of leaves and twigs of *C. dentata* against *B. subtilis E. coli, S. aureus, P. aerugenosa E. faecalis* as well as *C. albicans* has earlier been reported [26, 27]. This paper however, is the first report of the antimicrobial potency of *C. dentata* against verotoxic multidrug resistant (VDMR) bacteria pathogens.

Table 3. Phytochemical constituents of stem bark extracts of *Curtisia dentata.*

Phytoconstituent	Ethanol extract	Water extract
Saponins	+	+
Tannins	+	+
Alkaloids	-	-
Glycosides	-	+
Anthraquinones	+	+
Flavonoids	-	+

+ = present; - = absent

The study has shown that VMDR *E. coli* other than O157:H7 and *Acinetobacter* spp. are present in the South African water environment and that *C. dentata* has the potential to provide alternative source of antimicrobials that can be used in controlling these multi-drug resistant pathogenic bacteria. Authors are currently considering the effects of the plant extracts on the virulence as well as verocytotoxin production by the test bacteria.

Acknowledgement: The study was supported by Cape Peninsula University of Technology, Cape Town South Africa through the University Research Fund RP03.

References

[1] Sausa CP. The versatile strategies of *Escherichia coli* pathotypes: a mini review. *Journal of Venomous Animals and Toxins including Tropical Diseases.* 2006;12(3):363-373.
[2] Duffy G, Garvey P. Animal, food and biomedical aspects of verocytotoxigenic *E. coli*: a European concerted action project. *Research International.* 2000;33:267-272.
[3] Verweyen HM, Karch H, Brandis M, Zimmerhackl LB. Enterohemorrhagic *Escherichia coli* infections: following transmission routes. *Pediatrics Nephrology.* 2006;14:73-83.
[4] Dorsey CW, Tomaras AP, Connerly PL, Tolmasky ME, Crosa JH, Actis LA.The siderophore-mediated iron acquisition systems of *Acinetobacter baumannii* ATCC 19606 and *Vibrio anguillarum* 775 are structurally and functionally related. *Microbiology.* 2004;150:3657-3667.
[5] Grotiuz G, Sirok A, Gadea P, Varela G, Schelotto F. Shiga toxin 2-producing *Acinetobacter haemolyticus* associated with a case of bloody diarrhea. *Journal of Clinical Microbiology.* 2006;44(10):3838-3841.
[6] Castellanos ME, Asensio MT, Blanco VR, Suarez MR, Torrico AM, Llosa AC. Infective endocarditis of an intraventricular patch caused by *Acinetobacter haemolyticus. Infection.* 1995;23:243-245.
[7] Prashanth K, Badrinath S.. Epidemiological investigation of nosocomial *Acinetobacter* infections using arbitrarily primed PCR and pulse field gel electrophoresis. *Indian Journal of Medical Research.* 2005;122:408-418.
[8] Reinthaler FF, Feierl G, Galler H, Haas D, Leitner E, Mascher F, Melkes A, Posch J, Winter I, Zarfel G, Marth E. ESBL-producing *E. coli* in Austrian sewage sludge. *Water Research.* 2010;44:1981-1985.
[9] Abong'o BO, Momba MNB. Prevalence and characterization of *Escherichia coli* O157:H7 isolates from meat and meat products sold in Amathole District, Eastern Cape Province of South Africa. *Food Microbiology.* 2009; 26: 173-176.
[10] Browning NG, Botha J, Sacho H, Moore PJ. *Escherichia coli* O157:H7 haemorrhagic colitis. Report of the first South African case. *South African Surgical Journal.* 1990;28:28-29.
[11] Akinyemi KO, Oyefolu AO, Opere B, Otunba-Payne VA, Oworu AO.. *Escherichia coli* in patients with acute gastroenteritis in Lagos, Nigeria. *East African Medical Journal.* 1998;75:512-515.
[12] Doughari JH, Ndakidemi PA, Human IS, Bennade S. Verocytotoxic diarrhogenic bacteria and food and water contamination in developing countries: a challenge to the scientific and health community. *Reviews in Infection.* 2010;1(4):202-210.
[13] Notten A. *Curtisia dantata* (Burm.f.) C.A.Sm. South African National Biodiversity Institute's plant information website; 2004. Available at: http://www.plantzafrica.com/. Accessed September 28, 2010.
[14] Obire O, Tamuno DC, Wemedo SA. Bacteriological water quality of Elechi Creek in Port Harcourt, Nigeria. *Journal of Applied Sciences and Environmental Management.* 2005;9(1):79-84.
[15] Tarawneh KA, Al-Tawarah NM, Abdel-Ghani AH, Al-Majali, Ahmed M, Khleifat KM. Characterization of verotoxigenic *Escherichia coli* (VTEC) isolates from faeces of small ruminants and environmental samples in Southern *Jordan. Journal of Basic Microbiology.* 2009;49:310-317.
[16] Perilla G, Agello GMS, Bopp CMS, Elliot J, Facklam R, Knapp JS, Popovic T, Wells JMS, Dowell SF. Manual for Laboratory identification and antimicrobial susceptibility testing of bacterial pathogens of public health importance in the developing world. WHO.2003:103-139.
[17] National Committee for Clinical Laboratory Standards. Methods for dilution antimicrobial susceptibility tests for bacteria that grow aerobically; 15th ed. Approved standard (M100-S15), 25, No. 1. Wayne, PA: NCCLS; 2005.

[18] Chandran A, Hatha AAM,Varghese S, Sheeja KM. Prevalence of multidrug resistant *E. coli* serotypes in a Tropical Estuary, India. *Microbes and Environment.* 2008;23(2):153-158.
[19] Doughari JH, Obidah JS. Antibacterial potentials of stem bark extracts of *Leptadenia lancifoli.*against some pathogenic bacteria. *Pharmacologyonline.* 2008;3:172-180.
[20] Rojas JJ, Ochoa VJ, Ocampo SA, Muñoz JF. Screening for antimicrobial activity of ten medicinal plants used in Colombian folkloric medicine: A possible alternative in the treatment of non-nosocomial infections. *BMC Complimentary and Alternative Medicine.* 2006; 6:2-8.
[21] Vanbroekhoven K, Ryngaert A, Wattiau P, Demot R, Springael D. *Acinetobacter* diversity in environmental samples by 16S rRNA gene PCR-DGGE fingerprinting. *FEMS Microbial Ecology.* 2004;50:37-50.
[22] Joshi SG, Litake GM, Satpute MG, Telang NV, Ghole VS, Niphadkar KB. Clinical and demographic features of infection caused by *Acinetobacter* species. *Indian Journal of Medical Science.*2006;60(9):351-360.
[23] Barbe V, Vallene D, Fonknechten N, Kreimeyer A, Oztas S, Labarre L, Cruveiller S, Robert C, Duprat S, Wincker P, Ornston NL, Weissenbach J, Marlière P, Cohen GN, Médigue C. Unique features revealed by the genome sequence of *Acinetobacter* sp. ADP1, a versatile and naturally transformation competent bacterium. *Nucleic Acids Research.* 2004; 32(19):5766-5779.
[24] Bielaszewska M, Karsh H. Non-O157:H7 Shiga toxin (verocytotoxin)-producing *Escherchia coli* strains: epidemiological significance and microbiological diagnosis. *World Journal of Microbiology and Biotechnology.* 2000;16:711-718.
[25] Hawkey PM. The growing burden of antimicrobial resistance *Journal of Antimicrobial Chemotherapy.* 2008; 62(Suppl.)1: i1-i9.
[26] McGaw LJ, Eloff JN. Ethnoveterinary use of Southern African plants and scientific evaluation of their medicinal properties. *Journal of Ethnopharmacology.* 2008;119(3):559-574.
[27] Shai LJ, McGaw LJ, Eloff JN. Extracts of the leaves and twigs of the threatened tree *Curtisia dantata* (Cornaceae) aremore active against *Candida albicans* and other microorganisms than the stem bark extract. *South African Journal of Botany.* 2009;75:363-366.

Antimicrobial utilization in Intensive care units of a private tertiary care hospital

Pramil Tiwari[*]**, Vani Yadav and Shilpi Singh**

Department of Pharmacy Practice, National Institute of Pharmaceutical Education and Research (NIPER), S.A.S. Nagar-160062, Punjab, India

*Corresponding author email : ptiwari@niper.ac.in

OBJECTIVE - To evaluate the utilization of antimicrobials in ICUs of a private tertiary care hospital using RPM Plus indicators.

METHODOLOGY - A prospective observational study was conducted at two ICUs of the hospital-Coronary Care Unit (CCU) and Surgical ICU (SICU).

RESULTS & DISCUSSIONS -The number of patient records studied in SICU and CCU were 305 and 410, respectively. In SICU, the average number of total medications was 11.71±0.16, of which the average number of AMDs was found to be 1.32±0.04. For CCU, the average number of medications was 9.37±0.15 and out of this, the average number of AMDs prescribed was 1.79±0.05. 95% of AMDs were administered parenterally in SICU while it was 78.2% in CCU; however, this difference was statistically insignificant (p=0.546). The percentage of AMD prescribed by its generic name was 13.8% in SICU and 8.9% in CCU. AMDs prescribed from the National List of Essential Medicines (NLEM) in CCU were two times the numbers prescribed in SICU (42.5% vs 20.8%).

CONCLUSION – AMD consumption was found to be greater in CCU than SICU and the possible reasons could be that the patients admitted in CCU were relatively older and also because generally one antibiotic is used for surgical prophylaxis in SICU. There is a scope for improvement in the prescription of AMDs by generic name and also from the National List of Essential Medicines of India.

Keywords Antimicrobial utilization; Intensive care units

1. Introduction

Antimicrobial drugs (AMDs) are one of the most commonly utilized drugs in the hospital and their proper use and utilization is one of the major factors which ensures appropriate clinical outcomes.[1] In the past few years, there has been an increase in antimicrobial resistance to some of the commonly used antibiotics and it is more apparently encountered now.[2,3] This increase in the antimicrobial resistance over the years has not only challenged management of severe infections but has also impacted the way patient respond to treatment and the subsequent changes needed in empirical therapy.[4]

Antimicrobial resistance has emerged as important variable related to patient outcomes and overall resource use in the ICU. ICUs represent the most frequently identifiable source of nosocomial infections within the hospital, with infection rates and rates of antimicrobial resistance severalfold higher than in the general hospital setting.[2,5] The high prevalence of resistance in ICUs has been attributed to the severity of illness of the patients, prolonged hospital stays, and the widespread use of invasive devices and broad spectrum antimicrobials.[6,7] Besides treatment of infections, antimicrobials are administered as prophylaxis to prevent or limit major infections in such patients.[6] As a corollary, the total consumption is approximately ten times greater in ICU wards than in general hospital wards. Also a considerable part of total hospital antimicrobial consumption and cost are due to antimicrobials use in ICU.[8]

There are only limited data concerning the patterns of use of antimicrobial agents in developing countries especially in ICUs. This calls for an urgent need of medical auditing of prescribing trend which will help to optimize prescribing practice towards achieving cost effective therapeutics for patients. One of the mechanisms to ensure correct prescribing and use is the drug utilization review (DUR) process and regular evaluation of prescriptions. The assessment of drug utilization and monitoring of prescriptions is important for clinical, educational and pharmacoeconomic purposes and also restricting the emergence and spread of antimicrobial resistance.[9]

2. Methodology

AMD orders for patients from these ICUs were evaluated daily and the data was captured in the standard defined format from the treatment charts and case records of the inpatients. Patient related details: patient's name, age, sex, identification number, date of admission, date of discharge; drug related details: name, dosage, route of administration and duration of treatment, data related to antibiotic use(excluding antibacterials for TB and topical antimicrobials);diagnosis and progress of the patient. The patients will be followed from the date of admission till the time of discharge. All changes in diagnosis and progress will be considered for the evaluation. The data obtained was analyzed for selected prescribing indicators of the RPM plus criteria and the values were reported as average ± SEM or percentages, as applicable [10,11].

3. Results

The data was captured for 715 patients during the study period which was performed in Surgical ICU (SICU) and Coronary Care Unit (CCU) of a private tertiary care hospital. The data was then compiled into an excel sheet and the analysis was carried out to determine the various RPM indicators and the prescribing pattern of antimicrobials.

The average age of patients was 61.23±0.45 years. The number of males was higher with 73% while females accounted for 27% of the patient population. The average number of medications prescribed was 10.37±0.12. On an average 1.59±0.03 antimicrobials were prescribed to the patients. Considering both the ICUs, AMDs constituted about 15% of total medications used and about 63% of patients received one AMD. 23% of the patients received two AMDs while 14% of the patients received three or more AMDs. It was seen that parenteral route was preferred with 84.0% of the AMDs prescribed parenterally. Generic name prescription was seen in only 11% of the AMDs prescribed. The overall adherence of the prescribed AMDs to NLEM was 34.4% while it was observed that in both the ICUs, all AMDs were prescribed in accordance with the hospital formulary and showed no deviation. In the overall patient population, the average duration of the ICU stay was found to be 5.07±0.11 days. It was observed that cephalosporins was the most utilized drug class followed by floroquinolones and penicillins. In SICU, 1st generation cephalosporin was prescribed to most of the patients while in CCU, floroquinolone was the preferred class of AMDs.

The average number of medications was significantly higher in SICU when compared with CCU as seen in Table 1; 11.71±0.16 vs 9.37±0.15 (p<0.001). In SICU, as surgical prophylaxis single AMD use was preferred with average number of AMD utilized as 1.32±0.04 whereas it was greater in CCU with 1.79±0.05 (p<0.001). In SICU, 95% of AMDs were prescribed parenterally while in CCU it was 78.2%. Generic name prescription was 8.9% in CCU while it was13.8% in SICU. Adherence to NLEM was more than double in CCU than SICU; 42.5% vs 20.8% (p<0.001). Another indicator used was average duration of stay and it was 5.53±0.15 days in SICU and 4.72±0.17 days in CCU.

Table 1: RPM indicators used in the study.

	RPM Plus Indicator	**CCU**	**SICU**	**p value**
1.	Average no. of medications	9.37±0.15	11.71±0.16	<0.001
2.	Average no. of AMDs	1.79±0.05	1.32±0.04	<0.001
3.	% prescription by generic name	8.9	13.8	0.327
4.	% adherence to NLEM	42.5	20.8	<0.001
5.	% AMDs given parenterally	78.2	95	0.546
6.	Average duration of stay	4.72±0.17	5.53±0.15	<0.001

4. Discussion

Drug utilisation studies are undertaken with the objective of understanding the usage pattern of drugs which gives an insight to the prescribing pattern. It is known that ICUs are one of the most important sources of infections and this high risk of infections involve heavy consumption of antimicrobial agents which is about 10 times greater than in general wards. Thus they are often the target of surveillance programs and antimicrobial restrictions and the present study was also conducted to compute various indicators of AMD utilization in ICUs.

The average age of patients in the present study was found to be 61.23±0.45 years. This was found to be higher than a study [12] done in a tertiary care hospital in northern India (48.8years) and also the male proportion was higher as compared to the given study. But in another study done by Bergmans et al. [13] which was conducted in similar ICU setting, consisting surgical and non-surgical specialities with critically ill patients, showed median age of 60 years which is comparable to our study result. This implies a greater need of surveillance in these patients as they are more prone to drug interactions and adverse drug reactions.

In this study, the average number of medications prescribed was found to be 10.37±0.12. In a pilot study [14] conducted by Bista D et al in a teaching hospital in Nepal, the average number of drugs prescribed was 10.7.. Thus it is advisable to limit the practice of polypharmacy for optimization of therapy.

The present study documents an average of 1.59±0.03 antimicrobials prescribed to the patients. This is lower when compared to a study [15] done by Shankar RP et al in the Internal Medicine ward of a teaching hospital in Western Nepal (1.7). This may be because single antimicrobial agents are usually prescribed for both the ICUs considered in the present study- CCU and SICU.

AMDs constituted about 15% of total medications used with about 63% of patients receiving one AMD. A study [15] done by Shankar RP et al demonstrated that 50% of patients were prescribed only one AMD and another prevalence survey [16] done in Netherlands by Willemsen I et al found that 5% of the patients received two AMDs and only 1% were treated with three AMDs. The practice of prescribing minimum number of AMDs should be encouraged in the present setting also as it limits the emergence of resistance.

ICU settings demand high use of parenteral AMDs due to the critical condition of patients. In the present study also as high as 84% of AMDs were prescribed parenterally. This is very high as compared to a study [17] done in Israel hospital wards (64%) and a study done in a Critical Care Unit in Toronto [7] where the use was just 32%. This shows that there is a wide range seen in parenteral prescription. Nonetheless, use of oral form and also switching to oral AMDs as early as possible should be encouraged as it will reduce hospital stay and also the expenditures incurred.

Another indicator of drug utilization pattern which was found in the present study was generic name prescription. It was seen to be very low with only 11% of the AMDs prescribed by generic names. This was lower when compared to both, studies done in as well as outside India. In a study [18] done in our country in primary, secondary and tertiary level of health services by Ranjeeta K et al showed 27.1% generic drug prescription. Generic prescription should be encouraged to avoid confusion regarding multiple names of the same product and thereby simplifying dispensing of antimicrobials, facilitating generic substitution and improving hospital efficiency.

The overall adherence of the prescribed AMDs to NLEM was about 34.4% in the present study which is less when compared to a study [19] done in India by Mhetre NA et al (40%) as well as another study done in Nepal where 67% of prescribed AMDs complied with the NLEM of Nepal [15]. Use of essential drugs offer several advantages including cost, safety and effectiveness.

It is generally considered that the hospital formulary consists of drugs of choice to be used in a hospital and prescribing from it is one of the ways to optimize the use of medicines. In the present study, it was found that all AMDs were prescribed in accordance to the hospital formulary.

The average length of stay for patients receiving antimicrobials was found to be 5.07±0.11 days. This was shorter than studies done in other countries. An American study carried out by Yamashita SK et al [7] revealed average length of stay as 10.7 days and a study done in Greece [20] showed an average of 24.7 days. In the present study, the average length of stay may be shorter because patients admitted to the ICUs considered here require relatively shorter ICU stay.

Different drug classes were preferred in the two ICUs studied. It was noted that in CCU floroquinolone was the preferred class of drug whereas in SICU, 1st generation cephalosporin was prescribed to most of the patients as it was the drug of choice to be given as surgical prophylaxis. It is also documented in various studies and guidelines that cefazolin should be the drug of choice in cardiac surgeries. [21]

Thus, this study gives an account of prescribing and drug utilization pattern of AMDs in ICUs of a private tertiary care hospital in a developing country like India. The study points out the scope of improvement in certain areas like prescription of AMDs by generic name and prescription according to the NLEM of India. There was high AMD use in CCU which highlights the need of continuous monitoring of drug use and to develop certain guidelines for AMD prescription and ways to ensure adherence to them, especially in ICUs.

References

[1] Hariharana S, Pillaib G, McIntoshb D, Bhanjib Z, Culmerb L, McIntoshb KH. Prescribing patterns and utilization of antimicrobial drugs in a tertiary care teaching hospital of a Caribbean developing country. *Fundamental & Clinical Pharmacology*. 2009;23:609-615.

[2] Simor AE. Antimicrobial resistance in hospitals: How concerned should we be? *Canadian Medical Association Journal*. 2009;180(4):408-415.

[3] Erbay A, Çolpan A, Bodur H, Çevik MA, Samore MH, Ergönül Ö. Evaluation of antibiotic use in a hospital with an antibiotic restriction policy. *International Journal of Antimicrobial Agent*. 2003;21(4):308-312.

[4] Hunter PA, Reeves DS. The current status of surveillance of resistance to antimicrobial agents: report on a meeting. *Journal of Antimicrobial Chemotherapy*. 2002. 49;17-23.

[5] Vogelaersa D, Belsb DD, Forêtc F, Crand S, Gilberte E, Schoonheydtf K et al. Patterns of antimicrobial therapy in severe nosocomial infections: empiric choices, proportion of appropriate therapy, and adaptation rates—a multicentre, observational survey in critically ill patients. *International Journal of Antimicrobial Agent*. 2010;35:375-381.

[6] Allerberger F, Lechner A, Wechsler-Fordos A, Gareis R. Optimization of antibiotic use in hospitals-antimicrobial stewardship and the EU project ABS international. *Chemotherapy*. 2008;54(4):260-267.

[7] Yamashita SK, Louie M, Simor AE, Rachlis A. Microbiological surveillance and parenteral antibiotic use in a critical care unit. *Canadian Journal of Infectious Diseases*. 2000;11(2):107-111.

[8] Bergmans DC, Bonten MJ, Gaillard CA, Van Tiel FH, Van der Geest S, De Leeuw PW et al. Indications for antibiotic use in ICU patients: a one-year prospective surveillance. *Journal of Antimicrobial Chemotherapy*. 1997;39(4):527-535.

[9] WHO policy perspectives on medicines-promoting rational use of medicines: core components. Geneva: World Health Organization; 2002. Available at: http://apps.who.int/medicinedocs/pdf/h3011e /h3011e.pdf. Accessed June 17, 2010.

[10] Moore T, Bykov A, Zagorski A, Savelli T. Guidelines for implementing drug utilization review programs in hospitals-Rational Pharmaceutical Management Project (Russia). Moscow; 1997.

[14] How to Investigate Antimicrobial Drug Use in Hospitals: Selected Indicators. Draft published for the U.S. Agency for International Development by the Rational Pharmaceutical Management Plus Program. Arlington, VA: Management Sciences for Health.

[15] Habibi S, Wig N, Agarwal S, Sharma SK, Lodha R, Pandey RM,*et al*. Epidemiology of nosocomial infections in medicine intensive care unit at a tertiary care hospital in northern India. *Tropical Doctor*. 2008;38:233-235.

[16] Bergmans DC, Bonten MJ, Gaillard CA, Van Tiel FH, Van der Geest S, De Leeuw PW *et al*. Indications for antibiotic use in ICU patients: a one-year prospective surveillance. *Journal of Antimicrobial Chemotherapy*. 1997;39(4):527-35.

[17] Bista D, Saha A, Mishra P, Palaian S, Shankar PR. Pattern of Potential Drug-Drug Interactions in the Intensive Care Unit of a Teaching Hospital in Nepal: a Pilot Study. *Journal of Clinical and Diagnostic Research*. 2009;3:1713-1716.

[18] Shankar RP, Partha P, Shenoy NK, Easow MJ, Brahmadathan KN. Prescribing patterns of antibiotics and sensitivity patterns of common microorganisms in the Internal Medicine ward of a teaching hospital in Western Nepal: a prospective study. *Annals of Clinical Microbiology and Antimicrobials*. Available at: http://www.ann-clinmicrob.com/content/2/1/7. Accessed July 29, 2010.

[19] Willemsen I, Groenhuijzen A, Bogaers D, Stuurman A, Keulen P, Kluytmans J. Appropriateness of antimicrobial therapy measured by repeated prevalence surveys. *Antimicrobial Agents Chemotherapy*. 2007;51(3):864-867.

[20] Raveh D, Levy Y, Schlesinger Y, Greenberg A, Rudensky B, Yinnon AM. Longitudinal surveillance of antibiotic use in the hospital. QJM 2001;94(3):141-152.

[21] Ranjeeta K, Idris M, Vidya B, Anish K, Monika A, Shivendra S. Assessment of prescription pattern at the public health facilities of Lucknow district. *Indian Journal of Pharmacology*. 2008;40(6):243-247.

[22] Mhetre NA, Bodhankar SL, Pandit VA, Zambare GN. Study of pattern of drug usage in an urban area. *Indian Journal of Pharmacology*. 2003;35:316-317.

[23] Ntagiopoulos PG, Paramythiotou E, Antoniadou A, Giamarellou H, Karabinis A. Impact of an antibiotic restriction policy on the antibiotic resistance patterns of Gram-negative microorganisms in an Intensive Care Unit in Greece. *International Journal of Antimicrobial Agent*. 2007;30(4):360-5.

[24] Al-Momany NH, Al-Bakri AG, Makahleh ZM, Wazaify MMB. Adherence to international antimicrobial prophylaxis guidelines in cardiac surgery: a Jordanian study demonstrates need for quality improvement. *Journal of Managed Care Pharmacy*. 2009;15(3):262-267.

Bacterial clearance from blood in mice infected by *S. pneumoniae* (penicillin MIC=16 μg/ml) presenting specific IgG (non-protective levels) and treated with sub-therapeutic regimens of cefditoren (a highly bound cephalosporin)

Fabio Cafini[1], Jose Yuste[2], Maria-Jose Giménez[1], David Sevillano[1], Lorenzo Aguilar[1*], Luis Alou[1], Martha Torrico[1], Natalia González[1], Ernesto García[2], Jose Prieto[1], Pilar Coronel[3]

[1]Microbiology Department, School of Medicine, Universidad Complutense, Avda. Complutense s/n, 28040 Madrid, Spain;

[2]Molecular Microbiology and Infection Biology, Consejo Superior de Investigaciones Cientificas and CIBER-Enfermedades respiratorias, Ramiro de Maeztu 9, 28040 Madrid, Spain;

[3]Tedec-Meiji Farma, Ctra. M-300, Km. 30,500, 28802 Alcalá de Henares, Madrid, Spain

*Corresponding author: Fax no: +34 91 3941511; email: laguilar@med.ucm.es

A sepsis model using mice infected by a 23F *S. pneumoniae* was performed to explore the effect of the presence of non-protective levels of specific immunoglobulins on blood bacterial clearance by sub-therapeutic regimens of cefditoren (CDN). Intraperitoneal infecting inocula: $4.7x10^6$cfu/ml. The non-protective hyperimmune serum dilution (HS) and CDN regimen were administered to groups of 10 animals: a) Placebo group, b) HS group-received HS alone 1h pre-inoculation, c) CDN group- received CDN alone initiating the 48h tid treatment 1h post-inoculation, and d) HS+CDN group- received HS 1h pre-inoculation followed by CDN 48h tid treatment starting 1h post-inoculation. Survival and blood colony counts were recorded over 7 days. In the combined group (HS+CDN group) survival was 80% vs. 0% in other groups and bacteremia was significantly decreased.

Keywords: *S. pneumoniae*; serotype 23F; animal model; passive immunization; bacteremia; penicillin resistance.

INTRODUCTION

Colonisation is to some extend a B-cell immunising event [1,2]. Previous mice sepsis models (using a serotype 6B isolate non-susceptible to penicillin as infecting strain), where bacteremia with high colony counts was related with mortality, showed that both (bacteremia and mortality) decreased in a dose-related trend by administration of hyperimmune serum [3] or β-lactam treatment (amoxicillin or cefotaxime) [4]. When combined therapies (hyperimmune serum plus β-lactam) were explored, the strategy based on passive immunisation (prior to infection) followed by β-lactam administration post-infection was more effective than serotherapy plus antibiotherapy post-infection [5]. Pharmacodynamically, passive pre-immunisation (with non-protective dilutions of hyperimmune serum) of animals resulted in a significant reduction in the value of the parameter predicting efficacy for β-lactams (t>MIC) [6] when total drug was considered.

Taking into account the advantage of the same protein binding of cefditoren (CDN) in mice and humans, the aim of this study was to explore the antibacterial in vivo effect (bacteremia clearance) of non-protective CDN regimens and non-protective specific antibody serum concentrations, alone and in combination, in a mice sepsis model.

MATERIAL AND METHODS

A mice sepsis model was performed using a serotype 23F *S. pneumoniae* isolate ($MIC_{penicillin}$=16μg/ml; MIC_{CDN}=4μg/ml) as infecting strain. Groups of 20 mice were inoculated weekly (up to five weeks) by the intraperitoneal (ip) route with the heat-inactivated inoculum and exsanguinated by cardiac puncture to obtain hyperimmune serum (HS) as previously described [6]. IgG levels were measured as previously described [7]. A dose-ranging study with decreasing HS dilutions in one group of animals and cefditoren (CDN) doses in other group was performed to determine the HS dilution and the CDN tid regimen over 48h producing in separate ≤10% survival over 7 days after ip inoculation of $4.7x10^6$cfu/ml of the bacterial suspension. Once determined, these non-protective HS dilution (HS-np) and CDN regimen (CDN-np) were administered to four groups of 10 infected animals each, as follows:

a) Control group (K)- animals received placebo (PBS),

b) HS-np group- animals received the HS-np dilution ip 1h prior to inoculation (no CDN treatment),

c) CDN-np group- animals were ip treated with the tid CDN-np regimen for 48h, initiating the treatment 1h after bacterial ip inoculation,

d) HS-np +CDN-np group- animals received the HS-np dilution ip 1h prior to inoculation, and were ip treated with the tid CDN-np regimen for 48h, initiating the treatment 1h after bacterial ip inoculation.

Survival was recorded over 7 days. CDN serum concentrations were determined by biossay, protein binding was measured by an ultrafiltration method as previously described [8], and t>MIC (total and free) were calculated graphically from the semi-logarithmic plot of the concentration-time data. Colony counts were determined in blood samples collected at 2, 4, 6, 8, 24, 48, 72, 96, 120, 144 and 168h post-infection (only from alive animals). To collect blood samples, tails were disinfected, anaesthetized, and the terminal portion of the tail was eliminated with scissors. An 8-μl blood sample was collected by pressing the tail, resuspended in Todd-Hewitt broth containing 50 units of heparin, and plated onto blood agar for colony counting. To obtain the subsequent blood samples, the crust was removed and again, by pressing the anaesthetized tail, an 8-μl blood sample was collected and processed. The lower limit of detection was 10^2 cfu/ml.

RESULTS AND DISCUSSION

Mean IgG levels in the hyperimmune serum were 251 μg/ml. The non-protective dilution of hyperimmune serum was the 1/4 dilution. CDN protein binding determined in animals was 86.9%. The non-therapeutic regimen of CDN was 50 mg/kg (t>MIC of 22.4% for total concentrations and 12.7% for free concentrations). All animals receiving the non-therapeutic dose of CDN alone were dead within the first 24h, and those receiving placebo (control group) or the non-protective HS dilution alone were dead on day 2. The group with the combined strategy (pre-immunisation + CDN treatment) showed 80% survival at day 7.

Figure 1 shows the bacteremic profile in the four study groups and Table 1 shows colony counts in surviving animals over 7 days. In contrast to non-significant reductions in the control, HS-np and CDN-np groups, reductions in the HS-np + CDN-np group were >3 $\log_{10}$ (in vivo bactericidal activity in bacteremia clearance) at 168h versus time 0.

In this mice animal model with in vivo experimental protein binding for CDN of 86.9%, similar to the one determined in humans [9], pre-immunisation with a non-protective dilution of hyperimmune serum increased survival rates of a sub-therapeutic regimen of CDN from 0% to 80%, by obtaining an in vivo bactericidal activity highly decreasing blood colony counts.

This study shows that in the presence of non-protective levels of specific antibodies CDN t>MIC values (22.4% for total concentrations and 12.7% for free concentrations), lower than those classically considered predictive of efficacy, were related to in vivo efficacy (bacteremia clearance). As colonisation is an immunising event (producing non-protective levels of antibodies), the synergism between CDN and non-protective levels of specific antibodies in this study may represent a therapeutic advantage in infections caused by strains with high penicillin MIC.

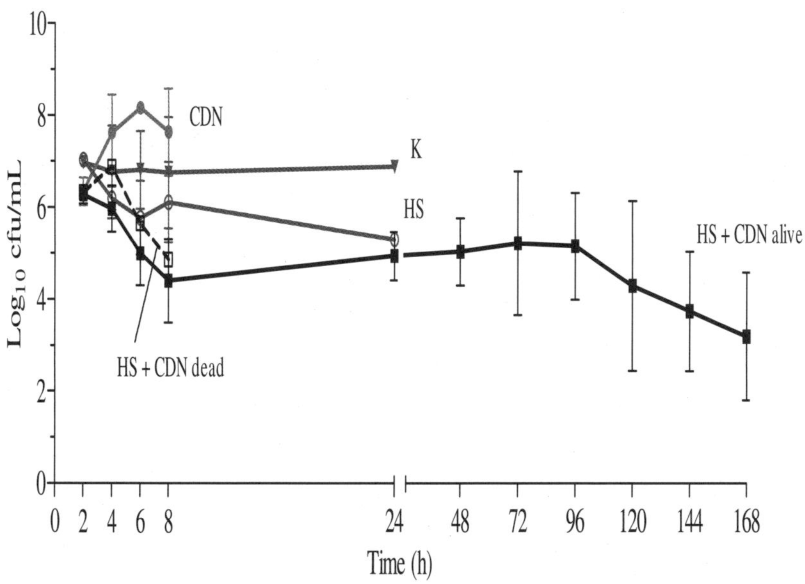

Figure 1. Colony counts ($\log_{10}$ cfu/ml) in blood over 168h for the four study groups (K: control; HS: non-protective hyperimmune serum dilution; CDN: sub-therapeutic cefditoren regimen)

Table 1. Mean ± SD of colony counts (log_{10} cfu/ml) in blood over 168h for surviving animals in the four study groups (HS-np: non-protective hyperimmune serum dilution; CDN-np: non-protective CDN regimen)

Time (hours)	Control	HS-np	CDN-np	HS-np + CDN-np
2	7.0 ± 0.1	7.0 ± 0.0	6.4 ± 0.3	6.3 ± 0.2
4	6.8 ± 1.0	6.2 ± 0.3	7.6 ± 0.8	6.2 ± 0.6
6	6.8 ± 0.8	5.8 ± 0.8	8.2 ± 0.1	5.2 ± 0.7
8	6.8 ± 1.2	5.8 ± 0.8	7.6 ± 1.0	4.5 ± 0.8
24	6.9 ± 0.0	5.3 ± 0.0	-	4.9 ± 0.5
48	-	-	-	5.0 ± 0.7
72	-	-	-	5.2 ± 1.6
96	-	-	-	5.2 ± 1.2
120	-	-	-	4.3 ± 1.6
144	-	-	-	3.7 ± 1.3
168	-	-	-	3.2 ± 1.4

REFERENCES

[1] Weinberger DM, Dagan R, Givon-Lavi N, Regev-Yochay G, Malley R, Lipsitch M. Epidemiologic evidence for serotype-specific acquired immunity to pneumococcal carriage. *J Infect Dis* 2008;197:1511-1518.

[2] Soininen A, Pursiainen H, Kilpi T, Käyhty H. Natural development of antibodies to pneumococcal capsular polysaccharides depends on the serotype: association with pneumococcal carriage and acute otitis media in young children. *J Infect Dis* 2001; 184:569-576.

[3] Yuste J, Jado I, Giménez MJ, Aguilar L, Molero F, Fenoll A, Casal J. Modification of bacteraemia by specific antibodies and relation with mortality in a pneumococcal mouse sepsis model. *Clin Exp Immunol* 2002;128:411-415.

[4] Yuste J, Jado I, Fenoll A, Aguilar L, Giménez MJ, Casal J. Beta-lactam modification of the bacteraemic profile and its relationship with mortality in a pneumococcal mouse sepsis model. *J Antimicrob Chemother* 2002;49:331-335.

[5] Yuste J, Fenoll A, Casal J, Giménez MJ, Aguilar L. Combined effect of specific antibodies (as serotherapy or preimmunization) and amoxicillin doses in treatment of *Streptococcus pneumoniae* sepsis in a mouse model. *Antimicrob Agents Chemother* 2002;46:4043-4044.

[6] Casal J, Aguilar L, Jado I, Yuste J, Giménez MJ, Prieto J, Fenoll A. Effects of specific antibodies against Streptococcus pneumoniae on pharmacodynamic parameters of beta-lactams in a mouse sepsis model. *Antimicrob Agents Chemother* 2002;46:1340-1344.

[7] Sha J, Erova TE, Alyea RA, Wang S, Olano JP, Pancholi V, Chopra AK. Surface-expressed enolase contributes to the pathogenesis of clinical isolate SSU of *Aeromonas hydrophila*. *J Bacteriol* 2009;191:3095-3107.

[8] Craig WA, Suh B. Protein binding and the antimicrobial effects: methods for the determination of protein binding. In: Lorian V, ed. *Antibiotics in laboratory medicine*, 3rd ed. Philadelphia: Williams and Wilkins, 1991:367-402.

[9] Wellington K, Curran MP. Cefditoren pivoxil: a review of its use in the treatment of bacterial infections. *Drugs* 2004;64:2597-2618.

Characterisation of Methicillin Resistant *Staphylococcus aureus* Isolates from Hospitalised Patients

Vladimir Kmet[1], Daniela Ohlasova[2] and Milan Niks[3]

[1]Institute of Animal Physiology, Slovak Academy of Sciences, Soltesovej 4, 040 01 Kosice, Slovakia, Email: kmetv@saske.sk
[2]Department of Microbiology, Medy, Marsala Koneva 1, 040 01 Kosice, Slovakia,
[3]Public Health Authority of the Slovak Republic, Trnavska 52, 826 45 Bratislava, Slovakia

The occurrence of Panton Valentin leukocidin in 99 invasive MRSA strains isolated from hospitalized patients was shown to be around 9,1 %. Five of nine PVL positive strains had SCCmec type IV casette. The predominating SCCmec type in PVL negative and SCCmec typable MRSA was II. Such strains were simultaneously resistant to erythromycin, clindamycin and ciprofloxacin. The linezolid resistance in invasive MRSA strains from Slovak hospital was not recorded. The MLVA characterization of 8 PVL positive isolates from various locations has demonstrated three different patterns. Except one hospital, isolates from over the country expressed the same MVLA pattern.

Keywords: MRSA, Panton-Valentin leukocidin, MLVA, hospital

1. Introduction

Methicillin-resistant *Staphylococcus aureus* continues to be a significant cause of both health care-associated and community-associated infections. In 2003 represented MRSA 3% of community and 6,3% of hospital *Staphylococcus aureus* isolates respectively. In 2009 the occurrence of community MRSA was already 13,8 %, in slovak intensive care units reached MRSA proportion as much as 17% [1]. The aim of the study was to characterize hospital MRSA isolates more accurately, namely to detect Panton-Valentine leukocidin genes and SCC*mec* cassette types in current invasive staphylococcal strains isolated from hospitalized patients.

2. Material and methods

During a period of 11 months 99 of methicillin resistant *Staphylococcus aureus* isolates from soft tissue abscesses and chirurgical wounds in inpatients (1 isolate per patient) from 7 county hospitals around the country were obtained and analyzed. For phenotypic methicillin resistance detection in *S. aureus* and antibiotic susceptibility determination (including linezolid), CLSI (2008) criteria were used. PVL gene (lukF-PV, lukS-PV), *mecA* gene [2] and SCC*mec* types [3-5] were determined by PCR. Additionally, multilocus variance analysis (MLVA) was applied to selected PVL positive strains to demonstrate relatedness of the isolates [6] .

3. Results and discussion

All of 99 methicillin resistant *Staphylococcus aureus* strains included in this study possessed the *mecA* gene. Of them, 9,1 % (9/99) were identified PVL positive. Five of nine PVL positive MRSA strains possessed SCC*mec* type IV cassette. One of them (No 1/7 BB), was obtained from a deep abscess in the patient on the second hospitalization day and therefore this strain should be considered rather a community acquired MRSA. Other PVL positive MRSA possessed also II and III types SCC*mec* casettes**,** while one was not typable.

Of 90 PVL negative MRSA strains, in 24 SCC*mec* cassettes were typable. The most common type was the II. All these isolates were typically resistant to erythromycin, clindamycin and ciprofloxacin in a susceptibility test. Similarly, Takano et al. [7] found that 90% of hospital MRSA were resistant to erythromycin, clindamycin and levofloxacin.

All of 99 tested MRSA strains were phenotypically susceptible to linesolid. It is difficult to cure staphylococal infections, because MRSA are multiple drug resistant. Linezolid resistance emerged in two isolates of ST36 MRSA colonizing the lungs of two paediatric cystic fibrosis patients [8]. Yanagihara et al. [9] (described that linezolid significantly reduced bacterial numbers and inflammation in a mouse model of PVL-positive *Staphylococcus aureus* haematogenous infection and improved the survival rate of infected mice.

The MLVA analysis of 8 PVL positive MRSA Slovak isolates has demonstrated 3 different patterns. All 3 patterns could be documented in 3 isolates from just one county hospital (BB). The other 5 PVL positive isolates, even from different geographical locations, expressed the identical MVLA pattern (A) and may be of the common origin. (Table 1). The simultaneous analysis of the variation in number of repeat units in seven individual genes (*sspA*, *spa*, *sdrC*, *sdrD*, *sdrE*, *clfA*, and *clfB*) endows MLVA with a discriminatory power comparable to PFGE [10].

Table 1 Geographical localisation and MLVA similarity of eight PVL positive MRSA strains

DateIsolate No.Region SCCmec typeMLVA pattern

VIII.7/37BB (middle)IVA
VIII.6/31BB (middle)IIIB
X.4/20KN (west)IVA
X.5/21KN (west)IVA
XI.3/19KE (east)NTA
XI.2/14KE (east)IVA
XI.1/7BB (middle)IVC
II.8/90CH (east)IIA

NT not typable

According to our study, the PVL positive MRSA represent about 9% of hospital MRSA isolates most of which seem to be of a common origin, spread over the country. The prevalence of PVL positivity in MRSA in Slovakia might be higher compared to other neighbour countries as Pantucek and colleagues [11] report in a recent study of 2336 *Staphylococcus aureus* wound isolates from Czech Republic as few as 20 PVL positive MRSA. PVL screening in austrian hospital revealed only 81 of 2,542 MRSA cases (3.2%) during the investigated period of 1999–2007 [12] .

Acknowledgments. The authors thank all the regional microbiology laboratories that submitted strains to the National Reference Laboratory for Antibiotics of Public Health Authority. This study was supported by Pfizer Luxembourg SARL, Slovakia and by slovak VEGA grant No. 2/0012/08

References

[1] Slovak National Antimicrobial Resistance Surveillance System: www.snars.sk

[2] McClure J.A., Conly J.M., Lau V., Elsayed S., Louie T., Hutchins W., Zhang K. Novel multiplex PCR assay for detection of the staphylococcal virulence marker Panton-Valentine leukocidin genes and simultaneous discrimination of methicillin-susceptible from -resistant staphylococci. *J Clin Microbiol* 2006, 44: 1141-1144

[3] Zhang K., McClure J.A., Elsayed S., Louie T, Conly J.M. Novel multiplex PCR assay for simultaneous identification of community-associated methicillin-resistant *Staphylococcus aureus* strains USA300 and USA400 and detection of mecA and Panton-Valentine leukocidin genes, with discrimination of *Staphylococcus aureus* from coagulase-negative staphylococci. *J Clin Microbiol.* 2008, **46:** 1118-1122

[4] Hanssen A.M., Kjeldsen G., Sollid J.U. Local variants of staphylococcal cassette chromosome mec in sporadic methicillin-resistant *Staphylococcus aureus* and methicillin-resistant coagulase-negative staphylococci: evidence of horizontal gene transfer? *Antimicrob. Agents Chemother.* 2004. 48: 285–296.

[5] Milheirico C., Oliveira D.C., de Lencastre H. Update to the Multiplex PCR Strategy for Assignment of *mec* Element Types in *Staphylococcus aureus. Antimicrob Agents Chemother* 2007, 251: 3374–3377

[6] Sabat A., Krzyszton-Russjan J., Strzalka W., Filipek R., Kosowska K., Hryniewicz W., Travis J., Potempa J. New method for typing *Staphylococcus aureus* strains: multiple-locus variable-number tandem repeat analysis of polymorphism and genetic relationships of clinical isolates. *J Clin Microbiol* 2003, 41: 1801-1804

[7] Takano T., Higuchi W., Yamamoto T. Superior in vitro activity of carbapenems over anti-methicillin-resistant *Staphylococcus aureus* (MRSA) and some related antimicrobial agents for community-acquired MRSA but not for hospital-acquired MRSA. *J. Infect. Chemother.* 2009, 15:54–57

[8] Hill R.L.R., Kearns A.M., Nash J., North S.E., Pike R., Newson T., Woodford N., Calver R., Livermore D.M. Linezolid-resistant ST36 methicillin-resistant *Staphylococcus aureus* associated with prolonged linezolid treatment in two paediatric cystic fibrosis patients. *J. Antimicrob. Chemother.* 2010, 65:442-445

[9] Yanagihara K., Kiharac R., Arakia N., Morinaga Y., Sekib M., Izumikaw K., Kakeya H., Yamamoto Y., Yamada Y. Kohno S., Tsukamoto K., Kamihira S. Efficacy of linezolid against Panton–Valentine leukocidin (PVL)-positive meticillin-resistant *Staphylococcus aureus* (MRSA) in a mouse model of haematogenous pulmonary infection. *Inter. J. Antimicrob. Agents* 2009, 34: 477-481

[10] Sabat A., Malachowa N., Miedzobrodzki J., Hryniewicz W. J. Comparison of PCR-based methods for typing *Staphylococcus aureus* isolates. *Clin. Microbiol.* 2006, 44: 3804–3807

[11] Pantucek R., Machova I., Kahankova J., Petras P., Kuntova L., Doskar J. Characteristics of Panton - Valentine leukocidin positive *Staphylococcus aureus* strains isolated in the Czech Republic. In 13th Int. Symposium on Staphylococci and Staphylococcal Infections. Cairns, Australia : Australian Society for Antimicrobials, Book of abstracts 2008, p. 66.

[12] Bauer C.C., Apfalter P., Daxboeck F., Bachhofner N., Stadler M., Blacky A., Diab-Elschahawi M.,Assadian O. Prevalence of Panton-Valentine leukocidin genes in methicillin-resistant *Staphylococcus aureus* isolates phenotypically consistent with community-acquired MRSA, 1999–2007, Vienna General Hospital. *Eur. J. Clin. Microbiol. Infect. Dis.* 2009, 28:909–912

Characterization of methicillin-resistant coagulase-negative Staphylococci Isolates from blood cultures in a Brazilian University Hospital

Valéria Cataneli Pereira[*,1] **and Maria de Lourdes Ribeiro de Souza da Cunha**[1]

[1] Department of Microbiology and Immunology, Biosciences Institute, UNESP - Univ Estadual Paulista, Distrito de Rubião Júnior, Botucatu, SP, Brazil.

*Corresponding author: e-mail: cataneli@ibb.unesp.br, Phone: +55 1438116240

Coagulase-negative staphylococci (CNS) are the microorganisms most frequently isolated from clinical materials and represent the leading cause of bacteremia in hospitals. The pathogenesis of infections caused by these bacteria can be aggravated due to antimicrobial resistance and virulence factors. Thus, the present study characterized the CNS species isolated from blood cultures from inpatients of the University Hospital of Botucatu School of Medicine for the detection of methicillin-resistant *Staphylococcus* (MRS) and staphylococcal enterotoxin genes. The gene *mecA* was detected in 83.7% CNS and the staphylococcal cassette chromosome *mec* was typed in MRS, indicating 45.8% type III, 25% type I, 11.1% type IV and 5.5% type II. The search for staphylococcal enterotoxin genes indicated that 77 samples were positive for at least one gene. The data obtained in this study highlight the importance of detecting and characterizing MRS and virulence factors to better understand the profile of the isolated CNS in order to appropriately treat infections caused by these microorganisms.

Keywords: Coagulase-negative staphylococci; *mecA*; enterotoxins; SCC*mec*; PCR.

1. Introduction

Coagulase-negative staphylococci (CNS) are members of the normal microflora of human skin, constituting the predominant microorganisms isolated from clinical materials [1]. Over the last two decades, CNS have been considered important pathogens, especially in immunocompromised patients, newborns and patients with prosthetic implant. CNS are recognized as opportunistic organisms prevailing in a large number of organic situations to produce severe infections [2-3]

In hospital settings, CNS are the leading cause of bacteremia, most frequently in patients at intensive care units (ICUs) [1]. Resistance to antibiotics by CNS has increased in recent years, impairing the treatment of infections caused by these microorganisms. Moreover, the pathogenesis of infections caused by these bacteria is complex and multifactorial, and several virulence factors are involved in infections caused by *Staphylococcus*, such as the production of toxins [4].

The increase in antimicrobial resistance in recent years has a fundamental importance for nosocomial infections. The problem is that infections caused by methicillin-resistant *Staphylococcus* (MRS) are difficult to treat and, in some cases, are susceptible only to glycopeptides and experimental drugs [5]. The intrinsic resistance of *Staphylococcus* to methicillin is due to the production of a supplemental penicillin-binding protein (PBP 2' or PBP 2a), which presents low affinity for semi-synthetic penicillins and is encoded by the chromosomal gene *mecA*. The gene *mecA* is identical in all *Staphylococcus* strains and is therefore a useful molecular marker of methicillin resistance [6]. This gene is carried on a mobile genetic element identified as a specific staphylococcal cassette chromosome *mec* (SCC*mec*). Thus, eleven types of SCC*mec* have been described [7] and types I, II and III are predominantly present in MRS isolates from hospitals, while other types are found in the community. SCC*mec* type III encodes the largest number of resistance genes and is an important pathogen in hospitals, causing serious infections [8]. SCC*mec* type IV is small in size (29.9-24.3 kb) and does not encode genes of resistance to different non-beta-lactam antibiotics, being susceptible to these agents [9].

Besides antimicrobial resistance, virulence factors of *Staphylococcus* such as the production of enterotoxins may aggravate infections caused by these microorganisms. Enterotoxins are also considered the major cause of food poisoning. They are associated with a form of gastroenteritis manifesting clinically with vomiting, with or without diarrhea, and with various immune system disorders [10]. Superantigens affect the immune system by binding to proteins of the class-II major histocompatibility complex (MHC II) and activating specific T-cell types through the variable regions of the T-cell receptors beta chains. The activation of such T-cells leads to the release of cytokines like Tumor Necrosis Factor (TNF), several interleukins and interferon-gamma [11]. The detection of staphylococcal enterotoxins is important to confirm an outbreak of food poisoning and to determine enterotoxigenic strains [12].

With the increasing number of infections caused by CNS, the study of virulence factors and their susceptibility to antimicrobial agents are extremely important to better understand the pathogenic potential of these microorganisms. Thus, the aim of the present study was to characterize the CNS species isolated from blood cultures from inpatients of the University Hospital of Botucatu School of Medicine (HC-FMB) in order to detect MRS and staphylococcal enterotoxin genes.

2. Material and Methods

2.1 Samples

A total of 86 CNS strains isolated from blood cultures from inpatients of HC-FMB were studied. The strains were isolated as described by Koneman et al [13] and species of the genus *Staphylococcus* were identify as described by Baker [14]. The phenotypic identification of CNS followed the criteria proposed by Kloos & Schleifer [15], and Kloos & Bannerman [16], according to the simplified scheme of biochemical tests [17], which includes the assessment of the following sugars: xylose, arabinose, sucrose, trehalose, mannitol, maltose, lactose, xylitol, ribose and fructose. In addition, the criteria also included characterization of hemolysins, nitrate reduction, urease, ornithine decarboxylase and novobiocin resistance.

The genotypic identification of CNS was performed by using primers of conserved sequences adjacent to 16S and 23S according to the technique of Internal Transcribed Spacer – PCR (ITS-PCR) described by Barry et al [18] and Couto et al [19]. To control the results, the following international reference strains were used: *S. auricularis* (ATCC 33753), *S. capitis* subsp. *capitis* (ATCC 27843), *S. capitis* subsp. *ureolyticus* (ATCC 49325), *S. caprae* (ATCC 35538), *S. cohnii* (ATCC 49330), *S. cohnii* subsp. *cohnii* (ATCC 29974), *S. epidermidis* (ATCC 12228), *S. epidermidis* (ATCC 35983), *S. haemolyticus* (ATCC 29970), *S. hominis* (ATCC 27844), *S. hominis* subsp. *novobiosepticus* (ATCC 700237), *S. lentus* (ATCC 700403), *S. lugdunensis* (ATCC 700328), *S. saprophyticus* (ATCC 15305), *S. schleiferi* subsp. *schleiferi* (ATCC 43808), *S. sciuri* subsp. *sciuri* (ATCC 29062), *S. simulans* (ATCC 27851), *S. xylosus* (ATCC 29979) and *S. warneri* (ATCC 10209).

2.2 DNA Extraction

Total nucleic acids were extracted from *Staphylococcus* strains cultured on blood agar, individually inoculated into BHI broth and incubated for 24 h at 37°C. DNA was extracted by using the Illustra kit (GE healthcare). Briefly, staphylococcal cells were first digested with 10 μg/ml lysozyme and 20 μg/ml proteinase K. Then, 500 μl of the extraction solution was added and the mixture was centrifuged at 5000 × *g* for 1 min. The supernatant was then transferred to a GFX column and centrifuged at 5000 × *g* for 1 min. The collected fluid was discarded and 500 μl of the extraction solution was again added to the column. After centrifugation and discarding of the collected fluid, 500 μl of the washing solution was added to the column which was centrifuged at 14,000 × rpm for 25 min. The column was then transferred to a 1.5-ml tube and 200 μl Milli-Q water heated to 70 °C was used for elution. The isolates were centrifuged at 5000 × *g* for 1 min and the GFX column was discarded. The extracted DNA was stored in a refrigerator at 4 °C.

2.3 Detection of the gene *mecA*

PCR reactions to detect the gene *mecA* were performed by following the parameters and primers described by Mukarami et al [20]. In all reactions, international reference strains were used as positive (*S. aureus* ATCC 33591) and negative (*S. aureus* ATCC 25923) controls.

2.4 Detection of staphylococcal enterotoxin genes

PCR reactions to detect enterotoxin genes were performed by following the parameters described by Johnson et al [21] and Cunha et al [22]. *S. aureus* international reference toxigenic strains were used as positive control, including ATCC 13565 (SEA), ATCC 14458 (SEB), ATCC 19095 (SEC), and ATCC 23235 (SED).

2.5 Detection of SCC*mec*

SCC*mec* was characterized in all MRS samples by using the method of Multiplex PCR reaction and following the parameters and primers described by Oliveira and Lencastre [23] and simplified by Machado et al [24].

3. Results and Discussion

3.1 Coagulase-Negative staphylococci

A total of 86 CNS strains isolated from blood cultures from inpatients of HC-FMB were studied. The phenotypic method identified 83.7% *S. epidermidis*, 8.2% *S. warneri*, 4.6% *S. haemolyticus* and 3.5% *S. capitis*. The genotypic method identified 70.9% *S. epidermidis*, 9.3% *S. warneri*, 7% *S. haemolyticus* and *S. capitis*, 4.6% *S. hominis*, and 1.2% *S. cohnii.*

ITS-PCR allowed the analysis of intergenic transcribed spaces between loci 16S and 23S of rRNA, a technique often used for the identification and discrimination of bacterial strains at species and subspecies level [25]. This technique was important to confirm the species identification and rule out possible errors during the phenotypic identification since these methods depend on the phenotypic expression of genes which, for various reasons, may not be expressed.

3.2 Detection of the gene *mecA* and typing of SCC*mec*

The gene *mecA* was detected in 83.7% CNS, being present in 85.2% *S. epidermidis*, 87.5% *S. warneri*, 83.3% *S. capitis*, 25% *S. hominis*, 100% *S. haemolyticus* and in the only strain of *S. cohnii* (Fig 1). Studies have reported that 60-85% of the *Staphylococcus* isolated from clinical materials are resistant to methicillin. [26]. The gene *mecA* was detected in all species isolated in the study, and *S. epidermidis* was most frequently isolated and had high percentage of methicillin resistance.

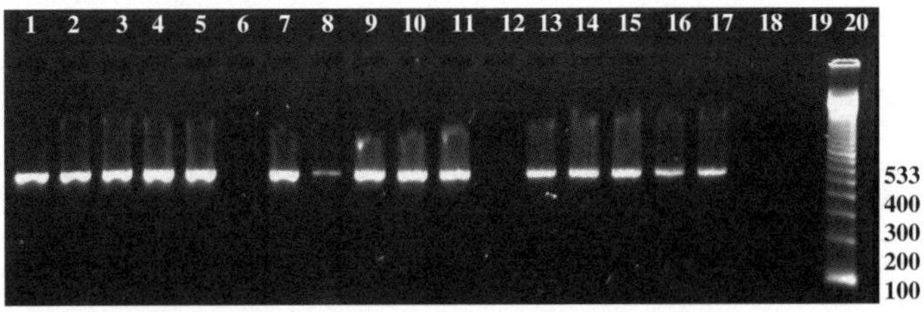

Fig 1. Agarose gel electrophoresis for the detection of the gene *mecA* (533 bp) in CNS strains by PCR. Lanes 1, 2, 3, 4, 5, 7, 8, 9, 10, 11, 13, 14, 15 and 16: positive strains; lanes 6 and 12: negative strains; lane 17: positive control (*S. aureus* ATCC 33591); lane 18: negative control (*S. aureus* ATCC 25923); lane 19: water; lane 20: molecular weight marker (100 bp).

All 72 samples positive for the gene *mecA* were subjected to multiplex PCR for SCC*mec* typing, which allowed the characterization of types I, II, III and IV (Fig 2). Among these samples, 45.8% were type III, 25% type I, 11.1% type IV, 5.6% type II and 12.5% were not typed. Similar results were found in the study of Machado et al [24], in which MRS characterization indicated 28% SCC*mec* type I, 3% type II, 52% type III and 13% type IV. Typing of SCC*mec* showed the prevalence of nosocomial MRS in samples isolated from blood culture. SCC*mec* types I, II and III mainly cause nosocomial infections and are significantly larger than other types; they can carry genes that provide resistance to various non-beta-lactam antibiotics [27].

The characterization of SCC*mec* type IV was important to detect the presence of community strains isolated from clinical materials. SCC*mec* type IV is smaller than hospital types and this probably increases their mobility and ability to move to the samples [28], suggesting that diseases caused by clones containing SCC*mec* type IV may increase in number [24].

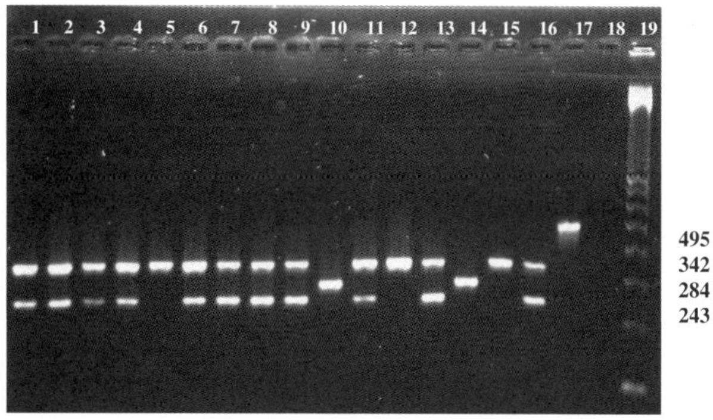

Fig 2. Agarose gel electrophoresis for the detection of genes SCC*mec* in CNS strains by Multiplex PCR. Lanes 1, 2, 3, 4, 6, 7, 8, 9, 11, 13 and 16: SCC*mec* type III (243pb and 342pb); lanes 5 and 15: SCC*mec* type IV (342pb); lanes 10 and 14: SCC*mec* type II (284pb); lane 17: SCC*mec* type I (495pb); lane 18: water; lane 20: molecular weight marker (100 bp).

3.3 Detection of enterotoxin genes

The genes *sea*, *seb*, *sec* and *sed* which encode the staphylococcal enterotoxins A, B, C and D, respectively, were searched in all CNS samples, indicating 77 samples positive for at least one gene, and 88.4% CNS positive for the gene *sea*, 29.1% for *seb*, 24.4% for *sec*, and 5.8% for *sed*. Of the 72 MRS samples, 94.4% were positive for one enterotoxin gene investigated in this study. Among these samples, the gene *sea* was detected in 94% MRS characterized as SCC*mec* type I, 100% type II, 91% type III and 88% type IV. The gene *seb* was detected in 33% MRS type I, 25% type II, and 41% type III. The gene *sec* was detected in 16.7% MRS type I, 50% type II, 19% type III and 25% type IV. On the other hand, the gene *sed* was detected in only 16.7% MRS type I (Fig 3).

The large number of staphylococcal enterotoxin genes in MRS samples is an important result because the expression of such genes may cause a variety of toxic effects that aggravate the pathogenesis of infections caused by these microorganisms. The data obtained in this study highlight the importance of detecting and characterizing MRS and virulence factors to better understand the profile of the isolated CNS in order to appropriately treat infections caused by these microorganisms.

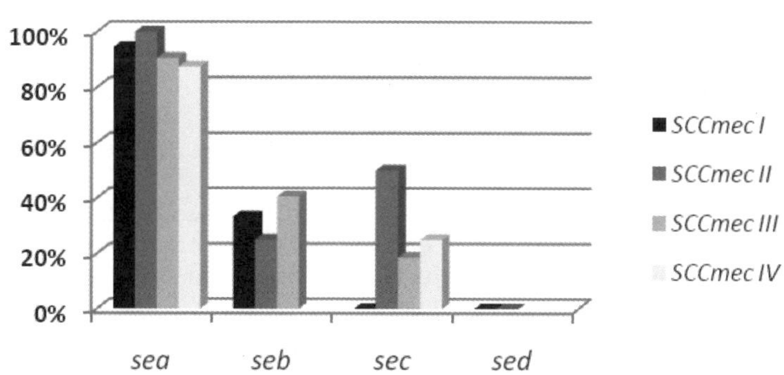

Fig 3. Characterization of SCC*mec* according to the detected staphylococcal enterotoxin genes.

Acknowledgements The support by FAPESP is gratefully acknowledged.

References

[1] Kloos WE, Bannerman TL. Update on clinical Significance of coagulase-negative staphylococci. *Clin Microbiol Rev.* 1994; 7: 117-140.

[2] Kloos WE, Bannerman TL. *Staphylococcus* and *Micrococcus.* In: Murray PR, Baron EJ, Pfaller MA, Tenover FC, Yolken RH, eds. *Manual of Clinical Microbiology.* 6 ed. Washington: American Society Microbiology.1999; 264-282.

[3] Rowlinson MC, LeBourgeois P, Ward K, Song Y, Finegold SM, Bruckner DA. Isolation of a Strictly Anaerobic Strain of *Staphylococcus epidermidis. J Clin Microbiol.* 2006; 44(3):857-860.

[4] Kein L S. Mapeamento dos estafilococos coagulase negativo no hospital universitário Antonio Pedro da Universidade Federal Fluminense, no período de 1998 a 2002. Niterói (RJ): Universidade Federal Fluminense; 2005.

[5] Stefani S, Varaldo PE. Epidemiology of Methicilin resistant staphylococci in Europe. *Clin Microbiol Infect.*2003; 9:1179-1186.

[6] Archer, G., Niemeyer, D.M. Origin and evolution of DNA associated with resistance to methicillin in staphylococci. *Trends in Microbiol.*1994;2:343-347.

[7] Classification of Staphylococcal Cassette Chromosome mec (SCCmec): Guidelines for Reporting Novel SCCmec Elements† International Working Group on the Classification of Staphylococcal Cassette Chromosome Elements (IWG-SCC). *Anticrobial Agents and Chemotherapy.* 2009; 53(12)4961–4967.

[8] Ito T, Katayama Y, Hiramatsu K. Cloning and nucleotide sequence determination of the entire *mec* DNA of pre methicillin-resistant *Staphylococcus aureus* N315. *Antimicrob. Agents Chemother.* 1999;43:1449-14.

[9] Lopes HV. CA-MRSA: um novo problema para o infectologista. *Rev Panam Infectol.* 2005;7(3):34-36.

[10] Dinges MM, Orwin PM, Schleivert PM. Exotoxins of *Staphylococcus aureus. Clin Microbiol Rev.* 2000;13(1):16-34.

[11] Vianello MA. Caracterização genotípica dos fatores de virulência e seu regulador *agr* em cepas de *Staphylococcus aureus* sensíveis à oxacilina. São Paulo: Faculdade de Ciências Farmacêuticas, Universidade de São Paulo; 2006.

[12] Bergdoll MS. Analytical methods for *Staphylococcus aureus.* Int *J Food Microbiol.*1990;10:91-100.

[13] Koneman EW, Allen SD, Janda WM, Schreckenberger PC, Winn Jr WC. *Color Atlas and Textbook of Diagnostic Microbiology.* 5th ed. Philadelphia: Lippincott. 1997.

[14] Baker JS. Comparison of various methods for differentiation of staphylococci and micrococci. *J Clin Microbiol.* 1984; 9:875-879.

[15] Kloos & Schleifer

[16] Kloos WE, Bannerman TL. *Staphylococcus* and *Micrococcus.* In: Murray PR, Baron EJ, Pfaller MA, Tenover FC, Yolken RH, editors. *Manual of Clinical Microbiology.* 6ed. Washington: American Society Microbiology; 1999;264-282.

[17] Cunha MLRS, Sinzato YK, Silveira LVA. Comparision of methods for identification of Coagulase-negative Staphylococci. *Mem Inst Oswaldo Cruz.*2004; 99:855-60.

[18] Barry T, Colleran G, Glennon M, Dunican LK, Gannon F. The 16S/23S ribosomal spacer region as a target for DNA probes to identify eubacteria. *PCR Methods Appl.* 1991; 1:51-56.

[19] Couto I, Pereira S, Miragaia M, Sanches IS, Lencastre H. Identifi cation of clinical staphylococcal isolates from humans by Internal Transcribed Spacer PCR. I. *Clin. Microbiol.*2001; 39:3099–3103.

[20] Murakami K, Minamide K, Wada K, Nakamura E, Teraoka H, Watanabe S. Identification of methicillin-resistant strains of staphylococci by polymerase chain reaction. *J. Clin. Microbiol.* 1991; 29:2240-44.

[21] Johnson WM, Tyler SD, Ewan EP, Ashton FE, Pollard DR, Rozee KR. Detection of genes for enterotoxins, exfoliative toxins, and toxic shock syndrome toxin 1 in Staphylococcus *aureus* by the polymerase chain reaction. *J Clin Microbiol.*1991; 29:426-430

[22] Cunha MLRS, Calsolari RAO, Araújo Jr JP. Detection of enterotoxin and toxic shock syndrome toxin 1 genes in *Staphylococcus*, with emphasis on coagulase-negative staphylococci. *Microbiology and Immunology.* 2007; 51:381–90.

[23] Oliveira DC, Lencastre H. Multiplex PCR strategy for rapid identification of structural types and variants of the *mec* element in Methicillin-Resistant *Staphylococcus aureus. Antimicrob.Agents Chemother.* 2002; 46(7):2155-61.

[24] Machado ABMP, Reiter KC, Paiva RM, Barth AL. Distribution of staphylococcal cassette chromosome *mec* (SCC*mec*) types I, II, III and IV in coagulase-negative staphylococci from patients attending a tertiary hospital in southern Brazil. *J Med Microbiol* .2007; 56:1328-1333

[25] Daffonchio D, Cherif A, Brusetti L, Rizzi A, Mora D, Boudabous A, Borin S. Nature of Polymorphisms in 16S-23S rRNA Gene Intergenic Transcribed Spacer Fingerprinting of Bacillus and Related Genera. *Appl Environ Microbiol.* 2003; 69:5128–5137.

[26] Kuehnert M J, Kruszon-Moran D, Hill HA, McQuillan G, McAllister K, Fosheim G, McDougal LK, Chaitram J, Jensen B, Fridkin SK, Killgore G, Tenover FC. Prevalence of *Staphylococcus aureus* nasal colonization in the United States, 2001–2002. *J. Infect. Dis.* 2006; 193:172–179.

[27] Shore A, Rossney AS, Keane CT, Enright MC, Coleman DC. Seven novel variants of the staphylococcal chromosomal cassette *mec* in methicillin-resistant *Staphylococcus aureus* isolates from Ireland. *Antimicrobs Agents Chemother.* 49; 2070-2083.

[28] Healy CM, Hulten KG, Palazzi DL, Cambell JR, Baker CJ. Emergence of New Strains of Methicillin-Resistant *Staphylococcus aureus* in a Neonatal Intensive Care Unit. Clinical Infectious Diseases. 2004;39:1460-1466.

Control of bacterial contamination in boar semen doses

J.M. Morrell [1*] and Margareta Wallgren [1, 2]

[1] Division of Reproduction, Swedish University of Agricultural Sciences (SLU), Box 7054, SE-75007 Uppsala, Sweden; [2] Quality Genetics, Råby, Hörby, Sweden

*Corresponding author: e-mail jane.morrell@kv.slu.se

The increase in antibiotic-resistant bacteria is thought to have been caused, at least partly, by the high levels of antibiotics used in human and veterinary medicine around the world. Antibiotic usage in the animal breeding industry varies according to species because of differences in semen storage and handling techniques, and is stipulated by an EU Directive for intra-community trade. The species in which liquid semen is stored for several days e.g. pigs and horses, have the most problem with bacterial contamination. Any reduction in antibiotic usage could help to reduce the spread of antibiotic resistance. Possible alternatives to antibiotics in semen extenders are considered, namely to reduce bacterial contamination by strict attention to hygiene, to remove bacteria by means of Single Layer Centrifugation, or to find alternatives microbicides, for example peptides from plants or from marine organisms. The potential removal of viruses by the same processing technique is considered. Lastly, possibilities for future research are considered.

Keywords antibiotics, semen extenders, boar semen, Single Layer Centrifugation (SLC), Androcoll-P.

1. Introduction

The purpose of this article is to consider the use of antibiotics to control bacterial contamination in semen doses for artificial insemination (AI), in particular in the boar semen industry. The problems associated with this usage will be reviewed and potential alternatives discussed. Finally, some speculations about the future are made.

2. Why use antibiotics in semen extenders?

It is difficult, although not impossible, to collect semen from domestic animals without some degree of microbial contamination (boars [1], 1996; stallions [2]; bulls [3]). Sources of contamination include the skin and hair of the male and of the semen collector, the respiratory tract and digestive tract of the male, the collection equipment and the environment in which the collection is made [4]. These bacteria are part of the normal flora of animals and do not arise because of ill-health. However, contamination with digestive tract bacteria should be kept to a minimum by strict attention to hygiene. Further sources of contamination arise in the laboratory where semen processing is carried out, for example, from receptacles, work surfaces and even the water used to make up semen extenders [5].

Natural mating is not an aseptic process. The female reproductive tract has developed its own physiological mechanisms to deal with microbial contamination during mating. However, for AI, it is standard practice to add extenders to semen to maintain sperm viability and permit lower sperm numbers to be used per female [6]. These extenders, containing salts, sugars and proteins, provide an excellent environment for supporting bacterial growth, which could cause disease in the female and provide a source of endotoxins that are toxic to spermatozoa. In addition, bacteria compete with spermatozoa for nutrients. A negative correlation has been found between bacterial contamination above a certain threshold and litter size in pigs [4], (Maroto Martin et al, 2010). Therefore, antibiotics are added, both to prevent disease in inseminated females and to maintain sperm viability. The addition of antibiotics is controlled by international, national or regional Directives. Within the European Union, Council Directive 90/429/EEC stipulates which antibiotics are to be added for intra-community trade and their minimum concentrations (Text Box 1). Thus, semen supplier must conform to this Directive.

Text Box 1: Extract from Council Directive 90/429/EEC, Annex C2 [7].

"An effective combination of antibiotics, in particular against leptospires and mycoplasmas, must be added to the semen after final dilution. This combination must produce an effect at least equivalent to the following dilutions:

- 500 IU per ml streptomycin
- 500 IU per ml penicillin
- 150 mg per ml lincomycin
- 300 mg per ml spectinomycin".

3. Level of use

The AI industry is enormous. In the European Union, for example, more than 90% of pigs, dairy cattle and turkeys are bred by AI, and the technique is also used to a lesser extent in beef cattle, sheep, horses and dogs. The benefits of AI include livestock improvement, disease control, facilitation of cross-breeding programmes convenience and economics [8]. However, the extensive use of AI means that there is also an attendant extensive use of antibiotics in semen extenders. A rough calculation of the amount of antibiotic usage for the pig breeding industry within the European Union is shown in Text Box 2.

Text Box 2: Calculation of usage of antibiotics per year in boar semen extenders within the European Union

- There are approximately 14 million sows in Europe [9], of which more than 90% are bred by AI;
- Each sow is expected to produce approximately 2.3 litters of piglets per year;
- Therefore, at least 50 million AI doses are used per year;
- Each dose is 80 ml semen + extender containing antibiotics;
- **Thus, 4 million L of boar semen extender containing antibiotics are used per year!**

This calculation does not include any doses prepared but not subsequently used, or any discarded semen extender.

By extrapolation from the antibiotic doses given in Text Box 1, the European pig industry uses **at least 160 billion IU streptomycin, 160 billion IU penicillin, 20000 kg lincomycin and 40000 kg spectinomycin per year** in semen doses for AI.

The figures vary between species because of differences between species in sperm number per AI dose and in the volume required for successful AI [10], as well as in the mode of semen preservation. In cattle AI, for example, frozen sperm doses are commonly used, whereas cooled semen doses are more frequent in pig breeding [11]. Therefore, the potential for bacterial growth is limited to the length of time during which the semen dose is cooled prior to freezing and any time post-thawing that the sperm dose is held above 0°C before AI. In contrast, it is customary to use cooled liquid semen for AI in pigs because pregnancy rates and litter sizes are higher than when frozen semen doses are utilised Thus, semen doses are stored at 16-18°C for 3-7 days before use. Although this temperature is not optimal for bacterial growth, some growth and multiplication do occur. It is thought that some contaminating bacteria even have the potential to adapt to these low environmental temperatures, leading to overgrowth in cooled extender [6]. Similarly, cooled stored semen doses are used for equine breeding, since cryopreservation protocols that are suitable for all individuals remain elusive. Thus, the total use of antibiotics in animal breeding worldwide is potentially very large.

4. Is the use of antibiotics a problem?

Globally there is increasing concern about the spread of antibiotic resistance in various strains of microbes [12]. Even a small amount of usage of antimicrobials against one particular strain of bacteria can cause the appearance of antibiotic resistance in a number of bacterial species [13]. Thus alternative strategies for controlling microbial contamination that can lead to a reduction in the use of antimicrobials should be welcomed and encouraged. The efficacy of any antimicrobial can be reduced by various physical conditions, such as the presence of organic matter, protein (which is a problem since seminal plasma is rich in protein), pH, and even water hardness. The microbes themselves may be able to circumvent exposure to antimicrobials, for example by the formation of biofilms or in association with protozoa in water.

Suitable alternatives to the use of antibiotics include reduction in environmental contamination, removing or reducing the contamination by physical means, or by controlling microbial growth using alternatives to antibiotics that have an antimicrobial effect. These options will now be considered individually.

5.1 Reduction in contamination

There is little that can be done about the normal flora of the skin, hair and respiratory tract of the male, although the collector can reduce his/her own potential contribution by strict attention to personal hygiene and the cleanliness of the collection equipment [14]. Collectors should wear gloves, preferably double, so that the outer, potentially contaminated, glove can be removed immediately before semen collection begins. Protective clothing and footwear should always be worn; this should be laundered regularly or disinfected as appropriate. People with respiratory infections should consider wearing masks or avoid the semen collection area altogether. Semen collection equipment and the collection area should be thoroughly cleaned and disinfected where appropriate. However, it should be borne in mind that all soaps and disinfectants are spermicidal to some degree: therefore, these substances should never be left on surfaces that will come in contact with semen. Most semen collection stations have strict standard operating procedures regarding cleanliness and hygiene, which should be strictly followed at all times [14].

5.2 Microbicidal alternatives to antibiotics

Alternatives to antibiotics such as peptides [15-16], and naturally occurring substances e.g. in plants or sea animals, have been identified and may be suitable for use in semen extenders in the future. Studies are progressing to identify candidate substances. The main problem with finding an alternative microbicide for use in semen extenders is that the substance must be microbicidal yet non-toxic to spermatozoa and must not interfere with sperm function. Thus substances that alter the pH, or are strongly oxidising, such as nitric oxide [17] cannot be used in the presence of spermatozoa, since they would kill or inactivate them [18]. The problem of finding substances that are toxic to microbes but non-toxic to mammalian cells is not unique to semen extenders: it occurs when preparing solutions for treating wound and soft tissue infections [17] as well as in solutions for disinfecting contact lenses [19].

5.3 Removal of bacteria from the semen before preparing the doses for AI

It may be possible to remove or reduce bacterial contamination from semen by centrifugation through a colloid. This method enables the best spermatozoa – the most motile, viable, morphologically normal spermatozoa with good chromatin integrity – to be selected from the rest of the ejaculate for use in assisted reproduction. Until recently, the method was restricted to density gradient centrifugation (DGC) for preparing semen for in vitro fertilization (IVF) or intracytoplasmic sperm injection (ICSI), usually for human fertility treatments [20]. IVF and ICSI are not widely used in animal assisted reproduction for practical reasons [21]. DGC was not used to prepare animal semen for AI because it was too impractical to deal with the high sperm concentration and/or large ejaculate volume of most animal ejaculates compared to human ejaculates. However, the colloid centrifugation technique has now been modified to enable only one layer of colloid to be used (Single Layer Centrifugation, SLC) and new species-specific colloid formulations, such as Androcoll, have been developed for use in SLC [20-21]. The combination of the two developments enables even large volumes of semen, such as are commonly encountered with boar ejaculates, to be processed easily in a practical manner [22].

6. The SLC method

The SLC methodology will now be considered in detail. The colloid, e.g. a species-specific formulation of Androcoll, is poured into a centrifuge tube and the extended semen (at a sperm concentration of 100×10^6/mL) is carefully layered on top [23] (Figure 1).

Figure 1: The Single Layer Centrifugation technique in practice.

Note: The 50-mL Falcon tube contains 15mL of Androcoll. Extended semen (in this case stallion semen) is being pipetted slowly on top of the colloid so that it forms a distinct layer. After centrifugation at 300 g for 20 min, the supernatant will be discarded and the sperm pellet transferred to a clean tube where it will be resuspended to the desired sperm concentration.

(Photograph courtesy of Marjet van Wienen)

After centrifugation at 300 g for 20 min, the supernatant, consisting of seminal plasma, extender, dead spermatozoa and colloid, is removed and the sperm pellet containing the most robust spermatozoa is resuspended in semen extender to the desired concentration for AI doses.

When the SLC-method was used with ten boar ejaculates, bacterial contamination was removed or substantially reduced [24]. Bacterial growth was absent from six SLC-samples, and the remaining four SLC-samples were only lightly contaminated, for example with Gram-negative bacteria. After storage for 24h at 16-18°C, the numbers of bacteria in the four SLC samples had not increased, whereas in the uncentrifuged controls, bacterial counts had nearly doubled. Therefore, it appeared that SLC could be a practical method for controlling bacterial contamination without the use of antibiotics, with the added advantage that sperm quality is improved at the same time. Similarly, studies have been conducted to investigate whether spermatozoa can be separated from other pathogens in the ejaculate, such as viruses, by SLC. Porcine circo virus 2 was removed, or substantially reduced, from spiked boar sperm samples using a combination of SLC and "swim-up" (Blomqvist et al, unpublished data). These results confirm earlier studies when stallion spermatozoa were obtained free of equine arteritis virus from a stallion known to be shedding the virus in semen [25]. Note that it is not the intention to use semen from diseased animals for AI, but rather to increase biosecurity in semen doses for AI to avoid disease-transmission. However, it should be noted that SLC is no substitute for good hygiene during semen collection, particularly since flagellated bacteria, such as enteric bacteria, may accompany the spermatozoa through the colloid and thus appear in the sperm pellet, albeit in reduced numbers, after centrifugation [24].

7. Future studies

The results of the preliminary study reported previously [24] showed considerable potential as a means of controlling bacterial growth without the use of antibiotics. However, further work is necessary before this method can be adopted by the breeding industries. The Directive refers particularly to the control of Leptospires and Mycoplasmas, neither of which was present in the ejaculates used in the study reported previously. Therefore, an attempt should be made to remove these organisms, either from ejaculates collected from naturally infected animals, or preferably by spiking normal ejaculates with known doses of these bacteria and quantifying their removal. Furthermore, it would be useful to repeat this spiking study with bacteria known to be pathogenic in other species, for example *Taylorella equigenitalia* and *Klebsiella pneumoniae* in horses.

In the preliminary experiment [24], the sperm pellet was harvested aseptically in a laminar air flow hood (LAF). However, few, if any, semen collection stations have access to such equipment. Therefore, it will be necessary to perform some SLCs in the same environment as normally provided at a semen collection station, to investigate potential re-contamination of the sperm pellet from the environment after SLC. Moreover, if some slight contamination is present, either as residual contamination in the sperm pellet or from re-contamination, how much bacterial multiplication occurs at the temperature and length of duration of semen storage in each species, bearing in mind that the storage temperatures and times vary between species.

Finally, it would be useful to scale-up the SLC method even further, particularly when dealing with boar ejaculates. Currently, both stallion and boar ejaculates have been processed as 25mL aliquots on top of 20 mL Androcoll in 100 mL glass tubes, or even as 50 mL aliquots on top of 40 mL colloid in 200 mL plastic tubes, although the latter caused problems with visualisation of the sperm pellet because they were not transparent [26, 22].

Further experiments are underway to see if even larger volumes of boar semen can be processed by the SLC-method.

An additional modification of the existing technique could potentially make the SLC more efficient at removing pathogens. Work with ejaculates from HIV patients has been carried out using double tubes [27] or "separable neck" tubes [28], to lessen the chance of recontaminating the sperm pellet after centrifugation. Similar modifications should be investigated for animal species, although the use of such specialised consumables may be too costly to permit its general adoption by the semen industry. Its use may be restricted to processing ejaculates from high-risk males. The SLC method for removing viruses should be tested with the semen of other species and other viruses of different densities.

Acknowledgements The support of the Swedish Farmers´ Foundation, Stockholm, Sweden, is gratefully acknowledged.

References

[1] Wallgren, M., Screening of mycoplasmal and bacterial content in semen from Swedish AI-boars. 14th Int. Pig. Vet. Congr., Bologna, Italy, 601.

[2] Malmgren L, Olsson Engvall E, Engvall A, Albihn A. Aerobic bacterial flora of semen and stallion reproductive tract and its relationship to fertility under field conditions. Acta vet. Scand. 1998;39:173-182.

[3] Smole I, Thomann A, Frey J, Perreten V. Repression of common bull sperm flora and *in vitro* impairment of sperm motility with *Pseudomonas aeruginosa* introduced by contaminated lubricant. Reprod. Dom. Anim. 2010;45:737-42.

[4] Maroto Martin LO, Cruz Munoz E, de Cupere F, van Driessche E, Echemendia-Blanco D, Machado Rodriguez J, Beeckmans S. Bacterial contamination of boar semen affects the litter size. Animal Reproduction Science 2010; 120 :95-104

[5] Althouse GC. & Lu KG. Bacteriospermia in extended porcine semen. Theriogenology 2005;63:537-584.

[6] Althouse GC. Sanitary procedures for the production of extended semen. Reprod Dom Anim 2008;43: 374-378.

[7] Council Directive 90/429/EEC. http://eur-lex.europa.eu

[8] Foote RH, The history of artificial insemination: Selected notes and notables. J. Anim Sci 2002;80:1-10.

[9] Marquer P Eurostat statistics in focus 8/2010. http:/epp.eurostat.ec.europa.eu/cache/ITY_OFFPUB/KS-SF-10-008/EN/KS-SF-10-008-EN.PDF.

[10] Revell S. Comparative AI in pigs and cattle. http://www.thecattlesite.com/articles/873/comparative-ai-in pigs-and-cattle accessed 25 November 2010.

[11] Gerrits RJ, Lunney JK, Johnson LA, Pursel VG, Kraeling RR, Rohrer GA, Dobrinsky JR. Perspectives for artificial insemination and genomics to improve global swine populations. Theriogenology 2005; 63:283-99.

[12] Catry B, van Duijkere E, Pomba MC, Greko C, Moreno MA, Pyörälä, S, Ruzauskas M, Sanders P, Threlfall EJ, Ungemach F, Törnele K, Munoz-Madero C, Torren-Edo J, Scientific Advisory Group on Antimicrobials (SAGAM). Reflection paper on MRSA in food-producing and companion animals: epidemiology and control options for human and animal health. Epidemiol. Infect. 2010;138:626-644.

[13] Johansson A, Greko C, Engström BE, Karlsson M. Antimicrobial susceptibility of Swedish, Norwegian and Danish isolates of Clostridium perfringens from poultry, and distribution of tetracycline resistance genes. Vet. Microbiol. 2004;99:251-257.

[14] Maes D, Nauwynck H, Rijsselaere T, Mateusen B, Vyt P, de Kruif A, Van Soom A. Diseases in swine transmitted by artificial insemination: An overview. Theriogenology 2008; 70: 1337-1345.

[15] Blondelle SE, Takahashi E, Dinh KT, Houghton RA. The antimicrobial activity of hexapeptides derived from synthetic combinatorial libraries J Appl. Bioteriol 1995;78:39-46 .

[16] Dathe M, Wieprecht T. Structural features of helical antimicrobial peptides: their potential to modulate activity on model membranes and biological cells. Biochim Biophys Acta 1999;1462:71-87.

[17] Englander L, Friedman A. Nitric Oxide Nanoparticle Technology. A Novel Antimicrobial Agent in the Contextof Current Treatment of Skin and Soft Tissue Infection. J Clin Aesthetic Dermatol. 2010;3:45–50.

[18] JB Weinberg, E Doty, J Bonaventura, AF Haney Nitric oxide inhibition of human sperm motility. Fertil Steril 1995; 64: 408-13.

[19] Willcox MDP, Hume EBH, AliwargaY, Kumar N and Cole N. A novel cationic-peptide coating for the prevention of microbial colonization on contact lenses 2008 doi:10.1111/j.1365-2672.2008.03942.

[20] Morrell JM, Rodriguez-Martinez, H. Biomimetic techniques for improving sperm quality in animal breeding: a review. The Open Andrology Journal 2009;1: 1-9.

[21] Morrell JM, Rodriguez-Martinez H. Practical applications of sperm selection techniques as a tool for improving reproductive efficiency. Veterinary Medicine International doi:10.4061/2011/2984767 .

[22] Van Wienen M, Johannisson A, Wallgren M, Parlevliet J, Morrell JM. Single Layer Centrifugation with Androcoll™-P can be scaled-up to process larger volumes of boar semen. Veterinary Science doi:10.5402/2011/548385.

[23] Morrell JM, Wallgren M. Androcoll™-P-Large selects boar spermatozoa with good membrane integrity from the sperm-rich fraction of the ejaculate. Animal Reproduction *in press*.

[24] Morrell JM, Wallgren M. Removal of bacteria from boar ejaculates by using Single Layer Centrifugation can reduce the use of antibiotics in semen extenders Animal Reproduction Science 2010 doi:10.1016/j.anireprosci2010.11.005.

[25] Morrell JM, Geraghty RJ. Effective removal of equine arteritis virus from stallion semen. Equine Veterinary Journal. 2006;38:224-229.

[26] Morrell JM, Johannisson A, Dalin A-M, Rodriguez-Martinez H. Single Layer Centrifugation with Androcoll™-E can be scaled-up to allow large volumes of stallion ejaculate to be processed easily. Theriogenology 2009;72:879-884.

[27] Politch JA, Chong Xu, Tucker L, Anderson DJ. Separation of human immunodeficiency virus type 1 from motile sperm by the double tube gradient method versus other methods Fertility Sterility 2004;81: 440-447.

[28] Kato S, Hanabusa H, Kaneko S, Takakuwa K, Suzuki M, Kuji N, Jinno Mm, Tanaka R, Kojima K, Iwashita M, YoshimuraY, Tanaka K. Complete removal of HIV-1 RNA and proviral DNA from semen by the swim-up method: assisted reproduction technique using spermatozoa free from HIV-1. AIDS 2006; 20:967–973.

Diffusion of extended-spectrum β-lactamase producing *Enterobacter cloacae* in a kidney transplantation unit

S. Hammami[1], I. Boutiba-Ben Boubaker[1,2*], A. Kamoun[1,2], R. Ghozzi[1,2], M. Saidani[1,2], A. Slim[2] and S. Ben Redjeb[1]

[1]Laboratoire de Recherche «Résistance aux Antimicrobiens» Faculté de Médecine de Tunis-Tunisia
[2] Laboratoire de Microbiologie Hôpital Charles Nicolle –Boulevard 9 Avril 1006 Tunis – Tunisia

[*]Corresponding author: ilhem.boutiba@rns.tn

From January to December 2009, 26 non-repetitive extended spectrum β-lactamases (ESBLs) producing *E. cloacae* were isolated in the kidney transplantation unit of Charles Nicolle hospital of Tunis (Tunisia). The present study was conducted in order to establish epidemiological relationship among them and to characterize the ESBLs produced. All isolates were analysed by pulsed-field gel electrophoresis (PFGE). Production of ESBL was detected by decreased susceptibility to extended spectrum cephalosporins and a positive double disk synergy test result. ESBLs were identified by PCR amplification and sequencing of $bla_{CTX\text{-}M}$, bla_{SHV} and bla_{TEM} genes. All strains were susceptible to carbapenems. They were resistant to fluoroquinolones, gentamicin, tobramycin and trimethoprim + sulfamethoxazole but variably resistante to netilmicin, amikacin and tetracyclines. PFGE revealed 24 unrelated profiles. All isolates carried a single ESBL (CTX-M-15: n=22 and SHV-12: n=4). Our results showed a multiclonal outbreak. Diffusion of *E. cloacae* producing CTX-M-15 ESBL in our kidney transplantation unit is the consequence of dissemination of identical or related plasmids harbouring CTX-M-15 gene among different clones. Early detection and prompt containment can limit the spread of these multiresistant pathogens.

Keywords: *E. cloacae*; ESBL; kidney transplantation unit

1. Introduction

Enterobacter cloacae is an increasingly important nosocomial pathogen. It has been associated with several outbreaks, usually involving strains that overproduce the chromosomal AmpC cephalosporinase or uncommonly, strains expressing extended spectrum β-lactamases (ESBLs). Until 2008, only sporadic cases of ESBL producing *E. cloacae* have been identified in our hospital. In 2009, 26 non repetitive ESBL producing *E. cloacae* were isolated in the kidney transplantation unit. The aim of this study was to establish epidemiological relationship among them and to characterize the ESBLs produced.

2. Material and methods

2.1 Bacterial strains and susceptibility testing

During 2009, 26 non-repetitive ESBL producing *E. cloacae* were collected in the kidney transplantation unit of Charles Nicolle Hospital. They were selected on the basis of the double-disk synergy test (DDST) result with or without cloxacillin (250mg/ml). These isolates were identified using the API20E system (bioMérieux, Marcy l'Étoile, France).

Susceptibility testing was performed by disk diffusion assays and MICs were determined by agar dilution technique and results were interpreted according to the guidelines of Clinical Laboratory Standards Institute (CLSI) [1].

2.2 PCR amplification and sequencing

Identification of bla_{SHV}, bla_{TEM} and $bla_{CTX\text{-}M}$ group 1 genes was performed as described previously using PCR-specific primers [2]. The PCR products were purified with a Gel Band Purification Kit (BioMatik) and sequenced on an ABI PRISM 310 DNA Sequencer (Applied Biosystems, Foster City, CA, USA). Similarity searches and alignments for both the nucleotide sequences were performed with the BLAST program (http://www.ncbi.nlm.nih.gov/BLAST).

2.3 Pulsed-field gel electrophoresis

Molecular characterization of ESBL-producing *E. cloacae* was done by macrorestriction analysis of genomic DNA with XbaI (Promega). DNA fragments were separated by pulsed-field gel electrophoresis (PFGE) in a CHEF-DR III system (Bio-Rad) as described previously [3]. Restriction patterns were analyzed by previously established criteria [4].

3. Results

E. cloacae isolates were recovered from urine (n=24), blood (n=1) and pus (n=1). Antibiotic susceptibility patterns and β-lactams MICs were shown in table 1. A higher level of resistance to cefotaxime (MIC_{50} 2048µg/ml) than to ceftazidime (MIC_{50} 256µg/ml) was recovered in 22 strains. Clavulanic acid partially or completely restored the activities of cefotaxime and ceftazidime. All isolates were resistant to cefepime (MIC_{50} 256µg/ml). PCR amplification and sequencing of *bla* genes showed that CTX-M-15 was carried in 22 strains and SHV-12 by 4 strains. TEM-1 was found in combination with ESBL in 19 strains. PFGE revealed 24 different pulsotypes (P1 to P24). Only pulsotypes (P1 and P11) were revealed in 2 strains respectively. The remaining strains were sporadic belonging to the 22 different pulsotypes (table 1).

4. Discussion

The prevalence of ESBL-producing *E. cloacae* was diverse worldwide [5-8], and the highest prevalence rate was 50%, reported by a tertiary university hospital in Korea [7]. A low rate of ESBL have been reported in Taiwan in 2007 (28%) [9]. However, the frequency of ESBL production can be underestimated in *E. cloacae* derepressed variants. In fact, the conventional phenotypic DDST is expected to be difficult and is dependent on the level of the chromosomal cephalosporinase. Performing the DDST on agar containing cloxacillin significantly improves the test sensitivity, by inhibition of the cephalosporinase activity [10].

As it was often described worldwide [11-13], many isolates were multidrug resistant. In fact, they additionally carried other antibiotic resistance genes. They often carried other β-lactamases genes (e.g TEM-1 found in 19 isolates). However, all isolates remained susceptible to carbapenems. This situation must be preserved because resistance to imipenem was recently reported in Spain [12], China [14] and France [15]. In our study, 24 of 26 ESBL producing *E. cloacae* strains were CTX-M-15 producers. CTX-M-15 is now found worldwide, mainly in *E. coli* and *K. pneumoniae* and since 2004, in *E. cloacae* [11, 16]. Reports regarding *E. cloacae* producing CTX-M β-lactamases in Africa were limited to Alger [17], Egypt [18] and recently in Tunisia [19].

SHV-12 producing *E. cloacae* was previously reported in Taiwan and Algeria [9, 20]. In Tunisia, this is the first report. Emergence of this ESBL represents a new alert and requires more intensive monitoring. In this study, SHV-12 was less frequent than the CTX-M–15 ESBLs. However, several studies have shown that SHV-12 was the most prevalent ESBL in isolates of *Enterobacteriaceae* from Korea [21, 22].

The different patterns revealed by PFGE suggest that diffusion of *E. cloacae* producing CTX-M-15 ESBL in this unit is the consequence of dissemination of identical or related plasmids harbouring CTX-M-15 gene among different clones. Early detection and prompt containment can limit the spread of these multiresistant pathogens.

Acknowledgements This work was partially funded by grants from the Ministry of Higher Education, Scientific Research and Technology, Tunisia.

Table 1. Characteristics of *E. cloacae* isolates

Strains	Date of isolation	Specimen	Antibiotic susceptibility patterns	MIC (μg/ml) TIC	TCC	CAZ	CAZ+ AC	CTX	CTX+ AC	FOX	FEP	β-lactamase production	PFGE
1	02/01/2009	blood	Gn, Tm, Net, Tet, C, NA, Ofx, Sxt	>2048	1024	128	2	>2048	1	512	256	CTX-M15	P1
2	05/01/2009	urine	Gn, Tm, Tet, C, NA, Ofx, Sxt	>2048	512	256	2	2048	1	256	512	CTX-M15	P2
3	22/01/2009	urine	Gn, Tm, Net, Am, Tet, C, NA, Ofx, Sxt	>2048	256	256	2	2048	1	256	256	CTX-M15	P3
4	23/01/2009	urine	Gn, Tm, Net, C, NA, Ofx, Sxt	>2048	256	512	0.5	512	1	256	512	TEM-1; CTX-M15	P4
5	02/02/2009	urine	Gn, Tm, Net, Tet, NA, Ofx, Sxt	>2048	1024	256	16	512	64	512	512	TEM-1; CTX-M15	P5
6	05/02/2009	urine	Gn, Tm, Tet, C, NA, Ofx, Sxt	>2048	1024	128	16	512	128	256	128	TEM-1; CTX-M15	P6
7	06/02/2009	urine	Gn, Tm, Tet, C, NA, Ofx, Sxt	>2048	1024	512	2	512	8	256	128	CTX-M15	P7
8	26/02/2009	urine	Gn, Tm, Tet, C, NA, Ofx, Sxt	>2048	256	256	2	2048	2	256	512	CTX-M15	P8
9	28/02/2009	urine	Gn, Tm, Tet, C, NA, Ofx, Sxt	>2048	256	256	1	2048	0.25	512	256	TEM-1; CTX-M15	P9
10	10/03/2009	urine	Gn, Tm, Tet, C, NA, Ofx, Sxt	>2048	>1024	256	16	2048	8	512	256	TEM-1; SHV-12	P10
11	13/03/2009	urine	Gn, Tm, Net, Tet, C, NA, Ofx, Sxt	>2048	256	64	4	2048	8	512	512	TEM-1; CTX-M15	P11
12	28/03/2009	urine	Gn, Tm, Tet, C, NA, Ofx, Sxt	>2048	512	512	2	>2048	8	512	2048	CTX-M15	P12
13	14/04/2009	urine	Gn, Tm, Tet, C, NA, Ofx, Sxt	>2048	128	256	0.5	256	2	512	512	TEM-1; SHV-12	P13
14	08/05/2009	urine	Gn, Tm, Net, C, NA, Ofx, Sxt	>2048	16	512	2	128	4	512	32	TEM-1; CTX-M15	P14
15	23/05/2009	pus	Gn, Tm, Net, Tet, C, NA, Ofx, Sxt	>2048	512	512	64	2048	128	512	128	TEM-1; CTX-M15	P15
16	28/05/2009	urine	Gn, Tm, C, NA, Ofx, Sxt	>2048	512	512	2	2048	0.5	512	64	TEM-1; CTX-M15	P16
17	02/06/2009	urine	Gn, Tm, Tet, C, NA, Ofx, Sxt	>2048	64	256	0.5	1024	8	512	32	TEM-1; SHV-12	P17
18	30/06/2009	urine	Gn, Tm, Net, Am, Tet, NA, Ofx, Sxt	>2048	256	64	2	2048	8	256	256	TEM-1; CTX-M15	P18
19	21/08/2009	urine	Gn, Tm, Net, Tet, C, NA, Ofx, Sxt	>2048	256	256	2	>2048	4	512	256	TEM-1; CTX-M15	P19
20	07/10/2009	urine	Gn, Tm, Tet, C, NA, Ofx, Sxt	>2048	256	128	0.5	2048	4	256	32	TEM-1; CTX-M15	P20
21	09/10/2009	urine	Gn, Tm, Net, Tet, C, NA, Ofx, Sxt	>2048	>1024	256	32	2048	8	512	2048	CTX-M15	P21
22	15/10/2009	urine	Gn, Tm, Net, Am, Tet, C, NA, Ofx, Sxt	>2048	>1024	256	4	2048	4	512	256	TEM-1; CTX-M15	P22
23	22/10/2009	urine	Gn, Tm, Net, Am, Tet, C, NA, Ofx, Sxt	>2048	256	256	1	512	0.5	512	256	TEM-1; SHV-12	P1
24	24/10/2009	urine	Gn, Tm, Tet, NA, Sxt	>2048	256	16	16	512	32	512	256	TEM-1; CTX-M15	P23
25	26/10/2009	urine	Gn, Tm, Net, C, NA, Ofx, Sxt	>2048	128	64	0.5	512	0.5	512	128	TEM-1; CTX-M15	P11
26	28/10/2009	urine	Gn, Tm, Net, Am, Tet, NA, Ofx, Sxt	>2048	256	64	0.5	2048	0.125	512	256	TEM-1; CTX-M15	P24

TIC: Ticarcillin; TCC: Ticarcillin- clavulanic acid; CAZ: Ceftazidime; CAZ+AC: Ceftazidime-clavulanic acid; CTX: Cefotaxime; CTX+AC: Cefotaxime-clavulanic acid; FEP: Cefepime; Gn: Gentamicin; Tm: Tobramycin; Net: Netilmicin; Am: Amikacin; Tet: Tetracyclin; C: Chloramphenicol; NA: Nalidixic acid; Ofx: Ofloxacin; Sxt: Trimethoprim- sulfamethoxazole

References

[1] Clinical and Laboratory Standards Institute. Performance Standards for Antimicrobial Susceptibility Testing. Sixteenth information supplement. M100-S16. Wayne, PA: CLSI; 2006.

[2] Ktari S, Arlet G, Mnif B, Gautier V, Mahjoubi F, Ben Jmeaa M, Bouaziz M, Hammami A. Emergence of multidrug-resistant *Klebsiella pneumoniae* isolates producing VIM-4 metallo-beta-lactamase, CTX-M-15 extended-spectrum beta-lactamase, and CMY-4 AmpC beta-lactamase in a Tunisian university hospital. *Antimicrob Agents Chemother.* 2006;50:4198-4201.

[3] Manzur A, Tubau F, Pujol M, Calatayud L, Dominguez MA, Pena C, Sora M, Gudiol F, Ariza J. Nosocomial outbreak due to extended-spectrum-beta-lactamase- producing *Enterobacter cloacae* in a cardiothoracic intensive care unit. *J Clin Microbiol.* 2007; 45:2365-2369.

[4] van Belkum A, Tassios PT, Dijkshoorn L, Haeggman S, Cookson B, Fry NK, Fussing V, Green J, Feil E, Gerner - Smidt P, Brisse S, Struelens M. Guidelines for the validation and application of typing methods for use in bacterial epidemiology. *Clin Microbiol Infect.* 2007;13 :1-46.

[5] Bell JM, Turnidge JD, Jones RN. Prevalence of extended-spectrum beta-lactamase-producing *Enterobacter cloacae* in the Asia-Pacific region: results from the SENTRY Antimicrobial Surveillance Program, 1998 to 2001. *Antimicrob Agents Chemother.* 2003;47:3989-3993.

[6] Canton R, Oliver A, Coque TM, Varela Mdel C, Perez-Diaz JC, Baquero F. Epidemiology of extended-spectrum beta-lactamase-producing *Enterobacter* isolates in a Spanish hospital during a 12-year period. *J Clin Microbiol.* 2002;40:1237-1243.

[7] Pai H, Hong JY, Byeon JH, Kim YK, Lee HJ. High prevalence of extended-spectrum beta-lactamase-producing strains among blood isolates of *Enterobacter spp.* collected in a tertiary hospital during an 8-year period and their antimicrobial susceptibility patterns. *Antimicrob Agents Chemother.* 2004;48:3159-3161.

[8] Schlesinger J, Navon-Venezia S, Chmelnitsky I, Hammer-Munz O, Leavitt A, Gold HS, Schwaber MJ, Carmeli Y. Extended-spectrum beta-lactamases among *Enterobacter* isolates obtained in Tel Aviv, Israel. *Antimicrob Agents Chemother.* 2005;49:1150-1156.

[9] Wu JJ, Ko WC, Tsai SH, Yan JJ. Prevalence of plasmid-mediated quinolone resistance determinants QnrA, QnrB, and QnrS among clinical isolates of *Enterobacter cloacae* in a Taiwanese hospital. *Antimicrob Agents Chemother.* 2007; 51:1223-1227.

[10] Drieux L, Brossier F, Sougakoff W, Jarlier V. Phenotypic detection of extended-spectrum beta-lactamase production in *Enterobacteriaceae*: review and bench guide. *Clin Microbiol Infect.* 2008;14:90-103.

[11] Paterson DL, Bonomo RA. Extended-spectrum beta-lactamases: a clinical update. *Clin Microbiol Rev.* 2005;18:657-686.

[12] Leotard S, Negrin N. Epidemiology of *Enterobacteriaceae* producing extended-spectrum beta-lactamase in Grasse Hospital (2005-2008). *Pathol Biol* 2010; 58:35-38.

[13] Wiegand I, Geiss HK, Mack D, Sturenburg E, Seifert H. Detection of extended-spectrum beta-lactamases among *Enterobacteriaceae* by use of semiautomated microbiology systems and manual detection procedures. *J Clin Microbiol.* 2007; 45:1167-1174.

[14] Miro E, Segura C, Navarro F, Sorli L, Coll P, Horcajada JP, Alvarez-Lerma F, Salvado M. Spread of plasmids containing the bla_{VIM-1} and bla_{CTX-M} genes and the qnr determinant in *Enterobacter cloacae, Klebsiella pneumoniae* and *Klebsiella oxytoca* isolates. *J Antimicrob Chemother.* 2010; 65:661-665.

[15] Wu Q, Liu Q, Han L, Sun J, Ni Y. Plasmid-mediated carbapenem-hydrolyzing enzyme KPC-2 and ArmA 16S rRNA methylase conferring high-level aminoglycoside resistance in carbapenem-resistant *Enterobacter cloacae* in China. *Diagn Microbiol Infect Dis.* 2010;66:326-328.

[16] Mamlouk K, Boutiba-Ben Boubaker I, Gautier V, Vimont S, Picard B, Ben Redjeb S, Arlet G. Emergence and outbreaks of CTX-M beta-lactamase-producing *Escherichia coli* and *Klebsiella pneumoniae* strains in a Tunisian hospital. *J Clin Microbiol.* 2006; 44:4049-4056.

[17] Touati A, Benallaoua S, Forte D, Madoux J, Brasme L, de Champs C. First report of CTX-M-15 and CTX-M-3 beta-lactamases among clinical isolates of *Enterobacteriaceae* in Bejaia, Algeria. *Int J Antimicrob Agents.* 2006;27:397-402.

[18] Khalaf NG, Eletreby MM, Hanson ND. Characterization of CTX-M ESBLs in *Enterobacter cloacae, Escherichia coli* and *Klebsiella pneumoniae* clinical isolates from Cairo, Egypt. *BMC Infect Dis.* 2009;9:84-88.

[19] Bourouis A, Dubois V, Coulange L, André C, Bejhadj C, Ben Moussa M, Quentin C, Belhadj O. First report of CTX-M-9 in a clinical isolate of *Enterobacter cloacae* in a Tunisian hospital. *Pathol Biol.* 2009; doi:10.1016/j.patbio.2009.03.008.

[20] Iabadene H, Messai Y, Ammari H, Ramdani-Bouguessa N, Lounes S, Bakour R, Arlet G. Dissemination of ESBL and Qnr determinants in *Enterobacter cloacae* in Algeria. *J Antimicrob Chemother.* 2008;62:133-136.

[21] Jeong SH, Bae IK, Lee JH, Sohn SG, Kang GH, Jeon GJ, Kim YH, Jeong BC, Lee SH. Molecular characterization of extended-spectrum beta-lactamases produced by clinical isolates of *Klebsiella pneumoniae* and *Escherichia coli* from a Korean nationwide survey. *J Clin Microbiol.* 2004;42:2902-2906.

[22] Ryoo NH, Kim EC, Hong SG, Park YJ, Lee K, Bae IK, Song EH, Jeong SH. Dissemination of SHV-12 and CTX-M-type extended-spectrum beta-lactamases among clinical isolates of *Escherichia coli* and *Klebsiella pneumoniae* and emergence of GES-3 in Korea. *J Antimicrob Chemother.* 2005;56:698-702.

Effect of antifungal agents on non-*Candida albicans Candida* species enzymatic activity

Negri M.[*,1], **Lorenço T.**[1], **Silva S.**[1], **Henriques M.**[1], **Azeredo J.**[1] **and Oliveira R.**[1]

[1] IBB, Centre of Biological Engineering, Universidade do Minho, Campus de Gualtar, 4710-057 Braga, Portugal

* Corresponding author: e-mail: melyssanegri@deb.uminho.pt, Phone: +351 253604408

To study the action of antifungal agents (fluconazole and amphotericin B) on proteinase, phospholipase, and haemolytic activity of several *Candida* species. The assays were performed with *C. albicans*, *C. glabrata*, *C. parapsilosis,* and *C. tropicalis* clinical isolates recovered from different body sites. The minimal inhibitory concentration (MIC) and the maximum antifungal concentration (MAC) were determined against both antifungal agents. Then, the enzymatic activity was measured for yeasts grown at MIC and MAC conditions. The results showed that all *Candida* species assayed were sensible to both antifungal agents. In general, all *Candida* species presented proteinase activity, with the exception of *C. glabrata,* and also presented haemolytic activity, with the exception of four *C. parapsilosis* strains. Fluconazole and amphotericin B, were causative agents of a decrease in proteinase and haemolytic activity. However, this effect was species and strain dependent for both antifungal agents. fluconazole and amphotericin B were able to influence the enzymatic activity of *Candida* species surviving cells. There are few works focusing on the effect of antifungals in proteinase and phospholipase activity of *C. albicans*, and to our knowledge, this is the first report on the effect of the two most important clinical antifungal agents on several other *Candida* species enzymatic activity.

Keywords: Candida; Fluconazole; Amphotericin B; Proteinase activity; Haemolytic activity

1. Introduction

Candida species are most frequently isolated from oral, vulvovaginal and urinary tracts and have been detected in approximately 31 to 55% of healthy individuals [1-2]. In addition *Candida* species are the most common fungi recovered from blood cultures in hospitalized patients [3]. Actually, an increasing incidence of fungal infections caused by *Candida* species has been recognized in immunocompromised patients, such as the intensive care, postsurgical, and cancer patients [4-6]. Additionally, management of fungal infections is markedly limited by problems of drug safety, resistance and effectiveness profile. Furthermore, current therapies for treatment of invasive candidosis is based on several antifungal drugs, but the most commonly used are amphotericin B, fluconazole and itraconazole [7].

Although *Candida albicans* is regarded as the most common causative agent involved in candidosis, its incidence has been decreasing and the number of infections caused by non-*Candida albicans Candida* (NCAC) species (such as: *C. glabrata*, *C. tropicalis* and *C. parapsilosis*) has become more prevalent [8-10]. Moreover, this fact may be related with some NCAC species that are intrinsically resistant to some triazoles, or may be sensitive only to high antifungal doses. Mechanisms of pathogenicity, such as human tissue invasion may be another phenomenon associated with NCAC prevalence [7, 11].

The secretion of enzymes by *Candida* (proteinases, phospholipases and haemolysins) is recognized as an important virulent factor and has been associated with its high pathogenicity [12-14]. The secretion of aspartic proteinases (Saps) facilitates colonization and invasion of host tissues by disruption of the host mucosal membranes [15]. In addition to proteinases, phospholipases are also often considered as relevant on *Candida* pathogenicity. Phospholipases are enzymes that hydrolyze phospholipids into fatty acids and the production of these enzymes could contribute to host cell membrane damage or to exposure of receptors facilitating adherence [16-17]. Furthermore, recent studies have demonstrated that haemolysin is used by *Candida* species for destroying the haemoglobin and extracting the elemental iron from the host cells [12-13].

In vitro studies have indicated that antifungal agents could be able to influence the proteinase and phospholipase expression of *Candida* species [18-21]. However, to the authors' knowledge, the correlation of the effect of antifungals on *Candida* proteinase, phospholipase and haemolysin activity has not yet been described. So, the aim of this work was to get insights on the effect of antifungal agents (fluconazole and amphotericin B) on *Candida* species enzymatic activity.

2. Materials and Methods

Yeast species and strain. This study was performed with *C. albicans* (1), *C. glabrata* (4), *C. parapsilosis* (5) and *C. tropicalis* (6) recovered from different body sites (blood, oral, vaginal and urinary tract). All isolates were obtained from the archive collection of the Biofilm Group of the Centre of Biological Engineering. Four reference strains of *C. albicans* (90028), *C. glabrata* (2001), *C. parapsilosis* (22019) and *C. tropicalis* (750), from the American Type Culture Collection (ATCC) were also examined. The identity of all isolates was confirmed using CHROMagar *Candida* (CHROMagar, Paris, France) and by PCR based sequencing using specific primers (ITS1 and ITS4) against the 5.8S subunit gene reference. Genomic DNA was extracted following previously described procedures [22]. The PCR products were sequenced using the ABI-PRISM Big Dye terminator cycle sequencing kit (Perkin Elmer, Applied Biosystems, Warrington, UK).

Antifungal susceptibility testing. The microdilution method, from the Clinical and Laboratory Standards Institute document M27-A2 [23], was performed in RPMI 1640 broth, to test fluconazole and amphotericin B,. The minimal inhibitory concentrations (MIC) and the maximum antifungal concentration (MAC) of the antifungal agents against each *Candida* isolate were determined. For fluconazole, the MIC was the lowest concentration of antifungal capable of inhibiting 50% of microbial growth compared with the positive control (growth without antifungal agent), and for amphotericin B the MIC was the lowest concentration that inhibited 90% of the growth, according to the standard. For fluconazole the MAC was determined as the highest concentration of antifungal capable of inhibiting 90% of microbial growth compared with the positive control. All the experiments were performed in triplicate, at least in three independent assays.

Proteinase, phospholipase and haemolysin production. The secretion of proteinases, phospholipases and haemolysins was detected by an opaque halo formation resulting from the degradation around the colonies grown in a specific agar plate, according to Ruchel et al. [15], Price et al. [24] and Negri et al. [12], respectively. These tests were performed using 25 μl of suspension collected from the well plate where MIC and MAC were determined. Samples were spread on BSA supplemented agar, egg-yolk agar and sheep blood agar for protease, phospholipase, and haemolysin activity determination, respectively. The enzymatic activity was measured by the ratio between the colony diameter and the colony diameter plus the precipitation zone. The results were presented as percentage of activity. All the experiments were performed in triplicate and in three independent assays.

3. Results

The susceptibility of *Candida* species was analyzed according to CLSI M27-A2, and thus classified into susceptible (S), susceptible-dose (S-DD) and resistant (R). All *Candida* clinical isolates were susceptible to fluconazole and amphotericin B. However, *C. glabrata* clinical isolates showed, in general, higher fluconazole MIC values, despite, being classified as susceptible according to CLSI M27-A2 parameters. Regarding MAC, only *C. parapsilosis* showed values lower than 64 μg/mL (Table 1).

In addition, Table 1 also presents the results of enzymatic activity. It was evident that all *Candida* species presented proteinase activity, with the exception of *C. glabrata*. However, differences were observed both among species and strains. Furthermore, *C. albicans* ATCC90028 and *C. parapsilosis* 534638, recovered from candiduria, were the strains with the highest values of proteinase activity (60%). Interestingly, all *Candida* species presented haemolytic activity (Table 1), with the exception of four *C. parapsilosis* strains (ATCC 22019, AM2, 513143, and 491861), highlighting strain variation. Moreover, the strain that produced the highest haemolytic activity was *C. glabrata* 554784 (82%) isolated from vagina. The totality of yeasts under study was negative for phospholipase activity, except *C. albicans* ATCC 90028 (data not shown).

The main goal of this study was to assess the enzymatic activity of *Candida* species after treatment with fluconazole and amphotericin B (Table 1). Regarding the effect of fluconazole (MIC and MAC) on proteinase activity, it was possible to observe a significant decrease for the majority of the strains under study. The exceptions were: *C. albicans* ATCC 90028 (MAC); and *C. tropicalis* 75 U (MIC and MAC). *Candida parapsilosis* presented the highest heterogeneity with 3 strains presenting increased and other 3 decreased values of activity for both MIC and MAC conditions. Furthermore, the highest increase (30%) was found for *Candida parapsilosis* ATCC 22019 (MAC conditions) and the highest decrease for *C. parapsilosis* 534638 (from urine) with 80% (MIC) and 69.1% (MAC) reductions.

The effect of amphotericin B was also evaluated and, again, a decrease on proteinase activity was found for all the strains under study after treatment at MIC conditions. The highest decrease was detected for *C. parapsilosis* ATCC 22019 corresponding to 81.8% of reduction. Curiously, the heterogeneity observed for *C. parapsilosis* after fluconazole treatment was not found in the presence of amphotericin B, where proteinase activity never increased.

Regarding haemolytic activity, it was observed that the antifungal effect was species, strain and antifungal dependent. Interestingly, it was possible to observe that few strains were able to maintain their activity profile after contact with both antifungal agents. In the case of fluconazole (MIC and MAC conditions), five strains (*C. glabrata* ATCC 2001, 562123 and 585626; *C. tropicalis* ATCC 750, 75U) out of 16 presented an increase in haemolytic activity. In presence of amphotericin B, four strains: *C. glabrata* 562123 and 585626; *C. tropicalis* ATCC 750 and 75U out of 16 presented an increase in haemolytic activity after treatment at MIC conditions.

Table 1 *Candida* species Minimum Inhibitory Concentration (MIC) and Maximum Antifungal Concentration (MAC) values in presence of fluconazole (FLU) and amphotericin B (AB) and the percentage of enzymatic activity (proteinase and haemolysin) after and before contact with antifungal agent

Strain	Site of isolation	Antifungal susceptibility (µg/mL)			Proteinase (%)				Haemolysin (%)			
					w/o*	With antifungal**			w/o*	With antifungal**		
		FLU		AB	control	FLU		AB	control	FLU		AB
		MIC	MAC	MIC		MIC	MAC	MIC		MIC	MAC	MIC
C. albicans	ATCC 90028	0.5	64	1	60.0	↓46.7	↑61.9	↑61.1	77.0	↓75.0	↓71.0	↓60.0
C. albicans 324LA/94	Oral	32	64	1	33.3	↓23.1	↓20.0	=33.3	29.0	↓24.0	↓25.0	↓24.0
C. glabrata	ATCC 2001	.32	64	1	0.0	0.0	0.0	0.0	70.0	↑81.0	↑79.0	=70.0
C. glabrata D1	Oral	8	64	1	0.0	0.0	0.0	0.0	77.0	↓72.0	↓73.0	↓70.0
C. glabrata 562123	Urine	4	64	1	0.0	0.0	0.0	0.0	65.0	↑75.0	↑72.0	↑70.0
C. glabrata 554784	Vaginal	2	64	1	0.0	0.0	0.0	0.0	82.0	↓77.0	↓73.0	↓22.0
C. glabrata 585626	Vaginal	2	64	1	0.0	0.0	0.0	0.0	69.0	↑82.0	↑79.0	↑76.0
C. parapsilosis	ATCC 22019	1	1	0.25	50.0	=50.0	↑65.0	↓9.1	0.0	0.0	0.0	0.0
C. parapsilosis AM2	Oral	2	8	0.5	40.0	↓12.5	↑53.9	=40.0	0.0	0.0	0.0	0.0
C. parapsilosis 553877	Urine	1	8	0.5	40.0	↓28.6	↓30.1	↓25.0	29.0	↓25.0	↓25.0	↓25.0
C. parapsilosis 534638	Urine	1	64	0.5	60.0	↓12.5	↓18.2	↓41.7	40.0	↓33.0	↓25.0	↓20.0
C. parapsilosis 513143	Vaginal	1	64	0.5	50.0	↓33.3	↓20.0	↓22.2	0.0	0.0	0.0	0.0
C. parapsilosis 491861	Vaginal	1	64	1	33.3	↓20.0	↑42.9	↓25.0	0.0	0.0	0.0	0.0
C. tropicalis	ATCC 750	0.5	64	1	28.6	↓15.4	↓11.1	↓0.0	17.0	↑18.0	↑24.0	↑25.0
C. tropicalis 66A	Blood	1	64	1	47.0	↓31.0	↓25.0	↓22.0	47.0	↓31.0	↓25.0	↓22.0
C. tropicalis 66B	Blood	2	64	1	42.5	↓33.0	↓29.0	↓33.0	42.5	↓33.0	↓29.0	↓33.0
C. tropicalis 69	Urine	1	64	1	45.5	↓33.0	↓30.0	↓33.0	45.5	↓33.0	↓30.0	↓33.0
C. tropicalis 75U	Urine	1	64	1	48.5	↑50.0	↑50.0	↑50.0	48.5	↑50.0	↑50.0	↑50.0
C. tropicalis 515468	Urine	0.5	64	1	51.5	↓33.0	↓31.0	↓38.0	51.5	↓33.0	↓31.0	↓38.0
C. tropicalis 12V	Vaginal	0.3	64	1	42.5	↓33.0	↓33.0	↓33.0	42.5	↓33.0	↓33.0	↓33.0

*Strains grown in solid medium supplemented (with blood or BSA) without contact with antifungal agent
**Strains grown in solid medium supplemented (with blood or BSA) after contact with antifungal agent
Equal (=), increase (↑) or decrease (↓) of enzymatic activity when compared with control (without contact with antifungal agent)

4. Discussion

In recent decades, the profiles of susceptibility to antifungal agents and virulence factors of *Candida* species have been changing [18]. This can be due to prolonged use of empiric antifungal treatments and the privileged prophylactic therapy. Furthermore, high-level of resistance to antifungal therapy has been observed and the virulence factors of *Candida* species altered [11, 18].

Epidemiological data suggests that NCAC species have been showing reduced susceptibility to azoles, namely in immunocompromised patients [7, 11]. However, amphotericin B has been a potent antifungal for *Candida* species [11, 25]. In this work, the 16 clinical isolates assayed, recovered from different body sites, were susceptible to both antifungal agents (Table 1), although, it is important to highlight that *C. glabrata* presented the highest MIC values, despite still being in the range of susceptibility, in agreement to the standard. The highest MIC values of *C. glabrata* for farnesol, comparatively to other *Candida* species, had been already reported [25]. Moreover, other authors also described that *C. glabrata* showed low susceptibility or even resistance to the recommended doses of azoles [7, 25-26].

The pathogenesis of invasive candidosis is related to several virulence factors, including the ability to produce hydrolytic enzymes such as proteinases, phospholipases and haemolysins. Furthermore, concerning the expression of enzymes, recent studies suggested that patients infected with *C. albicans* and subsequently treated with suboptimal doses of fluconazole presented higher virulence symptoms [27]. Moreover, NCAC species, despite presenting lower virulence *in vitro*, presented an equal or greater virulence *in vivo*, comparatively to *C. albicans* [11, 28-29]. It is also assumed that an important virulence factor of *C. albicans* is its higher ability to exhibit proteinase activity [14, 17, 29]. In addition, it was recently shown that proteinase activity of *C.*

parapsilosis strains has important implication on epithelium damage [30-31]. Accordingly, in the present study, it was possible to observe that *C. albicans*, *C. parapsilosis* and *C. tropicalis* were able to secret proteinases and that *C. albicans* reference strain and a clinical isolate of *C. parapsilosis* showed the highest values of proteinase production (60%). Bonassoli *et al.* [30] have reported that *C. parapsilosis* has high proteinase activity. Concerning *C. glabrata*, all strains were unable to produce proteinase activity, which is in accordance with other authors' observations [32].

Haemolysins are used by *Candida* species for degrading and extracting the elemental iron from host cells. Thus, haemolytic activity, despite being less studied, has been increasingly considered as an important *Candida* virulence factor. In the present work, it was possible to observe that all tested *Candida* species have the ability to produce haemolysins. Strain variation was particularly evident for *C. parapsilosis*. On the other hand, the highest haemolytic activity was observed for *C. glabrata* 554784 (a vaginal isolate) (82%). Luo *et al.* [13] also reported that *C. glabrata* presents higher haemolysin activity in opposition to *C. parapsilosis*. However, Bonassoli *et al.* [30] showed variable haemolytic activity among *C. parapsilosis* strains, although with high total haemolytic activity for all *C. parapsilosis* tested.

In relation to the phospholipase activity, only *C. albicans* presented positive results. In fact, there is some controversy concerning phospholipase activity in *Candida* species. It was believed that only *C. albicans* was able to produce extracellular phospholipases, as determined by egg yolk-based assay, comparatively to other *Candida* species [16]. However, recent works showed that NCAC species can be able to produce phospholipases but with lower activity relatively to *C. albicans* [17, 29, 33].

Regarding the antifungal effect on enzymatic activity, *Candida* species presented differences between both virulence factors (proteinase and haemolytic activity) and at both conditions tested (MIC and MAC). It is important to emphasize, that the effect of both antifungal agents were greatly species and strain dependent. Such findings undoubtedly reflect the inherent physiological differences between strains and could have significance with respect to pathogenic potential. Furthermore, for the great majority of strains, it was verified a decrease in the proteinase and haemolytic activities after contact with fluconazole and amphotericin B. These results corroborate the ones observed by Wu *et al.* [27], who studied the production of proteinase by *C. albicans,* isolated from the oral cavity of an AIDS patient, before and after the development of clinical and *in vitro* resistance to fluconazole. Those authors reported that extracellular proteinase decreased for the most fluconazole-susceptible isolate (MIC of 1 µg/mL) and increased in the case of the most fluconazole-resistant isolates (MIC > 64 µg/mL).

Nevertheless, in the present study some clinical isolates showed high enzymatic activity after contact with both antifungal agents (fluconazole and amphotericin B) at MIC and MAC conditions. The highest increase in proteinase activity (30%) was found in *C. parapsilosis* ATCC 22019 (fluconazole MAC conditions), while as haemolysin is concerned only few strains presented an increased activity after being challenged with the antifungal agents. There are few works focusing on the effect of antifungals in proteinase or phospholipase activity of *C. albicans* [19, 27] and, to our knowledge, this is the first report on the effect of amphotericin B and fluconazole on several *Candida* species enzymatic activity.

Summarizing, this work demonstrates that fluconazole and amphotericin B, two important antifungal agents currently used in clinical practice, were able to influence the enzymatic activity of NCAC species. Furthermore, this effect is species and strain dependent and no correlation was found between isolation site and the susceptibility and enzymatic activity profile. It must be emphasized, however, that further investigations with more isolates from specific environments are required to confirm these findings.

Acknowledgements The authors acknowledge 'Cordenação de Aperfeiçoamento de Pessoal de Nível Superior (CAPES)', Brazil, for supporting Melyssa Negri work through grant BEX 4642/06-6 and 'Fundação para a Ciência e Tecnologia (FCT)', Portugal, for supporting Sónia Silva work through grant SFRH/BD/28341/2006.

References

[1] Lai C-C, Tan C-K, Huang Y-T, Shao P-L & Hsueh P-R. Current challenges in the management of invasive fungal infections. *Journal of Infection and Chemotherapy.* 2008; 14: 77-85.

[2] Vazquez JA & Sobel JD. Mucosal candidiasis. *Infect Dis Clin North Am.* 2002; 16: 793-820, v.

[3] Pfaller MA, Jones RN, Messer SA, Edmond MB & Wenzel RP. National surveillance of nosocomial blood stream infection due to Candida albicans: frequency of occurrence and antifungal susceptibility in the SCOPE Program. *Diagn Microbiol Infect Dis.* 1998; 31: 327-332.

[4] Hagerty JA, Ortiz J, Reich D & Manzarbeitia C. Fungal infections in solid organ transplant patients. *Surg Infect (Larchmt).* 2003; 4: 263-271.

[5] Samaranayake LP, Fidel PL, Naglik JR, *et al.* Fungal infections associated with HIV infection. *Oral Dis.* 2002; 8 Suppl 2: 151-160.

[6] Kiehn TE, Edwards FF & Armstrong D. The prevalence of yeasts in clinical specimens from cancer patients. *Am J Clin Pathol.* 1980; 73: 518-521.
[7] Eggimann P, Garbino J & Pittet D. Epidemiology of Candida species infections in critically ill non-immunosuppressed patients. *Lancet Infect Dis* 2003; 3: 685-702.
[8] Bassetti M, Righi E, Costa A, *et al.* Epidemiological trends in nosocomial candidemia in intensive care. *BMC Infect Dis.* 2006; 6: 21.
[9] Chandra J, Kuhn DM, Mukherjee PK, *et al.* Biofilm Formation by the Fungal Pathogen Candida albicans: Development, Architecture, and Drug Resistance. *J Bacteriol.* 2001; 183: 5385-5394.
[10] Colombo AL, Perfect J, DiNubile M, *et al.* Global distribution and outcomes for Candida species causing invasive candidiasis: results from an international randomized double-blind study of caspofungin versus amphotericin B for the treatment of invasive candidiasis. *Eur J Clin Microbiol Infect Dis.* 2003; 22: 470-474.
[11] Krcmery V & Barnes AJ. Non-albicans Candida spp. causing fungaemia: pathogenicity and antifungal resistance. *J Hosp Infect.* 2002; 50: 243-260.
[12] Negri M, Martins M, Henriques M, *et al.* Examination of Potential Virulence Factors of Candida tropicalis Clinical Isolates From Hospitalized Patients. *Mycopathologia.* 2010; 169: 175-182.
[13] Luo G, Samaranayake LP & Yau JYY. Candida Species Exhibit Differential In Vitro Hemolytic Activities. *J Clin Microbiol.* 2001; 39: 2971-2974.
[14] Naglik JR, Albrecht A, Bader O & Hube B. Candida albicans proteinases and host/pathogen interactions. *Cell Microbiol.* 2004; 6: 915-926.
[15] Rüchel R. Proteinases of pathogenic fungi. *Mycoses.* 1999; 42: 48-52.
[16] Ghannoum MA. Potential Role of Phospholipases in Virulence and Fungal Pathogenesis. *Clin Microbiol Rev.* 2000; 13: 122-143.
[17] Kantarcio̓lu AS & Yücel A. Phospholipase and protease activities in clinical Candida isolates with reference to the sources of strains. *Mycoses.* 2002; 45: 160-165.
[18] Gauwerky K, Borelli C & Korting H. Targeting virulence: a new paradigm for antifungals. *Drug Discov Today* 2009; 14: 214-222.
[19] Kadir T, Gümrü B & Uygun-Can B. Phospholipase activity of Candida albicans isolates from patients with denture stomatitis: The influence of chlorhexidine gluconate on phospholipase production. *Archives of Oral Biology.* 2007; 52: 691-696.
[20] Ganendren R, Widmer F, Singhal V, *et al.* In vitro antifungal activities of inhibitors of phospholipases from the fungal pathogen Cryptococcus neoformans. *Antimicrob Agents Chemother.* 2004; 48: 1561-1569.
[21] Mata-Essayag S, Magaldi S, Hartung de Capriles C, *et al.* "In vitro" antifungal activity of protease inhibitors. *Mycopathologia.* 2001; 152: 135-142.
[22] Williams DW, Wilson MJ, Lewis MA & Potts AJ. Identification of Candida species by PCR and restriction fragment length polymorphism analysis of intergenic spacer regions of ribosomal DNA. *J Clin Microbiol.* 1995; 33: 2476-2479.
[23] CLSI (2002) Reference Method for broth dilution antifungal susceptibility testing of yeasts – Second Edition: Approved standard M27-A2. CLSI ed.^eds.), p.^pp. Clinical and Laboratory Standards Institute, Wayne, PA, USA.
[24] Price M, Wilkinson I & Gentry L. Plate method for detection of phospholipase activity in Candida albicans. *Sabouraudia.* 1982; 20: 7-14.
[25] Negri M, Henriques M, Svidzinski TIE, Paula CR & Oliveira R. Correlation between Etest®, disk diffusion, and microdilution methods for antifungal susceptibility testing of Candida species from infection and colonization. *Journal of Clinical Laboratory Analysis.* 2009; 23: 324-330.
[26] Fidel PL, Jr., Vazquez JA & Sobel JD. Candida glabrata: review of epidemiology, pathogenesis, and clinical disease with comparison to C. albicans. *Clin Microbiol Rev.* 1999; 12: 80-96.
[27] Wu T, Wright K, Hurst SF & Morrison CJ. Enhanced extracellular production of aspartyl proteinase, a virulence factor, by Candida albicans isolates following growth in subinhibitory concentrations of fluconazole. *Antimicrob Agents Chemother.* 2000; 44: 1200-1208.
[28] Biasoli MS, Tosello ME, Luque AG & Magaró HM. Adherence, colonization and dissemination of Candida dubliniensis and other Candida species. *Medical Mycology.* 2009; 22.
[29] Gokce G, Cerikcioglu N & Yagci A. Acid proteinase, phospholipase, and biofilm production of Candida species isolated from blood cultures. *Mycopathologia.* 2007; 164: 265-269.
[30] Bonassoli LA, Bertoli M & Svidzinski TIE. High frequency of Candida parapsilosis on the hands of healthy hosts. *J Hosp Infect.* 2005; 59: 159-162.
[31] Silva S, Henriques M, Oliveira R, *et al.* Characterization of Candida parapsilosis infection of an in vitro reconstituted human oral epithelium. *Eur J Oral Sci.* 2009; 117: 669-675.
[32] Kumar C, Kumar S & Menon T. Phospholipase and Proteinase Activities of Clinical Isolates of Candida from Immunocompromised Patients. *Mycopathologia.* 2006; 161: 213-218.
[33] Furlaneto-Maia L, Specian A, Bizerra F, de Oliveira M & Furlaneto M. In Vitro Evaluation of Putative Virulence Attributes of Oral Isolates of Candida spp. Obtained from Elderly Healthy Individuals. *Mycopathologia.* 2007; 166: 209-217.

Effect of chitosan, nisin and storage time on the growth of *Listeria innocua* and *Shewanella putrefaciens* in fish homogenates

L. I. Schelegueda[1,2], M. F. Gliemmo[1,2] and C. A. Campos[*,1,2]

[1]Departamento de Industrias, Facultad de Ciencias Exactas y Naturales, Universidad de Buenos Aires, Intendente Güiraldes 2160, Ciudad Universitaria, Capital Federal, 1428, Argentina.

[2]CONICET

*C. A. Campos: carmen@di.fcen.uba.ar

The aim of this study was to evaluate the effect of chitosan, nisin and storage time on the growth of *Listeria innocua* and *Shewanella putrefaciens* in fish homogenates at pH 5.5. For such purpose a factorial design in three blocks and a central point was performed. Fish homogenate was prepared processing Argentine hake fillets (*Merluccius hubbsi*) and distilled water in a ratio of 1:1; pH was adjusted to 5.5 using citric acid 10%w/w. Then, homogenates were sterilized for 15 minutes at 100ºC.
When antimicrobials were added, bacterial counts were significantly reduced. However, this effect depended on the microorganism being tested. After 48 hours of storage, both antimicrobials showed an additive effect reducing the development of both microorganisms between 4 and 5 log cycles. Furthermore, different trends with respect to storage time were noted. Nisin effect on counts was shown immediately after its application, while chitosan effect was noticed during storage.

Keywords nisin; chitosan

1. Introduction

Fish and other fishery products are highly perishable food. The process of degradation starts immediately after the fish slaughter and it often develops more quickly than in other meats. The major cause of the spoilage is microbial growth [1]. Therefore, the use of a preservation method is essential when it comes to the mentioned products. The use of antimicrobials together with other preservation measures has proved to be a good option to extend the shelf life of this kind of food [2].

In order to meet the requirements of consumers, it is necessary to minimize the use of chemical additives and replace them for natural antimicrobials.

Natural antimicrobials can be derived from plants, animals or microorganisms [3]. Usually they have a limited spectrum of activity, so that it is interesting to figure out how to optimize their use.

Natural antimicrobials such as nisin and chitosan are proposed to be used in fish products. It is interesting to explore the possibility of interactions when they are used together, because if there is synergism between them, the amount used could be decreased.

Chitosan is a natural biopolymer derived from deacetylation of chitin, a major component of the shells of crustacean. This natural preservative is known to be nontoxic, furthermore, it has anti-oxidative activity, biocompatibility and biodegradability. Its antibacterial activity is inversely affected by pH [4, 5].

Nisin is an antibacterial peptide produced by *Lactococcus lactis*. This bacteriocin effectively inhibits Gram-positive bacteria and also the outgrowth of spores of Bacilli and Clostridia. Nisin was the first antimicrobial peptide to obtain the "generally recognized as safe" status in the United States. Usualy, it is more effective at acidic pHs and it has its maximum activity at pH 5.5 [6, 7, 8].

There is a wide spectrum of bacteria able to contaminate and cause the spoilage of seafood. During its storage, the microflora changes according to their different abilities to tolerate the preservation conditions [1].

Listeria monocytogenes and *Shewanella putrefaciens* are two of the microorganisms usually isolated from fishery products. Both of them are able to grow at refrigeration temperatures. In addition, *L. monocytogenes* is able to tolerate low pH values and high salt concentrations [9]. In this study, *L. innocua* was selected as a representative microorganism of Gram-positive flora and as an alternative to the pathogen *L. monocytogenes* because of its similar response to stress factors [10].

S. putrefaciens, which is one of the typical spoilage microorganisms of seafood when it is stored in cold and aerobic conditions [1], was selected as a representative of the Gram-negative spoilage flora.

Based on the topics discussed, the aim of the present study was to evaluate the effect of chitosan, nisin and storage time and their interactions on the inhibition of *L. innocua* and *S. putrefaciens* growth in fish homogenates of pH 5.5 during storage at 30ºC.

2. Materials and methods

2.1 Test microorganisms

Both of the microorganisms used in this study, *Shewanella putrefaciens* ATCC 8071 and *Listeria innocua* 6a ATCC 33090, were stored at -20ºC in Mueller Hinton broth (Biokar Diagnostics, Beauvais, France) plus 10%w/w glycerol and 10%w/w skim milk. Before use they were grown twice in Mueller Hinton broth at 30ºC during 18 hours.

2.2 Test antimicrobials

Chitosan (Sigma, Germany) was dissolved in acetic acid 1%w/w and pH was adjusted to 5.5 using NaOH 0.4M. Nisin (Nisaplin, Danisco A/S DK, Denmark) was dissolved in distilled water acidified to pH 2.0 with HCl plus NaCl 0.75%w/w. Both solutions were autoclaved at 121ºC for 15 minutes. Before using the solution of nisin, its pH was adjusted to 5.5 using sterilized NaOH 0.4M.

2.3 Fish homogenates preparation

Fish homogenate was prepared processing Argentine hake fillets (*Merluccius hubbsi*) and distilled water in a ratio of 1:1; pH was adjusted to 5.5 using citric acid 10%w/w. Homogenate aliquots were put into screw-cap flasks and sterilized for 15 minutes at 100ºC.

The sterilized solutions of nisin and chitosan were added to the fish homogenates in order to achieve the selected concentrations, which are shown in Table 1. Then microorganisms were inoculated reaching a level of 10^5 CFU g^{-1}.

The inoculated homogenates were stored at 30ºC for 48 hours. *S. putrefaciens* and *L. innocua* populations were enumerated by pour-plating in Trypticase soy agar and Mueller-Hinton agar (Biokar Diagnostics, Beauvais, France), respectively. The plates were incubated for 48 hours at 30ºC.

Factors	Levels		
	-1	0	1
Chitosan (ppm)	0	1000	2000
Nisin (ppm)	0	2500	5000
Storage time (hours)	0	24	48

Table 1 Levels of studied factors and their codes

2.4 Design and statistical analysis

A factorial design in three blocks and a central point was performed, being the concentration of chitosan and nisin, and the storage time the factors studied. Table 1 shows the selected levels of the mentioned factors.
Data obtained were processed using the statistical program Statgraphics (Statgraphics Plus for Windows, version 5.0, 2001, Manugistics, Inc., Rockville, Maryland, USA). The significance level was 0.05%.

3. Results

Figure 1 shows the significance of each one of the factors studied and their respective interactions. As it was expected, in all cases, time exerted a significant effect on the growth of both microorganisms; the counts were higher with increasing storage time ($P < 0.05$). When nisin was added, counts were reduced compared to control samples ($P < 0.05$). A similar trend was observed by adding chitosan. A significant interaction ($P < 0.05$) between antimicrobials and time was noted; however, this effect depended on the antimicrobial and the microorganism being tested. There was no significant interaction between the two preservatives ($P > 0.05$).

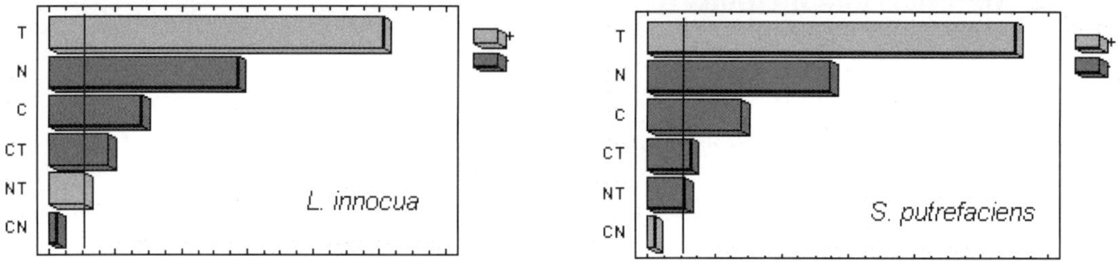

Fig. 1 Pareto charts showing the effect of chitosan, nisin and storage time on growth.
T: time; N: nisin; C:chitosan; CT: chitosan by time; NT: nisin by time; CN: chitosan by nisin.; light grey bar: positive effect; dark grey bar: negative effect

3.1 *Listeria innocua*

As it was mentioned above, bacterial counts increased with storage time. The development of *L. innocua* was affected by chitosan and nisin. However, different trends with respect to time are noted, as it is shown in Fig. 2. Nisin, alone or together with chitosan, reduced the inoculated level of *L. innocua* between 3.3 and 4.0 log cycles, immediately after being added to the fish homogenate. On the other hand, the ability of chitosan to reduce counts at time zero was negligible.

Nisin was not able to control the growth of *L. innocua* during the storage, since a regrowth of the microorganism was observed after 48 hours storage at 30ºC. This trend is linked to the fact that bacteriocins can induce a transitory inhibitory effect against *L. monocytogenes*, followed by a regrowth of cells [11]. Probably, this regrowth was caused by the appearance of resistant variants of *L. monocytogenes* to nisin [12] and the same behavior can be expected with *L innocua*.

The increase in bacterial population in the presence of chitosan and nisin was lower than in the presence of each preservative alone. However, microbial counts reached were high suggesting that another preservation factor must be included to control growth.

No synergism was shown between the antimicrobials, but the best results were observed in the homogenates containing both preservatives. In these samples the counts of *L. innocua* were lower than 6 log cycles after 48 hours of storage at 30ºC.

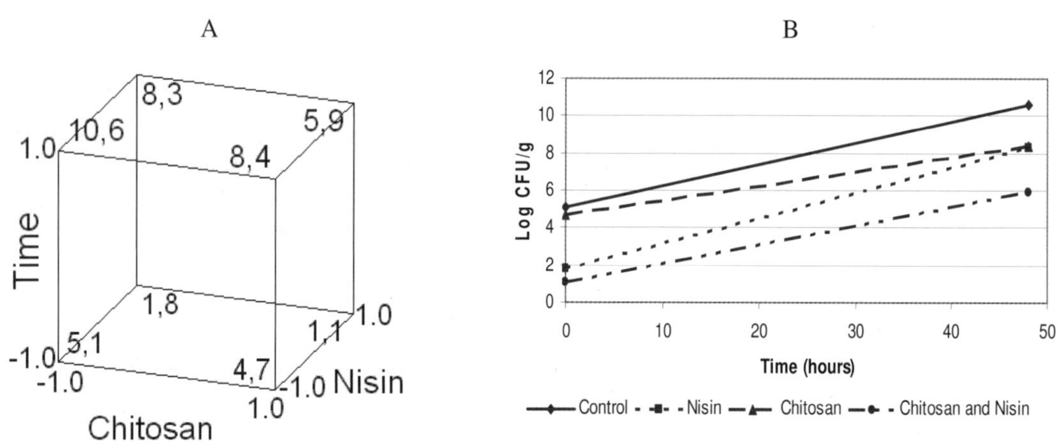

Fig. 2 A) Cubic graph showing the evolution of log counts of *L. innocua* as a function of codified levels of chitosan, nisin and storage time. B) Evolution of log counts of *L. innocua* during the storage.

3.2 *Shewanella putrefaciens*

The counts of *S. putrefaciens* increased over the time, as it happened with *L. innocua.* As it can be seen in Fig. 3, both preservatives reduced growth of *S. putrefaciens* during storage at 30ºC.

At time zero, both antimicrobials were effective. Chitosan reduced the inoculated level of S. *putrefaciens* approximately 1 log cycle, while nisin reduced it 2 log cycles. The lowest counts were found in presence of both preservatives, in this case they were 2.4 log cycles below the inoculated level.

After 48 hours of storage at 30ºC, the same tend was observed. Counts increased 7 log cycles in control samples, while the increase in chitosan fish homogenates was 5.6 log cycles. In the presence of nisin, the counts increased 5.8 log cycles. However, nisin showed to be more efficient than chitosan to control *S. putrefaciens* growth, since the level reached by the microorganism was lower.

Although there was no a statistical significant interaction that suggested a synergistic action between the two preservatives, results obtained showed that the development of the microorganism was lower when the antimicrobials were used in combination.

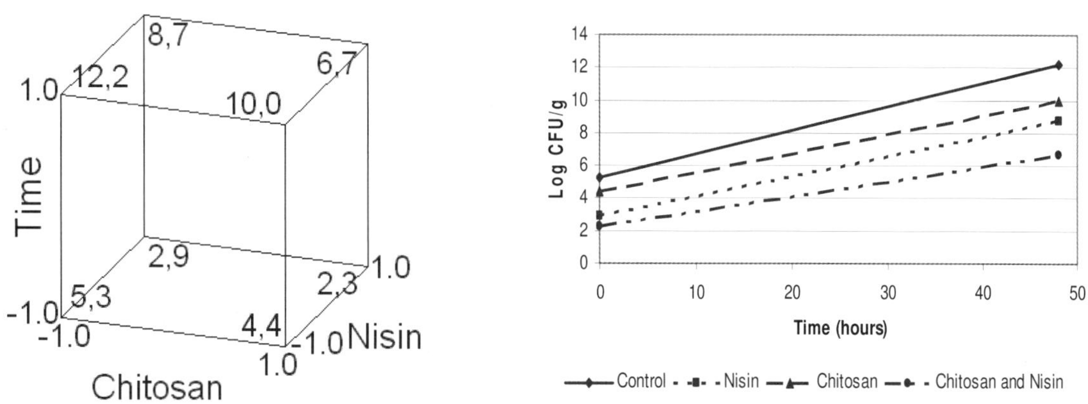

Fig. 3 A) Cubic graph showing the evolution of log counts of *S. putrefaciens* as a function of codified levels of chitosan, nisin and storage time. B) Evolution of log counts of *S. putrefaciens* during the storage.

4. Conclusion

Nisin and chitosan were able to inhibit the development of *L. innocua* and *S. putrefaciens* in fish homogenate at pH 5.5. In all cases, the counts were reduced by the presence of the test antimicrobials.

Regarding to *L. innocua*, nisin produced a decrease in the counts immediately after being applied, but a regrowth of the microorganism was observed after 48 hours storage at 30ºC. It must be highlighted that the regrowth was inhibited by the joint presence of nisin and chitosan.

It is known that nisin effectively inhibits Gram-positive bacterium. Despite this, it is clear that in this case, the preservative also controlled the growth of *S. putrefaciens*, Gram-negative bacterium. It should be noted that the inhibitory effect of nisin on *S. putrefaciens* during the storage was stronger than the effect of chitosan.

Although no synergism was observed between the two preservatives, the joint use of the studied antimicrobials could be promising, since the nisin effect on counts was shown immediately after its application, while chitosan action was noticed during the storage. Results obtained suggested that the joint use of both antimicrobials together with another stress factor such as refrigeration would be useful for extending shelf life of fish products.

Acknowledgements The support of Universidad de Buenos Aires and Consejo Nacional de Investigaciones Científicas y Técnicas de la República Argentina is gratefully acknowledged.

References

[1] Gram L, Daalgard P. Fish spoilage bacteria – problems and solutions. *Current Opinion in Biotechnology*. 2002; 13 (3): 262-266.

[2] Del Nobile, M.A, Corbo, M.R, Speranza , B, Sinigaglia, M, Conte, A and Caroprese, M. Combined effect of MAP and active compounds on fresh blue fish burger. *International Journal of Food Microbiology*. 2009;135: 281-287.

[3] Sofos JN, Beuchat LR, Davidson PM, Jonsons EA. Naturally occurring antimicrobials in food. *Regulatory Toxicology and Pharmacology*. 1998; 28: 71-72.

[4] No HK, Park NY, Lee SH, Meyers SP. Antibacterial activity of chitosans and chitosan oligomers with different molecular weights. *International Journal of Food Microbiology*. 2002; 74:65-72.

[5] Fan W, Sun J, Chen Y, Qiu J, Zhang Y, Chi Y. Effects of chitosan coating on quality and shelf life of silver carp during frozen storage. *Food Chemestry*. 2009; 115: 66-70.

[6] Cleveland J, Montville TJ, Nes IF, Chikindas ML. Bacteriocins: safe, natural antimicrobials for food preservation. *International Journal of Food Microbiology*. 2001; 71: 1-20.

[7] Food and Drug Administration. Nisin preparation: affirmation of GRAS status as a direct human food ingredient. *FDA, Federal Regulation*. 1998; 53: 11247.

[8] Dykes GA, Hancock REW, Hatings JW. Structural variations in nisin associated with different membrane mimicking and pH environments. *Biochemical and Biophysical Research Communications*. 1998; 247: 723-727.

[9] Himelbloom B, Nilsson L, Gram L. Factors affecting production of an antilisterial bacteriocin by *Carnobacterium piscicola* strain A9b in laboratory media and model fish sistems. *Journal of Applied Microbiology*. 2001; 91: 506-513.

[10] Friedly, E. C., Crandall, P. G., Ricke, S., O'bryan, C. A., Martin, Em., & Boyd, L. M. (2008). 721
Identification of Listeria innocua surrogates for *Listeria monocytogenes* in hamburger patties. *Journal of Food Science*, 73 (4), M174–178.B.

[11] Bouttefroy, A., Milliere, J., 2000. Nisin-curvaticin 13 combinations avoiding the regrowth of bacteriocin resistant cells of *Listeria monocytogenes* ATCC 15313. *International Journal of Food Microbiology*. 62, 65– 75.

[12] Harris, L., Fleming, H., Klaenhammer, T., 1991. Sensitivity and resistance of *Listeria monocytogenes* ATCC 19115, Scott A, and UAL500 to nisin. *Journal of. Food Protection*. 54, 836–840

ESBL-producing Enterobacteriaceae in the Northern Portugal – Antimicrobial Susceptibility and Molecular Epidemiology

R. Fernandes[*1,2] **and C. Prudêncio**[1,2]

[1]Ciências Químicas e das Biomoléculas, Escola Superior de Tecnologia da Saúde do Porto, Instituto Politécnico do Porto, 4400-330, Vila Nova de Gaia, Portugal
[2]Centro de Farmacologia e Biopatologia Química (U38-FCT), Laboratório de Bioquímica, Faculdade de Medicina, Universidade do Porto, 4200 - 319 Porto, Portugal

*Corresponding author: e-mail: rfernandes@eu.ipp.pt, Phone: +351 222061004

Extended-spectrum β-lactamases (ESBLs) prevalence was studied in the north of Portugal, among 193 clinical isolates belonging to citizens in a district in the boundaries between this country and Spain. It were recovered some members of Enterobacteriaceae family, producing ESBL enzymes, including *Escherichia coli* (67.9%), *Klebsiella pneumoniae* (30.6%), *Klebsiella oxytoca* (0.5%), *Enterobacter aerogenes* (0.5%) and *Citrobacter* freundii (0.5%). The β-lactamases genes, blaTEM, blaSHV and blaCTX-M were screened by Polymerase Chain Reaction (PCR) and sequencing approaches. TEM enzymes were among the most prevalent types (40.9%) followed by CTX-M (37.3%) and then SHV (23.3%) types. Almost all ESBL-producing isolates (98.9%) were resistant to the fourth-generation cephalosporin Cefepime. Of the 193 isolates 157 (81.3%) present transferable plasmids. Molecular typing studies were performed using PCR for *enterobacterial repetitive intragenic consensus* (ERIC) sequences. This study reports a high diversity of genetic patterns, A to J for *E. coli* isolates and K to O *for K. pneumoniae strains.*

Keywords Enterobacteriaceae; ESBL

1. Introduction

Extended-spectrum β-lactamases (ESBLs) are enzymes that confer resistance to Aztreonam, Cefotaxime, Ceftazidime, and related oxyimino-β-lactams as well as to older penicillins and cephalosporins but are inhibited by Clavulanic acid [1]. ESBL-producing Enterobacteriaceae were first reported in Europe in the 1980s and have since become a worldwide problem [2]. This has resulted in increased morbidity, mortality and cost in treating the infections they cause [3]. The first ESBL were mutants of the TEM and SHV plasmid-mediated penicillinases with one or more aminoacid substitutions. The mutations conferred resistance to all oxyimino-cephalosporins but not α-methoxy-cephalosporins (cephamycins) or carbapenems by causing enlargement of ESBL active site, which allowed the deflection of the oxyimino group diminishing the attack efficiency on the β-lactam ring [4]. TEM and SHV present to date over 200 members known (http://www.lahey.org/studies). Another ESBL group, include the CTX-M enzymes, that are organized in five major CTX-M groups: 1, 2, 8, 9, and 25 [5]. The CTX-M comprise a rapidly growing family distributed both over wide geographic areas among a wide range of bacteria of clinical significance and are becoming more prevalent than its ancestors TEM and SHV in several countries in Europe [6]. Several studies have been reported in Iberian Peninsula describing the genetic and clinical environments of ESBL occurrence [7-12]. Here we report the molecular and antimicrobial susceptibility profile of the ESBLs found in the Portuguese occidental cost in the boundaries between the two countries, Portugal and Spain. For this task it were used methods of molecular typing that have been developed for the identification of the β-lactamases *bla* genes in Enterobacteriaceae bacteria from clinical hospitalized and non-hospitalized patients for a period of two years.

2. Materials and Methods

2.1. Bacterial strains, identification and susceptibility

A total of 7529 clinical strains were included in the study. All isolates were gently provided from Clinical Pathology Laboratories and belong to patients samples recovered from September 2007 to August 2009 in the northern occidental coast of Portuguese territory (Minho). This area comprises several populations of the northern of Portugal and some at the boundaries with Galicia, Spain. Bacteria identification and preliminary antimicrobial susceptibility were determined in accordance with the guidelines of the Clinical and Laboratory

Standards Institute [13]. ESBL production was confirmed by the elipsoid method, using two E-test Strips (AB Biodisk, Sweden). Each strip contained several concentrations of a third generation cephalosporin, Ceftazidime (TZ) and Cefotaxime (CT) alone and with Clavulanic acid (TZL, CTL). The ESBL strip contained CT gradient at one end (0.25-16 mg/L) and a gradient of Cefotaxime (0.016-1 mg/L) plus 4 mg/L of Clavulanic acid (L), an inhibitor of ESBL, at the other end (CTL). The other ESBL strip contained TZ gradient at one end (0.5-32 mg/L) and a gradient of TZ (0.064-4 mg/mL) plus L (4 mg/L) at other end (TZL). According to manufacturer an ESBL is considered present whenever the E-test attends at least 8-fold MIC decrease on TZL or CTL.

2.2. Conjugation experiments

Transmissibility of resistance was test by matting clinical isolates to F^- strains of azide-resistant *E. coli* J53 Azi^R on trypticase soy broth (TSB) according to a method described previously [14].

2.3. Analytical isoelectric focusing (IEF)

Crude preparations of β-lactamases from clinical strains transconjugants were obtained by sonicating the cells in phosphate buffer, pH 7.0 as described previously [15]. Briefly crude extracts were concentrated and crude protein concentrated extract was added to a nitrocefin solution. The sample run in a IEF mini-gels, pH 3 to 10 for 30 min using both anode and cathode electrode buffers. β-lactamases isoelectric point (pI) was determined by pouring molten 3% agarose containing nitrocefin over the gel and comparing the bands to standards run on the same gel. The β-lactamase standards used were extracts of strains containing TEM-1, pI 5.4, SHV-5, pI 8.2 and CTX-M-14, pI 8.1.

2.4. Genetic molecular characterization of *bla* genes and typing

A single colony of each transconjugant was left to grow for 16h on MacConkey agar and was placed in 200 µL sterile water in a 1.5 mL microtube. Each tube was heated in a microwave oven at 600–700 W for 2 min to burst the cells and release their DNA. PCR was performed as described previously [15]. The recovered bands from agarose were subcloned for further sequencing. The nucleotide sequences of both ends of the insert were determined with M13 sequencing primers specific for the cloning vector [16]. ERIC profiles were analyzed with software, FPQuest™ version 4.5, Fingerprinting II (Bio-Rad Laboratories, CA, USA).

3. Results

The prevalence of ESBL-producing strains in the north Portuguese territory was 2.6% (n=193). The most frequent ESBL-producing organism was *E. coli* (67.9%, n=131), followed by *K. pneumoniae* (30.6%, n=59), *K. oxytoca* (0,5%, n=1), *E. aerogenes* (0,5%, n=1) and *C. freundii* (0,5%, n=1). The ESBL-producing strains were isolated from urine (n=127), sputum (n=42), bronchoalveolar lavage (n=14), blood (n=7) and ascitic fluid (n=3). The types of ESBL detected in the present study appear in the table 3 and were in percents as follows. TEM types were detected in 40.9%, CTX-M types in 37.3% and SHV types in 23.3%. Several isolates harboring also broad-spectrum β-lactamases, such as TEM-1 (pI=5.4), TEM-2 (pI=5.6) and SHV-1 (pI 7.6) types, but are not presented in this study.

MIC tests shown that isolates producing TEM, SHV and CTX-M ESBL types also appear in Table 1and were mostly resistant to Cefepime (98.9%) and susceptible to Carbapenems (100%) and Amikacin (99.5%). Nevertheless, some SHV-producing strains, shown an important reduced susceptibility to other member of Aminoglycosides, namely Gentamicin (11.1%). ESBL-producing strains from this sample also present reduced susceptibility to Quinolones (14.6%) in the generality, being more conspicuous in the members of the CTX-M family (1.9%) rather than TEM and SHV members (19.2% and 27.9% respectively).

Regarding interspecific genetic similarity, it was observed a high genetic diversity. It was possible to define 10 clusters for *E. coli* based on Pearson's correlation coefficient in PCR-ERIC based profile (data not shown). For *K. pneumoniae* it was defined 5 different clusters for the PCR-ERIC (data not shown).

4. Discussion

As reported in previous Portuguese [9-12; 14-15; 17-19], Spanish [8, 20-21] and other European [6] studies *E. coli* and *K. pneumonia* are the species where ESBL is the most frequently identified. In this study *E. coli* were the most frequent (n=131) organism expressing ESBL phenotypes, more than two fold of the *K. pneumonia* (n=59), the second most frequent.

TEM-52 and TEM-24 were the most frequent TEM types, 20.2% (n=39) and 12.9% (n=25) respectively. Members of the TEM-10 (n=8) and TEM-116 (n=4) were also detected. Within CTX-M family, the CTX-M-9 group is more prevalent than the CTX-M-1 group (58.3% against 41.6%). CTX-M-9 group was represented by CTX-M-9 (n=26, 36.1%) and CTX-M-14 (n=16, 22.2%). In the CTX-M-1 group, CTX-M-15 was most frequent type (n=24, 33.3%), followed by CTX-M-1 (n=4, 5.5%), CTX-M-3 (n=1, 1.3%) and CTX-M-32 (n=1, 1.3%). The SHV enzymes occurred only in 23.3% of all ESBL- producing organisms. Within this type, the most frequent was the SHV-12 (n=24, 53.3%), followed by SHV-5 (n=17, 37.8%) and finally SHV-2 (n=4, 8.8%). Some isolates co-produced more than one ESBL type: TEM-52/CTX-M-14 (n=1), TEM-116/CTX-M-14 (n=1) and TEM-116/CTX-M-15 (n=1).

Table 1 Characterization ESBL-producing strains

ESBL	IEF (pI)	Microorganism (no. of isolates)	Conjugation (% positive)	Resistance phenotype (% Non-susceptible: I+R)					ERIC types patterns (*)
				β-lactams		Non β-lactams			
				CEP	FOX	CIP	GEN	Bactrim	
TEM-4	5.9	*E. coli* (2)	100	100	50	100	50	100	A
		K. pneumoniae (1)	100	100	0	100	100	0	L
TEM-10	6.0	*E. coli* (8)	87.5	100	25	87,5	0	75	B,D,H
TEM-24	6.5	*E. coli* (12)	83.3	100	50	75	8.3	8.3	C,G,H,I
		K. pneumoniae (11)	81.8	100	20	72.7	50	63.6	K,L,O
		K. oxytoca (1)	100	100	100	100	0	100	n.a.
		E. aerogenes (1)	100	100	0	0	100	100	n.a.
TEM-52	6.0	*E. coli* (27)	81.4	100	36.4	84.6	53.8	42.3	A,B,E,F,J
		K. pneumoniae (12)	100	100	8.4	91.6	83.3	33.3	N,O
TEM-116	5.4	*E. coli* (n=4)	100	100	50	75	75	0	F
SHV-2	7.6	*K. pneumoniae* (3)	100	33.3	33.3	66.7	66.7	100	L
		E. coli (1)	100	0	0	100	100	100	A
SHV-5	8.2	*K. pneumoniae* (11)	72.7	100	27.3	63.6	72.7	27.3	L,N
		E. coli (6)	83.3	100	66.7	100	100	0	
SHV-12	8.2	*K. pneumoniae* (18)	83.3	100	27.7	61.1	94.4	77.8	K,M,N,O
		E. coli (6)	100	100	33.3	100	100	66.7	C,D,E
CTX-M-1	6.3	*E. coli* (1)	100	100	25	100	100	0	E
CTX-M-3	8.4	*C. freundii* (1)	0	100	0	0	100	0	n.a.
CTX-M-9	8.1	*E. coli* (26)	92.3	100	26.9	96.2	84.6	3.8	A,C,D,G,I,J
CTX-M-14	8.1	*E. coli* (16)	68.8	100	12.5	100	75	0	D,F,H, J
CTX-M-15	8.6	*E. coli* (21)	85.7	100	38.1	100	90.5	14.3	A,C,E,I,J
		K. pneumoniae (3)	100	100	100	100	100	100	N
CTX-M-32	9.0	*E. coli* (1)	100	100	0	100	100	0	D

ESBL: extended-spectrum β-lactamase; IEF: isoelectric focusing; pI: isoelectric point; I: Intermediate; R: Resistant; n.a.: Not applicable. Antimicrobials: CEP: Cefepime; FOX: Cefoxitin; CIP: Ciprofloxacin; Gen: Gentamicin; Bactrim: combination of Trimethoprim and Sulfamethoxazole; (*) Dendogram not shown.

Regarding the high diversity ESBL types obtained in our study, we find interesting to compare with neighbor regions, such as Douro Litoral located at south of Minho, Trás-os-Montes e Alto Douro, eastern and the Spanish province of Galiza, located at north of Minho (Figure 1). From this comparison and regarding to TEM types ESBL, TEM-10, TEM-20, TEM-26 and TEM-116 in *E. coli* and TEM-4 in *K. pneumoniae* were present only in this region but have been reported in other locations of Iberian Peninsula. Isolates of *E. coli* producing SHV-2 and SHV-5 were also found only in this region. Nevertheless, it has been described previously in the Iberian Peninsula (Hernandez et al., 2005). Regarding CTX-M types, it seems that CTX-M-14 is widespread among the northwestern Iberian Peninsula. *K. pneumoniae* harboring a CTX-M-15 was described for the first time in Portugal in 2005 [23] in Lisbon area but it is also found in the north of Portugal in this study and Douro [9]. Other ESBL-producing species non-*E.coli* and non-*K. pneumonia* were also found. It has been the first time that an *E. aerogenes* is described in this country as producer of a TEM-24 and the first time that is found a *C. freundii* as a producer of a CTX-M-32.

In what concerns to the clonally studies, it seems clear the high genetic diversity among the strains in study. Nevertheless, some considerations are needed. K cluster for *K. pneumoniae* was represented by TEM-24 producing members only. Nevertheless, it may be due to a coincidence. The two strains presenting the tightest resemblance were only 74% genetically related. The antibiogram for the four strains were also different between all them (data not shown). Moreover, they belong to four patients, two males and two females, interned in different clinical services. In the opposite situation, the CTX-M-15 producing *K. pneumoniae*, bellowing to N cluster, presented the exactly antibiogram profile and the strain less related to the other two, shared a 93% of genetic similarity and the most related two of them, presented a 98% of genetic similarity. The patients, three females with ages between 76 and 86 years old, shared the same clinical service, Medicine Females, during the same period, from October to November 2007.

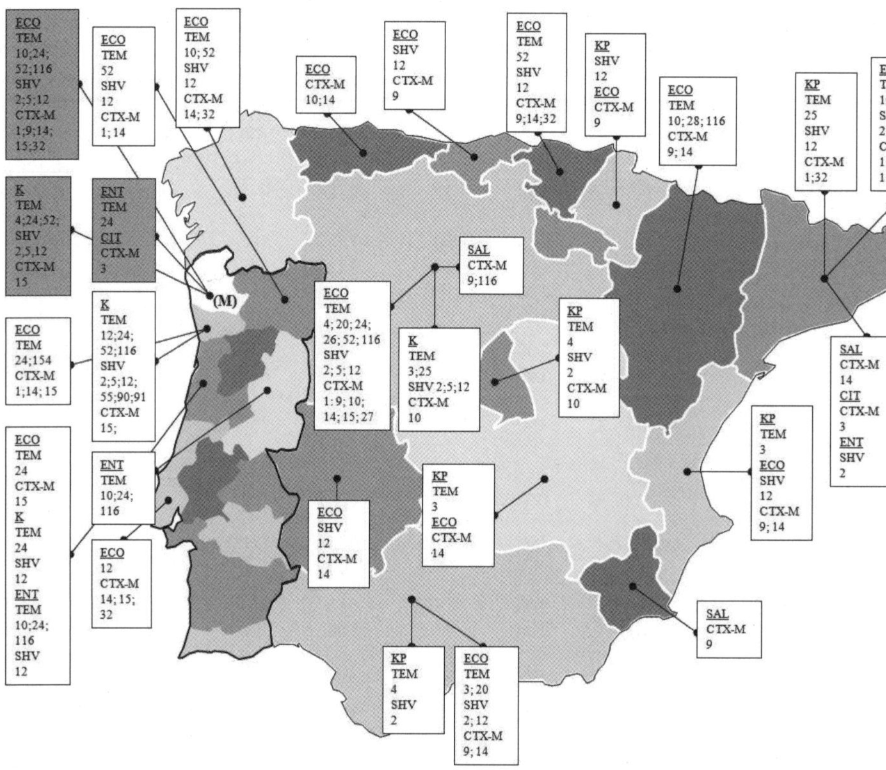

Figure 1. Occurrence of ESBL-TEM, ESBL-SHV and ESBL-CTX-M type in the north of Portugal. The present study territory, Minho (M), is represented at the map in white. Other regions either from Portugal or Spain, are represented with different gray tones. Data was collected from this study and from several studies in these two countries [8-12; 14,15;17-21; 23-27].

Cefepime presents, in this study a surprisingly low activity against ESBL-producing microrganisms. In our sample only two *K. pneumoniae* harboring SHV-2 ESBL was susceptible to Cefepime. All the other clinical isolates, 98.9% (n=191) expressing the ESBL phenotype were resistant to Cefepime. It seems interesting that a recent study showed that Cefepime was successfully administrated to three patients (two females and one male) with ages between 47 and 87 years old carrying a gram-negative ESBL positive strain [28]. Nevertheless Other studies worldwide start to describe the emergence of high resistance among ESBL gram-negative producers [29-30].

In summary, this work showed the high diversity of ESBL-enzymes occurring in Portugal. In this country, the most prevalent type is still the TEM-type but CTX-M is growing rapidly. The ERIC analysis shows to be efficient in the clonally relationships between the different *Enterobacteriaceae* isolates, and it was possible to detect an episode of nosocomial infection. Nevertheless, it was obtained indistinguishable patterns among ESBL-producing strains. The emergence of ESBL producers resistant to Cefepime in Portugal is a matter of concern. We believe that the uncontrolled use of cephalosporins may have an important role in the acquisition of resistance mechanisms, specially the production of ESBL enzymes. It seems to be urgent the establishment of policies to monitor drug delivery in hospital and ambulatory pharmacies and the implementation of public health defense strategies towards health promotion and drug resistance prevention.

References

[1] Bush, K., Jacoby, G.A., Medeiros, A.A. A functional classification scheme for □-lactamases and its correlation with molecular structure. *Antimicrob Agents Chemother.*1995, *39*: 1211–33.

[2] Paterson, D.L., Bonomo, R.A.. Extended-spectrum β-lactamases: a Clinical update. *Clin Microbiol Vet.* 2005,18:657–86.

[3] Paterson, D.L., Ko, W.C., Von Gottberg, A., Casellas, J.M., Mulazimoglu, L., Klugman K.P., *et al.* Outcome of cephalosporin treatment for serious infections due to apparently susceptible organisms producing extended-spectrum □-lactamases: Implications for the clinical microbiology laboratory. *J Clin Microbiol.* 2001, 39:2206–12.

[4] Livermore, D.M., Woodford, N. The □-lactamase threat in Enterobacteriaceae, Pseudomonas and Acinetobacter. *Trends Microbiol.* 2006,14:413–20.

[5] Bonnet, R. Growing group of extended-spectrum □-lactamases: the CTX-M enzymes. *Antimicrob Agents Chemother.* 2004, 48:1-14.

[6] Livermore, D.M. □-lactamases – the threat renews. *Curr Protein Pept Sci.* 2009, 10:397–400.

[7] Rodríguez-Baño, J., Navarro, M.D., Romero, L., Martinez-Martinez, L., Muniain, M.A., Perea, E.J. Epidemiology and clinical features of infections caused by extended-spectrum □-lactamase-producing *Escherichia coli* in non-hospitalized patients. *J Clin Microbiol.* 2004,42:1089–94.

[8] Hernández, J.R., Martínez-Martínez, L., Cantón, R., Coque, T.M., Pascual, A. Nationwide Study of *Escherichia coli* and *K. pneumoniae* producing extended-spectrum □-lactamases in Spain. *Antimicrob Agents Chemother.* 2005, 49:2122–25.

[9] Machado, E., Coque, T.M., Cantón, R., Baquero, F., Sousa, J.C., Peixe, L. Dissemination in Portugal of CTX-M-15-, OXA-1-, and TEM-1-producing Enterobacteriaceae strains containing the aac(6')-Ib-cr gene, which encodes an aminoglycoside- and fluoroquinolone-modifying enzyme. *Antimicrob Agents Chemother.* 2006, 50:3220–1.

[10] Mendonça, N., Leitão, J., Manageiro, V., Ferreira, E., Caniça, M. Spread of extended-spectrum □-lactamase CTX-M-producing *Escherichia coli* clinical isolates in community and nosocomial environments in Portugal. *Antimicrob Agents Chemother.* 2007, 51:1946–55.

[11] Amador, P., Fernandes, R., Prudêncio, C., Brito, L. Resistance to □-lactams in bacteria isolated from different types of Portuguese cheese. *Int J Mol Sci.* 2009, 10:1538–51.

[12] Fernandes, R., Gestoso, A., Freitas, J.M., Santos, P., Prudêncio, C. High resistance to fourth-generation cephalosporins among clinical isolates of Enterobacteriaceae producing extended-spectrum □-lactamases isolated in Portugal. *Int J Antimicrob Agents.* 2009, 33:184–5.

[13] Clinical and Laboratory Standards Institute. Methods for dilution antimicrobial susceptibility tests for bacteria that grow aerobically, 7th ed. Approved standard M7-A7. Clinical and Laboratory Standards Institute, Wayne, PA, 2007.

[14] Amador, P., Fernandes, R., Brito, L., Prudêncio, C. Antibiotic resistance in enterobacteriaceae isolated from Portuguese deli meats. *J Food Safety,* 2010 (In Press).

[15] Fernandes, R., Prudêncio, C. Post-surgical wound infections involving Enterobacteriaceae with reduced susceptibility to □-lactams in two Portuguese hospitals. *Int Wound J.* 2010, 7(6):508-14

[16] Naiemi, N.A., Duim, B., Savelkoul, P.H., Spanjaard, L., de Jonge, E., Bart, A. Widespread transfer of resistance genes between bacterial species in an intensive care unit: implications for hospital epidemiology. *J Clin Microbiol.* 2005, 43:4862-4.

[17] Costa, D., Poeta, P., Briñas, L., Sáenz, Y., Rodrigues, J., Torres, C. Detection of CTX-M-1 and TEM-52 □-lactamases in *Escherichia coli* strains from healthy pets in Portugal. *J Antimicrob Chemother.* 2004; 54:960–1

[18] Costa, D., Poeta, P., Sáenz, Y., Coelho, A.C., Matos, M., Vinué, L. Prevalence of antimicrobial resistance and resistance genes in faecal *Escherichia coli* isolates recovered from healthy pets. *Vet Microbiol.* 2008, 127:97–105.

[19] Fernandes, R., Vieira, M., Ferraz, R., Prudêncio, C. Bloodstream infections caused by multidrug-resistant Enterobacteriaceae: report from two Portuguese hospitals. *J Hosp Infect.* 2008, *70*: 93-5.

[20] Briñas, L., Zarazaga, M., Sáenz, Y., Ruiz-Larrea, F., Torres C. □-lactamases in Ampicillin-Resistant *Escherichia coli* Isolates from Foods, Humans, and Healthy Animals. *Antimicrob Agents Chemother.* 2002, 46:3156–63.

[21] Coque, T.M., Novais A., Carattoli, A., Poirel, L., Pitout, J., Peixe, L. Dissemination of clonally related *Escherichia coli* strains expressing extended-spectrum □-lactamase CTX-M-15. *Emerg Infect Dis.* 2008, 14:195–200.

[22] Miró, E., Mirelis, B., Navarro, F., Rivera, A., Mesa, R.J., Roig, M.C. Surveillance of extended-spectrum □-lactamases from clinical samples and faecal carriers in Barcelona, Spain. *J Antimicrob Chemother.* 2005, 56:1152–5.

[23] Conceição, T., Brízio, A., Duarte, A., Lito, L.M., Cristino, J.M., Salgado, M.J. First description of CTX-M-15-producing *K. pneumoniae* in Portugal. *Antimicrob Agents Chemother.* 2005, **49**:477–8.

[24] Briñas, L., Lantero, M., de Diego, I., Alvarez, M., Zarazaga, M., Torres C. Mechanisms of resistance to expanded-spectrum cephalosporins in *Escherichia coli* isolates recovered in a Spanish hospital. *J Antimicrob Chemother.* 2005, 56:1107–10.

[25] Coque, T.M., Oliver A, Pérez-Díaz JC, Baquero F, Cantón R. Genes Encoding TEM-4, SHV-2, and CTX-M-10 extended-spectrum □-lactamases are carried by multiple *Klebsiella pneumoniae* clones in a cingle hospital (Madrid, 1989 to 2000). *Antimicrob Agents Chemother.* 2000, 46:500–10.

[26] Costa, D., Vinué, L., Poeta, P., Coelho, A.C., Matos, M., Sáenz, Y. Prevalence of extended-spectrum □-lactamase-producing *Escherichia coli* isolates in faecal samples of broilers. *Vet Microbiol.* 2009, 138:339–44.

[27] Fernández, A., Gil, E., Cartelle, M., Pérez, A., Beceiro, A., Mallo, S. Interspecies spread of CTX-M-32 extended-spectrum □-lactamase and the role of the insertion sequence IS1 in down-regulating bla CTX-M gene expression. *J Antimicrob Chemother.* 2007, 59:841–7.

[28] Bhavnani, S.M., Ambrose, P.G., Craig, W.A., Dudley, M.N., Jones, R.N. Outcomes evaluation of patients with ESBL- and non-ESBL-producing *Escherichia coli* and *Klebsiella* species as defined by CLSI reference methods: report from the SENTRY Antimicrobial Surveillance Program. *Diagn Microbiol Infect Dis.*2006, 54:231–6.

[29] Grover, S.S., Sharma, M., Chattopadhya, D., Kapoor, H., Pasha, S.T. Singh G. Phenotypic and genotypic detection of ESBL mediated cephalosporin resistance in *Klebsiella pneumoniae*: Emergence of high resistance against cefepime, the fourth generation cephalosporin. *J Infect.* 2006, 53:279–88.

[30] Sader, H.S., Hsiung, A., Fritsche, T.R., Jones, R.N. Comparative activities of cefepime and piperacillin/tazobactam tested against a global collection of *Escherichia coli* and *Klebsiella* spp. with an ESBL phenotype. *Diagn Microbiol Infect Dis.* 2007, 57:341–4.

Observations on the antimicrobial susceptibility of *Staphylococcus pseudintermedius* following the introduction of cefovecin for clinical use in Europe

Y. Chaudhry[1], A. Robinson[2] and K.S. Godinho[3,4]

[1] Pfizer GmbH, Linkstr. 10, Postfach 610194, 10785 Berlin, Germany
[2] Pfizer Animal Health, 23-25 Avenue du Dr Lannelongue, F-75668 Paris, France
[3] Veterinary Research & Development, Pfizer Ltd., Sandwich, Kent CT13 9NJ, United Kingdom
[4] Corresponding author: Tel: +44 1304 645790. Fax: +44 1304 656257. E-mail: kevin.godinho@pfizer.com

In vitro susceptibility tests were conducted on field isolates of *Staphylococcus pseudintermedius* from canine clinical cases isolated from European countries during 2009-2010 for sensitivity to cefovecin and a panel of other commonly used antimicrobials by broth microdilution and oxacillin disc diffusion tests. Isolates suspected of multi-drug resistance were screened for the presence of the *mecA* gene. Eight isolates with zone diameters of less than 16 mm in the oxacillin disc diffusion test were deemed resistant to oxacillin and, following genotyping, seven were confirmed *mecA* positive. The MIC_{90} and MIC_{50} values for cefovecin were both 0.25 µg/mL and those isolates selected for genotyping had MIC values for cefovecin of ≥ 1 µg/mL. The results for cefovecin were compared with data from a similar survey conducted prior to launch in 1999-2003. The data indicate that to date for canine clinical isolates of *S. pseudintermedius*, there has been no apparent shift in susceptibility to cefovecin amongst isolates from first opinion cases across Europe but continued monitoring is recommended.

Keywords cefovecin; canine; *Staphylococcus pseudintermedius* susceptibility; methicillin resistance

1. Introduction

Following the introduction of a new antimicrobial agent onto the veterinary market, the instigation of a monitoring programme is an important requirement to detect changes in the pattern of bacterial sensitivity that might arise amongst the general population of the target species. This aspect has gained increasing significance with recently heightened concern over the possible selection for multi-drug resistant strains of potentially zoonotic bacteria, notably staphylococcal species. Taxonomic refinement has recently led to the most common staphylococcal species found in dogs as *Staphylococcus pseudintermedius* although previously described in many instances as *S. intermedius* [1,2]. Methicillin-resistant *Staphylococcal* (MRS) infections are identified through the presence and expression of the *mecA* gene coding for low-affinity penicillin binding protein 2A. Methicillin resistance is of particular importance as it commonly occurs in multi-drug resistant strains thus increasing the difficulty of achieving bacteriological cure in patients that may be immuno-compromised. Furthermore, homologues of the same gene are responsible for MRS in a range of staphylococci including isolates from both human and veterinary sources [3] and is found in methicillin-resistant *Staphylococcus aureus* (MRSA), methicillin-resistant *Staphylococcus intermedius* (MRSI) and methicillin-resistant *Staphylococcus pseudintermedius* (MRSP) [4].

Cefovecin is a cephalosporin that has only recently been developed for veterinary use in dogs and cats. It is administered as a single parenteral administration that may be repeated after an interval of 14 days. This authorised dose regimen allows for better control of the dose administered than repeated daily administration and minimises the risk of missed doses or inadvertent underdosage during a course of treatment. This paper reports the findings of a bacteriological survey to determine sensitivity to cefovecin of European strains of *S. pseudintermedius* isolated during 2009-2010.

2. Materials and Methods

2.1 Source of isolates

Field isolates of *S. pseudintermedius* were obtained from canine clinical cases, including pyoderma lesions, wounds and abscess cavities, by veterinarians across six European countries (France, Germany, Hungary, Poland, Spain and the United Kingdom) between 2009 – 2010 . All isolates were sourced from animals that were either untreated or had not received treatment with any antimicrobial for a minimum of 30 days prior to sampling. Only a single isolate was selected from each case. Bacteriological swabs were collected from wounds and skin infections and placed in AMES charcoal for transportation prior to identification. In cases of pyoderma, a swab was used to sample from the margin of the lesion, all superficial purulent exudate was removed and the area cleaned with physiological saline or water (without preservative) to reduce the number of contaminant organisms. Fresh exudate was then expressed and collected with the swab. Intact pustules were cleaned with 70% ethyl alcohol and allowed to dry prior to being ruptured with a sterile fine-gauge needle and a sample of pus collected on the swab. Samples from abscess cavities were collected from the abscess wall of recently formed abscesses. Species identification was performed using standard laboratory techniques including the use of API ID32-STAPH (bioMérieux, Marcy l'Etoile, France). Prior to sensitivity testing all bacterial isolates were stored on cryopreservative beads (Protect Beads™, Technical Services Consultants Ltd., Heywood, Lancashire, OL10 1DS, UK) at -75°C to -85°C.

2.2 Determination of minimum inhibitory concentrations

Susceptibility testing of the bacterial isolates to a panel of antimicrobials was performed in accordance with CLSI guidelines [5] at a single laboratory using a broth microdilution technique. In addition, susceptibility to oxacillin was tested by disc diffusion using 1 µg oxacillin disks (Oxoid, Hampshire, UK). The medium used for broth microdilution was cation-adjusted Mueller-Hinton broth (CAMHB) with TES buffer (Trek Diagnostic Systems, Cleveland, OH, USA) in standard round-bottomed 96-well microdilution plates (Alpha Laboratories Sterile Microtitre plates, Hampshire, UK). The dilution range for determination of minimum inhibitory concentrations (MICs) of the antimicrobials tested generally ranged from 0.06 to 64 µg/mL (see Table 1). The test system was validated and routinely monitored by concurrently determining the MIC of *Staphylococcus aureus* American Type Culture Collection (ATCC) 29213 (MIC to cefovecin 0.5-2 µg/mL) as a quality control organism.

Bacterial isolates were retrieved from storage and sub-cultured onto Columbia sheep blood agar, incubated overnight at 35±1°C and checked visually for morphology and purity. Cell suspensions of isolates were tested in accordance with CLSI methodologies including testing of appropriate standard control organisms. The MIC for each isolate was defined as the lowest concentration of drug to completely inhibit visible growth. The MIC_{90} and MIC_{50} values were defined as the lowest concentrations capable of inhibiting the growth of 90% and 50% of isolates, respectively.

2.3 PCR for presence of the *mecA* gene

The oxacillin disk diffusion test system was validated by concurrently determining the inhibition of *mecA* negative strain *S. aureus* ATCC 25923 (required inhibitory zone of 18-24 mm) and *mecA* positive strain *S. aureus* ATCC 43300. Those isolates suspected of resistance to oxacillin (inhibitory zone ≤17mm) were genotyped by polymerase chain reaction (PCR) test to determine the presence or absence of *mecA*. The PCR assay was based on a previously published methodology [6]. This method uses the primers 5′-CTCAGGTACTGCTATCCACC-3′ and 5′-CACTTGGTATATCTTCACC-3′ to amplify a 449 bp fragment of the *mecA* gene. Amplification was performed with *Taq* polymerase under suitable buffer conditions. An initial denaturation at 94°C for 5 min was followed by 30 cycles of 94°C for 6 sec, 52°C for 30 sec and 72°C for 30 sec, and final primer extension at 72°C for 10 min. The PCR products were separated by electrophoresis through 2% agarose gels in TRIS/Boric acid/EDTA buffer at 90V for 1.5 h. Gels were stained with ethidium bromide and viewed under UV light.

3. Results

A total of 153 *S. pseudintermedius* isolates of canine origin were obtained from veterinary clinics around Europe.Ten isolates were obtained from France (7%), 30 from Germany (20%), 10 from Hungary (7%), 70 from Poland (46%), 14 from Spain ((9%) and 19 from the United Kingdom (12%). Sixteen isolates were sourced from swabs taken of abscess cavities. The remainder were taken from swabs of pyoderma lesions. The distribution of MIC values and summary data following MIC testing against a standardized panel of antimicrobials are shown in Table 1. All isolates were resistant to metronidazole (>32 µg/mL). The MIC_{90} and MIC_{50} values for cefovecin were both 0.25 µg/mL (the mode was 0.25 µg/mL, and the range 0.12 to 32 µg/mL). Disk diffusion results for oxacillin are also shown in Table 1. Eight isolates with zone diameters of less than 16 mm were deemed resistant to oxacillin and, following genotyping, seven were confirmed *mecA* positive, as shown in Table 2. All eight isolates selected for genotyping had MIC values for cefovecin of ≥1 µg/mL.

Table 1 Distribution of sensitivity data for 153 isolates of *Staphylococcus pseudintermedius* isolated from canine clinical cases in Europe.

Antimicrobial concentration (μg/mL)	Antimicrobials included in panel for broth MIC test								OXA disc diffusion test	
	VEC n	LEX n	CEP N	CLI n	MAR n	ENR n	SXT† n	AUG† n	zone (mm)	n
≤0.06	-	-	33	3	-	6	5	5	>29	-
0.12	57	-	111	101	-	64	50	49	27-29	41
0.25	87	-	6	3	52	64	21	70	24-26	43
0.5	1	36	2	3	52	6	39	27	21-23	54
1	3	105	-	2	32	3	32	1	18-20	7
2	3	4	1	1	9	3	3	-	15-17	1
4	1	3	-	-	2	1	1	1	12-14	1
8	-	2	-	1	1	1	5	-	9-11	2
16	-	2	-	1	1	1	2	-	6-8	4
32	1	-	-	7	4	4	NT	-	3-5	-
≥64	NT	1	-	31	NT	NT	NT	-	0-2	-
Total number of isolates	153	153	153	153	153	153	153	153		153

VEC=cefovecin, LEX=cephalexin, CEP=cephalothin, CLI=clindamycin, MAR=marbofloxacin, ENR=enrofloxacin, SXT=trimethoprim/sulphamethoxazole (1:9 concentration ratio), AUG=amoxicillin/clavulanic acid (2:1 concentration ratio).† Drug concentrations for SXT and AUG are shown based on concentrations of trimethoprim and amoxicillin, respectively. OXA=oxacillin. NT=not tested.

Table 2 Antimicrobial susceptibility (MIC) of eight isolates suspected of being methicillin resistant and submitted for PCR analysis for the presence or absence of *mecA*.

Country of origin	Antimicrobials included in panel for broth MIC test for 8 isolates								OXA	
	VEC (μg/mL)	LEX (μg/mL)	CEP (μg/mL)	CLI (μg/mL)	MAR (μg/mL)	ENR (μg/mL)	STX (μg/mL)	AUG (μg/mL)	zone (mm)	*mecA*
Spain	4	16	0.5	>32	>16	>16	8/152	1/0.5	6	+
Spain	2	8	0.25	>32	>16	>16	8/152	0.5/0.25	6	+
Spain	1	4	0.25	>32	16	16	4/76	0.5/0.25	15	-
Germany	2	8	0.25	>32	>16	>16	16/304	0.5/0.25	13	+
Spain	1	4	0.25	>32	1	0.25	8/152	0.5/0.25	9	+
Spain	32	>64	2	>32	>16	>16	16/304	4/2	6	+
France	1	4	0.25	>32	0.25	0.12	1/19	0.5/0.25	9	+
France	2	16	0.5	0.12	8	8	0.12/2.38	0.5/0.25	8	+

VEC=cefovecin, LEX=cephalexin, CEP=cephalothin, CLI=clindamycin, MAR=marbofloxacin, ENR=enrofloxacin, SXT=trimethoprim/sulphamethoxazole, AUG=amoxicillin/clavulanic acid, OXA=oxacillin.

4. Discussion and Conclusion

Unfortunately recent refinements in species identification of staphylococci isolated from dogs and cats makes comparison with data from earlier surveys unreliable due to confusion over the speciation of the staphylococcal isolates. Indeed it is probable that the majority of isolates previously identified as *S. intermedius* are likely to have been *S. pseudintermedius*. For the purpose of the current study, based on standard laboratory techniques, it was assumed that all isolates were *S. pseudintermedius* but no additional confirmation, using phenotypic or molecular methods of identification was attempted. This does not reduce the value of the susceptibility data obtained as the methodology replicated that used in an earlier bacteriological survey undertaken prior to product launch [7]. In the earlier epidemiological survey the MIC_{90} for cefovecin against *S. (pseud)intermedius* using the same methodology for identification as the current survey was 0.25 μg/mL (mode 0.12 μg/mL, range ≤0.06 to 8 μg/mL). These results are similar to the present results (MIC_{90} 0.25 μg/mL, mode 0.25 μg/mL, range 0.12 to 32 μg/mL), in that values for each survey were within one dilution level of each other. It is acknowledged that the two surveys are not directly comparable, as the earlier survey included isolates from both canine and feline sources and from different locations across Europe. However, the earlier study found no differences in cefovecin susceptibility between European isolates and those collected in the United States, indicating a similar level of susceptibility across a wide geographic area. The current data would, therefore, indicate that to date susceptibility of *S. pseudintermedius* to cefovecin amongst clinical isolates has not diminished following the commercial launch of the branded product. No attempt was made in the earlier survey to determine the susceptibility of the isolates to methicillin or the presence of *mecA* in methicillin resistant isolates and it is not therefore possible to determine any change in the incidence of MRS. One isolate with an oxacillin inhibition zone of 15mm had an MIC to cefovecin of 1 μg/mL. All seven isolates positive for the *mecA* gene had MIC values for cefovecin of ≥1 μg/mL and hence did not affect the MIC_{50} or MIC_{90} values derived from the current study. In line with current CLSI recommendations, isolates resistant to oxacillin are also described as resistant to the cephalosporins tested here in the absence of *in vivo* data demonstrating efficacy. *In vitro,* however, only one isolate had an MIC indicating resistance to cefovecin (32 μg/mL) and of the remaining oxacillin-resistant isolates, one had an MIC of 4 μg/mL and the others had MIC values ≥1 μg/mL. A division of MRSP strains into two antibiogram types based on oxacillin susceptibility has been reported before [8] and also been noted for MRSA [6]. This may account for those isolates found to be carrying *mecA* but appearing to be susceptible to cephalosporins *in vitro*.

Current recommendations for testing susceptibility to methicillin suggest the use of disks containing 30 μg cefoxitin [5]. However, it has been found by others that the use of cefoxitin is unreliable for the prediction of oxacillin resistance in staphylococcal species found in dogs, and that 1 μg oxacillin disk diffusion with a ≤17 mm zone of inhibition is highly correlated with the presence of a *mecA* genotype [14,15]. For this reason oxacillin and not cefoxitin was used in the current survey to determine methicillin resistance.

The current survey detected a rate of 8/153 or 5.2% MRSP amongst the isolates tested which is lower than in some other surveys. The purpose of this study was to evaluate susceptibility amongst the wider population, not to identify the incidence of MRSP following previous high levels of mixed antimicrobial usage that may result in an increased incidence of multi-drug resistance in referral hospitals. Where clinicians have previously administered antimicrobials to treat specific conditions as this may induce a bias that is difficult to quantify [9] and care should be taken when interpreting the data from surveys. For example multiresistant *S. intermedius* was found to represent 23% of all *S. intermedius* isolates submitted to a dermatology referral clinic after antimicrobial therapy during an 18 month period [10] and 13.7% of isolates of *S. intermedius* were *mecA* positive in a survey in Poland [11]. The frequency of MRSI within 37 isolates of *S. intermedius* from dogs tested at a teaching hospital in the US was 15.6% [12]. However, a methicillin-resistance rate of 3% were reported in *S. intermedius* isolated from healthy dogs compared with 9% of dogs with inflammatory skin disease [9], indicating a lower recovery of methicillin-resistant isolates from less biased populations. In the present study isolates were derived from clinical cases that had not received antimicrobial therapy in the preceding 30 days. Only by excluding isolates from animals recently treated with antimicrobials can an unbiased estimate of the general population status be determined.

At present the data appear to indicate that there has been no shift in susceptibility of canine clinical isolates of *S. pseudintermedius* to cefovecin sourced from first opinion clinical cases. However, the situation needs continued monitoring and it is proposed that further surveys be conducted using similar or refined methodology to determine trends in sensitivity over time. In particular it is advised that future surveys should include determination of *mecA* status amongst isolates.

Acknowledgements The support of Pfizer Ltd for the funding of the study is gratefully acknowledged.

References

[1] Devriese LA, Vancanneyt M, Baele M, Vaneechoutte M, De Graef E, Snauwaert C, Cleenwerck I, Dawyndt P, Swings J, Decostere A, Haesebrouck F. *Staphylococcus pseudintermedius* sp. Nov., a coagulase-positive species from animals. *International Journal of Systemic and Evolutionary Microbiology*. 2005;55:1569-1573.

[2] Sasaki T, Kikuchi K, Tanaka Y, Takahashi N, Kamata S, Hiramatsu K. Reclassification of phenotypically-identified *Staphylococcus pseudintermedius* strains. *Journal of Clinical Microbiology*. 2007;45:2770-2778.

[3] Malik S, Peng H, Barton MD. Partial nucleotide sequencing of the mecA genes of Staphylococcus aureus isolates from cats and dogs. *Journal of Clinical Microbiology*. 2006;44: 413-416.

[4] Bannoehr J, Ben Zakour NL, Waller AS, Guardabassi L, Thoday KL, van den Broek AH, Fitzgerald JR. Population genetic structure of the *Staphylococcus intermedius* group: insights into *agr* diversification and the emergence of methicillin-resistant strains. *Journal of Bacteriology*. 2007;189:8685-8692.

[5] CLSI. Performance standards for antimicrobial disk and dilution susceptibility tests for bacteria isolated from animals; Approved standard – Third edition. M31-A3. Vol 28, No.8. Clinical and Laboratory Standards Institute, 940 West Valley Road, Suite 1400, Wayne, PA 19087, USA, 2008.

[6] Bignardi GE, Woodford N, Chapman A, Johnson AP, Speller DC. Detection of the *mec-A* gene and phenotypic detection of resistance in *Staphylococcus aureus* isolates with borderline or low-level methicillin resistance. *Journal of Antimicrobial Chemotherapy*. 1996;37:53-63.

[7] Stegemann MR, Passmore CA, Sherington J, Lindeman CJ, Papp G, Weigel DJ, Skogerboe TL. Antimicrobial activity and spectrum of cefovecin, a new extended spectrum cephalosporin, against pathogens collected from dogs and cats in Europe and North America. *Antimicrobial Agents and Chemotherapy*. 2006;50:2286-2292.

[8] Sasaki T, Kikuchi K, Tanaka Y, Takahashi N, Kamata S, Hiramatsu K. Methicillin-resistant *Staphylococcus pseudintermedius* in a veterinary teaching hospital. *Journal of Clinical Microbiology*. 2007;45:1118-1125.

[9] Griffeth, G.C., Morris, D.O., Abraham, J.L., Shofer, F.S., Rankin, S.C., 2008. Screening for skin carriage of methicillin-resistant coagulase-positive staphylococci and *Staphylococcus schleiferi* in dogs with healthy and inflamed skin. Vet. Dermatol. 19, 142-149.

[10] Loeffler A, Linek M, Moodley A, Guardabassi L, Sung JML, Winkler M, Weiss R, Lloyd DH. First report of multiresistant, *mecA*-positive *Staphylococcus intermedius* in Europe: 12 cases from a veterinary dermatology referral clinic in Germany. *Veterinary Dermatology*. 2007;18:412-421.

[11] Kizerwetter-Świda M, Chrobak D, Rzewuska M, Binek M. Antibiotic resistance patterns and occurrence of the *mecA* gene in *Staphylococcus intermedius* strains of canine origin. *Polish Journal of Veterinary Science*. 2009;12:9-13.

[12] Jones, R.D., Kania, S.A., Rohrbach, B.W., Frank, L.A., Bemis, D.A., 2007. Prevalence of oxacillin- and multidrug – resistant staphylococci in clinical samples in dogs: 1772 samples (2001-2005). J. Am. Vet. Med. Assoc. 230:221-227.

[13] Bemis DA, Jones RD, Frank LA, Kania SA. Evaluation of susceptibility test breakpoints used to predict *mecA*-mediated resistance in *Staphylococcus pseudintermedius* isolated from dogs. *Journal of Veterinary Diagnostic Investigation*. 2009;21:53-58.

[14] Schissler JR, Hillier A, Daniels JB, Cole LK, Gebreyes WA. Evaluation of Clinical Laboratory Standards Institute interpretative criteria for methicillin-resistant *Staphylococcus pseudintermedius* isolated from dogs. *Journal of Veterinary Diagostic Investigation*. 2009;21:684-688.

Oxacillin resistance among *Staphylococcus aureus* isolated from peritoneal dialysis related peritonitis

C. H. Camargo*[1,2], T. M. C. Moraes[2], M. L. R. S. Cunha[2], J. Caramori[1], A. L. Mondelli[1], A. C. Montelli[1,2] and P. Barretti[1]

[1] Departamento de Clínica Médica, Faculdade de Medicina de Botucatu, UNESP – Univ Estadual Paulista, Distrito de Rubiao Jr., s/n, CEP 18618-970, Botucatu, SP, Brazil.

[2] Departamento de Microbiologia e Imunologia, Instituto de Biociências de Botucatu, UNESP – Univ Estadual Paulista, Distrito de Rubiao Jr., s/n, CEP 18618-970, Botucatu, SP, Brazil.

* Corresponding author: e-mail: chcamargo@fmb.unesp.br, Phone: +55 14 38116240 extension 216

Peritonitis remains as the major cause of peritoneal dialysis failure, and *Staphylococcus* spp. are the main etiologic agents of these infections. Lower resolution rates are associated to *Staphylococcus aureus* episodes. Oxacillin resistance is an important factor in therapy definition, since it implies in resistance to all other β-lactam antimicrobials. In this study, we aimed to evaluate the oxacillin resistance, *mecA* gene detection, and staphylococcal cassette chromosome surrounding *mecA* gene (SCC*mec*) in *S. aureus* samples isolated from peritonitis in patients on peritoneal dialysis, from a single center from 1996 to 2008. Oxacillin resistance was found in 10/69 (14.5%) strains, while *mecA* gene was detected in 8/69 (11.6%). SCC*mec* typing was carried out on *mecA* gene-positive strains, but only three strains could be identified: two as type III, and one as type IA. Determining SCC*mec* alerts to the possibility of acquiring these infections on hospital environment, once that these cassette are traditionally associated to nosocomial infections.

Keywords Peritonitis; Peritoneal Dialysis; *Staphylococcus aureus*; Oxacillin resistance; *mecA* gene; SCC*mec*

1. Introduction

Peritoneal dialysis (PD) is an alternative to treatment for end-stage renal failure patients, and although the strong reduction in peritonitis rates over the last decades, these infections remains as an important cause of morbidity, technique failure, and mortalitity [1-3]. Coagulase-negative staphylococci (CoNS) are the main etiologic agents of PD-related peritonitis [4], while *Staphylococcus aureus* etiology are associated to worse outcome [5, 6]. The definition of antimicrobial susceptibility of peritonitis agents is crucial to adjust the initial antibiotic therapy [4, 7, 8]. Oxacillin (methicillin) - resistance is an important factor to define therapy of staphylococci infections, since it implies in resistance to all other β-lactam antimicrobials [9]. Although phenotypic methods have been employed to detect staphylococcal methicillin resistance, conflicting results are reported, particularly in heterogeneous resistant strains [10]. Molecular tests for *mecA* gene as polymerase chain reaction (PCR) or protein penicillin-binding 2a are the most accurate methods for oxacillin resistance detection [9]. Staphylococcal cassette chromosome surrounding *mecA* gene (SCC*mec*) is a mobile genetic element composed of variable structures [11] that allows its typing [10] and that can provide elements to elucidate the epidemiology of methicillin-resistant *S. aureus* (MRSA) infections [11]. For example, SCC*mec* types I to III are traditionally related to hospital environment whilst SCC*mec* types IV and V are commonly detected in MRSA community acquired infections [10, 12, 13]. We aimed to study the staphylococcal oxacillin resistance by phenotypic and genotypic tests, and the genetic *mecA* positive strains profile.

2. Material and Methods

This study was approved by the local Ethics Committee (Sept 14th 2009).

2.1 Peritonitis definition

Peritonitis was defined as the occurrence of cloudy peritoneal effluent with white blood cells (WBC) count more than 100/mL, with at least 50% polymorphonuclear neutrophilic cells [4].

2.2 Background

This study was carried out with strains isolated from peritonitis episodes occurred in PD patients at Botucatu Medical School, Sao Paulo, Brazil, from 1996 to 2008.

2.3 Specimen Processing

Peritoneal effluent (50 mL) was centrifuged at 3000g for 15 minutes, followed by resuspension of the sediment in 3 to 5 mL of sterile saline, and inoculation into a BACTEC® bottle. Primary isolation was carried out onto blood agar base and MacConkey agar plates.

2.4 *Staphylococcus aureus* identification

Following primary isolation, strains were identified by the macroscopic colony morphology on sheep blood agar medium, Gram stain, and catalase and coagulase production [14]. Established quality control strains were included in each one of this and the following the tests.

2.5 Phenotypic and genotypic oxacillin resistance detection

Oxacillin susceptibility testing was carried out by E-test strips, by streaking a 0.5 McFarland scale of log-phase growing staphylococci culture onto Mueller-Hinton agar plate. Lecture was achieved after 18 to 24h incubation at 35°C on aerobic conditions, and minimal inhibitory concentration (MIC) interpretation followed Clinical and Laboratory Standards Institute M100-S20 document [9]. The *mecA* gene was detected by PCR performed after DNA extraction (GE Healthcare Illustra GFX DNA kit) with specific primers [15]. A multiplex-PCR approach was employed to SCC*mec* typing, according to previous protocol proposed by Milheiriço and colleagues [16].

3. Results and Discussion

During a 13-year-period from 1996 and 2008, 72 *S. aureus* strains were isolated from peritonitis episodes in patients who were treated at Botucatu Medical School, Sao Paulo, Brazil; 69 of them were available to laboratory testing. Oxacillin resistance (MIC ≥ 4 µg/mL) was detected in 10/69 (14.5%) strains. The *mecA* gene investigation showed 8/69 (11.6%) positive-strains. In this study, 8 isolates were resistant to oxacillin by the phenotypic methods and did not contain *mecA* gene (Table 1). Resistance of *S. aureus* to oxacillin is mainly associated with the presence of the *mecA* gene; however, other types of resistance, called borderline, have also been identified in strains which do not have this gene. Borderline resistance may be due to inactivation of oxacillin mediated by hyperproduction of β-lactamase or a modified resistance, called MOD-SA, which is mediated by the production of modified penicillin-binding proteins (PBPs) with altered affinity to oxacillin [17]. A Korean single-center study identified 151 episodes of *S. aureus* peritonitis from 1992 through 2001, of which 35.1% showed methicillin resistance, using a disk-diffusion method [18]. In a multicenter study evaluating 12 PD units in United Kingdom region during 2002-2003, Davenport [2] reported oxacillin resistance in 14.2% of *S. aureus* causing peritonitis. Chiu et al. [19] reported the MRSA rate of 33% in peritonitis in pediatric patients treated by automated peritoneal dialysis between 2002 and 2006, from a single Chinese center. Analyzing 503 episodes of *S. aureus* peritonitis in Australia and New Zealand, from 2003 to 2006, methicillin resistance was found in 22% of the isolates [20]. Despite the different methods to detect methicillin resistance in those studies, MRSA rates seems to be similar to observed to strains isolated in our Service. It is remarkable to highlight that the results of bacterial resistance and *mecA* detection in this study showed agreement of 79.7% (Table 1).

Table 1 Comparison of methods to detect methicillin resistance in *S. aureus*. Bold numbers indicate concordant results.

Oxacillin	*mecA* gene			
	Positive	Negative		
			Sensibility	25.0%
			Specificity	86.9%
Resistant	**2**	8	Kappa index	0.1072
Sensible	6	**53**	Agreement	79.7%

SCC*mec* type could be determined in only 3 strains, despite 9 *S. aureus* isolates presented *mecA* gene. Of these strains, two were classified as SCC*mec* III and one as *SCCmec* Ia (Fig. 1).

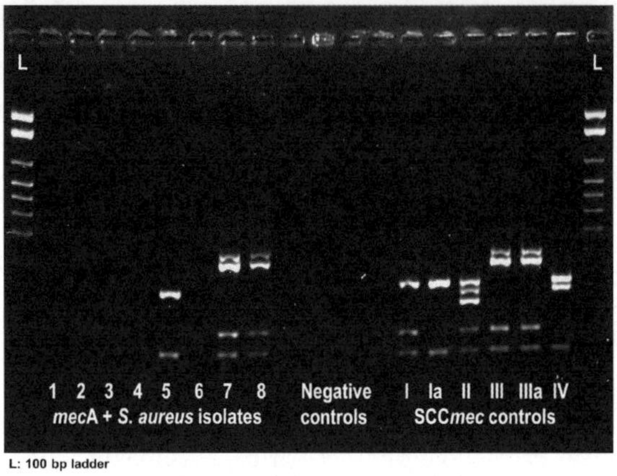

Fig. 1 *Agarose gel electrophoresis for SCC*mec *typing.* Multiplex-PCR was performed as Milheiriço and colleagues [16] protocol, and amplicons were applied on a 1.5% agarose gel with SYBR® Safe DNA gel stain.

Several studies suggest that *S. aureus* peritonitis is associated to carriage of these bacteria [21]. Identical strains causing peritonitis and the ones recovered from carriage sites suggest the role of endogenous source in the development of PD peritonitis [22]. The MRSA rate in this study was lower than those reported from nosocomial infections in Brazilian [17, 23] and in Latin American [24] hospitals, what can be explained by the community source of these strains. Some authors [25, 26], however, suggest that dialysis related infections, including haemodialysis and PD, should not be classified as community acquired infections inasmuch as patients are frequently in contact to hospital environment and its staff. According to this point of view, the results of SCC*mec* typing are logical since the SCC*mec* Ia and III are associated to hospital acquired infections [10]. This conclusion is limited due to the high amount of *mecA*-positive strains that could not have their SCC*mec* typed. Similar difficult to type SCC*mec* was reported previously [26].

To identify a strain with oxacillin resistance mediated by *mecA* gene is useful to define peritonitis management. Empirical therapy of PD peritonitis includes vancomycin or a first generation cephalosporin to cover Gram-positive infections [8]. Once culture results and bacterial susceptibility are known, antibiotic therapy should be adjusted, and vancomycin can be discontinued if a oxacillin-susceptible agent is identified. It is important to incentive the rationale use of vancomycin, considering that its prolonged use is an important risk factor to the development of vancomycin intermediate *S. aureus* (VISA) strains [27].

In summary, this study showed *S. aureus* oxacillin resistance mediated by *mecA* in PD-related peritonitis and the presence of SCC*mec* types, that in general are associated to hospital acquired infections. Further studies, including the use of molecular typing techniques are desirable to elucidate the MRSA PD peritonitis epidemiology.

Acknowledgements The authors are thankful to Fundação de Amparo à Pesquisa do Estado de São Paulo (FAPESP) for their support.

References

[1] Davenport A. Peritonitis remains the major clinical complication of peritoneal dialysis: The London, UK, Peritonitis Audit 2002–2003. *Peritoneal Dialysis International*. 2009;29:297-302.

[2] Heywood J, Bargman M. Coagulase-Negative Staphylococcal Peritonitis: Outcomes of Cephalosporin-Resistant Strains. *Advances in Peritoneal Dialysis*. 2010;26:34-36.

[3] Perez Fontan M, Rodriguez-Carmona A, Garcia-Naveiro R, Rosales M, Villaverde P, Valdes F. Peritonitis-related mortality in patients undergoing chronic peritoneal dialysis. *Peritoneal Dialysis International*. 2005;25:274-84.

[4] Piraino B, Bailie GR, Bernardini J, Boeschoten E, Gupta A, Holmes C, Kuijper EJ, Li PK, Lye WC, Mujais S, Paterson DL, Fontan MP, Ramos A, Schaefer F, Uttley L; ISPD Ad Hoc Advisory Committee. Peritoneal dialysis-related infections recommendations: 2005 update. *Peritoneal Dialysis International*. 2005;25:107-31.

[5] Bunke CM, Brier ME, Golper TA. Outcomes of single organism peritonitis in peritoneal dialysis: gram negatives versus gram positives in the Network 9 Peritonitis Study. *Kidney International*. 1997;52:524-9.

[6] Peacock SJ, Howe PA, Day NP, Crook DW, Winearls CG, Berendt AR. Outcome following staphylococcal peritonitis. *Peritoneal Dialysis International*. 2000;20:215-9

[7] Troidle L, Finkelstein F. Treatment and outcome of CPD-associated peritonitis. 2006. *Annals of Clinical Microbiology and Antimicrobials*. 2006;5:6.

[8] Li PK, Szeto CC, Piraino B, Bernardini J, Figueiredo AE, Gupta A, Johnson DW, Kuijper EJ, Lye WC, Salzer W, Schaefer F, Struijk DG. Peritoneal dialysis-related infections recommendations: 2010 update. *Peritoneal Dialysis International*. 2010;30:393-423.

[9] Clinical and Laboratory Standards Institute. Performance standards for antimicrobial susceptibility testing. CLSI approved standard M100-S20. Clinical and Laboratory Standards Institute, Wayne, PA. 2010.
[10] Martins A, Cunha ML. Methicillin resistance in *Staphylococcus aureus* and coagulase-negative staphylococci: epidemiological and molecular aspects. *Microbiology and Immunology*. 2007;51:787-95.
[11] Chongtrakool P, Ito T, Ma XX, Kondo Y, Trakulsomboon S, Tiensasitorn C, Jamklang M, Chavalit T, Song JH, Hiramatsu K. Staphylococcal cassette chromosome *mec* (SCC*mec*) typing of methicillin-resistant *Staphylococcus aureus* strains isolated in 11 Asian countries: a proposal for a new nomenclature for SCC*mec* elements. *Antimicrobial Agents and Chemotherapy*. 2006;50:1001-1012.
[12] Zhang K, McClure JA, Elsayed S, Louie T, Conly JM. Novel Multiplex PCR Assay for Characterization and Concomitant Subtyping of Staphylococcal Cassette Chromosome *mec* Types I to V in Methicillin-Resistant *Staphylococcus aureus*. *Journal of Clinical Microbiology*. 2005;43:5026-5033.
[13] Okuma K, Iwakawa K, Turnidge JD, Grubb WB, Bell JM, O'Brien FG, Coombs GW, Pearman JW, Tenover FC, Kapi M, Tiensasitorn C, Ito T, Hiramatsu K. Dissemination of New Methicillin-Resistant *Staphylococcus aureus* clones in the community. *Journal of Clinical Microbiology*. 2002;40:4289-4294.
[14] Gram-Positive Cocci: Part I: Staphylococci and Related Gram-Positive Cocci. In Winn WC, Allen SD, Allen S, Janda WM, Koneman EW, Schreckenberger PC, Procop GW, Woods GL. *Koneman's Color Atlas and Textbook of Diagnostic Microbiology*. 6th ed. Philadelphia, PA: Lippincott Williams & Wilkins; 2005:623-671
[15] Murakami K, Minamide K, Wada K, Nakamura E, Teraoka H, Watanabe S. Identification of methicillin-resistant strains of staphylococci by polymerase chain reaction. *Journal of Clinical Microbiology*. 1991;29:2240-2244.
[16] Milheiriço C, Oliveira DC, de Lencastre H. Update to the multiplex PCR strategy for assignment of *mec* element types in *Staphylococcus aureus*. *Antimicrobial Agents and Chemotherapy*. 2007;51:3374-3377.
[17] Martins A, Pereira VC, Cunha MLRS. Oxacillin Resistance of *Staphylococcus aureus* Isolated from the University Hospital of Botucatu Medical School in Brazil. *Chemotherapy*. 2010;56:112-119.
[18] Kim DK, Yoo TH, Ryu DR, Xu ZG, Kim HJ, Choi KH, Lee HY, Han DS, Kang SW. Changes in causative organisms and their antimicrobial susceptibilities in CAPD peritonitis: a single center's experience over one decade. *Peritoneal Dialysis International*. 2004;24:424-432.
[19] Chiu MC, Tong PC, Lai WM, Lau SC. Peritonitis and exit-site infection in pediatric automated peritoneal dialysis. *Peritoneal Dialysis International*. 2008;28 Suppl 3:S179-S182.
[20] Govindarajulu S, Hawley CM, McDonald SP, Brown FG, Rosman JB, Wiggins KJ, Bannister KM, Johnson DW. *Staphylococcus aureus* peritonitis in Australian peritoneal dialysis patients: predictors, treatment, and outcomes in 503 cases. *Peritoneal Dialysis International*. 2010;30:311-319.
[21] Kluytmans J, van Belkum A, Verbrugh H. Nasal carriage of *Staphylococcus aureus*: epidemiology, underlying mechanisms, and associated risks. *Clinical Microbiology Reviews*. 1997;10:505-520.
[22] Pignatari A, Pfaller M, Hollis R, Sesso R, Leme I, Herwald L. *Staphylococcus aureus* colonization and infection in patients on continuous ambulatory peritoneal dialysis. *Journal of Clinical Microbiology* 1990;28:1898-1902.
[23] Gales AC, Sader HS, Ribeiro J, Zoccoli C, Barth A, Pignatari AC. Antimicrobial susceptibility of gram-positive bacteria isolated in Brazilian hospitals participating in the SENTRY Program (2005-2008). *Brazilian Journal of Infectious Diseases*. 2009;13:90-98.
[24] Sader HS, Moet GJ, Jones RN. Antimicrobial resistance among Gram-positive bacteria isolated in Latin American hospitals. *Journal of Chemotherapy*. 2009;21:611-620.
[25] Calfee DP, Durbin LJ, Germanson TP, Toney DM, Smith EB, Farr BM. Spread of methicillin-resistant *Staphylococcus aureus* (MRSA) among household contacts of individuals with nosocomially acquired MRSA. *Infection Control and Hospital Epidemiology*. 2003;24:422-426.
[26] Lu PL, Tsai JC, Chiu YW, Chang FY, Chen YW, Hsiao CF, Siu LK. Methicillin-resistant *Staphylococcus aureus* carriage, infection and transmission in dialysis patients, healthcare workers and their family members. *Nephrology Dialysis Transplantation*. 2008;23(5):1659-1665.
[27] Howden BP, Davies JK, Johnson PD, Stinear TP, Grayson ML. Reduced vancomycin susceptibility in *Staphylococcus aureus*, including vancomycin-intermediate and heterogeneous vancomycin-intermediate strains: resistance mechanisms, laboratory detection, and clinical implications. *Clinical Microbiology Reviews*. 2010;23:99-139.

Resistance Detection and Susceptibility Profile in *Staphylococcus* spp. Isolated from Patients with Urinary Tract Infection (UTI)

Adriano Martison Ferreira[1,2]; Mariana Fávero Bonesso[1,2]; Alessandro Lia Mondelli[2] and Maria de Lourdes Ribeiro de Souza da Cunha[*,1]

[1] Laboratory of Bacteriology, Department of Microbiology and Immunology, Bioscience Institute, UNESP –Univ. Estadual Paulista, Distrito de Rubião Junior, Botucatu, SP, Brazil.
[2] Department of Tropical Diseases, Botucatu School of Medicine, UNESP – Univ. Estadual Paulista, Distrito de Rubião Junior, Botucatu, SP, Brazil.
[3] Department of Internal Medicine, Botucatu School of Medicine, UNESP – Univ. Estadual Paulista, Distrito de Rubião Junior, Botucatu, SP, Brazil.

*Corresponding author: e-mail: cunhamlr@ibb.unesp.br ; Distrito de Rubião Jr. CEP. 18618-970. Botucatu, São Paulo – Brasil.

UTI is one of the most common infections in clinical practice and *S. saprophyticus* is the second most frequent agent affecting the community. We identified *Staphylococcus* spp. and assessed its susceptibility profile in samples from UTI patients. The identification was done by means of the simplified method of biochemical tests and the highest percentage of resistance was to oxacillin (78.0%). Of the 18 *mecA*-positive staphylococci, 15 (83.3%) were resistant to cefoxitin and oxacillin disks simultaneously. Among 57 *S. saprophyticus*, 7 (12.3%) were resistant to the cefoxitin disk and none of them had the gene *mecA*, whereas 56 (98.2%) showed resistance to the oxacillin disk and 2 had the gene *mecA*, both with SCC*mec* type IV. Although cefoxitin disk and detection of the gene *mecA* are more sensitive to detect resistance to oxacillin, the oxacillin disk presented better results for *S. saprophyticus* probably due to a hyperproduction of β-lactamase, a change in a PBP other than PBP2a, or even a cutoff suggested by CLSI that overestimates the resistance to oxacillin for *S. saprophyticus*.

Keywords: Resistance; Urinary Tract Infection; *S. saprophyticus*; *mecA*.

1. Introduction

Urinary tract infection (UTI) is one of the most common diseases in internal medicine [1], constituting the second most common infection in humans [2], only less frequent than those of the respiratory tract.

Staphylococcus saprophyticus is the second most common agent of acute UTI affecting the population, following *E. coli* [3], and is mainly isolated from the urine of sexually active young women [4], leading to symptoms indistinguishable from those caused by *Escherichia coli*. There are also reports of septicemia and pyelonephritis by such a microorganism [5].

Production of PBP2a, a penicillin-binding protein with features of extreme low affinity to β-lactam antibiotics, is certainly the main mechanism responsible for the resistance to oxacillin, although other mechanisms have been described, such as change in other PBPs and hyperproduction of β-lactamases [6]. The gene *mecA* codifies the protein PBP2a and, together with its regulatory genes, is located in a mobile genetic element, the staphylococcal cassette chromosome *mec* (SCC*mec*) [7], which is composed of two genetic elements: *mec*, responsible for the resistance to methicillin, and the complex *ccr*, responsible for the integration and the excision of the chromosomal cassette in the bacterial genome [8].

Strains containing SCC*mec* types I, IV and V are usually more susceptible to antibiotics than the other *SCCmec* types because the former do not have resistance genes, except *mecA* [9]. SCC*mec* types I, II and III especially lead to nosocomial infections and are larger than types IV and V [10].

Resistance to methicillin has also been reported for species that not have the gene *mecA*, named borderline resistance. Two mechanisms have been proposed to explain this type of resistance: the first mechanism involves oxacillin inactivation mediated by hyperproduction of type A β-lactamase, which results in the partial hydrolysis of the β-lactam ring; these strains are known as borderline oxacillin-resistant *S. aureus* (BORSA). The second mechanism is named modified *S. aureus* (MODSA) due to the production of modified PBPs with low affinity to oxacillin [11].

Another important resistance mechanism in staphylococci is induced resistance to the antimicrobials belonging to the group MLS (macrolide, lincosamide and streptogramin). Resistance to macrolides and lincosamides results from three mechanisms: changes in the ribosome-binding site, codified by the gene erythromycin ribosomal methylase (*ermA* or *ermC*) and providing cross-resistance to macrolides, lincosamides and streptogramin B (MLS_B resistance); active efflux, codified by the specific gene methionine sulfoxide reductase (*msrA*), providing resistance to erythromycin, macrolides and streptogramin B; or drug inactivation

[12]. The present study aimed to evaluate the susceptibility profile and to identify *Staphylococcus* spp. in urine samples from patients with UTI sent to the Laboratory of Microbiology of the University Hospital of Botucatu School of Medicine.

2. Materials and Methods

2.1 Samples

This study used 101 *Staphylococcus* spp. samples which were isolated from the urine of different patients and considered compatible with UTI, showing colony count equal to or higher than 100.000 colony-forming units per milliliter of urine ($\geq 10^5$ CFU/mL) based on the evaluation criteria of Kass [13].

2.2 Identification and Antimicrobial Susceptibility Test

Staphylococci were identified by means of the simplified method of biochemical tests described by Cunha et al. [14] and antimicrobial susceptibility was tested for oxacillin 1μg, cefoxitin 30μg, cephalothin 30μg, gentamicin 10μg, linezolid 30μg, norfloxacin 10μg, penicillin G 10 UN, sulfamethoxazole/trimethoprim 25μg, vancomycin 30μg, nitrofurantoin 300μg, amoxicillin/clavulanic acid 30μg, erythromycin 15μg and clindamycin 2μg (Oxoid) by means of the disk-diffusion method according to the criteria recommended by CLSI [15].

2.3 Tests for the Detection of Resistance to Oxacillin and Induced Resistance to MLS Antimicrobials

Disks of oxacillin 1μg and cefoxitin 30μg (Oxoid) were used to predict resistance to oxacillin mediated by the gene *mecA*. The inhibition zone diameter was analyzed according to CLSI, document M100-S18 [15]. Induced resistance to MLS antimicrobials was detected by placing an erythromycin disk near the clindamycin disk (15 to 26mm) on the antibiogram plate. Erythromycin diffusion through agar induces resistance to clindamycin, resulting in a D-shaped flatness of the inhibition zone adjacent to the erythromycin disk [12].

2.4 Detection of gene *mecA* and Determination of Staphylococal Cassete Chromosome *mec* (SCC*mec*) Type for *S. aureus* and Coagulase-Negative Staphylococci

To detect the gene *mecA*, PCR was carried out according to the parameters described by Murakami et al [16] and the SCC*mec* type was determined by using the multiplex PCR method as described by Oliveira and Lencastre [17]. The amplification efficiency was monitored by electrophoresis using agarose gel 1.5% prepared in 1.0 X TBE buffer and stained with SYBR Green. The size of the amplified products was compared with the 100bp standard size and photographed under ultraviolet transillumination.

3. Results

The simplified method of biochemical tests led to the identification of 57 (56.5%) isolates as *S. saprophyticus*, 17 (16.9%) as *S. aureus*, 15 (14.9%) as *S. epidermidis*, 7 (6.9%) as *S. haemolyticus*, 4 (3.9%) as *S. warneri* and 1 (0.9%) as *S. lugdunensis*. The antimicrobial susceptibility percentages for *Staphylococcus* spp. according to the disk-diffusion method were: oxacillin 23.8%, penicillin 27.8%, erythromycin 59.4%, cefoxitin 73.2%, sulfamethoxazole/trimethoprim 78.2%, norfloxacin 84.2%, gentamicin 86.1%, amoxicillin/clavulanic acid 90.0%, clindamycin 93.0% and cephalothin 94.0%. In addition, 3.9% samples had intermediate results for sulfamethoxazole/trimethoprim, 1.9% for norfloxacin, 1.9% for erythromycin and 0.9% for cephalothin. All samples were susceptible to vancomycin, linezolid and nitrofurantoin.

Among the studied samples, 18 were *mecA*-positive staphylococci (Table 1), including: 6 *S. epidermidis*, 3 type III and 3 type IV SCC*mec*; 4 *S. aureus*, 2 type III and 2 type II; 5 *S. haemolyticus*, 2 type II and 3 nontypeable ones; 1 *S. warneri*, type III; and 2 *S. saprophyticus*, both type IV. As regards the tests to detect resistance to oxacillin, 15 (83.3%) out of the 18 *mecA*-positive staphylococci were resistant to cefoxitin and oxacillin disks simultaneously (Table 1). Of the 57 *S. saprophyticus* studied, 7 (12.2%) were resistant to the cefoxitin disk and none had the gene *mecA*, while 56 (98.2%) were resistant to the oxacillin disk and only 2 (3.5%) had the gene *mecA* (Table 2). The test to detect induced resistance to MLS antimicrobials (D-test) was positive for 10 isolates: 6 *S. saprophyticus*, 3 *S. aureus* and 1 *S. haemolyticus*.

Table 1. SCC*mec* classification, susceptibility profile according to the disk-diffusion method and origin of *Staphylococcus* spp. isolates.

SCC*mec* type	Species	Drugs													Origin	No
		FOX	OX	KF	P	SXT	NOR	CN	F	LZD	VA	DA	E	AMC		
II	*S. aureus*	R	R	R	R	S	R	S	S	S	S	R	R	R	WD	2
	S. haemolyticus	R	R	I	R	S	R	R	S	S	S	S	R	R	OC	1
	S. haemolyticus	R	R	S	R	S	S	S	S	S	S	S	S	S	WD	1
III	*S. aureus*	R	R	R	R	R	R	R	S	S	S	R	R	R	WD	2
	S. epidermidis	S	S	S	R	S	S	S	S	S	S	S	S	S	OC	1
	S. epidermidis	R	R	S	R	R	S	R	S	S	S	S	R	R	WD	1
	S. epidermidis	R	R	S	R	R	S	R	S	S	S	S	R	S	WD	1
	S. warneri	R	R	R	R	R	R	R	S	S	S	R	R	R	OC	1
IV	*S. saprophyticus*	S	R	S	S	S	S	S	S	S	S	S	S	S	HC	2
	S. epidermidis	R	R	S	R	R	R	R	S	S	S	S	S	R	ER	1
	S. epidermidis	R	R	S	R	R	R	R	S	S	S	S	S	S	HC	1
	S. epidermidis	R	R	S	R	R	R	S	S	S	S	S	S	R	OC	1
Nontypeable	*S. haemolyticus*	R	R	S	R	S	S	R	S	S	S	S	S	S	WD	1
	S. haemolyticus	R	R	S	R	R	R	R	S	S	S	R	R	S	OC	1
	S. haemolyticus	R	R	S	R	S	S	S	S	S	S	S	R	S	HC	1
Total																18

Origin: WD (ward), OC (outpatient clinics), HC (health center), ER (emergency room).
Drugs: FOX (cefoxitin), OX (oxacillin), KF (cephalothin), P (penicillin), SXT (sulfamethoxazole/trimethoprim), NOR (norfloxacin), CN (gentamicin), F (nitrofurantoin), LZD (linezolid), VA (vancomycin), DA (clindamycin), E (erythromycin) and AMC (amoxicillin/clavulanic acid).
R: resistant; S: susceptible.

Table 2. Detection of resistance to oxacillin using cefoxitin (30μg) and oxacillin (1μg) disks according to the method of agar disk-diffusion.

Species	*mecA*				Cefoxitin				Oxacillin			
	Positive		Negative		R		S		R		S	
	No	%	No	%	No	%	No	%	No	%	No	%
S. saprophyticus	2	3.5	55	96.5	7	12.2	50	87.8	56	98.2	1	1.8
S. aureus	4	25.0	12	75.0	5	31.2	11	68.8	3	18.8	13	81.2
S. epidermidis	6	37.5	10	62.5	6	37.5	10	62.5	6	37.5	10	62.5
S. haemolyticus	4	50.0	4	50.0	5	62.5	3	37.5	6	75.0	2	25.0
S. warneri	2	66.7	1	33.3	2	66.7	1	33.3	3	100	0	0.0
S. lugdunensis	0	0.0	1	100	0	0.0	1	100	0	0.0	1	100
Total	18	17.8	83	82.2	25	24.8	76	75.2	74	73.3	27	26.7

R: resistant; S: susceptible.

4. Discussion

Although UTIs by *S. saprophyticus* have been well described, the antimicrobial resistance and the dissemination of such a species have been scarcely studied. According to CLSI [18], routine tests involving urinary isolates of *S. saprophyticus* are not recommended since infections respond to the antimicrobials frequently used in the treatment. However, strains resistant to sulfamethoxazole/trimethoprim, with 82.4% susceptibility rate, were found in our evaluation.

In addition, resistance to antimicrobials has increased for *S. saprophyticus* and other species, requiring correct identification besides antimicrobial susceptibility test.

The antimicrobial susceptibility profile for patients with community UTI by *Staphylococcus* ssp. was similar to those reported by Gupta et al. [19] and Kahlmeter [20]. In Brazil, Camargo et al. [21] and Braoios et al. [22] evaluated the susceptibility profile of samples from São Paulo State and obtained results similar to those of the present study, except for *S. aureus*, which had lower rates of susceptibility to penicillin (17.6%) and erythromycin (35.2%); this can be related to 4 multiresistant *S. aureus* isolated from patients of HC-FMB.

As regards oxacillin, only 1.8% samples were susceptible, disagreeing from the data reported by Braoios et al. [22], who found susceptibility rates of 91.4% and did not note significant differences for other antimicrobials. Those authors also studied susceptibility to clindamycin and erythromycin, but without mentioning the D-test, which is essential to detect induced resistance to clindamycin. In our study, 10 isolates had positive D-test: 6 *S. saprophyticus*, 3 *S. aureus* and 1 *S. haemolyticus*. Although clindamycin is not frequently adopted to treat UTI, this antimicrobial can be eventually used in some specific cases, requiring thus the test.

Considering susceptibility to nitrofurantoin, several authors [20,21,22] have obtained results similar to those of the present study, with 100% susceptibility for both species, except for Gupta et al. [19] who reported 96.5% susceptibility for *S. saprophyticus* and 98.5% for *S. aureus*. Since nitrofurantoin is an antimicrobial specific for the urinary tract and has few side effects, low toxicity and low cost, in the present study it appears as a therapeutic option even against multiresistant *S. aureus*, which were susceptible only to vancomycin and linezolid among all tested antimicrobials.

Higashide et al. [23,24] investigated the susceptibility of 101 *S. saprophyticus* isolates, all of which were resistant to oxacillin; however, PCR for the detection of gene *mecA* indicated that 93 isolates were *mecA*-negative. As to the results of the disk-diffusion method with oxacillin (1μg), the inhibition zone diameter was ≤ 16mm and the isolates were considered methicillin-resistant coagulase-negative staphylococci (MRCoNS), suggesting that the oxacillin disk and the current cutoff for MRCoNS are not reliable for *S. saprophyticus* or resistance is mediated by a mechanism other than the presence of the gene *mecA*.

In addition, cefoxitin disk (30μg), also recommended to detect resistance to oxacillin, showed four cefoxitin-resistant isolates that were negative for *mecA*. Swenson & Tenover [25] and Frigatto [26] suggested that the cefoxitin disk has inconsistent results in the detection of *mecA*-positive isolates among CoNS including *S. saprophyticus*, which indicates that the correct cutoff using cefoxitin disk must be reviewed for such a species. Higashide et al. [23] stated that the diameter zone should be ≤ 19mm to correctly predict *mecA*-mediated resistance. In our study, however, the two *mecA*-positive *S. saprophyticus* had inhibition halo diameters equal to 27 and 28mm (susceptible) for cefoxitin and 12 and 17mm (resistant) for oxacillin, demonstrating that the results are inconsistent for some species such as *S. saprophyticus* even when the cutoff for cefoxitin is reconsidered, as proposed by Higashide et al. [23].

As regards the test for the detection of resistance to oxacillin, out of the 18 *mecA*-positive staphylococci, 15 (83.3%) were resistant to cefoxitin and oxacillin disks simultaneously, 1 *S. epidermidis* was susceptible to both antimicrobials and 2 *S. saprophyticus* were susceptible to cefoxitin but resistant to oxacillin, demonstrating again the difficulty in detecting the gene *mecA* in this species.

Prevalence of resistance to β-lactam antibiotics was first reported for *S. saprophyticus* isolates in Japão [16], then in Canada [27], United States [25] and Brazil [28]. However, there are few reports on *mecA*-mediated resistance in *S. saprophyticus* [23,29].

In a recent study, the prevalence of *mecA*-positive *S. saprophyticus* was 7.9% (8/101) isolates but the isolation frequency has not been reported so far; thus, the prevalence of resistance to oxacillin in *S. saprophyticus* is suggested to be increasing in clinical isolates and a rapid and correct detection of resistance is needed to treat UTI by this microorganism [23]. The incidence found in our study, however, was 3.5% (2/57), lower than that previously reported. Nevertheless, further studies are needed to assess the increase in resistance and the presence of *mecA*-positive isolates.

In the present study, the two *mecA*-positive *S. saprophyticus* were characterized as type IV and showed the same susceptibility profile, resistant only to oxacillin and susceptible to the remaining tested antimicrobials. Three *S. epidermidis* with SCC*mec* type IV were also isolated; however, these isolates were more resistant to the antimicrobials since all of them were resistant to cefoxitin, oxacillin, penicillin, sulfamethoxazole/trimethoprim and norfloxacin. Other two isolates were also resistant to gentamicin and amoxicillin/clavulanic acid but susceptible to the remaining antimicrobials.

The remaining *mecA*-positive isolates were identified and characterized as follows: 3 *S. epidermidis* with SCC*mec* type III; 4 *S. aureus*, 2 type III and 2 type II; 5 *S. haemolyticus*, 2 type II and 3 nontypeable ones; and 1 *S. warneri*, type III. According to the literature, isolates presenting SCC*mec* type III have a large number of resistance genes [30]. In our study, all isolates with SCC*mec* types II and III were resistant to cefoxitin and oxacillin, showing higher resistance to the antimicrobials, except for 1 *S. epidermidis* type III, resistant only to penicillin. Furthermore, 1 *S. haemolyticus*, type II, was resistant only to oxacillin, cefoxitin and penicillin but curiously susceptible to erythromycin, considering that SCC*mec* type II carries the transposon Tn554, responsible for the resistance to this drug.

The obtained results indicate that, although the cefoxitin disk and the detection of *mecA* gene are more sensitive in detecting resistance to oxacillin by different staphylococci species, oxacillin disk presented better results for *S. saprophyticus* probably due to a hyperproduction of β-lactamase, a change in a PBP other than PBP2a, or even a cutoff suggested by CLSI that overestimates the resistance to oxacillin in *S. saprophyticus*. Thus, some changes may be necessary to detect resistance.

Acknowledgements The support FAPESP is gratefully acknowledged.

References

[1] Warren JW, Abrutyn E, Hebel JR, Johnson JR, Schaeffer AJ, Stamm WE. Guidelines for antimicrobial treatment of uncomplicated acute bacterial cystitis and acute pyelonephritis in women. Infectious Diseases Society of América (IDSA). *Clin Infect Dis.* 1999;29:745-58.

[2] Valiquette L. Urinary tract infections in women. *Canadian Journal of Urology.* 2001;8:6-12.

[3] Gupta K, Scholes D, Stamm WE. Increasing prevalence of antimicrobial resistance among uropathogens causing acute uncomplicated cystitis in women. *JAMA.* 1999;;281:736–38.

[4] Svanborg C & Godaly G. Bacterial virulence in urinary tract infections. *Infect. Dis. Clin. N. Am.* 1997;11:513–29.

[5] Colledge C.L. *Staphylococcus* saprophyticus bacteremia. *J infect Dis.*1989;157:215.

[6] De Giusti M, et al. Phenotypic detection of nosocomial *mecA*-positive coagulase-negative staphylococci from neonates. *Journal of Antimicrobial Chemotherapy.* 1999;44:351-58.

[7] Enrigth MC, Robinson DA, Randle G, Feil EJ, Grundmann H, Spratt BG. The evolutionary history of methicillin-resistant *Staphylococcus aureus* (MRSA). *PNAS.* 2002;99(11): 7687-92.

[8] Zhang K, McClure JA, Elsayed S, Louie T, Conly JM. Novel multiplex PCR assay for characterization and concomitant subtyping of staphylococcal cassette cromossome mec types I to V in methicilin-resistant Staphylococcus aureus. *J. Clinical Microbiol.* 2005; 43:5026-5033.

[9] Hanssen A, Sollid JUE. *SCCmec* in staphylococci: genes on the move. *FEMS Immunol Med Microbiol.* 2006;46:08-20.

[10] Shore A, Rossney AS, Keane CT, Enrigth MC, Coleman DC. Seven novel variants of the staphylococcal chromosomal cassette mec in methicilin-resistant Staphylococcus aureus isolates from Ireland. *Antimicrob Agents Chemother.* 2005; 49:2070-2083.

[11] Caierão J, Musskopf M, Superti S, Dias CG, d'Azevedo PA. Automated systems in the identification and determination of methicillin resistance among coagulase negative staphylococci. *Mem Inst Oswaldo Cruz.* 2006;101(3): 277-79.

[12] Rossi F, Andreazzi DB. Resistência bacteriana: *Interpretando o antibiograma.* São Paulo: Editora Atheneu, 2005.

[13] Kass EH. Assymptomatic infections of the urinary tract. *Trans Assoc Am Phys.* 1956;69: 56-64.

[14] Cunha MLRS, Sinzato YK, Silveira LVA. Comparison of methods for the identification of coagulase-negative staphylococci. *Mem Inst Oswaldo Cruz.* 2004;99(8):855-60.

[15] Clinical and Laboratory Standards Institute. 2008. Performance standardsfor antimicrobial susceptibility testing, 18th information supplement. M100-S18. Clinical and Laboratory Standards Institute, Wayne, PA.

[16] Murakami K, Minamide K, Wada K, Nakamura E, Teraoka H, Watanabe S. Identification of methicillin-resistant strains of staphylococci by polymerase chain reaction. *J Clin Microbiol.* 1991;29:2240-44.

[17] Oliveira D.C., Lencastre H. Multiplex PCR strategy for rapid identification of structural types and variants of the *mec* element in Methicillin-Resistant *Staphylococcus aureus. Antimicrob.Agents Chemother.* 2002.;46 (7): 2155-2161.

[18] Clinical and Laboratory Standards Institute. 2010. Performance standardsfor antimicrobial susceptibility testing, 20th information supplement. M100-S20. *Clinical and Laboratory Standards Institute*, Wayne, PA.

[19] Gupta K, Sahm DF, Mayfield D, Stamm WE. Antimicrobial Resistance among Uropathogens That Cause Community-Acquired UrinaryTract Infections in Women: A Nationwide Analysis. *Clinical Infectious Diseases.* 2001;33: 89-94.

[20] Kahlmeter, G. An international survey of the antimicrobial susceptibility of pathogens from uncomplicated urinary tract infections: the ECO.SENS project. *J. Antimicrob. Chemother.* 2003;51:69–76.

[21] Camargo CBS, Pedro CC, Lourenço DS, Gironi RHA, Martinez R. Infecção de vias urinárias na comunidade de Ribeirão Preto-SP: Etiologia, sensibilidade bacteriana a antimicrobianos e implicações terapêuticas. *Medicina, Ribeirao Preto.* 2002; 35:173-8.

[22] Braoios A, Turatti TF, Meredija LCS, Campos TRS, Denadai FHM. Infecções do trato urinário em pacientes não hospitalizados: etiologia e padrão de resistência aos antimicrobianos. *J Bras Patol Med Lab.* 2009; 45: 6: 449-6.

[23] Higashide, M., M. Kuroda, S. Ohkawa, and T. Ohta. 2006. Evaluation of a cefoxitin disk diffusion test for the detection of *mecA*-positive methicillinresistant *Staphylococcus saprophyticus. Int. J. Antimicrob. Agents.* 2006; 27:500–504.

[24] Higashide M, Kuroda M, Omura CTN, Kumano M, Ohkawa S, Ichimura S, Ohta T. Methicillin-Resistant *Staphylococcus saprophyticus* Isolates Carrying Staphylococcal Cassette Chromosome *mec* Have Emerged in Urogenital Tract Infections. *Antimicrob Agents Chemother.* 2008; 2061–2068.

[25] Swenson JM. & Tenover FC. Results of disk diffusion testing with cefoxitin correlate with presence of *mecA* in *Staphylococcus* spp. *Journal of Clinical Microbiology.* 2005;43(8):3818-23.

[26] Frigatto EAM, et al. Is the cefoxitin disk test reliable enough to detect oxacillin resistance in coagulase-negative staphylococci? *J Clin Microbiol.* 2005;(4):2028-29.

[27] Hussain Z. et al. Correlation of oxacillin MIC with *mecA* gene carriage in coagulase-negative staphylococci. *Journal of Clinical Microbiology.* 2000;38(2):752-54.

[28] Souza Antunes AL, Secchi C, Reiter KC, Rodrigues Perez LR, Peixoto de Freitas AL, Alves d'Azevedo P. Evaluation of oxacillin and cefoxitin disks for detection of resistance in coagulase negative staphylococci. *Mem Inst Oswaldo Cruz* 2007; 102: 719–723.

[29] Soderquist B & C. Berglund C. Methicillin-resistant Staphylococcus saprophyticus in Sweden carries various types of staphylococcal cassette chromosome mec (SCCmec). *European Society of Clinical Microbiology and Infectious Diseases.* 2009; 15: 1170–1191

[30] Ito T, Katayama Y, Hiramatsu K. Cloning and nucleotide sequence determination of the entire mec DNA of pre-methicillin-resistant *Staphylococcus aureus* N315. *Antimicrob. Agents Chemother.* 1999; 43:1449-1458.

Resistance Distribution profile of MBL, ESBL and Multidrug resistant Gram negatives isolated at a tertiary care hospital in India

K. H. Bhutada[*,1] **and V. R. Shende**[2]

[1] Kamalnayan Bajaj Hospital, Aurangabad – 431 005, India
[2] Department of Biochemistry, Dr. B.A.M. University, Aurangabad – 431 004, India

* Corresponding author: e-mail: kanchan_bhutada@yahoo.com, Phone: +91 9422141480

Prevalence of extended spectrum β-lactamase (ESBLs) and metallo- β -lactamase (MBLs) constitute a serious threat to current β–lactam therapy leading to treatment failure and consequent escalation of cost. In a tertiary care hospital the patients are admitted for causes other than the underlying infection. Detection of infection and subsequent identification of resistance in the infectious agent puts additional burden of investigating effective drug for cure. To ease the process of drug selection to treat MBL and ESBL isolates a detailed study was initiated to identify the resistance distribution pattern in isolates. In a year long study a total of 390 isolates were found to contain 35.12% ESBLs, 9.48% MBLs, 32.56% multidrug resistants and the rest sensitive. Only a lone isolate was found to be both ESBL as well as MBL producer. In this study we also noted that all ESBLs were sensitive to cefoperazome+sulbactum, piperacilin+tazobactam and imipenem, MBLs were sensitive to polymixin-B and imipenem+cilastatin combination and multidrug resistants were found to respond well to carbapenems and polymixin-B.

Keywords ESBL; MBL; tertiary care hospital, multidrug resistance

1. Introduction

Microbial resistance is a growing major public health issue and a principle concern for the medical community. Production of β lactamase is a predominant means by which gram negative bacteria exhibit resistance to β lactam antibiotics. ESBLs (Extended spectrum β lactamases) are a group of enzymes that can hydrolyze a variety of β lactams including III generation cephalosporins (ceftazidime, cefotaxime and ceftriaxone) and IV generation cephalosporins (cefepime and cefpirome), monobactam (aztreonam) in addition to penicillins.[1] As a result of this bacteria that produce ESBLs become resistant to β lactams. ESBLs do not hydrolyze cephamycins and carbapenems and these may lead to increased prescription of more broad-spectrum and expensive drugs such as imipenem. Also strains producing metallo-β-lactamases (MBL) exhibit resistance to carbapenem (e.g., imipenem).[2] Acinetobactor species, pseudomonas and new *Klebsiella pneumoniae* are increasingly reported as resistant to carbapenems mainly due to either carbapenemases or impermability due to alterations in the outer membrane proteins.[3] If this happens a very narrow spectrum of antibiotics such as polymyxins (i.e. colistin & polymyxin B), sulbactum and possibly tigecycline would remain available for treatment.[4]

Similarly, multidrug resistance in *Enterobacteriaceae* is becoming an increasingly threatening public health problem due to resistance to first generation quinolones and aminoglycosides.[5] Three different mechanisms are active in providing resistance against quinolones and aminoglycosides.[6] In view of the above and also due to the isolations of many resistant strains at our hospital which is a tertiary care hospital we carried out a systematic study to determine the prevalence and nature of resistance in clinically important organisms. Our study also throws light on the susceptibility and thus other therapeutic options available to the physicians grappled with antibiotic resistance in this part.

2. Materials and Methods

2.1 ***Culture and Isolation***: The KBH is a 150 bedded hospital equipped with modern urban tertiary care facility that provides services to all of the srugical and medical subpecialities, including orthopedics, obstertrics and gynecology, oncology, cardiology, nephrology, radiology, ophthalmology with two intensive care units. A total of 390 gram negative organism were isolated from various samples mainly ETT, pus, urine and blood. All samples were processed for isolation and identification of bacterial pathogens according to standard microbiological techniques. No duplicate isolates were included in this study.

2.2 ***Antibiogram***: Antimicrobial sensitivity testing was performed on Mueller Hinton Agar (Hi-Media, India) plates by disk diffusion method according to CLSI guidelines. Zone of inhibition was recorded and interpreted as sensitive, intermediate or resistant based on CLSI standards. *E.coli* ATCC 25922 (ESBL negative) , *Pseudomonas aeruginosa* ATCC 27853 (ESBL negative) and *Klebsiella Pneumoniae* ATCC 700603 (ESBL positive) strains were used as control organisms for test performance.

2.3 ***ESBL and MBL detection***: All gram negative bacteria isolated from samples were tested for ESBL production by using four disks, ceftazidime (30mcg), ceftazidime/clavulanic acid (30/10mcg) cefotaxime (30mcg) and cefotaxime/clavulanic acid(30/10mcg) and interpreted according to CLSI guidelines (Fig.1). Imipenem resistant strains were tested for MBL using imipenem(10mcg), imipenem /EDTA(10/750mcg) discs. All ESBLs and MBLs were confirmed by MIC using e-test. All resistatnt strains (ESBL, MBL, Multi-drug) were screened for antibiotic susceptibility profile using additional higher drugs.

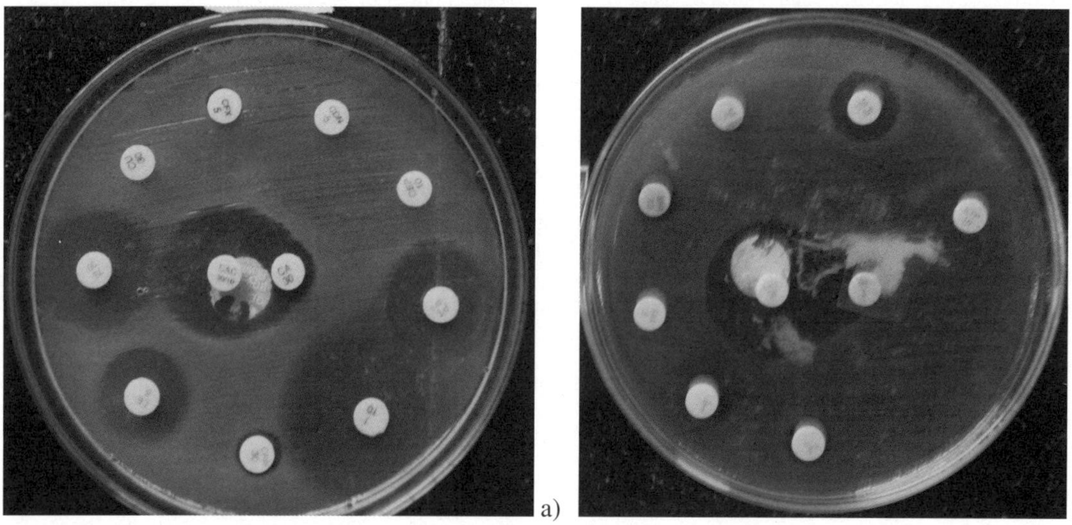

a) b)

Fig. 1 The increased zone read with a) ceftazidime+clavulanic acid and b) Imipenem+EDTA disks.

3. Result and Discussion

A total of 390 gram negative organisms were isolated from various patients admitted primarily for UTI, RTI, sepsis, bacterimia and diarrhoea. These were identified as *Escherichia coli* (174), *Pseudomonas aeruginosa* (71), *Klebsiella pneumoniae* (73), *Klebsiella* spp. (47) and *Acinetobacter baumanii* (26).

Eighty nine (22.83%) isolates were found sensitive to ceftazidime, cephotaxim, cefpirome, cefixime, cefidnir, levofloxacin, amikacin, cefoperazone+sulbactum, and piperacilin+tazobactam and imipenem. In these sensitive strains *E. coli* were 52, *K. pneumoniae* 12, *Klebsiella species* 13, *P. aeruginosa* 11 and a lone isolate of *A. baumanii*.

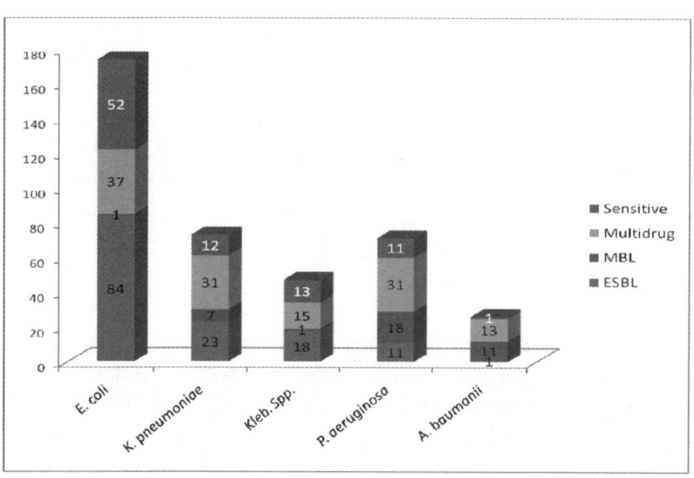

Fig. 2 Species wise distribution of ESBL, MBL and multi-drug resistance.

The 301 (77.17%) resistant isolates were grouped further into three categories. ESBL and MBL were identified based on their resistance to ceftazidime and imipenem and their sensitivity to ceftazidime+clavulanic acid and imipenem+EDTA combination, respectively. Multi-drug resistant strains were all those that were resistant to all except imipenem. There were 137 (35.12%) ESBL producers, 37 (9.48%) MBL producers and 127 (32.56%) multidrug resistant.

Of the 137 ESBL producers, 84 (61.31%) were *E. coli*, 23 (16.78%) *K. pneumoniae*, 18 (13.13%) *Klebsiella* spp, 11 (8.02%) *P. aeruginosa* and 1 (0.72%) *A. baumanii*. Also, in 37 MBL producing isolates 18 (48.65%) were *P. aeruginosa*, 11 (29.72%) *A. baumanii*, 7 (18.92%) *K. pneumoniae* and 1 (2.7%) each of *E. coli* and *Klebsiella* spp. Similarly, in the 127 multi-drug resistant isolates 37 (29.13%) were *E. coli*, 31 (24.40%) *K. pneumoniae*, 31 (24.40%) *P. aeruginosa*, 15 (11.81%) *Klebsiella* spp., and 13 (10.23%) were *A. baumanii*. Thus specieswise ESBLs production was found to be predominant in either *E. coli* or *K. pneumoniae* and that of MBL production was in *P. aeruginosa* and *A. baumanii*. A *K. pneumoniae* isolate was found both ESBL as well as MBL producer.

However, there was no predominance of multi-drug resistance among the groups of isolates identified. It was found to be 29.13% in *E. coli*, 24.4% each in *K. pneumoniae* and *P. Aeruginosa*, 11.8% in *K.* spp. and 10.23% in *A. baumanii*.

In a Korean study the *E. coli* isolates obtained from sputum showed higher ESBL production rates (37.5%) compared to that seen in urine isolates (8.3%).[7] Low ESBL production (9.3%) was also seen in *E. coli* isolates from a tertiary care hospital in Trinidad and Tobago.[8] Higher ESBL production rates (67.2%) were observed in *E. coli* isoated from tertiary care hospital in Iran,[9] and 24.4% in Tanzania.[10] *E. coli* isolates from different parts of India also showed very high prevalence of ESBL production in independent studies.[11, 12, 13]

ESBL production rates in K. pneumoniae varied in different regions but were high in Kuwait (37%), Tanzania (63.7%). [14, 15] However, the rates were comparable to those observed in the present study and with the isolates from Trinidad & Tobago (15.3%) and Korea (22.4%). [8,7] The rate of ESBL prevalence in Klebsiella spp. was found to be 13.13% in the current study. A very high rate (58%) of ESBL prevalence in K. spp. has been reported. [15]

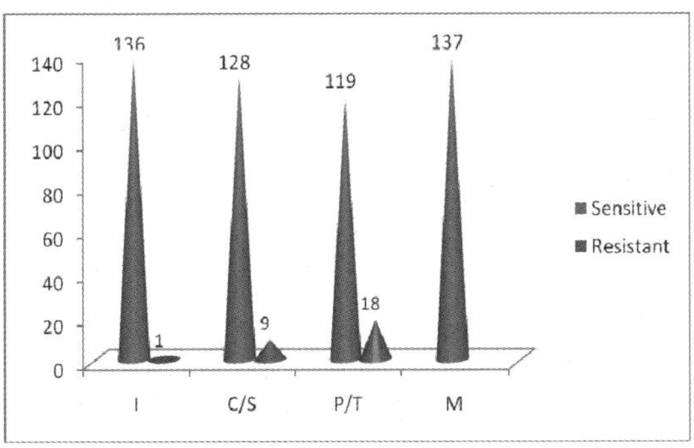

a)

Fig. 3 Susceptibility and resistance profile of a) ESBL, b) MBL and, c) multi-drug resistant isolates to higher and reserved antibiotics. I=imipenem, C/S=cephaperazome+sulbactum, P/T=piperacillin+ tazobactam, M=meropenem, P-B=polymixinB, I/C= imipenem+cilastatin, Ert=ertapenem, Far=faropenem, Gati=gatifloxacin. Value over first cone for sensitive isolate, on second cone depicts resistant isolates.

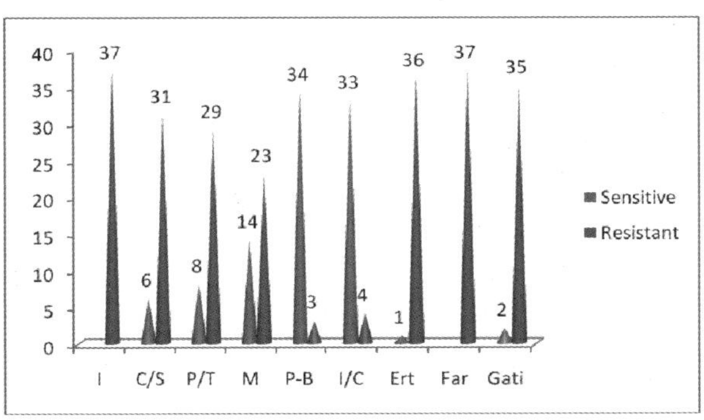

b)

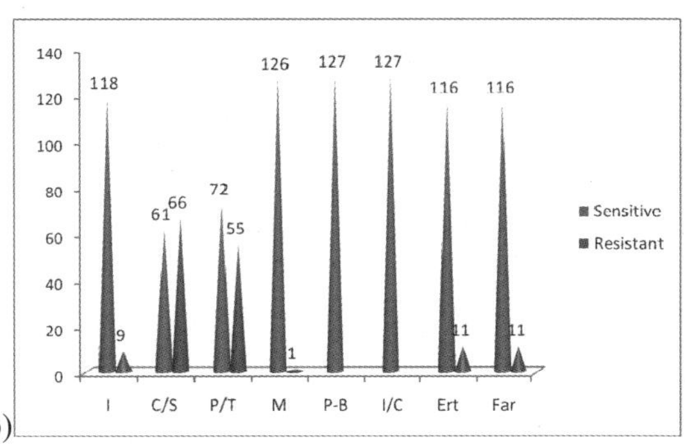

c)

Sensitivity of ESBL, MBL and multidrug resistants:

To test if the resistant isolates were sensitive to any single drug or combination further sensitivity test were carried out using additional drugs. In these studies all 137 ESBL producers, except one isolate, were sensitive to Imipenem, 128 (93.43%) to cefoperazone+ sulbactum and 119 (86.86%) were sensitive to piperacillin+ tazobactum. On the other hand, of the 37 MBL producers only 6 were found sensitive to cephoperazone+

sulbactum and 8 to piperacillin+tazobactum combination. However, most MBL producers were sensitive to meropenem (14), polymyxin-B (34), and imipenem+cilastatin (33) combination (Fig. 3).

Similarly, 127 multi-drug resistants were sensitive to meropenem (126), polymixin-B (127) and imipenem+cilastatin (127) combination as well as ertapenem (116) and faropenem (116). Thus, either imipenem alone or in combination with cilastatin appears to be the best combination faced with ESBL, MBL and multi-drug resistance challenge by the organism (Fig. 3).

The resistant group constituted nearly 77.17% of the total isolates. The high figures could be attributed partly to the patients' routing through the primary care and antibiotics prescriptions without cultuting or testing the sensitivity of the underlying infectious agent by consultant physians. Whatever be the cause very high resistance figures indicate and emphasize the need for a rational use policy for antimicrobials and strict adherence to the concept of reserve drugs in India.

References

[1] Gupta E, Mohanty S. et al. Emerging resistance to carbapenems in a tertiary care hospital in north India. *Ind. J. Med. Res.* 2006; 124: 95-98.

[2] Christian G Girke, et al., Clinical and economic impact of common multidrug resistant Gram-negative bacilli; *Antimicrol Agents & Chemotherapy.* 2008; 52: 813-821.

[3] Brown, S. and Amyes, S. Oxa (□)lactamasees in Acinetobacter: the story so far. *J. Antimicrob. Chemother.* 2006; 57: 1-3.

[4] Falagas M.E., Bliziotis, I.A. et al. Outcome of infectious due to pandrug resistant (PDR) Gram-negative bacteria. *BMC Infect. Dis.* 2005; 5: 24.

[5] Bonomo R.A. and D.Szabo (2006) Mechanism of multidrug resistance in Acinetobacter species and Pseudomonas aeruginosa. *Clin. Infect. Dis.* 2006; 43(Suppl. 2): 549-556.

[6] Park, C.H. et al. Prevalence in the United States of aac(6')-ib-cr encoding a ciprofloxacin modifying enzyme. *Antimicrob. Agents Chemother.* 2006; 50: 3953-3955.

[7] Ko KS, Lee MY, Song JH, Lee H, Jung DS, Jung SI, Kim SW, Chang HH, Yeom JS, Kim YS, Ki HK, Chung DR, Kwon KT, Peck KR, Lee NY. Prevalence and characterization of extended-spectrum beta-lactamase-producing Enterobacteriaceae isolated in Korean hospitals. *Diagn Microbiol Infect Dis.* 2008; 61:453-459.

[8] Akpaka PE, Swanston WH. Phenotypic detection and occurrence of extended-spectrum beta-lactamases in clinical isolates of Klebsiella pneumoniae and Escherichia coli at a tertiary hospital in Trinidad & Tobago. *Braz J Infect Dis.* 2008;12:516-520.

[9] Mehrgan H, Rahbar M. Prevalence of extended-spectrum beta-lactamase-producing Escherichia coli in a tertiary care hospital in Tehran, Iran. *Int J Antimicrob Agents.* 2008;31:147-151.

[10] Mshana SE, Kamugisha E, Mirambo M, Chakraborty T, Lyamuya EF. Prevalence of multiresistant gram-negative organisms in a tertiary hospital in Mwanza, Tanzania. *BMC Res Notes.* 2009; 2:49.

[11] Shobha KL, Ramchandra L, Rao G, Majumder S, Rao SP. Extended spectrum beta-lactamases (ESBL) in gram-negative bacilli at a tertiary care hospital. *J. Clin. Diag. Research* 2009 ; 3: 1307-1312

[12] Shiju MP, Yashavanth R, Narendra N. Detection of extended spectrum beta-lactamase production and multidrug resistance in clinical isolates of E. coli and K. pneumoniae in Mangalore. *J. Clin. Diagn. Research.* 2010 ; 4 : 2442-2445.

[13] Chaudhary U and Aggarwal R. Extended spectrum beta-lactamases (ESBL) – An emerging threat to clinical therapeutics. *Ind.J Med Microbiol*, 2004 ; 22: 75-80.

[14] Mokaddas EM, Abdulla AA, Shati S, Rotimi VO. The technical aspects and clinical significance of detecting extended-spectrum beta-lactamase-producing Enterobacteriaceae at a tertiary-care hospital in Kuwait. *J Chemother.* 2008;20:445-51.

[15] Jain A, Mondal R. Detection of extended spectrum beta-lactamase production in clinical isolates of Klebsiella spp. *Indian J Med Res.* 2008;127:344-346.

Study on selection and transfer of antimicrobial resistant *Escherichia coli* from broiler breeders to their progeny

G. Manfreda[1], **F. Pasquali** [*1], **A. Lucchi**[1], **A. De Cesare**[1], **D. Giovanardi**[2], **M. Stonfer**[3] and **A. Franchini**[1]

[1] Department of Food Science, University of Bologna, via del Florio 2, 40064 Ozzano dell'Emilia, Italy
[2] Laboratorio Tre Valli, Corte la Pellegrina 3, 37132, San Michele Extra, Verona, Italy
[3] Bayer Health Care, Animal Health Division, Viale Certosa 130, 20156 Milan, Italy

* Corresponding author: e-mail: frederique.pasquali3@unibo.it, Phone: +39 512097862

Escherichia coli is a normal inhabitant of the intestinal tract of chickens, but when an imbalance in bacterial flora of the intestinal tract occurs, *E. coli* may over grow and cause extraintestinal infections. The aim of this study was to evaluate the selection and transfer of *E. coli* clones and their antimicrobial resistant characters from broiler breeders to their progeny in absence or presence of an ongoing *E. coli* infection. Environmental samples and different organs of humanly euthanized broiler breeder were collected from group R1 and its sick progeny P1 (characterised by omphalitis symptoms) and from group R2 and its healthy progeny (P2). All samples were tested for the presence of *E. coli* and the isolates obtained were typed using *Xba*I Pulsed Field Gel Electrophoresis and *Eco*RI automated ribotyping. The 172 *E.coli* isolates were molecular discriminated into 90 *Xba*I PFGE types and 63 ribotypes. Five *E. coli* isolates from breeder R1and four from breeder R2 shared the same PFGE and/or ribotype with *E. coli* isolates from progeny P1 and P2. Interestingly *E. coli* strains isolated both in breeders and their progeny showed different antibiotic multiresistance profiles reinforcing the idea of a possible horizontal transfer of antibiotic resistance determinant genes to and from these isolates.

Keywords antimicrobial resistant *E. coli*; transfer; broiler breeders; progeny

1. Introduction

Escherichia coli is a normal inhabitant of the intestinal tract of chickens and is harmless as long as its growth and colonization is inhibited by other commensal intestinal microbial populations. When an imbalance in bacterial flora of the intestinal tract occurs, *E. coli* may over grow and cause extraintestinal infections. *E. coli* diseases, manifested as septicaemia in both parent flocks and their progeny, are often due to clonal strains [1]. The localisation of most antimicrobial resistance genes on mobile genetic elements rises concern on the possible spread of resistance determinants from *E. coli* to other intestinal bacteria circulating between birds thus enabling the persistence of antimicrobial resistance determinants also after the infection. Ampicillin, aminoglycoside, tetracycline, thrimethoprim and sulfonamide resistance genes are often transposon-plasmid located and can therefore be horizontally transferred [2, 3]. Moreover the exposure of hatching to resistant clones present on the surface of the egg can potentially result in dissemination of these strains from the breeder flock to progeny [4].

The aim of this study was to evaluate the selection and transfer of *E. coli* and its antimicrobial resistant characters from broiler breeders to their progeny in absence or presence of an ongoing *E. coli* infection.

2. Materials and methods

2.1 Sampling scheme

E. coli isolates were collected from a broiler breeder flock R1 and its sick progeny P1 (characterised by omphalitis) and from breeders belonging to flock R2 and its healthy progeny P2 (Table 1). Amoxicillin was administered in drinking water to P1 and P2 birds during the first week of live. Among each progeny group of healthy and sick chicks, 5-10 animals were picked out during the first week of life (1st sampling). Consistently, 5-10 animals of each progeny group were selected after the antibiotic therapy or at 21 days and at the end of the productive cycle (2nd and 3rd sampling). In parallel, three pools of environmental dust were sampled. Two weeks after the collection of healthy and sick chicks, 5-10 animals of each of the corresponding broiler breeder group (i.e., R1 or R2) were also selected.

All selected animals were humanely euthanized and from each of them different organs (liver, spleen, intestine, heart, lungs and tracheal mucus) were collected and cultured for the presence of *E. coli*.

2.2 *Escherichia coli* isolation and molecular characterisation

Five grams of dust were diluted 1:10 in Ringer's Solution (Oxoid, Hampshire, UK). An aliquot of 0,1 ml was plated on Eosin Methylene Blue agar (EMB) and incubated at 37°C for 24 h. One gram of viscera was plated on the same media and incubated at the same conditions as described before. From each plate, two colonies showing characteristic *E. coli* morphology were selected for confirmation through biochemical tests and one out of two positive was submitted for further analysis.

Overall 172 of *E. coli* isolates were molecular characterised by Pulsed Field Gel Electrophoresis (PFGE), using the standard PulseNet protocol with *Xba*I as restriction enzyme [5], and automated ribotyping according to the manufacturer instructions with *Eco*RI as restriction enzyme (Oxoid, Hampshire, UK). PFGE and ribotyping profiles were imported into Bionumerics 6.0 software (Applied Maths, Saint-Martens-Latem, Belgium) and the normalized profiles were compared using the discriminatory index (D) [6] and clustered by the Unweighted Pair Group Method with Arithmetic mean (UPGMA) with optimization 1% and tolerance 1%.

Table 1 Number of *E.coli* positive organs and dust.

	Sampling					
		Organs			Dust	
R1		13 (5)[§]			3	
R2		6 (6)			2	
	1st sampling		2nd sampling		3rd sampling	
	Organs	dust	Organs	dust	Organs	dust
P1	13(13)	-	39(20)	9(9)	13(8)	7
P2	-	-	33(18)	8(8)	17(12)	7

[§] numbers in brackets correspond to number of birds/dust sampled

2.3 Antibiotic susceptibility testing

The antibiotic sensibility of *E. coli* isolates toward amoxicillin (AMX) (Sigma, Milan, Italy), tetracycline (TET) (Sigma), spectinomycin (SP) (Sigma), enrofloxacin (ENR) (Bayer HealthCare, Milan, Italy) and ciprofloxacin (CIP) (Bayer HealthCare) was investigated by the determination of the Minimum Inhibitory Concentration (MIC) following standard procedures of the microdilution method reported in the document M31-A2 of the Clinical Laboratory Standard Institute (CLSI) [7].

3. Results

3.1 PFGE Analysis

The isolates possessed a number of recognition sites yielding between 12 and 24 DNA bands with a molecular mass between 28 and 790 kb. Considering a cut off of 90%, the 172 isolates were grouped into 91 clusters including 1 to 6 isolates. Seventeen isolates were untypable by PFGE and the D value achieved using this typing method was 0.992. Three out of 11 PFGE-types of breeder R1 were found in their respective progeny demonstrating a possible transfer of E. *coli* related isolates (i.e., isolates 169, 173, 183 from R1)(Fig. 1a).

3.2 Ribotyping analysis

A selection of 128 *E. coli* isolates was submitted to automated ribotyping. Considering a cut off of 90 %, the selected isolates were grouped into 63 clusters including 1 to 18 isolates. Two isolates were untypable by ribotyping showing a D value of 0.975. Two out of 12 ribotypes of breeder R1 as well as 2 out of 7 of breeder R2 were found in their respective progeny, P1 and P2, demonstrating a possible transfer of E. *coli* related isolates (i.e., isolates 173, 179 and 183 from R1 and 1,7, 2 and 4 from R2). (Fig. 1b).

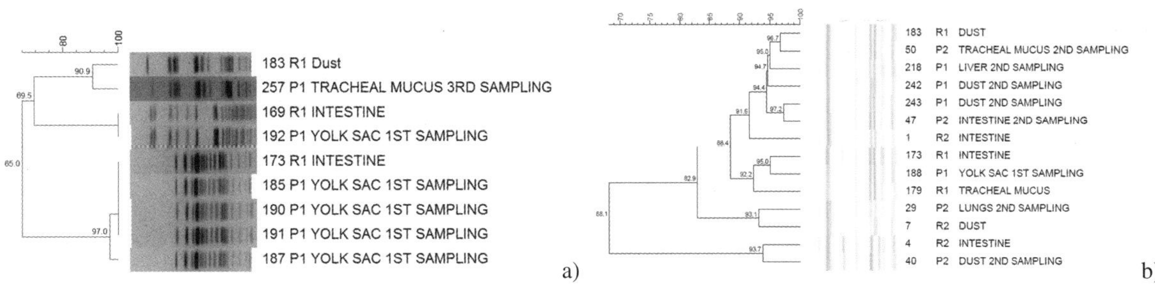

Fig. 1 PFGE-types (A) and RIBO-types (B) of *E. coli* isolates found in both breeders R1 and R2 and their progeny P1 and P2.

3.3 Antibiotic susceptibility

A total 17 out of 172 *E. coli* isolates were susceptible to all antibiotic tested, 12 and 143 were resistant to one and two or more antibiotics, respectively and 33 were fully resistant to all antibiotics. Table 2 summarises the antibiotic profiles associated to the *E. coli* isolates.

Table 2 Antibiotic resistance profiles of *E.coli* isolates.

Antibio-type	N° of *E. coli* isolates			
	R1	P1	R2	P2
Susceptible	3	5	3	6
TET	1	2	1	1
SP		3		
AMX		2		2
ENR, AMX	1			
ENR, TET	1			
TET, SP		4		5
TET, AMX	7	5	4	14
SP, AMX		9		5
ENR, TET, SP				1
ENR, TET, AMX	2	1		3
ENR, SP, AMX		1		1
TET, SP, AMX		5		19
ENR, TET, SP, AMX	1	14		5
ENR, CIP, SP, AMX		2		
ENR, CIP, TET, SP, AMX		28		5

Overall, 75-80 % of *E. coli* isolates were resistant to tetracycline and amoxicillin irrespective to the bird group of origin. Among sick birds, 81 % of both *E. coli* isolates from breeders and progeny were resistant or intermediate to enrofloxacin in comparison to only the 28% of isolates from healthy birds. Interestingly, in both sick and healthy progeny (P1 and P2) a higher percentage of *E. coli* isolates were resistant to ciprofloxacin and spectinomycin in comparison to the respective breeder groups R1 and R2 (Table 3).

Isolates belonging to breeders R1 and sharing the same PFGE-type with isolates of progeny P1 showed different antibiotic resistance profiles with a higher degree of multiresistance in the progeny than in the respective breeder group (Table 4).

Table 3 Antibiotic susceptibility of *E. coli* isolates from breeder R1 and its sick progeny P1 and breeder R2 and its healthy progeny P2.

		N(%)		
Antibiotic	bird group	Susceptible	Intermediate	Resistant
ENR	R1	3 (18,8)	8 (50)	5 (31,3)
	P1	15 (18,5)	20 (24,7)	46 (56,8)
	R2	6 (75,0)	2 (25)	0 (0)
	P2	48 (71,6)	4 (6)	15 (22,4)
CIP	R1	16 (100)	0 (0)	0 (0)
	P1	43 (53,1)	8 (9,9)	30 (37)
	R2	8 (100)	0 (0)	0 (0)
	P2	61 (91)	1 (1,5)	5 (7,5)
TET	R1	2 (12,5)	2 (12,5)	12 (75)
	P1	18 (22,2)	4 (4,9)	59 (72,8)
	R2	3 (37,5)	0 (0)	5 (62,5)
	P2	8 (11,9)	6 (9)	53 (79,1)
SP	R1	9 (56,3)	6 (37,5)	1 (6,3)
	P1	7 (8,6)	7 (8,6)	67 (82,7)
	R2	7 (87,5)	1 (12,5)	0 (0)
	P2	22 (32,8)	4 (6)	41 (61,2)
AMX	R1	3 (18,8)	2 (12,5)	11 (68,8)
	P1	9 (11,1)	5 (6,2)	67 (82,7)
	R2	4 (50,0)	0 (0)	4 (50)
	P2	13 (19,4)	0 (0)	54 (80,6)

Table 4 Antibiotic resistance profiles of isolates of breeders and progeny sharing the same PFGE-type.

PFGE-type	Isolate number	Antibio-type
88	183 R1	ENR, TET, SP, AMX
	257 P1	ENR, TET, SP, AMX
35	173 R1	susceptible
	185 P1	SP
	187 P1	TET, AMX
	190 P1	TET, AMX
	191 P1	TET, AMX
83	169 R1	TET, AMX
	192 P1	ENR, TET, SP, AMX

4. Discussion

In this study the transfer of *E. coli* clones and their antimicrobial resistant characters from broiler breeders R1 to their sick progeny P1 as well as from broiler breeder R2 to their healthy progeny P2 was investigated. Molecular typing by *Xba*I PFGE and *Eco*RI automated ribotyping discriminated the 172 *E. coli* isolates tested into 90 and 63 different clusters suggesting a high genetic diversity both within and between all the animal groups. This diversity might be due to different *E. coli* clones belonging to the intestinal microflora of the same bird. Five *E. coli* isolates from breeder R1 and 4 from breeder R2 showed the same PFGE and/or ribotype of *E. coli* isolates from progeny P1 and P2, respectively, demonstrating the transfer of these isolates from breeders to their progeny. Moreover, the isolate 183 was collected from dust thus suggesting the importance of the environment as a vehicle of transmission. Interestingly, *E. coli* strains isolated both in breeders and their progeny showed different antibiotic multiresistance profiles reinforcing the idea of a possible horizontal transfer of antibiotic resistance determinant genes to and from these isolates. Regarding antibiotic resistance, 90% of *E. coli* isolates were resistant to one or more antibiotic with high percentages of resistant isolates to tetracycline and amoxicillin regardless to the group of birds and the eventual antibiotic treatment applied, confirming how the presence of antibiotic resistance determinant genes is not strictly correlated to the antibiotic treatment.

References

[1] Christensen JP, Chadfield MS, Bojesen AM, Bisgaard M. (2003). Investigations on the clonality of *E. coli* infections in industrial poultry production. *Proceedings from the XIII Congress of the World Veterinary Poultry Association* Denver, USA, 2003:83.

[2] Mulvey MR, Boyd DA, Olson AB, Doublet B, Cloeckaert A. The genetics of Salmonella genomic island 1. *Microbes and Infections* 2006;8(7):1915-1922.

[3] Pasquali F, Kehrenberg C, Manfreda G, Schwarz S. Physical linkage of Tn*3* and part of Tn*1721* in a tetracycline and ampicillin resistance plasmid from *Salmonella* Typhimurium. *Journal of Antimicrobial Chemotherapy* 2005;55(4):562-565.

[4] Idris U, Lu J, Maier M, Sanchez S, Hofacre CL, Harmon BG, Maurer JJ, Lee MD. Dissemination of fluoroquinolone-resistant *Campylobacter spp.* within an integrated commercial poultry production system. *Applied and Environmental. Microbiology* 2006;72(5):3441-3447.

[5] Ribot EM, Fair MA, Gautom Ribot R, Cameron DN, Hunter SB, Swaminathan, B, Barrett TJ. Standardization of pulsed-field gel electrophoresis protocols for the subtyping of Escherichia coli O157:H7, Salmonella, and Shigella for PulseNet. *Foodborne Pathogens and Disease.*2006;3:59-67.

[6] Hunter PR, Gaston MA. Numerical index of the discriminatory ability of typing systems: an application of Simpson's index of diversity. *J. Clin. Microbiol.* 1988;26:2465-2466.

[7] National Committee for Clinical Laboratory Standards. Performance standards for antimicrobial disk and dilution susceptibility tests for bacteria isolated from animals-Second edition: Approved standard M31-A2. NCCLS, Wayne, PA, USA, 2002.

Survey of *Fusarium* species infecting potato and their resistance to fungicides used in Canada

K. Lugosch[1], R.D. Peters[2]*, T. Barasubiye[3] and K. Drake[2]

[1]Acadia University, Wolfville, NS Canada
[2]Agriculture and Agri-Food Canada, Crops and Livestock Research Station, Charlottetown, PE Canada
[3]Agriculture and Agri-Food Canada, Eastern Cereal and Oilseed Research Centre, Ottawa, ON Canada

* corresponding author: rick.peters@agr.gc.ca

Potato seed tubers with symptoms of decay were sampled from across Canada in spring 2010 to obtain a collection of *Fusarium* isolates. Isolates were identified to species level by microscopic examination of morphological features and molecular methods and then tested for sensitivity to thiabendazole and fludioxonil using a fungicide-amended agar assay. The major seed decay pathogens, *F. sambucinum, F. avenaceum* and *F. coeruleum* were frequently recovered. Isolates of *F. sambucinum* and *F. coeruleum* varied in their sensitivity to fungicides, with resistance to one or both products found in strains obtained from most Canadian provinces. Isolates of *F. avenaceum* with varying fungicide resistance were also identified, although most isolates of this species were sensitive to both products. Isolates of *F. oxysporum* recovered in the survey were always resistant to fludioxonil, but sensitive to thiabendazole. In summary, the survey confirms the presence of strains of *F. sambucinum* and *F. coeruleum* in Canada resistant to thiabendazole and/or fludioxonil. As well, fungicide resistance in other *Fusarium* spp. was also identified. Fungicide resistance has limited the effectiveness of the most common fungicides used against *Fusarium* spp. on potato seed in Canada. A re-evaluation of management practices for control of *Fusarium*-induced potato seed-piece decay is required.

Keywords: *Fusarium*; potato; *Solanum tuberosum*; fungicide resistance

Fusarium spp. are important pathogens of potato that cause yield losses at planting and in storage following harvest. Spores of the fungus are found in all soils where potatoes are grown and can survive for many years. Seed potatoes infected with *Fusarium* can rot after planting (seed-piece decay) causing "misses" in the field. Even if plants grown from infected seed do emerge, they often have reduced vigour and yield. The fungus can spread from infected to healthy seed during the cutting and handling process. After harvest, *Fusarium* spp. cause a dry rot in storage which reduces crop quality.

Potatoes infected with *Fusarium* spp. develop a spreading external decay that usually becomes shrunken and wrinkled in appearance. When diseased tubers are cut open, the brown decay can be seen spreading into the internal tissues of the tuber. The internal decay is usually marked by open cavities which contain the white mycelium of the fungus. *Fusarium* spp. can only infect potatoes through wounds. Thus, infection can occur when inoculum is spread from diseased to healthy seed during seed cutting and handling. As well, inoculum in soil attached to the surface of tubers can infect potatoes through wounds made during harvest and handling operations prior to storage.

Resistance of pathogens to pesticides is becoming a concern in the management of several potato diseases. Isolates of *Fusarium* spp. with resistance to thiophanate-methyl (Senator ® PSPT – a common potato seed-piece treatment) and thiabendazole (Mertect ® SC- a common post-harvest treatment) have been recovered in Canada [1-3]. More recently, isolates of *Fusarium* spp. with resistance to fludioxonil (Maxim® PSP - common potato seed piece treatments) have also been recovered in Canadian surveys [4]. In several instances, isolates of *F. sambucinum* with resistance to all of the afore-mentioned products have been found.

The current survey (2010) sought to assess the prevalence of different *Fusarium* species infecting seed potatoes across Canada and determine their sensitivity to commonly used seed-piece treatment and post-harvest fungicides. The survey was coordinated at Agriculture and Agri-Food Canada, Crops and Livestock Research Centre in Charlottetown, Prince Edward Island, Canada. Samples of infected potato seed tubers were sent by growers, provincial government personnel and potato industry representatives from different potato production regions across Canada. Each sample (representing a seed lot) contained 4 – 10 potato tubers with obvious signs of *Fusarium* infection. From these tubers, a 5mm x 10mm tissue sample was extracted from the border of a disease lesion. Tissue samples were surface-sterilized in 10% Javex (0.6% sodium hypochlorite) for 1 minute, followed by two rinses in sterile distilled water. Tissue pieces were then transferred to ½ strength potato dextrose agar (PDA) in Petri dishes. After 5 to 7 days, or when sufficient colony growth was observed, fungal colonies were hyphal-tipped and then transferred to fresh plates of ½ strength PDA to obtain pure cultures.

Isolates were identified to species level using micro-morphological features [5]. A sub-set of isolates representing the diversity of species in the collection was subjected to partial sequencing of the translation elongation factor 1-alpha for comparison with sequences of this gene in FUSARIUM-ID (http://fusarium.cbio.psu.edu; [6]) to confirm micro-morphological observations.

To determine the chemical sensitivity of isolates, agar plugs (5 mm diameter) were removed from the margins of 7-10 day-old cultures and transferred onto ½ strength PDA plates amended with 0 or 10 ppm of thiabendazole or fludioxonil. Chemicals were first dissolved in acetone and then added to the molten agar after autoclaving. Control plates (without fungicides) contained PDA with acetone only. Two replications were tested for each fungicide/isolate combination and the experiment was repeated.

After incubation at room temperature in the dark for 7 days, the diameter of hyphal growth in each plate (two observations per plate at right angles to each other) was recorded using digital calipers. The growth of isolates at 10 ppm of fungicide was compared with hyphal growth in the control plates (0 ppm). An isolate was considered to be resistant to the fungicide if growth in the fungicide-amended plate was >50% of the growth in the controls.

Samples were received from all provinces in Canada except Newfoundland. In total, 85 samples were received which yielded 171 individual isolates of *Fusarium* spp. The most common species identified were *F. sambucinum, F. avenaceum* and *F. coeruleum,* representing 47%, 30% and 13% of the isolates in the collection, respectively (Table 1). In addition, isolates of *F. oxysporum*, *F. cerealis*, *F. acuminatum* and *F. sporotrichioides* were also identified (Table 1).

Table 1. Frequency and chemical sensitivity of isolates of *Fusarium* spp. recovered from infected seed potato tubers in Canada in 2010.

Species	Total number of isolates	Thiabendazole		Fludioxonil	
		Sensitive	Resistant	Sensitive	Resistant
F. sambucinum	80	13	67	19	61
F. coeruleum	23	19	4	18	5
F. avenaceum	52	50	2	47	5
F. oxysporum	6	6	0	0	6
F. cerealis	2	2	0	1	1
F. acuminatum	1	1	0	1	0
F. sporotrichioides	7	7	0	7	0
Total	171	98	73	93	78

In vitro fungicide-amended agar assays clearly distinguished resistant from sensitive isolates for both fungicides. In most cases, growth of sensitive isolates in plates amended with 10 ppm fungicide was completely absent, whereas growth of resistant isolates in fungicide-amended plates was similar to that of the control, leading to unambiguous isolate classification for fungicide response.

The frequency of isolates of *Fusarium* spp. with resistance to thiabendazole and/or fludioxonil varied by species and geography (province of Canada). Most isolates of *F. sambucinum* were resistant to one or both fungicides (Table 1), whereas most isolates of other species were sensitive (Table 1). An exception was *F. oxysporum*, which was always sensitive to thiabendazole, but resistant to fludioxonil (Table 1).

The frequency of isolates of *F. sambucinum* with resistance to both fungicides was high and widespread across Canada (Table 2). However, in Alberta the frequency of resistance in *F. sambucinum* populations was closer to 50% (much lower than in other provinces). As well, all isolates of *F. sambucinum* recovered from British Columbia were sensitive to fludioxonil (Table 2).

Table 2. Chemical sensitivity of isolates of *F. sambucinum* collected from infected potato seed tubers from various Canadian provinces in 2010.

Province	Total number of isolates	Thiabendazole		Fludioxonil	
		Sensitive	Resistant	Sensitive	Resistant
Prince Edward Island	11	2	9	2	9
Nova Scotia	0	0	0	0	0
New Brunswick	18	4	14	5	13
Quebec	0	0	0	0	0
Ontario	8	0	8	0	8
Manitoba	7	0	7	1	6
Saskatchewan	20	0	20	0	20
Alberta	9	5	4	4	5
British Columbia	7	2	5	7	0
Total	80	13	67	19	61

As mentioned, most isolates of *F. avenaceum* were sensitive to both fungicides (Table 3). A small number of fungicide-resistant isolates were recovered from Prince Edward Island, Quebec and Manitoba (Table 3).

Table 3. Chemical sensitivity of isolates of *F. avenaceum* collected from infected potato seed tubers from various Canadian provinces in 2010.

Province	Total number of isolates	Thiabendazole		Fludioxonil	
		Sensitive	Resistant	Sensitive	Resistant
Prince Edward Island	29	28	1	27	2
Nova Scotia	0	0	0	0	0
New Brunswick	5	5	0	5	0
Quebec	10	9	1	8	2
Ontario	0	0	0	0	0
Manitoba	1	1	0	0	1
Saskatchewan	4	4	0	4	0
Alberta	2	2	0	2	0
British Columbia	1	1	0	1	0
Total	52	50	2	47	5

Most isolates of *F. coeruleum* recovered in the survey were sensitive to both fungicides (Table 4). However, strains with resistance to fludioxonil were found in Prince Edward Island and strains with resistance to thiabendazole were found in New Brunswick (Table 4).

Table 4. Chemical sensitivity of isolates of *F. coeruleum* collected from infected potato seed tubers from various Canadian provinces in 2010.

Province	Total number of isolates	Thiabendazole		Fludioxonil	
		Sensitive	Resistant	Sensitive	Resistant
Prince Edward Island	9	9	0	4	5
Nova Scotia	0	0	0	0	0
New Brunswick	5	1	4	5	0
Quebec	4	4	0	4	0
Ontario	0	0	0	0	0
Manitoba	0	0	0	0	0
Saskatchewan	0	0	0	0	0
Alberta	5	5	0	5	0
British Columbia	0	0	0	0	0
Total	23	19	4	18	5

In summary, the survey confirms the presence of strains of *F. sambucinum* and *F. coeruleum* in Canada resistant to thiabendazole and/or fludioxonil. As well, fungicide resistance in other *Fusarium* spp. was also identified. Fungicide resistance in F. sambucinum, the major dry rot and seed decay pathogen, is particularly widespread in Canada. Some partitioning of fungicide resistance along species and production zone lines was also evident. Fungicide resistance has limited the effectiveness of the most common fungicides used against *Fusarium* spp. on potato seed in Canada. A re-evaluation of management practices for control of *Fusarium*-induced potato seed-piece decay and dry rot is required.

Based on our research, knowing the predominant *Fusarium* spp. in a particular seedlot and their sensitivities to various chemical products would provide growers with important information to use to make disease management decisions. Thus, diagnostic testing of samples of tubers from seedlots could be a useful tool in the management of this important disease. Since fungicide resistance is a concern, alternating products from different chemical classes becomes an important strategy. In several field and storage trials (data not shown), mancozeb used as a seed treatment was able to control fungicide- resistant strains. Down the road, difenoconazole (another effective product in our trials) and other products may become available for seed treatment and post-harvest application.

Ultimately, the management of Fusarium dry rot and seed-piece decay depends upon an integrated approach that takes advantage of a number of control options and information generated by research studies. A summary of some of these control options would include:

At planting:

1. Use clean seed; store seed in a facility that has been properly disinfected
2. Warm seed tubers prior to cutting to promote rapid healing
3. Remove any diseased seed tubers prior to cutting
4. Disinfect seed cutting and handling equipment often and ensure that cutters are sharp to make a clean cut that heals quickly
5. Don't store cut seed for more than 10 days before planting
6. Determine the *Fusarium* spp. present by having seed tested by a diagnostic clinic if available
7. Use a registered fungicide seed treatment and access any available local information on pesticide resistance; rotate products of different chemical classes; mancozeb has been shown to control fungicide resistant strains; new seed treatment products are in development
8. Plant when soil and temperature conditions promote rapid sprout growth and emergence

At harvest and in storage:

1. Reduce tuber injury during harvest and handling operations
2. Provide conditions for rapid wound healing early in storage, then drop temperatures
3. Monitor storage conditions

4. Post-harvest treatments with thiabendazole will control most *Fusarium* spp., but not thiabendazole-resistant *Fusarium sambucinum* or *F. coeruleum*, therefore, access any available local information on pesticide resistance
5. New post-harvest treatments to manage dry rot in storage are in development and should be available in the next few years

Acknowledgements: The authors would like to acknowledge funding provided by Agriculture and Agri-Food Canada and Syngenta Canada Ltd. They would also like to thank participating potato growers, provincial government colleagues and industry representatives that contributed to sample submission.

References:

[1] Kawchuk LM, Holley JD, Lynch DR, Clear RM. Resistance to thiabendazole and thiophanate- methyl in Canadian isolates of *Fusarium sambucinum* and *Helminthosporium solani*. *American Potato Journal*. 1994;71:185-192.

[2] Peters RD, Macdonald IK, MacIsaac KA, Woodworth S. First report of thiabendazole-resistant isolates of *Fusarium sambucinum* infecting stored potatoes in Nova Scotia, Canada. *Plant Disease*. 2001;85:1030.

[3] Platt HW. Resistance to thiabendazole in *Fusarium* species and *Helminthosporium solani* in potato tubers treated commercially in eastern Canada. *Phytoprotection*. 1997;78:1-10.

[4] Peters RD, Platt HW, Drake KA, Coffin RH, Moorehead S, Al-Mughrabi KI, Howard RJ. First report of fludioxonil-resistant isolates of *Fusarium* spp. causing potato seed-piece decay. *Plant Disease*. 2008;92:172.

[5] Samson RA, Hoekstra ES, Frisvad JC, Filtenborg O, eds. *Introduction to Food- and Airborne Fungi.* 6th ed. Centraalbureau vor Schimmelcultures, Utrecht, The Netherlands; 2000.

[6] Geiser DM, Jiménez-Gasco M, Kang S, Makalowska I, Veeraraghavan N, Ward TJ, Zhang N, Kuldau GA, O'Donnell K. FUSARIUM-ID v. 1.0: A DNA sequence database for identifying *Fusarium*. *European Journal of Plant Pathology*. 2004;110:473-479.

Vancomycin Resistant Enterococci (VRE) in equine-faecal samples

Mohamed O. Ahmed[*,1,2], **Peter D. Clegg**[2], **Nicola J. Williams**[3], **Keith E. Baptiste**[4] **and Malcolm Bennett**[3]

[1]Department of Veterinary Microbiology, Faculty of Veterinary Medicine, Al-Fateh University P.O. Box 13662, Tripoli, Libya

[2]Department of Veterinary Clinical Sciences, Faculty of Veterinary Science, University of Liverpool, Leahurst, Chester High Road, Neston, CH64 7TE,UK

[3]Department of Veterinary Pathology, Faculty of Veterinary Science, University of Liverpool, Leahurst, Chester High Road, Neston, CH64 7TE,UK

[4]Department of Large Animal Sciences, Faculty of Life Sciences, University of Copenhagen, 2630 Taastrup, Denmark

*Corresponding author: email: mhmd73@lttnet.net, Phone: +218925400914

The occurrence and prevalence of vancomycin resistant enterococci (VRE) in equine-faecal samples of hospitalized (n=66) and non-hospitalized (n=72) horses were investigated in North West region in England in order to evaluate any potential role of horses to harbour this zoonotic pathogen. Modified culture methods with presumptive selection were used for *Enterococcus spp.* and VRE. A multiplex polymerase-chain reaction (PCR) method was used for genotyping of VRE. Overall 47 identified VRE isolates (from 264 faecal samples from 138 horses) were collected and only 9 isolates were confirmed and characterized by PCR. The *VanC* genotype (i.e. *vanC-1* gene) was identified in seven isolates and of these, six were from hospitalized horses. One positive isolate of each of *VanA* and *VanD* genotype were also detected both in non-hospitalized horses but were untypable by PCR. Furthermore, susceptibility testing by disc diffusion showed that all isolates (n=9) exhibited further resistance to other antibiotics as well as vancomycin (e.g. teicoplanin and streptogramins drugs). Unlike *VanA* genotype, which is reportedly associated with human infections, PCR revealed that genotypes of most isolates (i.e. VanC genotype) were not common to clinical human isolates. This study suggests that horses are unlikely to have a major role in the zoonotic transmission of VRE in this geographic area. Furthermore, there is an increased apparent prevalence of VRE in hospitalized horses suggesting a possible nosocomial transmission in hospital.

Keywords: Equine; Vancomycin resistant enterococci; Nosocomial infection; *VanC* genotype; *Enterococcus gallinarum*

1. Introduction

Vancomycin resistant enterococci (VRE) first reported in 1980s (1), is an emerging and re-emerged bacterial infectious disease, likely as a result of antibiotic selective pressures (2). VRE infections and colonization have been reported worldwide (3). Today, VRE has been associated with hospital-acquired infections in humans (4), and an important cause of nosocomial and community-acquired infections (5). Moreover VRE can act as a reservoir of antibiotic resistance genes to later donate resistance towards multiple antibiotics to other bacteria (6). VRE can be highly pathogenic and antibiotic treatments can be costly (7). VRE isolates have been reported in Europe with glycopeptides resistance in humans (8) and animals including horses (9).

Enterococci are part of the normal microflora of the gastrointestinal tracts of humans and many animals, as well as an important nosocomial pathogen frequently associated with human infections (10). Of the many enterococci species; *Enterococcus faecalis* and *Enterococcus faecium* are the most important causes of enterococcal humans infections. Furthermore, *E. faecium* and *E. faecalis* are human pathogens (11) that have been frequently isolated from the gastrointestinal tract of humans and animals. Up to now, six VAN resistance phenotypes have been described: *Van (A, B, C, D, E and G)*. These are distinguishable by the level of resistance to vancomycin and teicoplanin, ability to transfer as well as inducibility. Vancomycin resistance in these bacteria is associated mostly with the transferable and inducible *vanA* and *vanB* genes (12), which also have been reported widely among animals and environmental sources (13). Both genes (i.e. *vanA* and *vanB* genes), encode high levels of resistance to vancomycin and teicoplanin, except *VanB* phenotype that only represents a low to high-level resistance to vancomycin (14). *E. faeciuum* and *E. faecalis* are widely documented as *VanA* phenotype carrier strains in both human hospitals and animals (15). *Enterococcus faecium* is the major reservoir of acquired vancomycin resistance (16). The *VanC* phenotype, which is intrinsic and not transferable, are encoded by the *vanC-1, vanC-2* and *vanC-3* genes, and associated respectively with *E. gallinarum*, *E. casseliflavus*, and *E. flavescens*; these are rare causes of human disease, and associated with low-level resistance to vancomycin but not to teicoplanin (10). However, the *VanC* phenotype (i.e. *Enterococcus gallinarum*) is reported frequently from animals with different resistant phenotypes (15), including clinical human isolates (17).

The use of antibiotics has resulted in the emergence of antibiotic-resistant bacteria (18). Reservoirs of resistant bacteria in the environment and community (e.g. VRE) have been documented as consequences of use and abuse of antimicrobials (e.g. avoparcin) in animal husbandry (9) and growth promotors (i.e. virginiamycin) (19). There is little or no evidence of direct VRE transmission between animals and humans (20). However, contamination with VRE is assumed to exist in food of animal origin (20); therefore transmission could be possible via the food chain (21). VREs are widespread among enterococci_isolates from farm and pet animals. The aims described in this work were; to investigate and describe the prevalence of VRE; define possible unrecognized reservoirs of VRE in horses; and determine the susceptibility to different antimicrobials.

2. Materials & Methods

2.1 Source Of Desig*n* - Horses investigated were from two sources, hospitalized and non-hospitalized populations in the North-West of England. The non-hospitalized horses were randomly sampled at two different local premises (Livery yard). The hospitalized horses sampled in this study were from the Philip Leverhulme Equine Hospital (PLEH), University of Liverpool, a second opinion equine referral hospital. This study involved 264 horse faecal samples from 138 horses collected over a period of six months. From PLEH, a total of 109 faecal samples were collected from 66 horses. From the riding school ('premise A'), a total of 112 equine faecal samples were collected from 37 horses. From the Livery yard ('premise B'), a total of 43 faecal samples were collected from 35 horses. All samples were submitted into the laboratory and processed within the same day of collection.

2.2 Isolation & Identification of VRE – A modified method was adapted from Landman et al (22) and Novicki et al (23), based on the fact that selective media with a vancomycin impregnated disk was reported to be reliable, inexpensive method for VRE detection (22) and broth enrichment have been reported to improve identification rates of VRE (23). The selective media used was a bile-esculin azide agar supplemented with vancomycin and kanamycin aesculin azide agar based on reports of successful isolation of glycopeptide-resistant enterococci (23).

2.2.1 The method: Fresh faecal sample mixtures were prepared by mixing three grams of faeces with 5 ml brain-heart infusion broth (BHI broth) containing 5% glycerol in a 20ml universal container. The selective media kanamycin aesculin azide broth [KAB] (LABM, LAB107) supplemented with 0.5mg/l of bile salts (LABM) was inoculated with 1ml from BHI broth and incubated in aerobic conditions at 37°C for 24 hours. After incubation, kanamycin aesculin azide agar [KAA] (LABM, LAB106-UK) was streaked from KAB and a 5μg vancomycin disc (MAST-UK) was placed on the surface and incubated at 37°C for 48 hours. Presumptive enterococci colonies exhibiting growth to the vancomycin disc were identified as suspected VRE. Identification was based on colony morphology on the selective media. Suspected isolates were then sub-cultured on colombia agar containing 5% horse blood and further confirmed by multiplex polymerase-chain reaction (PCR). VRE-149 *E. faecalis* (provided by Department of Medical Microbiology-University of Liverpool) was used as a positive control. BHI broths were stored at –20°C.

2.3 Molecular Identification - Suspected VRE isolates were characterized at the species level and antibiotic resistance genes by PCR. DNA extraction was by a method adapted from Holmes & Quigley, 1981(24). PCR method was adapted from Dutka-Malen, 1995 (25) and designed to identify common genes responsible for the main reported species (i.e. *E. faecalis, E. faecium, E. gallinarium (vanC-1), E. casseliflavus (vanC-2)* and clinically important van genes (i.e. *vanA, vanB* (25)). Commercial primer (MWG, Ebersberg, Germany) sequences were used and positive controls for all genes were provided by the University of Liverpool (Dr. Nicola Williams) and used to confirm the accuracy of the PCR method. Confirmed VRE-PCR isolates were stored at –80°C for further susceptibility testing.

2.4 Antibiotic Susceptibility Tests - The complete protocol used for the disc diffusion method for antimicrobial susceptibility testing was in accordance with BSAC guidelines (26). An overnight culture of VRE on nutrient agar was suspended in 3ml sterile water and matched against 0.5 McFarlands standard. Isosensitest agar plates (ISA; Oxoid, Basingstoke, UK) were then inoculated and antibiotic discs (Mast Diagnostics) were placed on the inoculated ISA. Plates were incubated in aerobic conditions at 37ºC for 24 hours. Zone sizes were read and resistance was determined according to BSAC guidelines. Only confirmed VREs by PCR were tested to the following antibiotic discs; ampicillin (10μg), azithromycin (15μg), gentamicin (200μg), imipenem (10μg), linezolid (10μg), meropenem (10μg), synercid (quinupristin-dalfopristin) (15μg), teicoplanin (30μg), vancomycin (5μg).

3. Results & Discussion

A total of 47 suspected VRE isolates were found in 31 faecal samples from 26 horses. PCR confirmed nine out of 47 (i.e.19%) suspected VRE isolates from a total of nine faecal samples and eight individual horses from all locations detailed in Table 1: six horses were from the PLEH, one from premises A and two from premises B (i.e. 5.5%-hospital samples; 0.8%-premise A samples; 4.6%-premise B samples). PCR identified *vanC-1gene* (i.e. *E. gallinarum*) in seven isolates, all from PLEH except one isolate from premise A. The other two isolates from premise B were un-typable by PCR but respectively expressed *vanA* and *vanD genes* (Table 2). These un-typable isolates from premise B need further evaluation to characterize the vancomycin resistance. The remaining non-confirmed 38 suspected VRE isolates by PCR were not further studied and were excluded from any further analysis.

Table 1. Prevalence of VRE in equine hospital and non-hospital premises

	Numbers of samples assayed		Numbers of suspected VREs (from culture methods)			Numbers of confirmed VREs (from PCR analysis)	
Source	**Faecal samples**	**Horses**	**Isolates**	**Samples**	**Horses**	**Isolates**	**Horses**
Hospital	**109**	**66**	**25**	**12**	**11**	**6**	**5**
Premises A	**112**	**37**	**5**	**4**	**4**	**1**	**1**
Premises B	**43**	**35**	**17**	**15**	**11**	**2**	**2**
Total	**264**	**138**	**47**	**31**	**26**	**9**	**8**

Legend to table 1: Samples were collected from hospital (Philip Leverhulme Equine Hospital) and non-hospital (premises A and B) premises and assayed by culture methods, using selective media. DNA was extracted from 2-3 colonies /sample and analysed by multiplex PCR with VRE-149 *E. faecalis* as a positive control.

The direct use of PCR for identification of VREs in fecal samples is difficult due to inhibition by different substances and reported to be less sensitive (27). Thus, initial culture for the detection of VREs is still important, despite limited sensitivity (28). Disc diffusion remains widely used and recommended for laboratories especially in areas where VRE is persistent (29), because of its reliability for VREs (30) and an accurate tool for susceptibility testing for important zoonotic bacteria (31). However, susceptibility testing with disc diffusion identified two distinctive antibiotic resistant profiles among all isolates: *(Van, Teic, Qu, Azit, Mer) and (Van, Gent, Amp, Teic, Qu, Azitr, Imp, Mer)* (Table 2). The similarity between theses susceptibility profiles would suggest a nosocomial transmission within the hospital. Antibiotic resistance phenotypes could also possibly be used as an epidemiological tool.

Table 2. Characterization of VRE isolates from all sources

No	Source	Gene	Species	Antibiotic resistance profile
1	PLEH	*vanC-1*	*E. gallinarum*	Van,Tei,Qu,Azi,Mer
2		*vanC-1*	*E. gallinarum*	Van,Tei,Qu,Azi,Mer
3		*vanC-1*	*E. gallinarum*	Van,Gen,Amp,Tei,Qu,Azi,Imp,Mer
4		*vanC-1*	*E. gallinarum*	Van,Tei,Qu,Azi,Mer
5		*vanC-1*	*E. gallinarum*	Van,Gen,Amp,Tei,Qu,Azi,Imp,Mer
6		*vanC-1*	*E. gallinarum*	Van,Tei,Qu,Azi,Mer
7	Premise A	*vanC-1*	*E. gallinarum*	Van,Tei,Qu,Azi,Mer
8	Premise B	*vanA*	-	Van,Tei,Qu,Azi,Mer
9		*vanD*	-	Van,Tei,Qu,Azi,Mer

Legend to Table 2: Identification of *Enterococcus spp* (*E. faecalis*, *E. faecium*, *E. gallinarium* (*vanC-1*), *E. casseliflavus* (*vanC-2*) and vancomycin resistance genes *vanA, B, C1, C2* (Dutka-Malen and others 1995) and *vanD* (Perichon and others 1997) was by multiplex PCR, using commercially available primers (MWG, Ebersberg, Germany). Antibiotics resistance was tested using the following antibiotic discs; Amp, ampicillin (10μg); Azi, azithromycin (15μg); Gen, gentamicin (200μg); Imp, imipenem (10μg); linezolid (10μg); Mer, meropenem (10μg); Qu, synercid (quinupristin-dalfopristin) (15μg); Tei, teicoplanin (30μg),; Van, vancomycin (5μg). PLEH, Philip Leverhulme Equine Hospital.

The VRE were predominantly found in hospitalised horses in this study. However, PCR revealed that these isolates were distinct from those typically associated with human infections, namely either the *VanA* or *VanB* genotypes (12) which also have been widely reported among animal and environmental sources (32). Isolates were further tested and shown to be resistance to a range of antibiotics by disc diffusion test. Previously, *VanC* resistance phenotypes which are intrinsic and not transferable, were described with intrinsic low-level of resistance to vancomycin but susceptible to teicoplanin (10). However, the isolates in this study with *VanC*-encoded resistance exhibited resistant to vancomycin, teicoplanin and streptogramines (Table 2). The similarity between theses susceptibility profiles would suggest a nosocomial transmission within the hospital. Also, these antibiotic resistance profiles could also possibly be used as an epidemiological tool. However, this antibiotics resistant *VanC*-encoded phenotypes have also been reported in isolates from humans, from clinical cases and intestinal tract, identified as *E. gallinarum* carrying both *vanC1* and *vanA* genes, with a high-level of resistance to both vancomycin and teicoplanin (33). Moreover, the *VanC* genotypes have been reported with high levels of resistance to vancomycin and linzolid (30). *VanC*-type enterococci have been also described in clinical cases of human bacteremia caused by *E. gallinarum* and *E. casseliflavus* (34). The *VanC* phenotype is frequently reported from animals with different resistant phenotypes (15).

The *VanC* phenotypes is a constitutive, non-transmissible, chromosomal-based resistance and less potent phenotype than that expressed by the inducible *VanA* and *VanB* (10). Many enteroccocci spp are intrinsically resistant to vancomycin (14). Generally, resistance to vancomycin has been the result of either inactivation or sequestration of vancomycin, the production of intermediates that binds to and inactivates vancomycin, or target-site alteration (35). Chromosomal encoded resistance (i.e. low level of resistance to vancomycin) in the intrinsic enterococci (*VanC* phenotypes) appears distinct and evolved separately from the *VanA* and *VanB* types of resistance (i.e. high level of resistance to vancomycin) (35). *VanC* gene cluster analysis has shown a distinct variation from other *Van* phenotypes (35). This suggests the ability of *VanC* enterococcus genotypes to evolve, acquire and develop a distinct system for antimicrobial resistance. Such mechanisms requires further study for our collected isolates.

Resistance genes can be transferred and exchanged between bacterial species as well as genera. The *vanA* gene was reported in farm animals in different enterococci species including *E. gallinarum* with high resistance to teicoplanin (17,15). Therefore, commensal bacteria represent a reservoir of antibiotic resistance genes with the potential of transference to human and animal pathogens (36). Colonization by VRE of animal origin has been documented in the intestines of humans. However, little is known about the transfer of the *vanA* gene to human intestinal isolates (37). Nevertheless, there is a possibility of *vanA* gene transfer from animal enterococci to human enterococci phenotypes (38).

Increasingly, antimicrobial resistance of clinically important bacteria (e.g.VRE & MRSA) present difficult challenges for clinicians, and reflect the need for further newer therapy options (39). The use and misuse of antimicrobials can have great impact on therapy and public health of both humans and animals, since animals can constitute a reservoir for antimicrobial resistance and potentially zoonotic organisms (40).

To the authors knowledge this is first report of VRE in horses in the United Kingdom. This study indicates that horses are not a major reservoir for zoonotic VREs, in this geographic location, of widely reported and clinical significance to humans. Nevertheless, the *VanC* enterococci, especially *E. gallinarum* is of concern and should be further studied.

Acknowledgments Thanks also to stuff members at the Faculty of Veterinary Science at the University of Liverpool, Alfateh University and Biotechnology Research Centre in Tripoli/Libya for their help and support.

References

1. Uttley AH, C H Collins, J Naidoo, and RC George. Vancomycin-resistant enterococci. Lancet 1988; 1:57-58.
2. Cockerill, FR, and TF Smith. Response of the clinical microbiology laboratory to emerging (new) and reemerging infectious diseases. J.Clin.Microbiol. 2004; 42:2359-2365.
3. Courvalin P. Vancomycin resistance in gram-positive cocci. Clin.Infect.Dis. 2006; 42 Suppl 1:S25-S34.
4. Talebi M, Pourshafie MR, Oskouii M, Eshraghi SS. Molecular Analysis of vanHAX Element in Vancomycin Resistant Enterococci Isolated from Hospitalized Patients in Tehran. Iran Biomed. J. 2008; 12, 223-228.
5. Bates J, Jordens JZ, Griffiths DT. Farm animals as a putative reservoir for vancomycin-resistant enterococcal infection in man. J. Antimicrob. Chemother. 1994; 34, 507-514.
6. Nisbet DJ, Callaway TR, Edrington TS, Anderson RC, Poole TL. Effects of ionophores on Enterococcus faecalis and E. faecium growth in pure and mixed ruminal culture. Foodborne. Pathog. Dis 2008; 5, 193-198.
7. Brandl K, Plitas G, Mihu CN, Ubeda C, Jia T, Fleisher M, Schnabl B, DeMatteo RP, Pamer EG. Vancomycin-resistant enterococci exploit antibiotic-induced innate immune deficits. Nature 2008; 455, 804-807.
8. Woodford N, Chadwick PR, Morrison D, Cookson BD. Strains of glycopeptide-resistant Enterococcus faecium can alter their van genotypes during an outbreak. J. Clin. Microbiol. 1997; 35, 2966-2968

9. de NS, Sabia C, Messi P, Guerrieri E, Manicardi G, Bondi M. VanA-type vancomycin-resistant enterococci in equine and swine rectal swabs and in human clinical samples. Curr. Microbiol. 2007; 55 , 240-246.
10. Cetinkaya Y, Falk P, Mayhall CG. Vancomycin-resistant enterococci. Clin. Microbiol. Rev. 2000; 13, 686-707.
11. Gholizadeh Y, Courvalin P. Acquired and intrinsic glycopeptide resistance in enterococci. Int. J. Antimicrob. Agents 2000; 16 Suppl 1, S11-S17.
12. Jung WK, Lim JY, Kwon NH, Kim JM, Hong SK, Koo HC, Kim SH, Park YH. Vancomycin-resistant enterococci from animal sources in Korea. Int. J. Food Microbiol. 2007; 113, 102-107.
13. Eisner A, Feierl G, Gorkiewicz G, Dieber F, Kessler HH, Marth E, Kofer J. High prevalence of VanA-type vancomycin-resistant Enterococci in Austrian poultry. Appl. Environ. Microbiol. 2005; 71, 6407-6409.
14. Zirakzadeh A and R Patel. Vancomycin-resistant enterococci: colonization, infection, detection, and treatment. Mayo Clin.Proc. 2006; 81:529-536.
15. Chan YY, Abd Nasir MH, Yahaya MA, Salleh NM, Md Dan AD, Musa AM, Ravichandran M. Low prevalence of vancomycin- and bifunctional aminoglycoside-resistant enterococci isolated from poultry farms in Malaysia. Int. J. Food Microbiol. 2008; 122, 221-226.
16. Werner G, Coque TM, Hammerum AM, Hope R, Hryniewicz W, Johnson A, Klare I, Kristinsson KG, Leclercq R, Lester CH, Lillie M, Novais C, Olsson-Liljequist B, Peixe LV, Sadowy E, Simonsen GS, Top J, Vuopio-Varkila J, Willems RJ, Witte W, Woodford N. Emergence and spread of vancomycin resistance among enterococci in Europe. Euro. Surveill 2008; 13.
17. Togneri A, Lopardo H, Corso A. Bacteremia caused by Enterococcus gallinarum with a high level of glycopeptide resistance: 1st documented cases in Argentina. Rev. Argent Microbiol. 2003; 35, 96-99.
18. Hand WL. Current challenges in antibiotic resistance. Adolesc. Med. 2000; 11, 427-438.
19. Smith DL, Johnson JA, Harris AD, Furuno JP, Perencevich EN, Morris JG. Assessing risks for a pre-emergent pathogen: virginiamycin use and the emergence of streptogramin resistance in Enterococcus faecium. Lancet Infect. Dis. 2003; 3, 241-249.
20. Song JY, Hwang IS, Eom JS, Cheong HJ, Bae WK, Park YH, Kim WJ. Prevalence and molecular epidemiology of vancomycin-resistant enterococci (VRE) strains isolated from animals and humans in Korea. Korean J. Intern. Med. 2005; 20, 55-62.
21. Seong CN, Shim ES, Kim SM, Yoo JC. Prevalence and characterization of vancomycin-resistant enterococci in chicken intestines and humans of Korea. Arch. Pharm. Res. 2004; 27, 246-253.
22. Landman D, JM Quale, E Oydna, B Willey, V Ditore, M Zaman, K Patel, G Saurina, W. Huang. Comparison of five selective media for identifying fecal carriage of vancomycin-resistant enterococci. J.Clin.Microbiol. 1996; 34:751-752.
23. Novicki T J, J M Schapiro, B K Ulness, A Sebeste, L Busse-Johnston, K M Swanson, S R Swanzy, W Leisenring, AP Limaye. Convenient selective differential broth for isolation of vancomycin-resistant enterococcus from fecal material. J.Clin.Microbiol. 2004; 42:1637-1640.
24. Holmes DS, Quigley M. A rapid boiling method for the preparation of bacterial plasmids. Anal. Biochem. 1981; 114, 193-197.
25. Dutka-Malen S, Evers S, Courvalin P. Detection of glycopeptide resistance genotypes and identification to the species level of clinically relevant enterococci by PCR. J. Clin. Microbiol. 1995; 33, 24-27.
26. Andrews JM: BSAC standardized disc susceptibility testing method (version 7). *J Antimicrob Chemother* 2008, 62: 256-278
27. Satake S, N Clark, D Rimland, FS Nolte, FC Tenover. Detection of vancomycin-resistant enterococci in fecal samples by PCR. J.Clin.Microbiol. 1997; 35:2325-2330.
28. D'Agata EM, S Gautam, WK Green, YW Tang. High rate of false-negative results of the rectal swab culture method in detection of gastrointestinal colonization with vancomycin-resistant enterococci. Clin.Infect.Dis. 2002; 34:167-172.
29. Swenson JM, MJ Ferraro, DF Sahm, P Charache, FC Tenover. New vancomycin disk diffusion breakpoints for enterococci. The National Committee for Clinical Laboratory Standards Working Group on Enterococci. J.Clin.Microbiol. 1992; 30:2525-2528.
30. Yasliani S, AM Mobarez, RH Doust, M Satari, O Teymornejad. Linezolid vancomycin resistant Enterococcus isolated from clinical samples in Tehran hospitals. Indian J.Med.Sci. 2009; 63:297-302.
31. Senok A, Yousif A, Mazi W, Sharaf E, Bindayna K, Elnima e, Botta G. Pattern of antibiotic susceptibility in Campylobacter jejuni isolates of human and poultry origin. Jpn. J. Infect. Dis. 2007; 60, 1-4.
32. Mallon DJ, Corkill JE, Hazel SM, Wilson JS, French NP, Bennett M, Hart CA. Excretion of vancomycin-resistant enterococci by wild mammals. Emerg. Infect. Dis. 2002; 8, 636-638.
33. Neves FP, Ribeiro RL, Duarte RS, Teixeira LM, Merquior VL. Emergence of the vanA genotype among Enterococcus gallinarum isolates colonising the intestinal tract of patients in a university hospital in Rio de Janeiro, Brazil. Int. J. Antimicrob. Agents 2009; 33, 211-215.
34. Koganemaru H, Hitomi S. Bacteremia caused by VanC-type enterococci in a university hospital in Japan: a 6-year survey. J. Infect. Chemother. 2008; 14, 413-417.
35. Reynolds PE and P Courvalin. Vancomycin resistance in enterococci due to synthesis of precursors terminating in D-alanyl-D-serine. Antimicrob.Agents Chemother. 2005; 49:21-25.
36. Phillips I, Casewell M, Cox T, De GB, Friis C, Jones R, Nightingale C, Preston R, Waddell J. Does the use of antibiotics in food animals pose a risk to human health? A critical review of published data. J. Antimicrob. Chemother. 2004; 53, 28-52.

37. Lester CH, Frimodt-Moller N, Sorensen TL, Monnet DL, Hammerum AM. In vivo transfer of the vanA resistance gene from an Enterococcus faecium isolate of animal origin to an E. faecium isolate of human origin in the intestines of human volunteers. Antimicrob. Agents Chemother. 2006; 50, 596-599.
38. Bourgeois-Nicolaos N, Moubareck C, Mangeney N, Butel MJ, Doucet-Populaire F. Comparative study of vanA gene transfer from Enterococcus faecium to Enterococcus faecalis and to Enterococcus faecium in the intestine of mice. FEMS Microbiol. Lett. 2006; 254, 27-33.
39. Rossi F, Andreazzi D. Overview of tigecycline and its role in the era of antibiotic resistance. Braz. J. Infect. Dis. 2006; 10, 203-216.
40. Wegener HC. Antibiotics in animal feed and their role in resistance development. Curr. Opin. Microbiol. 2003; 6, 439-4

Chemistry

Antimicrobial Cyclic Pseudopeptides including Aza-β^3-Amino Acids

M. Laurencin[1], B. Legrand[2], L. Mouret[2], A. Bondon[2], Y. Fleury[3] and M. Baudy Floc'h[1]

[1] *ICMV, UMR CNRS 6226 Sciences Chimiques de Rennes*, 263 Avenue du Général Leclerc, 35042 Rennes Cedex, France

[2] *Université de Rennes 1. RMN-ILP, UMR CNRS 6026, CS 34317, Campus de Villejean, 35043 Rennes cedex, France*

[3] Laboratoire Universitaire de Biodiversité et d'Ecologie Microbienne, EA 3882, Université de Bretagne Occidentale

Keywords cyclic pseudoeptides, antibacterial aza-β^3 peptides

1. Introduction

Antibiotic resistance of pathogens against conventional antibiotics is increasing at a rate that far exceeds the pace of new development of drugs. So, antimicrobial peptides, both synthetic and from natural sources, have raised interest as potential useful drugs in the future. Small antimicrobial cyclic hexapeptides completely composed of L-aminoacids[1] or alterning L and D aminoacids[2] had already been described with both bacteria and fungi activities.

However, due to proteolytic degradation, peptides are not ideal candidates for pharmaceutical development. That is why numerous researches try to develop non natural peptidic analogues for enhancing metabolic stability, bioavailability, and biological absorption. In this class of peptidomimetics, pseudopeptides consisting exclusively or including aza β^3-amino acids have emerged as a promising new class of compounds that favour hydrogen bond formation and can enhance biological activities when compared to natural parent peptides[3].

We have designed "mixed" cylic pseudopeptides composed of α- and aza-β^3-amino acids that target bacterial cell wall and induce the death of the pathogens.

2. Results and Discussion

In our laboratory, we have developed the synthesis of Fmoc-aza-β^3-amino acids those bear natural proteinogenic side chains (like aza-β^3-lysine) or hydrophobic non proteinogenic side chains specially designed for antimicrobial application (like aza-β^3-1-naphthylalanine).

Table: Structure and biological activities of cyclopseudopeptides **P1** and **P2**
(MIC: Minimum Inhibitory Concentration in μM; ND : Not Determined)

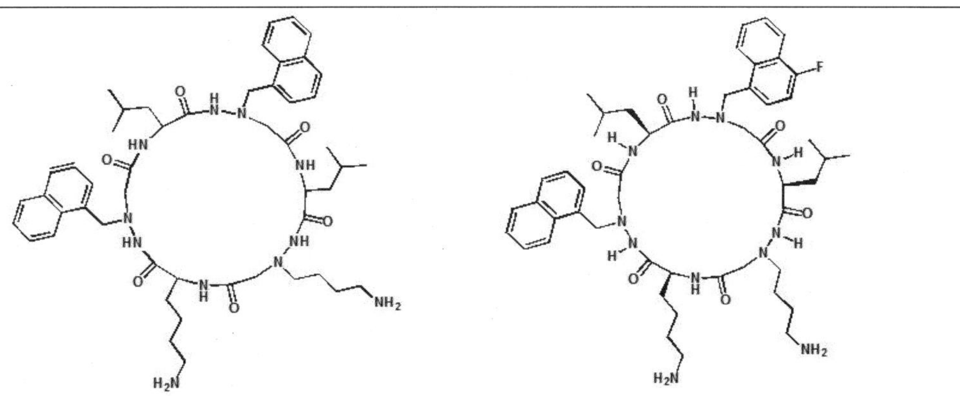

Pseudopeptide P1 Pseudopeptide P2

	MIC 48 h on *Staphylococcus aureus*	MIC 48 h on *Enterococcus faecalis*	MIC 48 h on *Streptococcus euqinus*	MIC 48 h on *Escherichia coli*	MIC 48 h on *Pseudomonas aeruginosa*	Hemolysis at 100μM	MIC 72h on *Cryptococcus neoformans*
P1	25	50	6,25	50	50	1,5	25
P2	12,5	25	6,25	25	25	5	12,5
melittin	25	12,5	3,12	6,25	50	100	ND

The synthesis of these cyclic pseudopeptides is very suitable with automated solid phase synthesis in Fmoc chemistry. So, we can build and test a lot of different cyclic antimicrobial pseudopeptides.

Among the twenty first synthesised cyclopseudopeptides, some of them have broad spectrum antibactericidal activities on gram positive and gram negative bacteria with low minimum inhibitory concentrations (MIC). On the other hand, this type of molecules is not hemolytic and cytotoxic at antimicrobial activity levels

To explain the mechanism of action of our pseudopeptides that act on the microbial membranes we performed NMR studies. We have demonstrated that in solution cycles auto-associate at high concentrations and we investigate their behaviour in presence of small unilamellar lipidic vesicules (SUV)[5].

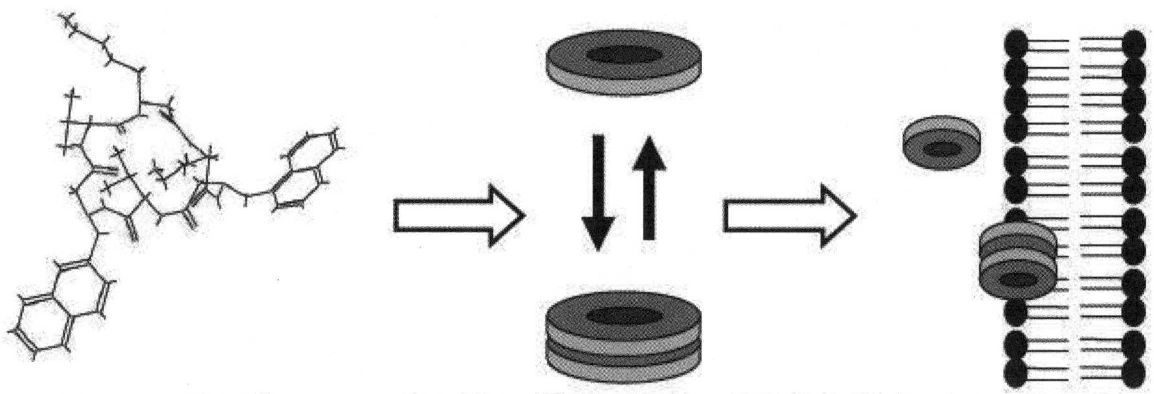

Scheme: Schematic representation of possible interactions of antimicrobial cyclopseudopeptides

CD spectrum and fluorescence studies have revealed that cyclopseudopeptide P1 adopt a well defined structuration in presence of SDS micelles or anionic SUV and a stacking of bulky aromatic side chains in these conditions. Therefore this phenomenon of self-association seems to be facilitated at the lipid interface.

3. Conclusion

Cyclopseudopeptides including aza-β^3-aminoacids display a broad spectrum of antimicrobial activities without significant cytotoxicity.

Now we try to elucidate the mechanism of action of this type of molecules. We already generate the structure of the free form but for the associated form it's difficult to discriminate inter- and intramolecular interactions. That's why we developed aza-β^3-Fluoro-naphtylalanine residues. Due to the chemical inertness and extreme hydrophobicity of fluorocarbons, P2 is more active on bacteria and fungi than P1 without augmentation of the cytotoxicity

References

[1] Dathe M., Nikolenko H., Klose J., Bienert M., Biochemistry, **43** (2004), 9140-9150.

[2] Ferandez-Lopez S., Kim HS., Choi E., Delgado M., Granja JR., Khazanov A., Kraehenbuehl K., Long G., Weinberger DA, Wilcoxen KM., Ghadiri MR., Nature, **412** (2001), 452-455.

[3] Dali H., Busnel O., Hoebeke J., Bi L., Decker P., Briand J.P., Baudy-Floc'h M., Muller S., *Mol. Immunol.,* **44** (2007), 3024-3036.

[4] Baudy-Floc'h M., Laurencin M., Zatylnyj-Gaudin C., Henry J., CNRS patent, **N° 07/07108** (2007)

[5] Da Costa G., Mouret L., Chevance S., Le Rumeur E., Bondon A., ***Eur. Biophys. J.***, **36** (2007), 933-942

Effect of Paracetamol on the Pharmacokinetics of Cephalexin in Dogs

N. A. Afifi, M. Atef, K. Abo-El-Sooud and N. El-Mokadem

Pharmacology Department, Faculty of Veterinary Medicine, Cairo University, PO Box 12211 Giza, Egypt

The pharmacokinetics of single oral dose (20 mg kg^{-1} b.wt.) of cephalexin alone and concomitantly with paracetamol was studied in normal healthy dogs. Moreover, the effect of simultaneous co-administration during five continuous days of treatment on hepato-renal functions was also evaluated. Cephalexin was rapidly absorbed from the gastrointestinal tract and the calculated peak serum concentration of cephalexin C_{max} was 16.47 µg ml^{-1} attained at 1.96 h (t_{max}). Paracetamol treatment significantly decreased the cephalexin concentrations from 30 minutes till 4 hours post administration. Paracetamol significantly lowered the peak serum concentration and the area under the concentration curve AUC of cephalexin. The relative bioavailability (F_r) of cephalexin with paracetamol was 76.65%. The percents of cephalexin protein binding in normal dog's serum either alone or in combination with paracetamol were 15.60, and 14.24 % respectively, indicating that the co-administration were significantly decreased the binding tendency. Following multiple doses of cephalexin alone, AP, ALT and AST activities, urea and creatinine were significantly increased till the fifth day then returning to the normal level. While concomitant administration of cephalexin with paracetamol induced transient significantly increases in the concentrations of urea and creatinine during the treatment period. Therefore, concomitant use of paracetamol with cephalexin may require close monitoring for clinical consequence of potential drug interaction.

Introduction

Co-administration of non steroidal anti-inflammatory drugs (NSAIDs) with antibiotics are commonly prescribed in veterinary medicine **(Milanova and Lashev 2002)** during the inflammatory conditions associated with bacterial infections. Some pharmacokinetic interactions between antibiotics and NSAIDs have been described **(Carbon, 1990)**, for instance phenylbutazone and indomethacin alter the pharmacokinetics of aminoglycosides and penicillins **(Hekman *et al.*,1982 and Firth *et al.*, 1990)**.

Cephalexin is a bactericidal first generation cephalosporin interferes with cell-wall formation (particularly with transpeptidation) and effective against both Gram-positive and Gram-negative bacteria **(Brander *et al.*, 1991)**. It is used particularly for treatment of pyodermatitis, folliculitis and frunculosis in dogs **(Kietzmann *et al.*, 1990; Frank and Kunkle, 1993)**. Paracetamol is SAIDs commonly used in dogs. Its proposed that inhibit prostaglandin synthesis in the central nervous system to interfere with nociceptive mediators in the brain, **(Shimada *et al.*,1994).** The present study was done to study the effect of the co-administration of paracetamol on the pharmacokinetics and the *in-vitro* serum protein binding of cephalexin in dogs. Moreover to investigate the effect of the multiple combined therapy on some hepatic and renal functions.

Materials and Methods

Animals: 10 clinically healthy both sex, mongrel dogs of 2-2.5 years old and weighting 15 - 20 kg each were used. Animals were kept indoors under good hygienic conditions, fed on antibiotic free diet and were given water *ad libitum*. Animals were kept under observation for one month before the beginning of experiment to ensure complete withdrawal of any resident drugs.

Drugs:
1.Cephalexin monohydrate was supplied from Amoun Pharmaceutical Industry Company, El Obour city, Cairo, Egypt.
2. Acetaminophen (Paracetamol)®: was supplied from Misr for medical products Company, Cairo, Egypt.

Pharmacokinetic study of cephalexin:
In this experiment; the animals (10 dogs) were randomly divided into 2 groups of five dogs each. Animals of the first group received a single dose of 20 mg kg^{-1} b.wt. of cephalexin monohydrate orally by stomach tube .Those in the second group was given paracetamol orally in dose of 20 mg.kg^{-1} b.wt. simultaneously with cephalexin dose.
Blood samples (3 ml) were collected in clean sterile centrifuge tubes, from the left femoral vein just before and at 30 minutes, 1, 2, 4, 6, 8, 10,12, 24 hours after antibiotics administration in all groups. Blood samples were

allowed to clot at room temperature, and the serum were separated by centrifugation at 3000 r.p.m. for 15 minutes and stored at -20 ^{0}C until assessment.

Cephalexin Assay:
The serum concentrations of cephalexin was estimated by agar plate diffusion technique **(Tsai and Kondo, 2001)** using Bacillus *subtilis* ATCC MB32SDR as the test organism

Pharmacokinetic analysis:
Following oral administrations, data were analyzed by both compartmental (Rstrip, Micromath Scientific Software, version 5.0, Salt Lake City, UT, USA) and non-compartmental analysis based on the statistical moment theory (Yamaoka *et al.*, 1978) to determine peak concentration (C_{max}), time to peak concentration (t_{max}), elimination half-life ($t_{1/2el}$), area under the curve (AUC) from zero to infinity by the trapezoidal rule and mean residence time (MRT) in serum.
The relative bioavailability was calculated from equation of **Rowland and Tozer ,1995:**

$F_r = AUC_{\text{Co-administration}} / AUC_{\text{cephalexin alone}} \times 100$

Estimation of protein binding percent:
In-vitro studies were made for estimation of blood protein binding percent of cephalexin either alone or in combination with paracetamol in normal dog's serum according to the method of **Craig and Suh,(1980)**.

The effect of multiple doses of Cephalexin on liver and kidneys:
Cephalexin alone or with paracetamol were administered once daily for 5 consecutive days . Blood samples were taken just before and after daily administration of the drugs till fifth day and on the fifteenth day from the last doses. The activity of Alkaline Phosphatase (AP), Aspartate Aminotransferase (AST) and Alanine Aminotransferase (ALT), urea and creatinine concentrations were estimated using Bio Merieux Kits (Vitek, Inc., USA).

Result

The serum concentrations of cephalexin after a single oral dose of 20 mg kg^{-1} body weight alone and with paracetamol in dogs were illustrated in figure (1). The serum levels of cephalexin against time were adequately described by one compartment open model. After 30 minutes, serum concentration was 5.44 µg ml^{-1}, it continued to increase until it reached its maximum concentration 17.26 µg ml^{-1} 2 hours after administration, then it decreased gradually till reach its lowest level 1.2 µg ml^{-1} after 12 hours. Paracetamol significantly decreased cephalexin concentrations from 30 minutes till 4 hours post administration.

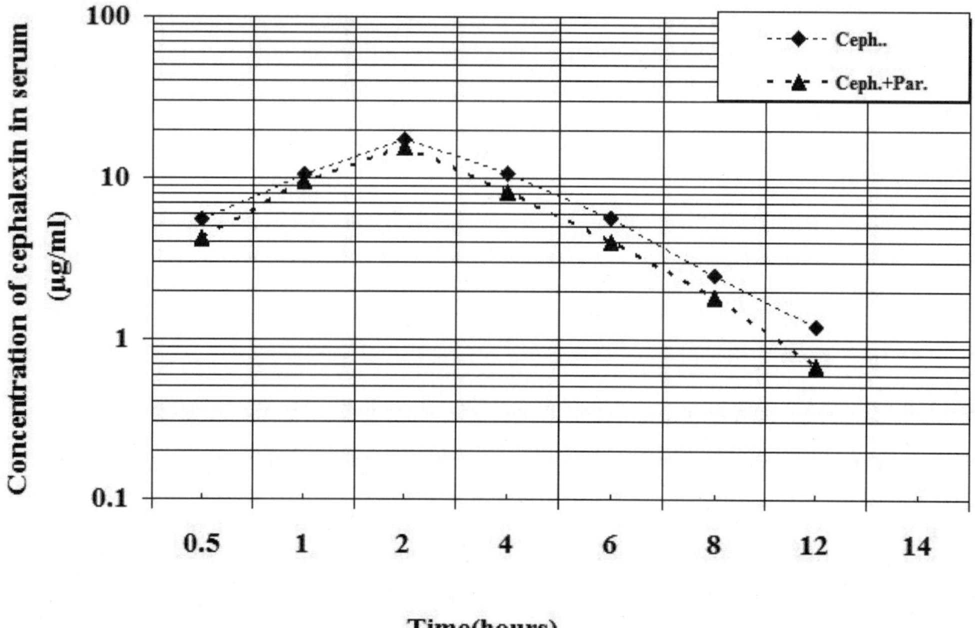

Figure 1: Serum concentrations of cephalexin (µg/ml) following a single oral administration alone and with paracetmol in dogs

Pharmacokinetic parameters of cephalexin alone and with paracetamol in dogs were showed in **Table 1**. Cephalexin was rapidly absorbed from the gastrointestinal tract and the calculated peak serum concentration C_{max} was 16.47 μg ml^{-1} and was attained at 1.96 h (t_{max}). Paracetamol significantly decreased the peak serum concentration, and the area under the concentration curve AUC of cephalexin.

Table (1): Mean ± SE of Pharmacokinetic parameters of cephalexin after single oral administration alone and with paracetamol in dogs *(n=5)*

Parameter	Unit	Cephalexin	Cephalexin + Paracetamol
K_{ab}	h^{-1}	0.84±0.04	0.97±0.08
$t_{0.5ab}$	h	0.83±0.04	0.74±0.06
C_{max}	μg ml^{-1}	16.47±0.37	14.17±0.29***
t_{max}	h	1.96±0.06	1.86±0.05
K_{el}	h^{-1}	0.414±0.02	0.423±0.03
$t_{0.5\ el}$	h	1.69±0.09	1.65±0.12
A.U.C.	μg.h ml^{-1}	86.61±3.5	66.39±2.58**
MRT	h	3.93±0.15	3.78±0.13
F_r	%	---	76.65

K_{ab} absorption rate constant, $t_{1/2ab}$ absorption half-life, C_{max} peak drug concentration, t_{max} time to peak concentration, K_{el} elimination rate constant, $t_{1/2el}$ elimination half-life, AUC: area under the curve from zero to infinity MRT mean residence time and Fr is the relative bioavailability.

*** Significant at (P < 0.05) in comparison with Cephalexin alone**
**** Significant at (P < 0.01) in comparison with Cephalexin alone**
***** Significant at (P < 0.001) in comparison with Cephalexin alone**

The percents of cephalexin protein binding in normal dog's serum either alone or in combination with paracetamol were 15.60, and 14.24 % respectively **(Table 2)**. Paracetamol was significantly decreased the *in vitro* cephalexin protein binding tendency.

Following multiple doses of cephalexin alone for 5 consecutive days, AP, ALT and AST activities, urea and creatinine were significantly increased till the fifth day then returning to the normal level **(Table 3)**.

Table (2): The *in-vitro* protein binding percent of cephalexin alone and with paracetamol

concentration (**µg ml^{-1}**)	**Cephalexin**	**Cephalexin +Paracetamol**
0.1	14.4	13.6
0.5	14.6	13.9
1	16.3	14.8
2.5	16.9	14.1
5	15.8	14.8
$\overline{X} \pm S.E$	15.60±0.48	14.24±0.24*

*** Significant at (P < 0.05) in comparison with Cephalexin alone**
**** Significant at (P < 0.01) in comparison with Cephalexin alone**

Discussion

The rapid absorption rate of cephalexin from the gastrointestinal tract suggesting the hypothesis of absorption through carrier-mediated transport in the gastrointestinal tract of the dogs rather than diffusion across membranes **(Carli *et al.*, 1999)**. In the clinical practice, the persistence of antibiotic concentration in serum and tissues above the minimal inhibitory concentration **MIC** should be achieved. The predicted average serum concentration of cephalexin in dogs after oral administration, suggested that the 20 mg kg $^{-1}$ body weight should be administered every 6-8 hours in order to attain serum concentration greater than 5 µg ml^{-1} . Longer time intervals between dosing cause a drop in serum concentrations below the MIC values (<2 µg ml^{-1}) for cephalexin sensitive bacteria **(Campbell and Rosin,1998)**. The experimental data of observed serum concentrations revealed that pharmacokinetics of cephalexin in dogs following single oral administration was adequately described by one compartment open model. This model had also been used by various workers to describe the pharmacokinetics of cephalexin in human volunteers **(Lode *et al.*, 1979** and **Liu *et al.*,1997)** and dogs **(Carli *et al.*, 1999)** after oral route.

The peak serum concentration C_{max} values of cephalexin after its oral administration alone and following co-administration with paracetamol were 16.47, and 14.17 µg ml^{-1}, respectively. These results are lower than that reported by **Campbell and Rosin, 1998** (46.2 µg ml^{-1}) and **Carli *et al.*, 1999** (20.3 µg ml^{-1}) in dogs. This difference may be due to the use of different dosages and formulations. Paracetamol treated animals had significantly lower area under curve (AUC) and peak serum concentration C_{max} were indicative of poorer availability. The relative bioavailability (F_r) with value paracetamol was 76.65% suggested that the absorption is significantly decreased compared to oral administration of cephalexin alone.

In-Vitro plasma protein binding of cephalexin in the present study showed that it were lower after the combination of cephalexin with paracetamol indicating thaparacetamol altered the plasma protein binding of cephalexin. In this respect, **Tsivou *et al.*, 2005** stated that NSAIDs showed a greater binding affinity for tissue proteins compared to the cephalosporins, and probably play an antagonistic role in protein binding, resulting in higher concentrations of antibiotics. **In contrary, Ramadan, 1990** found that the serum protein binding tendency of epicillin alone or combined with diclofenac as calculated *in vitro* was 14.30 and 14.00%, respectively.

The result showed that the use of cephalexin alone or following there concomitant administrations with paracetamol provoked a transient alterations of hepato-renal functions. The results are in agreement with that reported **Krause *et al.*, (2005)** who found an elevation of serum creatinine within 1 to 4 days after treatment with a variety of non-steroidal anti-inflammatory drugs (NSAID) but the renal functions completely normalized within 7 to 16 days. Consequently, concomitant use of paracetamol with cephalexin may require close monitoring for clinical consequence of prospective drug interaction.

Table (3): Mean ± SE values of serum enzymatic activities and urea and creatinine concentrations following daily doses of cephalexin alone and with paracetamol for 5 consecutive days in normal healthy dogs (n=5)

Parameter	Days	Control	Cephalexin	cephalexin + paracetamol
(AP) ($\mu g\ L^{-1}$)	1st day	24.8±0.6	26.2±0.45	24.5±0.51
	5th day	24.2±0.96	26.5±1.02*	26.3±0.32
	15th day	24.6±0.51	24.9±0.51	25.2±0.51
(AST) ($\mu g\ L^{-1}$)	1st day	11.2±0.55	13.2±0.58*	11.8±0.46
	5th day	10.8±0.75	14.8±0.45**	13.8±0.73*
	15 th day	10.6±0.53	10.21±0.32	10.4±0.51
(ALT) ($\mu g\ L^{-1}$)	1st day	14.3±0.82	16.6±0.93	15.4±0.98
	5th day	13.9±0.71	16.4±0.45*	16.2±0.66*
	15th day	14.42±0.25	14.2±0.55	14.3±0.45
Urea ($mg 100\ ml^{-1}$)	1st day	20.2±0.43	22.4±0.75*	20.8±0.66
	5th day	21.4±0.52	23.4±0.37*	22.4±0.51
	15 th day	20.8±0.51	21.2±0.75	20.7±0.75
Creatinine ($mg\ 100\ ml^{-1}$)	1st day	1.44±0.18	1.97±0.10*	1.50±0.10
	5th day	1.37±0.24	1.98±0.1*	1.67±0.07
	15th day	1.25±0.08	1.24±0.05	1.29±0.09

***Significant at (P < 0.05) in comparison with Control**
**** Significant at (P < 0.01) in comparison with Control**

References

Brander, G. C.; Pugh, D. M.; Bywater, R. J. and Lenkin, W.L.(1991): Veterinary applied pharmacology and therapeutics. 5th ed.Boiller Trindal London. pp. 430--473

Campbell, B. G. and Rosin, E. (1998): Effect of food on absorption of cefadroxil and cephalexin in dogs. J. Vet. Pharmacol. Therap., 21(5): 418-420.

Carbon, C. (1990): Antibiotics and anti-inflammatory agents: drug interactions. Pathol. Biol. (Paris), 38: 255–260

Carli, S.; Anfossi, P.; Villa, R.; Castellani, G.; Mengozzi, G. and Montesissa, C.(1999): Absorption kinetics and bioavailability of cephalexin in the dogs after oral and intramuscular administration. J. Vet. Pharmacol. Therap., (22): 308-313.

Craig, A. W. and Suh, B.(1980): Protein binding and the antibacterial effects, "Antibiotics in laboratory Medicine" Bailtmore, Maryland, Williams and Wilkins, pp. 265-297.

Firth, E.C.; Nouws, J.F.; Klein, W. R. and Driessens, F.(1990): The effect of phenylbutazone on the plasma disposition of penicillin G in the horse. J. Vet. Pharmacol. Therap., 13(2):179-185.

Frank, L. A. and Kunkle, G. A.(1993): Comparison of the efficacy of cefadroxil and generic and proprietary cephalexin in the treatment of pyoderma in dogs. J. A. V. M. Ass., 203: 530-533.

Hekman, P.; Nouws, F.M and Van Ginneken, C. A. M. (1982): Effect of tomanol on the pharmacokinetics and tissue distribution of penicillin G in dairy cows. Vet. Q., (4): 12-18.

Kietzmann ,M.; Mischke, R.; Albrecht, N. and Nolte, L.(1990): Vertraglichkeit und Pharmacokinetik von Cefalexin (Cefaseptin® Dragees) beim Hund. Kleintiepraxis, 35: 390-398.

Krause I, Cleper R, Eisenstein B, Davidovits M. (2005): Acute renal failure, associated with non-steroidal anti-inflammatory drugs in healthy children. Pediatr. Nephrol., 20(9):1295-8

Liu, X. D.; Xie, L.; Gao, J. P.; Lat, L. S. and Liu, G. Q.(1997): Cefixime absorption Kinetics after oral administration to human . Eur J Drug Metab Pharmacokinet., 22: 185-188.

Lode, H.; Stahlmann. R. and Koeppe. P. (1979): Comparative pharmacokinetics of cephalexin, cephaclor, cephadroxil and CGP9000. Antimicrob Agents Chemother., 16(1): 1-6.

Milanova, A. and Lashev, L. (2002): Pharmacokinetics of oleandomycin in dogs after intravenous or oral administration alone and after pretreatment with metamizole or dexamethasone. Vet. Res. Commun., 26(1): 61-71.

Ramadan, A. (1990): Effect of Diclofenac on pharmacokinetic aspects of epicillin in experimental *Streptococcus faecalis* infection in dog. J. Drug Res. Egypt, 19: 137-149.

Rowland, M. and Tozer, T., (1995): Clinical Pharmacokinetics: Concepts and applications, Lea & Febiger, Philadelphia, PA pp. 9-48.

Shimada, S.G.; Otterness, I.G. and Stiff, J.T. (1994): A study of the mechanism of action of the mild analgesic dipyrone. Agents Actions, 41(3-4): 188-192.

Tsai, C., and Kondo, F (2001): Improved agar diffusion method for detecting residual antimicrobial agents. J. Food Protect., 64: 361-366.

Tsivou E, Melakopoulos I, Kotsiou A, Anagnostopoulou S, Tesseromatis C. (2005): Alterations in cefalosporin levels in the serum and mandible of hyperlipaedemic rats after co-administration of ibuprofen. Eur. J. Drug Metab. Pharmacokinet, 30(3): 171-174.

Yamaoka, K., Nakagawa, T. and Uno, T., (1978): Statistical moments in pharmacokinetics. J. Pharmacokinet. Biopharma., 6: 547-558.

Importance of the C9 absolute configuration for the antifungal activity of natural and semisynthetic sesquiterpenes

M. Derita[*1], **M. Di Liberto**[1] **and S. Zacchino**[1]

[1]Pharmacognosy, Faculty of Biochemical and Pharmaceutical Sciences, National University of Rosario, Suipacha 531, Rosario, 2000, Argentine.

Keywords: Sesquiterpenes, antifungal, semisynthesis.

1. Introduction

Since the early 1980s, fungal infections have emerged as major causes of morbi-mortality, mainly among immunocompromised patients. The majority of deaths were associated with species of *Candida*, *Aspergillus* and *Cryptococcus*.[1] Instead, dermatophytes such as *Trichophyton* and *Microsporum* spp. produce superficial infections (tineas) which are usually not threatening but dramatically diminish the quality of life of human beings.[2] Although it appears to be an array of antifungal agents (polyenes, azoles, allylamines and the recent echinocandins) there are, in fact, few therapeutic options. Decreased susceptibilities of yeasts to the currently available antifungal agents[3] added to the increase in the number of reported cases of resistance,[4] have led to a general consensus that new efforts for detecting novel antifungal entities remain a priority. In this context, the study of plants with history of ethnopharmacological use for ailments related to fungal infections, can serve two goals: validation of the use of traditional medicines and finding new leads.[5]

The genus *Polygonum* (Polygonaceae) is represented in Argentina by 21 species which are divided into five sections: *Amblygonum*, *Echinocaulon*, *Tiniaria*, *Polygonum* and *Persicaria*. Most of these species are reported to have interesting biological activities.[6] In a previous work we reported the isolation of two sesquiterpene dialdehydes [polygodial (**1**) and isopolygodial (**2**)] and a related alcohol [drimenol (**3**)] from *Polygonum acuminatum* Kunth, a species belonging to *Persicaria* section.[7] In this paper we describe the obtainment of isodrimenol (**4**) by semisynthetic means, and the evaluation of the antifungal activities of the four compounds in order to establish the importance of C9 absolute configuration of these sesquiterpene-drimanes for its antifungal activities.

2. Material and Methods

Compounds **1-3** were isolated from dichloromethane extracts of leaves of *P. acuminatum* collected in summer 2008 and compound **4** was prepared by catalytic hydrogenolysis of isodrimendiol (**5**) which, in turn, was obtained by reduction of **2** with sodium borohydride (1:4). Figure 1.

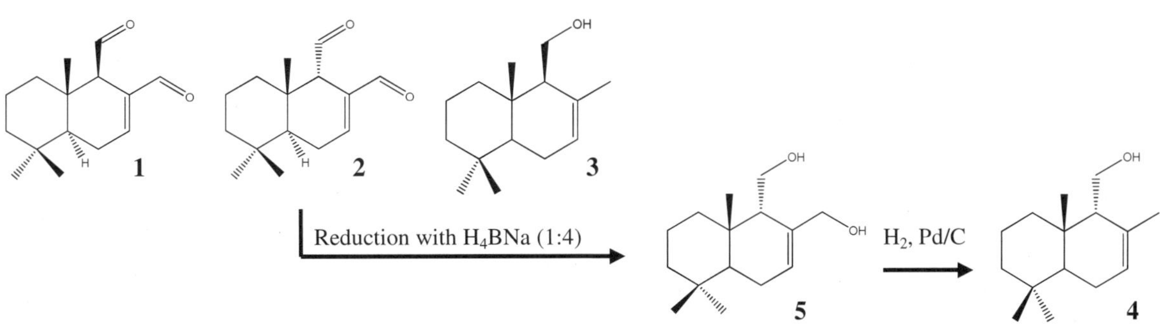

Figure 1: Structures of naturally occurrence polygodial (**1**), isopoligodial (**2**), drimenol (**3**); and isodrimenol (**4**) obtained by catalytic hydrogenolysis of isodrimendiol (**5**) prepared by reduction of **2** with sodium borohydride (1:4).

Compounds **1-4** were evaluated for their antifungal activity against a panel of human pathogenic fungi by the microbroth dilution assay recommended by CLSI (Clinical and Laboratory Standards Institute).[8] For the antifungal evaluation, strains from the American Type Culture Collection (ATCC, Rockville, MD, USA) and Centro de Referencia en Micología, CEREMIC [C, Faculty of Biochemical and Pharmaceutical Sciences,

Suipacha 531 (2000)-Rosario, Argentina] were used: *Candida albicans* (*Ca*) ATCC 10231, *Saccharomyces cerevisiae* (*Sc*) ATCC 9763, *Cryptococcus neoformans* (*Cn*) ATCC 32264, *Aspergillus flavus* (*Afl*) ATCC 9170, *Aspergillus fumigatus* (*Afu*) ATTC 26934, *Aspergillus niger* (*An*) ATCC 9029, *Trichophyton rubrum* (*Tr*) C 110, *Trichophyton mentagrophytes* (*Tm*) ATCC 9972 and *Microsporum gypseum* (*Mg*) C 115. Strains were grown on Sabouraud-chloramphenicol agar slants for 48 h at 30 °C, maintained on slopes of Sabouraud-dextrose agar (SDA, Oxoid) and subcultured every 15 days to prevent pleomorphic transformations. Inocula of cell or spore suspensions were obtained and quantified following reported procedures (CLSI).[8] Minimum Inhibitory Concentration (MIC) of each compound was determined by using broth microdilution techniques according to the guidelines of CLSI for yeasts: document M27-A2 and for filamentous fungi, M38A.[8] For the assay, stock solutions of pure compounds (100 µL) were two-fold diluted with the culture medium. A volumen of 100 µL of inoculum suspension [adjusted to 1–5 × 10^4 cells/spores as Colony Forming Units (CFU/mL)] was added to each well with the exception of the sterility control where sterile water was added to the well instead. Concentrations from 250 to 0.98 µg/mL were obtained for compounds **1-4** in the different wells. Ketoconazole (Sigma Chem. Co., St. Louis, MO), Terbinafine (Novartis) and Amphotericin B (Sigma) were used as positive controls.

3. Results and Discussion

Results showed (table 1) that compound **1** was active against 6/9 tested fungi with MICs between 3.9 to 62.5 µg/mL, meanwhile compound **2** was active against 4/9 tested fungi (MICs 62.5 µg/mL), indicating that the aldehyde function in C-9β makes compound **1** more active and with broader spectrum of action than **2**. Otherwise, compound **3** was active against 4/9 tested fungi with MICs between 62.5 to 125 µg/mL, while compound **4** was active against 6/9 tested fungi with MICs between 31.2 to 250 µg/mL, indicating that the alcohol function in C-9α makes compound **4** more active and with broader spectrum of action than **3**. These results confirm the importance of C9 absolute configuration for the antifungal activity of sesquiterpene dialdehydes and monoalcohols.

Table 1: Antifungal activity for compounds **1-4** and standards drugs (MICs in µg/mL). I: inactive (MIC > 250 µg/mL).

Compounds	MICs (µg/mL)								
	Ca	*Sc*	*Cn*	*Afu*	*Afl*	*An*	*Mg*	*Tr*	*Tm*
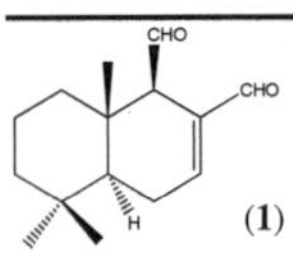 (1)	3.9	15.6	7.8	I	I	I	62.5	7.8	7.8
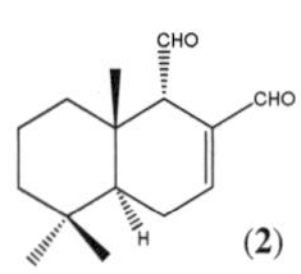 (2)	I	I	62.5	I	I	I	62.5	62.5	62.5
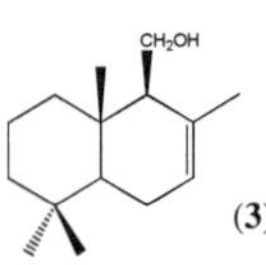 (3)	I	I	125	I	I	I	62.5	62.5	62.5
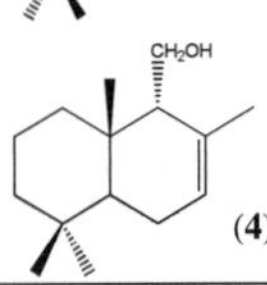 (4)	125	125	31.2	I	I	I	125	62.5	250
Standars drugs									
Amphotericin	1.0	0.50	0.25	0.50	0.50	0.50	0.12	0.07	0.07
Ketoconazole	0.50	0.50	0.25	0.12	0.50	0.25	0.04	0.02	0.02
Terbinafine	-	-	-	-	-	-	0.04	0.01	0.04

To go on with this approach, we are preparing others derivatives of **1** in order to enlarge the number of tested compounds and find other features that make them more actives.

Acknowledgments CONICET, ANPCyT, UNR, ERASMUS MUNDUS, UNIBO.

References

[1] Pfaller M, Diekema D. Epidemiology of invasive candidiasis: a persistent public health problem. *Clinical Microbiology Reviews*. 2007; 20: 133-163.
[2] Weitzman I, Summerbell R. The dermatophytes. *Clinical Microbiology Reviews*. 1995; 8: 240-259.
[3] Hsueh P, Lau Y, Chuang Y, Wan J, Huang W, Shyr J, Yan J, Yu K, Wu J, Ko W, Yang Y, Liu Y, Teng L, Liu Ch, Luh K. Antifungal susceptibilities of clinical isolates of *Candida* species, *Cryptococcus neoformans*, and *Aspergillus* species from Taiwan: Surveillance of multicenter antimicrobial resistance on Taiwan program data from 2003. *Antimicrobial Agents and Chemotherapy*. 2005; 49: 512-517.
[4] White T, Holleman S, Dy F, Mirels L, Stevens D. Resistance mechanisms in clinical isolates of *Candida albicans*. *Antimicrobial Agents and Chemotherapy*. 2002; 46: 1704-1713.
[5] Verpoorte R. Pharmacognosy in the new millennium: leadfinding and biotechnology. *Journal of Pharmacy and Pharmacology*. 2000; 52: 253-262.
[6] Natural Products Alert (NaPralert). Program for Collaborative Research in the Pharmaceutical Sciences. (2002). College of Pharmacy, University of Illinois. Chicago, III, USA.
[7] Derita M, Leiva M, Zacchino S. Influence of plant part, season of collection and content of the main active constituent, on the antifungal properties of *Polygonum acuminatum* Kunth. *Journal of Ethnopharmacology*. 2009; 124: 377-383.
[8] CLSI (Clinical and Laboratory Standards Institute). Methods M 27-A2, Vol. 22 (15): 1-29 y M 38-A, Vol. 22 (16): 1-27. Wayne Ed. (2002).

Specificity determinants dramatically reduce hemolytic activity in amphipathic α-helical antimicrobial peptides: antimicrobial activity against Gram-negative pathogens, *Acinetobacter baumannii* and *Pseudomonas aeruginosa*

Ziqing Jiang[1], Adriana I. Vasil[2], Lajos Gera[1], Michael L. Vasil[2] and Robert S. Hodges[*,1]

[1]Department of Biochemistry & Molecular Genetics and [2]Department of Microbiology, University of Colorado, School of Medicine, Anschutz Medical Campus, Aurora, CO, 80045, USA

*Corresponding author: e-mail: robert.hodges@ucdenver.edu, Phone: +1 3037243268, Fax: +1 3037243249

Keywords specificity determinants; cyclic β-sheet and α-helical antimicrobial peptides; Gram-negative pathogens; *Acinetobacter baumannii*; *Pseudomonas aeruginosa*

1. Introduction

It is urgent to develop new classes of antimicrobial agents due to the rapidly growing problem of increased resistance to classical antibiotics. *Acinetobacter baumannii* and *Pseudomonas aeruginosa* are two increasingly problematic hospital-associated Gram-negative pathogens due to dramatic increases in the incidence of antibiotic-resistant species. In a recent study in Mexico, 550 clinical isolates of *A. baumannii* and 250 clinical isolates of *P. aeruginosa* were analyzed for the prevalence of multi-drug resistance [1]. 74% of *A. baumannii* and 34% of *P. aeruginosa* were multi-drug resistant.

Amphipathic cationic antimicrobial peptides (AMPs) have been proposed as a potential new class of antibiotics with the ability to kill target cells rapidly, with broad spectrum activity and effectiveness against some of the most serious antibiotic-resistant pathogens isolated in clinics. AMPs are found in organisms that are evolutionarily quite distant. Overall, native AMPs are structurally diverse including peptides with no secondary structure to peptides in a linear or cyclic conformation containing α-helical and/or β-sheet structure frequently stabilized by disulfide bonds.

Although the exact mode of action of AMPs has not been established, it is generally accepted that the both the inner and outer membranes of gram negatives can be the main target of AMPs, whereby peptide accumulation in the membrane causes increased permeability and a loss of barrier function, resulting in the leakage of cytoplasmic or periplasmic components and cell death. In addition, AMPs have also been shown to target outer membrane components [2-4]. However, because their sole or main targets are cell membranes, hemolytic activity or toxicity to mammalian cells is always a potential issue preventing them from being used as systemic therapeutics. The goal in the development of antimicrobial peptides is to optimize hydrophobicity, to minimize eukaryotic cell toxicity and maximize antimicrobial activity, which in turn optimizes the therapeutic index. The vast majority of native AMPs are very hemolytic. Thus, the introduction of the "specificity determinant" design concept was developed in our laboratory as a biophysical mechanism to remove toxicity to eukaryotic cells (as measured by hemolytic activity against human red blood cells) of amphipathic α-helical or cyclic β-sheet antimicrobial peptides [5-9].

Our definition of a "specificity determinant" is any amino acid substitution on the non-polar face of amphipathic α-helical or amphipathic cyclic β-sheet antimicrobial peptides that dramatically reduces toxicity (i.e. hemolytic activity). This has been generally accomplished by substituting a positively charged lysine residue for a hydrophobic residue in the center of the non-polar face.

We also demonstrated that the sole target of our AMPs is the cytoplasmic membrane and their interactions with the membrane did not involve a stereoselective interaction with a chiral enzyme, lipid or protein receptor, since the all-L and all-D conformations had similar biological and biophysical properties [9]. Thus, antimicrobial peptides that meet these criteria can be prepared in the all-D-conformation, and are completely resistant to proteolytic enzyme degradation, which in turn enhances their potential as clinical therapeutics.

2. Experimental procedures

Gram-negative bacteria strains used in this study - *A. baumannii* strains were originally isolated from different patients and organs/tissues: strains 649 (blood), 689 (groin), 759 (gluteus), 821 (urine), 884 (axilla), 899 (perineum), 964 (throat), 985 (pleural fluid) and 1012 (sputum) or were purchased from the American Type Culture Collection, ATCC 17978 (fatal meningitis) and ATCC 19606 (urine). *P. aeruginosa* strains PAO1, WR5, PAK, PA14, M2 and CP204 are described [10]. All *A. baumannii* and *P. aeruginosa* strains have been maintained at -80 °C in the laboratory of Michael Vasil.

Measurement of Antimicrobial Activity (MIC) - MICs were determined by a standard microtiter dilution method in Mueller Hinton (MH) medium. The plates were incubated at 37 °C for 24 hours. MIC is given based on three sets of determinations. The MICs were determined as the lowest peptide concentration that inhibited growth [10].

Measurement of Hemolytic Activity (HC_{50}) - Peptide samples (concentration determined by amino acid analysis) were added to 1% human erythrocytes in phosphate-buffered saline (100 mM NaCl, 80 mM Na_2HPO_4, 20 mM NaH_2PO_4, pH 7.4) and the reaction mixtures were incubated at 37°C for 18 h in microtiter plates up to a peptide concentration of 1000 μg/mL as recently described [10].

Calculation of Therapeutic Index (HC_{50}/MIC Ratio) - Represents the specificity of antimicrobial peptides for prokaryotic versus eukaryotic cells. The larger values of therapeutic index indicate greater specificity for prokaryotic cells.

3. Results

Our definition of a "specificity determinant" is any amino acid substitution on the non-polar face of amphipathic α-helical or amphipathic cyclic β-sheet antimicrobial peptide that dramatically reduces toxicity to human cells. That is, a specificity determinant results in peptide specificity for prokaryotic membranes in preference to eukaryotic membranes. Toxicity is typically measured as hemolysis of human red blood cells (hRBC). Although, in the examples shown in this study, a positively charged lysine residue was substituted in the center of the non-polar face, usually replacing a large hydrophobic residue like Val or Leu, the positively charged residue can be any positively charge residue (e.g. Lys, Arg, Orn or others). In fact, we showed in the amphipathic cyclic β-sheet peptides that changing the stereochemistry of a positively charged residue on the polar face of the molecule from L-Lys to D-Lys put the positively charged side-chain on the non-polar face of the molecule with a dramatic effect on hemolytic activity [5, 6].

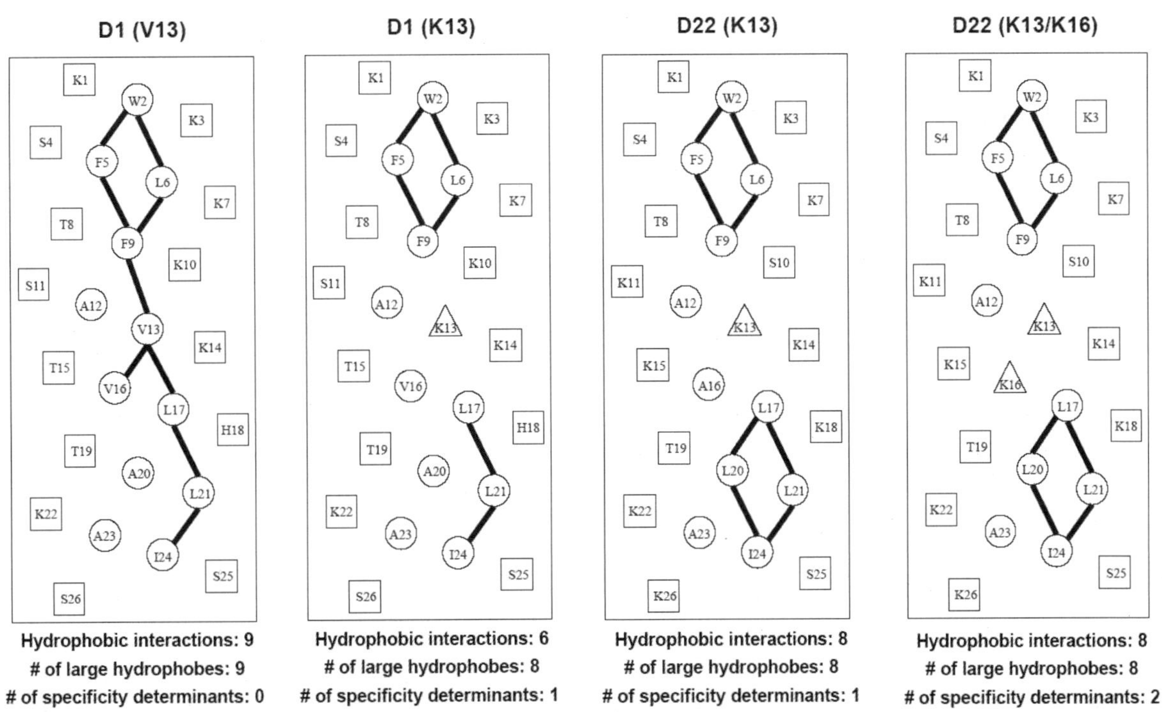

Fig. 1 Helical net representations of the sequences of peptides D1 (V13), D1 (K13), D22 (K13) and D22 (K13/K16). The one letter code is used for amino acid residues displayed in the helical net. The prefix D denotes that all residues in the peptides are in the D-conformation. The "specificity determinant(s)" lysine residue(s) in the center of the non-polar face are denoted by triangles. The polar and positively charged residues on the polar face are boxed. The hydrophobic residues on the non-polar face are circled. The *i* to *i*+3 / *i* to *i*+4 hydrophobic interactions between large hydrophobes (Val, Ile, Leu, Phe and Trp) along the helix are shown as black bars.

Similarly, substituting a Val residue on the non-polar face with a Lys residue (maintaining the same stereochemistry of the substituting residue) in the amphipathic cyclic β-sheet peptides accomplished the same effect. In the amphipathic α-helical peptides, Chen et al, 2005 [8], substituted residues varying in a wide range of hydrophilicity/hydrophobicity in the center of the non-polar face in both the L- and D- conformation in an all-L-antimicrobial peptide. Their results showed that the positively charged substitution in the center of the non-polar face disrupts the α-helical structure and that changing the stereochemistry of the substituting residue even if not a positively charged residue can accomplish a similar effect [8].

Table 1 Summary of the biological activity of D1 (V13) analogs against 11 diverse strains of *Acinetobacter baumannii* and 6 diverse strains of *Pseudomonas aeruginosa*.

Peptide Name	# of S.D.[a]	Hemolytic activity		Antimicrobial activity					
				Acinetobacter baumannii			*Pseudomonas aeruginosa*		
		HC_{50}[b] (μM)	Fold[c]	MIC_{GM}[d] (μM)	Therapeutic index[e] HC_{50}/MIC	Fold[f]	MIC_{GM}[d] (μM)	Therapeutic index[e] HC_{50}/MIC	Fold[f]
D1 (V13)	0	1.8	1.0	0.7	2.57	1.0	1.8	1.0	1.0
D1 (K13)	1	140.9	78.3	1.1	128.1	49.8	4.1	34.4	34.4
D22 (K13)	1	81.3	45.2	0.8	101.6	39.5	2.3	35.3	35.3
D22 (K13/K16)	2	351.5	195.3	0.8	439.4	171.0	2.5	140.6	140.6

a. S.D., number of "Specificity Determinants".
b. HC_{50} is the peptide concentration that produces 50% hemolysis of human red blood cells after 18 h at 37 °C in the standard microtiter dilution method.
c. Fold increase in HC_{50} (less hemolytic) compared to that of D1 (V13) (Fig.1).
d. MIC is minimal inhibitory concentration that inhibited growth at 37°C after 24h. GM is the geometric mean of the MIC values for 11 diverse strains of *Acinetobacter baumannii* and 6 diverse strains of *Pseudomonas aeruginosa.*
e. Therapeutic index is the ratio of the HC_{50} value (μM) over the geometric mean MIC value (μM). Large values indicate greater antimicrobial specificity.
f. The fold improvement in therapeutic index compared to that of D1 (V13) (Fig.1).

Though specificity determinants create the greatest reduction in toxicity of amphipathic α-helical peptides, the sequence and amino acid composition of both the polar and non-polar face influence the effectiveness of the specificity determinant. For example, if we compare peptide D1 (K13) with D22 (K13) both have a single specificity determinant K13, but the environment or context surrounding the specificity determinant is different (Fig. 1). The improvement in hemolytic activity from D1 (V13) to D1 (K13) is 78-fold (HC_{50} 1.8 μM to 140.9 μM, respectively), while the improvement in hemolytic activity from D1 (V13) to D22 (K13) is 45-fold (HC_{50} 1.8 μM to 81.3 μM, respectively) (Table 1).

Table 2 Effect of a single specificity determinant in amphipathic cyclic β-sheet and other α-helical antimicrobial peptides on hemolytic activity.

Peptide Name	Number of specificity determinant	Amphipathic peptide type	Hemolytic activity[a]	Fold[b]	Reference
GS14	0	cyclic β-sheet	100% lysis at 1.5 μg/mL	1.0	[5]
GS14K_D4	1	cyclic β-sheet	100% lysis at 200 μg/mL	133.3	[5]
GS14	0	cyclic β-sheet	100% lysis at 2 μg/mL	1.0	[7]
GS14K_D4	1	cyclic β-sheet	100% lysis at 125 μg/mL	62.5	[7]
GS14 (K3L4)	1	cyclic β-sheet	37% lysis at 100 μg/mL	–	[6]
RTA3 (F4W)	1	α-helical	4% lysis at 80 μg/mL	20.0	[11]
RTA3 (F4W/R5L)	0	α-helical	80% lysis at 80 μg/mL	1.0	[11]
B2RP	1	α-helical	50% lysis at 90 μM	5.0	[12]
B2RP (K16L)	0	α-helical	50% lysis at 18 μM	1.0	[12]

a. Hemolytic activity is measured as follows: cyclic β-sheet peptides, lysis of human erythrocytes at 24 hours and 37 °C; RTA3 (F4W), lysis of horse erythrocytes at 1 hour and 37 °C; B2RP, lysis of human erythrocytes at 1 hours and 37 °C.
b. Fold improvement in hemolytic activity (less hemolytic) with Lys on the non-polar face compared to Leu in different antimicrobial peptides, GS14K_D involved a conformation change. The value 1.0 is given to the most hemolytic peptide in each series.

The reason K13, in the context of D22 (K13), is less effective is that D22 (K13) is more hydrophobic on the non-polar face and more hydrophilic /positively charged on the polar face than D1 (K13) (Fig. 1). This context obviously lowers the effectiveness of the specificity determinant but less than a factor of 2 (Table 1). However, the substitution of a second specificity determinant at position 16 to generate peptide D22 (K13/K16) results in a substantial further improvement of the HC_{50} value relative to D1 (V13) of 195-fold (Table 1). The antimicrobial activity is generally unaffected by the substitution of the specificity determinant in the non-polar face. The effect of the specificity determinants on the therapeutic indices is dramatic with improvements from 40 to 171-fold against *A. baumannii* and 34 to 141-fold against *P. aeruginosa* (Table 1).

Though this concept of specificity determinants (placing a Lys residue in the center of the non-polar face of amphipathic antimicrobial peptides) was first introduced by our laboratory in 1999 with cyclic β-sheet peptides [5] and then extensively documented with amphipathic α-helical antimicrobial peptides from 2005 to present [8-10, 13-16], other laboratories are now validating this concept in other antimicrobial peptides [11, 12]. The improvement in hemolytic activity were between 5 and 20-fold for α-helical antimicrobial peptides (Table 2). The improvements in hemolytic activity with the cyclic β-sheet peptides were even more dramatic than the α-helical peptides where hemolytic activity decreased from 62.5 to 133.3–fold (Table 2), but also resulted in large unexpected improvements in antimicrobial activity of 65-fold and 100-fold for particular Gram-negative and Gram-positive organisms, respectively [5].

Since toxicity is so important in the development of antimicrobial peptides as clinical useful therapeutics, there must be a gold standard developed for determining hemolytic activity. Many variations of the assay exist and inconsistencies in the red blood source, peptide exposure time and reporting of peptide concentrations impede comparison of AMP toxicity across structural classes and research groups. Non-human red blood cell results should not be compared to human red blood cell (hRBC) hemolytic data. Membrane lipid composition varies among mammalian species and vastly different hemolysis results have been obtained across species. Thus, if the desire is to develop AMPs for human clinical use we strongly recommend using hRBC for hemolytic assays. The length of time erythrocytes are exposed to AMPs during the hemolytic assay is the least standardized parameter. Reported protocols have exposure times varying from 10 min to 24 hours. The most commonly used times are 30 minutes and 1 hour. Although such short exposure times provide valuable information about relative acute toxicity across a peptide series, longer exposure times are necessary to evaluate the longer term toxicity that could result if AMPs are not fully metabolized and cleared within 1 hour *in vivo*. We have shown in time course studies substantially greater hemolysis when exposure time is increased. Therefore we suggest hemolysis should be measured using a time course approach extending to 18 hours with the use of peptide concentrations from 1 to 1000 μg/mL and reported as HC_{50} values in μM as the gold standard (see our recent review [17] and references therein for a detailed discussion of this topic).

4. Conclusions

Specificity determinants, in general, that is, placing a positively charged residue in the center of the non-polar face of amphipathic cyclic β-sheet or α-helical antimicrobial peptides affect both their biological and biophysical properties. The specificity determinants in the example shown here reduced toxicity as measured by hemolytic activity against human red blood cells by 78-fold, for a single specificity determinant, and 195-fold for the combined effect of two specificity determinants. Biophysical properties of the peptides are also affected by specificity determinants as follows: (i) disrupts α-helical structure in aqueous conditions; (ii) decreases peptide self-association in aqueous conditions; (iii) the random coil monomer can more easily pass through the capsule and cell wall to reach the cytoplasmic membrane; (iv) α-helical structure is readily induced by the hydrophobic environment of the membrane for antimicrobial activity.

We have clearly documented our discovery of specificity determinants and their utility to dramatically lower toxicity to human red blood cells, a necessary step in the commercialization of amphipathic antimicrobial peptides as potential therapeutics.

Acknowledgements This research was supported by a NIH grant from the National Institute of Allergy and Infectious Diseases (NIAID) R01 AI067296 (R.S.H.), the John Stewart Chair in Peptide Chemistry to R.S.H and the Department of Defense, Office of Naval Research through a STTR grant to BioAmps International Inc. The content is solely the responsibility of the authors and does not necessarily represent the official views of NIAID or NIH.

References

[1] Garza-Gonzalez E, Llaca-Diaz JM, Bosques-Padilla FJ, Gonzalez GM. Prevalence of multidrug-resistant bacteria at a tertiary-care teaching hospital in Mexico: special focus on Acinetobacter baumannii. *Chemotherapy*, 2010; 56: 275-9.

[2] Eckert R, Qi F, Yarbrough DK, He J, Anderson MH, Shi W. Adding selectivity to antimicrobial peptides: rational design of a multidomain peptide against Pseudomonas spp. *Antimicrob Agents Chemother*, 2006; 50: 1480-8.

[3] Lin YM, Wu SJ, Chang TW, Wang CF, Suen CS, Hwang MJ, Chang MD, Chen YT, Liao YD. Outer membrane protein I of Pseudomonas aeruginosa is a target of cationic antimicrobial peptide/protein. *J Biol Chem*, 2010; 285: 8985-94.

[4] Srinivas N, Jetter P, Ueberbacher BJ, Werneburg M, Zerbe K, Steinmann J, Van der Meijden B, Bernardini F, Lederer A, Dias RL, Misson PE, Henze H, Zumbrunn J, Gombert FO, Obrecht D, Hunziker P, Schauer S, Ziegler U, Kach A, Eberl L, Riedel K, DeMarco SJ, Robinson JA. Peptidomimetic antibiotics target outer-membrane biogenesis in Pseudomonas aeruginosa. *Science*, 2010; 327: 1010-3.

[5] Kondejewski LH, Jelokhani-Niaraki M, Farmer SW, Lix B, Kay CM, Sykes BD, Hancock RE, Hodges RS. Dissociation of antimicrobial and hemolytic activities in cyclic peptide diastereomers by systematic alterations in amphipathicity. *J Biol Chem*, 1999; 274: 13181-92.

[6] McInnes C, Kondejewski LH, Hodges RS, Sykes BD. Development of the structural basis for antimicrobial and hemolytic activities of peptides based on gramicidin S and design of novel analogs using NMR spectroscopy. *J Biol Chem*, 2000; 275: 14287-94.

[7] Lee DL, Powers JP, Pflegerl K, Vasil ML, Hancock RE, Hodges RS. Effects of single D-amino acid substitutions on disruption of beta-sheet structure and hydrophobicity in cyclic 14-residue antimicrobial peptide analogs related to gramicidin S. *J Pept Res*, 2004; 63: 69-84.

[8] Chen Y, Mant CT, Farmer SW, Hancock RE, Vasil ML, Hodges RS. Rational design of alpha-helical antimicrobial peptides with enhanced activities and specificity/therapeutic index. *J Biol Chem*, 2005; 280: 12316-29.

[9] Chen Y, Vasil AI, Rehaume L, Mant CT, Burns JL, Vasil ML, Hancock RE, Hodges RS. Comparison of biophysical and biologic properties of alpha-helical enantiomeric antimicrobial peptides. *Chem Biol Drug Des*, 2006; 67: 162-73.

[10] Jiang Z, Vasil AI, Gera L, Vasil ML, Hodges RS. Rational design of a-helical antimicrobial peptides to target Gram-negative pathogens, *Acinetobacter baumannii* and *Pseudomonas aeruginosa*: utilization of charge, "specificity determinants", total hydrophobicity, hydrophobe type and location as design parameters to improve the therapeutic ratio. *Chem Biol Drug Des* (submitted).

[11] Hawrani A, Howe RA, Walsh TR, Dempsey CE. Origin of low mammalian cell toxicity in a class of highly active antimicrobial amphipathic helical peptides. *J Biol Chem*, 2008; 283: 18636-45.

[12] Conlon JM, Ahmed E, Condamine E. Antimicrobial properties of brevinin-2-related peptide and its analogs: Efficacy against multidrug-resistant Acinetobacter baumannii. *Chem Biol Drug Des*, 2009; 74: 488-93.

[13] Chen Y, Guarnieri MT, Vasil AI, Vasil ML, Mant CT, Hodges RS. Role of peptide hydrophobicity in the mechanism of action of alpha-helical antimicrobial peptides. *Antimicrob Agents Chemother*, 2007; 51: 1398-406.

[14] Jiang Z, Kullberg BJ, van der Lee H, Vasil AI, Hale JD, Mant CT, Hancock REW, Vasil ML, Netea MG, Hodges RS. Effects of hydrophobicity on the antifungal activity of a-helical antimicrobial peptides. *Chem Biol Drug Des*, 2008; 72: 483-495.

[15] Jiang Z, Vasil AI, Hale JD, Hancock RE, Vasil ML, Hodges RS. Effects of net charge and the number of positively charged residues on the biological activity of amphipathic alpha-helical cationic antimicrobial peptides. *Biopolymers (Peptide Science)*, 2008; 90: 369-83.

[16] Jiang Z, Higgins MP, Whitehurst J, Kisich KO, Voskuil MI, Hodges RS. Anti-tuberculosis activity of alpha-helical antimicrobial peptides: de novo designed L- and D-enantiomers versus L- and D-LL-37. *Protein Pept Lett*, 2010 (in press).

[17] Hodges RS, Jiang Z, Whitehurst J, Mant CT, “Development of antimicrobial peptides as therapeutic agents” In Development of Therapeutic Agents, Handbook in Pharmaceutical Sciences (Editor-in-Chief, Shayne Gad and Managing Editor, Michael Eventhal) John Wiley and Sons (in press)

Studies to find a better high level disinfectant for heat-sensitive endoscopes

Norman Miner*[1], Valerie Harris[1], Towanda Ebron[1] and Natalie Lukomski[1]

[1]MicroChem Laboratory, Inc., 1107-C S. Airport Circle, Euless, TX 76040, USA

*Corresponding author: email: nminer@microchemlab.net, Phone: (817) 283-9100

Heat-sensitive, fiber optic, flexible endoscopes have made great contributions for the non-invasive diagnosis and treatment of disease. These heat-sensitive endoscopes need to be carefully cleaned, rinsed, high level disinfected, and rinsed to remove the disinfectant between each patient. Alkaline glutaraldehyde, ortho-phthalaldehyde, and peracetic acid are widely used high level disinfectants, but they each have limitations. We combined various concentrations of alcohols, salts, surfactants, buffers, and pH values with glutaraldehyde to find a chemically stable high level disinfectant that quickly kills all species of mycobacteria in suspension or dried onto a substrate, and can kill bacterial spores in a practical exposure time, and be easily rinsed from endoscope surfaces.

Keywords glutaraldehyde; Aldahol High Level Disinfectant;

At this time it is estimated that there are about fifty million endoscopy procedures done world-wide with heat-sensitive, semi-critical endoscopes (e.g. gastroscopes, colonoscopes, bronchoscopes, cystoscopes, and urethroscopes) annually. Each of these endoscopes need to be thoroughly cleaned, rinsed to remove the cleaning agent, disinfected with a high-level disinfectant, rinsed to remove the disinfectant, and dried prior to each new patient procedure.

The incidence of disease transmission from one patient to another by means of an endoscope is very rare. However, it can and does happen. In 2004 Brazil reported more than 2000 cases of serious wound infections during medical procedures of liposuction. The causative agent for these infections was traced to glutaraldehyde-resistant *Mycobacterium massiliense*. Alkaline glutaraldehyde was the high level disinfectant being used during this outbreak [1].

Commonly used high level disinfectants are alkaline glutaraldehyde (e.g. Cidex Solution), ortho-phthalaldehyde (e.g. Cidex OPA), and peracetic acid, which is produced by hydrogen peroxide mixed with acetic acid. Each of these high level disinfectants has various limitations that compromise their use for the disinfection of endoscopes.

Alkaline glutaraldehyde has slow mycobactericidal activity requiring 45 min at 25°C to kill 6 $\log_{10}$ of *Mycobacterium bovis* var. BCG in suspension. By consensus alkaline glutaraldehyde is often used for 20.0 min rather than 45.0 min on the theory that the endoscopes are cleaned, and thus probably carry less than 6 $\log_{10}$ of mycobacteria. There is also often an assumption that 20°C, or ambient temperature, is "about as good" as 25°C, and the alkaline glutaraldehyde is often used at ambient temperature. Actually the extra 5.0 degrees C makes a large difference in the degree of microbial kill (see Figure 6).

Sporicidal activity of alkaline glutaraldehyde is also slow, requiring about 120.0 to 180.0 min at 25°C to kill 6 $\log_{10}$ of *Bacillus subtilis* spores in suspension. This limitation has also been accepted for decades on the basis that the medical requirement is to disinfect semi-critical endoscopes, not sterilize them. With the possible exception of bronchoscopes, endoscopes do not penetrate sterile areas of the body. However, when viable microbes remain within the channels of endoscopes, some microbes that are capable of dividing about every 30 to 60 minutes can begin to form a biofilm, even though this might take several days or longer to occur. Biofilms are capable of encasing other microbes such as bacteria and viruses, and sheltering them from disinfectants. Work has been published to show that endoscopes containing a biofilm cannot be disinfected [2, 3]. We have had that experience in our laboratory.

Another limitation of glutaraldehyde solutions is that they have irritating and sensitizing vapors of glutaraldehyde. Medical personnel and patients must be protected from these vapors.

Ortho-phthalaldehyde (OPA), also known as 1, 2 benzene dialdehyde, has become increasingly popular because there is little odor in the air from ortho-phthalaldehyde due to its very low vapor pressure. While glutaraldehyde kills the spores of spore-forming bacteria slowly, OPA does not kill *B. subtilis* spores at all in any practical exposure time, even at 25°C, and OPA does not claim to be a sterilant. The kill of mycobacteria might be limited to certain test conditions such as mycobacteria in suspension, and not dried onto any surface.

OPA is very insoluble in water (0.6%) or in isopropanol (11%), and this physical fact might make it difficult to rinse OPA away from endoscopes. We attempted to rinse OPA away from the surfaces of small pieces of endoscope insertion tubes and bending rubbers carrying at most 1.0 ml of 0.55% OPA by rinsing them in 100 ml of rinse water with up to fifteen serial rinses, a dilution factor of 10^{30}. Even after fifteen rinses, we were still able to demonstrate the presence of OPA adsorbed to the endoscope surfaces. Solutions of OPA are contra-indicated for use in the surveillance of bladder cancers with cystoscopes due to at least thirteen published episodes of anaphylactic shock [4-6]. Allergic reactions have also happened with the use of trans-esophageal echocardiogram probes disinfected with OPA, and with a colonoscope and laryngoscopes disinfected with OPA [7-9]. Both glutaraldehyde and OPA are known allergens, but glutaraldehyde, soluble as 67 g/100 ml of water, was able to be rinsed away from the insertion tube sections, while OPA was not able to be rinsed away even after fifteen serial rinses as described above.

Peracetic acid (PAA) is a broad-spectrum and fast-acting antimicrobial chemistry. Its aggressive oxidative chemistry might be damaging to some endoscopy materials. The mode of action for peracetic acid is to break down into a hydroxyl radical, which combines with and denatures biochemicals. This breakdown limits the use and re-use life of PAA, thus limiting the number of endoscopes that can be disinfected with one "loading" of PAA. This limitation makes PAA the most expensive high level disinfectant per endoscope disinfected.

We began studies to find a better high level disinfectant that did not have the above-mentioned limitations. We chose glutaraldehyde as the base chemistry because of its long history of compatibility with endoscope materials, its chemical stability, and its high solubility, and thus rinsability in water and alcohol. We combined glutaraldehyde with various concentrations of alcohols, salts, surfactants, pH values, and buffers. Each combination was tested against mycobacteria in suspension, mycobacteria dried onto surfaces, and spore-forming bacteria.

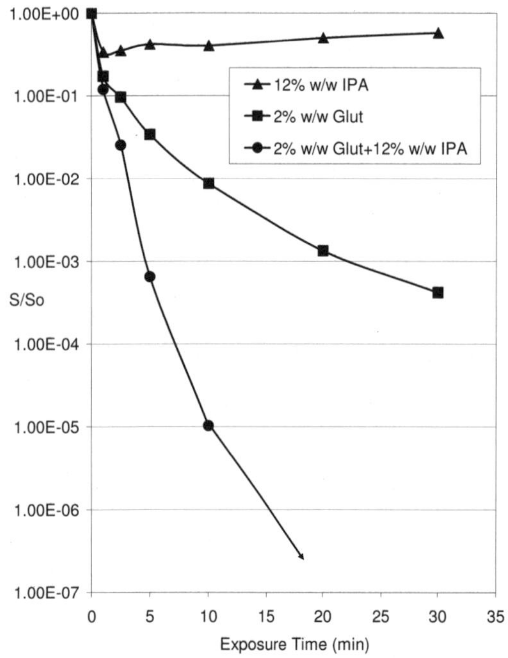

Fig. 1 shows that a small concentration of isopropanol in the range of 12% w/w greatly accelerates the rate-of-kill of mycobacteria by 2% w/w glutaraldehyde. (Glut = glutaraldehyde, IPA = Isopropanol)

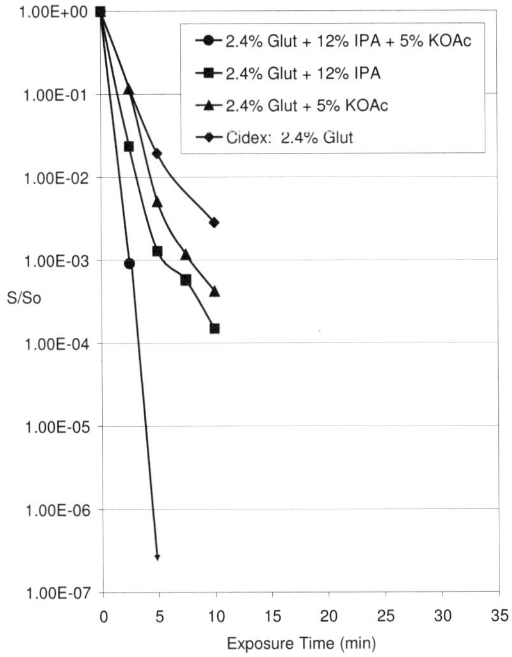

Fig. 2 shows that 5% w/w potassium acetate also accelerates the rate-of-kill of mycobacteria by glutaraldehyde, and the combination of 2.4% w/w alkaline glutaraldehyde with 12% w/w isopropanol and 5% w/w potassium acetate kills more than 6 $\log_{10}$ of mycobacteria within 5.0 min at 20ºC. (Glut = glutaraldehyde, IPA = Isopropanol, KOAc = potassium acetate)

The rate-of-kill of mycobacteria is greatly affected by whether or not the mycobacteria are dried onto an inert surface. In the USA the standard tests for determining the kill of mycobacteria, and thus the label exposure time claim, are usually done with the mycobacteria in suspension. Figure 3 and 4 show the kill of atypical mycobacteria when the mycobacteria are dried onto a braided polyester suture surface.

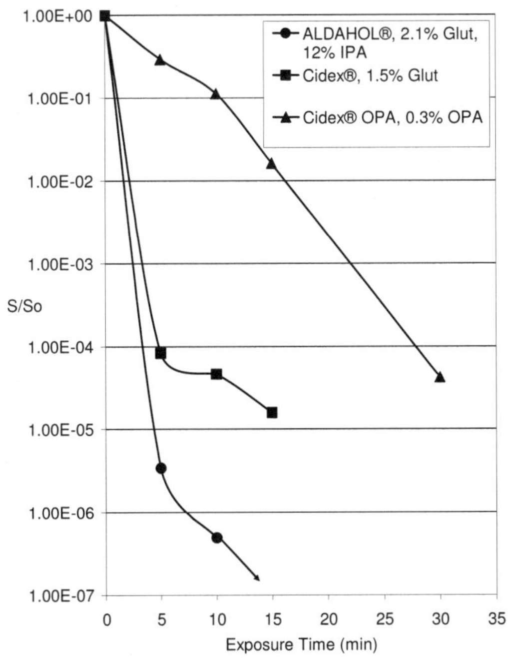

3)

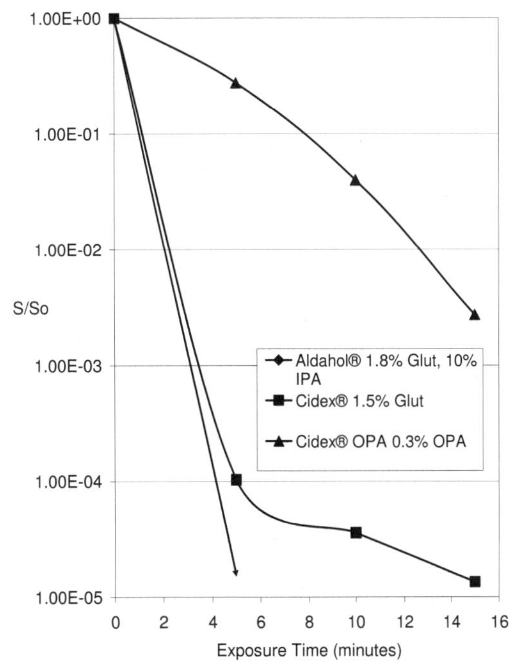

4)

Figs. 3 and 4 show that the kill of *Mycobacterium chelonae* (3) and *M. massiliense* (4) by ortho-phthalaldehyde is inhibited when the mycobacteria are dried onto an inert surface such as braided polyester suture material. This inhibition was not seen for Aldahol. (Glut = glutaraldehyde, IPA = Isopropanol, KOAc = potassium acetate, OPA = ortho-phthalaldehyde). Figure 3 test temperatures = Aldahol and Cidex OPA at 20°C and Cidex at 25°C. Figure 4 test temperature = 25°C for all three test chemistries.

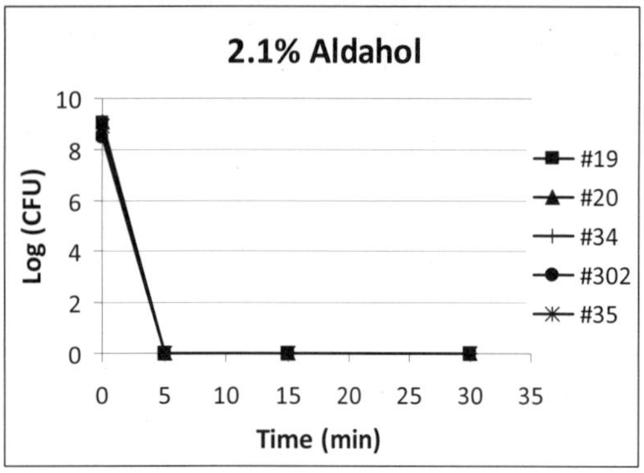

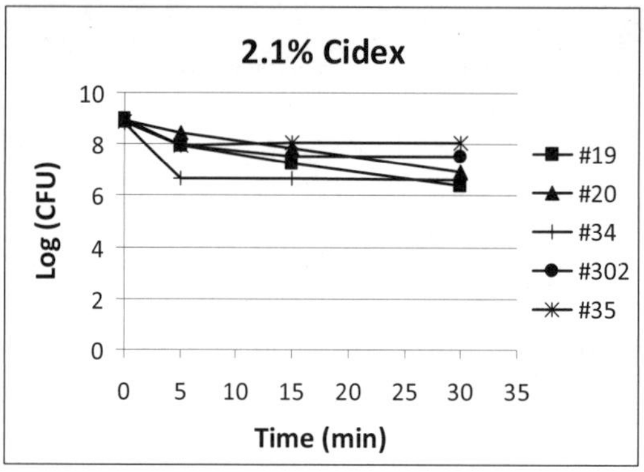

Fig. 5 In tests done by Dr. Mary Jackson of Colorado State University, and reproduced here by permission, a solution of 2.1% w/w alkaline glutaraldehyde in combination with 12% w/w isopropanol and 5% w/w potassium acetate (2.1% Aldahol (a)) killed more than 8 $\log_{10}$ of glutaraldehyde-resistant *M. massiliense* in suspension within 5.0 min at 25°C while Cidex Solution (b) did not kill the glutaraldehyde resistant *M. massiliense* even in 30 minutes. The numbers on the right of the graphs identify various strains of *M. massiliense*.

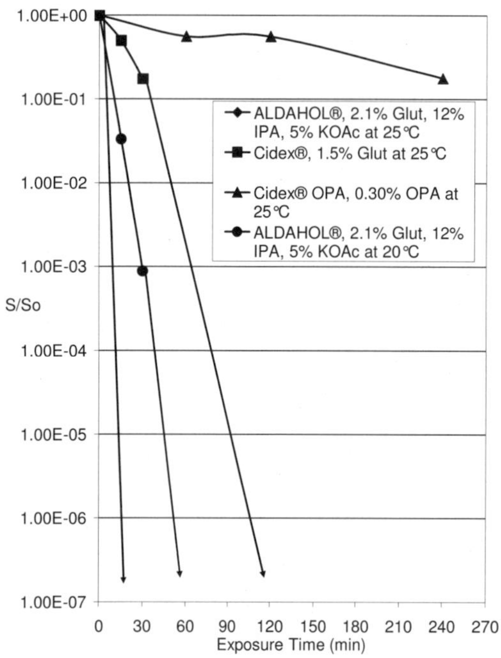

Fig. 6 compares the rate-of-kill of spores of *Bacillus subtilis* by Aldahol High Level Disinfectant (2.1% w/w alkaline glutaraldehyde in combination with 12% w/w isopropanol and 5.0% potassium acetate) at 20°C and 25°C, Cidex Solution (1.5% w/w alkaline glutaraldehyde) at 25°C, and an OPA solution at 0.3% w/w OPA at 25°C. (Glut = glutaraldehyde, IPA = Isopropanol, OPA = ortho-phthalaldehyde)

Small sections of colonoscope insertion tubes (1.0 cm x 2.0 cm) as cut from the full insertion tube and various types of bending rubbers, were soaked in full-strength Aldahol High Level Disinfectant (3.4% w/w glutaraldehyde + 20% w/w isopropanol + 5% w/w potassium acetate), Cidex Solution (2.4% w/w alkaline glutaraldehyde), and Cidex OPA Solution (0.55% w/w OPA) for 10.0 min at 20°C, and then triple rinsed at least three times (a one million-fold dilution) in 100 ml of rinse water per rinse. Each 1.0 x 2.0 cm piece of insertion tube and bending rubber was then placed on a lawn of *Staphylococcus aureus* on nutrient agar and incubated for 24 to 48 hrs at 35°C.

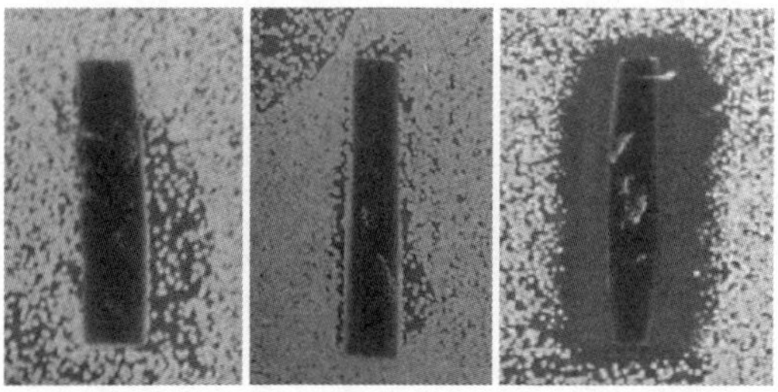

Fig. 7 shows the zone of inhibition (ZOI) surrounding the piece of Viton bending rubber as soaked in OPA, where the OPA has not rinsed away from the bending rubber surface, and remains in enough strength to inhibit the growth of *S. aureus* in the ZOI. The rinse procedure did rinse away the Cidex Solution and the Aldahol High Level Disinfectant. (From left to right, Aldahol – no zone, Cidex – no zone, Cidex OPA - ZOI).

In summary, by combining alkaline glutaraldehyde with low concentrations of isopropanol, and potassium acetate salts, Aldahol High Level Disinfectant was developed that killed at least 6 $\log_{10}$ of mycobacteria in suspension or as dried onto an inert surface within 10.0 min at 20°C or 25°C, and killed spores of *B. subtilis* in suspension within 60 min at 20°C or within 30 min at 25°C [10]. This formulation is robust and may be used and re-used for 14 days or until the glutaraldehyde concentration declines to 2.1% w/w. Aldahol is completely soluble in water, and rinses easily from endoscope surfaces. Careful attention must be given to protect medical personnel from the vapors of glutaraldehyde.

References

[1] Duarte,RS, Lorenco,MCS, de Souza Fonseca, LS, et.al. An epidemic of postsurgical infections caused by *Mycobacterium massiliense* . *J. Clin. Microbiol.* 2004; 42:5493-5501.

[2] Pajkos,A, Vickery, K, and Cossart,Y. Is biofilm accumulation on endoscope tubing a contributor to the failure of cleaning and decontamination? *J. Hosp. Infec.* 2004; 58:224-229.

[3] Bisset,L, Cossart,YE, Selby,W, et.al. A prospective study of the efficacy of routine decontamination for gastrointestinal endoscopes and the risk factors for failure. *Am. J. Infect. Control.* 2006; 34:274-280.

[4] Joshi,SN, Rosenfeld,S. Two cases of ortho-phthalaldehyde induced allergic reactions in patients undergoing surveillance cystoscopy. *J. Allergy Clin. Immunol.* 2004; 113:S311.

[5] Sokol,WN. Nine episodes of anaphylaxis following cystoscopy caused by Cidex OPA (ortho-phthalaldeyde) high-level disinfectant in four patients after cystoscopy. *J. Allergy Clin. Immunol.* 2004; 114:392-397.

[6] Cooper,DE, White,AA, Werkema,AN, Auge,BK. Anaphylaxis following cystoscopy with equipment sterilized with Cidex OPA (ortho-phthalaldehyde): a review of two cases. *J. Endourol.* 2008; 9:2181-2184.

[7] Streckenbach, SC, Alston, TA. Perioral stains after ortho-phthalaldehyde disinfection of ECHO probes. *Anesthesiology.* 2003; 99:1032.

[8] Fusasaki,T., Nishahara,T., and Iwakoshi,K. A case of anaphylaxis following colonoscopy caused by OPA (ortho-phthalaldehyde). Jap. *J. Occup. Med. and Trauma.* 2007; 55:201-205.

[9] Suzukawa,M, Komiya,A, Koketsu,R, et.al. Three cases of ortho-phthalaldehyde-induced anaphylaxis after laryngoscopy: detection of specific IgE in serum. *Allergology Int'l.* 2007; 56:313-316.

[10] Miner,N., Harris, V., Cao,TD, et.al. Aldahol high level disinfectant. *Am. J. Infect. Control.* 2010; 38:205-211.

Towards the rational design of antimicrobial peptides: Recent developments in computational tools

M. Torrent[*,1,2], **D. Andreu**[1], **M.V. Nogués**[2] **and E. Boix**[2]

[1] Department of Experimental and Health Sciences, Universitat Pompeu Fabra, Barcelona Biomedical Research Park (PRBB), Dr. Aiguader 80, 08003, Barcelona, Spain.
[2] Department of Biochemistry and Molecular Biology, Universitat Autònoma de Barcelona, Biosciences Faculty, 08193, Cerdanyola del Vallès, Spain.

*Corresponding author: e-mail: marc.torrent@upf.edu; marc.torrent@uab.cat, Phone: +34 935811707

Host defence antimicrobial proteins and peptides (AMPs) are effectors of the innate immune system and play a vital role in the prevention of bacterial infections. Some computational methods have been developed in order to find AMPs with potential application in the pharmacological industry.
On one hand, we have designed a protein scanning system that is able to perform a fast screening analysis over large protein sets in order to identify potential active peptides. The method differs from previously reported predictive algorithms as it is based on high throughput screening experimental results obtained from a synthetic peptide library. On the other hand, we have analyzed AMP databases and showed that by only considering few peptide descriptors it is possible to successfully predict active compounds and assess their antimicrobial activity. This system, based only in sequence-derived parameters, is a powerful tool to discover and optimize new AMPs.
In summary, the currently available computational tools can successfully predict antimicrobial sequences in both proteins and peptides. Notwithstanding, more research is needed in order to efficiently estimate their antimicrobial activity.

Keywords: antimicrobial peptides; bioinformatics; artificial neural networks; peptide drug design.

1. A new generation of antibiotics: the antimicrobial peptides

Antibiotics are defined as chemical substances produced by living organisms that are detrimental to microorganisms. Antibiotics emerged in the mid of 19th century with the discovery of sulfonamides and penicillins. Following, new antibiotics were discovered, such as macrolides, tetracyclins and cephalosporins, which were proved effective in the treatment of bacterial infections [1]. Unfortunately, the widespread use and abuse of antibiotics have caused a clinical warning due to the emergence of resistant microorganisms, mainly in hospitals, such as the resistant *Streptococus pneumonia* strains found in Europe and Asia [2]

Most antibiotics used for the treatment of bacterial infections can be classified in four major classes, attending to its mechanism of action: (1) cell wall synthesis disruption, (2) protein synthesis inhibition, (3) nucleic acid synthesis disturbance and (4) metabolic pathway inhibition (Table 1). The mechanisms of resistance adopted by bacteria include: (a) the development of new genes encoding enzymes, such as β-lactamases, that destroy antibiotics before they can promote damage in the cell wall, (b) the exclusion of antibiotics from the cell before they can reach their active site by efflux pumps, (c) alterations in some metabolic pathways that lead to altered bacterial cell walls that remove the antibiotic binding site and (d) downregulation of porin genes with the aim to constraint the access of antibiotics to the intracellular targets [3-6].

Antimicrobial peptides (AMPs) have been studied since 1981 when Hans Boman and coworkers discovered the cecropins [7, 8]. Although the mechanism of action of antimicrobial peptides can be diverse, most AMPs share general traits. Mainly, these peptides can interact with the membrane and the cell wall of bacteria where they exert their action. AMPs can bind to the negatively charged phospholipids of the membrane and insert, partially or totally, in order to promote the bacteria cell depolarization that eventually conclude in the cell death [9-12].

The development of resistance mechanisms against AMPs are, however, more intricate than antibiotics because they entail changes in the overall cell structure. Mainly, these mechanisms imply: (a) degradation by proteases, (b) cell wall chemical modification and (c) modification of the cell membrane phospholipid composition (Figure 1) [12, 13]. Despite the existence of several resilient strains, it is not easy to induce AMP resistance in the laboratory. As an example, protegrins induce low resistance in bacteria after multiple passages at a half minimal inhibitory concentration (MIC). Some systems have been associated to resistance against AMPs, as the PhoPQ or the PmrAB systems, but there is still a lack of evidence in this issue [14-16].

In this context, AMPs have been proposed as promising candidates in the fight of bacterial infections because they generally display a broad activity spectrum and, more interestingly, very few AMPs are associated with intrinsically resistant bacteria [17]. Thus, many efforts have been focussed in the development of AMP templates that can help in the treatment of bacterial infections resistant to classical antibiotics. In addition, other potential applications have been described for AMPs. Some of them include the modulation of the immune response and the use of AMPs as adjuvants together with classical antibiotics. These strategies are currently under evaluation and regarded as potential effective treatments [18, 19].

For all these purposes a large amount of AMP candidates have to be considered and tested. In this context, bioinformatics may play an important role in order to filter and select the most promising templates. An essential step in this effort requires an organized and freely available repository of AMPs containing its sequences, properties and biological effects, together with supporting tools able to analyze specific peptide properties. There is a growing need for versatile computational tools for the prediction of new AMPs, with the aim to design novel potent peptides against bacterial pathogens [20-22]. Notwithstanding, the low sequence homology displayed by AMPs and the non-conserved structure patterns makes the prediction of peptides with antimicrobial activity a difficult task.

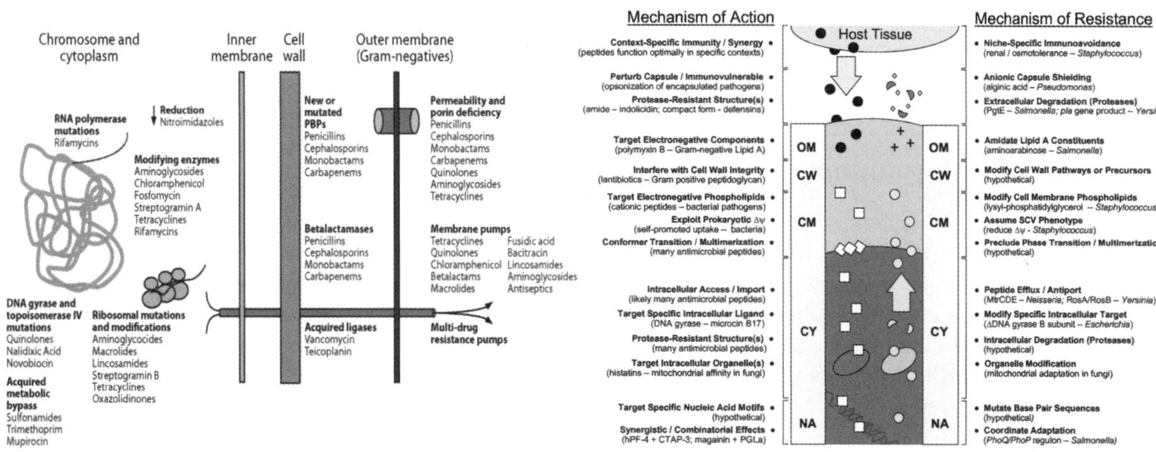

Fig. 1 *Antimicrobial mechanisms of action and resistance*. Comparison between the antimicrobial mechanism of action and resistance of antibiotics (A) and AMPs (B). Some similarities can be observed as the extracellular degradation of antimicrobials and the reinforcement of the cell wall. However, AMPs resistance mechanisms mainly introduce structural changes rather than blocking strategies. Reproduced with permission from [6] (A) and [12] (B).

2. Scanning of protein sequences

Proteins that belong to the innate immune system are a good source of AMPs, and represent therefore promising targets for the development of new active peptides. We have designed a method able to scan protein sequences in order to identify the antimicrobial domains that direct the protein function [23]. This system assigns an antimicrobial index to each amino acid based on high-throughput results obtained previously. Then, a sliding window system scans the protein sequence and builds up an antimicrobial profile where defined-regions below a threshold are identified as antimicrobial domains. The algorithm uses a 7-residue window. For a protein to be predicted as antimicrobial, the algorithm must detect at least one antimicrobial region, defined as a sequence of >12 residues with an over-threshold mean propensity value (PV) with at most 2 intervening residues with under-threshold PVs. The results show that this method can identify not only the antimicrobial proteins but also predict antimicrobial domains with an overall accuracy of 85% (Fig. 2).

This approach has been successfully applied to the Eosinophil Cationic Protein [24]. The sequence of this protein has been analyzed by our method and the region detected has been synthesized. The results obtained show that the derived peptide displays equipotent activity if compared with the whole protein (Fig. 2).

Thus, this method can give a first approach in spotting the key active regions of bactericidal proteins. The selected regions may provide a useful starting point in the development of new antimicrobial peptide derived drugs.

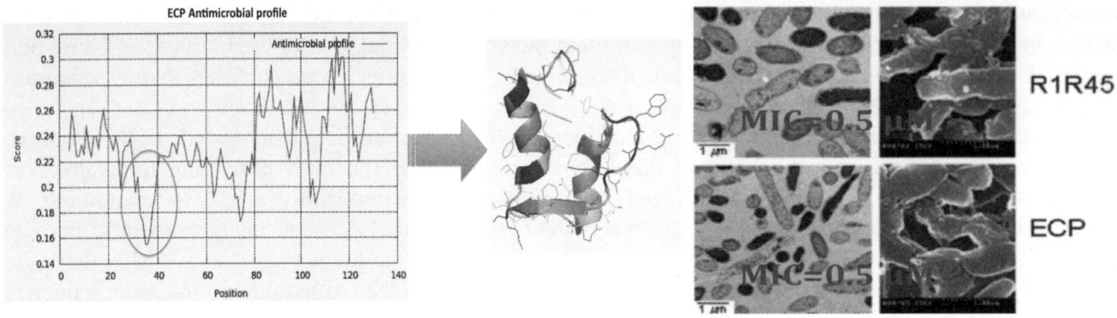

Fig. 2 *Scanning proteins for antimicrobial activity.* A sliding window system scans the protein sequence in order to find antimicrobial domains. An average propensity value (AvPV) is assigned to each amino acid to build up an antimicrobial profile. 12-residue minimum length sequences below the threshold are considered as putative antimicrobial domains. The figure shows the antimicrobial profile of Eosinophil Cationic Protein (ECP) and the peptide designed encompassing the detected antimicrobial domain (R1R45). Antimicrobial activity, calculated as Minimal Inhibitory Concentration (MIC) value and electron micrographs reported for the corresponding synthetic peptide show its equipotency when compared with the whole protein. Adapted with permission from [24].

3. Predicting antimicrobial peptides

By using computational techniques we can identify and obtain new antimicrobial peptide templates and derive optimization processes in order to improve activity. To develop this methodology a first learning phase step is required in which peptide descriptors combined with current antimicrobial peptide databases are used to train the system and derive correlations between the descriptors and the desired biological activities. The most used systems in antimicrobial peptide analysis are artificial neural networks (ANNs) and support vector machines (SVMs). Once a significant accuracy has been achieved, data can be validated experimentally and used later to find drugs with improved activity [25].

With this goal we have developed a prediction system based on the antimicrobial mechanism of action of peptides. We have assigned to each step of the model a single descriptor and then computed the corresponding parameters for over 1000 of experimentally tested peptides reported in the CAMP database [26]. The descriptors used comprise: isoelectric point (pI), chain length, *in-vivo* aggregation, *in-vitro* aggregation, peptide structure and hydrophobicity. By using an artificial neural network we have tested these descriptors and assessed the suitability of these classifiers to categorize the peptides as antimicrobials or non-antimicrobials.

The receiver-operating characteristics (ROC) plots indicate that good results are obtained by this approach, which displays high specificity and selectivity and an overall accuracy of 90%. Moreover, we have tested if our descriptors are able to correlate with antimicrobial activity. Thus we have used external peptide databases with experimentally determined activity and shown that our descriptors correlated well with antimicrobial activity.

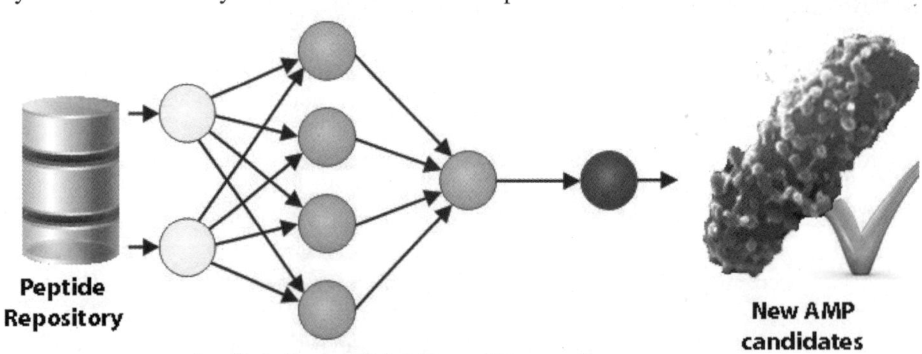

Fig. 3 *Detection and optimization of new peptide templates.* Using ANNs a huge amount of information contained in the nowadays-available AMPs repositories can be scanned. By using only sequence-based descriptors a prediction system can be constructed and new templates generated "*in-silico*". The combination of "*in-silico*" prediction systems with high-throughput experimental techniques can be a valuable methodology in the discovery of highly active new AMPs.

4. Concluding remarks

Although the antimicrobial prediction systems described above are useful tools to derive new AMPs, more information is still required in order to develop acceptable candidates for peptide drug design. Peptides with a low cytotoxicity and improved pharmacokinetics and pharmacodynamics are necessary [27]. Thus, antimicrobial peptide research has to overcome some barriers in order to be broadly applicable in the pharmaceutical industry. However, from an optimistic point of view, some antimicrobial peptides are now under development or in preclinical and clinical trials and some peptides for topical use are currently commercialized [10].

The bioinformatics strategies presented here can help in the development of ncw pcptidc-based drugs that can be useful in the treatment of bacterial infections, in particular in those infections produced by bacteria resistant to classical antibiotics.

Acknowledgements M.T. is the recipient of a postdoctoral grant from Alianza Cuatro Universidades (Spain). Work supported by the European Union (HEALTH-F3-2008-223414), the Spanish Ministry of Science and Innovation (BIO2008-04487-CO3-02, BFU2009- 09371) and the Generalitat de Catalunya (SGR2009-494, SGR2009-795).

References

[1] Khardori N. Antibiotics--past, present, and future. *Med Clin North Am.* 2006, 90:1049-1076

[2] Linares J, Ardanuy C, Pallares R, Fenoll A. Changes in antimicrobial resistance, serotypes and genotypes in Streptococcus pneumoniae over a 30-year period. *Clin Microbiol Infect.* 2010, 16:402-410

[3] Savjani JK, Gajjar AK, Savjani KT, Mechanisms of resistance: useful tool to design antibacterial agents for drug - resistant bacteria. *Mini Rev Med Chem.* 2009, 9:194-205

[4] Perez-Llarena FJ, Bou G, Beta-lactamase inhibitors: the story so far. *Curr Med Chem.* 2009, 16:3740-65

[5] Martinez M, Silley P. Antimicrobial drug resistance. *Handb Exp Pharmacol.* 2010, 199:227-264

[6] Sundsfjord A, Simonsen GS, Haldorsen BC, Haaheim H, Hjelmevoll SO, Littauer P, Dahl KH. Genetic methods for detection of antimicrobial resistance. *APMIS.* 2004, 112:815-837

[7] Boman HG, Steiner H. Humoral immunity in Cecropia pupae. *Curr Top Microbiol Immunol.* 1981, 94-95: 75-91

[8] Steiner H, Hultmark D, Engstrom A, Bennich H, Boman HG. Sequence and specificity of two antibacterial proteins involved in insect immunity. *Nature.* 1981, 292:246-248

[9] Kindrachuk J, Napper S. Structure-activity relationships of multifunctional host defence peptides. *Mini Rev Med Chem.* 2010, 10:596-614

[10] Guani-Guerra E, Santos-Mendoza T, Lugo-Reyes SO, Teran LM. Antimicrobial peptides: general overview and clinical implications in human health and disease. 2010, *Clin Immunol.* 135:1-11

[11] Yeaman MR, Yount NY. Unifying themes in host defence effector polypeptides *Nat Rev Microbiol.* 2007, 5:727-740

[12] Yeaman MR, Yount NY. Mechanisms of antimicrobial peptide action and resistance. *Pharmacol Rev.* 2003, 55:27-55

[13] Peschel A. How do bacteria resist human antimicrobial peptides? *Trends Microbiol.* 2002, 10:179-186

[14] Devine DA, Hancock, RE. Cationic peptides: distribution and mechanisms of resistance. *Curr Pharm Des.* 2002, 8:703-714

[15] Prost LR, Miller SI. The Salmonellae PhoQ sensor: mechanisms of detection of phagosome signals. *Cell Microbiol.* 2008, 10:576-582

[16] Gunn JS. The Salmonella PmrAB regulon: lipopolysaccharide modifications, antimicrobial peptide resistance and more. *Trends Microbiol.* 2008, 16:284-290

[17] Harder J, Glaser R, Schroder JM. The role and potential therapeutical applications of antimicrobial proteins in infectious and inflammatory diseases. *Endocr Metab Immune Disord Drug Targets.* 2007, 7:75-82

[18] Marr AK, Gooderham WJ, Hancock RE. Antibacterial peptides for therapeutic use: obstacles and realistic outlook. *Curr Opin Pharmacol.* 2006, 6:468-472

[19] Bommarius B, Kalman D. Antimicrobial and host defense peptides for therapeutic use against multidrug-resistant pathogens: new hope on the horizon. *IDrugs.* 2009, 12:376-380

[20] Juretic D, Vukicevic D, Ilic N, Antcheva N, Tossi A. Computational design of highly selective antimicrobial peptides. *J Chem Inf Model.* 2009, 49:2873-2882

[21] Lata S, Mishra N K, Raghava G P. AntiBP2: improved version of antibacterial peptide prediction *BMC Bioinformatics. 2010,* 11 Suppl 1, S19

[22] Hilpert K, Fjell CD, Cherkasov A. Short linear cationic antimicrobial peptides: screening, optimizing, and prediction. *Methods Mol Biol.* 2008, 494:127-159

[23] Torrent M, Nogues VM, Boix E. A theoretical approach to spot active regions in antimicrobial proteins. *BMC Bioinformatics.* 2009, 10:373-379

[24] Torrent M, de la Torre BG, Nogues VM, Andreu D, Boix E. Bactericidal and membrane disruption activities of the eosinophil cationic protein are largely retained in an N-terminal fragment. *Biochem J.* 2009, 421:425-434

[25] Fjell CD, Jenssen H, Hilpert K, Cheung WA, Pante N, Hancock RE, Cherkasov A. Identification of novel antibacterial peptides by chemoinformatics and machine learning. *J Med Chem.* 2009, 52:2006-2015

[26] Thomas S, Karnik S, Barai RS, Jayaraman VK, Idicula-Thomas S. CAMP: a useful resource for research on antimicrobial peptides *Nucleic Acids Res.* 2010, 38:D774-D780

[27] Czock D, Markert C, Hartman B, Keller F. Pharmacokinetics and pharmacodynamics of antimicrobial drugs. *Expert Opin Drug Metab Toxicol.* 2009, 5:475-487

Antimicrobial Microbes and Viruses.

Biosynthesis of Antibiotics.

Antimicrobial properties of *Lactobacillus plantarum* Tensia (DSM 21380) and Inducia (DSM 21379)

M. Rätsep [*] [1,2], P. Hütt [1,2], R. Avi [2], M. Utt [3] and E. Songisepp [1]

[1] Bio-Competence Centre of Healthy Dairy Products LLC, Kreutzwaldi 1, 51014, Tartu, Estonia
[2] Institute of Microbiology, Faculty of Medicine, University of Tartu, 50411, Estonia
[3] Institute of General and Molecular Pathology, Department of Immunology, Faculty of Medicine, University of Tartu, 50411, Estonia

* Corresponding author: e-mail: merlera@ut.ee, Phone: +372 7 374173

The objective of present work was to characterize the antimicrobial properties of probiotic strains *Lactobacillus plantarum* Tensia and Inducia. The strains were screened for antimicrobial activity against *L. monocytogenes* assessed in different experimental conditions, for putative bacteriocin production and the presence of plantaricin precursor peptides encoding genes. Heating reduced the antimicrobial activity of the supernatants of both strains indicating the proteinous nature of the antimicrobial compounds. Tensia`s supernatant kept its activity in wide pH range, referring also to the possibility, that the strain may produce other antimicrobial substances than organic acids. Besides, the strain harbored all the plantaricin precursor peptides EF and JK encoding genes while Inducia had only *pln*E and *pln*F. As Tensia`s supernatant had stronger antimicrobial activity and kept its antimicrobial activity in different experimental conditions, it refers to the possibility, that Tensia may produce both plantaricin EF and JK which belong to class II bacteriocins. Weaker performance of Inducia in the same experimental conditions indicates that Inducia may produce only plantaricin EF.

Keywords: *Lactobacillus plantarum*; antimicrobial properties; bacteriocins.

Foodborne illnesses associated with *Campylobacter jejuni*, *Clostridium perfringens*, *Listeria monocytogenes*, *Salmonella* spp and *Staphylococcus aureus* are still problem worldwide. The consumption of more food that has been formulated with chemical preservatives has also increased consumer concern and created a demand for more "natural" and "minimally processed" food. As a result, there has been a great interest in naturally produced antimicrobial agents.

Bacterial fermentation of perishable raw materials has been used for centuries to preserve the nutritive value of food and beverages over an extended shelf-life. Use of lactic acid bacteria (LAB) and their metabolites (organic acids, hydrogen peroxide, diacetyl etc.) are the acting antimicrobial components in foodstuff. Most important step in preserving food by fermentation is to convert sugars to organic acid and by acidification of food environment preventing spoilage organism to grow.

Since food safety has become an increasingly important international concern, the application of antimicrobial peptides (e.g. bacteriocins) from LAB that target food pathogens without toxic or other adverse effects has received great attention.

Bacteriocins of Gram-positive bacteria can be classified into four broad groups: (1) lantibiotics, (2) small non-modified peptides, (3) large proteins, and (4) cyclic peptides [1]. *Lactobacillus plantarum* strains are producing several antimicrobial peptides which could be called as plantaricins. Most plantaricins described today belonging to class II according to classification by Heng 2007. Plantaricins are divided into subgroups according to structure and characteristics of antimicrobial action.

The objective of this work was to detect and characterize the antimicrobial properties of novel probiotic strains *Lactobacillus plantarum* Tensia and Inducia.

1. Material and methods

1.1 Bacterial strains and culture conditions

1.1.1 Lactobacilli

The *Lactobacillus plantarum* strains Tensia (DSM 21380) and Inducia (DSM 21379) have been isolated from healthy Estonian children [2,3]. Both strains belong to the Bio-Competence Centre of Healthy Dairy Products LLC (www.tptak.ee).

The strains were stored as stocks at -80°C in skimmed milk and propagated twice in Man-Rogosa-Sharpe (MRS, Oxoid, United Kingdom) broth before use.

1.1.2 Pathogens

The lactobacillus strains were tested against *Listeria monocytogenes* ATCC 51774 and ATCC 13932. Pathogens were propagated twice in Brain Heart Infusion (BHI, Oxoid) broth and before use grown on blood agar.

1.2 Antimicrobial activity

The *L. plantarum* strains were screened for antimicrobial activity and putative bacteriocin production. The antimicrobial properties were assessed in different experimental conditions with agar spot diffusion test [4,5].

The agar drop diffusion test was carried out with supernatant of *L. plantarum* strains. Aliquot of 5 ml of supernatant was adjusted with 6N NaOH to pH 6.0±0.15 and the other 5 ml remained as original (pH 3.65-3.85), microbial cells were removed by filtration through a 0.22-μm-pore-size filter (Orange Scientific GyroDisc Syr. CA-PC 30 mm, Belgium). Lawns of each indicator strain were prepared by inoculating ca. 10^6 cells into 4 ml of soft-overlay of BHI (0.7% agar) poured onto the surface of BHI agar plates (1.5% agar); thereafter the plates were allowed to dry for 15 min in a drying hood. Five microliters of the lactobacilli supernatants was spotted onto the indicator lawns, and then the plates were allowed to dry for 10 min. The plates were incubated at 37°C for 1 day and they were examined for growth inhibition zones. The indicator strains used were *L. monocytogenes* ATCC 51774 and *L. monocytogenes* ATCC 13932. Assessment criteria for pathogen growth inhibition zone (mm) were given as follows: weak ≤0.5; medium 0.5-5.7; strong ≥ 5.

1.2.1 The impact of growth medium and incubation temperature on supernatant antimicrobial activity

Overnight cultures of lactobacilli strains were inoculated (1 %) into MRS broth and standardized cheese-milk. The incubation was performed at 15°C for 30 days (as a cheese model), at 30°C and 37°C for 20-22 hours without agitation. Culture supernatants were used in parallel as neutralized (pH 6.0±0.15) and original (pH 3.65-3.85) samples.

1.2.2 Effect of pH and temperature on supernatants antimicrobial activity

Cell-free supernatants of Tensia and Inducia were adjusted to pH 6.0. Supernatant was heated 10 minutes at different temperatures (60°C, 70°C, 80°C, 90°C, and 100°C).
To assess the effect of pH on antimicrobial activity, the pH of the supernatant was adjusted to different values from pH 4.0 to 10.0.

1.3 Design of primers

The *L. plantarum* strains Tensia and Inducia were tested for presence of bacteriocin precursor peptide encoding genes in genomic DNA. *L. plantarum* WCFS1 was used as a positive control [6].
Forward and reverse RT-PCR primers (Table 1) were designed based on sequences of genes *plnE*, *plnF*, *plnJ* and *plnK* from *Lactobacillus plantarum* WCFS1 genome using Primer Express software http://www.niehs.nih.gov/training/manuals/primerex.

Genomic DNA from 24h old lactobacilli cultures were isolated with QIAamp Mini Kit according to producer protocol (QIAGEN GmbH, Germany) and stored at -20°C.

PCR was carried out as follows: final volume of reaction mixture 50 μl, contained 1.5 mM Mg^{2+} (Fermentas, Lithuania), 1x reaction buffer (Fermentas), 0.2 mM dNTP (Deoxynucleoside triphosphates) (Fermentas, Lithuania), 1 μM both (*forward* and *reverse*) primers (DNA Technology A/S, Denmark), 1 U Taq polymerase (Fermentas, Lithuania) and 2 μl DNA. Termocycler Eppendorf Master Cycler Gradient was used for PCR reaction. Termocycler program is presented in Table 2.

Presence of detectable genes were determined by anlayzing PCR products on 2% agarose gel in UV light after gel electrophoresis and staining DNA with ethidium bromide.

Table 1. Sequence of plantaricin precursor peptides encoding genes.

Tested gene	Primer	Sequence
plnE	E 1F	5` - ATCGGCAGGCCCAACAG - 3`
	E 1R	3` - GCGTGACCGTGAATTAAATGC – 5`
plnF	F 1F	5` - GACAGCGCTAATGACCCAATC - 3`
	F 1R	3` - GCGTGACCGTGAATTAAATGC – 5`
plnJ	J 1F	5` - CCCAACAACCCCGTTCAAC - 3`
	J 1R	3` - TGCTACAGTAAAGGGAGAGGTATGAC – 5`
plnK	K 2F	5` - ACTTATTATAATCCCTTGAACCACCAA - 3`
	K 2R	3` - GCTGACGCTGAAAAGAATATTTCT – 5`

Table 2. PCR termocycler program.

1. Denaturation	95°C 5 min.
2. Denaturation	95°C 30 sec.
3. Annealing	54°C 30 sec.
4. Polymerisation	72°C 35 sec.
5. Repeat from 2x	44x
6. Extra polymerisation	72°C 10 min.

2. Results and discussion

2.1 Supernatant's antimicrobial activity

The antimicrobial activity of *L. plantarum* Inducia appeared to be weak at tested conditions. The neutralised (pH 6.0±0.15) supernatant of the strain Tensia sustained its antilisterial activity at both temperatures, though significantly weaker than that of the original (pH 3.65-3.85) supernatant (Table 3). This could be explained by synergistic action of organic acids and putative bacteriocin [7-9]. As neutralisation excludes the antimicrobial effect of organic acids produced by the strain, the antimicrobial activity of neutralised supernatant of Tensia could be caused by the presence of other antimicrobial compounds than organic acids.

Table 3. Impact of lactobacilli incubation temperature (grown in MRS broth) and supernatant`s neutralisation to antimicrobial activity against indicator organism *Listeria monocytogenes* shown in growth inhibition zone diameter (mm).

Strain	Incubation temperature			
	30°C		37°C	
	original pH	neutral pH	original pH	neutral pH
Tensia	5.1±1.4[#]	3.7±2.6[#]	5.4±0.5[§]	4.2±2.0[§]
Inducia	1.1±2.1	1.1±2.1	1.4±2.2	1.4±2.3

Pathogen growth inhibition diameter (mm): weak ≤0.5; medium 0.5-5.7; strong ≥ 5.7; [#]p=0.03; [§]p=0.04

2.2 Impact of low incubation temperature on antimicrobial activity

It's known that stress situation (e.g. suboptimal growth conditions such as low temperature, deficit of nutrients etc.) increases bacteriocin synthesis. At lower temperatures the metabolism of microbes slows down meaning lower production of antimicrobial organic acids. In order to protect its ecological niche, microbes start to produce other antimicrobial compounds incl. bacteriocins [8,9]. Though facultatively heterofermentative species *Lactobacillus plantarum* is able to grow at 15°C, although temperature is still suboptimal for the species.

Total antilisterial activity of the supernatant from lactobacilli grown in MRS broth at low temperature (15°C) was found to be significantly higher, compared to growth in milk at the same temperture ($p<0.01$).

Tensia supernatant's antilisterial activity in MRS broth and in milk strengthened significantly towards the end of incubation period (Figure 1).

The same tendency was noted by cultivation of Inducia in MRS broth. The antilisterial activity of the supernatant of this strain grown in milk occurred on the twentieth day of incubation. However, on thirtieth day of incubation no antilisterial activity was registered. This phenomenon could be explained by the decomposition of the antimicrobial compounds.

Type and concentration of carbon and nitrogen sources present in the growth environment, as well as NaCl are, in general, reported as the most influencing factors in bacteriocin production [10]. MRS medium is an optimal medium for cultivation of lactobacilli as well as for bacteriocin synthesis. In milk nutrient contents and availability is poorer than in MRS. Some of LAB are able to grow and produce bacteriocins in milk, however others need induction through co-incubation with starter cultures, to get necessary nutrients in suitable form [11]. Our results show, that strains Tensia and Inducia are able to reorganise their metabolism, grow and produce antimicrobial compounds in poorer nutrient environment.

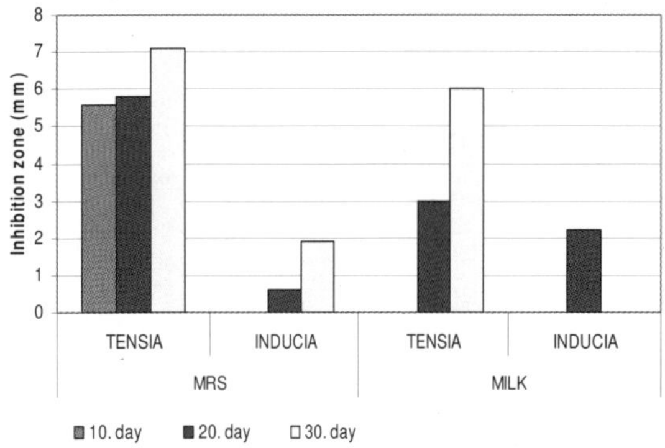

Figure 1. Impact of low incubation temperature (15°C) to antimicrobial activity of *L. plantarum* Tensia and Inducia against indicator organism *L. monocytogenes* shown in growth inhibition zone diameter (mm).

2.3 Effect of pH and temperature on supernatants antimicrobial activity

Heating reduced the antimicrobial activity of the supernatants of Tensia and Inducia (Table 4). 10-minute heating at 100°C abolished the antimicrobial activity of Tensia's supernatant, whereas Inducia`s supernatant lost its antimicrobial activity already at 80°C. These results are showing proteinous nature of antimicrobial compounds produced by Tensia.

Table 4. Impact of heating on antilisterial activity of lactobacilli supernatants shown in growth inhibition zone diameter (mm).

Heating temperature	Tensia	Inducia
60°C	3.9±1.5	1.3±1.1
70°C	3.8±0.8	1.1±0.9
80°C	4.0±1.5	0
90°C	3.2±2.5	0
≥100°C	0	0

Pathogen growth inhibition zone diameter (mm): weak ≤0.5; medium 0.5-5.7; strong ≥ 5.7

Increasing the pH value of Tensia`s supernatant to 9 had no impact on its antimicrobial activity. Antimicrobial activity of Tensia`s supernatant decreased significantly at pH 10 ($p<0.001$), indicating that potential cationic antimicrobial peptides are loosing their charge and biological effect.

The fact, that Tensia`s supernatant keeps its activity in wide pH range, refers to the possibility, that Tensia may produce some other antimicrobial substances than organic acids.

2.4 Plantaricin precursor peptides encoding genes

Many *L. plantarum* strains isolated from different environments are known to produce more than one bacteriocin [13]. Bacteriocin production in this species may partly contribute to its success in colonizing a wide variety of niches such as fermenting wine and olives, fermented cheeses, vegetables and sausages, as well as the human saliva and gastrointestinal tract [12-15].

Our results are showing, that the strain Tensia harbored all the plantaricin precursor peptides EF and JK encoding genes and Inducia has *pln*E and *pln*F (Table 5). The results indicate that *L. plantarum* Inducia antimicrobial activity may be somehow weaker due to only one gene pair of plantaricin precursor peptides encoding genes.

Table 5. Presence of amplified PCR products for bacteriocin precursor peptides *plnE, plnF, plnJ* ja *plnK* in *L. plantarum* Tensia and Inducia.

	Bacteriocin genes			
	plnE	*plnF*	*plnJ*	*plnK*
L. plantarum WCFS1	+	+	+	+
L. plantarum Tensia (DSM 21380)	+	+	+	+
L. plantarum Inducia (DSM 21379)	+	+	-	-

3. Conclusion

In conclusion, Tensia`s supernatant had stronger antimicrobial activity and it kept the antimicrobial activity in different experimental conditions, it refers to the possibility, that Tensia may produce both plantaricin EF and JK which belong to class II bacteriocins. Weaker performance of Inducia in the same experimental conditions indicates that Inducia may produce only plantaricin EF.

Acknowledgements: The support EU Structural Funds (PROJECT EU22868; EU27789; EU28662; EU30002) and Estonian Science Foundation grant 5525 is gratefully acknowledged.

References

[1] Heng NCK, Wescombe PA, Burtons JP, Jack RW, Tagg JR. The Diversity of Bacteriocins in Gram-Positive Bacteria. In:Riley MA and Chavan MA, eds. *Bacteriocins: Ecology and Evolution.* Berlin: Springer-Verlag Heidelberg, 2007:45-92.

[2] Mikelsaar M, Annuk H, Shchepetova J, Mändar R, Sepp E, Björkstén B. Intestinal *Lactobacilli* of Estonian and Swedish children. *Microb Ecol Health Dis* 2002;14:75-80.

[3] Annuk H, Shchepetova J, Kullisaar T, Songisepp E, Zilmer M, Mikelsaar M. Characterization of intestinal lactobacilli as putative probiotic candidates. *J. Appl Microbiol* 2003;94:403-412.

[4] Jimenez-Diaz R, Rios-Sanchez RM, Desmazeaud M, Ruiz-Barba JL, Piard JC. Plantaricins S and T, Two New Bacteriocins Produced by *Lactobacillus plantarum* LPCO10 Isolated from a Green Olive Fermentation. *Appl. Environ. Microbiol.* 1993;59:1416-1424.

[5] Todorov SD, Dicks LMT. Effect of Growth on Bacteriocin Production by Lactobacillus plantarum ST194BZ, a Strain Isolated from Boza. *Food Tecnol. Biotecnol.* 2005;43(2);165-173.

[6] Kleerebezem M, Boekhorst J, van Kranenburg R, Molenaar D, Kuipers OP, Leer R, Tarchini R, Peters SA, Sandbrink HM, Fiers MWEJ, Stiekema W, Klein Lankhorst RM, Bron PA, Hoffer SM, Nierop Groot M, Kerkhoven R, de Vries M, Ursing B, de Vos WM, Siezen RJ. Complete genome sequence of *Lactobacillus plantarum* WCFS1. *PNAS.* 2003;100:1990-1995.

[7] Mezaini A, Chihib NE, Bouras AD, Nedjar-Arroume N, Hornez JP. Antibacterial Activity of Some Lactic Acid Bacteria Isolated from an Algerian Dairy Product. *J. of Environ. And Public Health.* 2009.

[8] Neysens P, Messens W, Gevers D, Swings J, De Vuyst L. Biphasic kinetics of growth and bacteriocin production with *Lactobacillus amylovorus* DCE 471 occur under stress conditions. *Microbiol.* 2003;149:1073-1082.

[9] Zamfir M, Grosu-Tudor S. Impact of stress conditions on the growth of *Lactobacillus acidophilus* IBB 801 and production of acidophilin 801. *J. Gen. Appl. Microbiol.* 2009;55:277-282.

[10] Delgado A, Arroyo López FN, Brito D, Peres C, Fevereiro P, Garrido-Fernández A. Optimum bacteriocin production by *Lactobacillus plantarum* 17.2b requires absence of NaCl and apaprently follows a mixed metabolite kinetics. *J. of Biotecnol.* 2007;130:193-201.

[11] Somkuti GA, Steinberg DH. Pediocin production in milk by *Pediococcus acidilactici* in co-culture with *Streptococcus theermophilus* and *Lactobacillus delbrueckii* subsp. bulgaricus. *J. Ind. Microbiol. Biotechnol.* 2010;37:65-69.

[12] Rojo-Bezares B, Sáenz Y, Navarro L, Jiménez-Díaz R, Zarazagga M, Ruiz-Larrea F, Torres C. Characterization of a new organization of the plantaricin locus in the inducible bacteriocin-producing Lactobacillus plantarum J23 of grape must origin. Arch. Microbiol. 2008;189:491-499.

[13] Diep DB, Straume D, Kjos M, Torres C, Nes IF. An overview of the mosaic bacteriocin *pln* loci from *Lactobacillus plantarum. Peptides.* 2009;1562-1574.

[14] Maldonado A, Ruiz-Baraba JL, Jiménez-Díaz R. Purification an Genetic Characterization of Plantaricin NC8, a Novel Coculture-Inducible Two-Peptide Bacteriocin from *Lactobacillus plantarum* NC8. *Appl. and Environmental Microbiol.* 2003;383-389.

[15] Sáenz Y, Rojo-Bezares B, Navarro L, Dí́iez L, Somalo S, Zarazaga M, Ruiz-Larrea F, Torrs C. Genetic diversity of the *pln* locus among oenological *Lactobacillus plantarum* strains. *Int. J of Food Microbiol.* 2009;134:176-183.

Cell Growth Control by tRNase Ribotoxins from Bacteria and Yeast

Eyemen Kheir[1,2], Christian Bär[1], Daniel Jablonowski[1,3] and Raffael Schaffrath[1,3*]

[1] Department of Genetics, University of Leicester, Leicester LE1 7RH, UK
[2] Department of Zoology, Downing Street, University of Cambridge, Cambridge CB2 3EJ, UK
[3] Institut für Biologie, Mikrobiologie, Heinrich-Plett-Str., Universität Kassel, D-34109 Kassel, Germany

*email: rs240@le.ac.uk; phone: +44-116-252-5233 or +49-561-8044-175

Microbial endoribonuclease toxins (ribotoxins) ensure survival of their producers against other microbial competitors. Often, they cleave tRNA anticodons leading to tRNA depletion and cell death. The tRNases PrrC and zymocin from *E. coli* and dairy yeast *Kluyveromyces lactis*, respectively, attack tRNA anticodons that possess specific nucleobase modifications at their wobble position. Intriguingly, these modifications are similar and functionally conserved among prokaryotes and eukaryotes. Therefore, our idea was to take the basic biology of tRNase toxins and apply this to cell systems, including HeLa tumour cells, whose proliferation heavily relies on tRNA functioning for protein synthesis. Our pilot findings indicate that expression of the zymocin tRNase not only inhibits growth of the yeast *Saccharomyces cerevisiae* but also affects the viability of higher eukaryotic cells. Similarly, we observe that *E. coli* tRNase PrrC is lethal when expressed in yeast cells. Hence, microbial tRNase ribotoxins may be invoked as novel anti-proliferative factors for biomedical intervention schemes.

Keywords: *K. lactis*; zymocin; ribotoxin; anticodon nuclease; tRNase; tRNA modification

1. Introduction

Transfer RNAs (tRNAs) are not only essential for messenger RNA (mRNA) translation but also play roles in translation-unrelated processes including, but not limited to, protein degradation, retroviral replication and RNA polymerase switches during chloroplast biogenesis [1]. Moreover, recent studies have shown that tRNAs are also targets for bacterial and fungal endoribonuclease toxins (ribotoxins) which specifically cleave tRNA anticodons and thereby cause cells to cease growth (Fig. 1 A, B, D).

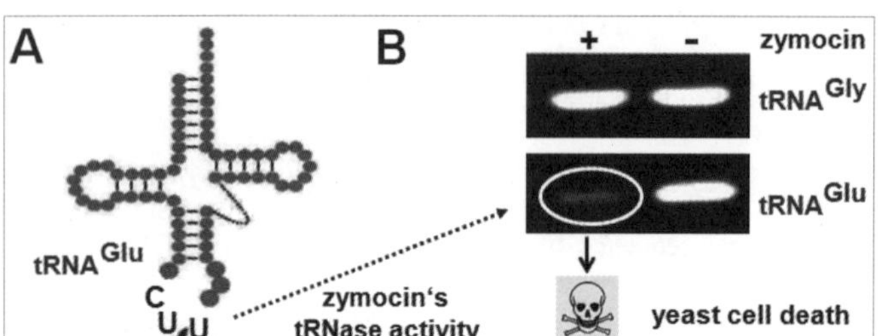

Fig. 1. The zymocin tRNase complex from dairy yeast *K. lactis* kills baker's yeast *S. cerevisiae*. **A**. Zymocin's tRNase subunit (*scissors*) cleaves $tRNA^{Glu}$ carrying a uridine in the anticodon wobble position (CU<u>U</u>). **B**. RT-PCR assays have shown that zymocin treatment leads to selective depletion of $tRNA^{Glu}$ (*circled*) while the $tRNA^{Gly}$ pool remains intact. **C**. The tRNase specificity is determined by wobble uridine modifications in position 2 (*dotted circle*: thiolation) and 5 (*solid circle*: methoxy-carbonylmethyl side chain with methyl group shown as an inner *dotted circle*) of the pyrimidine ring. Protein complexes and their subunits involved in these modifications are labelled by arrows. **D**. Yeast mutants lacking thiolase (*ncs2*Δ, *ncs6*Δ), wobble methylase (*trm9*Δ, *trm112*Δ) or Elongator (*elp1*Δ-*elp3*Δ) function resist tagainst the tRNase (zym^R) whereas a wild-type (wt), zymocin-sensitive (zym^S) strain forms an eclipse-shaped zone of cell death in *K. lactis* killer bioassays.

Therefore, secreted bacterial ribotoxins such as colicins D and E5 as well as multi-subunit yeast tRNase complexes from *Kluyveromyces lactis (Fig. 1 D)* and *Pichia acaciae,* which kill other yeasts including *Saccharomyces cerevisiae,* provide a means to discriminate self from non-self species and ensure survival of their producers against other microbial competitors [2-5]. In addition to these secreted killer toxins, some bacteria maintain endogenous tRNA anticodon nucleases (PrrC; RloC) which usually are dormant, unless being activated under conditions of stress [6-9]. Such programmed tRNA damage is a common bacterial feature and particularly deployed in response towards infection and cell lysis by phages [6]. Moreover, recently a growing body of evidence has demonstrated that in eukaryotes, tRNA damage is involved in stress responses, too. Intriguingly, tRNA fragments produced by tRNA cleavage can be a signal for downstream processes required for rescue and recovery from cellular stress [10-12] in a fashion that is compatible with tRNA pools required for functioning in mRNA translation and other tRNA processes (see above).

Often, yeast tRNase ribotoxins including the toxin complexes zymocin and PaT from *K. lactis* and *P. acaciae* attack tRNA anticodons that possess, in their at the wobble positions, uridine nucleobases with specific posttranscriptional modifications [3-6, 13] (Fig. 1C). Intriguingly, the pathways required for generating these wobble modifications have been shown to be conserved among lower and higher eukaryotes. In addition, among prokaryotic and eukaryotic microorganisms, these wobble modifications appear to play related roles for tRNA functioning during mRNA decoding, although they are not necessarily chemically identical [6, 13] (Fig. 2A). Therefore, our idea was to take the basic biology of tRNase ribotoxins and apply this to clinically relevant settings using HeLa tumuor cells whose proliferation heavily relies on tRNA functioning. To assess growth-related tRNase effects, HeLa cell proliferation was monitored in response to conditional expression of the ribotoxin from yeast. Our pilot findings indicate that the tRNase does not solely inhibit yeast growth but also affects the viability of higher eukaryotic HeLa cells. Similarly, we observe that the *E. coli* tRNase PrrC efficiently kills yeast cells. Our data therefore suggest that microbial tRNases may be invoked as novel anti-proliferative factors for biomedical intervention schemes.

2. Materials and Methods

A clone of the *E. coli prrC* gene, kindly provided by Gabriel Kaufmann, served as a template for PCR amplification of a construct compatible for cloning under *GAL1* promoter control in a yeast single-copy vector. The resulting *prrC* expression vector, an empty *GAL1* vector control and in parallel, a similar plasmid engineered for *GAL1*-driven expression of zymocin's tRNase subunit (γ-toxin) gene, were transformed into *S. cerevisiae* wild-type strain W303A and a *trm9*Δ deletion mutant [4, 14] (Fig. 2B). Toxicity of the microbial tRNases was assessed essentially as described [4, 14] using serial dilutions of cells spotted onto glucose-repressing and galactose-inducing media. As a source for zymocin's tRNase (γ-toxin) subunit gene served plasmid pABY1643 [3] which was used to PCR amplify a DNA fragment encoding a tRNase-GST fusion gene. The fusion was cloned under *GAL1* promoter control into yeast vector pYES2 and under Tet operator control into HeLa expression vector pTREX (Invitrogen). Upon confirmation that the pYES2-based tRNase-GST fusion was lethal in yeast strain W303A (Fig. 3A), HeLa cells were transfected with the pTREX vector carrying the tRNase fusion gene. As a control served empty vector pTREX. The antibiotics zeocin and blasticidin were used to select for maintenance of pTREX and the Tet repressor gene on a stable plasmid, respectively, required for the tetracycline-dependent expression strategy (Invitrogen). Both cell populations were seeded into 6-well plates at a density of 1.5×10^5 cells per well and incubated in the presence of the above antibiotics at 37°C. The following day, fresh media was used prior to addition of tetracycline (1μg/ml) and induction of the tRNase-GST fusion gene. Next, using three independent expression experiments, cells were incubated and counted at 24 h time intervals (Fig. 3B).

3. Results and Discussion

3.1 *E.coli* tRNase PrrC induces cell death in *S. cerevisiae*

We and others have been exploiting baker's yeast *S. cerevisiae* as a model system to study the effects of intracellular expression of tRNases including γ-toxin, the anticodon nuclease (ACNase) subunit of the zymocin complex [4, 13-16] and bacterial, colicin-type tRNases [17, 18]. Here, we show that this strategy is also applicable to the ACNase activity of the tRNase PrrC from *E. coli. Under control of the galactose-inducible GAL1 promoter, conditional expression of the bacterial prrC* gene efficiently arrests the proliferation of yeast cells on galactose media (Fig. 2B). The observed toxicity is mirrored in a recent study on PrrC homologs from other bacteria which demonstrates that a PrrC homolog from *Streptococcus mutans* also displays 'killer' activity against yeast in a fashion that is comparable to the potent PrrC prototype from *E. coli* [19]. Intriguingly, the same study also revealed that conditional expression of PrrC homologs from *Neisseria meningitidis* and *Xanthomonas campestris* had no such antifungal effects [19]. Whether their failure to induce yeast cell death is due to PrrC expression levels too low for efficient ACNase activity and tRNA restriction or due to tRNA recognition requirements distinct from their bacterial hosts are questions that remain to be elucidated. As for the yeast tRNA recognised by *E. coli* PrrC, however, the target appears to be $tRNA^{Lys}$ as is the case within the bacterial context [19-21]. This is particularly noteworthy, since prokaryotic and eukaryotic $tRNA^{Lys}$ species differ with regards to their anticodon wobble base modifications (Fig. 2A) which govern the tRNase specificities of zymocin and PrrC [3, 4, 20, 21]. Our observation that loss of the Trm9-Trm112 wobble methlyase activity, which is required for cell death induction by zymocin [4, 22], cannot protect against the tRNA attack of PrrC (Fig. 2B) suggests that the yeast and bacterial tRNases differ in their ACNase capacities or in the way they recognize their modified target tRNAs.

High-copy tRNA suppression studies have recently indicated that overexpression of the wobble uridine containing $tRNA^{Lys(UUU)}$, but not the isoacceptor $tRNA^{Lys(CUU)}$, protected yeast cells from toxicity of PrrC variants that are less lethal than wild-type PrrC tRNase [19]. Together with non-protection by higher-than-normal-levels of $tRNA^{Glu(UUC)}$ [19], the major wobble uridine containing tRNA target of the zymocin tRNase [3, 4], this strongly suggests that albeit lethal, the PrrC and zymocin tRNases attack different tRNA anticodons in yeast and in *E. coli* [20, 21]. Our observations and recent data by Meinecke *et al.* [19] showing that bacterial PrrC-type ACNases from *E. coli* and *S. mutants* (and potentially from other bacterial sources) do block yeast cell proliferation, provide an attractive experimental means to further examine whether yeast and fungal pathogens of plant and mammalian cells such as *Candida, Cryptococcus, Phytophthora, Ustilago etc* may be growth-inhibited by microbial tRNases, too. Because budding yeast appears to lack an endogenous repair system able to efficiently heal and seal the tRNA breaks created by tRNases [16], other yeast or fungi may also be vulnerable to the use of ribotoxins as antifungals.

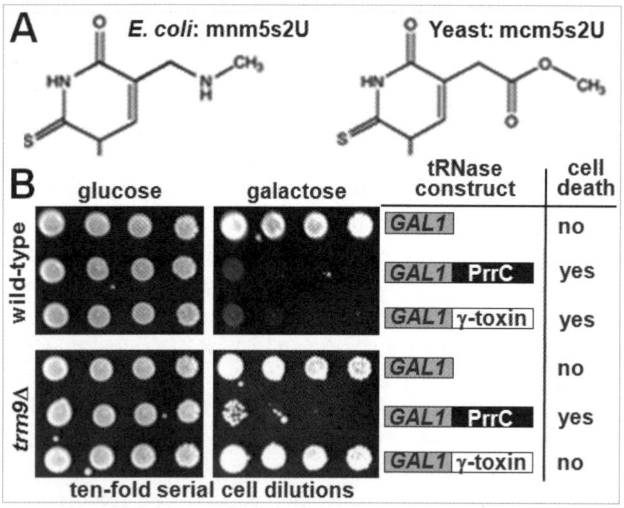

Fig. 2. Despite distinct wobble modifications in $tRNA^{Lys}$ between bacteria and yeast, expression of the PrrC tRNase from *E. coli* leads to *S. cerevisiae* cell death. **A**. Structures of modified wobble uridines in $tRNA^{Lys}$ from yeast (right: mcm5s2U) and *E. coli* (left: mnm5s2U, the PrrC target). **B**. Expression of the PrrC tRNase and the zymocin tRNase (γ-toxin subunit) in yeast cells. Both tRNase genes were fused under *GAL1*-promoter control and induced by galactose. Empty vector served as internal control and glucose was used to repress the *GAL1*-promoter and shut-off tRNase (γ-toxin) gene expression. Expression studies were performed in wild-type yeast strain W303A, known to be growth-inhibited by intracellular expression of the zymocin tRNase γ-toxin subunit, and in a *trm9Δ* mutant reported to resist against zymocin's tRNase attack [4, 13, 22].

3.2 Zymocin's tRNase subunit reduces the viability of HeLa tumour cells

Recent studies demonstrated that the wobble uridine modification function of the six-subunit Elongator complex, which is essential for the tRNase activity of zymocin (Fig. 1C, D), is conserved between *S. cerevisiae* and *Arabidopsis thaliana* [23]. Thus, zymocin resistance of yeast Elongator mutants can be fully reversed by expressing the respective plant Elongator subunit gene [23, 24]. Together with our report that tRNAs isolated from an Elongator-minus *Arabidopsis* mutant lack the wobble uridine modification (mcm5s2U) (Fig. 1C) crucial for tRNA cleavage by zymocin in yeast cells [3, 4, 23], there is the option that cells from higher eukaryotes, including plants and mammals, might maintain pools of wobble uridine-modified tRNAs that in principle, are attackable by tRNase ribotoxins. In agreement with this notion, we recently observed that *in planta*, expression of zymocin's tRNase subunit induces a hypersensitive reaction. Again, this suggests therapeutic niches for ribotoxins to inactivate proliferation-relevant tRNA functions in cells or organisms that are of potential biomedical or agrobiological significance. In this regard, the anti-retroviral use of tRNases proposed by Kaufmann and co-workers [21, 25] is particularly noteworthy as cleavage and depletion of $tRNA^{Lys}$, the primer for HIV reverse transcriptase, by the tRNase PrrC may block HIV replication and infection.

In addition, tRNase ribotoxins from bacteria and yeast may prove suitable in anti-cancer therapy [26]. For example, there are extensive studies on the anti-cancer activity of Onconase, a frog ribonuclease that is responsible for tRNA damage induction in human tumour cells and readily gains entry into mammalian cells where it appears to evade endogenous RNase inhibitors [27, 28]. Obviously, in contrast to Onconase, the yeast zymocin tRNase complex employs a much more elaborate docking and uptake system involving cell wall chitin and plasma membrane sphingolipids (for review see [13]) and bacterial PrrC-type ACNases apparently have a higher selectivity for specific tRNA target sites than does the frog RNase Onconase [21]. Nonetheless, the demonstration that the tRNase activities of PrrC (Fig. 2B), colicin D and colicin E5 are growth inhibitory to yeast cells [17-19] together with our findings that the zymocin tRNase is active *in planta*, does justify further biomedical pilot trials with tRNase ribotoxins.

We therefore decided to study the effect of conditional expression of zymocin's tRNase gene in the mammalian HeLa tumour cell line using Tet 'on/off', the regulated pTREX expression system that operates through tetracycline-mediated Tet repressor inactivation (Invitrogen). Prior to expressing the tRNase (γ-toxin-GST) fusion gene from pTREX, the toxicity of the PCR construct was confirmed in yeast; we observed that under *GAL1* promoter control, the tRNase-GST fusion efficiently arrested yeast growth (Fig. 3A) on galactose medium suggesting the PCR construct is suitable for pTREX-based expression in HeLa cells. Upon transfection of HeLa cells with empty vector pTREX as a control and pTREX harbouring the tRNase-GST gene, tetracycline was added to the growth medium for 24 hours to allow for Tet repressor inactivation by the antibiotic and induction of the yeast gene (Fig. 3B). We found that over a further growth period of 72 hours post-induction, HeLa cells lacking the tRNase gene had their cell count increased by 430% (Fig. 3B). Contrast this to HeLa cells carrying the tRNase expression construct in pTREX: here, following induction of the tRNase gene, the cell titer increased by some 215% in the same time window corresponding to an overall ~2-fold decrease in viability compared to the control group (Fig. 3B). Even more drastic was the effect of tetracycline addition and tRNase gene induction on HeLa cell viability during the first 24 hours post-induction: here, the control group acquired a cell number increase of 113%, whereas tRNase expressing HeLa cells increased their numbers by a mere ~10% (Fig. 3B). This implies that immediately after tRNase gene induction, cell viability becomes compromised, which is reflected in an overall ~2-fold reduced proliferation compared to the control group at the end of the expression study. Thus, in the beginning, HeLa cells hardly duplicated with the population as a whole remaining static rather than proliferative. Nonetheless, in contrast to yeast which ceases cell growth in the presence of the tRNase (Fig. 3A), HeLa cells are not entirely knocked-out by the ribotoxin (Fig. 3B). Yet, their performance and viability are clearly affected by induction of the yeast tRNase.

These observations suggest that the yeast ribotoxin, albeit induced and produced, is only partially active, perhaps due to reduced induction kinetics of the tetracycline-based pTREX expression system. Alternatively, the tRNase may be less potent in the context of HeLa cells since its tRNase activity targets mammalian tRNAs less efficiently than yeast tRNAs, probably as result of variations in tRNA wobble base modifications. Also, HeLa cells may maintain tRNA repair enzymes that efficiently counter anticodon cleavage by the tRNase and heal the broken ends, similar to plant tRNA ligase which is a potent overexpression suppressor of zymocin [16]. Finally, delayed HeLa cell recovery may also indicate that a subpopulation no longer expresses the yeast tRNase under inducing conditions. For reasons inherent to the protocol to maintain stable pTREX transfectants by random genomic integration of the plasmid, it is likely that not all integration events are identical with some of them having the structural integrity of their tRNase expression cartridge affected. So a mixed population may arise from different integration events that are selected for by the pTREX borne zeocin-resistance marker gene. Potentially, this leads to loss of the tRNase expression cassette and lack of tRNase production. As a result, long-term selection for zeocin-resistance will propagate cells that seem to progressively recover from 'tRNase expression' when, in fact, they are cured off the cytotoxic tRNase construct in pTREX. So to select HeLa cells that stably maintain the pTREX-based tRNase construct, transgene targeting to a unique genomic locus is highly recommended. In combination with anti-GST Western blots to monitor the production levels of the tRNase-GST fusion, this should ultimately allow to comprehensively assess the effects of tRNA cleavage by tRNase ribotoxins on the proliferation of the HeLa model tumour cell line.

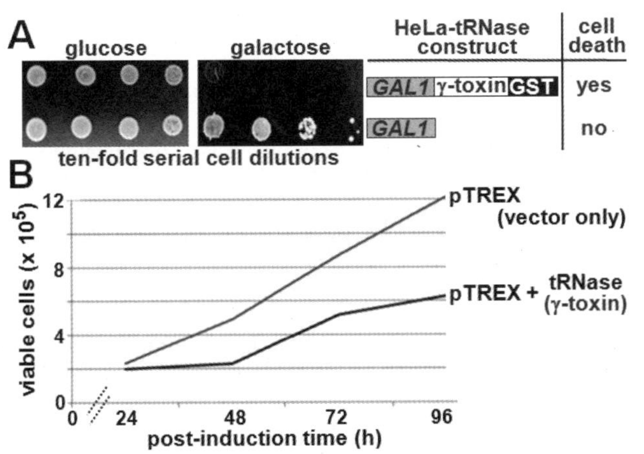

Fig. 3. Zymocin's tRNase subunit significantly reduces the viability of HeLa tumour cells. **A.** Prior to HeLa expression studies, the tRNase fusion (γ-toxin-GST) was tested in yeast for lethality using *GAL1*-promoter induction (see also Fig. 2B). **B.** Upon confirming that the fusion was as lethal as the wild-type tRNase ribotoxin, the construct was used for tetracycline-dependent 'Tet on/off' expression in HeLa cells using vector pTREX. Together with an empty vector control, HeLa cells were transfected and assayed for growth inhibition by the tRNase-GST fusion starting with 1.5 x 10^5 cells. This involved tetracycline addition, Tet repressor inactivation and gene induction over 24 h followed by a further 72 h growth period post-induction. The values plotted represent mean values of triplicates.

Acknowledgements. We thank G. Kaufmann for the *prrC* clone and S. Zink and P. Studte for technical assistance. Support by the Biotechnology and Biological Sciences Research Council (BBSRC: BB/F019106/1) to DJ & RS is greatly acknowledged. EK and CB received MSc and PhD studentship support from Leicester University, UK.

References

[1] Phizicky EM, Hopper AK. tRNA biology charges to the front. *Genes Dev*. 2010; 24: 1832-1860.

[2] Tomita K, Ogawa T, Uozumi T, Watanabe K, Masaki H. A cytotoxic ribonuclease which specifically cleaves four isoaccepting arginine tRNAs at their anticodon loops. *Proc Natl Acad Sci USA*. 2000; 97: 8278-8283.

[3] Lu J, Huang B, Esberg A, Johansson MJ, Byström AS. The *Kluyveromyces lactis* gamma-toxin targets tRNA anticodons. *RNA*. 2005; 1:1648-1654.

[4] Jablonowski D, Zink S, Mehlgarten C, Daum G, Schaffrath R. $tRNA^{Glu}$ wobble uridine methylation by Trm9 identifies Elongator's key role for zymocin-induced cell death in yeast. *Mol. Microbiol*. 2006; 59: 677-688.

[5] Klassen R, Paluszynski JP, Wemhoff S, Pfeiffer A, Fricke J, Meinhardt F. The primary target of the killer toxin from *Pichia acaciae* is $tRNA^{Gln}$. *Mol Microbiol*. 2008; 69: 681-697.

[6] Kaufmann G. Anticodon nucleases. Trends Biochem. Sci. 2000; 25: 70-74.

[7] Condon C. Shutdown decay of mRNA. *Mol Microbiol*. 2006; 61: 573-583.

[8] Nariya H, Inouye M. MazF, an mRNA interferase, mediates programmed cell death during multicellular *Myxococcus* development. *Cell*. 2008; 132: 55-66.

[9] Davidov E, Kaufmann G. RloC: a wobble nucleotide-excising and zinc-responsive bacterial tRNase. *Mol Microbiol*. 2008; 69: 1560-1574.

[10] Thompson DM, Lu C, Green PJ, Parker R. tRNA cleavage is a conserved response to oxidative stress in eukaryotes. *RNA*. 2008; 14: 2095-2103.

[11] Thompson DM, Parker R. The RNase Rny1p cleaves tRNAs and promotes cell death during oxidative stress in *Saccharomyces cerevisiae*. *J Cell Biol*. 2009; 185: 43-50.

[12] Thompson DM, Parker R. Stressing out over tRNA cleavage. *Cell*. 2009; 138: 215-219.

[13] Jablonowski D, Schaffrath R. Zymocin, a composite chitinase and tRNase killer toxin from yeast. *Biochem. Soc. Trans*. 2007; 35: 1533-1537.

[14] Frohloff F, Fichtner L, Jablonowski D, Breunig KD, Schaffrath R. *Saccharomyces cerevisiae* Elongator mutations confer resistance to the *Kluyveromyces lactis* zymocin. *EMBO J*. 2001; 20: 1993-2003.

[15] Keppetipola N, Jain R, Meineke B, Diver M, Shuman,S. Structure-activity relationships in *Kluyveromyces lactis* γ-toxin, a eukaryal tRNA anticodon nuclease. *RNA*. 2009; 15: 1036-1044.

[16] Nandakumar J, Schwer B, Schaffrath R, Shuman S. RNA repair: an antidote to cytotoxic eukaryal RNA damage. *Mol Cell*. 2008; 31: 278-286.

[17] Ogawa T, Hidaka M, Kohno K. Masaki H. Colicin E5 ribonuclease domain cleaves *Saccharomyces cerevisiae* tRNAs leading to impairment of the cell growth. *J Biochem*. 2009; 145: 461-466.

[18] Shigematsu M, Ogawa T, Kido A, Kitamoto HK, Hidaka M, Masaki H. Cellular and transcriptional responses of yeast to the cleavage of cytosolic tRNAs by colicin D. *Yeast*. 209; 26: 663-673.

[19] Meineke B, Schwer B, Schaffrath R, Shuman S. Determinants of eukaryal cell killing by the bacterial ribotoxin PrrC. Nucleic Acids Res. 2010 Sep 19. [Epub ahead of print]

[20] Jiang Y, Meidler R, Amitsur M, Kaufmann G. Specific interaction between anticodon nuclease and the tRNA(Lys) wobble base. *J Mol Biol*. 2001; 305: 377-88.

[21] Jiang Y, Blanga S, Amitsur M, Meidler R, Krivosheyev E, Sundaram M, Bajji AC, Davis DR, Kaufmann G. Structural features of tRNALys favored by anticodon nuclease as inferred from reactivities of anticodon stem and loop substrate analogs. *J Biol Chem*. 2002; 277: 3836-3841.

[22] Studte P, Zink S, Jablonowski D, Bär C, von der Haar T, Tuite MF, Schaffrath R. tRNA and protein methylase complexes mediate zymocin toxicity in yeast. *Mol Microbiol*. 2008; 69:1266-1277.

[23] Mehlgarten C, Jablonowski D, Wrackmeyer U, Tschitschmann S, Sondermann D, Jäger G, Gong Z, Byström AS, Schaffrath R, Breunig KD. Elongator function in tRNA wobble uridine modification is conserved between yeast and plants. *Mol Microbiol*. 2010; 76: 1082-1094.

[24] Chen Z, Zhang H, Jablonowski D, Zhou X, Ren X, Hong X, Schaffrath R, Zhu JK, Gong Z. Mutations in ABO1/ELO2, a subunit of holo-Elongator, increase abscisic acid sensitivity and drought tolerance in *Arabidopsis thaliana*. *Mol Cell Biol*. 2006; 26:6902-6912.

[25] Shterman N, Elroy-Stein O, Morad I, Amitsur M, Kaufmann G. Cleavage of the HIV replication primer tRNALys,3 in human cells expressing bacterial anticodon nuclease. *Nucleic Acids Res*. 1995; 23: 1744-1749.

[26] Makarov AA, Kolchinsky A, Ilinskaya ON. Binase and other microbial RNases as potential anticancer agents. *Bioessays*. 2008; 30:781-790.

[27] Ardelt W, Ardelt B, Darzynkiewicz Z. Ribonucleases as potential modalities in anticancer therapy. *Eur J Pharmacol*. 2009; 625: 181-189.

[28] Saxena SK, Sirdeshmukh R, Ardelt W, Mikulski SM, Shogen K, Youle RJ. Entry into cells and selective degradation of tRNAs by a cytotoxic member of the RNase A family. *J Biol Chem*. 2002; 277: 15142-15146.

Comparison of Anti-Listerial Effect Spectrum of Bacteriocins

Selin Kalkan[1], Emel Ünal[2], Zerrin Erginkaya[2]

[1] Osmaniye Korkut Ata University, Bahce Vocational School of Higher Education, Department of Food Technology, Osmaniye, Turkey

[2] Cukurova University, Faculty of Agriculture, Department of Food Engineering, Adana, Turkey

Listeria spp. is one of the important responsibilities of food-borne outbreaks. Therefore, studies of *Listeria* spp. and *Listeria* spp. inactivation treatments have gained great importance and intensity in worldwide. *L. monocytogenes* is the most important and most studied were made on *Listeria* spp. strain. *L. monocytogenes* is a gram positive pathogen and it can be found in a wide variety such as raw and processed animal or plant foods. It can be adapt and grow a wide range of environmental conditions such as low temperature (2-4 ^oC), acidic pH and high salt concentration. Due to fatal listeriosis, it was definitely not allowed finding in processed food. In recent years, minimally processed and ready to eat foods are frequently consumed, increased the risk of food-borne listeriosis. As a protection strategy, bacteriocins or bacteriocins producing lactic acid bacteria has created a remarkable effect about food-borne pathogen *Listeria monocytogenes* bio-control. Many studies on this issue are emphasized, using of nisin, pediocin PA-1/AcH and enterocin A high anti-listerial activity. However, effect of nisin is usually against Gram negative. For high anti-listerial activities with using nisin, to be necessary low pH, NaCl concentration, EDTA and other protective factors such as pulsed electric field and high hydrostatic pressure, different processing method with combined. According to recent studies, the Class IIa bacteriocins provide the most effective anti-listerial effect, including pediocin PA-1/Ach, enterocin A, sakacin P and curvacin A. In this manner, using of Class IIa bacteriocins in variety food such as particularly nisin wasn't very active in meat products, with their strong anti-listerial efficiency, suggested that important potential as bio-preservative.

In this review, anti-listerial effect spectrums of bacteriocins produced by different groups of microorganism were compared and potential use of these bacteriocins as bio-preservative in food was evaluated.

Keywords: Bacteriocin, anti-listerial, *Listeria* spp.

1. Introduction

Following outbreaks of food borne *Listeria*, all over the world *Listeria* spp. and inactivation of *Listeria* species in foods has gained great importance and density [1]. *Listeria monocytogenes* is most important and most work done on the strain of *Listeria* genus. It is Gram positive pathogenic bacteria which can be found in many animal and vegetable foods. It's capable of adapting to various environmental conditions such as refrigerator conditions, high acid and salt concentration. It's associated with many food poisoning occur worldwide and it's important in ensuring food safety [2]. Nowadays it's recognized as an important human pathogen, infecting especially immune depressed patients, elderly, pregnant and newborns [3].

Consumers increasing demands for minimally processed foods with less chemical additives [4] make the control of food borne pathogens such as *Listeria* spp. a more difficult task. FDA applies to ready to eat foods; *L. monocytogenes* is a zero-tolerance policy [2]. Within this context, the use of bio-preservation methods in the control of *Listeria* spp. has received great attention in recent years. As a new bio-control method, *L. monocytogenes* bio-control with bacteriocins extract or bacteriocins-producing lactic acid bacteria has created a remarkable effect [4].

2. Anti-listerial activities of bacteriocins

Bacteriocins are extracellular active proteins synthesized as ribosomal by some of Gram positive and Gram negative bacteria and have antibacterial effect. Biochemical properties, domain of the spectra, mechanisms of action, molecular weight and genetic structure quite different from each other [5]. The classification currently in use, taking into account the classification of Gram positive bacteria of Klaenhammer. This classification is located in 4 different class and the anti-listeria activities of Class I and Class II bacteriocins are notable.

Nisin is the most widely used in food preservation, including Class I bacteriocins. However, its application on meat is clearly limited, especially if pH is above 5.0 [6]. Thus researchers have led to new searches for a more efficient use of nisin, especially increases of the anti-listerial effect. Gandhi and Chikindas (2007) stated that when was used reuterin synthesized by *Lactobacillus reuteri* in combination with nisin, it acted synergistically to inhibit *L. monocytogenes*. Branen and Davidson (2004) investigated the effect of ethylenediaminetetraacetic acid (EDTA) and lactoferrin on the antimicrobial activity of nisin. Low levels of EDTA used in the study synergistically enhanced the activity of nisin against *L. monocytogenes* [7]. In another similar research, Bhatti *et al.* (2004) stated that Tween 80, a nonionic emulsifier, partially counteracted the loss of the anti-listerial activity of nisin. In addition, according to experiments with fat-free milk *L. monocytogenes* was the rather more sensitive to the nisin compared to in pasteurized and homogenized milk [8] In addition, new food processing methods, which highlighted the electric field, high hydrostatic pressure, high pressure homogenization, pulsed light an ionizing radiation can be used in combination with nisin and anti-listerial increase in activity can be achieved [9]. These results demonstrate that the chemical composition and treatment of foods may play an important role in the anti-listerial effects of nisin.

Class II bacteriocins are dominant anti-listerial was determined by many of research. These bacteriocins have significant potential as a bio-preservative in a very large number of foods. These class bacteriocins known as pediocin-like bacteriocins. With the discovery pediocin AcH/PA-1, these bacteriocins have become the most noticeable and most working on of the LAB bacteriocins class. Anti-listerial activity works best on the intensifying Class IIa bacteriocins are pediocin AcH/PA-1, enterocin A ve B, sakacin A ve P ve curvacin A [10]. In vitro studies have shown that pediocin AcH/PA-1 produced by *Pediococcus acidilactisi*, is a powerful anti-listerial activity against different types of *L. monocytogenes* [11]. But it's not yet been approved by the FDA for use as a food additive [12]. Enterococcal bacteriocins potential evaluated in many studies for control food pathogens like *L. monocytogenes* and anti-listerial activity is strongly underlined. A few of these enterococcal bacteriocins are enterocin AS-48, enterocin 416K1, enterosin A and B, sakacin A and P, enterocin L50A and B with enterocin EJ97 [13]. Activity of enterocin EJ97 against *L. monocytogenes* was greatly enhanced by nitrate and nitrite salts. This compound is widely used in meat and also in fish preservation. According to these results, the combination of low nitrite concentrations together with enterocin EJ97 to control growth of *L. monocytogenes* in foods should be investigated [4]. Enterocin 416K1 was studied and partially characterized with the aim of assessing its use as ''natural preservative'' for application in food production. Taking into account its bactericidal activity, proteinaceous nature, heat resistance, and low molecular weight, enterocin 416K1 can be classified as a small, heat-stable listeria-active [14].

3. Comparison of spectra of anti-listerial bacteriocins

Bacteriocins binding to cytoplasm membrane to form pores in cell. They enter into the cell and therefore the anti-listerial activity takes place by intracellular compounds to leak out of the cell. In particular, result in loss of the K^+ ions which duty as the balance of intracellular pH and ATP can cause cell destruction. These effects carry out by Class I bacteriocins after a certain number of bacteriocins molecules interacting with the phospholipids groups to form the lipid- protein pore (wedge model). These take place in Class II bacteriocins, the bacteriocins molecule binding to the membranes vertical manner carry out of a in the membrane by forming an ion channel (Barrel-Stave model) [15].

Class I bacteriocins, including nisin, are needed use of high concentrations bacteriocins to be effective in the inhibition of *L. monocytogenes*. In addition, usage levels depending on pH, salt and phosphate content with varies process conditions.

Recently a lot of studies demonstrated that Class II bacteriocins applications are provided high anti-listerial activity than Class I bacteriocins. In addition, in studies using model food systems, Class IIa bacteriocins has been proven effective with high antibacterial activity against *L. monocytogenes* when nisin in the absence of effective. Class II bacteriocins were compared among themselves in term of anti-listerial activity, pediocin PA-l and enterocin A have been shown to exhibit spectra of activity broader than those of sakacin P and curvacin A, which has been ascribed to the presence of an extra disulfide bond in the two former bacteriocins. It has also been speculated that bacteriocins with fewer amino acid residues would tend to have a relatively broader antibacterial spectrum than those with larger numbers of residues. In addition, both specificity and potency of the activity of Class IIa bacteriocins are apparently influenced by the lipid composition of target membranes which would determine the occurrence and the degree of interactions with bacteriocins molecules. Therefore, the comparable activities of Class IIa bacteriocins against strains of *Listeria* spp. could be due to the similarity in membrane lipid composition within this genus [10].

4. Conclusion

Consequently, due to this association of a narrow inhibitory spectrum and high anti-listerial activity, Class IIa bacteriocins are perhaps the most promising of the tested LAB bacteriocins for use in foods that are prone to *Listeri*a spp. contaminations. However, the bacteriocins extract or bacteriocins producer species are well characterized and are subject to approval when used in foods. In this context, the promising result obtained in studies show that a high potential for inhibiting pathogenic *Listeria* spp. as an alternative bio-control method.

References

[1] Erol İ, Çelik T H, Şireli U T, Özdemir H. Bakteriyosin oluşturan starter kültürlerin fermente Türk sucuklarında *Listeria monocytogenes* üzerine etkileri. *Tr. J. of Veterinary and Animal Sciences.* 1999; 23: 793-802.

[2] Gandhi M , Chikindas M L. *Listeria*: A foodborne pathogen that knows how to survive *International Journal of Food Microbiology*. 2007; 113:1–15.

[3] Martinez R C R, De Martinis E C P. Anti-listerial activity of a crude preparation of *Lactobacillus sakei* 1 bacteriocin and its lack of influence on *Listeria monocytogenes* haemolytic activity. *Food Control.* 2005; 16: 429–433.

[4] Garcı´a M T, Can˜amero M M, Lucas R, Omar N B, Pulido R P, Ga´lvez A. Inhibition of *Listeria monocytogenes* by enterocin EJ97 produced by *Enterococcus faecalis* EJ97. *International Journal of Food Microbiology.* 2004; 90: 161–170.

[5] Dinçer E, Kıvanç M, Karaca H. Biyokoruyucu olarak Laktik asit bakterileri ve bakteriyosinler. *Gıda.* 2009; 3:1-8.

[6] Ananou S, Garriga M, Hugas M, Maqueda M, Martínez-Bueno M, Ga´lvez A, Valdivia E. Control of *Listeria monocytogenes* in model sausages by enterocin AS-48. *International Journal of Food Microbiology.* 2005; 103: 179–190.

[7] Branen J K, Davidson P M. Enhancement of nisin, lysozyme, and monolaurin antimicrobial activities by ethylenediaminetetraacetic acid and lactoferrin. *Int. J. Food Microbiol.* 2004; 90 (1): 63–74.

[8] Bhatti M, Veeramachaneni A, Shelef L A. Factors affecting the anti-listerial effects of nisin in milk. *International Journal of Food Microbiology.* 2004; 97: 215– 219.

[9] Ga´ lvez A, Abriouel H, Benomar N, Lucas R. Microbial antagonists to food-borne pathogens and biocontrol. *Current Opinion in Biotechnology.* 2010; 2:142–148.

[10] Ennahar S, Sonomoto K, Ishızakı A. Class IIa bacteriocins from lactic acid bacteria: antibacterial activity and food preservation. *Journal of Bıoscıence and Bıoengıneerıng.* 1999; 87 (6): 705-716.

[11] Dabour N, Zihler A, Kheadr E, Lacroix C, Fliss İ. In vivo study on the effectiveness of pediocin PA-1 and *Pediococcus acidilactici* UL5 at inhibiting *Listeria monocytogenes.* International Journal of Food Microbiology. 2009; 133: 225–233.

[12] Cleveland J, Montville T J, Nes I F, Chikindas, M L. Bacteriocins: safe, natural antimicrobials for food preservation. *International journal of food microbiology.* 2001; 71: 1-20.

[13] Khan H, Flint S, Yu P. Enterocins in food preservation. *International journal of food Microbiology.* 2010; 1-10.

[14] Sabia C, Manicardi G, Messi P, de Niederha¨usern S, Bondi M. Enterocin 416K1, an anti-listerial bacteriocin produced by *Enterococcus casselif lavus* IM 416K1 isolated from Italian sausages. *International Journal of Food Microbiology.* 2002; 75:163–170.

[15] Kurt Ş and Zorba Ö. Bakteriyosinler ve gıdalarda kullanım olanakları. *YYÜ Vet Fak. Derg.* 2005; 16(1):77-83.

Isolation and screening of novel antibiotic producing *Streptomyces* from southwest Turkey soils

A. Ugur[*,1] and O. Ceylan [2]

[1]Department of Biology, Faculty of Science, Mugla University, Kotekli, 48170, Mugla, Turkey
[2]Department of Nursing, Health College, Usak University, 64000,Usak, Turkey

*Corresponding author: e-mail: aysel ugur@hotmail.com, Phone: +90 2762272072

The purpose of this study is to explore secondary metabolites belonging to *Streptomyces* which show antimicrobial activities on bacteria. For this aim, a total of 234 different *Streptomyces* isolates were recovered from soil samples taken from areas with various vegetation types in Mugla and its surrounding. All the isolates were screened with spektra-plak method regarding the production of antibacterial metabolites. In this screening process, the target bacteria chosen from pathogens which are 10 Gram positive and 9 Gram negative clinically-significant bacterial pathogens including *Stenotrophomonas maltophilia* and *Staphylococcus* which are multi-resistant to antibiotics. Out of these, 61 isolates exhibited inhibitory activity against at least one of the tested microorganisms. 73.9% isolates produced antibacterial substances against Gram positive bacteria, 47.4% against Gram negative bacteria. The results indicated that 26 isolates were highly active against at least one clinically significant multi-resistant *Staphylococcus spp.* with an inhibition zone at ≥20 mm.

Keywords: Streptomyces, antibiotic producing, Staphylococcus, Stenotrophomonas

1. Introduction

Today the antibiotic resistance problem has grown to include all of the major bacterial pathogens and all classes of antibiotic compounds [1-2]. The appearance of multiresistant pathogenic strains has caused a therapeutic problem of enormous proportions. For instance, they cause substantial morbidity and mortality specially among the elderly and immunocompromised patients [3]. For this reason, search for new antibiotics effective against multi-drug resistant pathogenic bacteria is presently an important area of antibiotic research [4].

Soil is a natural reservoir for microorganisms and their antimicrobial products [4]. Filamentous soil bacteria belonging to the genus *Streptomyces* are widely recognized as industrially important microorganisms because of their ability to produce many kinds of novel secondary metabolites including antibiotics [5, 6, 7, 8].

The purpose of this study is to explore secondary metabolites belonging to *Streptomyces* which show antimicrobial activities on many Gram negative and Gram positive bacterial pathogens including *Stenotrophomonas maltophilia* and *Staphylococcus* which are multi-resistant to antibiotics.

2. Material and Methods

2.1. Isolation of microorganisms

Soil samples were collected from various locations in Mugla province. Several diverse habitats in different areas were selected for the isolation of *Streptomyces* strains. These habitats included the rhizosphere of plants, agricultural soil, preserved areas and forest soils. The samples were taken up to a depth of 20 cm after removing approximately 3 cm of the soil surface. The samples were placed in sterile glass bags, closed tightly and stored in a refrigerator. The following screening procedure was adopted for the isolation of *Streptomyces* [8]. The soil was pretreated with $CaCO_3$ (10:1 w/w) and incubated at 37°C for 4 days. It was then suspended in sterile Ringer solution (1/4 strength). Test tubes containing a 10^{-2} dilution of the samples were placed in a water bath at 45°C for 16 h so that the spores would separate from the vegetative cells and the dilutions were inoculated on the surface of the Actinomycete Isolation Agar (AIA)(Difco 0957) plates. The plates were incubated at 28°C until the sporulation of *Streptomyces* colonies occurred. *Streptomyces* colonies (where the mycelium remained intact and the aerial mycelium and long spore chains were abundant) were then removed and transferred to the Yeast Extract-Malt Extract Agar (ISP2) slants. Pure cultures were obtained from selected colonies for repeated sub culturing. After antimicrobial activity screening, the isolated *Streptomyces* strains were maintained as suspensions of spores and mycelial fragments in 10% glycerol (v/v) at -20°C in the Mugla University Collection of Microorganisms (MU).

2.2. Characterization of the isolates

Streptomyces colonies were characterized morphologically and physiologically following the directions given for the International *Streptomyces* Project (ISP) [9]. All isolates putatively assigned to the genus *Streptomyces* on the basis of colony morphology, notably aerial spore mass colour, substrate mycelial pigmentation and the colour of any diffusible pigments were subcultured onto ISP2 agar [5, 7]. Aerial mycelium, substrate mycelium and diffuse pigment colours of all isolates were grouped by NBS/ISCC colour system. The NBS/ISCC system is a standarized set of color terms. It defines a set of 276 color centroids. This system considers ten base terms (pink, red, orange, brown, yellow, olive, green, blue, violet and purple) and some combination of them (for instance, reddish orange or bluish green), building twenty-eight hues [10].

2.3. Test microorganisms

For determination the antimicrobial activity spectrum of *Streptomyces* isolates following test strains were used: *Bacillus subtilis* ATCC 6633, *Bacillus cereus* RSKK 863, *Micrococcus luteus* NRRL B-4375, *Staphylococcus aureus* ATCC 25923, *Streptococcus mutans* ATCC 35668, *Escherichia coli* ATCC 25922, *Pseudomonas aeruginosa* ATCC 27853, *Enterobacter aerogenes* RSKK 720 and *Salmonella typhimurium* CCM 5445 were used as test microorganisms.

Also, *Staphylococcus aureus* MU 38, MU 40, MU 46, *Staphylococcus xylosus* MU 34, *Staphylococcus epidermidis* MU 30 and *Strenotrophomonas maltophilia* MU 52, MU 64, MU 69, MU 99, MU 137, multiple antibiotic resistant bacteria were used. The strains of coded MU were obtained from the Mugla University Culture Collection.

The above mentioned bacteria, except *S. mutans*, were cultured in Nutrient Broth (NB) (Difco); *S. mutans* was cultured in Brain Heart Infusion Broth (BHIB)(Difco). *P. aeruginosa* and *S. maltophilia* strains were incubated at 30±0.1°C for 18–24 h and 24–48 h, respectively. Other bacteria strains were incubated at 37±0.1°C for 24–48 h. Inocula were prepared by adjusting the turbidity of the medium to match the 0.5 Mcfarland Standard Dilutions of this suspension, in 0.1% peptone (w/v) solution in sterile water inoculated on NB, BHIB to check the viability of the preparation. The cultures of the microorganisms were maintained in their appropriate agar slants at 4°C throughout the study and used as stock cultures.

2.4. In vitro screening of isolates for antagonism

Müeller-Hinton agar (MHA, Merck) plates were prepared and inoculated with *Streptomyces* isolate by a single streak of inoculum in the center of the petri dish. After 4 days of incubation at 28°C the plates were seeded with test organisms by a single streak at a 90° angle to *Streptomyces* strains. The microbial interactions were analyzed by the determination of the size of the inhibition zone [11].

3. Results and Discussion

A total of 234 different *Streptomyces* isolates were recovered from 117 soil samples, collected from different locations in Mugla region, were screening for their potential as a source of antibiotics active against antibiotic-resistant bacteria. According to colony morphology, notably aerial spore mass colour, substrate mycelial pigmentation and the colour of any diffusible pigments, these strains belonged the genus *Streptomyces*. Colour grouping of *Streptomyces* isolates were represented in Table 1.

Table 1. Colour grouping of *Streptomyces* isolates.

Colour	Brown-yellow	White	Grey	Yellow	Red	Blue	Violet	Orange	Pink	Total
	Number of isolate									
Aerial mycelium colour	-	27	161	20	8	2	5	4	7	234
Reverse side colour	37	-	92	58	6	16	13	6	6	234
Soluble pigment colour	-	-	9	14	20	6	-	19	1	69

As shown in Table 1, isolates varies within each colour series, with production of such compounds being recorded as a soluble pigment in the colours of red (28.9%), orange (27.5%), yellow (20.2%), grey (13%), blue (8.7%) and pink (1.4%). The reverse side colour was grey (39.3%), yellow ((24.8%), Brown-yellow (15.8%), blue (6.8%), violet (5.5%), and red, orange, pink(2.6%).

All of the isolates were tested for their ability to produce inhibitory substances against nineteen test microorganisms (Table 2). 73.9% isolates produced antibacterial substances against Gram positive bacteria, 47.4% against Gram negative bacteria. Out of these 61 isolates exhibited inhibitory activity against at least one of the tested microorganisms. One hundred eleven isolates produced antibacterial substances against both Gram negative and Gram positive bacteria. 26 of the isolates were shown to have a very potent in vitro antimicrobial activity against the Gram positive test organisms and clinically significant multi-resistant *Staphylococcus spp.*(Table 3). The isolates produced 11-34 mm, 7-35 mm, 8-32 mm inhibition zones on *S.aureus* ATCC 25923, MU 38 and MU 40, respectively. Two of the isolates (452-1 and 539-1) were active against *S.aureus* MU 46 with 34 mm and 21 mm inhibition zones, respectively. The isolates produced 7-38 mm and 9-30 mm inhibition zones on *S.epidermidis* MU 30 and *S. xylosus* MU 34, respectively. Especially, the isolate 15-H-1 inhibited the growth of *S.aureus* MU 38, MU 40, *S.epidermidis* MU 30 and *S.xylosus* MU 34 strains with >30 mm inhibition zones. In addition, the isolates 192-2, 350-1, 426-1 and 452-1 inhibited the growth of tested *S.aureus* MU 38 with >30 mm inhibition zones. It appears that the inhibitory substances from 49X-2 and 328-4 are more effective against *S.epidermidis* MU 30. *S. aureus* infections have increased in the past 20 years primarily due to the increased use of invasive devices in both hospital and home care settings, as well as the increase in the number of elderly and/or immune compromised patients. Antibiotic-resistant strains of *S. aureus* also have increased, limiting viable therapies to treat and prevent infections. These antibiotic-resistant strains –known as methicillin-resistant *S.aureus* (MRSA)- can result in serious medical complications and death [12]. *S. aureus* MU38 and MU40 are Methicillin-Resistant *Staphylococcus aureus* (MRSA). MRSA is one of the most significant causes of healthcare-associated (HA) infections worldwide; it is associated with significant morbidity and mortality and imposes a huge burden on healthcare resources [13, 14]. MRSA is responsible for the largest outbreak of hospital-acquired infection (HAI) that the world has ever seen [15]. Worldwide, many strains of *S. aureus* are already resistant to all antibiotics; thus, the organism has progressed one step toward becoming an unstoppable killer [16].

Table 2. Antimicrobial activity of *Streptomyces* isolates.

Test Microorganisms	Inhibitory effect			
	0	1	2	3
	Number of isolate			
Gram-positive bacteria				
M.luteus NRRL B–4375	107	51	53	23
B.subtilis ATCC 6633	218	6	5	5
B.cereus RSKK 863	210	8	14	2
S.mutans ATCC 35668	219	8	2	5
S.aureus ATCC 25923	147	40	37	10
S.aureus MU 38	218	7	4	5
S.aureus MU 40	217	12	4	1
S.aureus MU46	231	1	1	1
S.epidermidis MU 30	199	22	10	3
S.xylosus MU 34	226	6	1	1
Gram-negative bacteria				
E.coli ATCC 25922	154	37	39	4
P.aeruginosa ATCC 27853	198	31	5	-
E.aerogenes RSKK 720	218	5	10	1
S.typhimurium CCM 5445	223	6	3	2
S.maltophilia MU 52	234	-	-	-
S.maltophilia MU 64	234	-	-	-
S.maltophilia MU 69	234	-	-	-
S.maltophilia MU 99	234	-	-	-
S.maltophilia MU 137	234	-	-	-

The inhibitory effect of isolates was divided into four groups according to the size of the inhibition zone and as follows: 0 (passive group, ≤ 10 mm); group 1 (11-20 mm, slightly active); group 2 (21-30 mm, moderately active); and group 3 (≥31 mm, highly active).

Table 3. Antimicrobial activity against Gram-positive bacteria selected *Streptomyces* isolates.

Strain number	*M.luteus* NRRL B–4375	*B.subtilis* ATCC 6633	*B.cereus* RSKK 863	*S.mutans* ATCC 35668	*S.aureus* ATCC 25923	*S.aureus* MU 38	*S.aureus* MU 40	*S.aureus* MU 46	*S.epidermidis* MU 30	*S.xylosus* *MU 34*
	Inhibition zone (mm)									
10-H(2)	17	17	17	5	17	-	-	-	21	-
15-H-1	27	-	4	14	23	35	32	-	38	*>30*
17-H(2)	27	28	27	-	34	-	-	-	24	-
18-2	18	15	10	14	>30	22	8	-	7	*9*
26-1	20	-	-	-	10	-	-	-	25	-
27-3	22	-	>30	-	18	-	-	-	20	-
29A(1)	22	-	27	-	17	-	-	-	21	-
49X-2	-	>30	-	>30	-	-	20	-	35	-
67A-3	29	-	-	-	27	-	8	-	24	-
73-1	28	27	26	-	33	-	-	-	20	-
78A-2	25	-	22	16	21	16	20	-	-	*21*
90-3	25	-	5	14	-	29	15	-	-	*16*
97-2	23	-	30	3	27	24	-	-	16	-
192-2	27	-	22	-	24	32	-	-	19	-
204-1	32	>30	-	>30	20	-	17	-	25	-
309-1	10	-	30	4	22	15	21	-	20	*20*
328-4	27	>30	17	>30	26	-	25	-	38	-
329-1	15	-	7	19	-	20	-	-	-	-
330-2	15	12	22	-	16	-	22	-	24	-
350-1	12	-	29	-	-	33	-	-	22	-
380-2	26	25	-	14	21	20	18	-	12	*20*
426-1	21	-	3	>30	27	31	18	-	-	*20*
450-1E	35	22	23	-	33	-	-	-	25	-
450-3	13	-	19	-	11	27	-	-	17	-
452-1	27	37	29	-	22	31	29	34	26	-
539-1	*32*	*17*	*25*	*9*	*16*	*7*	*18*	*21*	-	-

One hundred eleven isolates produced antibacterial substances against one or more Gram negative bacteria tested. The isolates produced 4-33 mm, 2-25 mm, 2-45 mm and 4-39 mm inhibition zones on *E.coli* ATCC 25922, *P.aeruginosa* ATCC 27853, *E.aerogenes* RSKK 720 and *S.typhimurium* CCM 5445, respectively. Six isolates inhibited the growth of all the tested Gram negative bacteria. Five isolates (25-2, 26-1, 49X-3, 54-2, 427-2) produced >30 mm inhibition zones for *E.coli* ATCC 25922. The isolate (26-1) inhibited the growth of *P.aeruginosa* ATCC 27853 with 25 mm inhibition zone. The highest inhibition zone (45 mm) was on *E.aerogenes* RSKK 720. Twentyone isolates showed antibacterial activity on *E.aerogenes* RSKK 720. The inhibition zones of eleven of them were higher than 20 mm. Twenty isolates showed antibacterial activity on *S.typhimurium* CCM 5445. Three isolates (49X-2, 328-4 and 350-2) produced >30 mm inhibition zones for *S.typhimurium* CCM 5445. One isolate (12-4-12(2)) were only effective *S.maltophilia* MU 64 and MU 69 with inhibition zones 10 mm and 9 mm, respectively. Other isolates were not active against all *S.maltophilia* strains. *S. maltophilia* has also been described to be an important nosocomial pathogen [17]. The treatment of infections caused by this microorganism is difficult because *S. maltophilia* is frequently resistant to most of the widely used antibiotics [18, 19, 20, 21].

In this work, we have shown that a total 234 different Streptomycetes isolates associate with soil have the ability to produce antimicrobial compounds against microorganisms, especially multiple antibiotic resistant Gram positive and Gram negative bacteria. This study have shown that most of our isolates have the ability to produce antimicrobial compounds against clinically significant multi-resistant *Staphylococcus spp.* For this reason, further investigations are needed in order to determine the active metabolites of these isolates.

Acknowledgements Work supported by Turkish State Planning Organization (Grant No:DPT 2003K-1208500)

References

[1] Guerrant RL, Blackwood BL. Threats to a global survival: the growing crisis of infectious diseases — our ''unfinished agenda''. *Clin Inf Dis.* 1999; 28:966-986.

[2] Johnson AP. Antibiotic resistance in the intensive care unit setting. *Expert Rev Anti Infect Ther.* 2002; 1:253-260.

[3] Barsby T, Kelly MT, Gagne SM, Andersen RJ. Bogorol A produced in culture by a marine *Bacillus* sp. reveals a novel template for cationic peptide antibiotics. *Org. Lett.* 2001; 3:437-40.

[4] Dancer SJ. How antibiotics can make us sick: the less obvious adverse effects of antimicrobial chemotherapy. *The Lancet Infectious Diseases.* 2004; 4:611-619.

[5] Williams ST, Goodfellow M, Alderson G, Wellington EMH, Sneath PHA, Sackin MJ Numerical classification of *Streptomyces* and related genera. *Journal of General Microbiology.* 1983a; 129, 1743–1813.

[6] Crandall LW, Hamil RL. Antibiotics produced by *Streptomyces*: major structural classes. In: Queener SW, Day LE (ed.). *The bacteria*, Vol. 9, Academic Press, Orlando, Fla, 1986:355–401.

[7] Williams ST, Goodfellow M, Alderson G. Genus *Streptomyces* Waksman and Henrici 1943, 339AL. In: Williams ST, Sharpe ME, Holt JG (ed.) *Bergey's Manual of Systematic Bacteriology*, Vol.4. Williams & Wilkins, Baltimore. 1989:2452–2492.

[8] Korn-Wendisch F, Kutzner HJ. The family *Streptomycetaceae*. In: Balows A, Truper HG, Dworkin M, Harder W, Schleifer KH (ed.) *The prokaryotes*, Springer-Verlag, 2nd edn. New York, 1992:921–995.

[9] Shirling EB, Gottlieb D. Methods, classification, identification and description of genera and species. Vol. 2. The Williams and Wilkins Company, Baltimore. 1966:61- 292.

[10] David Mundie. Available at: http://www.anthus.com/colors/nbs.html

[11] Madigan MT, Martinko JM, Parker J. Antibiotics: isolation and characterization. In: *Brock Biology of Microorganisms*, 8th edn. Prentice-Hall International Inc. New Jersey; 1997:440–442.

[12] Chenoworth, C. Four-year prospective evaluation of community-acquired bacteremia: Epidemiology, microbiology and patient outcome. *Diagnostic Microbiology and Infectious Disease.* 2001;41:15–22.

[13] Grundmann H, Aires-de-Sousa M, Boyce J, Tiemersma E. Emergence and resurgence of meticillin-resistant *Staphylococcus aureus* as a public-health threat. *Lancet* 2006;368:874-885.

[14] Boyce JM, Cookson B, Christiansen K, Hori S, Vuopio-Varkila J, Kogagoz S, Öztop AY, Vanden-Grauls CM, Harbarth S, Pittet D. Meticillin-resistant *Staphylococcus aureus*. *Lancet Infect Dis* 2005;5:653-663.

[15] Gould IM. The clinical significance of methicillin-resistant *Staphylococcus aureus*. *Journal of Hospital Infection.* 2005;61:277–282.

[16] Ugur A, Ceylan O. Occurrence of resistance to antibiotics, metals, and plasmids in clinical strains of *Staphylococcus* spp. *Archives of Medical Research.* 2003;34:130–136.

[17] Denton M, Kerr KG. Microbiological and clinical aspects of infection associated with *Stenotrophomonas maltophilia*. *Clinical Microbiology Reviews.* 1998;11:57–80.

[18] Vartivarian S, Anaissie E, Bodey G, Sprigg H, Rolston K. A changing pattern of susceptibility of *Xanthomonas maltophilia* to antimicrobial agents: implications for therapy. *Antimicrobial Agents and Chemotherapy.* 1994;38:624-627.

[19] Liu PL, Lau YJ, Hu BS, Shyr JM, Shi ZY, Tsai WS, Lin YH, Tseng CY. Comparison of susceptibility to extended-spectrum □-lactam antibiotics and ciprofloxacin among Gram-negative bacilli isolated from intensive care units. *Diagnostic Microbiology and Infectious Disease.* 1995;22:285-291.

[20] Skaehill P. Management of *Stenotrophomonas maltophilia* infections. *The Consult Pharmacist.* 2000;15:74- 76.

[21] Krueger TS, Clark EA, Nix DE. In vitro susceptibility of *Stenotrophomonas maltophilia* to various antimicrobial combinations. *Diagnostic Microbiology and Infectious Disease* 2001;41:71-78.

Production and characterization of enterocin from *Enterococcus faecium* IJ-06 and IJ-21 isolated from indigenous dairy products

Imran Javed[*,1,2] **and Safia Ahmed**[2]

[1] Department of Dairy Technology, University of Veterinary and Animal Sciences, Lahore, 54000 Pakistan
[2] Department of Microbiology, Quaid-i-Azam University, Islamabad, 45320 Pakistan

*Corresponding author: e-mail: imranjaved@uvasedu.pk Phone: +34 924258615

The study was conducted to investigate previously isolated *Enterococcus faecium* IJ-06 from local processed cheese and *Enterococcus faecium* IJ-21 from yoghurt for their bacteriocin production potential, antimicrobial activity and relevant virulence traits. Isolated stains were non hemolytic, susceptible to vancomycin and both strains possess genes for enterocin A (*entA*) and enterocin P (*entP*). Assessment of *vanA, vanB* genes and virulence determinants *cyl* were done through PCR. The enterocin was inhibitory towards *Listeria monocytogenes, Staphylococcus aureus, Bacillus subtlis, Bacillus cereus* and other enterococci. Characterization of enteriocins with specific treatment followed by residual activity assay showed its thermo-stability, with a wide pH range of activity and sensitivity to proteases. The molecular weight of partially purified enteocins was ~4.5-4.7 kDa as estimated by SDS-PAGE. It is promising for further investigation of the safe exploitation of these strains or their enterocins for microbial food safety.

Keywords: Bacteriocin; Enterocin; Antimicrobial; Enterococci

1. INTRODUCTION

Bacteriocins are peptides or polypeptides with an antibacterial activity usually against their closely related strains and genera [1]. Since the last decade the number of characterized bacteriocins produced by enterococci (enterocins) is increasing. Interestingly, enterocins are particularly active against pathogenic bacteria, such as Listeria, Clostridium and Staphylococcus [2-3]. Most bacteriocins from LAB are synthesized as biologically inactive pre-peptides containing an N-terminal extension. The mature peptides are often cationic, amphiphilic, membrane permeabilizing molecules [4]. Class IIa of bacteriocins is comprised of nonmodified, heat-stable pediocin-like, with strong antilisterial bacteriocins for example enterocin A and P etc. [5-6].

Enterococcus is an important genus belonging to the group of lactic acid bacteria (LAB), which play a significant role in the ripening of traditional Mediterranean country products like sausages, cheeses etc. [3]. Enterococci is also much controversial group which shows favorable metabolic activities (lipolytic and esterolytic activity, citrate utilization, etc.) contributing to the typical taste and flavour [3,7]. However, there is a potential health risk in the use of enterococci. In addition, there is a clear increasing antibiotic resistance development among enterococci [8-9].

In this work, the enterocin of *Enterococcus faecium* IJ-06 and IJ-21 previously isolated from indigenous processed cheese and yoghurt sample respectively were studied. Their virulence determinants such as resistance to glycopeptides antibiotics and cytolysin were investigated. Finally enterocins produced were partially purified and characterization was done for thermo-stability, pH range of activity and inactivation with proteolytic enzymes in view of their technological potential and practical use.

2. MATERIALS AND METHODS

2.1 Bacterial strains and media: *Enterococcus* spp. were propagated at 37°C in de Man-Rogosa-Sharpe (MRS) while *Listeria* spp. were propagated at 35°C in Brain Heart Infusion (BHI) broth, respectively. All the media were from Oxoid (Oxoid, Basingstoke, United Kingdom). Stock cultures of all strains were kept in the appropriate medium with 10% (v/v) glycerol and stored at -20°C. *Enterococcus faecium* ATCC 27273 and LMG 11423^T were used as reference strains.

2.2 Qualitative test for haemolysis and vancomycin resistance: The *Enterococcus faecium* IJ-06 an IJ-21 strains were grown overnight in MRS medium at 37°C, and then transferred onto Blood Agar Base plates containing 5-7% (v/v) of sterile sheep and human blood (University Medical center). The plates were incubated overnight at 37°C. The haemolytic reaction was recorded by observation of zone of hydrolysis around the colonies. Glycopeptide antibiotic resistance was determined by standard broth dilution technique resulting tubes with a final concentration of 64, 32, 16, 8, 4, and 2 ug/ml of the antibiotic. Overnight culture was inoculated to each tube and incubated at 37°C for 24 hrs. The minimum inhibitory concentration (MIC) values were determined as the lowest antibiotic concentration that did not allow growth of the strains.

2.3 Identification of bacteriocin structural genes and quantitative determination of bacteriocin titer: Polymerase Chain Reaction was performed to see whether structural genes already reported enterocins from *E. faecium* were present in the bacteriocinogenic isolates investigated. Genomic DNA isolated by DNeasy kit (Qiagen, Biosciences, Maryland, USA) and used as template. The PCR reactions were carried out in a 50 µl volume that consisted of 39.25 µl water, 5 µl of a 10x reaction buffer, 1.5 µl $MgCl_2$ (50 mM), 1 µl dNTPs (10 mM), 1 µl of primer (10mM), 0.25 µl of Taq DNA polymerase (1 U) and 2.5ng of template. The reaction conditions were 30 sec denaturation at 95°C, followed by 30 cycles of 2min at 72°C, followed by 5 min at 72°C and a cool down to 4°C. Secondary PCR was conducted by same conditions by taking 2µl from primary PCR as template. DNA from these reactions was separated by 2% (w/v) agarose gel electrophoresis; gels were stained with 0.1 µg/ml of ethidium bromide and visualized by UV light. Quantitative determination of bacteriocin activity was performed by agar well diffusion and spot on lawn assay [10]. To examine the inhibitory spectrum, both cell-free culture supernatant and partially purified enterocin were used without dilution.

2.4 Partial purification of enterocins: Enterocin was isolated from a 1 liter culture of the *Enterococcus faecium* IJ-06 and *Enterococcus faecium* IJ-21, grown in MRS medium at 37°C for 24 h. The pH of the cell-free culture supernatant was adjusted to 6.5 by addition of 1N NaOH. The cells were removed by centrifugation (5 500 g, 4°C 20 min) and this cell-free culture supernatant was brought to a final ammonium sulphate concentration of 40% saturation by slow addition of the salt, and was stirred overnight at 4°C. Then, the mixture was centrifuged (5 500 g, 4°C, 30 min) and the surface pellicles and bottom pellets were harvested and resuspended in 1 ml of sodium phosphate buffer (pH 6.5). To one volume of the resuspended product, 10 volumes of a methanol–chloroform mixture (1:2 v/v) were added, and the mixture was extracted at 4°C for 2 h. The sample was centrifuged (5 500 g, 4°C, 30 min), the supernatant fraction discarded and the pellet air-dried. The pellet was resuspended in 2 ml of ultrapure water (MilliQ; Millipore N.V., Brussels, Belgium). This partially purified enterocin was stored at -20°C for further assay. To quantify bacteriocin activity, enterocin sensitive strains were used as indicator. Protein concentrations were determined by a modification of the method of Bradford [11] using a Protein Assay Kit from BioRad Laboratories (California, USA). Bovine serum albumin was used as a standard.

2.5 Sensitivity of partially purified bacteriocins to different enzyme, pH and heat treatment: Sensitivity of the enterocin to protease, lipase, lysozyme, amylase, and catalase was tested using partially purified enterocin samples. The residual activity (AU/ml) was measured by bioassay, using enterocins sensitive strains as indicator. Thermal stability of the partially purified enterocin was determined by incubation of partially purified enterocin solutions at 60°C, 80°C for 30 min, at 100°C and 121°C for 15 min. The pH stability was assayed at pH 4.0, 5.0, 6.0, 7.0, 8.0, and 9.0, at room temperature (25°C) after 1 hr of incubation of partially purified enterocin solutions. The remaining activity (AU ml) was measured by bioassay.

2.6 Molecular weight estimation by SDS-PAGE: Sodium dodecyl sulphate-polyacrylamide gel electrophoresis (SDS-PAGE) was used for the approximation of molecular weight of produced bacteriocin. Polyacrylamide concentration in the stacking and the resolving gels was 7.5 and 15.0%, respectively. Electrophoresis was conducted at a constant voltage of 60 V for 1 h and 90 V for 2 h. The molecular weight marker for SDS-PAGE, ultra-low range (Sigma) with six polypeptides of known molecular weights (range 1060–26 600 kDa), was used. The gel stained and photo-digitalized then was washed with distilled water overnight. The gel was overlaid with MRS soft agar seeded with actively growing cells of *E. faecalis* IJ-11 and incubated at 37 °C for 18 h and zone of inhibition was assayed.

3. RESULTS

3.1. Identification and safety investigation of isolated strain

The strain IJ-06 and IJ-21 isolated previously from indigenous dairy products were identified as *Enterococcus faecium* IJ-06 and *Enterococcus faecium* IJ-21 by molecular characterization. Nucleotide sequences for 16S rRNA were deposited with GenBank at NCBI with accession numbers EU547774 and EU547778 respectively. Two of the major factors involved in a potential health risk associated with the use of enterococci were evaluated, namely vancomycin resistance and cytolysin production. Both strains under investigation showed γ-haemolytic activity on animal and human blood, indicating absence of haemolysis. The vancomycin and teicoplanin MIC value was 4µg/ml, indicating sensitivity of these strains for these antibiotics.

3.2 Bacteriocin structural genes analysis by PCR: After agarose gel electrophoresis, a PCR fragment with a size of 108 and 155 bp were amplified from genomic DNA of both *E. faecium* IJ-06 and *E faecium* IJ-21, which corresponds to the enterocin P and A respectively (Fig. 1). Investigation for other enterocins was also done with different sets of respective primers but no signals were obtained that corresponded with the enterocins A, B, 31,

AS-48, L50A, L50B, 1071A, 1071B. Both strains also lacked the structural genes for the production of cytolysin (*cyl*), and vancomycin resistance genes (*vanA* and *vanB*).

3.3 Isolation and partial purification of enterocins: Bacteriocin production from *E. faecium* IJ-06 and IJ-21 in batch fermentation occurred in early exponential phase after 8 h of incubation in MRS broth (37°C, 1% inoculum, 7.0 pH) and reached maximum (upto 6400 AU/ml) after 16 h, and it remained stable until 48 h. Bacteriocin then started decreasing and low levels were detected at 120 h.

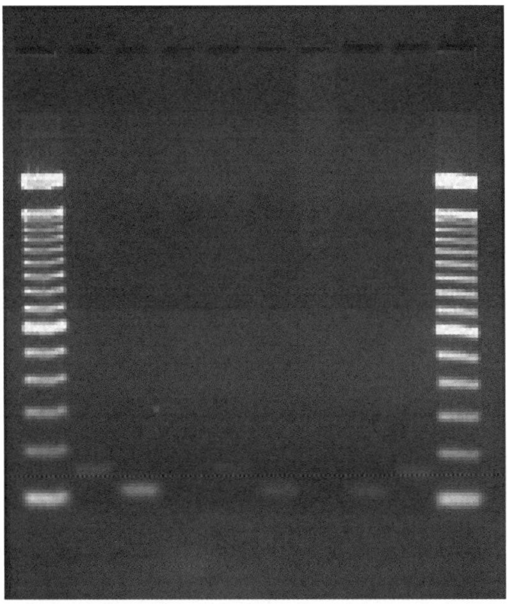

Fig. 1: PCR amplification of bacteriocin structural genes from isolated strain *Enterococcus faecium* IJ-06 and IJ-12
M: DNA marker, L2: Ent. faecium IJ-06 entA, L3: Ent. faecium IJ-06 entP, L4: Ent. faecalis IJ-12 entA, L5: Ent. faecium IJ-21 entA, L6: Ent. faecium IJ-21 entP, L7: Ent. faecium IJ-31 entA, L8: Ent. faecium IJ-31 entP, L9: Ent. faecium ATCC 27273 entA

3.4 Inhibitory spectrum: The cell-free culture supernatant and the partially purified enterocins were tested against a broad range of Gram positive bacteria (35 strains) and *E. coli* ATCC10536 as Gram-negative control (Table 1). A narrow inhibitory spectrum was found for both cell-free culture supernatant and partially purified enterocins, when only lactic acid bacteria other than enterococci were considered as indicator strains. The antimicrobial activity was detected against a wide range of microorganisms like *Listeria monocytogenes, Bacillus cereus*, *Enterococcus faecium* and *Enterococcus faecalis.* An increase in inhibitory spectrum was observed in case of partially purified enterocin preparation.

Table 1: Antimicrobial activity of isolated enterococcal strains

Producer Strain	Indicator strain	Tested strains[a]	Inhibited strains[b]	Activity[c]	Average Zone of inhibition[d] (mm)
IJ-06	*Listeria* spp.	5	4	6400	25
	Enterococcus spp.	11	11	6400	28
IJ-21	*Listeria* spp.	5	4	6400	25
	Enterococcus spp.	11	11	6400	28
	Bacillus cereus	3	2	3200	24
	Bacillus subtlis	4	2	6400	26
Control	*Listeria* spp.	5	4	6400	23
	Enterococcus spp.	11	11	6400	25

[a] Total No. of strains tested
[b] No. of inhibited strains
[c] Activity in Arbitrary units per ml (AU/mL)
[d] Activity expressed as average zone of inhibition in mm

3.5 Sensitivity of partially purified enterocins to enzyme, pH and heat treatment: Antibacterial activity of the partially purified enterocins studied was completely destroyed upon treatment, enterocin activities were not affected by lipase, lysozyme, and catalase. The temperature stability of the enterocins depended on the pH. At acidic pH values the enterocins studied were more heat-stable than at neutral or alkaline pH. The activity was higher at low pH.

3.6 Molecular weight estimation by Tricine-SDS-PAGE: The molecular weight of the enterocin studied was approximately 4.5-47 kDa when observed by SDS-PAGE. Antimicrobial assay of gel after electrophoresis and extensive washing showed a single inhibitory zone for the enterocin from *E. faecium* IJ-06 and IJ-21 that means only one type of enterocin was produced showing activity against *Bacillus subtilis* ATCC27142 (Fig. 2).

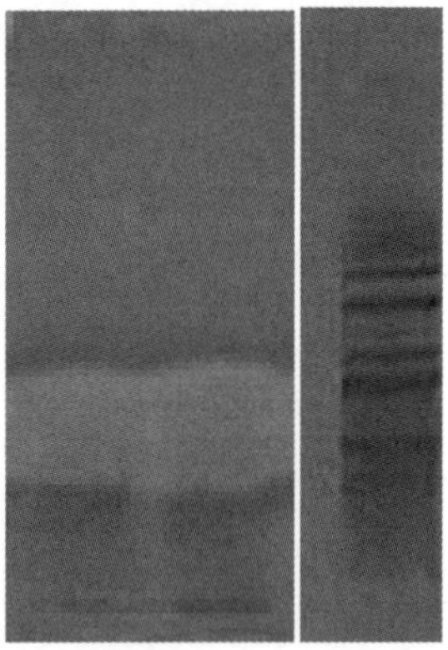

Fig. 2: Molecular weight of enterocin produced by *Ent. faecium* IJ-06 and IJ-21 estimated by SDS-PAGE M: Molecular weight marker (Sigma); 1: Zone of inhibition of enterocin from *Ent. faecium* IJ-06; 2: Zone of inhibition of enterocin from *Ent. faecium* IJ-21 against *Ent. faecalis* IJ-11.

4. DISCUSSION

The enterococci used in the present study were sensitive to the glycopeptide antibiotics vancomycin and teicoplanin, in addition, strains did not show haemolytic activity. Furthermore, the absence of the glycopeptide resistance genotypes and the genes involved in the production of the lantibiotic cytolysin was confirmed by PCR. The incidence of isolating vancomycin resistant enterococci has been increasing rapidly in the last few years but its rare in food isolates [13]. Also, cytolysin is involved in the pathogenesis of enterococal diseases, but the haemolytic capacity is restricted to haemolysin/bacteriocin producers [8]. The enterocins studied in this paper consisted of pure peptide, as their activities were inactivated by treatment with proteolytic enzymes, while lipolytic and glycolytic enzymes did not affect them. Moreover, they shared the typical characteristics of class II bacteriocins, being small (<5 kDa), hydrophobic, and heat-stable. They displayed a narrow inhibitory spectrum towards other lactic acid bacteria; Listeria and Clostridium were more sensitive to a concentrated enterocin preparation. Although SDS-PAGE is not an accurate technique to calculate the molecular weight of small, hydrophobic bacteriocins like enterocins [5], but it gives valuable information about the presence of either one or two peptides [3,10,12]. In this study, a single inhibition zone was seen for all the partially purified enterocins, indicating the production of one enterocin. Whereas the enterocin A and P genes were present in the genome of both *E. faecium* IJ-06 and IJ-21. In summary, it is clear that both the examination of the inhibitory spectrum and the characterization of the biochemical properties do not give sufficient information to distinguish between different enterocins. On the other hand, purification of bacteriocins to homogeneity is tedious and cumbersome. Therefore, it may be interesting to include PCR as a first step to detect Enterocin genes. In addition, both bacteriocin production and resistance will certainly play an important physiological role in the diverse ecosystems characteristic for lactic acid bacteria such as milk, meat, vegetables, the gastrointestinal tract, etc.

Acknowledgements The support to this project by Higher Education Commission of Pakistan is gratefully acknowledged.

REFERENCES

[1] De Vuyst L, Vandamme EJ. *Antimicrobial potential of lactic acid bacteria.* In Bacteriocins of Lactic Acid Bacteria: Microbiology, Genetics and Applications ed. De Vuyst L, Vandamme EJ. London: Blackie Academic & Professional.1994: 91–142.

[2] Aymerich T, Garriga M, Ylla J, Vallier J, Monfort JM Hugas M. Application of enterocins as biopreservatives against *Listeria innocua* in meat products. *Journal of Food Protection.* 2000;63:721–726.

[3] Moreno MRF, Sarantinopoulos P, Tsakalidou E, De Vuyst, L. The role and application of enterococci in food and health. *International Journal of Food Microbiology.* 2005;106:1–24.
[4] Cotter PD, Hill C, Ross RP. Bacteriocins: developing innate immunity for food. *Nature Review Microbiology.* 2005;3:777–788.
[5] Aymerich T, Holo H, Havarstein LS, Hugas M, Garriga M, Nes IF. Biochemical and genetic characterization of enterocin A from *Enterococcus faecium*, a new antilisterial bacteriocin in the pediocin family of bacteriocins. *Applied and Environmental Microbiology.* 1996;62:1676–1682.
[6] Cintas LM, Casaus P, Havarstein LS, Hernandez PE Nes IF. Biochemical and genetic characterization of enterocin P, a novel sec-dependent bacteriocin from *Enterococcus faecium* P13 with a broad antimicrobial spectrum. *Applied and Environmental Microbiology.* 1997;63:4321–4330.
[7] Sarantinopoulos, P., Leroy, F., Leontopoulou, E., Georgalaki, M.D., Kalantzopoulos, G., Tsakalidou, E. and De Vuyst, L. Bacteriocin production by *Enterococcus faecium* FAIR-E 198 in view of its application as adjunct starter in Greek Feta cheese making. *International Journal of Food Microbiology.* 2002;72:125–136.
[8] Booth MC, Bogie CP, Sahl HG, Siezen RL, Hatter KL, Gilmore MS. Structural analysis and proteolytic activation of *Enterococcus faecalis* cytolysin, a novel lantibiotic. *Molecular Microbiology.* 1996;21: 1175–1184.
[9] McDonald LC, Kuehnert M, Tenover FC, Jarvis WR. Vancomycin-resistant enterococci outside the health-care setting: prevalence, sources, and public health implications. *Emerging Infectious Diseases.* 1997;3:311–317.
[10] De Vuyst L, Callewaert R, Pot B. Characterization of the antagonistic activity of *Lactobacillus amylovorus* DCE 471 and large scale isolation of its bacteriocin amylovorin L471. *Systematic and Applied Microbiology.* 1996;19:9–20.
[11] Bradford MM. A rapid and sensitive method for the quantification of microgram quantities of protein using the principle of protein-dye binding. *Analytical Biochemistry.* 1976;72:248–254.
[12] Cleveland J, Montville TJ, Nes IF, Chikindas ML. Bacteriocins: safe, natural antimicrobials for food preservation. *International Journal of Food Microbiology.* 2001;71:1–20.
[13] Ginzburg E, Namias N, Brown M, Ball S, Hameed SM, Cohn SM. Gram-positive infection in trauma patients: new strategies to decrease emerging Gram-positive resistance and vancomycin toxicity. *International Journal of Antimicrobial Agents.* 2000;16:39–42.

Structural Model for a Holin-like Protein Tmp3 Exhibiting Broad Spectrum Antibacterial Activity

Thangamani Rajesh and Paramasamy Gunasekaran*

Department of Genetics, Centre for Excellence in Genomic Sciences, School of Biological Sciences, Madurai Kamaraj University, Madurai, 625 021. India.

*Corresponding author: e-mail: gunagenomics@gmail.com, Phone: +91-452-2458478; Fax: +91-452-245-9181

Holins are phage encoded small membrane proteins whose expression is lethal to bacterial host cells. Using metagenomics based approaches; we have identified a small ORF of 105 bp that encodes a protein Tmp3 with broad spectrum antibacterial activity. Tmp3 shared sequence similarity with holin-like proteins and as 3D structural information has not been available for any of the existing holins, the structure of Tmp3 was predicted using LOMETS server. The stability of the predicted structure was assessed by molecular dynamics simulations both in water and membrane environment. The proposed structural model for Tmp3 was found to be highly stable and reliable.

Keywords Holin-like protein; Antibacterial activity; Structural Model; Molecular Dynamics Simulations

1. Introduction

Due to the rapid emergence of antibiotic resistance pathogens, alternate therapeutics have been extensively studied in recent years. One among the alternatives to antibiotics are virolysins or bacteriophage encoded lytic proteins. dsNucleic acid phages employ a holin-endolysin system to accomplish host cell lysis for the release of progeny particles. In this system, endolysins act on cell-walls of bacterial cells and cleave them for the release of progeny viruses. As in certain cases, the endolysins lack secretory signals; they accumulate within the cytoplasmic membrane. In such a condition, holins are responsible for the formation of holes in the cytoplasmic membrane of bacterial cell, facilitating the endolysins to act on bacterial membrane [1]. The result of the net process is the lysis of bacterial cells. As endolysins by nature cleave the bacterial cell wall, they have been widely studied as antibacterials [2]. However, the applicability of holins as antibacterial agents remains unexplored. In a metagenomics based study, we identified a holin-like protein coding gene *tmp3* (Genbank acession number: HM748329). The expressed recombinant protein Tmp3 had broad spectrum antibacterial activity against both Gram-positive and Gram-negative bacteria. As structural information for holin-like proteins does not exist, we evaluated the 3D structure of Tmp3 by *in silico* analysis.

Computational approaches for protein structure prediction have made significant development in recent years [3]. The availability of reliable prediction tools has provided necessary information on protein structures for targets without good templates in protein data bank. One of the recently available structure prediction tool is LOMETS, a local threading meta-server, for quick and automated predictions of protein tertiary structures and spatial constraints. Through this server, threading alignments are made rapidly when compared with traditional meta-servers. Models generated through this server have been shown to be 7% more accurate than that were obtained employing the best individual servers and are also statistically significant. In the present study, a structural model for Tmp3 was obtained employing LOMETS server and was validated using molecular dynamics simulations [4].

2. Materials and Methods

2.1 3D structure modeling

The Tmp3 structure was predicted using the automatic threading and homology modeling service LOMETS [4]. High confidence models were selected and evaluated for stability using molecular dynamics simulations.

2.2 Molecular Dynamics (MD) Simulations

VMD (Visual Molecular Dynamics) and NAMD in conjunction with CHARMM (Chemistry at HARvard Molecular Mechanics) force field were used for MD simulation [5]. The initial coordinates of the protein model was used to begin with.

2.3 Generation of Protein Structure File (PSF)

PSF files were generated from the initial PDB file of the protein using VMD with the help of Automatic PSF builder. The generated PDB file together with the PSF file contains guessed coordinates for hydrogen atoms of the structure.

2.4 Solvating the protein

The protein (Tmp3) to be analysed was solvated by enclosing it in water to more closely resemble the cellular environment. This is performed by placing the protein either in a water sphere in vaccum, in preparation for minimization and equilibration without periodic boundary conditions or in a water box in preparation for minimization and equilibration with periodic boundary conditions. For Tmp3, the protein was placed in a cubic water box of thickness 12 Å, large enough that the protein does not interact with its image in the next cell if periodic boundary conditions were used. The TIP3 model was used and the water box was generated with the help of TCL script from the PSF and PDB files of the protein.

Energy minimization and MD simulations were carried out using the CHARMM27 force field as implemented in NAMD. The simulation system was energy minimized using the steepest descent and conjugate gradient algorithms. The system was first minimized with positional restraints on protein atoms and later without any positional restraints. In order to remove unfavorable contacts, the system equilibrated at 250 K, first with positional restraints on protein atoms for 25 ps and the next 25 ps with unrestrained system. This was followed by productive molecular dynamics run at 300 K for 1 ns with a time step of 2 fs by the use of Leap Frog integrator. All bonds involving hydrogen atoms were constrained to their equilibrium values by means of the SHAKE algorithm. The coordinates of the system were recorded at every 0.4 ps interval during the simulations. For all simulations, periodic boundary conditions were employed to eliminate surface effects. A short-range cutoff of 10 Å was used for all nonbonded interactions while the smooth Particle Mesh Ewald (PME) method was applied for long-range electrostatic forces. The Berendson method with Langevin piston was employed to control temperature and pressure. The group pressure was maintained at 1 atm for 0.1 ps. All simulations were carried in dual core workstation.

2.5 Trajectory analysis and model evaluation

CARMA, EUCB tools and TCL scripts implemented in VMD were used for the analysis of MD trajectory. Prior to analysis, water molecules were removed and all frames were superimposed on to the first frame to remove global (rotational and translational) movements. Root Mean Square Deviations (RMSD) and Root Mean Square Fluctuations (RMSF) of the backbone atoms were calculated relative to the initial MD structure. VMD timeline plug-in employing STRIDE algorithm was used to determine the percentage of secondary structure throughout the simulations. Backbone hydrogen bonds were calculated with an acceptor-donor angle cut-off of 45° and distance cut-off at 3.6 Å.

3. Results

3.1 Three dimensional model building for Tmp3

Homology modeling always produces accurate models as it depends on experimentally determined templates for model building. When homology modeling fails, threading approach is employed to build model for proteins and peptides. To model the 3-D structure of Tmp3, threading based server, LOMETS was employed. In this method the target sequences are threaded through a representative PDB structure library to search for possible folds and the fragments are rearranged to obtain a full-length model. The quality of the model generated for Tmp3 (Figure 1) was assessed by performing a validation with LOMETS server. The results obtained indicate that the structural assignment to be highly significant (Table1).

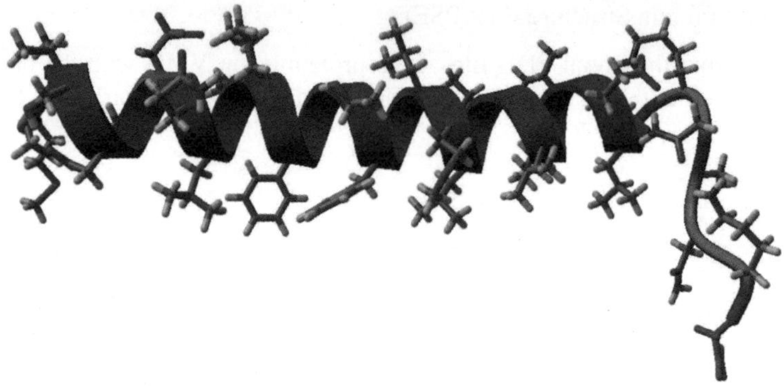

Figure 1. Structural model for Tmp3 obtained from LOMETS server.

Table 1. LOMETS server validation for modeled structure of Tmp3

S. No	Parameter	Values
1	Structure Z-score	Better than average
1	1st generation packing quality	-1.985
2	2nd generation packing quality	0.445
4	Bond lengths	0.760
5	Bond angles	1.248
6	Omega angle restraints	0.256 (tight)
7	Side chain planarity	1.795
8	Improper dihedral	1.136
9	Inside/Outside distribution	1.392

3.2 Molecular dynamics simulations

Lack of templates with significant sequence similarity and the use of fragments based ab-initio modeling to model a structure may incorporate errors on packing of secondary structures and loop conformations in the model. Hence, it is necessary to critically assess the quality of the model and refine the model by MD simulations. The 3D model of Tmp3 was subjected to 0.6 ns molecular dynamics simulations in explicit water to access the stability of the structure. Tmp3 model was solvated with TIP3P water models and net-charge of the system was neutralized by adding counter ions. The system was first energy minimized with positional restraints on protein atoms and followed by unrestrained energy minimization. After energy minimization, the system was equilibrated for 50 ps MD at 250 K with positional restraints on protein atoms and 100 ps unrestrained MD at 250 K to gently relax the system. Productive MD was carried out for 0.6 ns. CHARMM27 force field in conjunction with NAMD/VMD was used for all the energy minimization and MD simulation studies. Protein stability for the model was determined by assessing the following quality metrics:

1. Root Mean Square Deviation (RMSD)

RMSD is an indicator of conformational change along the trajectory with respect to the initial structure. Figure 2a. shows the RMS positional deviations as functions of simulation time for Cα atoms of Tmp3. Absence of a sharp rise or downfall in the RMSD indicates that there is no structural instability or unfolding for the model of Tmp3 in water environment.

2. Trajectory analysis

CARMA, EUCB tools and tcl scripts implemented in VMD were used for the analysis of MD trajectory. Prior to analysis, water molecules were removed and all frames were super imposed on to the first frame to remove global (rotational and translational) movements. VMD timeline plug-in employing STRIDE algorithm was used to determine the percentage of secondary structure throughout the simulations.

Trajectory images obtained with Tmp3 during various stages of MD simulations also indicate that minimal structural deviations in the modeled structure of Tmp3 (Figure 2b).

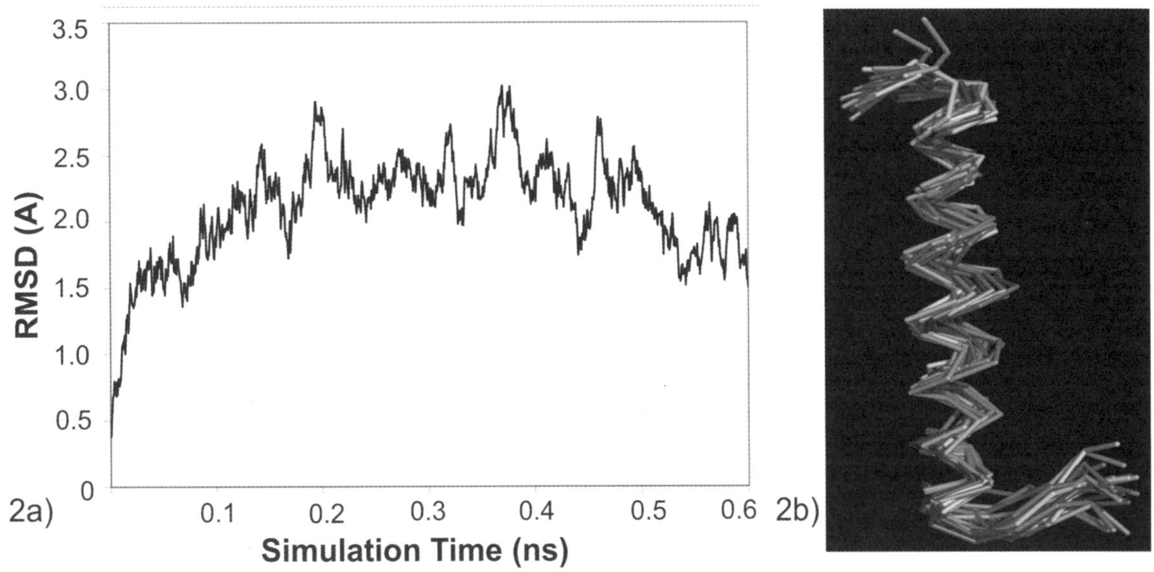

Figure 2. a) RMSD plot of Tmp3 over 0.6 ns. b) Trajectory snap shot indicating various structural conformations of Tmp3 in water environment during different stages of MD simulations.

3.3 Molecular dynamics simulations of Tmp3 in membrane

As Tmp3 interacts with bacterial cell walls, MD simulations was performed with modeled structure of Tmp3 in a membrane environment. The structural model of Tmp3 was further evaluated with molecular dynamics simulation by embedding in POPC membrane. Peptide with membrane was fully solvated and the system was first energy minimized with positional restraints on protein atoms, followed by unrestrained energy minimization. After energy minimization, the system was equilibrated for 100 ps MD at 300 K with positional restraints on protein atoms to relax lipid atoms and 50 ps unrestrained MD at 250 K to gently relax the system. Productive MD was carried out for 0.6 ns. Figure 3.a indicates the conformation of Tmp3 in a membrane environment.

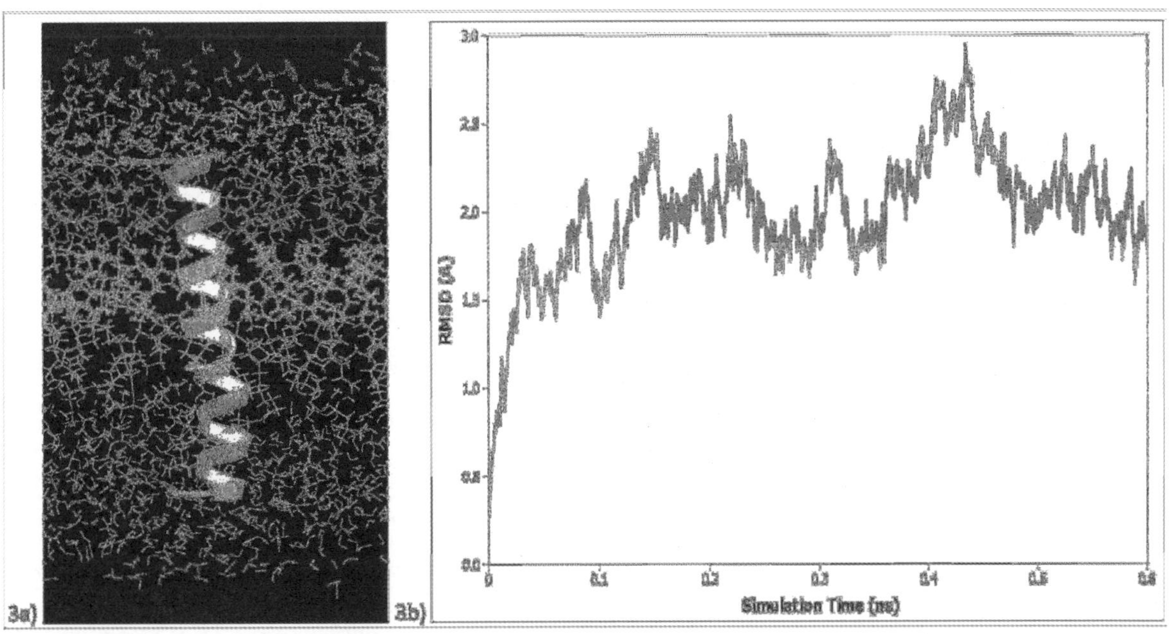

Figure 3. a) Tmp3 in POPC-membrane system. b) RMSD plot of Tmp3 over 0.6 ns.

RMS positional deviations as functions of simulation time for Cα atoms of Tmp3 was studied and an absence of a sharp rise or downfall in the RMSD indicates that there is no structural instability or unfolding for the model of Tmp3 in membrane environment.

4. Discussion

Although the three-dimensional structure of Tmp3 has not been determined experimentally, the structure prediction algorithms suggest that Tmp3 is likely to consist of a single α-helix with an extended loop at the C-terminal. The overall topology of the 3D structure of Tmp3 with a single α-helix is similar to that of those described for maganins. Magainins are cationic, amphipathic α-helical antimicrobial peptides identified in the skin extracts of frogs *Phyllomedusa sauvagei* and *Xenopus laevis* [6]. The presence of a single α-helix in Tmp3 suggest that the hydrophobic transmembrane within Tmp3 may be related to a membrane permeation. The recent models for membrane permeation proposed to proteins structurally similar to Tmp3 suggest that, these kinds of peptides follow 'barrel-stave' mechanism where the hydrophobic transmembrane domain with α-helices form bundles and produce a transmembrane pore leading to bacterial lysis [7]. At present, there is no evidence of holin-like proteins act on cell membranes of bacteria when provided externally. Our current experiments are underway to determine whether Tmp3 functions either in a monomeric or multimeric manner over bacterial membranes.

Acknowledgements Authors thank Department of Biotechnology, New Delhi, India for financial support through research grant (BT/PR-8346/BCE/08/489/2006). TR thanks UGC, India for providing meritorious student research fellowship in Biosciences (No. F. 4-1/2006(BSR)/5-67/2007). The Centre for Advanced Studies in Functional Genomics, Centre for Excellence in Genomic Sciences and Networking Resource Centre in Biological Sciences are gratefully acknowledged for support facilities.

References

[1] Young R. Bacteriophage lysis: mechanism and regulation. *Microbiol Rev.* 1992;56:430-481.

[2] Borysowski J, Weber-Dabrowka B, Gorski A. Bacteriophage endolysins as a novel class of antibacterial agents. *Exp Bio Med.* 2006;231:366-377.

[3] Wiese CK. Review of "The Ten Most Wanted Solutions in Protein Bioinformatics", by Anna Tramontano. *BioMedical Engineering OnLine* 2006; 5:1doi:10.1186/1475-925X-5-1.

[4] Wu S, Zhang Y. LOMETS: A local meta-threading-server for protein structure prediction. *Nucleic Acids Res.* 2007;35:3375-3382.

[5] Humphrey W, Dalke A Schulten K. "VMD - Visual Molecular Dynamics". *J. Molec. Graphics.* 1996;14:33-38.

[6] Zairi A, Tangy F, Bouassida K, Hani K. Dermaseptins and magainins: antimicrobial peptides from frogs' skin-new sources for a promising spermicides microbicides-a mini review. *J Biomed Biotechnol.* 2009. doi:10.1155/2009/452567.

[7] Oren Z, Shai Y. Mode of action of linear amphipathic alpha-helical antimicrobial peptides. *Biopolymers.* 1998;47:451-463.

The hospital environment as a source of new antimicrobial substances: antimicrobial activity of *Bacillus amyloliquefaciens* strain HNA3 isolated in a surgery room

M.Guida[*], R.A. Nastro[§#], M.Inglese[°*], M.Trifuoggi[§], S.Scherillo[*], R. Gesuele[*], V. Di Onofrio[+] and G.Liguori[+]

*Department of Biological Sciences – University of Naples "Federico II"
°Laboratory of environmental microbiology and ecotoxicology "Inglese Mafalda &co. s.a.s"
§Department of Chemical Sciences "Paolo Corradini", University of Naples "Federico II"
+Chair of Epidemiology and Hygiene, University of Naples "Parthenope"

[#]Corresponding author: Dr. Nastro Rosa Anna-e-mail address: r.nastro@alice.it

In the course of a survey on the environmental quality in eight different private hospitals, a *Bacillus amyloliquefaciens strain HNA3* producing an inhibition halo on a *Bacillus subtilis* strain, growing on the same surface. An inhibition zone was evident against *Bacillus subtilis* isolate on the same wall, on *Pseudomonas aeruginosa* ATCC 15442, *Aeromonas hydrophyla* (environmental isolate and ATCC 7966), *Bacillus cereus* ATTC 11778 and a *Vibrio harvey* strain from thermal water. No effects were detected on *B.subtilis* ATTC 6633, *Staphylococcus aureus* (ATTC 6538 and ATTC 25923), *Streptococcus faecalis* (ATCC 33186), *Escherichia coli* (ATTC 10536), *Salmonella tiphymurium* (ATCC 14028), *Shigella sonnei* (ATCC 25931), and *Klebsiella pneumoniae* (ATCC 13883). The susceptibility at both antibiotics and disinfectants of *Bacillus subtilis* strain isolated on the surgery wall were tested. According to our preliminary findings, the environmental hospital could be considered as the candidate place for searching new antibiotic producers.

Keywords: *Bacillus amyloliquefaciens strain HNA3,* antibiotic compounds; clinical environment.

1. Introduction

According to the available literature, many species belonging to the genus Bacillus are known as antibiotics and enzymes producers (1;2;3). *Bacillus amyloliquefaciens* was isolated in 1943 by Fukomoto, but it was renamed and confirmed as a distinct species in 1987 (4). Recently, the European Food Safety Agency (EFSA) approved the use of *B.amyloliquefaciens* spores as probiotic in poultry food because of its ability to produce α-amilase (3;5). Moreover, its ability to produce antimicrobial compounds had been estabilished by different researchers against protozoa (6) and fungi (7;8). But, till now, there is no evidence of an antimicrobial activity against Gram negative bacteria.

2. Methods

A RODAC contact plate containing Plate Count Agar (PCA) by Oxoid ltd. (as all the media used in this survey) was used to evaluate the microbial contamination of a wall in a surgery room of an hospital, in Naples district. After an incubation at 37°C for 24 hours, an uniform, white patina developed on the surface of the agar due to the growth of a *Bacillus subtilis* strain. The patina was interrupted by an inhibition halo surrounding a small, white colony of *Bacillus amyloliquefaciens* strain HNA3. Both strains were identified genetically (NCBI nucleotide Genbank). As first step, we tested the in-vitro susceptibility (9) of the environmental isolate of *B. subtilis* to 13 antibiotics chosen among the most commonly used in medical practices in Italy (10), and to 6 disinfectants (table 1). This way, we tried a first characterization of the "target" bacterium.

Bacillus amyloliquefaciens HNA3 isolate was suspended in 6 ml of Buffered Peptone Water (BPW) (pH=7,0), incubated at 37°C for 6-8 hours and then plated by using a 5μl loop on Tryptic Soy Agar (TSA). This procedure was repeated several times to obtain pure cultures. Then, we tested in-vitro ability of *B. amyloliquefaciens HNA3* to inhibit the growth of several bacterial species belonging both to ATCC and environmental gram-positive and gram-negative strains. Isolated colonies of test strains were transferred to Tryptic Soy Broth (TSB) to a turbidity of 0.5 McFarland units and 100 μl of suspension was seeded on the dried surface of a plate containing Mueller Hinton Agar (MHA). Then, a 1-μl loop was used to transfer a uniform amount of an isolated colony of *B. amyloliquefaciens HNA3* colony to the centre of each plate previously seeded with test bacteria (Table 2 and 3).

Aminoglicosides	Pennicilines	Fluoroquinolones	Macrolides	Cephalosporins	Quinolones	Disinfectants
Tobramicin	Imipenem	Ciprofloxacin	Eritromicin	Cefalotin	Pipemic acid	Ethanol 96%
Chloramphenicol	Ampicillin			Cefotaxime		Sodium hypocloride 5%
Gentamicin	Aztreonam					Hydrogen peroxide
Amikacin						Iodopovidone 10%
						Benzalchonium chloride 1,5 g/l
						Amuchina 2%

Table 1: List of antibiotics and disinfectants used in *Bacillus subtilis* isolate in-vitro susceptibility tests.

Table 2: List of the ATTC strains whose susceptibility against *B. amyloliquefaciens* strain HNA3 was tested.

Gram positive bacteria	***Gram negative bacteria***
Bacillus cereus ATTC 11778	*Escherichia coli ATTC 10536*
Bacillus subtilis ATTC 6633	*Shigella sonnei ATTC 25931*
Staphylococcus aureus ATTC 6538	*Klebsiella pneumoniae ATTC 13883*
Staphylococcus aureus ATTC 25923	*Pseudomonas aeruginosa ATTC 15442*
Streptococcus faecalis ATTC 33186	*Aeromonas hydrophila ATTC 7966*
	Salmonella typhimurium ATTC 14028

Table 3: List of *Vibrionaceae* isolated from thermal water and used to check the antibiotic activity of *B. amiloliquefaciens* strain HNA3.

Vibrio cholerae	*Vibrio vulnifuicus*
Vibrio alginoliticus	*Vibrio pomeroy*
Vibrio fluvialis	*Vibrio harvey*
Photobacterium damsela	*Vibrio hyspanicus*
Aeromonas hydrophila	

3. Results

In table 4 are shown *Bacillus subtilis* pattern of susceptibility to the chosen antibiotics and disinfectants. *B. subtilis* was susceptible to all the antibiotics except for cefotaxime, inhibiting bacterial cell wall synthesis by binding to one or more of the penicillin-binding proteins (PBPs) thus inhibiting cell wall biosynthesis, and aztreonam, inhibiting mucopeptide synthesis in the bacterial cell wall thereby blocking peptidoglycan crosslinking. Penicillines and cephalosporines, inihibiting transpeptidase activity, have been both effective together with macrolids (inihibiting the 50S subunity), aminoglicosides (binding the 30S subunity) and quinolones (blocking DNA biosynthesis) (11;12). It is possible to hypothesize that *B. amyloliquefaciens* antimicrobial compounds mechanism of action is different from the cefotaxime and aztreonam ones. Among the disinfectants, sodium hypocloride 5% was the most effective against *B. subtilis*.

Table 4: *Bacillus subtilis* isolate susceptibility to disinfectants and antibiotics.

Disinfectant	***Inihibition halo diameter (mm)***	***Antibiotics***	***Susceptibility/resistance***	***Antibiotics***	***Susceptibility/resistance***
Ethanol 96%	*0*	*Tobramicin*	*Susceptible*	*Ciprofloxacin*	*Susceptible*
H2O2 3%	*0*	*Cloramphenicol*	*Susceptible*	*Eritromicin*	*Susceptible*
H2O2 10%	*Bacteriostatic effect*	*Gentamicin*	*Susceptible*	*Cephalotin*	*Susceptible*
NaClO 5%	*40 mm*	*Amikacin*	*Susceptible*	*Cefotaxime*	*Resistant*
Quaternary salts of ammonium	*20 mm*	*Imipenem*	*Susceptible*	*Pipemic acid*	*Susceptible*
Iodopovidone (10% iodine)	*15 mm*	*Aztreonam*	*Resistant*	*Rifampicin*	*Susceptible*

As to the susceptibility tests to *B. amyloliquefaciens* strain HNA3, after 24 hours of incubation at 37°C, all seeded plates were examined. Of the gram positive test strains, the environmental isolate of *B. subtilis* was inhibited while no zone of inhibition was observed for *B. subtilis* ATTC strain, *Staphylococcus aureus* (both ATTC strains), and *Streptococcus faecalis*. Of the gram negative test strains, *Aeromonas hydrophyla* (environmental isolate and ATCC) and *Pseudomonas aeruginosa* had zones of inhibition, while no halo of inhibition was detected in *Escherichia coli*, *Salmonella tiphymurium*, *Shigella sonnei*, and *Klebsiella pneumoniae* cultures. Additional testing of *Vibrio* species resulted in a zone of inhibition for an environmental isolate of *Vibrio harvey* but not for environmental isolates of *Vibrio cholerae* and *Vibrio vulnificus.*

Table 5: Susceptibility of test bacteria to *B. amyloliquefaciens* antibiotic molecules.

Bacterial strain	***Inhibition halo (mm)***
Bacillus cereus (ATTC 17778)	*13 mm*
Pseudomonas aeruginosa (ATTC 15442)	*13 mm*
Aeromonas hydrophila (ATTC 7966)	*15 mm*
Aeromonas hydrophila (thermal water)	*14 mm)*
Vibrio harvey (thermal water)	*57 mm*
Bacillus subtilis (environmental strain)	*18 mm*

4. Conclusions

The preliminary findings here reported, shed light to the ability of *Bacillus amyloliquefaciens* strain HNA3 to produce antibiotics against potential pathogenic bacteria as *Pseudomonas aeruginosa* (responsible of a great number of nosocomial infections), *Aeromonas hydrophyla* (an opportunistic pathogen) and *Vibrio harvey* (associated to some diseases in aquatic organisms) . The production of antibiotics plays an important role in the competition among different microbial strains for food and space (quorum sensing). It is possible, then, that *B. amyloliquefaciens strain HNA3* in a hospital environment (like a surgery room) had to compete with the above mentioned gram negative bacteria. Further studies will be needed to define the nature of the antimicrobial compounds produced by *B. amyloliquefaciens strain HNA3*, in the order to evaluate possible biotechnological applications.

Acknowledgements The support of Dr. Pina Tirimperio, Dr.Daniela Scudiero and Dr. Amelia di Costanzo are gratefully acknowledged.

References

[1] Mannanov R. N. and Sattarova R. K.. Antiobitics produced by Bacillus bacteria. *Chemistry of Natural Compounds,* 2001.Vol. 37, No. 2.

[2] Lee K.-L., Jun K.-D., Kim W-S., Paik H.-D. Partial cahacterization of polyfermenticin SCD, a newly identified bacteriocin of *Bacillus polyfermenticus*.(2008). *Lett. in Appl. Microbiol*. Vol.32, Iss.146-151.

[3] Gangadharan D., Sivaramakrishnan S., Madhavan Nampoothiri K. & Pandey A. Solid Culturing of *Bacillus amyloliquefaciens* for Alpha Amylase Production.(2006) Food Technol. Biotechnol. 44 (2) 269–274

[4] Priest F. G., Googfellow M., Shute L.A., Berkley R. C. W. *Bacillus* amyloliquefaciens sp. nov. norn. rev.Intern. Journal of Systematic Bacteriol. Jan,. 1987, Vol. 37, No. 1; p. 69-71

[5] European Food Safety Agency. Safety and efficacy of Ecobiol® (*Bacillus amyloliquefaciens*) as feed additive for chickens for fattening - Scientific Opinion of the Panel on Additives and Products or Substances used in Animal Feed. http://www.efsa.europa.eu/EFSA/efsa_locale-1178620753812_1211902039267.htm. Accessed November 29, 2010.

[6] Brittes Benitez L., Caumo K., Brandelli A., Brittes Rott M.. Bacteriocin-like substance from *Bacillus myloliquefaciens* shows remarkable inhibition of *Acanthamoeba polyphaga. Parasitol Res*, 2010;Oct.22 Epub ahead of print.

[7] Caldeira A.T., Savluchinske Feio S., Arteiro J.M. S., Roseiro J. C. Bacillus amyloliquefaciens CCMI 1051 in vitro activity against wood contaminant fungi. *Annals of Microbiology*, 2007; 57 (1) 29-33.

[8] Arguelles-Arias A., Ongena M., Halimi B., Lara Y., Brans A., Joris B., Fickers P. *Bacillus amyloliquefaciens* G1 as a source of potent antibiotics and other secondary metabolites for biocontrol of plant pathogens. *Microbial Cell Factories* 2009, **8**:63.

[9] Lalitha M.K. (2004). Manual on Antimicrobial Susceptibility Testing. Indian Journal of Medical Microbiology. Special Issue available online at www.ijmm.org

[10] Regione Emilia Romagna. Sorveglianza dell'antibioticoresistenza e uso di antibiotici sistemici in Emilia-Romagna Rapporto 2007. http://asr.regione.emilia-romagna.it/wcm/asr/collana_dossier/doss173.htm accessed on 30 November.

[11] Marne Gaynor and Alexander S. Mankin. Macrolide Antibiotics: Binding Site, Mechanism of Action, Resistance (2005) *Frontiers in Medicinal Chemistry, ,2*, 21-35.

[12] Blondeau J.M. Fluoroquinolones: mechanism of action, classification, and development of resistance. (2004). Survey of Ophtalmology Vol 42, Issue 2, Suppl.2, Pages S73-S78.

The Inhibitor activity of *Lactobacillus rhamnosus* against rope-forming *Bacillus* strains

Selçuk Arslan[1], Zerrin Erginkaya[1] and Emel Ünal[*,1]

[1] University of Çukurova, Faculty of Agriculture, Department of Food Engineering, Adana, Turkey

* Corresponding author: e-mail: unale@cu.edu.tr, Phone: +34 905365785022

In this study, it was investigated inhibitor activity of *Lactobacillus rhamnosus* used as biopreservative against rope-forming *Bacillus* strains as in vitro (agar diffusion method and in broth medium) and in vivo (in dough from bread flour).
Supernatants and cells of *Lactobacillus rhamnosus* were evaluated with regard to inhibitor activity after incubation for 24 hours. After incubation for 24 hours in broth medium, it was found that 6 isolate from 20 isolate (*Bacillus spp.*) were inhibited by both supernatants and cells. According to agar diffusion method, while 5 isolate are resistant to supernatant, 4 isolate are resistant to cells. Also in dough added only *Bacillus subtilis* ($1,3x10^4$cfu/g), counts of *Bacillus subtilis* increased, whereas an important change didn't occur in counts of *Bacillus subtilis* ($2,9x10^4$cfu/g) in dough added mix of *Bacillus subtilis* ($1x10^4$cfu/g) and *Lactobacillus rhamnosus*. As a result, it was determined that while *Lactobacillus rhamnosus* has inhibitor activity against some *Bacillus subtilis* strains, it didn't have any inhibitor activity against some of them.

Keywords: *Lactobacillus rhamnosus,* ropy spoilage, dough

1. Introduction

Flour with many physical, chemical and biological problems causes significant economic losses. Mold contamination and rope formation are the problems often encountered in these products [1, 2].

Ropy spoilage is caused mainly by *Bacillus subtilis* and *Bacillus licheniformis*, the spores of which contaminate raw materials such as flour, bread improvers, yeast, etc., and survive baking temperatures [3, 4, 5].

As an increasing number of consumers prefer natural foods without chemical preservatives. So nowadays, new preservation methods are focused on the use of lactic acid bacteria (LAB) or of their metabolites. Metabolites of lactic acid bacteria such as organic acid, diacetil, acetoin, hydrogen peroxide, reuterine, antifungal peptide and bacteriocin have been used commonly as natural preservative [6, 7].

In previous studies, it was determined that lactic acid bacteria have inhibitor effect on *Bacillus* spp. as much as chemical preservatives have [8, 9, 10, 11].

The aim of this study is to determine inhibitor activity of *Lactobacillus rhamnosus* against rope-forming *Bacillus* strains as in vitro (agar diffusion method and in broth medium) and in vivo (in dough from bread flour).

2. Material and Method

2.1. Material

In this study, wheat flour samples were supplied from some flour and bread factory in Gaziantep. Also, as reference strains, *Bacillus subtilis* strains were provided by University of Ankara and *Lactobacillus rhamnosus* strains were bought from Denisco USA.

2.2. Method

2.2.1. Isolation of *Bacillus* spp. from flour samples

For *Bacillus* isolation, vegetative form of *Bacillus* spp. in diluted samples were inhibited at 65°C, 60 minute and then diluted samples were cultivated on Nutrient Agar (NA) and incubated at 30^0C, for 5 days [12, 13].

2.2.2. The determination of inhibitor activity of *Lactobacillus rhamnosus* against Bacillus strains

Lactobacillus rhamnosus was grown in MRS broth and was diluted to 10^5cfu/ml. The suspensions of *L. rhamnosus* were centrifuged (5000g, 10 min). After centrifugation, it was separated as cell and supernatant.

Bacillus isolates from flour were grown in Nutrient Broth (NB) at 30 °C, for 24 hours. *Bacillus* isolates were diluted to 10^6cfu/ml. Suspension of *Bacillus* spp. was used to determine inhibitor activity of *L. rhamnosus* as in vitro (agar well diffusion method and in broth medium) and in vivo (in dough from bread flour) [12, 13].

2.2.2.1. Use of agar well diffusion method for the determination of inhibitor activity

10^6cfu/ml from *Bacillus* isolate was cultivated on PCA and then 100 µl from supernatant or cell of *L. rhamnosus* were placed in wells in plates seeded with bioassay strains. The plates were incubated at 37 °C, for 24-48 hours and the diameters of the inhibition zone were taken. [12, 13].

2.2.2.2. Use of broth medium for the determination of inhibitor activity

1 ml from cell or supernatant of *Lactobacillus rhamnosus* (10^5cfu/ml) was added in nutrient broth medium (10 ml) that contain *Bacillus* spp. (10^6cfu/ml). Mix was cultivated on PCA for determining counts of Bacillus spp. in this mix and incubated at 37 °C, for 24 hours [12, 13].

2.2.2.3. Use of bread dough for the determination of inhibitor activity

25 g from flour and 5 ml from *Bacillus subtilis* (10^7cfu/ml) or 5 ml from *L. rhamnosus* (10^5cfu/ml) were mixed and bread dough was prepared. Diluted each dough sample was cultivated on PCA and incubated at 37 °C, for 24 hours and *Bacillus* strain was counted [12, 13].

3. Results and Discussion

As a result of this study, after incubation for 24 hours in broth medium, it was found that 6 isolate from 20 isolate (*Bacillus spp.*) were inhibited by both supernatants and cells. According to agar diffusion method, while 5 isolate are resistant to supernatant, 4 isolate are resistant to cells. Also in dough added only *Bacillus subtilis* ($1,3x10^4$cfu/g), counts of *Bacillus subtilis* increased, whereas an important change didn't occurred in counts of *Bacillus subtilis* ($2,9x10^4$ cfu/g) in dough added mix of *Bacillus subtilis* ($1x10^4$cfu/g) and *Lactobacillus rhamnosus*. It has been shown the inhibitor activity of *Lactobacillus rhamnosus* against rope-forming *Bacillus* strains in picture 1.

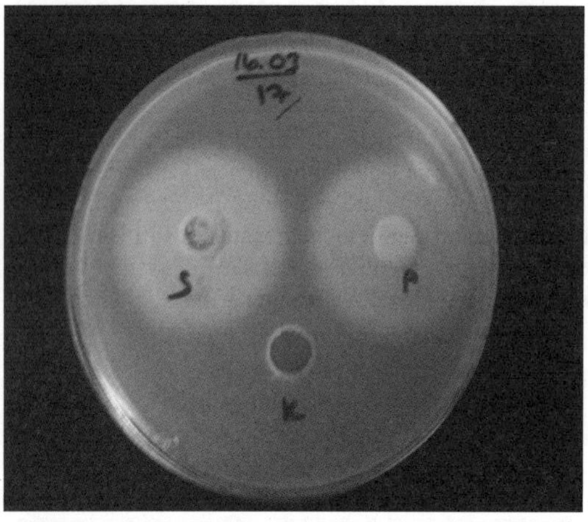

Picture 1. The inhibitor activity of *Lactobacillus rhamnosus* against rope-forming *Bacillus* strains

In conclusion, it was determined that while *Lactobacillus rhamnosus* has inhibitor activity against some *Bacillus subtilis* strains, it didn't have any inhibitor activity against some of them. Also, while addition of *Lactobacillus rhamnosus* to both broth medium and agar medium led to decreasing in counts of *Bacillus* spp., *L. rhamnosus* in bread dough prevent growth of *Bacillus subtilis* but it didn't occurred any decreasing in counts of *Bacillus subtilis*.

References

[1] Daniel, AR. Bakery Materials and Methods. Mc Laren & Sons,Ltd.London, England, 1963

[2] Demir, Y. Trakya bölgesinde üretilen ekmeklik buğday unlarında rop sporu varlığı ve bazı kalite kriterlerinin belirlenmesi, Yüksek Lisans Tezi, *Trakya Üniversitesi Fen Bilimleri Enstitüsü Gıda Mühendisliği Ana Bilim Dalı*, Tekirdağ, 2006.

[3] Palop, A, Mana AS, P, Condo AN, S. Sporulation temperature and heat resistance of *Bacillus* spores: a review. *Journal of Food Safety*, 1999: 19, 57-72.

[4] Simmonds, P, Mossel, BL, Intaraophan, T, Deeth, HC. Heat resistance of *Bacillus* spores when adhered to stainless steel and its relationship to spore hydrophobicity, *J Food Prot*, 2003:66,2070-2075.

[5] Iurlina, MO, Saiz, AM, Fuselli, SR, Fritz, R. Prevalence of *Bacillus* spp. in different food products collected in Argentina. *LWT*, 2006:(39) 105–110.

[6] Adams, MR. Safety of Industrial Lactic Acid Bacteria. *Journal of Biotechnology*, 1999:68,171-178.

[7] Ross, RP, Desmond, C, Fitzgerald, GF, Stanton, C. Overcoming the technological hurdles in the development of probiotic foods, *J. of Appl. Microbiol.*,2005: 98, 1410-1417.

[8] Spelhaug, SR, Harlander, SK. Inhibition of foodborne bacterial pathogens by bacteriocins from *Lactococcus lactis* and *Pediococcus pentosaceous. Journal of Food Prot.*, 1989:52(12);856-862.

[9] Rosenquist, H, Hansen, A. The antimicrobial effect of organic acids, sourdough and nisin against *Bacillus subtilis* and *B. licheniformis* isolated from wheat bread. *Journal of Applied Microbiology*. 1998:85(3), 621–631.

[10] Olimpia, P, Giuseppe, B, Giancarlo, M, Greco, T, Francesso, V. Sünme-Producing Strains of *Bacillus* spp. from Wheat Bread and Strategy for Their Control by Lactic Acid Bacteria, Dipartimento di Scienza degli Alimenti, *Universita` degli Studi di Napoli Federico II*, 80055 Portici, Italy, 2002.

[11] Adams, MR. Safety of Industrial Lactic Acid Bacteria. *Journal of Biotechnology*, 1999:68,171-178.

[12] Halkman, K. Gıda Mikrobiyolojisi Uygulamaları. *Merck.Başak matb.*, Ankara. 2005:358 s.

[13] Sübmuth, R, Eberspaecher, J, Haag, R, Springer, W. Biochemisch Mikrobioogisches Prakticum. *Thieme Verlag-Stuttgart*, 1987:409 s.

Uncovering the mechanism of resistance against type III bacteriocin, enterolysin A

L. Maliničová, M. Piknová, P. Pristaš[*] and P. Javorský

Institute of Animal Physiology, Slovak Academy of Sciences, Šoltésovej 4-6, 04001 Košice, Slovakia

*Corresponding author: e-mail: pristas@saske.sk, Phone: +421556783120

Enterolysin A (EnlA) is a peptidoglycan hydrolase produced by some enterococcal strains with a relatively broad inhibitory spectrum. This enzyme is being classified as a class III bacteriocin (large heat-labile bacteriocin) and it is interesting due to its domain structure resembling the structure of bacteriophage peptidoglycan-degrading enzymes (endolysins). N-terminal catalytic domain is responsible for cleaving chemical bonds in peptidoglycan and C-terminal domain has been thought to be responsible for EnlA binding to its cell wall substrate. Using EnlA C-terminal domain fused with green fluorescent protein (GFP-CWB) followed by fluorescence assay we demonstrated that this part of EnlA represents true cell wall binding (CWB) domain. Although resistance to bacteriocins often results from the expression of so-called "immunity protein" (in bacteriocin producing strain encoded by the same operon as bacteriocin itself), we suppose that in the case of EnlA, resistance results from the absence of specific receptor needed for EnlA binding in resistant strains.

Keywords enterolysin A; CWB domain; resistance

1. Introduction

1.1 Peptidoglycan hydrolases

Peptidoglycan hydrolases (PGHs) represent a broad group of enzymes of different origin, but similar structure. They are able to cleave chemical bonds in peptidoglycan, which is the main constituent of bacterial cell wall. Activity of these enzymes results in a cell wall disruption and subsequent bacterial cell lysis.

Peptidoglycan is a polymer of alternating units of the disaccharide *N*-acetylmuramic acid-(β-1,4)-*N*-acetylglucosamine (NAM-NAG), that are cross-linked by short peptides. Wall peptides are cross-linked with other peptides that are attached to a neighbouring glycan strand, thereby generating a three-dimensional molecular network that surrounds the cell [1]. Except peptidoglycan, Gram-positive cell wall contains another important component, teichoic acids. These anionic polymers occur in two distinct forms depending on whether they are linked to the peptidoglycan (teichoic acid) or to the head groups of cytoplasmic membrane lipids (lipoteichoic acid) [2].

According to the specific bond that is split down in the peptidoglycan, distinct classes of PGHs are defined. As Figure 1 shows, *N*-acetylmuramoyl-L-alanine amidases cleave the amide bond between the NAM and L-alanine of the peptide bridge, endopeptidases and carboxypeptidases cleave short peptide bridges, and *N*-acetylglucosaminidases and *N*-acetylmuramidases hydrolyze β-1,4-bonds between NAM and NAG in glycan strands.

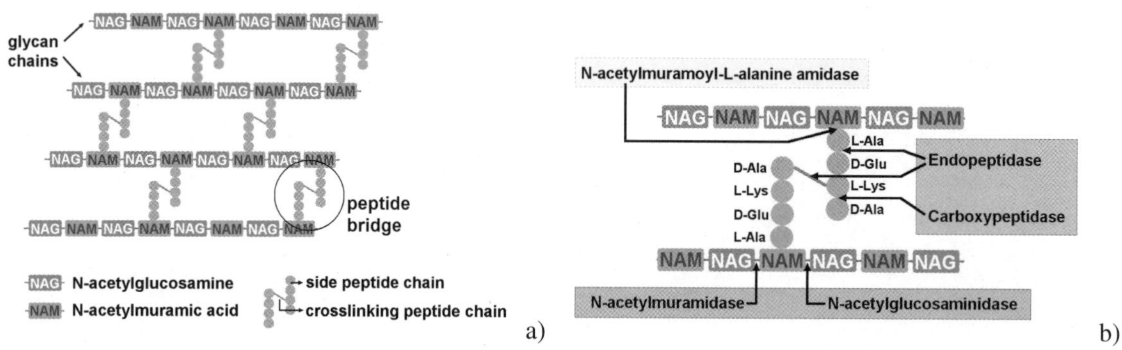

Fig. 1 Substrate specificity of peptidoglycan hydrolases. a) Schematic structure of the peptidoglycan and b) classification of PGHs according to the specific bond in the peptidoglycan that they cleave.

There are many sources of PGHs including animals, plants, insects, bacteria, and bacteriophages. Although enzymes from different sources may have similar structures and functionalities, they have important differences with significant impacts on their applications [3]. According to their origin PGHs may be classified into lysozymes, bacteriocins, autolysins, and endolysins.

1.1.1 Modular structure of peptidoglycan hydrolases

One of the common features of PGHs is their modular structure, that is, they are composed of at least two distinct functional domains (modules) [4]. Generally, the N-terminal domain contains the catalytic activity of the enzyme, which cleaves specific bond in the peptidoglycan as described above. The C-terminal domain of enzyme has specificity for a cell wall substrate. Thus, unless the binding domain binds to its substrate, the catalytic domain will not cleave, offering specificity to most enzymes studied [5].

1.2 Enterolysin A

Enterolysin A was first described in 2003 as the first bacteriocin from genus *Enterococcus* belonging to class III, the large and heat-labile bacteriocins, with a broad inhibitory spectrum including selected enterococci, streptococci, pediococci, lactococci, and lactobacilli. The N-terminal region of the protein contains a domain found in the M37 family of metallopeptidases (catalytic domain). The C-terminal part of EnlA is distantly related to the known hydrolases only, but it shows significant sequence similarity to two bacteriophage endolysins [6]. Until now this part of the protein has been speculated to possess a cell wall binding function (putative CWB domain).

Despite being classified as a bacteriocin, EnlA thanks to its modular structure resembles the bacteriophage peptidoglycan-degrading enzymes (endolysins).

1.3 Resistance to bacteriocins

All bacteriocin producers are insensitive to the bacteriocin(s) they produce. Resistance to bacteriocins in most cases results from expression of so-called bacteriocin immunity protein. The genetic determinants proposed or confirmed to confer immunity are frequently found downstream of the bacteriocin structural gene(s) in the same operon as bacteriocin itself [7]. Bacteriocin immunity proteins may act by two ways: (1) they may impact and enzymatically modify molecules of bacteriocin, or (2) they may interact with bacteriocin receptor and change its structure.

However, the mechanisms by which bacteria are resistant to bacteriocins vary, depending on the mechanism of resistance apparent in each particular strain. It is speculated that resistance may be mediated by alterations in components of the membrane or cell wall preventing adherence or functioning of the bacteriocin [8]. Some bacteriocins (mainly those with domain structure) require specific receptors in sensitive bacterial cell surface, therefore absence of such receptors results in bacteriocin resistance.

1.4 Aims

The aims of our work were: (1) to investigate the function of C-terminal domain of enterolysin A and its participation in cell wall targeting, (2) to examine the relationship between EnlA sensitivity/resistance and binding ability of its C-terminal domain, (3) to analyse the regions flanking *enl*A gene (are there any genes typical either for bacteriocin operons or for phage lytic systems?), (4) to uncover the mechanism of resistance against EnlA, and (5) to identify the nature of EnlA receptor on bacterial cell surfaces.

2. Experimental procedures

2.1 Chimeric protein GFP-CWBenlA fluorescence binding assay

Fusion of EnlA C-terminal domain with green fluorescent protein (GFP-CWBenlA) followed by fluorescence assay was used to study species-specific directing and binding of EnlA to the bacterial cell wall envelopes.

Both *gfp* gene (encoding Green Fluorescent Protein) and C-terminal domain of *enl*A gene were amplified using primers with specific restriction sites. Gfp amplicon was cloned into pQE-30 UA expression vector (generating plasmid pQE-GFP). To construct GFP-CWBenlA, SacI/SalI digested pQE-GFP vector and C-terminal domain of *enl*A were ligated together and the resultant plasmid construct was designated pGEnlCWB. Fusion gene *gfp*-CWB*enl*A was heterologously expressed in the host strain *E. coli* ER2566 (New England Biolabs) and recombinant clones were identified by fluorescence test followed by SDS-PAGE analysis.

To study the cell wall binding potential of C-terminal domain of EnlA, the following conditions were used: recombinant clone expressing GFP-CWBenlA was grown on LB agar plates overnight, colonies were washed off with 0.9% NaCl, harvested by centrifugation and disrupted by homogenization. The cell debris was removed by centrifugation and the supernatant was used in binding assays.

For binding assays, supernatant containing GFP-CWBenlA was mixed with various strains of enterococci and streptococci, reaction mixtures were incubated for 40 minutes, bacteria were sedimented by centrifugation, washed and suspended in physiological saline, and GFP-CWBenlA binding was immediately visualised by fluorescence microscopy.

All strains used in previous binding ability test were consecutively tested on sensitivity/resistance against purified EnlA by agar diffusion test.

2.2 Analysis of the flanking DNA regions of *enl*A gene

To elucidate if there are any genes typical either for bacteriocin operons or for phage lytic systems, neighbouring DNA regions of *enl*A gene were amplified using inverse PCR and vectorette PCR (Universal Vectorette™ System; Sigma-Aldrich).

As a template total DNA of *Enterococcus faecalis* II/1 (natural producent of EnlA) was used [9]. All obtained amplicons were sequenced and sequences compared against http://www.ncbi.nlm.nih.gov/ GenBank database using the Blastx algorithm.

2.3 Identifying the nature of EnlA receptor

To identify a chemical nature of the receptor, a competitive binding assay for galactose and rhamnose (frequent saccharides in enterococcal cell wall) was used. These saccharides were added parallel to the mixture of GFP-CWBenlA and tested cells, and a prospective influence on binding ability of GFP-CWBenlA was monitored.

The second approach was to treat tested cells with proteinase (10 and 100 x diluted proteinase K, 934 U/ml, Fermentas) or lysozyme (4,5 mg/ml and 0,45 mg/ml; Lysozyme from chicken egg white, 185000 U/mg, SERVA) followed by GFP-CWBenlA binding assay (described above).

In both approaches two sensitive and two resistant enterococcal strains were tested. After incubation, mixture containing GFP-CWBenlA and tested cells was centrifugated and fluorescence in supernatant was measured by Synergy 2 Multi-Mode Microplate Reader (BioTek).

3. Results and discussion

3.1 Confirmation of presumptive function of EnlA C-terminal domain

As Figure 2 shows, the results of fluorescence assay clearly demonstrated that the C-terminal domain of enterolysin A targets protein specifically to the cell walls of sensitive bacteria and represents thus a true cell wall binding (CWB) domain. Besides, data obtained using fluorescence binding assay correlated with sensitivity/resistance of bacterial species to EnlA determined by agar diffusion test (see Table 1).

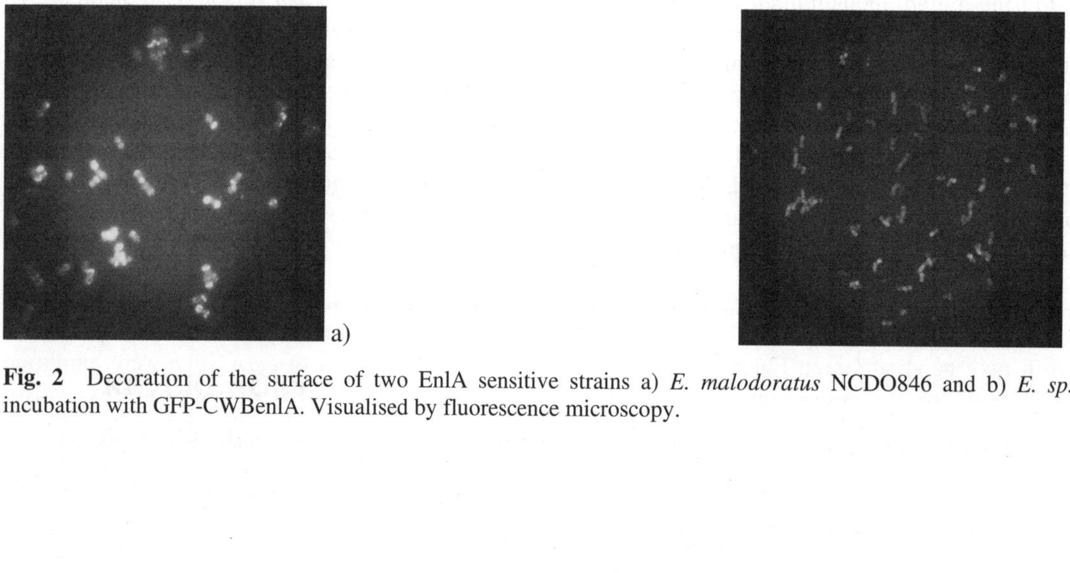

a) b)

Fig. 2 Decoration of the surface of two EnlA sensitive strains a) *E. malodoratus* NCDO846 and b) *E. sp.* 3A after incubation with GFP-CWBenlA. Visualised by fluorescence microscopy.

Table 1 Sensitivity/resistance of bacterial strains to EnlA vs. ability to bind GFP-CWBenlA. Sensitivity/resistance of bacterial species to EnlA hydrolase determined by agar diffusion test correlated with data obtained using fluorescence binding assay.

Bacterial strain	Sensitivity to EnlA	GFP-CWBenlA binding	Bacterial strain	Sensitivity to EnlA	GFP-CWBenlA binding
E. faecium 1A	+	+	*S. bovis* 46/2	-	-
E. faecium 1C	+	+	*S. bovis* 47/3	-	-
E. faecium 1D	+	+	*S. bovis* 59/2	-	-
E. faecium 1F	+	+	*S. gallolyticus* LRC0253	+	+
E. faecium 1H	+	+	*S. gallolyticus* LRC0255	-	-
E. durans 2C	+	+	*E. malodoratus* NCDO 846	+	+
E. sp. 3A	+	+	*E. faecalis* II/1	-	-
E. mundtii 3B	+	+	*E. hirae* 3B-1	+	+
E. mundtii 3E	+	+	*E. faecium* CTC492	+	+
E. casseliflavus 3F	+	+	*E. faecium* AO	+	+
E. durans 5A	+	+	*E. faecium* LMG2771	+	+
E. hirae 6A	+	+	*E. faecalis* 7827	+	+
E. hirae 8E	+	+	*E. faecalis* 7790	-	-
E. durans 8H	+	+	*E. faecium* AL41	+	+
E. faecium 51	-	-	*E. faecalis* 7563	+	+
S. bovis 44/9	+	+	*E. sp.* 8B-1	-	-

3.2 EnlA – bacteriocin or phage endolysin?

Sequence analysis of amplicons obtained by inverse and vectorette PCR revealed the presence of gene encoding C39 family endopeptidase, located downstream of *enl*A. These enzymes participate in bacteriocin maturation and secretion. Region upstream of enlA contains 3 genes coding for hypothetical proteins (with still unknown function), which sequences and organisation are nearly identical as in case of *E. faecalis* JH1 and *E. faecalis* D6 strains. No genes typical for phage lytic systems were found (data not shown).

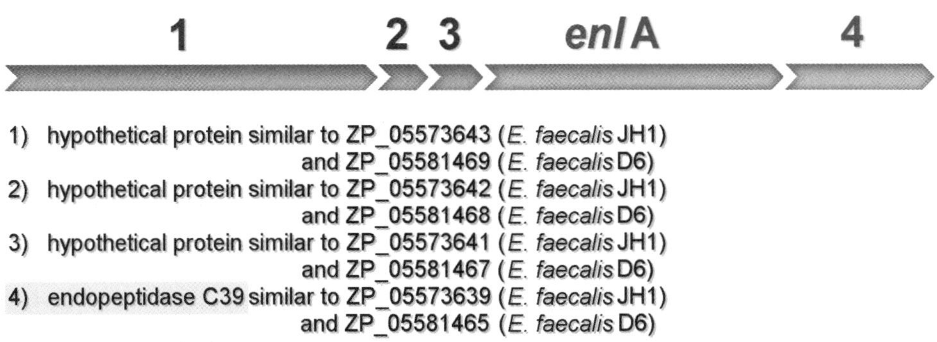

Fig. 3 Flanking DNA regions of *enl*A gene.

3.3 The nature of EnlA receptor

Any compounds present on bacterial cell surface (peptidoglycan subunits, saccharides, proteins, lipoteichoic acid, etc.) could serve as a PGHs receptor. In the series of experiments the chemical nature of EnlA receptor was elucidated using competitive binding assay.

According to type of receptor at least seven different proven or hypothetical CWB domains are reported in the literature: (a) the cysteine and histidine-dependent amidohydrolases/peptidases (CHAP) domain, (b) the GW domain for noncovalent attachment to lipoteichoic acid, (c) the SH3 domain, (d) the PlyPSA CWB domain, (e) the FtsN-peptidoglycan-binding domain, (f) the choline-binding domain, and (g) the LysM CWB domain [10]. Sequence analysis of *enl*A gene did not comfirm the presence of such domains and thus *enl*A C-terminal domain is probably a representative of a novel type of CWB domains.

Adding galactose or rhamnose to the mixture of GFP-CWBenlA and sensitive cells, and also treatment with proteinase had no significant effect in GFP-CWBenlA binding assay (compared with control). On the other hand, treatment with lysozyme significantly increased the binding ability of GFP-CWBenlA to the surface of sensitive cells (see Figure 4).

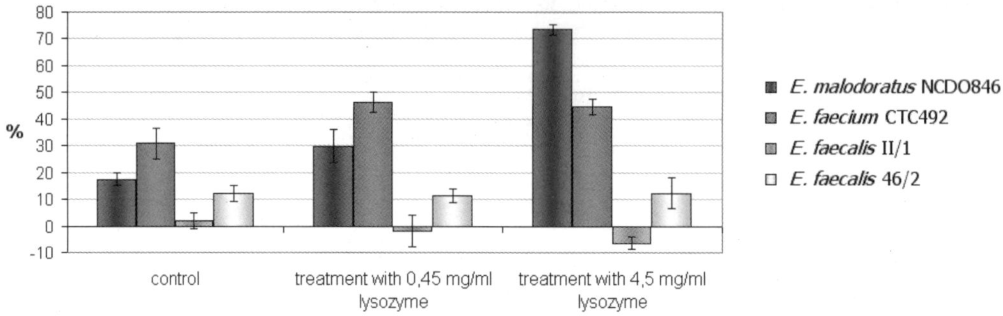

Fig. 4 Binding ability of GFP-CWBenlA to the cell surfaces of sensitive (*E. malodoratus* NCDO846, *E. faecium* CTC492) and resistant (*E. faecalis* II/1, *E. faecalis* 46/2) enterococcal strains. The graph illustrates the percentage of total amount of GFP-CWBenlA in solution, which bound on the bacterial cell surfaces after treatment the cells with lysozyme, compared with control.

4. Conclusions

Using fusion with GFP we demonstrated that the C-terminal domain of enterolysin A is involved in specific recognition and binding to the target cell envelopes and represents true cell wall binding domain. Sensitivity/resistance to EnlA is dependent on binding ability/disability of its CWB domain.

Neighbouring DNA region downstream of *enl*A contains gene coding for endopeptidase C39 (protein important in bacteriocin maturation and secretion). No genes similar to the known bacteriocin immunity proteins or genes typical for phage lytic modules were found.

We suppose that real mechanism of resistance against enterolysin A is absence of specific receptor on the cell surface, which is necessary for binding of the bacteriocin molecule.

Since the treatment of sensitive cells with lysozyme (which totally desintegrates peptidoglycan) increased binding ability of GFP-CWBenlA to the cell surface, receptor is probably not embedded only in peptidoglycan layer. We suppose that EnlA receptor could be the lipoteichoic acid (LTA), which is the specific antigen of Lancefield group D streptococci and enterococci [11]. LTA extends as far as to phospholipid bilayer and therefore, it remains anchored to the cell surface after lysozyme treatment. Removing of peptidoglycan layer by lysozyme may even improve the access of EnlA to its receptor.

Acknowledgements The support by APVV Grant No. 0586-07 and VEGA Grant No. 2/0006/08 is gratefully acknowledged.

References

[1] Layec S, Decaris B, Leblond-Bourget N. Diversity of Firmicutes peptidoglycan hydrolases and specificities of those involved in daughter cell separation. *Research in Microbiology.* 2008;159:507-515.

[2] Schirner K, Marles-Wright J, Lewis RJ, Errington J. Distinct and essential morphogenic functions for wall- and lipoteichoic acids in *Bacillus subtilis. The EMBO Journal.* 2009;28:830-842.

[3] Parisien A, Allain B, Zhang J, Mandeville R, Lan CQ. Novel alternatives to antibiotics: bacteriophages, bacterial cell wall hydrolases, and antimicrobial peptides. *Journal of Applied Microbiology.* 2008;104:1-13.

[4] Borysowski J, Weber-Dabrowska B, Górski A. Bacteriophage endolysins as a novel class of antibacterial agents. *Experimental Biology and Medicine.* 2006;231:366-377.

[5] Fischetti VA. Using phage lytic enzymes to control pathogenic bacteria. *BMC Oral Health.* 2006; 6(Suppl 1):S16.

[6] Nilsen T, Nes IF, Holo H. Enterolysin A, a cell wall-degrading bacteriocin from *Enterococcus faecalis* LMG 2333. *Applied and Environmental Microbiology.* 2003;69:2975-2984.

[7] Gajic O, Buist G, Kojic M, Topisirovic L, Kuipers OP, Kok J. Novel mechanism of bacteriocin secretion and immunity carried out by lactococcal multidrug resistance proteins. *The Journal of Biological Chemistry.* 2003;278:34291-34298.

[8] Dykes GA. Bacteriocins: ecological and evolutionary significance. Trends in Ecology & Evolution. 1995;10:186-189.

[9] Nigutová K, Serenčová L, Piknová M, Javorský P, Pristaš P. Heterologous expression of functionally active enterolysin A, class III bacteriocin from *Enterococcus faecalis*, in *Escherichia coli. Protein Expression and Purification.* 2008;60:20-24.

[10] Vollmer W, Joris B, Charlier P, Foster S. Bacterial peptidoglycan (murein) hydrolases. *FEMS Microbiology Reviews.* 2008;32:259-286.

[11] Ruoff KL. Recent taxonomic changes in the genus *Enterococcus. European Journal of Clinical Microbiology & Infectious Diseases: official publication of the European Society of Clinical Microbiology.* 1990;9:75-79.